Handbook of Energy

Handbook of Energy

VOLUME I: DIAGRAMS, CHARTS, AND TABLES

Edited by

CUTLER J. CLEVELAND

Boston University, Boston, Massachusetts, USA

CHRISTOPHER MORRIS

Morris Books, Escondido, California, USA

ELSEVIER

AMSTERDAM • BOSTON • HEIDELBERG • LONDON • NEW YORK
OXFORD • PARIS • SAN DIEGO • SAN FRANCISCO • SYDNEY • TOKYO

Elsevier
225 Wyman Street, Waltham, MA 02451, USA
Radarweg 29, PO Box 211, 1000 AE Amsterdam, The Netherlands
The Boulevard, Langford Lane, Kidlington, Oxford OX5 1GB, UK

British Library Cataloguing in Publication Data
A catalogue record for this book is available from the British Library

Library of Congress Cataloging-in-Publication Data
A catalog record for this book is available from the Library of Congress

For information on all Elsevier publications
visit our web site at store.elsevier.com

ISBN: 978-0-08-046405-3

Images courtesy of iStockphoto:
Santosha, Image #14058272; Globestock, Image #6124616; Klubovy, Image #15475299;
Falun, Image #17038819; RASimon, Image #4139892

Working together
to grow libraries in
developing countries

www.elsevier.com • www.bookaid.org

CONTENTS

Energy is central to human existence. Indeed, human history can be told in terms of the history of energy. The discovery of fire, the domestication of animals, the discovery of fossil fuels, the electrification of cities, advances in nuclear physics, and the recent oil wars in the Middle East are all pivotal points in human history.

Energy is now central to one of the preeminent challenges facing humanity: a sustainable human existence on the planet. Energy is a key driver of macroeconomic growth. In the environmental dimension, conventional energy sources are major drivers of environmental stress at global as well as local levels. In terms of the social dimension, energy is a prerequisite for the fulfillment of many basic human needs and services, and inequities in energy provision and quality often manifest themselves as issues of social justice.

The *Handbook of Energy* was conceived to provide up-to-date and essential information about energy to a diverse audience. Its distinguishing features are its integration of the social, physical, biological, and engineering sciences, its breadth of coverage, its diversity of content types, and its reliance on authoritative, peer-reviewed information.

Information is organized by five broad themes:

1. *Sources:* The story of energy begins with the generation of the stocks and flows of energy in the environment that humans ultimately tap. This theme includes content related to all the sources of energy that humans use: solar energy, hydrogen, coal, electricity, wind, and so on.
2. *Foundations:* Energy is the common link between the living and non-living realms of the universe and thus is an integrator across all fields in science, technology, engineering, and mathematics education and research. This theme includes content related to the physical foundations of energy that have been established in the natural sciences and engineering fields.
3. *Applications.* Humans do not value coal or oil *per se.* Rather, they seek the services that energy provides: warmth, illumination, and mobility. This theme includes content related to the applications of energy, such as energy conversion, lighting, end use efficiency, and transportation by land, water, and air.
4. *Impacts.* The extraction, processing, transport, and use of energy have wide-ranging impacts. This theme includes content related to climate change, air and water pollution and other environmental effects, and health and safety.
5. *Correlations.* This theme includes content related to the economics and business of energy.

The diagrams, charts, and tables within this organizational scheme are grouped to supply the reader with logical and ready access to information. Thus, all the diagrams, charts, and tables related to energy conversion are grouped together, as are all those related to oil and gas, all those related to climate change, and so on. This approach exposes the reader to the broad, interdisciplinary nature of the *Handbook,* while at the same time enabling the reader to easily "drill down" and find information on a specific topic.

The information in the *Handbook* is drawn predominantly from peer-reviewed sources in academia and government: journals, textbooks, reference handbooks, and technical manuals. Less frequent sources include research institutes, corporations, and nongovernmental organizations.

Cutler J. Cleveland
Boston University, Boston, Massachusetts, USA

Christopher Morris
Morris Books, Escondido, California, USA

Mallory Nomack and Amanda Winchester contributed significant technical expertise in the construction of databases and in the production of the charts using MATLAB. Rebecca Bar assisted in data collection and table construction.

The following individuals and organizations assisted in making data available: Jeongwoo Han, Michael Wang, Argonne National Laboratory; Mark Friedl, Lucy Hutrya, Robert Kaufmann, Ranga Myneni, Boston University; Christopher Weber, Carnegie Mellon University; Maurizio Cocchi, etaflorence; Ida Arabshahi, Global Methane Initiative; Jean-Paul Rodrigue, Hofstra University; Daniel Couvida, International Commission on Large Dams; Steve Usher, International Journal on Hydropower and Dams; Larry Sherwood, Interstate Renewable Energy Council; Ben McLellan, Kyoto University; Alix Anne Bluhm, McKinsey & Company; Kevin E. Trenberth, National Center for Atmospheric Research; Garvin Heath, Jordan Macknick, Ethan Warner, National Renewable Energy Laboratory; Steven Smith, Pacific Northwest National Laboratory; Tim Colton, Shipbuildinghistory.com; John Helliwell, Shun Wang, University of British Columbia; Mike Ashby, University of Cambridge; Lado Kurdgelashvili, University of Delaware; Gregory F. Nemet, University of Wisconsin; Louise Guey-Lee, Erik Kreil, Peter Wong, U.S. Energy Information Administration; Colin Williams, U.S. Geological Survey; Kirk Hamilton, Govinda Timilsina, World Bank; Elena Nekhaev, World Energy Council.

The team at Elsevier contributed considerable editorial, technical, and administrative support: Candice Janco, Jill Cetel, and Mohanapriyan Rajendran.

PART I

SOURCES

Bioenergetics

Figures

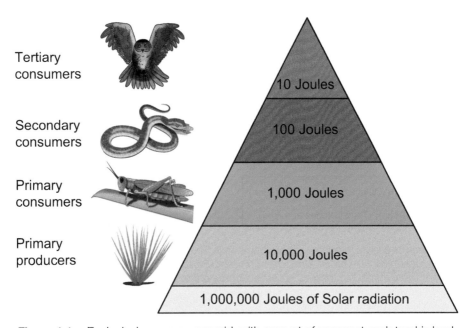

Tertiary consumers

Secondary consumers

Primary consumers

Primary producers

10 Joules

100 Joules

1,000 Joules

10,000 Joules

1,000,000 Joules of Solar radiation

Figure 1.1. Ecological or energy pyramid, with amount of energy at each trophic level.

Handbook of Energy. http://dx.doi.org/10.1016/B978-0-08-046405-3.00001-2

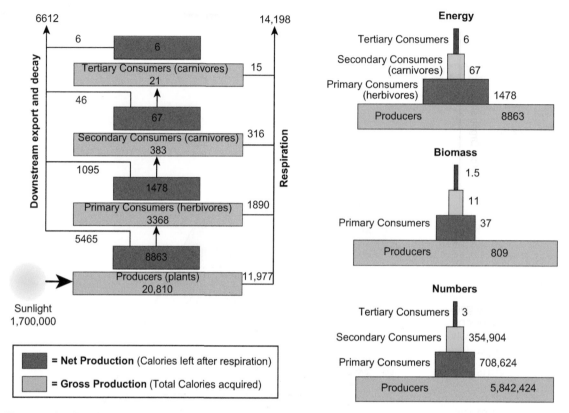

Figure 1.2. The flow of energy through a river ecosystem in Silver Springs, Florida. Energy units are kilocalories per square meter per year (kcal/m²/yr). Biomass units represent the dry weight of organic matter (per square meter). The pyramid of numbers is derived from a census of the populations of autotrophs, herbivores, and two levels of carnivores on an acre (0.4 hectare) of a typical grassland. The figures represent number of individuals counted at each trophic level.

Source: Energy data from Odum, Howard T. 1957. Trophic Structure and Productivity of Silver Springs, Florida. Ecological Monographs 27:55–112; adapted from Kimball, John W., Kimball's Biology Pages, <http://users.rcn.com/jkimball.ma.ultranet/BiologyPages/W/Welcome.html>.

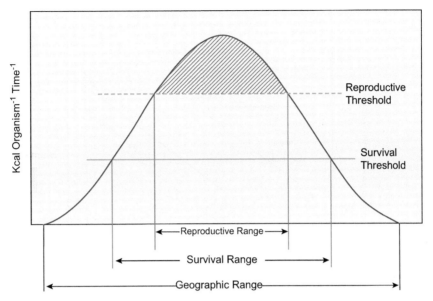

Figure 1.3. Organisms have ranges of tolerance for environmental factors (sunlight, temperature, pH, ec.). Optimum conditions are those that are most favorable for an organism to survive, grow and reproduce. This optimum is somewhere within the range of tolerance for that organism.
Source: Adapted from Hall, C.A.S., J.A. Stanford and R. Hauer. 1992. The distribution and abundance of organisms as a consequence of energy balances along multiple environmental gradients. Oikos 65: 377–390.

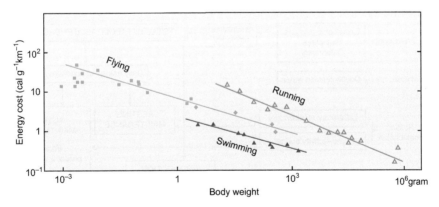

Figure 1.4. Energy cost of locomotion for swimming, flying, and running animals, as a function of body size.
Source: Data from Schmidt-Nielsen, K. 1972. Locomotion: energy cost of swimming, flying and running. Science 177, 222–228.

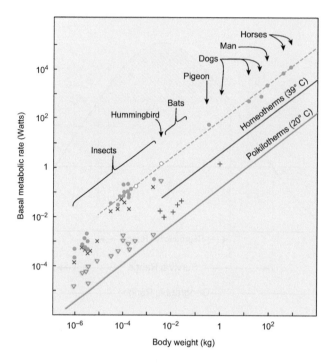

Figure 1.5. The relation between basal metabolic rate and body weight.
Source: Data from Monteith, J.L. and M.H. Unsworth. 1990. Principles of Environmental Physics (Second Edition), (London, Edward Arnold).

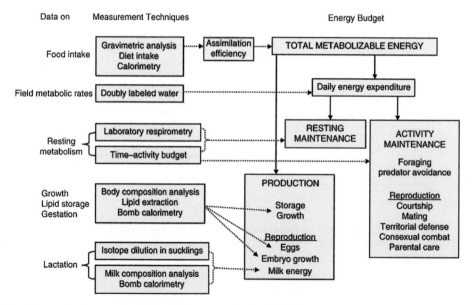

Figure 1.6. Generalized energy budget used to examine reproductive energetics.
Source: Kunz, Thomas H. and Kimberly S. Orrell. 2004. Reproduction, Energy Costs of, In: Cutler J. Cleveland, Editor-in-Chief, Encyclopedia of Energy, (New York, Elsevier), Pages 423-442.

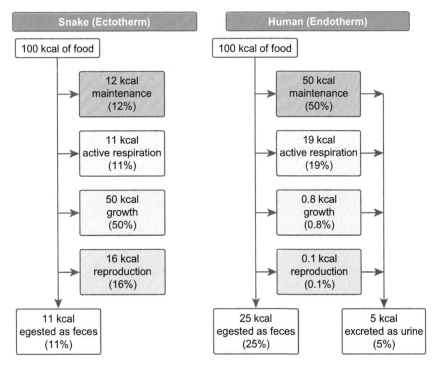

Figure 1.7. Comparative energetics of a snake (ectotherm) and human (endotherm).
Source: Data from Kaufmann, Robert K. and Cleveland, Cutler J. 2007. Environmental Science (McGraw-Hill, Dubuque, IA).

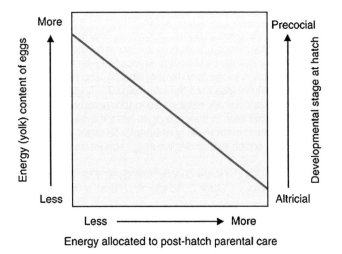

Figure 1.8. General relationship among energy content of eggs, energy allocated to post-hatch parental care, and development stage of offspring at hatch.
Source: Kunz, Thomas H. and Kimberly S. Orrell. 2004. Reproduction, Energy Costs of, In: Cutler J. Cleveland, Editor-in-Chief, Encyclopedia of Energy, (New York, Elsevier), Pages 423-442.

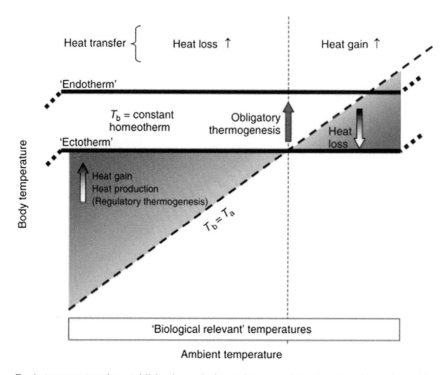

Figure 1.9.   Body temperature is established as a balance between heat input and heat loss. Heat input occurs through heat transfer or from obligatory or regulatory thermogenesis. Heat transfer, either loss or gain, between the animal and its environment can occur via conduction, convection, radiation, or evaporation/ condensation. The rate of heat transfer for each mode is proportional to the surface area and, except for evaporation/conduction, is proportional to the temperature gradient between the animal and the environment. For an 'ectotherm' at an established body temperature (T_b) in equilibrium with a given ambient temperature (T_a) heat gain equals heat loss (intersection with $T_b = T_a$). At temperatures on either side of this single T_a, for T_b to remain constant changes in heat transfer need to occur; at lower T_a heat loss will increase and heat needs to be added (or heat loss reduced) and for higher T_a heat gain will increase and heat loss needs to increase (and/ or heat input reduced). The increase in cellular metabolism associated with leakier membranes in 'endotherms' is coupled with increased heat production (obligatory thermogenesis). The production of heat will result in an increase in T_b to a temperature where heat loss from the increasing T_b–T_a gradient will reestablish equilibrium at the single T_a in which equilibrium was originally established in the 'ectotherm.' To maintain T_b with changing T_a again requires adjustments in heat transfer. In the case of an 'endotherm,' a decrease in T_a is initially met over a narrow temperature range through changes in heat transfer to offset the increasing heat loss (known as the 'thermal neutral zone', TNZ), after which further decline in T_a is countered with increased heat production (regulatory thermogenesis).

Source: Frappell, P., K. Cummings. 2008 Homeotherms, In: Sven Erik Jorgensen and Brian Fath, Editors-in-Chief, Encyclopedia of Ecology, (Oxford, Academic Press), Pages 1884-1893.

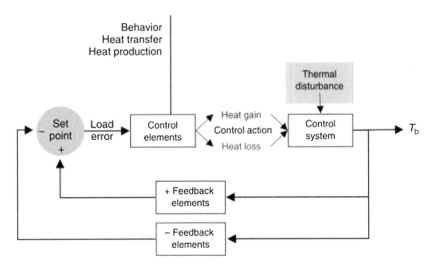

Figure 1.10. Dual controller model for the regulation of body temperature (T_b) with a thermal set-point. One group of feedback elements (sensors) responds with increased activity to rising temperature while the other group responds with increased activity to falling temperature. By comparing the activity of both groups of sensors the load error is generated (the difference between set-point and T_b) and the control elements are activated in proportion to the load error. An increase in T_b results in dominance of warm sensor activity and the control elements (e.g., changes in behavior, heat transfer properties, or heat production) restore equilibrium by increasing heat loss. When the activity from warm and cold sensors is equal, the load error is zero and T_b is at its set-point. For T_b to be regulated the system requires T_b to be displaced from set-point. How far the system can be displaced from equilibrium establishes the load error that is tolerated for the system.
Source: Frappell, P., K. Cummings. 2008 Homeotherms, In: Sven Erik Jorgensen and Brian Fath, Editors-in-Chief, Encyclopedia of Ecology, (Oxford, Academic Press), Pages 1884-1893.

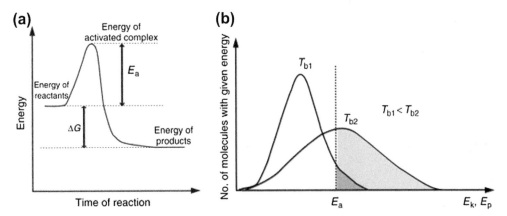

Figure 1.11. (a) Activation energy (E_a) of a chemical reaction. ΔG – Gibbs free energy change of the chemical reaction. For the reaction to occur, the energy of reactants must be increased by the value of E_a, after which the reaction proceeds spontaneously. A key function of enzymes as biological catalysts is to reduce the activation energy barrier of activated complex and thus to facilitate biochemical reactions. (b) The rate of biochemical reactions increases with increasing body temperature (T_b) due to increase in the fraction of molecules with energy levels exceeding E_a. E_p, E_k – Kinetic and/or potential energy of the reactant molecules.
Source: Sokolova, I. 2008. Temperature Regulation, In: Sven Erik Jorgensen and Brian Fath, Editors-in-Chief, Encyclopedia of Ecology, (Oxford, Academic Press), Pages 3509-3516.

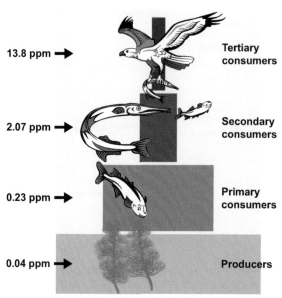

13.8 ppm ➡ Tertiary consumers

2.07 ppm ➡ Secondary consumers

0.23 ppm ➡ Primary consumers

0.04 ppm ➡ Producers

The numbers are representative values of the concentration in the tissues of **DDT** and its derivatives (in parts per million, ppm)

Figure 1.12. Principle of biomagnification. The numbers are representative values (ppm) of the concentration of the pesticide DDT and its derivatives. The horizontal bars describe the amount of energy at each trophic level. *Source: Data from Kimball's Biology Pages, <http://biology-pages.info>.*

Charts

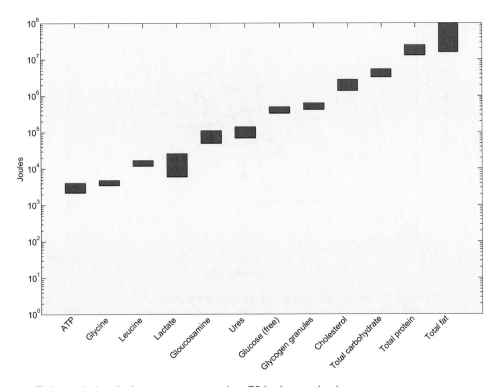

Chart 1.1. Estimated chemical energy resources in a 70 kg human body.
Source: Data from Freitas, Robert A. 1999. Nanomedicine, Volume I: Basic Capabilties (Austin, TX, Landes Bioscience).

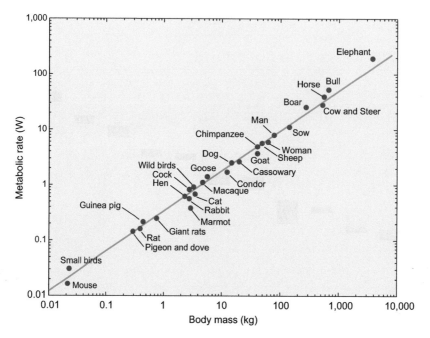

Chart 1.2. Metabolic rate (W) versus body mass (kg) for selected animals.
Source: Data from West, Geoffrey B. 2012. The importance of quantitative systemic thinking in medicine, The Lancet, Volume 379, Issue 9825, Pages 1551-1559.

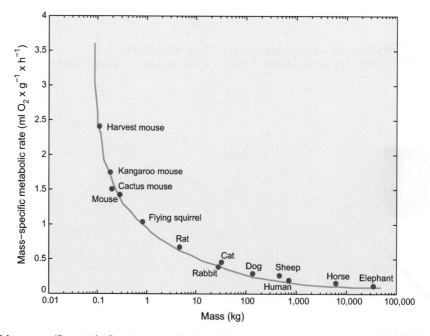

Chart 1.3. Mass-specific metabolic rate versus body mass.
Source: Data from Eckert, Roger and D.J. Randall. 1983. Animal Physiology, (San Francisco, W. H. Freeman).

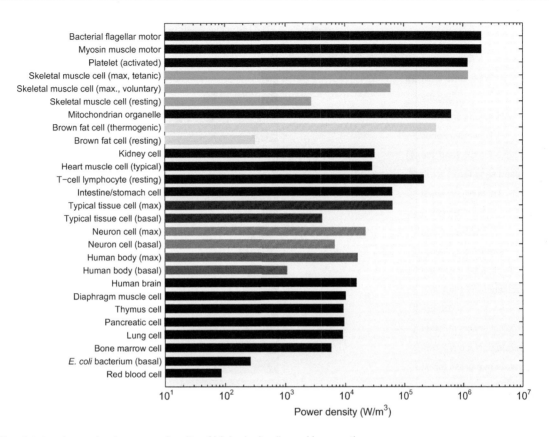

Chart 1.4. Approximate power density of biological cells and human tissues.
Data from Freitas, Robert A. 1999. Nanomedicine, Volume I: Basic Capabilties (Austin, TX, Landes Bioscience).

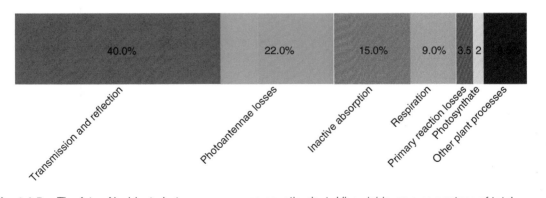

Chart 1.5. The fate of incident photon energy on an aquatic plant, *Ulva rigida*, as a percentage of total.
Source: Data from Gordillo FJL, Figueroa FL, Niell FX. 2003. Photon- and carbon-use efficiency in Ulva, rigida at different CO2 and N levels. Planta; 218(2):315-322.

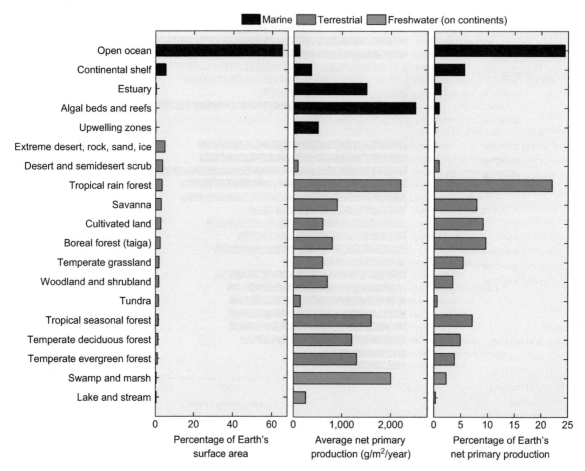

Chart 1.6. Net primary production of biomes.
Source: Data from Campbell, Neil A. and Jane B. Reece. Biology 7th Edition. San Francisco: Pearson Benjamin Cummings, 2005.

Tables

Table 1.1. Energy costs of human activity

Activity	METS[a]	Description
Bicycling	8	Bicycling, general
Bicycling	16	Bicycling, >20 mph, racing, not drafting
Conditioning exercise	7	Bicycling, stationary, general
Conditioning exercise	12.5	Bicycling, stationary, 250 watts, very vigorous effort
Conditioning exercise	8	Calisthenics, heavy, vigorous effort
Conditioning exercise	8	Circuit training
Conditioning exercise	6	Weight lifting
Conditioning exercise	9	Stair-treadmill ergometer, general
Conditioning exercise	3.5	Rowing, stationary, 50 watts, light effort
Conditioning exercise	12	Rowing, stationary, 200 watts, very vigorous effort
Conditioning exercise	6	Demonstrating aerobics exercises to class
Conditioning exercise	4	Water aerobics, water calisthenics
Dancing	4.8	Ballet or modern, twist, jazz, tap, jitterbug
Dancing	7	Aerobic, high impact
Dancing	5.5	Ballroom, dancing fast
Fishing and hunting	3	Fishing, general
Fishing and hunting	6	Fishing in stream, in waders
Fishing and hunting	2	Fishing, ice, siting
Fishing and hunting	5	Hunting, general
Home activities	3	Cleaning, heavy or major
Home activities	3.5	Vacuuming
Home activities	2	Cooking or food preparation
Home activities	7.5	Carrying groceries upstairs
Home activities	2.3	Ironing
Home activities	1.5	Sitting, knitting, sewing, light wrapping (presents)
Home activities	4	Walk/run, playing with child(ren), moderate
Home repair	3	Automobile repair
Home repair	3	Carpentry, general workshop
Home repair	4.5	Painting
Home repair	4.5	Washing and waxing hull of sailboat, car, powerboat, airplane
Inactivity quiet	0.9	Sleeping
Inactivity light	1	Reclining, reading
Inactivity light	1	Meditating
Lawn and garden	2.5	Mowing lawn, riding mower
Lawn and garden	6	Mowing lawn, walking, hand mower
Lawn and garden	4.3	Raking lawn
Lawn and garden	4	Gardening, general
Miscellaneous	1.3	Sitting, reading, book, newspaper, etc.
Miscellaneous	1.5	Sitting, talking or talking on the phone
Music playing	4	Drums
Music playing	2.5	Piano or organ
Music playing	2.5	Violin

Continued

Table 1.1. Continued

Activity	METS[a]	Description
Music playing	3	Guitar, rock and roll band (standing)
Occupation	6	Coal mining, general
Occupation	5.5	Construction, outside, remodeling
Occupation	8	Farming, bailing hay, cleaning barn, poultry work
Occupation	12	Firefighter, general
Occupation	2.5	Police, directing traffic (standing)
Occupation	1.3	Police, riding in squad car (sitting)
Occupation	1.5	Sitting, light office work, general
Running	7	Jogging, general
Running	8	Running, 5 mph (12 min/mile)
Running	18	Running, 10.9 mph (5.5 min/mile)
Running	15	Running, stairs, up
Sexual activity	1.5	Active, vigorous effort
Sexual activity	1.3	General, moderate effort
Sexual activity	1	Passive, light effort, kissing, hugging
Sports	8	Basketball, game
Sports	2.5	Billiards
Sports	3	Bowling
Sports	9	Football, competitive
Sports	8	Frisbee, ultimate
Sports	4.5	Golf, walking and carrying clubs
Sports	3.5	Golf, using power cart
Sports	8	Hockey, field
Sports	8	Hockey, ice
Sports	8	Lacrosse
Sports	11	Rock climbing, ascending rock
Sports	5	Skateboarding
Sports	12	Roller blading (in-line skating)
Sports	10	Soccer, competitive
Sports	7	Soccer, casual, general
Sports	5	Softball, fast or slow pitch, general
Sports	5	Tennis, doubles
Sports	8	Tennis, singles
Transportation	1	Riding in a car or truck
Walking	7	Backpacking
Walking	6	Hiking, cross country
Walking	2.5	Bird watching
Walking	6.5	Race walking
Walking	2.5	Walking, 2.0 mph, level, slow pace, firm surface
Walking	6.3	Walking, 4.5 mph, level, firm surface, very brisk
Water activities	10	Swimming laps, freestyle, fast, vigorous effort
Water activities	7	Swimming laps, slow, moderate or light effort
Winter activities	7	Skating, ice, general
Winter activities	7	Skiing, general

Table 1.1.   Continued

Activity	METS[a]	Description
Winter activities	7	Skiing, cross country, 2.5 mph, slow or light effort, ski walking
Winter activities	8	Skiing, downhill, vigorous effort

[a]The term MET (metabolic equivalent) is the ratio of a work metabolic rate to the resting metabolic rate. One MET is defined as 1 kcal/kg/hour and is roughly equivalent to the energy cost of sitting quietly. A MET also is defined as oxygen uptake in ml/kg/min, with one MET roughly equivalent to 3.5 ml/kg/min. Multiples of 1 MET indicate a higher energy cost for a specific activity. For example, a 2 MET activity requires twice the energy cost of sitting quietly.

Source: Adapted from Ainsworth, B.E. 2002. The Compendium of Physical Activities Tracking Guide. Prevention Research Center, Norman J. Arnold School of Public Health, University of South Carolina.

Table 1.2. Daily average energy requirement for women aged 30 to 59.9 years

Mean Weight kg	BMR/kg		1.45 × BMR				1.75 × BMR				2.05 × BMR				2.20 × BMR				Height (m) for BMI values		
	kJ	kcal	MJ	kJ/kg	kcal	kcal/kg	MJ	kJ/kg	kcal	kcal/kg	MJ	kJ/kg	kcal	kcal/kg	MJ	kJ/kg	kcal	kcal/kg	24.9	21	18.5
45	113	27	7.3	165	1750	39	8.9	195	2100	47	10.4	230	2500	56	11.1	250	2650	59	1.34	1.46	1.58
50	105	25	7.6	150	1800	36	9.2	185	2200	44	10.7	215	2550	51	11.5	230	2750	55	1.42	1.54	1.64
55	96	24	7.8	145	1850	34	9.5	170	2250	41	11.1	200	2650	48	11.9	215	2850	52	1.49	1.62	1.72
60	93	22	8.1	135	1950	33	9.8	165	2350	39	11.4	190	2750	46	12.3	205	2950	49	1.55	1.69	1.8
65	88	21	8.3	130	2000	31	10.1	155	2400	37	11.8	180	2800	43	12.6	195	3000	46	1.62	1.76	1.87
70	85	20	8.6	125	2050	29	10.4	150	2500	36	12.1	175	2900	41	13	185	3100	44	1.68	1.83	1.95
75	81	19	8.8	120	2100	28	10.7	140	2550	34	12.5	165	3000	40	13.4	180	3200	43	1.74	1.89	2.01
80	78	19	9.1	115	2150	27	11	135	2600	33	12.8	160	3050	38	13.8	170	3300	41	1.79	1.95	2.08
85	76	18	9.3	110	2250	26	11.2	130	2700	32	13.2	155	3150	37	14.1	165	3400	40	1.85	2.01	2.14

BMR (basal metabolic rate): the amount of daily energy expended by humans and other animals at rest.
BMI (body mass index): heuristic proxy for human body fat based on an individual's weight and height.
Source: Adapted from Food and Agriculture Organization of the United Nations, United Nations University, World Health Organization. 2004. Human Energy Requirements: Report of a Joint FAO/WHO/UNU Expert Consultation, (Rome, Food and Agriculture Organization of the United Nations).

Table 1.3. Daily average energy requirement for men aged 30 to 59.9 years

Mean Weight kg	BMR/kg		1.45 × BMR				1.75 × BMR				2.05 × BMR				2.20 × BMR				Height (m) for BMI values		
	kJ	kcal	MJ	kJ/kg	kcal	kcal/kg	MJ	kJ/kg	kcal	kcal/kg	MJ	kJ/kg	kcal	kcal/kg	MJ	kJ/kg	kcal	kcal/kg	24.9	21	18.5
50	121	29	8.8	175	2100	42	10.6	210	2550	51	12.4	250	2950	59	13.3	265	3200	64	1.42	1.54	1.64
55	114	27	9.1	165	2200	40	11	200	2650	48	12.9	235	3100	56	13.8	250	3300	60	1.49	1.62	1.72
60	109	26	9.5	160	2250	38	11.4	190	2750	46	13.4	225	3200	53	14.4	240	3450	57	1.55	1.69	1.8
65	104	25	9.8	150	2350	36	11.9	180	2850	44	13.9	215	3300	51	14.9	230	3550	55	1.62	1.76	1.87
70	100	24	10.2	145	2450	35	12.3	175	2950	42	14.4	205	3450	49	15.4	220	3700	53	1.68	1.83	1.95
75	97	23	10.5	140	2500	34	12.7	170	3050	40	14.9	200	3550	47	16	215	3800	51	1.74	1.89	2.01
80	94	22	10.9	135	2600	32	13.1	165	3150	39	15.4	190	3650	46	16.5	205	3950	49	1.79	1.95	2.08
85	91	22	11.2	130	2700	32	13.5	160	3250	38	15.9	185	3800	45	17	200	4050	48	1.85	2.01	2.14
90	89	21	11.6	130	2750	31	14	155	3350	37	16.3	180	3900	43	17.5	195	4200	47	1.9	2.07	2.21

BMR (basal metabolic rate): the amount of daily energy expended by humans and other animals at rest.
BMI (body mass index): heuristic proxy for human body fat based on an individual's weight and height.
Source: Adapted from Food and Agriculture Organization of the United Nations, United Nations University, World Health Organization. 2004. Human Energy Requirements: Report of a Joint FAO/WHO/UNU Expert Consultation, (Rome, Food and Agriculture Organization of the United Nations).

Table 1.4.   Activities accounting for 90% of human energy expenditure in the United States for males and females

	Male		Female
Rank	Activity description	Rank	Activity description
1	Job: Office work, typing	1	Driving car
2	Driving car	2	Taking care of child/baby (feeding, bathing)
3	Watching TV/movie, home or theater	3	Watching TV/movie, home or theater
4	Taking care of child/baby (feeding, bathing, etc.)	4	Job: Office work, typing
5	Job: Industrial plant/factory	5	Cleaning house, general
6	Job: Construction site	6	Activities performed while sitting
7	Activities performed while sitting	7	Eating (sitting)
8	Eating (sitting)	8	Food preparation (cooking, baking, etc.)
9	Yardwork-general: mowing lawn, trim	9	Talking/ Visiting, in person or on phone
10	Talking/ Visiting, in person or on phone	10	Job: Light intensity, stand/walking
11	Job: Farm hand	11	Shopping for non-foods (e.g. clothing)
12	Attending event (social) involving talking	12	Attending event (social) involving talking
13	Job: Store clerk, bartender, hair stylist	13	Laundry
14	Job: Light intensity, stand/walking	14	Job: Industrial plant/factory
15	Cleaning house, general	15	Job: Store clerk, bartender, hair stylist
16	Fishing and Hunting	16	Job: Restaurant staff (e.g. waiter, chef)
17	Shopping for non-foods (e.g. clothing)	17	Yardwork-general: mowing lawn, etc.
18	Job: Mechanic	18	Job: Teaching class
19	Food preparation (cooking, baking, etc.)	19	Cleaning kitchen (sweeping)
20	Job: Restaurant staff (e.g. waiter, chef)	20	Shopping for food, putting groceries away
21	Job: Driving (e.g. truck driver, bus driver, etc.)	21	Walking, moderately, (e.g. doing errands, etc.)
22	Remodeling, repairing house, workshop	22	Swimming, exercise
23	Walking, moderately, (e.g. doing errands, etc.)	23	Home projects (sewing, wrapping presents, etc.)[a]
24	Sports: Golf[a]	24	Gardening: Weeding, landscaping, etc.
25	Dancing/ Heavy Partying (all styles)	25	Moving, packing items[2]
26	Job: Teaching class		
27	Car maintenance, repair[a]		
28	Gardening: Weeding, landscaping, etc.		
29	Swimming, exercise		
30	Exercise, aerobics		
31	Yardwork-hard: chopping firewood, digging, etc.[a]		

[a]Activities that were not present on list for entire sample.

Source: Dong, Linda, Gladys Block and Shelly Mandel. 2004. Activities Contributing to Total Energy Expenditure in the United States: Results from the NHAPS Study, International Journal of Behavioral Nutrition and Physical Activity, 1:4.

Table 1.5. Activities that account for 90% of human energy expenditure in the United States, not including sleeping[a]

Rank	Activity Description	MET[b]	% of non-sleeping activity	Cumulative Percentage (%)
1	Driving car	2.3	10.9	10.9
2	Job: Office work, typing	1.5	9.2	20.1
3	Watching TV/movie, home or theater	1	8.7	28.8
4	Taking care of child/baby, (feeding, bathing, dressing)	3	8.4	37.2
5	Activities performed while sitting quietly	1.3	5.7	42.9
6	Eating (sitting)	1.5	5.3	48.2
7	Cleaning house, general	3	4	52.2
8	Talking/ Visiting, in person or on phone	1.5	3.8	56
9	Job: Industrial plant/factory (e.g. assembly line)	3	3.7	59.7
10	Food preparation (e.g. cooking, baking, setting table)	2	3	62.7
11	Job: Construction site	5.5	2.7	65.4
12	Job: Light intensity, stand/walking (e.g. hospital staff, real estate)	3	2.7	68.1
13	Yard work-general (e.g. mowing lawn, trimming hedges)	4.3	2.6	70.7
14	Attending event (social) involving talking while sitting	1.6	2.3	73
15	Shopping for non-foods (e.g. clothing)	2.3	2	75
16	Job: Light standing (e.g. store clerk, bartender, hair stylist, lab work)	2	1.8	76.8
17	Job: Farm hand (chores: baling hay, cleaning barn)	8	1.6	78.4
18	Job: Restaurant staff (e.g. waiter, chef)	3	1.5	79.9
19	Job: Teaching class	1.8	1.1	81
20	Laundry	2.2	1	82
21	Walking, moderately, (e.g. doing errands, walking to school)	2.8	1.1	83.1
22	Fishing and Hunting	3.3	0.9	84
23	Cleaning kitchen (sweeping)	3.3	0.9	84.9
24	Shopping for food, putting groceries away	2.4	0.8	85.7
25	Swimming	7	0.9	86.6
26	Job: Mechanic	3	0.7	87.3
27	Gardening: Weeding, landscaping, picking vegetables	3.7	0.6	87.9
28	Remodeling, repairing house, workshop, concrete work	3.6	0.6	88.5
29	Exercise, aerobics	6.5	0.6	89.1
30	Job: Driving (e.g. truck driver, bus, ambulance, tractor)	2.7	0.6	89.7
31	Dancing/ Heavy Partying (all styles of dancing)	4	0.5	90.2

[a]Sleeping or napping is the most common activity, averaging about eight hours in the past 24-hour period, for both males and females. Sleeping or napping contributes 19% of the overall energy expenditure.

[b]The term MET (metabolic equivalent) is the ratio of a work metabolic rate to the resting metabolic rate. One MET is defined as 1 kcal/kg/hour and is roughly equivalent to the energy cost of sitting quietly. A MET also is defined as oxygen uptake in ml/kg/min, with one MET roughly equivalent to 3.5 ml/kg/min. Multiples of 1 MET indicate a higher energy cost for a specific activity. For example, a 2 MET activity requires twice the energy cost of sitting quietly.

Source: Dong, Linda, Gladys Block and Shelly Mandel. 2004. Activities Contributing to Total Energy Expenditure in the United States: Results from the NHAPS Study, International Journal of Behavioral Nutrition and Physical Activity, 1:4.

Table 1.6. Girls' energy requirements

Age (years)	Weight (kg)	TEE[a] MJ/d	TEE[a] kcal/d	E_g[b] MJ/d	E_g[b] kcal/d	BMR_est[c] MJ/d	BMR_est[c] kcal/d	Daily energy requirement MJ/d	Daily energy requirement kcal/d	Daily energy requirement kJ/kg/d	Daily energy requirement kcal/kg/d	PAL[d] Tee/BMR
1-2[e]	10.8	3.561	851	0.057	14	2.505	599	3.618	865	335	80.1	1.42
2-3	13.0	4.330	1035	0.052	12	3.042	727	4.982	1047	337	80.6	1.42
3-4	15.1	4.791	1145	0.045	11	3.317	793	4.836	1156	320	76.5	1.44
4-5	16.8	5.152	1231	0.040	10	3.461	827	5.192	1241	309	73.9	1.49
5-6	18.6	5.522	1320	0.042	10	3.614	864	5.564	1330	299	71.5	1.53
6-7	20.6	5.920	1415	0.054	13	3.784	904	5.974	1428	290	69.3	1.56
7-8	23.3	6.431	1537	0.071	17	4.014	959	6.502	1554	279	66.7	1.60
8-9	26.6	7.019	1678	0.087	21	4.294	1025	7.106	1698	267	63.8	1.63
9-10	30.5	7.661	1831	0.094	23	4.626	1105	7.755	1854	254	60.8	1.66
10-11	34.7	8.287	1981	0.106	25	4.841	1157	8.393	2006	242	57.8	1.71
11-12	39.2	8.884	2123	0.106	25	5.093	1217	8.990	2149	229	54.8	1.74
12-13	43.8	9.414	2250	0.108	26	5.351	1279	9.523	2276	217	52.0	1.76
13-14	48.3	9.855	2355	0.99	24	5.603	1339	9.954	2379	206	49.3	1.76
14-15	52.1	10.168	2430	0.80	19	5.816	1390	10.248	2449	1974	47.0	1.75
15-16	55.0	10.370	2478	0.052	12	5.978	1429	10.421	2491	189	45.3	1.73
16-17	56.4	10.455	2499	0.019	5	6.056	1447	10.474	2503	186	44.4	1.73
17-18	56.7	10.473	2503	0.000	0	6.073	1451	10.473	2503	185	44.1	1.72

Girls' energy requirements calculated by quadratic regression analysis of TEE on weight, plus allowance for energy deposition in tissue during growth (E_g).

[a]TEE (total energy expenditure): The energy spent, on average, in a 24-hour period by an individual or a group of individuals. TEE (MJ/d) $1.102 + 0.273$ kg $- 0.0019$ kg^2.

[b]E_g = energy deposition in tissue during growth = 8.6 kJ or 2 kcal/g weight gain.

[c]BMR (basal metabolic rate): The minimal rate of energy expenditure compatible with life. It is measured in the supine position under standard conditions of rest, fasting, immobility, thermoneutrality and mental relaxation; BMR$_{est}$: basal metabolic rate estimated with predictive equations on body weight.

[d]PAL (physical activity level) = TEE/BMR. To calculate requirements, add Eg, or multiply by 1.01.

[e]Requirements for 1 to 2 years adjusted by 7 percent to fit with energy requirements of infants.

Source: Adapted from Food and Agriculture Organization of the United Nations, United Nations University, World Health Organization. 2004. Human Energy Requirements: Report of a Joint FAO/WHO/UNU Expert Consultation, (Rome, Food and Agriculture Organization of the United Nations).

Table 1.7. Boys' energy requirements

Age (years)	Weight (kg)	TEE[a]		E_g[b]		BMR_est[c]		Daily energy requirement				PAL[d]
		MJ/d	kcal/d	MJ/d	kcal/d	MJ/d	kcal/d	MJ/d	kcal/d	kJ/kg/d	kcal/kg/d	TEE/BMR
1-2[e]	11.5	3.906	934	0.057	14	2.737	654	3.963	948	345	82.4	1.43
2-3	13.5	4.675	1117	0.047	11	3.235	773	4.722	1129	350	83.6	1.45
3-4	15.7	4.187	1240	0.049	12	3.602	861	5.236	1252	334	79.7	1.44
4-5	17.7	5.644	1349	0.047	11	3.792	906	5.691	1360	322	76.8	1.49
5-6	19.7	6.092	1456	0.047	11	3.982	952	6.139	1467	312	74.5	1.53
6-7	21.7	6.531	1561	0.052	12	4.172	997	6.583	1573	303	72.5	1.57
7-8	24.0	7.024	1679	0.057	14	4.390	1049	7.081	1692	295	70.5	1.60
8-9	26.7	7.589	1814	0.066	16	4.647	1111	7.655	1830	287	68.5	1.63
9-10	29.7	8.198	1959	0.078	19	4.932	1179	8.276	1978	279	66.6	1.66
10-11	33.3	8.903	2128	0.092	22	5.218	1247	8.995	2150	270	64.6	1.71
11-12	37.5	9.689	2316	0.106	25	5.529	1321	9.795	2341	261	62.4	1.75
12-13	42.3	10.539	2519	0.123	29	5.884	1406	10.662	2548	252	60.2	1.79
13-14	47.8	11.452	2737	0.137	33	6.291	1504	11.588	2770	242	57.9	1.82
14-15	53.8	12.371	2957	0.139	33	6.735	1610	12.510	2990	233	55.6	1.84
15-16	59.5	13.171	3148	0.127	30	7.157	1711	13.298	3178	224	53.4	1.84
16-17	64.4	13.802	3299	0.099	24	7.520	1797	13.901	3322	216	51.6	1.84
17-18	67.8	14.208	3396	0.061	15	7.771	1857	14.270	3410	210	50.3	1.83

Boys' energy requirements calculated by quadratic regression analysis of TEE on weight, plus allowance for energy deposition in tissue during growth (E_g).
[a]TEE (total energy expenditure): The energy spent, on average, in a 24-hour period by an individual or a group of individuals. TEE (MJ/d) = 1.298 + 0.265 kg- 0.0011kg^2.
[b]E_g (energy deposition in tissue during growth) = 8.6 kJ or 2 kcal/g weight gain.
[c]BMR (basal metabolic rate): The minimal rate of energy expenditure compatible with life. It is measured in the supine position under standard conditions of rest, fasting, immobility, thermoneutrality and mental relaxation; BMR_est: basal metabolic rate estimated with predictive equations on body weight.
[d]PAL (physical activity level) = TEE/BMR. To calculate requirements, add Eg, or multiply by 1.01.
[e]Requirements for 1 to 2 years adjusted by 7 percent to fit with energy requirements of infants.
Source: Adapted from Food and Agriculture Organization of the United Nations, United Nations University, World Health Organization. 2004. Human Energy Requirements: Report of a Joint FAO/WHO/UNU Expert Consultation, (Rome, Food and Agriculture Organization of the United Nations).

Table 1.8. Energy requirements of breastfed, formula-fed and all infants

Age (months)	Breastfed[a]			Formula-Fed[b]			All (breast and formula-fed)[c]		
	Boys	Girls	Mean	Boys	Girls	Mean	Boys	Girls	Mean
kJ/kg/d									
1	445	415	430	510	490	500	475	445	460
2	410	395	405	460	455	460	435	420	430
3	380	375	380	420	420	420	395	395	395
4	330	335	330	360	370	365	345	350	345
5	330	330	330	355	365	360	340	345	345
6	325	330	330	350	355	355	335	340	340
7	320	315	320	340	340	340	330	330	330
8	320	320	320	340	340	340	330	330	330
9	325	320	320	340	340	340	330	330	330
10	330	325	325	340	340	340	335	330	335
11	330	325	325	340	340	340	335	330	335
12	330	325	330	345	340	340	335	330	335
kcal/kg/d									
1	106	99	102	122	117	120	113	107	110
2	98	95	97	110	108	109	104	101	102
3	91	90	90	100	101	100	95	94	95
4	79	80	79	86	89	87	82	84	83
5	79	79	79	85	87	86	81	82	82
6	78	79	78	83	85	84	81	81	81
7	76	76	76	81	81	81	79	78	79
8	77	76	76	81	81	81	79	78	79
9	77	76	77	81	81	81	79	78	79
10	79	77	78	82	81	81	80	79	80
11	79	77	78	82	81	81	80	79	80
12	79	77	78	82	81	81	81	79	80

Numbers are rounded to the closest 5 kJ/kg/d, and 1 kcal/kg/d, using the following predictive equations for total energy expenditure (TEE):
[a]TEE (MJ/kg/d) = (−0.635 +0.388 weight)/weight.
[b]TEE (MJ/kg/d) = (−0.122 + 0.346 weight) /weight.
[c]TEE (MJ/kg/d) = (−0.416 + 0/371 weight)/weight.
Source: Adapted from Food and Agriculture Organization of the United Nations, United Nations University, World Health Organization. 2004. Human Energy Requirements: Report of a Joint FAO/WHO/UNU Expert Consultation, (Rome, Food and Agriculture Organization of the United Nations).

Table 1.9. Total energy expenditure measured in well-nourished non-pregnant and pregnant women

Country	No.	Measurement, week of gestation	Weight (kg)	TEE (MJ/d)	BMR (MJ/d)	AEE (MJ/d)	PAL	Preg TEE/NP TEE[a] (%)	Preg AEE/NP AE[b] (%)	TEE (kJ/kg/d)
Sweden	19	NP	60.7	10.1	5.6	4.5	1.80			166
	19	36	72.7	12.2	7.3	4.9	1.67	20.8	8.9	168
UK	12	NP	61.7	9.5	6.1	3.5	1.57			154
	12	36	73.6	11.3	7.6	3.7	1.49	18.2	6.6	153
USA	10	NP	63.5	9.2	5.5	3.7	1.68			147
	10	34-36	75.1	11.4	7.1	4.4	1.61	23.7	16.6	153
Mean non-pregnant			60.6	9.9	5.7	4.2	1.74			164
sd[c]			2.2	0.4	0.2	0.5	0.11			11
Mean 30-36 weeks			72.1	11.5	7.2	4.3	1.60	16.5	3.0	160
sd			2.2	0.8	0.2	0.9	0.14	6.9	15.4	11

TEE (total energy expenditure): The energy spent, on average, in a 24-hour period by an individual or a group of individuals.

BMR (basal metabolic rate): The minimal rate of energy expenditure compatible with life. It is measured in the supine position under standard conditions of rest, fasting, immobility, thermoneutrality and mental relaxation.

AEE (activity energy expenditure): Thermodifiable component of total energy expenditure (TEE) derived from all activities, both volitional and nonvolitional.

[a] preg = pregnant

[b] NP = non-pregnant

[c] sd = standard deviation of the mean

Source: Adapted from Food and Agriculture Organization of the United Nations, United Nations University, World Health Organization. 2004. Human Energy Requirements: Report of a Joint FAO/WHO/UNU Expert Consultation, (Rome, Food and Agriculture Organization of the United Nations).

Table 1.10.   Additional energy cost of pregnancy in women with an average gestational weight of 12 kg

A. Rates of tissue deposition

	1st trimester g/d	2nd trimester g/d	3rd trimester g/d	Total deposition g/280 d
Weight gain	17	60	54	12000
Protein deposition[a]	0	1.3	5.1	597
Fat deposition[a]	5.2	18.9	16.9	3741

B. Energy cost of pregnancy estimated from the increment in BMR and energy deposition

				Total energy cost	
	1st trimester kJ/d	2nd trimester kJ/d	3rd trimester kJ/d	MJ	kcal
Protein deposition[a]	0	30	121	14.1	3370
Fat deposition[a]	202	732	654	144.8	34600
Efficiency of energy utilization[b]	20	76	77	15.9	3800
Basal metabolic rate	199	397	993	147.8	35130
Total energy cost of pregnancy (kJ/d)	**421**	**1235**	**1845**	**322.6**	**77100**

C. Energy cost of pregnancy estimated from the increment in TEE and energy deposition

				Total energy cost	
	1st trimester kJ/d	2nd trimester kJ/d	3rd trimester kJ/d	MJ	kcal
Protein deposition[a]	0	30	121	14.1	3370
Fat deposition[a]	202	732	654	144.8	34600
Total energy expenditure[c]	85	350	1300	161.4	38560
Total energy cost of prengancy (kJ/d)	**287**	**1112**	**2075**	**320.2**	**76530**

Weight gain and tissue deposition in first trimester computed from last menstrual period (i.e. an interval of 79 days). Second and third trimesters computed as 280/3=93 days each. Basal metabolic rate (BMR): The minimal rate of energy expenditure compatible with life. It is measured in the supine position under standard conditions of rest, fasting, immobility, thermoneutrality and mental relaxation.
[a]Protein and fat deposition estimated from longitudinal studies of body composition during pregnancy, and an energy value of 23.6kJ (5.65 kcal)/g protein depositied, and 28.7 kJ (9.25 kcal)/g fat deposited.
[b]Efficiency of food energy utilization for protein and fat deposition taken as 0.90.
[c]Efficiency of energy utilization not included in this calculation, as the energy cost of synthesis is included in the measurement of total energy expenditure (TEE) by doubly labelled water (DLW).

Table 1.11.   Energy cost of human milk production by women who practice exclusive breastfeeding

Months postpartum	Mean milk intake (g/day)	Human milk intake, corrected for insensible water losses (g/day[a])	Gross energy content (kJ/g[b])	Daily gross energy secreted (kJ/day)	Energy cost of milk production (kJ/day[c])
1	699	734	2.8	2,055	2,569
2	731	768	2.8	2,149	2,686
3	751	789	2.8	2,208	2,760
4	780	819	2.8	2,293	2,867
5	796	836	2.8	2,340	2,925
6	854	897	2.8	2,511	3,138
Mean	769	807	2.8	2,259	2,824

[a]Insensible water losses assumed to be equal to 5 percent milk intake.
[b]Gross energy content measured by adiabatic bomb calorimetry or macronutrient analysis.
[c]Based on energetic efficiency of 80 percent.
Source: Adapted from Food and Agriculture Organization of the United Nations, United Nations University, World Health Organization. 2004. Human Energy Requirements: Report of a Joint FAO/WHO/UNU Expert Consultation, (Rome, Food and Agriculture Organization of the United Nations).

Table 1.12.   Protein, fat, and energy deposition during growth in the first year of life

Age (months)	Protein gain (g/d)	Fat mass gain (g/d)	Weight gain (g/d)	Energy accrued in normal growth	
				kJ/g	kcal/g
Boys					
0–3	2.6	19.6	32.7	24.1	6.0
3–6	2.3	3.9	17.7	11.6	2.8
6–9	2.3	0.5	11.8	6.2	1.5
9–12	1.6	1.7	9.1	11.4	2.7
Girls					
0–3	2.2	19.7	31.1	26.2	6.3
3–6	1.9	5.8	17.3	15.6	3.7
6–9	2.0	0.8	10.6	7.4	1.8
9–12	1.8	1.1	8.7	9.8	2.3

Energy equivalents: 1 g protein = 23.6 kJ (5.65 kcal); 1 g fat = 38.7 kJ (9.25 kcal).
Source: Adapted from Food and Agriculture Organization of the United Nations, United Nations University, World Health Organization. 2004. Human Energy Requirements: Report of a Joint FAO/WHO/UNU Expert Consultation, (Rome, Food and Agriculture Organization of the United Nations).

Table 1.13. Equations for estimating BMR from body weight[a]

	BMR: MJ/day	see[b]	BMR: kcal/day
Age (years)			
Males	0.249 kg − 0.127	0.292	59.512 kg − 30.4
< 3	0.095 kg + 2.110	0.280	22.706 kg + 504.3
3–10	0.074 kg + 2.754	0.441	17.686 kg + 658.2
10–18	0.063 kg + 2.896	0.641	15.057 kg + 692.2
18–30	0.048 kg + 3.653	0.700	11.472 kg + 873.1
30–60	0.049 kg + 2.459	0.686	11.711 kg + 587.7
≥ 60			
Females	0.244 kg − 0.130	0.246	58.317 kg − 31.1
< 3	0.085 kg + 2.033	0.292	20.315 kg + 485.9
3–10	0.056 kg + 2.898	0.466	13.384 kg + 692.6
10–18	0.062 kg + 2.036	0.497	14.818 kg + 486.6
18–30	0.034 kg + 3.538	0.465	8.126 kg + 845.6
30–60	0.038 kg + 2.755	0.451	9.082 kg + 658.5
≥ 60			

[a]Weight is expressed in kg. Predictive equations for children and adolescents are presented for the sake of completeness. BMR (basal metabolic rate): The minimal rate of energy expenditure compatible with life. It is measured in the supine position under standard conditions of rest, fasting, immobility, thermoneutrality and mental relaxation.
[b]standard error of estimate.
Source: Adapted from Food and Agriculture Organization of the United Nations, United Nations University, World Health Organization. 2004. Human Energy Requirements: Report of a Joint FAO/WHO/UNU Expert Consultation, (Rome, Food and Agriculture Organization of the United Nations).

Table 1.14. The energetics of a Northern Eem Neanderthal society 125,000 years ago[a]

Example of Time Use and Corresponding Rate of Monthly Average Energy Use (W/cap)	4 Males	1 Male	1 Female	4 Females
Hunt: tracking down prey (8 h, 1 day in month)	2.04	2.04	1.71	
Hunt: prey killing (1 h, 1 day in month)	0.45	0.45	0.37	
Hunt: parting mammoth, drying, (3 h, 1 day in month)	0.96	0.96	0.8	
Hunt: eat, sleep, rest at hunt site (12 h, 1 day in month)	1.69	1.69	1.41	
Hunt: eat, watch, cut, scrape, sleep (24 h, 12 days in month)		44.16	36.96	
Hunt: carrying meat back (10 h, 7 or 1 days in month)	26.83	3.83	3.21	
Hunt: sleep, rest at home (14 h, 7 or 1 days in month)	13.77	1.97	1.65	
Hunt: returning to hunt site from home (8 h, 6 days in month)	12.27			
Hunt: eat, sleep, rest at hunt site (16 h, 6 days in month)	14.72			
Home: wood cutting (8 h, 5 days in month)	15.33			
Home: stone flaking, tools construction, clothes making (8 h, 11, 16, 30 d)	22.49	32.71	27.38	61.33
Home: fire attention, child rearing, food prep., leisure, eat (8 h, 16, 30 d)	22.9	22.9	19.16	42.93
Home: sleep (8 h, 16 or 30 days in month)	16.36	16.36	13.69	25.67
Monthly average energy expenditure, adult humans (W/cap)	149.81	127.06	106.35	129.93

Continued

Table 1.14.   Continued

Summary:	Male	Female		
Average adult minimum energy requirement (W/cap)	145.26	125.22		
Total average adult energy requirement, W for whole group of 10 adults	1352.37			
Children's average energy requirement (W for whole group of 15 children)	400			
Equivalent meat intake (loss 30%), W and GJ per month for whole group of 25 people	2503 W or 6.5 GJ/month			
Equivalent meat intake (loss 30%), in kg per month for whole group of 25 people	813 kg/month			
Fires: 5 cooking fires 8 h, 30 d (346 kg dry wood), average over month	1667 W			
Fires: large outdoor fire 8 h, 30 d (622 kg dry wood)	3000 W			
Fires: Possible fire at hunt site 12 h, 14 days in month (168 kg)	810 W			

[a]Based on a location in Northern Europe with an average temperature of about 8° C. These are etimates of activities requiring energy use beyond the basic metabolic rate (which for Neanderthal males averaged 92 W, for females 77 W), based on a group with 10 adult members and 15 children. Activities include hunting, wood provision, and tool making; fires were used for cooking and heating the cave or hut used for dwelling; but without woolen covers and some clothes and footwear, survival at Northern latitudes would not be possible.
Source: Adapted from Sørensen, Bent. 2011. Renewable Energy (Fourth Edition), (Boston, Academic Press); based on data from B. Sørensen, Energy use by Eem Neanderthals, Journal of Archaeological Science, 36 (2009), pp. 2201–2205.

Table 1.15.   Sustainable power of individual animals in good condition

Animal	Typical weight kN (kgf)	Pull-weight ratio	Typical pull N (kgf)	Typical working speed m/s	Power output W	Working hours per day	Energy output per day MJ
Ox	4.5 (450)	0.11	500 (50)	0.9	450	6	10
Buffalo	5.5 (50)	0.12	650 (65)	0.8	520	5	9.5
Horse	4.0 (400)	0.13	500 (50)	1	500	10	18
Donkey	1.5 (150)	0.13	200 (20)	1	200	4	3
Mule	3.0 (300)	0.13	400 (40)	1	400	6	8.5
Camel	5.0 (500)	0.13	650 (65)	1	650	6	14

Source: Adapted from Carruthers, I. Rodriquez, M. 1992. Tools for Agriculture, a guide to appropriate equipment for small holder farmers. I.T., C.T.A., Intermediate Technology Publication, U.K.

Table 1.16. Comparison of metabolic rates of endotherms and ectotherms of similar body mass

Species	Common name	Endotherm (ENDO)/ ectotherm (ECTO)	Temperature (°C)	Body mass (g)	Metabolic rate(ml O_2 h^{-1})
Dipsosaurus dorsalisa	Desert iguana	ECTO	25	35	1.68
Dipsosaurus dorsalisa	Desert iguana	ECTO	30	35	2.45
Dipsosaurus dorsalisa	Desert iguana	ECTO	35	35	5.25
Dipsosaurus dorsalisa	Desert iguana	ECTO	40	35	6.3
Notoryctes caurinusb	North-western marsupial mole	ENDO	30.8	34	21.4
Phyllostomus discolorc	Pale spear-nosed bat	ENDO	34.6	33.5	11.1
Gerbillus allenbyid	Allenby's gerbil	ENDO	36.3	35.3	38.8
Acanthodactylus erythruruse	Fringe-toed lizard	ECTO	20	9	1.17
Acanthodactylus erythruruse	Fringe-toed lizard	ECTO	25	9	1.62
Acanthodactylus erythruruse	Fringe-toed lizard	ECTO	30	9	2.25
Acanthodactylus erythruruse	Fringe-toed lizard	ECTO	35	9	3.15
Crocidura crosseif	Crosse's shrew	ENDO	34.3	10.2	22.4
Perognathus longimembrisg	Little pocket mouse	ENDO	34.7	8.9	9.5
Pteronotus davyih	Davy's naked-backed bat	ENDO	38.8	9.4	24.4

For endotherms, basal metabolic rate is shown. For ectotherms, standard metabolic rate over a range of temperatures is shown.
Source: Adapted from Labocha, M.K., J.P. Hayes, Endotherm, In: Sven Erik Jorgensen and Brian Fath, Editor(s)-in-Chief, Encyclopedia of Ecology, (Oxford, Academic Press), Pages 1270-1276.

Table 1.17.   Basal metabolic rate (BMR) of some mammals over a large range of body mass

Species	Common name	Body mass(g)	BMR (ml O_2 h^{-1})	Mass-specific BMR (ml O_2 g^{-1}h^{-1})
Sorex minutus	Pygmy shrew	3.3	28.4	8.6
Peromyscus maniculatus	Deer mouse	14.9	44.4	3
Clethrionomys glareolus	Bank vole	20.9	52.5	2.5
Petrodromus tetradactylus	Four-toed elephant shrew	206.1	179.5	0.87
Erinaceus europaeus	European hedgehog	750	337.5	0.45
Fossa fossa	Fanaloka	2 260.0	904	0.4
Felis pardalis	Ocelot	10 416.0	3 229.0	0.31
Antilocapra americana	Pronghorn	37 800.0	9 318.0	0.25
Panthera onca	Jaguar	68 900.0	12 402.0	0.18
Alces alces	Moose	325 000.0	51 419.0	0.16

BMR is the metabolic rate in the absence of physical activity in post-absorptive animals, within the zone of thermal neutrality, and during the inactive phase of the normal circadian cycle.
Source: Adapted from Labocha, M.K., J.P. Hayes, Endotherm, In: Sven Erik Jorgensen and Brian Fath, Editor(s)-in-Chief, Encyclopedia of Ecology, (Oxford, Academic Press), Pages 1270-1276.

Table 1.18.   Maximum energy reserve size and fasting endurance of food- and fat-storing hibernators

Storage Form and Species	Body mass (g)	Resting Metabolic Rate (kJ/day)	Maximum Reserve Size (g)/Body Mass g	(g)	kJ	Euthermic Endurance (days)
Fat:						
Little brown bat, Myotis lucifugus	10	6	3.8	0.39	151	25
Jumping mouse, Zapus princeps	36	26	15	0.4	580	23
Arctic ground squirrel, Spermophilus parryii	985	251	473	0.48	18,800	75
Woodchuck, Marmota monax	4,900	578	2,840	0.4	112,884	195
Brown bear, Ursus arctos	237,400	22,408	135,600	0.44	5,389,829	241
Food:						
Pocket mouse, Perognathus parvus	24	19	4,400	1342	32,217	1,703
Long-tailed hamster, Cricetulus triton	73	43	35,000	3511	256,270	5,955
Eastern chipmunk, Tamias striatus	100	70	12,200	893	89,328	1,276
European hamster, Cricetus cricetus	362	109	90,000	1092	395,388	3,616

Source: Hadapted from Humphries, M. M., Thomas, D. W., & Kramer, D. L. 2003. The Role of Energy Availability in Mammalian Cost-Benefit Approach. Physiological & Biochemical Zoology, 76(2), 165.

Table 1.19. Thermal resistance, *s*, of animals[a]

Tissue	s cm⁻¹	
	Vaso-constricted	Dialated
Steer	1.7	0.5
Man	1.2	0.3
Calf	1.1	0.5
Pig (3 months)	1.0	0.6
Down sheep	0.9	0.3
Coats	s cm⁻¹ per cm depth	percent of still air
Still air	4.7	100
Red fox	3.3	70
Lynx	3.4	65
Skunk	3	64
Husky dog	2.9	62
Merino sheep	2.8	60
Down sheep	1.9	40
Blackfaced sheep	1.5	32
Cheviot sheep	1.5	32
Ayrshire cattle: flat coat	1.2	26
Ayrshire cattle: erect coat	0.8	
Galloway cattle	0.9	19

[a]s is the resistance to sensible heat loss. A resistance of of 100 s m⁻¹ (or 1 S cm⁻¹) is equivalent to insulation of 0.078 K m² W⁻¹.
Source: Adapted from Monteith, J.L. and M.H. Unsworth. 1990. Principles of Environmental Physics (Second Edition), (London, Edward Arnold).

Table 1.20. Energy Content of Food

Food	Kcal/100 g
Fruits	
Apples	52
Apricots	48
Bananas	89
Figs	74
Grapefruit	32
Guavas	51
Mangos	65
Melons, cantaloupe	34
Oranges	47
Strawberries, raw	32
Tomatoes, red	18
Vegetables	
Asparagus, raw	20
Broccoli, cooked	35
Carrots, raw	41
Chickpeas, cooked	164

Continued

Table 1.20. Continued

Food	Kcal/100 g
Corn, sweet, yellow, raw	86
Cucumber, peeled, raw	12
Lentils, cooked	101
Lettuce, romaine, raw	17
Mushrooms, raw	22
Onions, raw	42
Peas, edible-podded, raw	42
Peppers, sweet, red, raw	26
Pickles, cucumber, dill	18
Potato, flesh and skin, raw	77
Potatoes, russet, raw	79
Meats and proteins	
Beans, black, cooked	132
Beans, navy	67
Beef	230–400
Catfish, channel, wild, raw	95
Caviar	252
Chicken	200-500
Cod, atlantic, raw	82
Crab, blue, raw	87
Flatfish (flounder and sole), raw	91
Herring, atlantic, raw	158
Lamb	150–500
Mixed nuts, dry roasted	594
Peanut butter	635
Pork, cooked	418
Salmon, Atlantic, wild, raw	142
Shrimp, raw	106
Soybeans, cooked	141
Tofu, firm	77
Turkey	205
Dairy	
Cheese, brie	334
Cheese, cheddar	403
Cheese, swiss	380
Egg, whole, raw, fresh	147
Milk, whole, 3.25% milkfat	60
Yogurt, plain whole milk	61
Fats and Oils	
Butter	717
Margarine	719
Olive oil	884
Grains	
Bread, french	274
Bread, rye	259
Bread, whole-wheat	246

Table 1.20. Continued

Food	Kcal/100 g
Cereal, breakfast	300–450
Cookies, chocolate chip	497
Crackers, matzo	351
Rice noodles, cooked	109
Rice, white, cooked	130
Spaghetti, cooked	141
Alcohol	
Beer	33
Whiskey, 86 proof	250
Wine	77

Source: U.S. Department of Agriculture.

Table 1.21. Digestability, Heat of Combustion, and net physiologic energy value of proteins, lipids, and carbohydrates

Food Group	Digestability[a] (%)	Heat of Combustion (kCal $\times$ g^{-1})	Net Energy[b] (kCal $\times$ g^{-1})
Protein			
Animal food	97	5.65	4.27
Meats, fish	97	5.65	4.27
Eggs	97	5.75	4.37
Dairy products	97	5.65	4.27
Vegetable food	85	5.65	3.74
Cereals	85	5.8	3.87
Legumes	78	5.7	3.47
Vegetables	83	5	3.11
Fruits	85	5.2	3.36
Average protein	92	5.65	4.05
Lipid			
Meat and eggs	95	9.5	9.03
Dairy products	95	9.25	8.79
Animal food	95	9.4	8.93
Vegetable food	90	9.3	8.37
Average lipid	95	9.4	8.93
Carbohydrate			
Animal food	98	3.9	3.82
Cereals	98	4.2	4.11
Legumes	97	4.2	4.07
Vegetables	95	4.2	3.99
Fruits	90	4	3.6
Sugars	98	3.95	3.87
Vegetable food	97	4.15	4.03
Average carbohydrate	97	4.15	4.03

[a]the percentage of a foodstuff taken into the digestive tract that is absorbed into the body.
[b]net physiologic energy values are computed as the coefficient of digestibility x heat of combustion adjusted for energy loss in urine.
Source: Adapted from McArdle,William D, Frank I. Katch, Victor L. Katch. 2009. Exercise Physiology: Nutrition, Energy, and Human Performance (7th Edition), Lippincott Williams & Wilkins.

Biomass

Figures

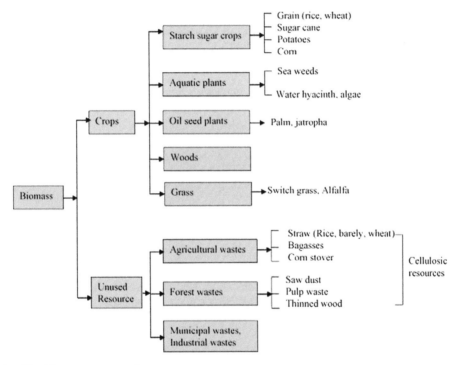

Figure 2.1. The biomass resource base.
Source: Naik, S.N., Vaibhav V. Goud, Prasant K. Rout, Ajay K. Dalai. 2010. Production of first and second generation biofuels: A comprehensive review, Renewable and Sustainable Energy Reviews, Volume 14, Issue 2, Pages 578-597.

Handbook of Energy. http://dx.doi.org/10.1016/B978-0-08-046405-3.00002-4

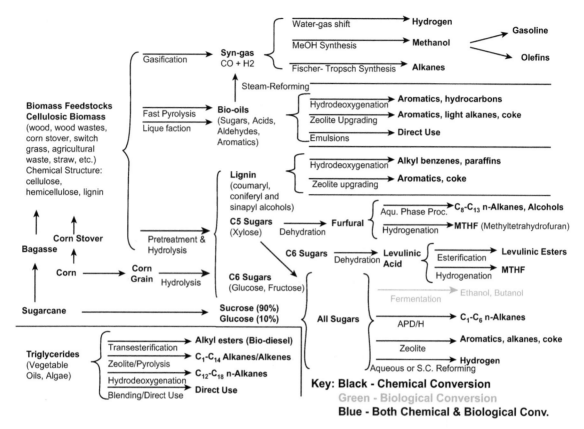

Figure 2.2. Pathways for biomass energy.
Source: Adapted from National Science Foundation. 2008. Breaking the Chemical and Engineering Barriers to Lignocellulosic Biofuels: Next Generation Hydrocarbon Biorefineries, Ed. George Huber. University of Massachusetts Amherst. National Science Foundation. Bioengineering, Environmental, and Transport Systems Division. Washington D.C.

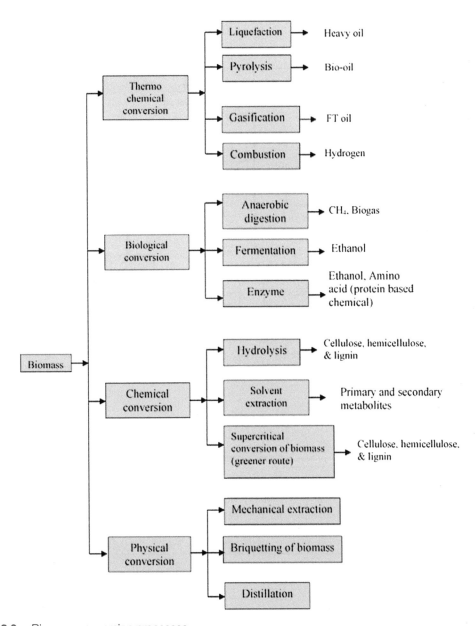

Figure 2.3. Biomass conversion processes.
Source: Naik, S.N., Vaibhav V. Goud, Prasant K. Rout, Ajay K. Dalai. 2010. Production of first and second generation biofuels: A comprehensive review, Renewable and Sustainable Energy Reviews, Volume 14, Issue 2, Pages 578-597.

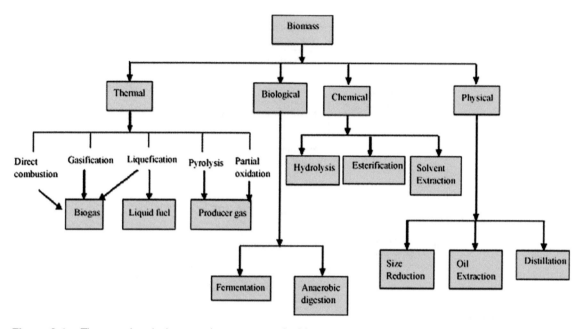

Figure 2.4. Thermo-chemical conversion processes for biomass.
Source: Naik, S.N., Vaibhav V. Goud, Prasant K. Rout, Ajay K. Dalai. 2010. Production of first and second generation biofuels: A comprehensive review, Renewable and Sustainable Energy Reviews, Volume 14, Issue 2, Pages 578-597.

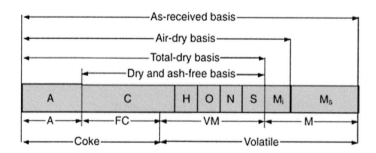

A ash O oxygen M_i inherent moisture H hydrogen
N nitrogen M_s surface moisture C carbon S sulfur

Figure 2.5. Bases for expressing fuel composition of biomass.
Source: Basu, Prabir. 2010. Biomass Gasification and Pyrolysis, (Boston, Academic Press).

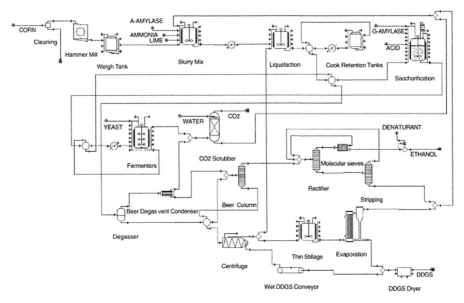

Figure 2.6. Simplified flow diagram of the dry-grind ethanol from corn process, the most widely used method in the U.S. for generating fuel ethanol by fermentation of grain.
Source: Kwiatkowski, Jason R., Andrew J. McAloon, Frank Taylor, David B. Johnston. 2006. Modeling the process and costs of fuel ethanol production by the corn dry-grind process, Industrial Crops and Products, Volume 23, Issue 3, Pages 288-296.

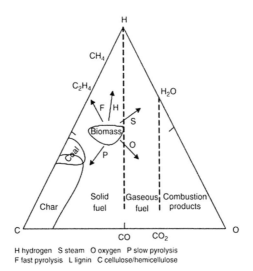

Figure 2.7. C-H-O ternary diagram of biomass showing the gasification process. The three corners of the triangle represent pure carbon, oxygen, and hydrogen—that is, 100% concentration. Points within the triangle represent ternary mixtures of these three substances. The side opposite to a corner with a pure component (C, O, or H) represents zero concentration of that component. For example, the horizontal base opposite to the hydrogen corner represents zero hydrogen—that is, binary mixtures of C and O.
Source: Basu, Prabir. 2010. Biomass Gasification and Pyrolysis, (Boston, Academic Press).

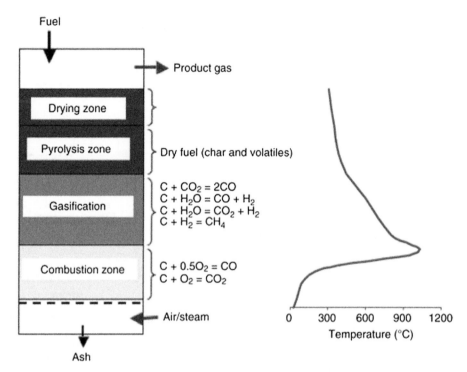

Figure 2.8.   Stages of gasification in an updraft gasifier.
Source: Basu, Prabir. 2010. Biomass Gasification and Pyrolysis, (Boston, Academic Press).

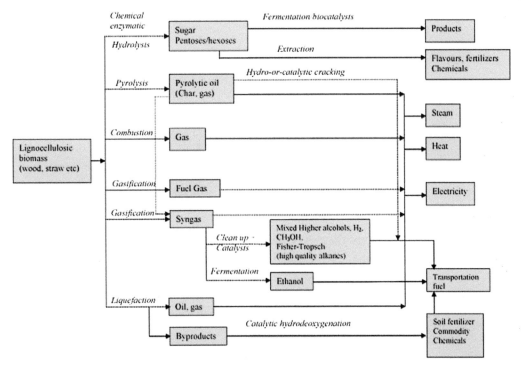

Figure 2.9. Forest based and lignocellulosic biorefinery.
Source: Naik, S.N., Vaibhav V. Goud, Prasant K. Rout, Ajay K. Dalai. 2010. Production of first and second generation biofuels: A comprehensive review, Renewable and Sustainable Energy Reviews, Volume 14, Issue 2, Pages 578-597.

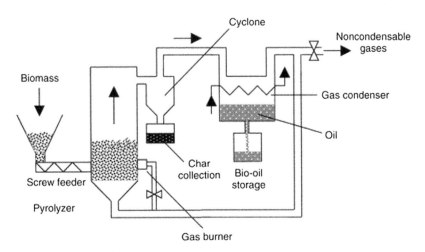

Figure 2.10. Simplified layout of a biomass pyrolysis plant.
Source: Basu, Prabir. 2010. Biomass Gasification and Pyrolysis, (Boston, Academic Press).

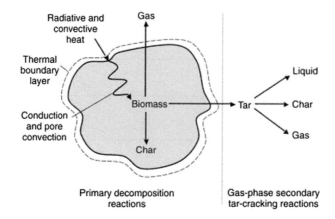

Figure 2.11. Pyrolysis in a biomass particle.
Source: Basu, Prabir. 2010. Biomass Gasification and Pyrolysis, (Boston, Academic Press).

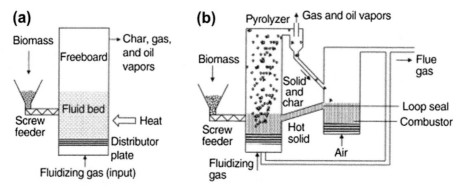

Figure 2.12. Biomass pyrolyzer designs: (a) bubbling fluidized bed, (b) circulating fluidized bed.
Source: Basu, Prabir. 2010. Biomass Gasification and Pyrolysis, (Boston, Academic Press).

Charts

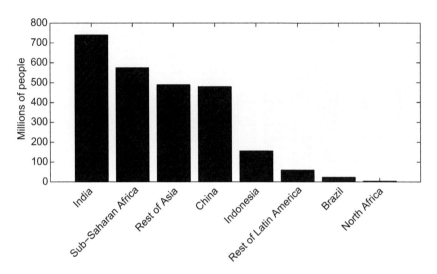

Chart 2.1. Number of people using traditional biomass, by region, 2009.
Source: Data from International Energy Agency (IEA), Energy statistics database, <http://www.iea.org/stats/index.asp>.

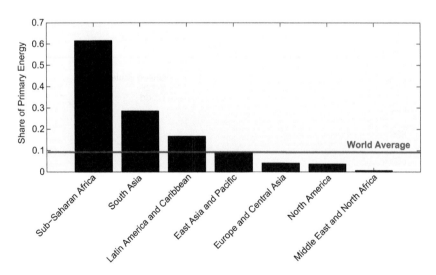

Chart 2.2. Biomass share of total primary energy supply, by region, 2009.
Source: Data from International Energy Agency (IEA), Energy statistics database, <http://www.iea.org/stats/index.asp>.

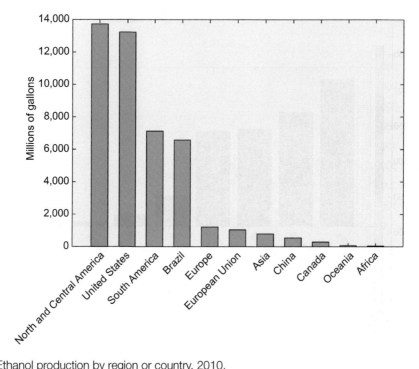

Chart 2.3. Ethanol production by region or country, 2010.
Source: Data from Oak Ridge National Laboratory, Transportation Energy Data Book: Edition 30, *<http://cta.ornl.gov/data/index.shtml>.*

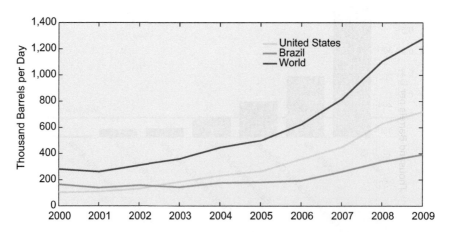

Chart 2.4. Fuel ethanol consumption in the United States, Brazil, and the world, 2000-2009.
Source: Data from International Energy Agency (IEA), Energy statistics database, <http://www.iea.org/stats/ *index.asp>.*

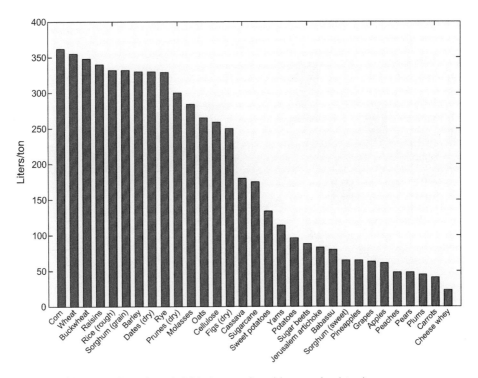

Chart 2.5. Potential fermentation ethanol yields from various biomass feedstocks.
Source: Data from Klass, Donald L. 1998. Biomass for Renewable Energy, Fuels, and Chemicals, (San Diego, Academic Press).

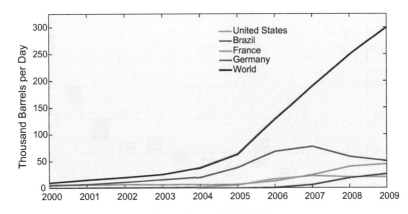

Chart 2.6. Biodiesel use in leading consuming nations, 2000-2009.
Source: Data from International Energy Agency (IEA), Energy statistics database, <http://www.iea.org/stats/index.asp>.

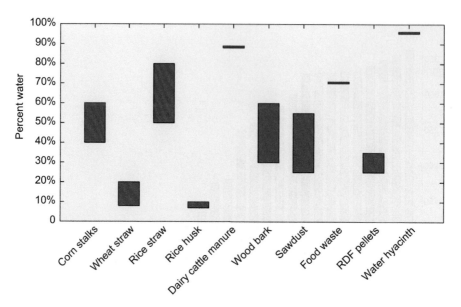

Chart 2.7. Moisture content of some biomass feedstocks.
Source: Data from Klass, Donald L. 1998. Biomass for Renewable Energy, Fuels, and Chemicals, (San Diego, Academic Press).

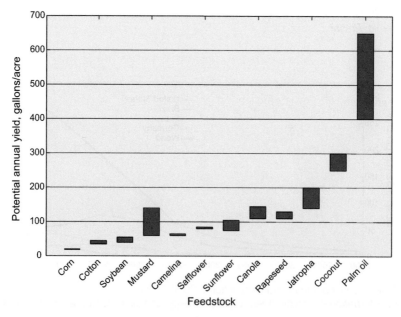

Chart 2.8. Potential biodiesel yield from selected triglyceride feedstocks.
Source: Data from Hoekman, S. Kent, Amber Broch, Curtis Robbins, Eric Ceniceros, Mani Natarajan. 2012. Review of biodiesel composition, properties, and specifications, Renewable and Sustainable Energy Reviews, Volume 16, Issue 1, Pages 143-169.

Tables

Table 2.1. Categories of biomass resources

Feedstock type	Definitions	Resources
Sugars/starches	Traditional agricultural crops suitable for fermentation using first generation technologies, some food processing residues are sugar and starch materials	Agricultural crops (sugars/starches), food processing residues containing residual sugars
Lignocellulosic biomass	Clean woody and herbaceous materials from a variety of sources, including clean urban biomass that is generally collected separately from the municipal waste stream (wood from the urban forest, yard waste, used pallets)	Agricultural residues, cellulosic energy crops, food processing residues, forest residues, mill residues, urban wood wastes, yard wastes
Bio-oils	Traditional edible and non-edible oil crops and waste oils suitable for conversion to biodiesel	Agricultural and forestry oil bearing crops and trees, waste oils/fats/grease
Solid wastes	Primarily lignocellulosic biomass, but that may be contaminated (e.g., construction and demolition woods) or co-mingled with other biomass types	Municipal solid waste, construction and demolition wood, food wastes, non-recycled paper, recycled materials
Other wastes	Other biomass wastes that are generally separate from the solid waste stream which include biogas and landfill gas	Animal waste, waste from wastewater treatment biogas and landfill gas

Source: Adapted from Chandra, R., H. Takeuchi, T. Hasegawa. 2012. Methane production from lignocellulosic agricultural crop wastes: A review in context to second generation of biofuel production, Renewable and Sustainable Energy Reviews, Volume 16, Issue 3, Pages 1462-1476.

Table 2.2. Proximate properties of common agricultural lignocellulosic biomass

Biomass	Volatile fraction in dry matter, %	Fixed carbon, %	Ash, %	Heating value, MJ/kg dry wt.
Maize stover	75.2–93.2	19.3	6.9–10.3	16.2–16.5
Sugarcane bagasse	70.9	7	14.7–22.1	10.0–14.3
Sugarcane leaves	77.4	14.9	7.7	17.4
Wheat straw	79.6–91.3	11.7–16.8	4.1–10.5	16.8–18.9
Rice straw	69.3–95	11.8–17.3	14.7–16.2	14.5–15.3
Rice husk	59.5–75.7		17.1–24.3	12.3–16.5

Source: Adapted from Chandra, R., H. Takeuchi, T. Hasegawa. 2012. Methane production from lignocellulosic agricultural crop wastes: A review in context to second generation of biofuel production, Renewable and Sustainable Energy Reviews, Volume 16, Issue 3, Pages 1462-1476.

Table 2.3.   Area, total living biomass, and mean biomass of the world's major terrestrial ecosystems

Ecosystem type	Area (10^6 ha)	Total biomass (Pg)	Mean biomass (Mg ha^{-1})
Tropical forests	1850	350	190
Temperate forests	2450	185	75
Arctic tundra	560	4	7
Mediterranean shrublands	280	34	120
Croplands	1350	8	6
Tropical savannas and grasslands	2760	160	57
Temperate grasslands	1500	12	8
Deserts	2770	20	7
Ice	1550	0	0
Total	15070	770	50

Source: Houghton, R.A. 2008. Biomass, In: Sven Erik Jorgensen and Brian Fath, Editor(s)-in-Chief, Encyclopedia of Ecology, (Oxford, Academic Press), Pages 448-453.

Table 2.4.   Primary production and biomass estimates for the world

Ecosystem	Area (10^{12} m^2)	Mean plant biomass (kg C/m^2)	Carbon in vegetation (10^{15}g)	Mean net primary production (g C/ m^2/yr)	Net primary productivity (10^{15}g C/yr)
Tropical wet and moist forest	10.4	15	156	800	8.3
Tropical dry forest	7.7	6.5	49.7	620	4.8
Temperate forest	9.2	8	73.3	650	6
Boreal forest	15	9.5	143	430	6.4
Tropical woodland and savanna	24.6	2	48.8	450	11.1
Temperate steppe	15.1	3	43.8	320	4.9
Desert	18.2	0.3	5.9	80	1.4
Tundra	11	0.8	9	130	1.4
Wetland	2.9	2.7	7.8	1300	3.8
Cultivated land	16	1	21.5	760	12.1
Rock and ice	15.2	0	0	0	0
Global total	145.2		558.8		60.2

Source: Adapted from Amthor, J.S. and members of the Ecosystems Working Group (1998) Terrestrial Ecosystem Responses to Global Change: a research strategy. ORNL Technical Memorandum 1998/27, Oak Ridge National Laboratory, Oak Ridge, Tennessee. 37 pp.

Table 2.5. **Estimates of global bioenergy potentials**

Study	Energy Flow (EJ/year)	Year
Bioenergy crops and residues, excluding forestry[a]	58–61	2050
World Energy Assessment mid-term potential[b]	94–280	2050
Berndes et al. review of mid-term potential[c]	35–450	2050
Fischer/Schrattenholzer mid-term potential[d]	370–450	2050
Hoogwijk et al., 2003[e]	33–1135	2050
IPCC-SRES scenarios mid-term[f]	52–193	2050
Bioenergy potential on abandoned farmland[g]	27–41	2050
Bioenergy potentials in forests[h]	0–71	2050
Surplus agricultural land (not needed for food & feed)[i]	215–1272	2050
Bioenergy crops (second generation)[j]	34–120	2050

[a]Erb, Karl-Heinz, Helmut Haberl, Fridolin Krausmann, Christian Lauk, Christoph Plutzar, Julia K. Steinberger, Christoph Müller, Alberte Bondeau, Katharina Waha, Gudrun Pollack. Eating the Planet: Feeding and fuelling the world sustainably, fairly and humanely – a scoping study, Working Paper 116, Institute of Social Ecology, Klagenfurt University, November 2009.

[b]Turkenburg, W.C., 2000. Renewable Energy Technology. In: Goldemberg, J. (Eds.), World Energy Assessment: Energy and the challenge of sustainability. United Nations Development Programme (UNDP), United Nations Department of Economic and Social Affairs, World Energy Council (WEC), New York, pp. 219–272.

[c]Berndes, G., Hoogwijk, M., van den Broek, R., 2003. The contribution of biomass in the future global energy supply: a review of 17 studies. Biomass and Bioenergy, 25, 1–28.

[d]Fischer, G., Schrattenholzer, L., 2001. Global bioenergy potentials through 2050. Biomass and Bioenergy, 20, 151-159.

[e]Hoogwijk, M., Faaij, A., Broek, R.v.d., Berndes, G., Gielen, D., Turkenburg, W., 2003. Exploration of the ranges of the global potential of biomass for energy. Biomass and Bioenergy, 25, 119–133.

[f]Nakicenovic, N., Swart, R., 2000. Special Report on Emission Scenarios. Intergovernmental Panel on Climate Change (IPCC), Cambridge University Press, Cambridge.

[g]Campbell, J.E., Lobell, D.B., Genova, R.C., Field, C.B., 2008. The Global Potential of Bioenergy on Abandoned Agriculture Lands. Environmental Science & Technology, 42, 5791–5795.

[g]Field, C.B., Campbell, J.E., Lobell, D.B., 2008. Biomass energy: the scale of the potential resource. Trends in Ecology & Evolution, 23 (2), 65–72.

[h]Smeets, E.M.W., Faaij, A.P.C., 2007. Bioenergy potentials from forestry in 2050. Climatic Change, 81 (3), 353–390.

[i]Smeets, E.M.W., Faaij, A.P.C., Lewandowski, I.M., Turkenburg, W.C., 2007. A bottom-up assessment and review of global bio-energy potentials to 2050. Progress in Energy and Combustion Science, 33 (1), 56–106.

[j]WBGU, 2008. Welt im Wandel. Zukunftsfähige Bioenergie und nachhaltige Landnutzung. Wissenschaftlicher Beirat der Bundesregierung Globale Umweltveränderungen (WBGU), Berlin.

Table 2.6. Energy potential from biofuel crops using current technologies and future cellulosic technologies

Feedstock type	Feedstock mass 2002	Gross biofuel conversion[a]	Gross biofuel energy[b]	Net energy balance ratio[c]	Net biofuel energy[d]
	(Mt y⁻¹)	(GJ/ton)	(EJ y⁻¹)	(Output/Input)	(EJ y⁻¹)
Corn kernel	696	8	5.8	1.25	1.2
Sugar cane	1324	2	2.8	8	2.4
Cellulosic biomass	–	6	–	5.44[e]	–
Soy oil	35	30	1	1.93	0.5
Palm oil	36	30	1.1	9	1
Rape oil	17	30	0.5	2.5	0.3

[a]Useful biofuel energy per ton of crop for conversion into biofuel (1GJ = 10⁹J).
[b]Product of feedstock mass and gross biofuel conversion (1EJ = 10¹⁸J).
[c]Ratio of the energy captured in biomass fuel to the fossil energy input.
[d]Energy yield above the fossil energy invested in growing, transporting and manufacturing, calculated as gross biofuel energy × (net energy-balance ratio −1)/net energybalance ratio.
[e]Not yet achievable at the industrial scale. Calculated assuming energy for biorefining does not come from fossil fuels.
Source: Adapted from Field, Christopher B., J. Elliott Campbell, David B. Lobell. 2008 Biomass energy: the scale of the potential resource, Trends in Ecology & Evolution, Volume 23, Issue 2, Pages 65-72.

Table 2.7. Production of biofuels, top 20 nations, 2010 (thousand barrels/day)

Country	2000	2005	2010	Change 2000-10
United States	105.5	260.6	887.6	741.0%
Brazil	183.9	276.4	527.3	186.8%
Europe	17.1	76.8	248.3	1352.1%
Germany	4.3	35.8	62.0	1341.9%
France	7.9	10.9	55.0	596.2%
China	0.0	21.5	43.0	--
Argentina	0.1	0.2	38.1	38000.0%
Canada	3.7	4.6	26.4	613.5%
Spain	1.6	8.2	24.0	1400.0%
Thailand	0.0	1.6	18.5	--
Italy	1.6	7.8	16.5	931.3%
Belgium	0.0	0.0	13.5	--
Colombia	0.0	0.5	12.0	--
Poland	0.0	3.2	11.0	--
Netherlands	0.0	0.1	9.5	--
Indonesia	0.0	0.2	8.0	--
Australia	0.0	0.6	7.9	--
Austria	0.4	1.6	7.7	1825.0%
Sweden	0.0	1.6	7.5	--
India	2.9	3.9	7.0	141.4%
World	314.6	656.3	1,855.6	489.9%

Source: United States Department of Energy, Energy Information Administration, International Energy Statistics,
<http://www.eia.gov/countries/data.cfm>, accessed 21 June 2012.

Table 2.8. Proximate analysis and pyrolysis characteristics of selected biomass plant species

Biomass	Fixed carbon (dry wt %)	Volatile matter (dry wt %)	Ash (dry wt %)	Range (°C)	Devolatilzation Max rate at (°C)	Max rate (wt %/°C)	Emitted at 320-500°C (wt%)
Bagasse	16.9	75.1	8	280–510	385	0.53	54
Bamboo dust	15.6	75.3	9.1	240–600	270	0.395	37.8
Cotton stalks	22.4	70.9	6.7	280–520	390	0.7	47.5
Corn cobs	16.2	80.2	3.6	370–710	400	0.49	45.6
Groundnut shell	25	68.1	6.9	300–720	505	0.67	49.3
Jute sticks	19	75.3	5.7	220–500	390	1.06	57
Mustard shells	14.5	70.1	15.4	300–550	370	0.67	42.5
Pigeon pea	14.8	83.5	1.8	290–650	390	0.48	54
Pine needles	26.1	72.4	1.5	320–680	410	0.38	44.5
Prickly acacia	22.3	77	0.6	270–680	340	0.35	44.1
Rice husks	19.9	60.6	19.5	340–510	390	0.66	48
Sal seed leaves	20.2	60	19.7	200–650	440	0.14	22.5

Source: Adapted from Klass, Donald L. 1998. Biomass for Renewable Energy, Fuels, and Chemicals, (San Diego, Academic Press).

Table 2.9. Characteristics of gaseous biofuels

Property	Methane	Propane	Syngas	Dimethyl ether (DME)	Hydrogen
Chemical formula	CH_4	C_3H_8	$CH_{2.68} O_{1.26}$	CH_3OCH_3	H_2
Density (kg/m³) at 15 °C	717	490	765–785	670	70.8
Boiling point (°C)	−161	−42	−192	−25	−252.8
Vapor pressure at 0 °C (bar)	246	9.3	-	6.1	-
Flammable limits % in air	5–15	2.1–9.4	5–32	3.4–17	4–75
Ignition temperature (°C)	537	470	-	235	500
Max. burning velocity (cm/s)	37	43	-	50	289
Stoichiometric air/fuel ratio	16.9	15.7	9.1	9	34.1
Lower calorific value (MJ/kg)	49	46.3	10–18	28.8	120
Cetane no.	0	5	70	55–60	-
Carbon (%, w/w)	75	82	21.8	52.2	-
N_2 (%, w/w)	-	-	1–3	0	-
O_2 (%, w/w)	-	-	73	34.8	-
H_2 (%, w/w)	25	18	4.87	13	100
Sulfur (%, w/w)	-	-	-	-	-

Source: Adapted from Gupta, K.K., A. Rehman, R.M. Sarviya. 2010. Bio-fuels for the gas turbine: A review, Renewable and Sustainable Energy Reviews, Volume 14, Issue 9, Pages 2946-2955.

Table 2.10. Characteristics of liquid biofuels

Properties	Diesel	JET-A	Straight vegetable oils	Biodiesel	Bioethanol	Biomethanol	Pyrolysis oil
Density (kg/m³)	827.4	807	900–940	860–900	794–810	796	984–1250
Kinetic viscosity (cSt at 40 °C)	1.7283	0.88	30–40	3.5–5	1.4–1.7	1.4–1.7	32–45
Flash point (°C)	44	38	230–280	120–180	13	11	56–130
Cloud point (°C)	–6	-	–4 to 12	–3 to –12	-	-	-
Pour point (°C)	–16	–47	–12 to 10	–15 to 5	–117	–161	–35 to –10
Lower calorific value (MJ/kg)	43	43.23	38–39	39–41	25–26	20	13–18
Ignition temperature	250	220	325–370	177	423	463	580
Cetane no.	45–55	55	37–42	48–60	8	5	10
Stoichiometric air/fuel ratio	14.6	14	13.8	13.8	9.79	6:01	34:01:00
Carbon (%, w/w)	80.33	80–83	76.11	77–81	52.2	37.5	32–48
H_2 (%, w/w)	14	10–14	-	12	13.1	12.6	7-8.5
N_2 (%, w/w)	1.76	-	0	0.03	-	-	<0.4
O_2 (%, w/w)	1.19	-	11	9–11	34.8	49.9	44–60
Sulfur (%, w/w)	<0.4	<0.4	0	<0.03	-	-	<0.05

Source: Adapted from Gupta, K.K., A. Rehman, R.M. Sarviya. 2010. Bio-fuels for the gas turbine: A review, Renewable and Sustainable Energy Reviews, Volume 14, Issue 9, Pages 2946-2955.

Table 2.11. Characteristics of selected biofuel feedstocks and fuels

		Cellulose (Percent)	Hemi-cellulose (Percent)	Lignin (Percent)	Extractives (Percent)	Ash (Percent)	Sulfur (Percent)	Potassium (Percent)	Ash melting temperature [some ash sintering observed] (C)	Cellulose fiber length (mm)	Chopped density at harvest (kg/m3)	Baled density [compacted bales] (kg/m3)
Bioenergy Feedstocks	Corn stover	30-38	19-25	17-21	3.3-11.9	9.8-13.5	0.06-0.1	b	b	1.5	b	b
	Sweet sorghum	27	25	11		5.5	b	b	b	b	b	b
	Sugarcane bagasse[a]	32-43	19-25	23-28	1.5-5.5	2.8-9.4	0.02-0.03	0.73-0.97	b	1.7	50-75	b
	Sugarcane leaves	b	b	b		7.7	b	b	b	b	25-40	b
	Hardwood	45	30	20		0.45	0.009	0.04	[900]	1.2	b	b
	Softwood	42	21	26		0.3	0.01	b	b	b	b	b
	Hybrid poplar	39-46	17-23	21-8	1.6-6.9	0.4-2.4	0.02-0.03	0.3	1,350	1-1.4	150 (chips)	b
	Bamboo	41-49	24-28	24-26	4.9-24.0	0.8-2.5	0.03-0.05	0.15-0.50	b	1.5-3.2	108	105-133
	Switchgrass	31-34	24-29	17-22		2.8-7.5	0.07-0.11	b	1,016	b	70-100	130-150 [300]
	Miscanthus	44	24	17		1.5-4.5	0.1	0.37-1.12	1,090 [600]	b	b	b
	Giant Reed	31	30	21	N/A	5-6	0.07	b	b	1.2	b	b
Liquid Biofuels	Bioethanol	N/A	N/A	N/A	N/A	b	<0.01	b	N/A	N/A	N/A	790[a]
	Biodiesel	N/A	N/A	N/A	N/A	<0.02	<0.05	<0.0001	N/A	N/A	N/A	875[a]
Fossil Fuels	Coal (low rank; lignite/sub-bituminous)	N/A	N/A	N/A	N/A	5-20	1.0-3.0	0.02-0.3	~1,300	N/A	N/A	700[a]
	Coal (high rank bituminous/anthracite)	N/A	N/A	N/A	N/A	1-10	0.5-1.5	0.06-0.15	~1,300	N/A	N/A	850[a]
	Oil (typical distillate)	N/A	N/A	N/A	N/A	0.5-1.5	0.2-1.2	b	N/A	N/A	N/A	700-900[a]

Notes:
N/A = Not Applicable.
[a]Bulk density.
[b]Data not available.
Source: Wright, Lynn, Bob Boundy, Bob Perlack, Stacy Davis, Bo Saulsbury. 2006. Biomass Energy Data Book, Edition 1, Oak Ridge National Laboratory (Oak Ridge, TN).

Table 2.12.   Common biomass densification technology feedstocks, and products

Machine	Feedstock Type	Moisture (wt %)	Densified bulk pressure Size (cm)	Density (kg/m³)	Moisture (wt %)
Impact press	Wood residues	15–17	Briquettes	990–1200	8–15
Extrusion press	Wood residues	10–20	Briquettes	1300–1400	8–15
Hydraulic press	Wood residues		Briquettes	590–800	8–15
Briquetting machine	Wood residues		Briquettes	>990	8–15
Pelleting machine	Wood residues	8–15	0.5–2.5 dia.	800	8–15
Biotruck 2000	Hay and straws		6 × 1.4 × 4	800–1200	22
Extruder	Hogged bark, some wood	56.5	5.7 dia.	1070	34.8
Extruder	Western hemlock sawdust	64.2	5.7 dia.	1100	36.5
Flat die press	Fine straws	10–20	0.6–2.0 dia.	450–650	10–15

Source: Klass, Donald L. 1998. Biomass for Renewable Energy, Fuels, and Chemicals, (San Diego, Academic Press).

Table 2.13.   Typical oil yields of various biomass sources

Crop	Oil yield (l/ha)
Corn	172
Soybean	446
Peanut	1,059
Canola	1,190
Rapeseed	1,190
Jatropha	1,892
Karanj (Pongamia pinnata)	2,590
Coconut	2,689
Oil palm	5,950
Microalgae (70% oil by wt.)	136,900
Microalgae (30% oil by wt.)	58,700

Source: Adapted from Singh, Jasvinder, Sai Gu. 2010. Commercialization potential of microalgae for biofuels production, Renewable and Sustainable Energy Reviews, Volume 14, Issue 9, Pages 2596-2610.

Table 2.14. Stages in charcoal formation

Temperature	Transformation
20 to 110° C	The wood absorbs heat as it is dried, giving off its moisture as water vapour (steam).
	The temperature remains at or slightly above 100° C until the wood is bone dry.
110 to 270° C	Wood starts to decompose giving off some CO, CO_2, acetic acid and methanol.
	Heat is absorbed.
270 to 290° C	Heat is evolved and breakdown continues spontaneously providing the wood is not cooled below this decomposition temperature.
	Mixed gases and vapour continue to be given off together with some tar.
290 to 400° C	The combustible gases CO, H_2 and CH_4 together with CO_2 and the condensible vapours: water, acetic acid, methanol, acetone, etc. and tars which begin to predominate as the temperature rises.
400 to 500° C	The transformation of the wood to charcoal is practically complete. The charcoal at this temperature still contains appreciable amounts of tar, perhaps 30% by weight trapped in the structure.
	To drive off this tar the charcoal is subject to further heat inputs to raise its temperature to about 500° C, thus completeing the carbonisation stage.

Source: Adapted from Panwar, N.L., Richa Kothari, V.V. Tyagi. 2012. Thermo chemical conversion of biomass – Eco friendly energy routes, Renewable and Sustainable Energy Reviews, Volume 16, Issue 4, Pages 1801-1816.

Table 2.15. Combustion data for some solid fuels[a]

Parameter	Pine wood	Kentucky bluegrass	Feedlot manure	Bituminous coal	Anthracite coal	Coke
Moisture, wt %	15	15	15	3.1	5.2	0.8
Higher heating value, MJ/kg	18.05	15.92	11.36	32.61	29.47	29.5
C/H wt ratio	8.2	7.8	6.6	16	33.6	106
Air/fuel wt ratio	5.37	5.51	3.97	10.81	9.92	10.09
Product CO_2, wt/wt fuel	1.9	1.68	1.29	2.94	2.96	3.12
Product H_2O, wt/wt fuel	0.56	0.53	0.47	0.49	0.22	0.07
N_2 from air, wt/wt fuel	4.85	4.97	3.58	3.83	7.58	7.73
CO_2 in dry flue gas, mol %	19.9	17.4	18.4	20	19.9	20.4
NO_2 in dry flue gas, mol %	0.032	1.55	1.13	0.21	0	0.266
SO_2 in dry flue gas, mol %	0	0.054	0.075	0.035	0.101	0.089

[a]Each biomass fuel is assumed to contain 15.0 wt % moisture. The overall assumptions are that combustion is complete, the ash and nitrogen air are inert, and all organic nitrogen and sulfur are oxidized to NO_2 or SO_2.
Source: Adapted from Klass, Donald L. 1998. Biomass for Renewable Energy, Fuels, and Chemicals, (San Diego, Academic Press).

Table 2.16.   Comparison of ultimate analysis (dry basis) of some biomass and other fossil fuels

Fuel	C (%)	H (%)	N (%)	S (%)	O (%)	Ash (%)	HHV[a] (kJ/kg)
Maple	50.6	6	0.3	0	41.7	1.4	19,958
Douglas fir	52.3	6.3	9.1	0	40.5	0.8	21,051
Douglas fir (bark)	56.2	5.9	0	0	36.7	1.2	22,098
Redwood	53.5	5.9	0.1	0	40.3	0.2	21,028
Redwood (waste)	53.4	6	0.1	39.9	0.1	0.6	21,314
Sewage sludge	29.2	3.8	4.1	0.7	19.9	42.1	16,000
Rice straw	39.2	5.1	0.6	0.1	35.8	19.2	15,213
Rice husk	38.5	5.7	0.5	0	39.8	15.5	15,376
Sawdust	47.2	6.5	0	0	45.4	1	20,502
Paper	43.4	5.8	0.3	0.2	44.3	6	17,613
MSW	47.6	6	1.2	0.3	32.9	12	19,879
Animal waste	42.7	5.5	2.4	0.3	31.3	17.8	17,167
Peat	54.5	5.1	1.65	0.45	33.09	5.2	21,230
Lignite	62.5	4.38	0.94	1.41	17.2	13.4	24,451
PRB coal	65.8	4.88	0.86	1	16.2	11.2	26,436
Anthracite	90.7	2.1	1	7.6	11.4	2.5	29,963
Petcoke	86.3	0.5	0.7	0.8	10.5	6.3	29,865

[a]HHV (high heating value): the amount of heat produced by the complete combustion of a unit quantity of fuel. The HHV is realized when all products of the combustion are cooled down to the temperature before the combustion, and when the water vapor formed during combustion is condensed.
Source: Adapted from Basu, Prabir. 2010. Chapter 2 - Biomass Characteristics, Biomass Gasification and Pyrolysis, (Boston, Academic Press), Pages 27-63.

Table 2.17.   Comparison of physical and chemical properties of bio-oil and three fossil fuels

Property	Bio-Oil	Heating Oil	Gasoline	Diesel
Heating value (MJ/kg)	18–20	45.5	44	42
Density at 15 °C (kg/m³)	1200	865	737	820–950
Flash point (°C)	48–55	38	40	42
Pour point (°C)	−15	−6	−60	−29
Viscosity at 40 °C (cP)	40–100 (25% water)	1.8–3.4 cSt	0.37–0.44	2.4
pH	2.0–3.0	–		
Solids (% wt)	0.2–1.0	–	0	0
Elemental Analysis (% weight)				
Carbon	42–47	86.4	84.9	87.4
Hydrogen	6.0–8.0	12.7	14.76	12.1
Nitrogen	<0.1	0.006	0.08	392 ppm
Sulfur	<0.02	0.2–0.7		1.39
Oxygen	46–51	0.04		
Ash	<0.02	<0.01		

cP: centipoise; cSt: centipoise. Values for gasoline and diesel are for a representative sample and can vary.
Source: Adapted from Basu, Prabir. 2010. Biomass Gasification and Pyrolysis, (Boston, Academic Press).

Table 2.18. Heat content ranges for various biomass fuels (dry weight basis[a])

Fuel type & source	English Higher heating value		Metric Higher heating value		Metric Lower heating value	
	Btu/lb	MBtu/ton	kJ/kg	MJ/kg	kJ/kg	MJ/kg
Agricultural Residues						
Corn stalks/stover	7,587–7,967	15.2–15.9	17,636–18,519	17.6–18.5	16,849–17,690	16.8–18.1
Sugarcane bagasse	7,450–8,349	14.9–16.7	17,317–19,407	17.3–19.4	17,713–17,860	17.7–17.9
Wheat straw	6,964–8,148	13.9–16.3	16,188–18,940	16.1–18.9	15,082–17,659	15.1–17.7
Hulls, shells, prunings	6,811–8,838	13.6–17.7	15,831–20,543	15.8–20.5		
Fruit pits	8,950–10,000	17.9–20.0				
Herbaceous Crops						
Miscanthus			18,100–19,580	18.1–19.6	17,818–18,097	17.8–18.1
Switchgrass	7,754–8,233	15.5–16.5	18,024–19,137	18.0–19.1	16,767–17,294	16.8–18.6
Other grasses			18,185–18,570	18.2–18.6	16,909–17,348	16.9–17.3
Bamboo			19,000–19,750	19.0–19.8		
Woody Crops						
Black locust	8,409–8,582	16.8–17.2	19,546–19,948	19.5–19.9	18,464	18.5
Eucalyptus	8,174–8,432	16.3–16.9	19,000–19,599	19.0–19.6	17,963	18.0
Hybrid poplar	8,183–8,491	16.4–17.0	19,022–19,737	19.0–19.7	17,700	17.7
Willow	7,983–8,497	16.0–17.0	18,556–19,750	18.6–19.7	16,734–18,419	16.7–18.4
Forest Residues						
Hardwood wood	8,017–8,920	16.0–17.5	18,635–20,734	18.6–20.7		
Softwood wood	8,000–9,120	16.0–18.24	18,595–21,119	18.6–21.1	17,514–20,768	17.5–20.8
Urban Residues						
Muncipal solid waste	5,644–8,542	11.2–17.0	13,119–19,855	13.1–19.9	11,990–18,561	12.0–18.6
Refuse-derived fuel	6,683–8,563	13.4–17.1	15,535–19,904	15.5–19.9	14,274–18,609	14.3–18.6
Newspaper	8,477–9,550	17–19.1	19,704–22,199	19.7–22.2	18,389–20,702	18.4–20.7
Corrugated paper	7,428–7,939	14.9–15.9	17,265–18,453	17.3–18.5	17,012	
Waxed cartons	11,727–11,736	23.5–23.5	27,258–27,280	27.3	25,261	

[a]This table attempts to capture the variation in reported heat content values (on a dry weight basis) in the US and European literature based on values in the Phyllis database, the US DOE/EERE feedstock database, and selected literature sources.

Source; Adapted from Wright, Lynn, Bob Boundy, Bob Perlack, Stacy Davis, Bo Saulsbury, 2006. Biomass Energy Data Book, Edition 1, Oak Ridge National Laborator,(Oak Ridge, TN).

Table 2.19. Typical proximate analyses and high heating values of selected biomass, coal, & peat

Category	Name	Type	Moisture range (wt%)	Orgnaic matter (dry wt %)	Ash (dry wt %)	High heating vale (MJ/ dry kg)
Waste	Cattle manure	Feedlot	20–70	76.5	23.5	13.4
	Actived biosolids	Sewage	90–97	76.5	23.5	18.3
	Primary biosolids	Sewage	90–98	73.5	26.5	19.9
	Refuse-derived fuel (RDF)	Municipal	15–30	86.1	13.9	12.7
	Sawdust	Woody	15–60	99	1	20.5
Herbaceous	Cassava	Tropical	20–60	96.1	3.9	17.5
	Euphoria lathyris	Warm season	20–60	92.7	7.3	19
	Kentucky bluegrass	Cool season	10–70	86.5	13.5	18.7
	Sweet sorghum	Warm season	20–70	91	9	17.6
	Switchgrass	Warm season	30–70	89.9	10.1	18
Aquatic	Griant brown kelp	Marine	85–97	54.2	45.8	10.3
	Water hyacinth	Fresh water	85–97	77.3	22.7	16
Woody	Black alder	Hardwood	30–60	99	1.0	20.1
	Cottonwood	Hardwood	30–60	98.9	1.1	19.5
	Eucalyptus	Hardwood	30–60	97.9	2.4	18.7
	Hybrid poplar	Hardwood	30–60	99	1.0	19.5
	Loblolly pine	Softwood	30–60	99.5	.5	20.3
	Redwood	Hardwood	30–60	99.8	.2	21
	Sycamore	Hardwood	30–60	98.9	1.1	19.4
Biomass Derivatives	Paper		3–12	94.0	6	17.6
	Pine bark	Softwood	5–30	97.1	2.9	20.4
	Rice straw		5–15	80.8	19.2	15.2
	Redwood chat		2–6	95.9	4.1	30.5
Coal	Illinois bituminous	Soft	5–10	91.3	8.7	28.3
	North Dakota lignite	Soft	5–15	89.6	10.4	14
Peat	Reed sedge	Young coal	70–90	92.3	7.7	20.8

Source: Adapted from Klass, Donald L. 1998. Biomass for Renewable Energy, Fuels, and Chemicals, (San Diego, Academic Press).

Table 2.20. Specific heat of biomass and related materials

Type	Specific Heat (kJ/kg.K)	Temperature (K)
Carbon	0.7	299–349
	1.6	329–1723
Cellulose	1.34	
Graphite	0.843	273–373
	1.621	329–1723
Wood (Oven dry, avg. 20 species)	1.374	273–379
Wood charcoal	0.843	273–273

Source: Adapted from Basu, Prabir. 2010. Biomass Gasification and Pyrolysis, (Boston, Academic Press).

Table 2.21. Heating values of stoichiometric air-fuel mixtures of pure alcohols, MTBE, isooctane, and gasoline

Fuel	Mol. wt.	Higher heating value[a]		Heating value of air-fuel mixtures[b]
		(MJ/kg)	(MJ/L)	(MJ/m³)
Methanol	32.04	22.33	17.7	3.643
Ethanol	46.07	29.77	23.5	3.724
1-Propanol	60.1	33.48	26.93	3.722
2-Propanol	60.1	33.08	25.98	3.678
1-Butanol	74.12	36.07	29.21	3.746
2-Methyl-1-propanol	74.12	36.05	28.88	3.743
2-Methyl-2-propanol	74.12	35.55	28.08	3.692
1-Pentanol	88.15	37.7	30.7	3.758
2-Methyl-2-butanol	88.15	37.27	30.15	3.715
Cyclohexanol	100.16	37.23	35.83	3.736
3-Methy-3-pentanol	102.18	37.99	31.48	3.663
Benzyl alcohol	108.14	34.62	36.07	3.751
1-Heptanol	116.2	39.81	32.72	3.769
3-Ethyl-3-pentanol	116.2	38.91	32.64	3.684
MTBE	88.15	38.12	28.23	3.8
Isooctane	114.23	47.79	33.07	3.745
Gasoline	112.21	47.2	34.0–36.8	3.787

[a]HHV (high heating value): the amount of heat produced by the complete combustion of a unit quantity of fuel. The HHV is realized when all products of the combustion are cooled down to the temperature before the combustion, and when the water vapor formed during combustion is condensed. The conditions for combustion are atmospheric pressure, 20° C, and product water in the liquid state. The higher heating value per unit volume is calculated from the density (20° C/4° C) and the higher heating value per unit mass.
[b]Calculated from the heats of combustion at 20° C and the mole percent of fuel in the stoichiometric air-fuel mixture.
Source: Adapted from Klass, Donald L. 1998. Biomass for Renewable Energy, Fuels, and Chemicals, (San Diego, Academic Press).

Table 2.22.   Standard heating value of constituents of typical product gas from biomass gasification

Gases	H_2	CO	CO_2	CH_4	C_2H_6	C_2H_4	C_2H_2	C_3H_8
HHV[a] (MJ/Nm³)	12.74	12.63		39.82	70.29	63.41	58.06	101.24
LHV[a] (MJ/Nm³)	10.78	12.63		35.88	64.34	59.45	56.07	99.09
Viscosity (µP)	90	182	150	112	94	103	104	82
Thermal conductivity (W/m.K)	0.182	0.0251	0.0166	0.0343	0.0218	0.0214	0.0213	0.0183
Specific heat (kJ/kg.K)	3.467	1.05	0.85	2.226	1.926	1.691	1.775	1.708

Gases	C_3H_6	$i\text{-}C_4H_8$	$i\text{-}C_4H_{10}$	$n\text{-}C_4H_{10}$	C_6H_6	N_2	NH_3	H_2S
HHV[a] (MJ/Nm³)	93.57	125.08	133.12	134.06	142.89		13.07	25.1
LHV[a] (MJ/Nm³)	87.57	116.93	122.91	123.81	141.41		10.13	23.14
Viscosity (µP)						180		
Thermal conductivity (W/m.K)						0.026		
Specific heat (kJ/kg.K)						1.05		

[a]High heating value (HHV) = the amount of heat produced by the complete combustion of a unit quantity of fuel. The HHV is realized when all products of the combustion are cooled down to the temperature before the combustion, and when the water vapor formed during combustion is condensed.

LHV (low heating value): the high heating value (HHV) minus the latent heat of vaporization of the water vapor formed by combustion. The LHV thus assumes that the latent heat of vaporization of water in the fuel and the reaction products is not recovered.

Source: Adapted from Basu, Prabir. 2010. Biomass Gasification and Pyrolysis, (Boston, Academic Press).

Table 2.23.   Properties of vegetable oils

Vegetable oil	Kinematic viscosity at 38°C (mm²/S)	Cetane no.	Heating value (MJ/kg)	Cloud point (°C)	Pour point (°C)	Flash point (°C)	Density (kg/l)
Babassu	30.3	38	-	20	-	150	0.946
Corn	34.9	37.6	39.5	-1.1	-40	277	0.9095
Cottonseed	33.5	41.8	39.5	1.7	-15	234	0.9148
Crambe	53.6	44.6	40.5	10	-12.2	274	0.9048
Diesel	3.06	50	43.8	-	-16	76	0.855
Linseed	27.2	34.6	39.3	1.7	-15	241	0.9236
Palm	39.6	42	-	31	-	267	0.918
Peanut	39.6	41.8	39.8	12.8	-6.7	271	0.9026
Rapeseed	37	37.6	39.7	-3.9	-31.7	246	0.9115
Safflower	31.3	41.3	39.5	18.3	-6.7	260	0.9144
Sesame	35.5	40.2	39.3	-3.9	-9.4	260	0.9133
Soya bean	32.6	37.9	39.6	-3.9	-12.2	254	0.9138
Sunflower	33.9	37.1	39.6	7.2	-15	274	0.9161

Source: Adapted from Misra, R.D., M.S. Murthy. 2010. Straight vegetable oils usage in a compression ignition engine—A review, Renewable and Sustainable Energy Reviews, Volume 14, Issue 9, Pages 3005-3013.

Table 2.24. Early history of ethanol in select countries

Country	Year	
Angola	1933	A decree from Lisbon, Portugal mandates two types of alcohol-gasoline blending. One is "Alcoolaina" with 25% alcohol, 75% gasoline and a blending agent. "Gasalcool" has a 75% alcohol, 25% gasoline blend.
Argentina	1931	A government report recommends a 30% blend of alcohol in motor gasoline because it can be produced at less cost than importing gasoline.
Australia	1927	The Australian National Power Alcohol Company starts with a capacity of two million gallons per year. The ethanol plant primarily uses molasses, along with cassava (manioc) and sweet potatos grown in rotation with sugarcane. Shellkol" fuel is marketed by Shell Oil Co. at 15%-35% concentrations.
	1933	Competition from gasoline suppliers leads to the Motor Spirits Vendors Act that mandates that ethanol be blended with petrol. The blending program lasts through World War II, when ethanol is used in chemicals and plastics.
Austria	1931	The Minister of Finance mandate 25% blending; in 1934 gasoline importers are required to buy alcohol at 2% of their sales volume while domestic gasoline producers are required to buy at 3.75% of their sales volumes. The alcohol is sold in 20—40% blends.
Brazil	1922	The Brazilian Congress of Coal and other National Fuels (Congresso de Carvac e Otres Combustiveis Nacionaes) recommendeds establishment of alcohol cooperative societies equipped with fermenting, distilling, denaturing, and carburizing plants and tank wagon for distributing; organization of alcohol demonstrating and distributing agencies throughout Brazil; government vehicle use of ethanol; and reduced taxes.
	1933	The Instituto do Assucar e do Alcool is established to promote alcohol fuels and provide technical assistance.
	1975	The Brazilian government launches the National Alcohol Program -Pró-Álcool- (Portuguese: 'Programa Nacional do Álcool'), a nation-wide program financed by the government to phase out automobile fuels derived from fossil fuels, such as gasoline, in favor of ethanol produced from sugar cane.
Chile	1933	A law requires blending of more than 10% and less than 25% alcohol in all gasoline, but is dropped to 9% and later 6% as the law became effective. Most alcohol is produced from imported Peruvian molasses.
China	1930s	"Benzolite," a mixture of 55% alcohol, 40% benzene and 5% kerosene is sold as a transportation fuel.
Cuba	1920s	A blend of 80% gasoline and 20% alcohol called "Espiritu" is sold as a transportation fuel.
Czechoslovakia	1932	A law is passed making 20% alcohol blending compulsory. "Dynakol," a mixture of about 50% alcohol, 20% benzene and 30% gasoline, is used by government agencies.
France	1915–1916	Concerned about wartime oil supplies, the government creates a state alcohol monopoly to encourage more alcohol fuel use from a tax on beverage alcohol; another law encourages industrial alcohol production from sugar beets, but grape producers were protected from industrial alcohol demands.
	1923	A government report recommends a 40%-50% blending level for a "national fuel," and a new law requires gasoline importers to buy alcohol for 10% blends from the State Alcohol Service.
Germany	1899	Government opens the Centrale fur Spiritus Verwerthung (office of alcohol sales) which supports alcohol through subsidies to alcohol producers and a tariff on imported oil.
	1915	During World War I, with when oil shortages threaten to likely to paralyze Germany's transportation system, thousands of engines are quickly modified to burn alcohol.
	1932	A 10% blend is compulsory. About two thirds comes from potatoes, the rest from grain wood sulfite liquors and beets.
Hungary	1929	"Moltaco," a blend of 20% ethanol and 80% gasoline, is made compulsory by royal decree.
India	1940	Uttar Pradesh passes a 20% power alcohol blending law; several other provinces follow suit.
	1948	The "Indian Alcohol Act" mandates 20% blending where feasible, but is not widely adopted.
Italy	1924	Tax incentives given to alcohol production for fuel.

Continued

Table 2.24. continued

	Year	Description
	1926	A royal decree in 1926 sets motor fuel composition at 30% alcohol and 70% gasoline; eleven alcohol manufacturers are producing for the fuel market by 1931.
Japan	1942	Compulsory blending of 5% ethanol becomes effective.
	1941-1945	The Army, Navy and Air Corps use alcohol during WWII due to the scarcity of oil supplies.
Latvia	1931	Compulsory mixtures of 25% ethanol; increased to 50% and 66% by 1936, when the legislation was lifted.
Lithuania	1936	A law sets national fuel levels at 25% alcohol; ethanol is sold in a blend called "Motorin".
Panama	1925	A law mandates a 10% alcohol blend in all gasoline sold nationwide.
Philippines	1922	The Philippean Motor Alcohol Corp. forms, and by 1931, "Gasonol" (spelled with an "n"), blend of 20% ethanol and 5% kerosene, is used on a commercial scale.
	1937	The Philippines leads the world in the use of pure alcohol fuels. Studebaker, McCormick, GM, and International Harvester sell pure alcohol-fueled cars and trucks. Three large bus companies, including Manila's Batangas Transportation Co., run their buses on 100% ethanol.
Poland	1927	A large alcohol factory is built at the Kutno Chemical Works; by 1939 it produces 54,000 liters per day.
	1932	A 25% alcohol blending mandate is decreed.
Sweden	1915	A reduction in oil imports during WWI leads to a quick expansion of paper mill sulfite alcohol production; from four to 22 plants with a capacity of about 22 million liters per year. The Government orders a return to gasoline after the war.
United Kingdom	1928	National Distillers Company markets an alcohol motor fuel blend with the Cleveland Oil Co. known as "Cleveland Discol." Other blends included "Koolmotor" marketed by the Cities Service Corp. and "National Benzol" marketed by the National Benzol Association. National Distillers continues to sell Cleveland Discol until 1968, when the alcohol fuels unit is purchased by British Petroleum and converted to chemical feedstock production. Various acts of Parliament encourage use of power alcohol, although blending is never made compulsory.
United States	1862	A heavy tax on both industrial and beverage alcohol cripples the "camphene" illuminating fuel industry with sales around 90 million gallons per year. Progressives led by President Theodore Roosevelt end\ the tax on industrial alcohol in 1906.
	1900	On the eve of the automotive era, the widespread availability of ethyl alcohol and concerns about access to oil reserves leads manufacturers to produce engine designs that can be adapted to both fuels. Henry's Ford's first vehicles—and those being produced in Europe—are designed to run on multiple fuels, including alcohol.
	1919-1933	During the Prohibition era in the United States, it is illegal to sell, manufacture, and transport alcohol, posing a significant barrier to the use of ethanol as a fuel. Industrial ethanol distillers and fuel retailers are accused of being allied with moonshiners, and ethanol can only be sold when mixed with gasoline.
	1930	In an attempt to help fight the 25% farmer unemployment rate that followed corn's fall from 45 cents to just 10 cents per bushel during the Great Depression, nearly three dozen bills to subsidize ethanol production are introduced in a broad effort to develop industrial uses for farm crops. Called Farm Chemurgy, the movement inspired campy ethanol slogans like "Try a tankfull. You'll be thankful." A 10% alcohol motor fuel blend is marketed by the Agrol Company of Atchison, Kansas as an experiment into broadening markets for surplus crops. The experiment is supported by Henry Ford and the Chemical Foundation, and at its peak in 1938, some 2,000 service stations in 8 different states sold Agrol. Agrol suspends business in 1939.
	1942	The United States uses ethyl alcohol for aviation fuel and synthetic "Buna-S" rubber during World War II. By 1944, petroleum based synthetic rubber production lags, and three quarters of all tires, raincoats, engine gaskets and other rubber products for the war effort come from ethanol

Source: Adapted from Kovarik, William, Ethanol's first century: Blending programs in Europe, Asia, Africa and Latin America, paper to the 30th International Symposium on Alcohol Fuel, Rio de Janeiro, Brazil, November 2006.

Table 2.25. Comparison of ethanol, gasoline, blended gasoline fuel properties

Property	Ethanol	Gasoline	E_{10}	E_{20}
Specific gravity	0.79	0.72–0.75	0.73–0.76	0.74–0.77
Heating value (Btu/gal)	76,000	117,000	112,900	109,000
Vapor pressure	17	59.5	64	63.4
Oxygen content	35	0	3.5	7

Source: Adapted from Naik, S.N., Vaibhav V. Goud, Prasant K. Rout, Ajay K. Dalai. 2010. Production of first and second generation biofuels: A comprehensive review, Renewable and Sustainable Energy Reviews, Volume 14, Issue 2, Pages 578–597.

Table 2.26. Comparison of diesel, biodiesel, and green diesel fuel properties

	Petroleum ULSD[a]	Biodiesel (FAME)[b]	Green diesel[c]
% oxygen	0	11	0
Specific gravity	0.84	0.88	0.78
Sulfur (ppm)	<10	<1	<1
Heating value (MJ/kg)	43	38	44
Cloud point (°C)	−5	−5 to +15	−10 to 20
Distillation (°C)	200-350	340-355	265-320
Cetane	40	50-65	70-90
Stability	Good	Marginal	Good

[a]ULSD: Ultra low sulfur diesel.
[b]FAME: Fatty acid methyl esters.
[c]Green diesel is an isoparaffin-rich diesel produced from renewable feedstock containing triglycerides and fatty acids by process of catalytic saturation, hydrodeoxygenation, decarboxylation and hydroisomerization. This technology can be used for any type of oil feedstock to produce an isoparaffin-rich diesel substitute. The product is an aromatic and low sulfur diesel fuel with a high cetane blending value.
Source: Adapted from Naik, S.N., Vaibhav V. Goud, Prasant K. Rout, Ajay K. Dalai. 2010. Production of first and second generation biofuels: A comprehensive review, Renewable and Sustainable Energy Reviews, Volume 14, Issue 2, Pages 578–597.

Table 2.27. Fuel properties of ethyl esters from various feedstocks

Fuel property	Kinematic viscosity (40 °C)	Density (15 °C)	HHV	LHV	Cetane number	Flash point	CFFP	CP	PP
Unit	mm²/s	kg/m³	MJ/kg	MJ/kg	-	°C	°C	°C	°C
Candlenut	5.6	878				194			−10
Cardoon	3.4	870	40		49	188		−3	−6
Castor Oil	4.6–17.4	912[b]–919				151			
Coconut	2.6	880			51	112	5		
Cotton	5.2	877		40.6	52	175			
Groundnut		906	38.3					23	
Jatropha	4.0–5.5[a]	883–886	41.7		59	117–190			−5
Linseed		884	40					−2	−9
Mustard	4.4–5.7	810–830							
Olive			38.2						
Palm kernel	4.8	883	40.6	37.3		167		6	2
Palm	3.5[a]–5.8	870–880				126–178		6–8	4–6
Radish	5	872[b]				182–184	−2		
Rapeseed/canola	4.4	869–902	38.7–40.3					−1 to 8	−15 to 12
Soybean	4.4–4.9	876[b]–925[b]	40		48	189–195	−7 to −5	0-1	−4 to −2
Sunflower	4.6–4.9	850–873	38.6–39.7		49	187		−1	−8
Grease	6						0	−3	9
Tallow	6.2						8	15	12
Waste cooking oil	2.7–9.1	855–887	37.8–40.7		38–49	119–196	−6 to −1	−3 to 11	−7 to 10

HHV: Higher Heating Value; LHV: Lower Heating Value; CFPP: Cold Filter Plugging Point; CP: Cloud Point; PP: Pour Point.
[a]Kinematic viscosity (30 °C).
[b]Density (20 °C).
Source: Adapted from Brunschwig, C., W. Moussavou, J. Blin. 2012. Use of bioethanol for biodiesel production, Progress in Energy and Combustion Science, Volume 38, Issue 2, Pages 283–301.

Table 2.28. Fuel properties of ethyl esters compared to methyl esters, crude oil, and diesel fuel

Fuel property	Unit	Oil	FAME	FAEE	Diesel fuel
Composition	-	TG C12–C22	FAME C12–C22	FAEE C12–C22	HC C10–C21
Kinematic viscosity (40° C)	mm^2/s	30–106	1.9–6.0	2.6–6.2	1.9–4.1
Density (15° C)	kg/m^3	915–940	860–894	810–919	750–850
HHV	MJ/kg	37.5–40.6	39–41	38.2–41.7	43.0–46.5
LHV	MJ/kg	35.0–39.5	37.4–39.5	37.3–40.6	36.6–43.8
Cetane number	-	35–45	48–65	48–59	40–55
Flash Point	°C	240-330	100–170	112–196	60–80
Cloud Point	°C	-	–3 to 12	–3 to 23	–15 to 5
Pour Point	°C	–1 to 31	–15 to 10	–15 to 12	–35 to –15

TG: triacylglycerides; FAME: fatty acid methyl ester; FAEE: fatty acid ethyl ester; HC: hydrocarbons; HHV: Higher Heating Value; LHV: Lower Heating Value; CFPP: Cold Filter Plugging Point; CP: Cloud Point; PP: Pour Point.
Source: Adapted from Brunschwig, C., W. Moussavou, J. Blin. 2012. Use of bioethanol for biodiesel production, Progress in Energy and Combustion Science, Volume 38, Issue 2, Pages 283–301.

Table 2.29. Potential biodiesel yield from triglyceride feedstocks

Source	Potential annual yield, gallons/acre
Corn	18–20
Cotton	35–45
Soybean	40–55
Mustard	60–140
Camelina	60–65
Safflower	80–85
Sunflower	75–105
Canola	110–145
Rapeseed	110–130
Jatropha	140–200
Coconut	250–300
Palm oil	400–650
Algae	>5000[a]

[a]Figure for algae is based upon extrapolations from small scale operations, and is quite speculative.
Source: Adapted from Hoekman, S. Kent, Amber Broch, Curtis Robbins, Eric Ceniceros, Mani Natarajan. 2012. Review of biodiesel composition, properties, and specifications, Renewable and Sustainable Energy Reviews, Volume 16, Issue 1, Pages 143–169.

Table 2.30. Typical properties of diesel, soybean oil, rapseed oil, and ester fuels

Property	No. 2 Diesel	Soybean Oil			Rapeseed Oil		
		Oil	Methyl ester	Ethyl ester	Oil	Methyl ester	Ethyl ester
Specific gravity	0.8495	0.92	0.886	0.881	0.91	0.88	0.876
Viscosity at 40°C, mm²/s	2.98	33	3.891	4.493	51	5.65	6.17
Cloud point, °C	-12	-4	3	0	-21	0	-2
Pour point, °C	-23	-12	-3	-3		-15	-10
Flash point, °C	74		188	171		179	124
Boiling point, °C	191		339	357		347	273
Water & sediment, vol %	<0.005		<0.005	<0.005		<0.005	<0.005
Carbon residue, wt %	0.16		0.068	0.071		0.08	0.06
Ash, wt %	0.002		0	0		0.002	0.002
Sulfur, wt %	0.036	0.01	0.012	0.008	0.01	0.012	0.014
Cetane number	49	38	55	53	32	62	65
Copper corrosion	1A		1A	1A		1A	1A
Higher heating value, MJ/kg	45.42	39.3	39.77	39.96	40.17	40.54	40.51
MJ/L	38.58	36.2	35.24	35.2	36.6	35.68	38
Fatty acid composition, wt %							
Palmitic (16:0)		9.8	9.9	10	1	2.2	2.6
Stearic (18:0)		2.4	3.8	3.8		0.9	0.9
Oleic (18:1)		28.9	19.1	18.9	32	12.6	12.8
Linoleic (18:2)		50.7	55.6	55.7		12.1	11.9
Linolenic (18:3)		6.5	10.2	10.2	15	8	11.9
Eicosenoic (20:1)			0.2	0.2		7.4	7.3
Behenic (22:0)			0.3	0.3		0.7	0.7
Erucic (22:1)			0	0	50	49.8	49.5

The figures in parentheses after each fatty acid denote the number of carbon atoms and double bonds in the fatty acid.
Source: Adapted from Klass, Donald L. 1998. Biomass for Renewable Energy, Fuels, and Chemicals, (San Diego, Academic Press).

Table 2.31. Constituents of producer gas

Compounds	Symbol	Gas (vol.%)	Dry gas (vol.%)
Carbon monoxide	CO	21	22.1
Carbon dioxide	CO_2	9.7	10.2
Hydrogen	H_2	14.5	15.2
Water vapor	H_2O	4.8	-
Methane	CH_4	1.6	1.7
Nitrogen	N_2	48.4	50.8
Gas high heating value:			
Generator gas (wet basis)	5506 kJ/Nm^3		
Generator gas (dry basis)	5800 kJ/Nm^3		
Air ratio required for gasification	2.38 kg wood/kg air		
Air ratio required for gas combustion	1.15 kg wood/kg air		

Source: Adapted from Panwar, N.L., Richa Kothari, V.V. Tyagi. 2012. Thermo chemical conversion of biomass – Eco friendly energy routes, Renewable and Sustainable Energy Reviews, Volume 16, Issue 4, Pages 1801-1816.

Table 2.32. Typical producer-gas composition and operating conditions for atmospheric bubbling fluidized-bed gasifiers

	Air	Steam (pure)	Steam-O_2 mixtures
Operating conditions			
ER	0.18–0.45	0	0.24–0.51
S/B (kg/kg daf)	0.08–0.66	0.53–1.10	0.48–1.11
T (°C)	780–830	750–780	785–830
Gas composition			
H_2 (vol%, dry basis)	5.0–16.3	38–56	13.8–31.7
CO (vol%, dry basis)	9.9–22.4	17–32	42.5–52.0
CO_2 (vol%, dry basis)	9.0–19.4	13–17	14.4–36.3
CH_4 (vol%, dry basis)	2.2–6.2	7–12	6.0–7.5
C_2H_n (vol%, dry basis)	0.2–3.3	2.1–2.3	2.5–3.6
N_2 (vol%, dry basis)	41.6–61.6	0	0
Steam (vol%, wet basis)	11–34	52–60	38–61
Yields			
Tars (g/kg daf)	3.7–61.9	60–95	2.2–46
Char (g/kg daf)	na	95–110	5–20
Gas (Nm^3/kg daf)	1.25–2.45	1.3–1.6	0.86–1.14
LHV (MJ/Nm^3)	3.7–8.4	12.2–13.8	10.3–13.5

na: not available; daf: dry ash-free basis; ER: equivalence ratio; S/B: steam-to-biomass ratio (H_2O (kg/h)/biomass (kg daf/h)); LHV: low heating value.
Source: Adapted from Puig-Arnavat, Maria, Joan Carles Bruno, Alberto Coronas. 2010. Review and analysis of biomass gasification models, Renewable and Sustainable Energy Reviews, Volume 14, Issue 9, Pages 2841-2851.

Table 2.33. Typical biomass gasification reactions at 25 °C

Reaction Type	Reaction
Carbon Reactions	
R1 (Boudouard)	$C + CO_2 \leftrightarrow 2CO + 172$ kJ/mol
R2 (water-gas or steam)	$C + H_2O \leftrightarrow CO + H_2 + 131$ kJ/mol
R3 (hydrogasification)	$C + 2H_2 \leftrightarrow CH_4 - 74.8$ kJ/mol
R4	$C + 0.5 O_2 \rightarrow CO - 111$ kJ/mol
Oxidation Reactions	
R5	$C + O_2 \rightarrow CO_2 - 394$ kJ/mol
R6	$CO + 0.5O_2 \rightarrow CO_2 - 284$ kJ/mol
R7	$CH_4 + 2O_2 \leftrightarrow CO_2 + 2H_2O - 803$ kJ/mol
R8	$H_2 + 0.5 O_2 \rightarrow H_2O - 242$ kJ/mol
Shift Reaction	
R9	$CO + H_2O \leftrightarrow CO_2 + H_2 - 41.2$ kJ/mol
Methanation Reactions	
R10	$2CO + 2H_2 \rightarrow CH_4 + CO_2 - 247$ kJ/mol
R11	$CO + 3H_2 \leftrightarrow CH_4 + H_2O - 206$ kJ/mol
R14	$CO_2 + 4H_2 \rightarrow CH_4 + 2H_2O - 165$ kJ/mol
Steam-Reforming Reactions	
R12	$CH_4 + H_2O \leftrightarrow CO + 3H_2 + 206$ kJ/mol
R13	$CH_4 + 0.5 O_2 \rightarrow CO + 2H_2 - 36$ kJ/mol

Basu, Prabir. 2010. Chapter 5 - Gasification Theory and Modeling of Gasifiers, Biomass Gasification and Pyrolysis, (Boston, Academic Press), Pages 117–165.

Table 2.34. Attributes of some commercial biomass gasifiers

Parameters	Fixed/Moving Bed	Fluidized Bed	Entrained Bed
Feed size	<51 mm	<6 mm	<0.15 mm
Tolerance for fines	Limited	Good	Excellent
Tolerance for coarse	Very good	Good	Poor
Exit gas temperature	450–650 °C	800–1000 °C	>1260 °C
Feedstock tolerance	Low-rank coal	Low-rank coal and excellent for biomass	Any coal including caking but unsuitable for biomass
Oxidant requirements	Low	Moderate	High
Reaction zone temperature	1090 °C	800–1000 °C	1990 °C
Steam requirement	High	Moderate	Low
Nature of ash produced	Dry	Dry	Slagging
Cold-gas efficiency	80%	89%	80%
Application	Small capacities	Medium-size units	Large capacities
Problem areas	Tar production and utilization of fines	Carbon conversion	Raw-gas cooling

Source: Adapted from Basu, Prabir. 2010. Chapter 6 - Design of Biomass Gasifiers, Biomass Gasification and Pyrolysis, (Boston, Academic Press), Pages 167–228.

Table 2.35. Ultimate and proximate analysis of various lignocellulosic biomass

Biomass type	Ultimate analysis (db, % w/w)						Proximate analysis (% w/w)				
	C	H	O	N	S	Ash	VM	FC	M	LHV (MJ/kg)
Cedar wood	51.1	5.9	42.5	0.12	0.02	0.3	80–82	18–20	a	19.26
Wood sawdust	46.2	5.1	35.4	1.5	0.06	1.3	70.4	17.9	10.4	18.81
Olive oil residue	50.7	5.89	36.97	1.36	0.3	4.6	76	19.4	9.5	21.2
Rice husk	45.8	6	47.9	0.3	–	0.8	73.8	13.1	12.3	13.36
Rice straw	38.61	4.28	37.16	1.08	0.65	12.64	65.23	16.55	5.58	14.4
Pine sawdust	50.54	7.08	41.11	0.15	0.57	0.55	82.29	17.16	a	20.54
Spruce wood pellet	49.3	5.9	44.4	0.1	–	0.3	74.2	17.1	8.4	18.5
Coffee husk	46.8	4.9	47.1	0.6	0.6	1	74.3	14.3	10.4	16.54
Coffee ground	52.97	6.51	36.62	2.8	0.05	1	71.8	16.7	10.5	22
Larch wood	44.18	6.38	49.32	0.12	–	0.12	76.86	14.86	8.16	19.45
Grapevine pruning waste	46.97	5.8	44.49	0.67	0.01	2.06	78.16	19.78	a	17.91
Jute stick	49.79	6.02	41.37	0.19	0.05	0.62	76–78	21.4–23.4	a	19.66
Sugar-cane bagasse	48.58	5.97	38.94	0.2	0.05	1.26	67–70	28.74–30.74	a	19.05
Corn cob	40.22	4.11	42.56	0.39	0.04	2.97	71.21	16.11	9.71	16.65
Peach stone	51.95	5.76	40.7	0.79	0.01	0.65	81.3	18.1	8.53	21.6
Wheat straw	46.1	5.6	41.7	0.5	0.08	6.1	75.8	18.1	a	17.2
Cotton stem	42.8	5.3	38.5	1	0.2	4.3	72.3	15.5	7.9	15.2
Straw	36.57	4.91	40.7	0.57	0.14	8.61	64.98	17.91	8.5	14.6
Camphor wood	43.43	4.84	38.53	0.32	0.1	0.49	72.47	14.75	12.29	17.48
Beech wood	48.27	6.36	45.2	0.14	–	0.8	81	18	a	19.2
Switchgrass	47	5.3	41.4	0.5	0.1	4.6	58.4	17.1	20	18.7

VM: volatile matter; FC: fixed carbon; M: moisture; LHV: low heating value.
[a]dry basis.

Source: Adapted from Alauddin, Zainal Alimuddin Bin Zainal, Pooya Lahijani, Maedeh Mohammadi, Abdul Rahman Mohamed. 2010. Gasification of lignocellulosic biomass in fluidized beds for renewable energy development: A review, Renewable and Sustainable Energy Reviews, Volume 14, Issue 9, Pages 2852–2862.

Table 2.36.   Composition of representative lignocellulosic feedstocks

Feedstocks	Carbohydrate composition (% dry wt)		
	Cellulose	Hemicellulose	Lignin
Barley hull	34	36	19
Barley straw	36–43	24–33	6.3–9.8
Bamboo	49–50	18–20	23
Banana waste	13	15	14
Corn cob	32.3–45.6	39.8	6.7–13.9
Corn stover	35.1–39.5	20.7–24.6	11.0–19.1
Cotton	85–95	5–15	0
Cotton seed hairs	80–95	5–20	0
Cotton stalk	31	11	30
Coffee pulp	33.7–36.9	44.2–47.5	15.6–19.1
Douglas fir	35–48	20–22	15–21
Eucalyptus	45–51	11–18	29
Hardwood stems	40–55	24–40	18–25
Herbaceous energy crops	45	30	15
Rice straw	29.2–34.7	23–25.9	17–19
Rice husk	28.7–35.6	11.96–29.3	15.4–20
Wheat straw	35–39	22–30	12–16
Wheat bran	10.5–14.8	35.5–39.2	8.3–12.5
Grasses	25–40	25–50	10–30
Newspaper	40–55	24–39	18–30
Sugarcane bagasse	25–45	28–32	15–25
Sugarcane tops	35	32	14
Pine	42–49	13–25	23–29
Poplar wood	45–51	25–28	10–21
Olive tree biomass	25.2	15.8	19.1
Jute fibres	45–53	18–21	21–26
Switchgrass	35–40	25–30	15–20
Grasses	25–40	25–50	10–30
Winter rye	29–30	22–26	16.1
Oilseed rape	27.3	20.5	14.2
Softwood stem	45–50	24–40	18–25
Oat straw	31–35	20–26	10–15
Nut shells	25–30	22–28	30–40
Sorghum straw	32–35	24–27	15–21
Tamarind kernel powder	10–15	55–65	–
Waste paper	76	13	11
Water hyacinth	18.2–22.1	48.7–50.1	3.5–5.4

Source: Adapted from Menon, Vishnu, Mala Rao. 2012. Trends in bioconversion of lignocellulose: Biofuels, platform chemicals and biorefinery concept, Progress in Energy and Combustion Science, Available online 16 March 2012.

Table 2.37. Wood pellet standards established by the Pellet Fuels Institute (PFI)[a]

Fuel Property	PFI Premium	PFI Standard	PFI Utility
Bulk Density, lb./cubic foot	40.0–46.0	38.0–46.0	38.0–46.0
Diameter, inches	0.230–0.285	0.230–0.285	0.230–0.285
Diameter, mm	5.84–7.25	5.84–7.25	5.84–7.25
Pellet Durability Index	≥96.5	≥95.0	≥95.0
Fines, % (at the mill gate)	≤0.50	≤1.0	≤1.0
Inorganic Ash, %	≤1.0	≤2.0	≤6.0
Length, % greater than 1.50	≤1.0	≤1.0	≤1.0
Moisture, %	≤8.0	≤10.0	≤10.0
Chloride, ppm	<300	<300	<300
Other Properties:			
Energy Content: 16.9 - 18 MJ/kg			
Space requirements: 1.5 m^3/ton			

[a]The Pellet Fuels, PFI Standards Program is a third-party accreditation program providing specifications for residential and commercial-grade fuel. The American Lumber Standard Committee (ALSC) serves as the program's accreditation body.
Source: Adapted from the Pellet Fuels Institute, <http://pelletheat.org/pfi-standards/pfi-standards-program/>, accessed 8 May 2012; Karkania,V., E. Fanara, A. Zabaniotou. 2012. Review of sustainable biomass pellets production – A study for agricultural residues pellets' market in Greece, Renewable and Sustainable Energy Reviews, Volume 16, Issue 3, Pages 1426–1436

Table 2.38. Comparison of biomass, biomass pellets, and torrefied pellets[a] (TOP) characteristics

Characteristic	Biomass	Pellets	TOP pellets
Moisture content (%)	35	10	3
Calorific value (MJ/kg)	10.5	16	21
Mass density (bulk) (kg/m^3)	550	600	800
Energy density (GJ/m^3)	5.8	9	16.7

[a]Torrefaction removes moisture and low energy volatiles from the roasted wood, producing a product that is more energy dense (more energy per unit of weight) than wood.
Source: Adapted from Karkania,V., E. Fanara, A. Zabaniotou. 2012. Review of sustainable biomass pellets production – A study for agricultural residues pellets' market in Greece, Renewable and Sustainable Energy Reviews, Volume 16, Issue 3, Pages 1426–1436

Table 2.39.   Oil content of some microalgae

Microalga	Oil content (% dry wt)
Botryococcus braunii	25–75
Chlorella sp.	28–32
Crypthecodinium cohnii	20
Cylindrotheca sp.	16–37
Dunaliella primolecta	23
Isochrysis sp.	25–33
Monallanthus salina	> 20
Nannochloris sp.	20–35
Nannochloropsis sp.	31–68
Neochloris oleoabundans	35–54
Nitzschia sp.	45–47
Phaeodactylum tricornutum	20–30
Schizochytrium sp.	50–77
Tetraselmis sueica	15–23

Source: Adapted from Chisti, Yusuf. 2007. Biodiesel from microalgae, Biotechnology Advances, Volume 25, Issue 3, Pages 294–306.

Table 2.40.   Advantages and limitations of various extraction methods for algae oil

Extraction methods	Advantages	Limitations
Oil press	Easy to use, no solvent involved	Large amount of sample required, slow process
Solvent extraction	Solvents used are relatively inexpensive; reproducible	Most organic solvents are highly flammable and/or toxic; solvent recovery is expensive and energy intensive; large volume of solvent needed
Supercritical fluid extraction	Non-toxicity (absence of organic solvent in residue or extracts), "green solvent" used; non-flammable, and simple in operation	Often fails in quantitative extraction of polar analytes from solid matrices, insufficient interaction between supercritical CO_2 and the samples
Ultrasound	Reduced extraction time; reduced solvent consumption; greater penetration of solvent into cellular materials; improves the release of cell contents into the bulk medium	High power consumption; difficult to scale up

Source: Adapted from Singh, Jasvinder, Sai Gu. 2010. Commercialization potential of microalgae for biofuels production, Renewable and Sustainable Energy Reviews, Volume 14, Issue 9, Pages 2596–2610.

Table 2.41. Microalgae commonly used for bioenergy and the organic chemicals industry

Algal species	Algal class	Product(s)	Culture technique	Advantages or drawbacks
Botryococcus braunii	Chlorophyceae	Triterpene oils	Photobioreactor	Whole genome available
Chlorella sp	Chlorophyceae	Carbohydrates, protein	Ponds, photobioreactor	Widespread and adaptable
Chlamydomonas reinhardtii	Chlorophyceae	Oils, carbohydrates, hydrogen and methane	Photobioreactor	Transformable
Dunaliella salina	Chlorophyceae	b-Carotene	Brackish seawater ponds	Needs high salinity
Nannochloropsis	Eustigmatophyceae	Polyunsaturated fatty acids	Seawater ponds	Lipid bodies produced under nitrogen stress
Ostreococcus tauri	Prasinophyceae- (Chlorophyceae)	Oils	Photobioreactor	Smallest known microalga, whole genome available
Pavlova lutheri	Prymnesiophyceae	Fatty acids, aquaculture feedstock	Photobioreactor	Little used
Arthrospira platensis (Spirulina)	Cyanobacteria	Health food	Ponds, photobioreactor	Filamentous morphology
Synechocystis and *Synecococcus*	Cyanobacteria	Isoprenes, oils	Photobioreactor	Whole genomes available, transformable

Source: Adapted from Larkum, Anthony W.D., Ian L. Ross, Olaf Kruse, Ben Hankamer. 2012. Selection, breeding and engineering of microalgae for bioenergy and biofuel production, Trends in Biotechnology, Volume 30, Issue 4, Pages 198–205.

Table 2.42. Methane yield from different algae strains

Biomass	Methane yield (m^3 kg^{-1})
Laminaria sp.	0.26–0.28
Gracilaria sp.	0.28–0.4
Macrocystis	0.39–0.41
L. Digitata	0.5
Ulva sp.	0.2

Source: Adapted from Singh, Jasvinder, Sai Gu. 2010. Commercialization potential of microalgae for biofuels production, Renewable and Sustainable Energy Reviews, Volume 14, Issue 9, Pages 2596–2610.

Table 2.43.   Examples of biomass productivity and estimated solar energy capture efficiency

Location	Biomass community	Annual yield dry Matter (t/ha-year)	Aveage insolation (W/m^2)	Solar energy capture effiency (%)
Alabama	Johnsongrass	5.9	186	0.19
Sweden	Enthropic lake angiosperm	7.2	106	0.138
Denmark	Phytoplankton	8.6	133	0.36
Minnesota	Willow and hyprid poplar	8–11	159	.30–.41
Mississippi	Water hyacinth	11.0–33.0	194	0.31–0.94
California	*Euphorbia Lathyris*	16.3–19.3	212	0.45–0.54
Texas	Switchgrass	8–20	212	0.22–.56
Alabama	Switchgrass	8.2	186	.26
Texas	Sweet sorghum	22.2–40.0	239	0.55–.99
Minnesota	Maize	24	169	.72
New Zealand	Temperate grassland	29.1	159	1.02
West Indies	Tropical marine angiosperm	30.3	212	.79
Nova Scotia	Sublittoral seaweed	32.1	133	1.34
Georgia	Subtropical saltmarsh	32.1	194	0.92
England	Coniferous forest, 0–21 years	34.1	106	1.79
Israel	Maize	34.1	239	0.79
New South Wales	Rice	35	186	1.04
Congo	Tree plantation	36.1	212	0.95
Netherlands	Maize, rye, two harvest	37.0	106	1.94
Marshall Islands	Green algae	39.0	212	0.95
Germany	Temperate reedswamp	46.0	133	1.92
Puerto Rico	*Panicum maximum*	48.9	212	1.28
California	Algea, sewage pond	49.3–74.2	218	1.26–1.89
Colombia	Pangola grass	50.2	186	1.50
West Indies	Tropical forest, mixed ages	59	212	1.55
Hawaii	Sugarcane	74.9	186	2.24
Puerto Rico	*Pennistum purpurcum*	84.5	212	2.21
Java	Sugarcane	86.8	186	2.59
Puerto Rico	Napier grass	106	212	2.78
Thailand	Green algae	164	186	4.9

Source: Adapted from Klass, Donald L. 1998. Biomass for Renewable Energy, Fuels, and Chemicals, (San Diego, Academic Press).

Table 2.44. Biomass productivity figures for heterotrophic and mixotrophic microalgae cultures[a]

Species	Product	Culture	Organic carbon source	X_{max} (g l^{-1})	Total lipid (%)	P_{volume} (g l^{-1} day^{-1})
Heterotrophic						
Galdieria sulphuraria	C-phycocyanin	Continuous	-	83.3	-	50
Galdieria sulphuraria	C-phycocyanin	Fed-batch	-	109	-	17.5
Chlorella protothecoides	Biodiesel	Fed-batch	-	3.2	57.8	-
Chlorella protothecoides	Biodiesel	Fed-batch	-	16.8	55.2	-
Chlorella protothecoides	Biodiesel	Fed-batch	-	51.2	50.3	-
Chlorella	Docosahexaenoic acid	Fed-batch	-	116.2	-	1.02
Crypthecodinium cohnii	Docosahexaenoic acid	Fed-batch	-	109	56	-
Crypthecodinium cohnii	Docosahexaenoic acid	Fed-batch	-	83	42	-
Chlorella	-	Fed-batch	-	104.9	-	14.71
Chlorella protothecoides	Biodiesel	Fed-batch	-	15.5	46.1	-
Chlorella protothecoides	Biodiesel	Fed-batch	-	12.8	48.7	-
Chlorella protothecoides	Biodiesel	Fed-batch	-	14.2	44.3	-
Mixotrophic						
Spirulina platensis	-	-	Glucose	2.66	-	-
Spirulina platensis	-	-	Acetate	1.81	-	-
Spirulina sp.	-	-	Glucose	2.5	-	-
Spirulina platensis	-	-	Molasses	2.94	-	0.32

[a]Heterotrophic microalgae are grown on organic carbon substrates such as glucose in stirred tank bioreactors or fermenters. Mixotrophic microalgage can operate as either autotrophic or heterotrophic.

X = biomass concentrations; P = productivity per unit area or volume.

Source: Adapted from Brennan, Liam, and Philip Owende. 2010. Biofuels from microalgae—A review of technologies for production, processing, and extractions of biofuels and co-products, Renewable and Sustainable Energy Reviews, Volume 14, Issue 2, Pages 557–577.

Table 2.45.   Biomass productivity in open pond and closed photobioreactor systems to produce microalgae

Algae species	Reactor type	Volume (l)	X_{max} (g l⁻¹)	P_{aerial} (g m⁻² day⁻¹)	P_{volume} (g l⁻¹ day⁻¹)	PE (%)
Open pond production systems						
Chlorella sp.	NA	NA	10	25	-	-
N/A	NA	NA	0.14	35	0.117	-
Spirulina platensis	NA	NA	-	-	0.18	-
Spirulina platensis	NA	NA	0.47	14	0.05	-
Haematococcus pluvialis	NA	NA	0.202	15.1	-	-
Spirulina	NA	NA	1.24	69.16	-	-
Various	NA	NA	-	19	-	-
Spirulina platensis	NA	NA	0.9	12.2	0.15	-
Spirulina platensis	NA	NA	1.6	19.4	0.32	-
Anabaena sp.	NA	NA	0.23	23.5	0.24	>2
Chlorella sp.	NA	NA	40	23.5	-	6.48
Chlorella sp.	NA	NA	40	11.1	-	5.98
Chlorella sp.	NA	NA	40	32.2	-	5.42
Chlorella sp.	NA	NA	40	18.1	-	6.07
Closed photobioreactor systems						
Porphyridium cruentum	Airlift tubular	200	3	-	1.5	-
Phaeodactylum tricornutum	Airlift tubular	200	-	20	1.2	-
Phaeodactylum tricornutum	Airlift tubular	200	-	32	1.9	2.3
Chlorella sorokiniana	Inclined tubular	6	1.5	-	1.47	-
Arthrospira platensis	Undular row tubular	11	6	47.7	2.7	-
Phaeodactylum tricornutum	Outdoor helical tubular	75	-	-	1.4	15
Haematococcus pluvialis	Parallel tubular (AGM)	25,000	-	13	0.05	-
Haematococcus pluvialis	Bubble column	55	1.4	-	0.06	-
Haematococcus pluvialis	Airlift tubular	55	7	-	0.41	-
Nannochloropsis sp.	Flat plate	440	-	-	0.27	-
Haematococcus pluvialis	Flat plate	25,000	-	10.2	-	-
Spirulina platensis	Tubular	5.5	-	-	0.42	8.1
Arthrospira	Tubular	146	2.37	25.4	1.15	4.7
Chlorella	Flat plate	400	-	22.8	3.8	5.6
Chlorella	Flat plate	400	-	19.4	3.2	6.9
Tetraselmis	Column	ca. 1,000	1.7	38.2	0.42	9.6
Chlorococcum	Parabola	70	1.5	14.9	0.09	-
Chlorococcum	Dome	130	1.5	11	0.1	-

X = biomass concentrations; P = productivity per unit area or volume; PE = photosynthetic efficiency = the fraction of light energy that is fixed as chemical energy during photoautrophic growth.

Source: Adapted from Brennan, Liam, and Philip Owende. 2010. Biofuels from microalgae—A review of technologies for production, processing, and extractions of biofuels and co-products, Renewable and Sustainable Energy Reviews, Volume 14, Issue 2, Pages 557–577.

Table 2.46.   Energetic performance parameters of mesophilic[a] digestion of sewage sludge

	VS[b] destruction (% VS-input)	Specific biogas production (m³ biogas/kg VS-destroyed)	Calorific value of biogas (MJ/m³)	Energy yield (MJ/kg VS-input)
Range of value	40–50	0.8–1.2	15.9–27.8	
Typical value	45	1	25.8	11.6

[a]30–38° C.
[b]volatile solids content.
Source: Adapted from Cao,Yucheng, Artur Pawłowski. 2012. Sewage sludge-to-energy approaches based on anaerobic digestion and pyrolysis: Brief overview and energy efficiency assessment, Renewable and Sustainable Energy Reviews, Volume 16, Issue 3, Pages 1657–1665.

Table 2.47.   Energetic performances of moderate-temperature fast pyrolysis (400–550° C) in sewage sludge-to-energy system

Sludge feedstock			Pyrolysis conditions				Bio-oil product		AEE[b] (%)
Type[a]	VS[a] (%)	CV[a] (MJ/kg)	Heating source	Operating mode	Temperature (°C)	Time (min)	Yield (%)	CV[a] (MJ/kg)	
PS	84	23	Electric	Batch	500	20	42	37	67.6
WAS	69	19	Electric	Batch	500	20	31	37	60.1
ADS	59	17	Electric	Batch	500	20	26	37	56.6
ADS	47	12.3	Electric	Continuous	550		36	32.1	94
ADS	46.6	11.9	Electric	Continuous	550		27.9	31.2	73
ADS	38.3	8.9	Electric	Continuous	550		24.3	30.6	83.5
SS	75.5		Microwave	Batch	490	10	40	35	
SS	75.5		Electric	Batch	500	30	37	30	

[a]VS = volatile solids content, CV = calorific value, PS = primary sludge, WAS = waste activated sludge, ADS = anaerobically-digested sludge, SS = sewage sludge.
[b]AEE = apparent energy efficiency, calculated according to the ratio of the energy content of bio-oil product to the energy content of its feedstock.
Source: Adapted from Cao,Yucheng, Artur Pawłowski. 2012. Sewage sludge-to-energy approaches based on anaerobic digestion and pyrolysis: Brief overview and energy efficiency assessment, Renewable and Sustainable Energy Reviews, Volume 16, Issue 3, Pages 1657–1665.

Hydropower

Figures

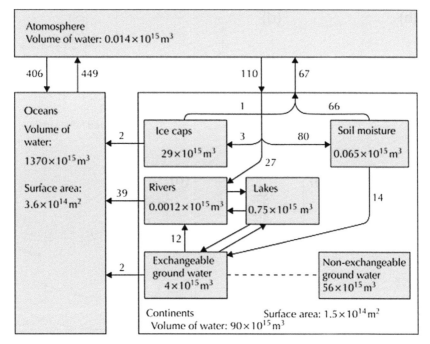

Figure 3.1.   Schematic summary of the water cycle, including free water to a depth of about 5 km below the Earth's surface.
Source: Sørensen, Bent. 2011. Renewable Energy (Fourth Edition), (Boston, Academic Press).

Handbook of Energy. http://dx.doi.org/10.1016/B978-0-08-046405-3.00003-6

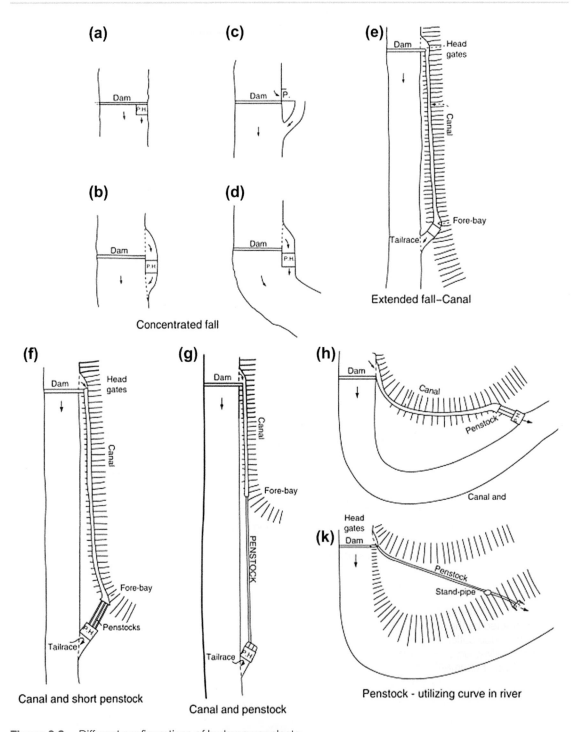

Figure 3.2.   Different configurations of hydropower plants.
Source: Lejeune, A. and S.L. Hu. 2012. Hydro Power: A Multi Benefit Solution for Renewable Energy, In: Ali Sayigh, Editor-in-Chief, Comprehensive Renewable Energy, (Oxford, Elsevier), Pages 15-47.

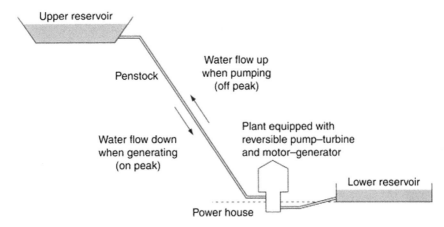

Figure 3.3. Typical layout of a storage hydroelectricity power plant.
Source: Hino, T. and A. Lejeune. 2012. Pumped Storage Hydropower Developments, In: Ali Sayigh, Editor-in-Chief, Comprehensive Renewable Energy, (Oxford, Elsevier), Pages 405-434.

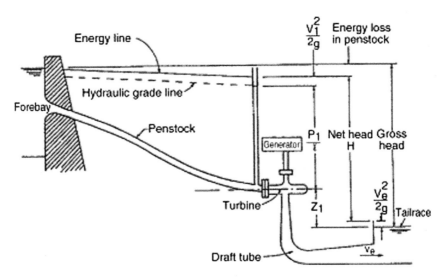

Figure 3.4. Schematic of a hydropower installation. V_1^2 is head lost before the turbine, P_i/γ is the positive head extracted by the turbine, Z_1 is the suction head on the turbine, and $V_c^2/2g$ is the velocity head remaining at the draft tube outlet.
Source: Gulliver, John S. and Roger E.A. Arndt. 2004. Hydropower, History and Technology of, In: Cutler J. Cleveland, Editor-in-Chief, Encyclopedia of Energy, (New York, Elsevier), Pages 301-314.

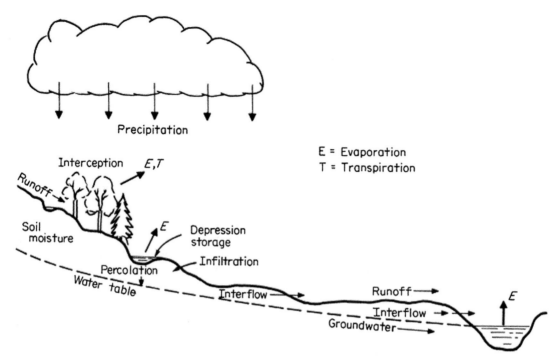

Figure 3.5.   Processes affecting the relationship between precipitation and runoff.
Source: Gulliver, John S. and Roger E.A. Arndt, Hydroelectric Power Stations, In: Robert A. Meyers, Editor-in-Chief, Encyclopedia of Physical Science and Technology (Third Edition), (New York, Academic Press), Pages 489-504.

Charts

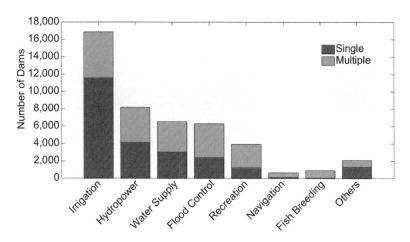

Chart 3.1.   Number and purpose of registered dams in the world.
Source: Data from International Commission on Large Dams, Registry of Dams,
<http://www.icold-cigb.org/>.

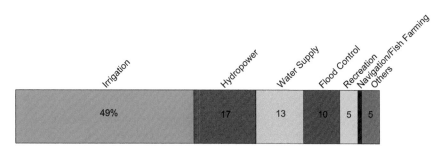

Chart 3.2.   Distribution of single purpose dams in the world.
Source: Data from International Commission on Large Dams, Registry of Dams,
<http://www.icold-cigb.org/>.

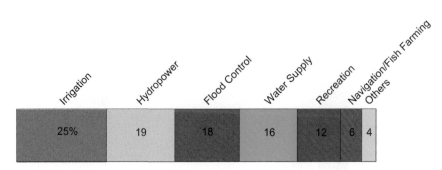

Chart 3.3. Distribution of multipurpose dams in the world.
Source: Data from International Commission on Large Dams, Registry of Dams,
<http://www.icold-cigb.org/>.

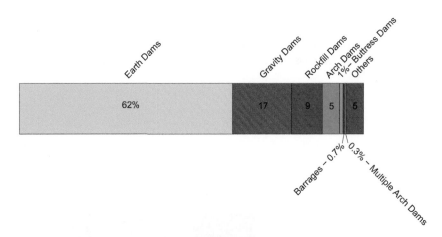

Chart 3.4. Distribution of dams in the world by type of construction.
Source: Data from International Commission on Large Dams, Registry of Dams,
<http://www.icold-cigb.org/>.

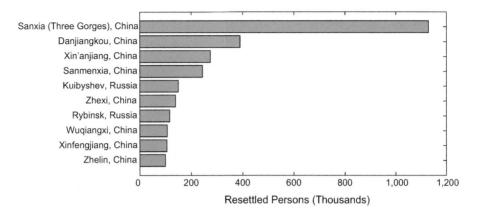

Chart 3.5. Largest resettlements of people caused by dam construction.
Source: Data from International Commission on Large Dams, Registry of Dams,
<http://www.icold-cigb.org/>.

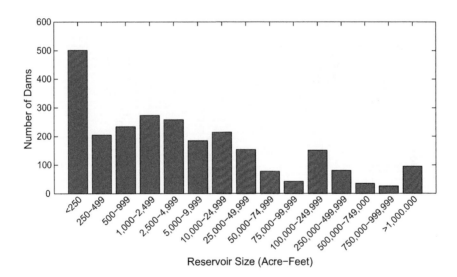

Chart 3.6. Reservoir size distribution for United Sates hydroelectric plants.
Source: Data from National Performance of Dams Program, Dams Directory, Department of Civil &
Environmental Engineering, Stanford University, <http://npdp.stanford.edu/npdphome/damdir.htm#>.

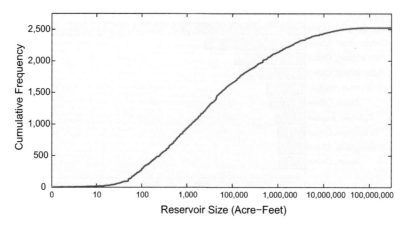

Chart 3.7.   Cumulative frequency plot of storage area for reservoirs in the United States.
Source: Data from National Performance of Dams Program, Dams Directory, Department of Civil & Environmental Engineering, Stanford University, <http://npdp.stanford.edu/npdphome/damdir.htm#>.

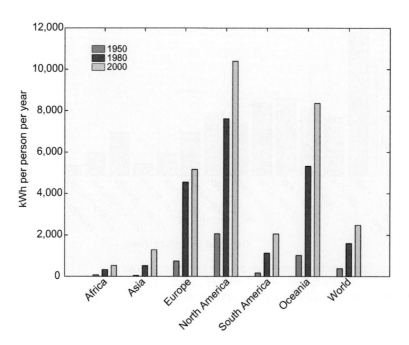

Chart 3.8.   Historical role of hydropower in the world electricity system, 1950-2000.
Source: Data from Sternberg, R. 2010. Hydropower's future, the environment, and global electricity systems, Renewable and Sustainable Energy Reviews, Volume 14, Issue 2, Pages 713-723.

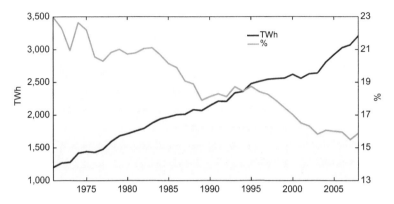

Chart 3.9.   Global electricity generation from hydropower.
Source: Data from International Energy Agency (IEA), Energy statistics database, <http://www.iea.org/stats/index.asp>.

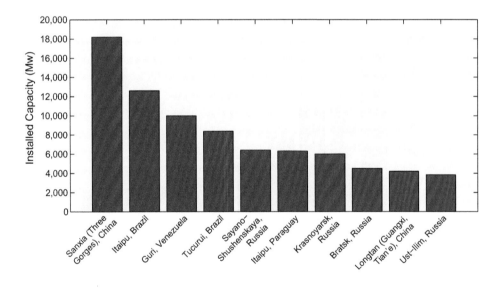

Chart 3.10.   World's largest hydroelectric plants, 2010.
Source: Data from International Commission on Large Dams, Registry of Dams, <http://www.icold-cigb.org/>.

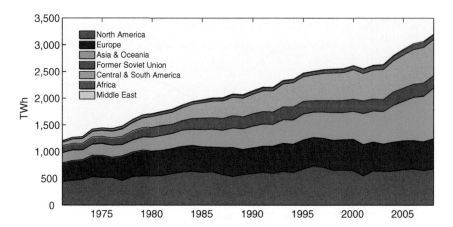

Chart 3.11. Electricity generation from hydropower energy by region, 1970-2009.
Source: Data from International Energy Agency (IEA), Energy statistics database, <http://www.iea.org/stats/index.asp>.

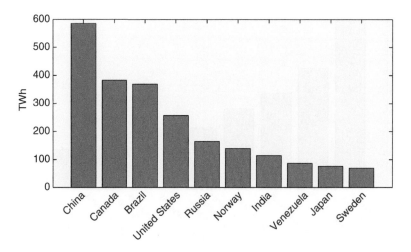

Chart 3.12. Net generation of electricity from hydropower by country, top 10 nations, 2008.
Source: Data from International Energy Agency (IEA), Energy statistics database, <http://www.iea.org/stats/index.asp>.

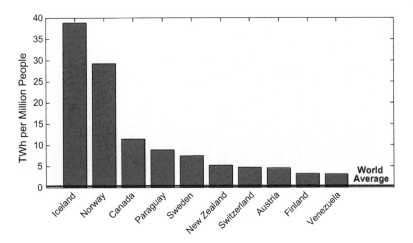

Chart 3.13. Per capita net generation of electricity from hydropower, top 10 nations, 2008.
Source: Data from International Energy Agency (IEA), Energy statistics database, <http://www.iea.org/stats/index.asp>.

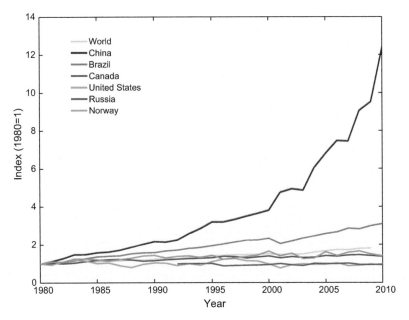

Chart 3.14. Index of net generation of electricity by hydropower by top producing nations, 1980-2010.
Source: Data from BP, Statistical Review of World Energy 2011.

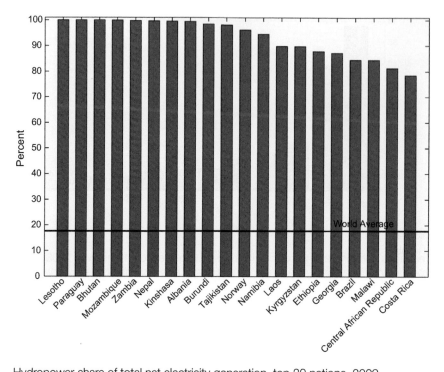

Chart 3.15. Hydropower share of total net electricity generation, top 20 nations, 2009.
Source: Data from United States Department of Energy, Energy Information Administration, International Energy Statistics, <http://www.eia.gov/countries/data.cfm>.

Tables

Table 3.1.   **Global Hydropower Potential**

Geographical region	Technically feasible potential (~GWh/year)	Economically feasible potential (~GWh/year)
Africa	1,750,000	1,100,000
Countries with major resources:		
Angola	90,000	65,000
Cameroon	115,000	103,000
Congo, Democratic Republic of	774,000	419,000
Ethiopia		260,000
Madagascar	180,000	49,000
Asia (including Russia and Turkey)	6,800,000	3,600,000
Countries with major resources:		
China, People's Republic of	2,200,000	1,270,000
India	660,000	
Japan	135,000	114,000
Russian Federation	1,670,000	852,000
Turkey	215,000	123,000
Vietnam	100,000	80,000
Australia/Oceania	270,000	107,000
Countries with major resources:		
Australia		30,000
New Zealand	77,000	40,000
Europe	1,035,000	791,000
Countries with major resources:		
Austria	54,000	50,000
France	72,000	70,000
Iceland	64,000	44,000
Italy	69,000	54,000
Norway	200,000	187,000
Spain	70,000	41,000
Sweden	200,000	130,000
North and Central America	1,663,000	1,000,000
Countries with major resources:		
Canada	981,000	536,000
Mexico	49,000	32,000
United States	528,000	376,000
South America	2,700,000	1,600,000
Countries with major resources:		
Argentina	130,000	
Bolivia	126,000	50,000
Brazil	1,300,000	764,000
Chile	162,000	
Colombia	200,000	140,000
Ecuador	133,000	106,000
Venezuela	261,000	100,000
World Totals	14,218,000	8,198,000

Source; Adapted from Sommers, Garold L. 2004. Hydropower Resources, In: Cutler J. Cleveland, Editor-in-Chief, Encyclopedia of Energy, (Elsevier, New York), Pages 325–332; based on data from The International Journal on Hydropower and Dams: World Atlas and Industry Guide, (Aqua~Media International Westemead House, Sutton, UK.

Table 3.2. Historical role of hydropower in the global electrcity system, 1950–2000[a]

1950

	Thermal	Hydropower	Total	kWh/p/y
World	102,473	57,360	154,092	382
Africa	3,062	485	3,521	66
Asia	7,328	7,584	14,912	41
Europe	55,665	27,355	83,259	733
N. America	N/A	N/A	N/A	2,061
S. America	2,892	2,164	5,056	167
Oceania	2,045	944	2,089	1,019

1960

	Thermal	Hydropower	Total	kWh/p/y
World	369,957	149,571	520,755	763
Africa	7,285	1,979	9,264	128
Asia	20,119	16,512	36,631	122
Europe	117,262	53,879	171,885	1,607
N. America	161,375	53,569	215,253	3,643
S. America	7,225	6,096	13,321	349
Oceania	4,856	2,755	7,650	1,960

1970

	Thermal	Hydropower	Total	kWh/p/y
World	817,945	290,652	1,125,392	1,347
Africa	16,608	7,508	24,116	244
Asia	93,995	42,775	138,537	279
Europe	223,934	99,365	330,832	3,077
N. America	321,766	88,114	416,711	5,934
S. America	14,416	14,654	29,070	564
Oceania	12,896	6,888	19,976	3,620

1980

	Thermal	Hydropower	Total	kWh/p/y
World	1,403,416	465,396	2,013,006	1,601
Africa	29,728	13,170	42,898	321
Asia	226,923	718,218	324,294	510
Europe	350,619	137,169	534,763	4,551
N. America	547,203	132,180	742,989	7,611
S. America	26,099	41,914	68,383	1,131
Oceania	22,598	10,234	32,992	5,322

1990

	Thermal	Hydropower	Total	kWh/p/y
World	1,771,591	628,429	2,746,127	2,207
Africa	52,370	19,332	72,742	486
Asia	404,973	120,592	572,942	817
Europe	376,809	171,588	678,338	5,710
N. America	638,100	160,186	926,446	8,658
S. America	34,759	80,375	116,809	1,504
Oceania	32,890	12,256	45,750	10,600

2000

	Thermal	Hydropower	Total	kWh/p/y
World	2,230,616	763,808	3,377,828	2,475
Africa	77,451	22,104	101,463	524
Asia	750,720	192,269	1,020,013	1,288
Europe	627,616	241,317	1,055,070	5,168
N. America	682,143	178,957	981,638	10,388
S. America	52,127	119,149	165,274	2,055
Oceania	40,559	13,284	54,370	8,362

[a]Units are installed MW unless otherwise specified. Thermal = derived from a facility in which the prime mover is steam driven. kWh/p/y = per capita electricity consumption.
Source: Adapted from Sternberg, R. 2010. Hydropower's future, the environment, and global electricity systems, Renewable and Sustainable Energy Reviews, Volume 14, Issue 2, Pages 713-723.

Table 3.3. Regional hydropower technical potential and installed capacity, 2009

Region	Technical potential, annual generation TWh/yr (EJ/yr)	Technical potential, installed capacity (GW)	2009 Total generation TWh/yr (EJ/yr)	2009 Installed capacity (GW)	Undeveloped potential (%)	Average regional capacity factor (%)
North America	1,659 (5.971)	338	628 (2.261)	153	61	47
Latin America	2,856 (10.283)	608	732 (2.635)	156	74	54
Europe	1,021 (3.675)	338	542 (1.951)	179	47	35
Africa	1,174 (4.226)	283	98 (0.351)	23	92	47
Asia	7,681 (27.651)	2,037	1,514 (5.451)	402	80	43
Australasia/ Oceania	185 (0.666)	67	37 (0.134)	13	80	32
World	**14,576 (52.470)**	**3721**	**3,551 (12.783)**	**926**	**75**	**44**

Source: Adapted from Kumar, A., T. Schei, A. Ahenkorah, R. Caceres Rodriguez, J.-M. Devernay, M. Freitas, D. Hall, Å. Killingtveit, Z. Liu, 2011: Hydropower. In IPCC Special Report on Renewable Energy Sources and Climate Change Mitigation [O. Edenhofer, R. Pichs-Madruga, Y. Sokona, K. Seyboth, P. Matschoss, S. Kadner, T. Zwickel, P. Eickemeier, G. Hansen, S. Schlömer, C. von Stechow (eds)], Cambridge University Press, Cambridge, United Kingdom and New York, NY, USA; based on data from International Journal of Hydropower and Dams. 2010. World Atlas & Industry Guide , Wallington, Surrey, UK, 405 pp.

Table 3.4. Net generation of hydroelectricity by top 20 nations, 1980-2010

Country	1980	1990	2000	2010	Rank 2010	Rank 1980	Change 2000–10
China	57.6	125.1	220.2	713.8	1	8	224.2%
Brazil	128.4	204.6	301.4	401.0	2	3	33.1%
Canada	251.0	293.9	354.9	348.0	3	2	–2.0%
United States	279.2	292.9	275.6	257.1	4	1	–6.7%
Russia	n/a	n/a	162.4	164.5	5	n/a	1.3%
Norway	82.7	119.9	137.5	116.4	6	5	–15.4%
India	46.5	70.9	73.7	110.3	7	9	49.6%
Venezuela	14.4	36.6	62.2	76.0	8	16	22.2%
Japan	87.8	88.4	86.4	73.4	9	4	–15.0%
Sweden	58.1	71.8	77.8	70.6	10	7	–9.2%
France	68.3	52.8	66.5	62.2	11	6	–6.5%
Turkey	11.2	22.9	30.6	51.3	12	20	67.7%
Italy	45.0	31.3	43.8	50.1	13	10	14.4%
Spain	29.2	25.2	29.3	41.8	14	12	42.8%
Colombia	14.3	27.2	31.8	39.8	15	17	25.5%
Mexico	16.7	23.2	32.8	36.7	16	15	12.0%
Switzerland	32.5	29.5	36.5	35.7	17	11	–2.1%
Austria	28.5	31.2	41.4	34.4	18	13	–16.9%
Pakistan	8.6	16.8	17.0	28.0	19	23	64.4%
Vietnam	1.2	5.3	14.4	27.4	20	38	90.0%
World	1,722.9	2,144.5	2,619.5	3145.2[a]			20.1%

[a]Value is for 2009

Source: United States Department of Energy, Energy Information Administration, International Energy Statistics, <http://www.eia.gov/countries/data.cfm>.

Table 3.5.   Cost estimates for hydropower

Study	LCOE[a] (cents/kWh)	Comments
WEA 2004	2–10	Large Hydro
	2–12	Small Hydro (<10 MW) (Not explicitly stated as levelized cost in report)
IEA-WEO 2008	7.1	
IEA-ETP 2008	3–12	Large Hydro
	5.6–14	Small Hydro
EREC/Greenpeace	10.4	
BMU Lead Study 2008	7.3	Study applies to Germany only
Krewitt et al 2009	9.8	Indicative average LCOE year 2000
IEA-2010	2.3–45.9	Range for 13 projects from 0.3 to 18,000 MW
	4.8	Weighted average for all projects
REN21	5–12	Small Hydro (<10 MW)
	3–5	Large Hydro (>10 MW)
	5–40	Off-Grid (<1 MW)

[a]The levelized cost of energy (LCOE) is defined as the constant price per unit of energy that causes the investment to just break even. Put another way, it is the unique cost price of the outputs of a project that makes the present value of the revenues (benefits) equal to the present value of the costs over the lifetime of the project.

Studies cited:

Hall, D.G., Hunt, R.T., Reeves, K.S. and Carroll, G.R., (2003). Estimation of Economic Parameters of U.S. Hydropower Resources. INEEL/EXT-03-00662, U.S. Department of Energy Idaho Operations Office, Idaho Falls, ID, USA, 25 pp.

Lako, P., H. Eder, M. de Noord, and H. Reisinger (2003). Hydropower Development with a Focus on Asia and Western Europe. Overview in the Framework of VLEEM 2. ECN-C--03-027, Verbundplan, Vienna, Austria, 96 pp.

UNDP/UNDESA/WEC (2004). World Energy Assessment Overview: 2004 Update. J. Goldemberg, and T.B. Johansson (eds.), United Nations Development Programme, United Nations Department of Economic and Social Affairs, and World Energy Council, New York, New York, USA, 85 pp.

IEA (2008a). World Energy Outlook 2008. International Energy Agency, Paris, France, 578 pp.

IEA (2008b). Energy Technology Perspectives 2008. Scenarios and Strategies to 2050. International Energy Agency, Paris, France, 646 pp.

Teske, S., T. Pregger, S. Simon, T. Naegler, W. Graus, and C. Lins (2010). Energy [R]evolution 2010—a sustainable world energy outlook. Energy Efficiency, doi:10.1007/s12053-010-9098-y.

BMU (2008). Lead Study 2008 - Further development of the Strategy to increase the use of renewable energies within the context of the current climate protection goals of Germany and Europe. German Federal Ministry for the Environment, Nature Conservation and Nuclear Safety (BMU), Berlin, Germany, 118 pp.

Krewitt, W., K. Nienhaus, C. Kleßmann, C. Capone, E. Stricker, W. Graus, M. Hoogwijk, N. Supersberger, U. von Winterfeld, and S. Samadi (2009). Role and Potential of Renewable Energy and Energy Efficiency for Global Energy Supply. Climate Change 18/2009. ISSN 1862-4359. Federal Environment Agency, Dessau-Roßlau, 336 pp.

IEA (2010b). Projected Costs of Generating Electricity. International Energy Agency, Paris, France, 218 pp.

REN21 (2010). Renewables 2010: Global Status Report. Renewable Energy Policy Network for the 21st Century Secretariat, Paris, France, 80 pp.

Table 3.6.   World's tallest dams, 2010

Dam Name	Height(m)	Purposes	Country
ROGUN	335	H,I	Tadjikistan
NUREK	300	I,H	Tadjikistan
XIAOWAN (YUNNAN)	292	HCIN	China
GRANDE DIXENCE	285	H	Switzerland
INGURI	272	H,I	Georgia
VAJONT	262	H	Italy
MANUEL M. TORRES	261	H	Mexico
TEHRI	261	IS	India
ALVARO OBREGON	260	IS	Mexico
MAUVOISIN	250	H	Switzerland
ALBERTO LLERAS C.	243	H	Colombia
MICA	243	H	Canada
SAYANO -SHUSHENSKAYA	242	NH	Russia
ERTAN	240	H,C,I	China
LA ESMERALDA (CHIVOR)	237	H	Colombia
KISHAU	236	IH	India
OROVILLE	235	CISHR	United States
EL CAJON	234	HICR	Honduras
CHIRKEY	233	H,I,S	Russia
SHUIBUYA	233	HCN	China

H = hydroelectricity; I = irrigation; N = navigation; R = recreation; S = water supply; C = flood control.
Source: International Commission on Large Dams, Register of Dams, <http://www.icold-cigb.org/GB/World_register/world_register.asp>.

Table 3.7.   Dams with largest generation of hydroelectricity, 2010

Dam Name	Year Completed	Installed Capacity (Mw)	Energy (GWh/year)	Country
SANXIA (THREE GORGES)	2009	18,200	84,000	China
ITAIPU	1991	12,600	90,000	Brazil
GURI	1986	10,000	52,000	Venezuela
TUCURUI	1984	8,370		Brazil
SAYANO -SHUSHENSKAYA	1990	6,400	22,800	Russia
ITAIPU	1983	6,300		Paraguay
KRASNOYARSK	1967	6,000	19,600	Russia
BRATSK	1964	4,500	22,500	Russia
LONGTAN(GUANGXI,TIAN'E)	2001	4,200		China
UST-ILIM	1977	3,840	21,200	Russia
ILHA SOLTEIRA	1973	3,444	3,230	Brazil
ERTAN	1999	3,300	17,000	China
YACYRETA	1994	3,100	20,000	Argentina
XINGO	1994	3,000	19,277	Brazil
MACAGUA II	1996	2,940	16,600	Venezuela
GEZHOUBA	1988	2,715	15,700	China
MINAMIAIKI	2004	2,700	2,160	Japan
VOLGOGRAD	1958	2,541	10,800	Russia
PAULO AFONSO IV	1979	2,460	8,743	Brazil
ATATURK	1992	2,400	8,900	Turkey

Source: International Commission on Large Dams, Register of Dams, <http://www.icold-cigb.org/GB/World_register/world_register.asp>.

Table 3.8.   World's largest pumped-storage hydroelectricity facilities

Station	Country	Capacity (MW)
Bath County Pumped Storage Station	United States	2,772
Kannagawa Pumped Storage Power Station	Japan	2,700
Robert Moses Niagara Power Plant	United States	2,515
Guangzhou Pumped Storage Power Station	China	2,400
Huizhou Hydroelectric Power Station	China	2,400
Dniester Hydroelectric Power Station	Ukraine	2,268
Grande Dixence Dam	Switzerland	2,069
Mt. Hope Dam	United States	2,000
Okutataragi Pumped Storage Power Station	Japan	1,932
Ludington Pumped Storage Power Plant	United States	1,872
Tianhuangping Pumped Storage Power Station	China	1,800
Zhuhai Pumped Storage Station	China	1,800
Dinorwig Power Station	United Kingdom	1,728
Mingtan Dam	Taiwan	1,602
Kazunogawa Dam	Japan	1,600
Kruonis Pumped Storage Plant	Lithuania	1,600
Sir Adam Beck Hydroelectric Power Stations	Canada	1,600
Castaic Dam	United States	1,566
Raccoon Mountain Pumped Storage Plant	United States	1,530
Tumut-3	Australia	1,500

Source: Adapted from Hino, T. and A. Lejeune. 2012. Pumped Storage Hydropower Developments, In: Ali Sayigh, Editor-in-Chief, Comprehensive Renewable Energy, (Oxford, Elsevier), Pages 405-434.

Table 3.9.   Classification of hydropower generation[a]

Hydro generator	Capacity	Feeding
Large	More than 100 MW	National power grid
Small	Up to 25 MW	National power grid
Mini	Below 1 MW	Micro power grid
Micro	Between 6 and 100 kW	Small community or remote industrial areas
Pico	Up to 5 kW	Domestic and small commercial loads

[a]Capacity ranges are approaximate and vary by country.
Source: Adapted from Haidar, Ahmed M.A., Mohd F.M. Senan, Abdulhakim Noman, Taha Radman. 2012. Utilization of pico hydro generation in domestic and commercial loads, Renewable and Sustainable Energy Reviews, Volume 16, Issue 1, Pages 518-524.

Table 3.10.   Applications of energy storage in power systems with different discharge times

	Application	Power rating	Discharge duration	Storage capacity	Response time	System location
Fast discharge	Transit and end use ride-through	<1 MW	Seconds	~ 2 kWh	<1/4 cycle	End use
	Transmission and distribution stabilization	up to 100 s MVA	Seconds	20–50 kV Ah	<1/4 cycle	Transmission and distribution
Short to long discharge	Voltage regulation	up to 10 MVAR	Minutes	250–2 500 kV Arh	<1/4 cycle	Transmission
	Fast response spinning reserve	10–100 MW	<30 m	5 000–500 000 kWh	<3 s	Generation
	Conventional spinning reserve	10–100 MW	<30 m	5 000–500 000 kWh	<10 min	Generation
	Uninterruptible power supply	<2 MW	~ 2 h	100–4 000 kWh	Seconds	End use
	End use and transmission peak shaving	<5 MW	1–3 h	1 000–150 000 kWh	Seconds	End use and distribution
	Transmission upgrade deferral	up to 100 s MW	1–3 h	1 000–500 000 kWh	Seconds	Transmission
	Renewable matching (short discharge)	<100 MW	Min–1 h	10–100 000 kWh	<1 cycle	Generation
Long discharge	Renewable matching (long discharge)	<100 MW	1–10 h	1 000–100 000 kWh	Seconds	Generation
	Load levelling	100 s MW	6–10 h	100–10 000 MWh	Minutes	Generation
	Load following	10–100 s MW	Several hours	10–1 000 MWh	< cycle	Generation and distribution
	Emergency back-up	<1 MW	24 h	24 MWh	Seconds	End use
	Renewables back-up	100 kW–1 MW	Days	20–200 MWh	Seconds–Minutes	Generation and end-use

MVA = megavolt-ampere; MVAR = megavolt-ampere reactive.
Source: Adapted from Hino, T. and A. Lejeune. 2012. Pumped Storage Hydropower Developments, In: Ali Sayigh, Editor-in-Chief, Comprehensive Renewable Energy, (Oxford, Elsevier), Pages 405-434.

Table 3.11. **Turbine type and head classification used in hydropower**

Turbine	Runner type	Head pressure	Approximate Height
Reaction	Propeller, Kaplan	Ultra Low	Below 3 m
Reaction	Propeller, Kaplan		
Impulse	Crossflow	Low	Above 3 m
Reaction	Francis, Pump as Turbine		Above 40 m
Impulse	Crossflow, Turgo, Multi-jet Pelton	Medium	
Impulse	Pilton, Turgo, Multi-jet Pelton	High	Above 100 m

Source: Adapted from Haidar, Ahmed M.A., Mohd F.M. Senan, Abdulhakim Noman, Taha Radman. 2012. Utilization of pico hydro generation in domestic and commercial loads, Renewable and Sustainable Energy Reviews, Volume 16, Issue 1, Pages 518-524.

Table 3.12. **Annual water availability in select river basins**

Region	River basin	Catchment Area km^2	Annual Precipitation (km^3)	km^2/km^3 [a]	Mean Flow (m^3/s)
Asia	Yangtze	1,810,000	1,003	1,805	35,000
	Ganges/Brahmaputra	1,750,000	1,389	1,260	20,000
	Indus	960,000	220	4,364	3,850
	Mekong	646,000	459	1,407	15,900
	Armur	1,860,000	355	5,240	12,500
Europe	Danube	578,300	176	3,286	6,450
	Volga	1,360,000	252	5,397	8,000
	Pechora	317,000	137	2,314	4,060
	Rhine	103,700	51	2,050	2,200
Africa	Niger	2,090,000	302	6,920	5,700
	Congo	3,680,000	1,320	2,788	42,000
	Nile	2,870,000	161	17,826	1,584
North America	Mississippi	2,980,000	515	5,786	17,545
	Columbia	668,000	237	2,819	6,650
	Colorado	637,000	16	38,813	168
	St. Lawrence	1,026,000	320	3,206	10,400
South America	Amazon	6,920,000	6,920	1,000	180,000
	Paraná	3,100,000	811	3,822	19,500
	Orinoco	1,000,000	1,010	990	28,000
	Bio Bio	21,220	36	596	1,230

[a]The larger the catchment area per km^3, the more uncertain the streamflow regime.
Source: Adapted from Sternberg, R. 2010. Hydropower's future, the environment, and global electricity systems, Renewable and Sustainable Energy Reviews, Volume 14, Issue 2, Pages 713-723.

Table 3.13.   Hydropower 80% depletion dates for various regions based on current rates of sedimentation[a]

Region	Hydropower dams: Date by which 80% is filled with sediment
Africa	2100
Australasia	2035
Europe and Russia	2070
North America	2060
Asia	2080
Central America	2060
Middle East	2060
South America	2080

[a]The ability of hydropower dams to generate power may be severely impacted when they reach a level of sedimentation of 80%.
Source: Adapted from Horlacher, H., T. Heyer, C.M. Ramos, M.C. da Silva. 2012. Management of Hydropower Impacts through
Construction and Operation, In: Ali Sayigh, Editor-in-Chief, Comprehensive Renewable Energy, (Oxford,Elsevier), Pages 49-91.

Table 3.14.   Annual reduction in reservoir volume due to sedimentation in various regions

Region	Number of dams studied	Annual reduction in reservoir volume due to sedimentation (%)
North America	7,205	0.2
South America	1,498	0.1
North Europe	2,277	0.2
South Europe	3,220	0.17
North Africa	280	0.08
Sub-Saharan Africa	966	0.23
China	1,851	2.3
South Asia	4,131	0.52
Central Asia	44	1
Southeast Asia	277	0.3
Pacific Border	2,778	0.27
Middle East	895	1.5
World total	25,422	0.5–1.0 (average)

Source: Adapted from Horlacher, H., T. Heyer, C.M. Ramos, M.C. da Silva. 2012. Management of Hydropower Impacts through Construction
and Operation, In: Ali Sayigh, Editor-in-Chief, Comprehensive Renewable Energy, (Oxford,Elsevier), Pages 49-91; based on data Alves
E (2008) Sedimentation in Reservoirs by Turbidity Currents (in Portuguese). PhD Thesis, Laboratório Nacional de Engenharia Civil.

Table 3.15. Range of gross CO_2 and CH_4 emissions from hydropower freshwater reservoirs; numbers of studied reservoirs are given in parentheses

	Boreal and temperate		Tropical	
	CO_2	CH_4	CO_2	CH_4
	(mmol/m²/d)	(mmol/m²/d)	(mmol/m²/d)	(mmol/m²/d)
Diffusive fluxes	−23 to 145 (107)	−0.3 to 8 (56)	−19 to 432 (15)	0.3 to 51 (14)
Bubbling	0	0 to 18 (4)	0	0 to 88 (12)
Degassing	~0.2 (2) to 0.1 (2)	n.a.	4 to 23 (1)	4 to 30 (2)
River below the dam	n.a.	n.a.	500 to 2500 (3)	2 to 350 (3)

[a]The degassing (generally in mg/d) is attributed to the surface of the reservoir and is expressed in the same units as the other fluxes (mmol/m²/d).
Source: Adapted form Kumar, A., T. Schei, A. Ahenkorah, R. Caceres Rodriguez, J.-M. Devernay, M. Freitas, D. Hall, Å. Killingtveit, Z. Liu, 2011: Hydropower. In IPCC Special Report on Renewable Energy Sources and Climate Change Mitigation [O. Edenhofer, R. Pichs-Madruga, Y. Sokona, K. Seyboth, P. Matschoss, S. Kadner, T. Zwickel, P. Eickemeier, G. Hansen, S. Schlömer, C. von Stechow (eds)], Cambridge University Press, Cambridge, United Kingdom and New York, NY, USA.

Table 3.16. Types of hydropower projects (HPP), their main services and distinctive environmental and social impacts

Type	Energy and water management services	Main environmental and social impacts
All	Renewable electricity generation Increased water management options	Barrier for fish migration and navigation, and sediment transport; Physical modification of riverbed and shorelines
Run-of-river	Limited flexibility and increased variability in electricity generation output profile; Water quality (but no water quantity) management	Unchanged river flow when powerhouse in dam toe; when localized further downstream reduced flow between intake and powerhouse
Reservoir (Storage)	Storage capacity for energy and water; Flexible electricity generation output; Water quantity and quality management; groundwater stabilization; water supply and flood management	Alteration of natural and human environment by impoundment, resulting in impacts on ecosystems and biodiversity and communities; dislocation of human settlements; Modification of volume and seasonal patterns of river flow, changes in water temperature and quality, land use change-related GHG emissions
Multipurpose	As for reservoir HPPs; Dependent on water consumption of other uses	As for reservoir HPP; Possible water use conflicts; Driver for regional development
Pumped storage	Storage capacity for energy and water; net consumer of electricity due to pumping; No water management options	Impacts confined to a small area; often operated outside the river basin as a separate system that only exchanges the water from a nearby river from time to time

Source: Adapted from Kumar, A., T. Schei, A. Ahenkorah, R. Caceres Rodriguez, J.-M. Devernay, M. Freitas, D. Hall, Å. Killingtveit, Z. Liu, 2011: Hydropower. In IPCC Special Report on Renewable Energy Sources and Climate Change Mitigation [O. Edenhofer, R. Pichs-Madruga, Y. Sokona, K. Seyboth, P. Matschoss, S. Kadner, T. Zwickel, P. Eickemeier, G. Hansen, S. Schlömer, C. von Stechow (eds)], Cambridge University Press, Cambridge, United Kingdom and New York, NY, USA.

Table 3.17.   Dams that caused the largest resettlement of people

Dam Name	Reservoir Capacity (10^3 m^3)	Resettled persons	Country
SANXIA (THREE GORGES)	39,300,000	1,130,000	China
DANJIANGKOU	20,890,000	390,000	China
XIN'ANJIANG	17,860,000	274,237	China
SANMENXIA	35,400,000	243,700	China
KUIBYSHEV	58,000,000	150,000	Russia
ZHEXI	3,570,000	139,500	China
RYBINSK	25,400,000	116,700	Russia
WUQIANGXI	2,990,000	107,148	China
XINFENGJIANG	13,896,000	106,000	China
ZHELIN	7,920,000	99,845	China
XIJIN	3,000,000	89,323	China
HUALIANGTING	2,398,000	61,124	China
VOTKINSK	9,400,000	61,000	Russia
SHUIKOU (FUJIAN)	2,340,000	60,000	China
GAOZHOU	1,151,100	58,867	China
WAN'AN	2,216,000	57,216	China
MIYUN	4,375,000	56,908	China
ANKANG	2,580,000	56,520	China
ITAPARICA	10,780,000	56,000	Brazil
GUANTING	4,160,000	53,000	China

Source: International Commission on Large Dams, Register of Dams, <http://www.icold-cigb.org/GB/World_register/world_register.asp>.

Table 3.18.   Key features of the Three Gorges Project[a]

Normal pool level	175 m (156 m at the initial stage)
Limit level for flood control	145 m (135 m at the initial stage)
Low level in dry season	155 m (140 m at the initial stage)
Pool level with 1% flood	166.9 m
Design peak flow (0.1% flood)	98 800 m^3 s^{-1}
Design flood level (0.1% flood)	175 m
Check peak flow (0.01% flood plus 10%)	124 000 m^3 s^{-1}
Check flood level	180.4 m
Total storage capacity (below the normal pool level)	39.3 billion m^3
Storage capacity for flood control (145–175 m)	22.15 billion m^3
Regulating storage (155–175 m)	16.5 billion m^3
Crest elevation	185 m
Installed gross capacity/guaranteed output	22.4 GW/4.99 GW
Average annual output	90.0 billion kWh
Single unit capacity/number of units	700 MW/32 units

[a]The Three Gorges Project is one of he world's largest hydrelectric facilities. It is located at the boundary between the middle and downstream of the Yangtze River with its dam situated in Sandouping of Yichang City, Hubei Province, about 40 km upstream of the Gezhouba Project completed in the 1980s.
Source: Adapted from Suo, L., X. Niu, H. Xie. 2012. The Three Gorges Project in China, In: Ali Sayigh, Editor-in-Chief, Comprehensive Renewable Energy, (Oxford, Elsevier), Pages 179-226.

Wind

Figures

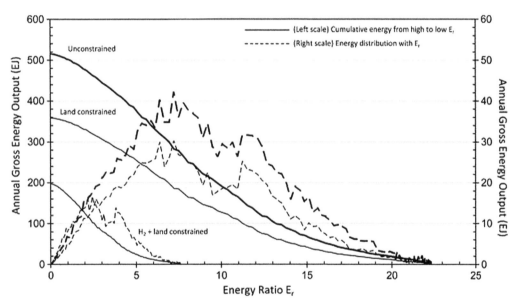

Figure 4.1. Global annual gross wind energy output against energy ratio (E_r). E_r is the ratio of energy produced to the direct and indirect input energy. Energy is shown as both a cumulative sum for the marginal E_r (left axis) and distribution in E_r (right axis). In the unconstrained case, all land area is considered; in the land constrained case, sensitive areas such as forests and wetlands, as well as urban areas, irrigated cropland and pasture are removed. Removing these areas has the effect of reducing the gross amount of energy produced. *Source: Moriarty, Patrick and Damon Honnery. 2012. What is the global potential for renewable energy?, Renewable and Sustainable Energy Reviews, Volume 16, Issue 1, Pages 244-252.*

Handbook of Energy. http://dx.doi.org/10.1016/B978-0-08-046405-3.00004-8

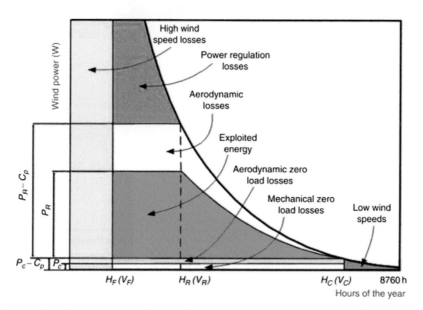

Figure 4.2. Distribution of energy losses through the process of conversion from available wind power to useful electrical over the course of one year. P=power (Watts); P_R=nominal power (Watts); V=velocity (meters/sec); C_p=power coefficient (dimensionless); C_p represents the energy generated by the turbine relative to the energy available from the wind.
Source: Zafirakis, D.P., A.G. Paliatsos, J.K. Kaldellis. 2012. 2.06 - Energy Yield of Contemporary Wind Turbines, In: Editor-in-Chief: Ali Sayigh, Editor(s)-in-Chief, Comprehensive Renewable Energy, (Oxford, Elsevier), Pages 113-168.

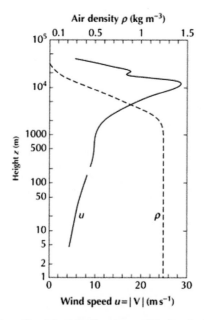

Figure 4.3. Annual average height profile of the kinetic energy density of wind at latitudes 50°–56° N.
Source: Sørensen, Bent. 2011. Renewable Energy (Fourth Edition), (Boston, Academic Press).

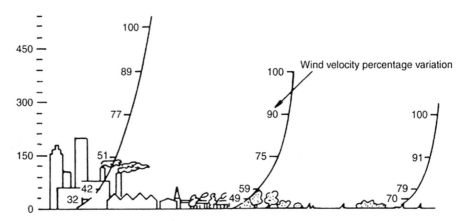

Figure 4.4. The effect of ground roughness on the atmospheric boundary layer development, showing how wind speed varies with altitude. The vertical axis is height (meters).
Source: Katsaprakakis, D. Al and D.G. Christakis. 2012. 2.07 - Wind Parks Design, Including Represen-tative Case Studies, In: Editor-in-Chief: Ali Sayigh, Editor(s)-in-Chief, Comprehensive Renewable Energy, (Oxford, Elsevier), Pages 169-223.

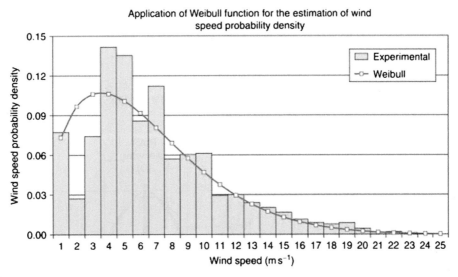

Figure 4.5. Use of Weibull distribution (a continuous probability distribution) to describe wind speed probability density.
Source: Zafirakis, D.P., A.G. Paliatsos, J.K. Kaldellis. 2012. 2.06 - Energy Yield of Contemporary Wind Turbines, In: Editor-in-Chief: Ali Sayigh, Editor(s)-in-Chief, Comprehensive Renewable Energy, (Oxford, Elsevier), Pages 113-168.

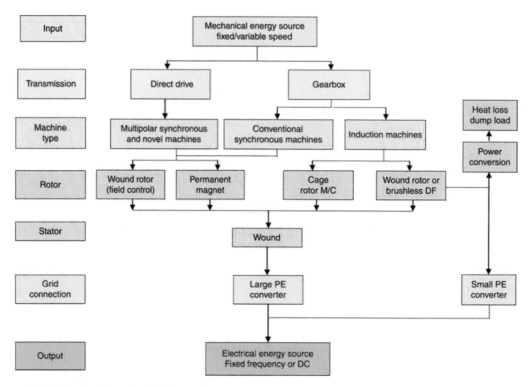

Figure 4.6. Wind turbine classification system.
Source: Stavrakakis,G.S. 2012. 2.10 - Electrical Parts of Wind Turbines, In: Editor-in-Chief: Ali Sayigh, Editor(s)-in-Chief, Comprehensive Renewable Energy, (Oxford, Elsevier), Pages 269-328.

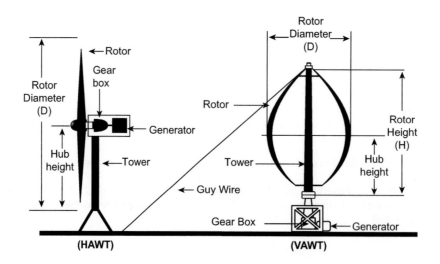

Figure 4.7. Horizontal (HAWT) and vertical axis (VAWT) wind turbines.
Source: Adapted from United Sates Department of Energy, National Renewable Energy Laboratory.

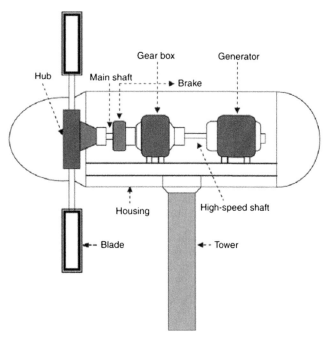

Figure 4.8. Cross-sectional view of a horizontal axis wind turbine.
Source: Mathew, S. and G.S. Philip. 2012. 2.05 - Wind Turbines: Evolution, Basic Principles, and Classifications, In: Editor-in-Chief: Ali Sayigh, Editor(s)-in-Chief, Comprehensive Renewable Energy, (Oxford, Elsevier), Pages 93-111.

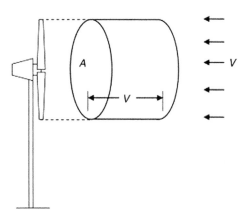

Figure 4.9. A wind turbine interacting with wind stream. The power of the wind *(P)* is given by $P = 1/2 \ \varrho A V^3$, where *V* is velocity, *A* is the swept area exposed to the wind stream of velocity *V* normal to the rotor plane, and ϱ is the density of air. The condition for maximum power extraction *(P)* from the wind is given by $P_{max} = (16/27) \times 1/2 \ \varrho A V^3$; the factor 16/27 is called the Betz coefficient which represents the maximum fraction that an ideal wind rotor under the given conditions can extract from the flow. The fraction of extracted power is frequently the power coefficient C_p, which in practice seldom exceeds 40% if measured as the mechanical power of a real wind rotor.
Source: Mathew, S. and G.S. Philip. 2012. 2.05 - Wind Turbines: Evolution, Basic Principles, and Classifications, In: Editor-in-Chief: Ali Sayigh, Editor(s)-in-Chief, Comprehensive Renewable Energy, (Oxford, Elsevier), Pages 93-111.

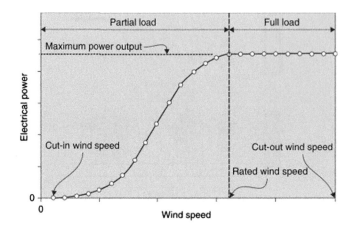

Figure 4.10. Typical power curve for a wind turbine.
Source: Carta, J.A. 2012. 2.18 - Wind Power Integration, In: Editor-in-Chief: Ali Sayigh, Editor(s)-in-Chief, Comprehensive Renewable Energy, (Oxford, Elsevier), Pages 569-622.

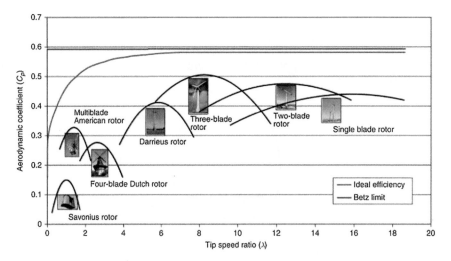

Figure 4.11. The relation between power coefficient (C_P) with the tip speed ratio (λ) for a wind turbine. C_p represents the energy generated by the turbine relative to the energy available in the wind. λ is the ratio between the rotational speed of the tip of a blade and the actual velocity of the wind. Both C_P and λ are dimensionless parameters. Hence, the CP–λ relationship for a particular rotor is valid for dimensionally similar rotors of any size.
Source: Zafirakis, D.P., A.G. Paliatsos, J.K. Kaldellis. 2012. 2.06 - Energy Yield of Contemporary Wind Turbines, In: Editor-in-Chief: Ali Sayigh, Editor(s)-in-Chief, Comprehensive Renewable Energy, (Oxford, Elsevier), Pages 113-168.

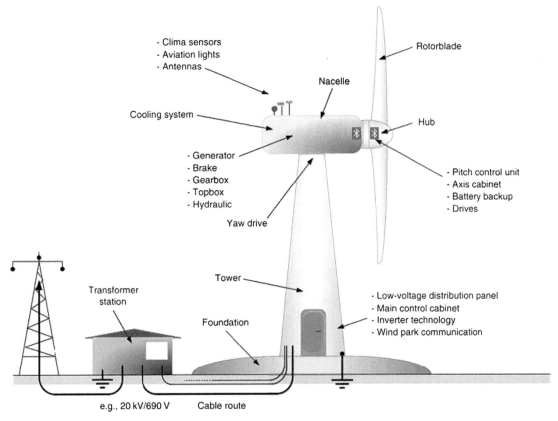

Figure 4.12. A grid-connected onshore wind turbine

Source: Stavrakakis,G.S. 2012. 2.10 - Electrical Parts of Wind Turbines, In: Editor-in-Chief: Ali Sayigh, Editor(s)-in-Chief, Comprehensive Renewable Energy, (Oxford, Elsevier), Pages 269-328.

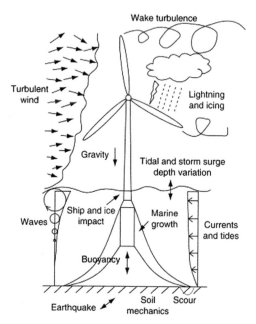

Figure 4.13. Loading sources for an offshore wind turbine.
Source: Kapsali, M. and J.K. Kaldellis. 2012. 2.14 - Offshore Wind Power Basics, In: Editor-in-Chief: Ali Sayigh, Editor(s)-in-Chief, Comprehensive Renewable Energy, (Oxford, Elsevier), Pages 431-468.

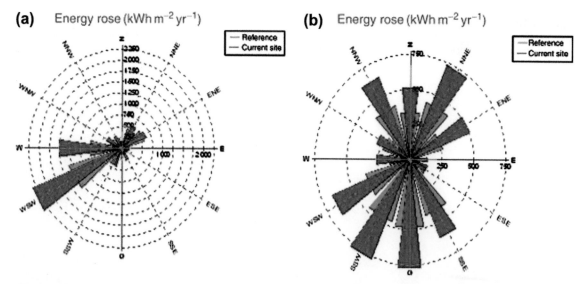

Figure 4.14. A wind energy rose for the west coast of Sweden, summarizing the strength, direction, and frequency of wind.
Source: Wizelius, T. 2012. 2.13 - Design and Implementation of a Wind Power Project, In: Editor-in-Chief: Ali Sayigh, Editor(s)-in-Chief, Comprehensive Renewable Energy, (Oxford, Elsevier), Pages 391-430.

Charts

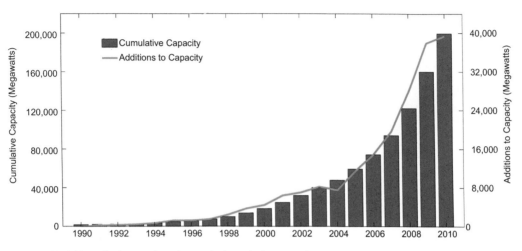

Chart 4.1. World installed annual and cumulative wind generation capacity, 1990-2010.
Source: Data from European Wind Energy Association.

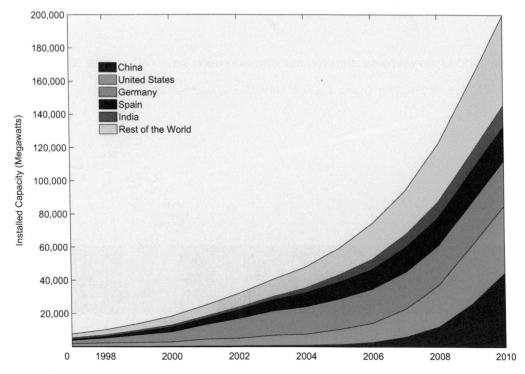

Chart 4.2. Installed cumulative wind generation capacity, top 5 nations and the rest of the world, 1997-2010.
Source: Data from Global Wind Energy Council, <http://www.gwec.net/>, accessed 2 February 2012.
European Wind Energy Association, <http://www.ewea.org/>.

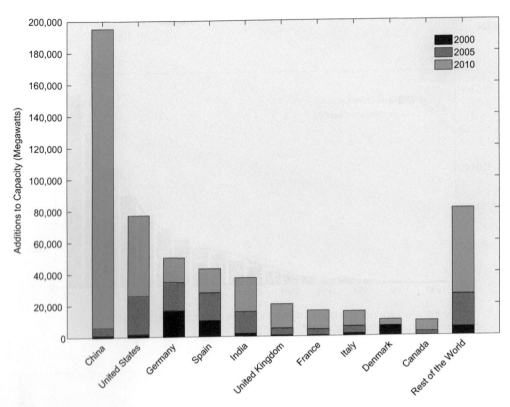

Chart 4.3. Additions to wind generating capacity for top nations and the rest of the world, 2000, 2005 and 2010.
Source: Data from Global Wind Energy Council, <http://www.gwec.net/>.

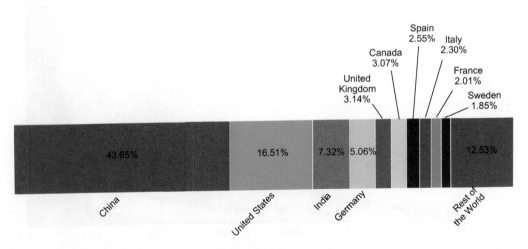

Chart 4.4. Top 10 nations share of new installed wind generating capacity, 2011.
Source: Data from Global Wind Energy Council, <http://www.gwec.net/>.

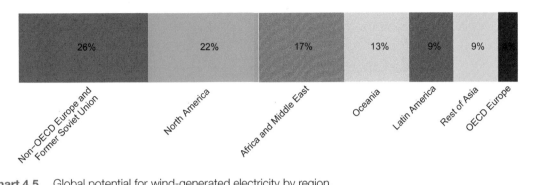

Chart 4.5. Global potential for wind-generated electricity by region.
Source: Data from Lu, Xi, Michael B. McElroy and Juha Kiviluoma. 2009. Global potential for wind-generated electricity, Proceedings of the National Academy of Sciences, 106 (27), pp. 10933–10938.

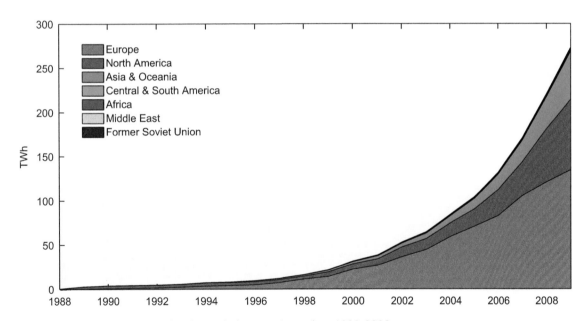

Chart 4.6. Electricity generation from wind energy, by region, 1988-2009.
Source: Data from International Energy Agency (IEA), Energy statistics database, <http://www.iea.org/stats/index.asp>.

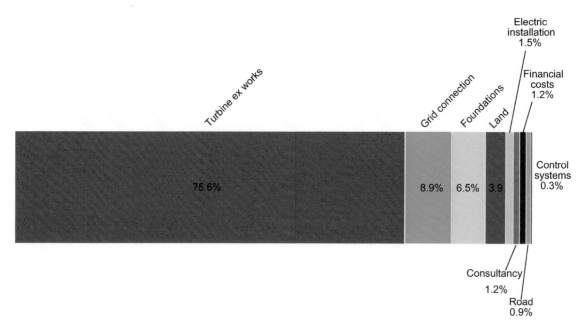

Chart 4.7. Typical investment costs for an onshore wind farm.
Source: Data from Kaldellis, J.K., D.P. Zafirakis. 2012. Trends, Prospects, and R&D Directions in Wind Turbine Technology, In: Ali Sayigh, Editor-in-Chief, Comprehensive Renewable Energy, (Oxford, Elsevier), Pages 671-724.

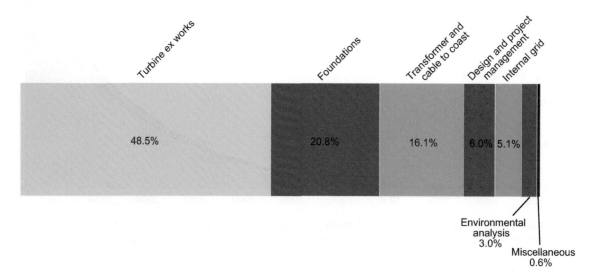

Chart 4.8. Typical investment costs for an offshore wind farm.
Source: Data from Kaldellis, J.K., D.P. Zafirakis. 2012. Trends, Prospects, and R&D Directions in Wind Turbine Technology, In: Ali Sayigh, Editor-in-Chief, Comprehensive Renewable Energy, (Oxford, Elsevier), Pages 671-724.

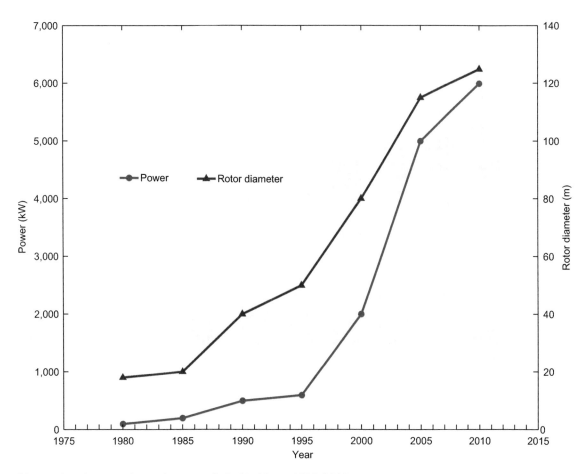

Chart 4.9. Average size and power of wind turbines, 1980-2010.
Source: Data from United States Department of Energy, National Renewable Energy Laboratory, National Wind Technology Center, <http://www.nrel.gov/wind/>.

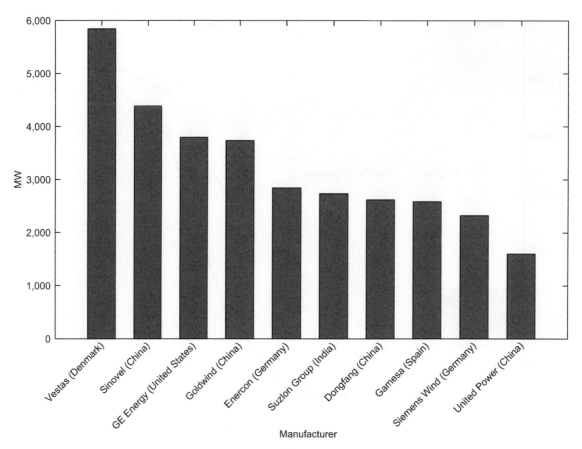

Chart 4.10. Top 10 wind turbine manufacturers by installed capacity, 2010.
Source: Data from Global Wind Energy Council, <http://www.gwec.net/>.

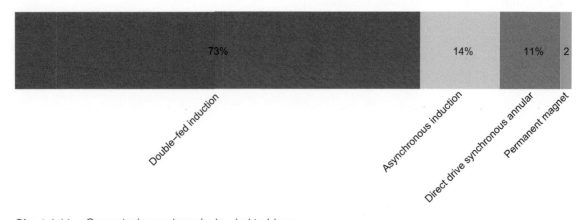

Chart 4.11. Generator/converter pairs in wind turbines.
Source: Data from Stavrakakis,G.S. 2012. 2.10 - Electrical Parts of Wind Turbines, In: Editor-in-Chief: Ali Sayigh, Editor(s)-in-Chief, Comprehensive Renewable Energy, (Oxford, Elsevier), Pages 269-328.

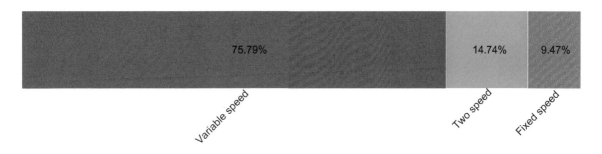

Chart 4.12. Market shares of variable speed, two speed, and fixed speed wind turbines.
Source: Data from Mathew, S. and G.S. Philip. 2012. 2.05 - Wind Turbines: Evolution, Basic Principles, and Classifications, In: Editor-in-Chief: Ali Sayigh, Editor(s)-in-Chief, Comprehensive Renewable Energy,(Oxford, Elsevier), Pages 93-111.

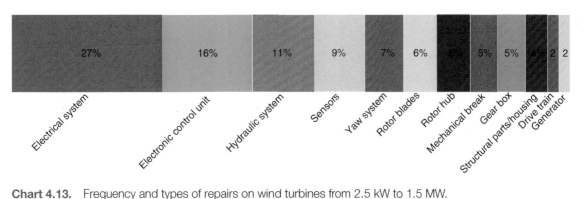

Chart 4.13. Frequency and types of repairs on wind turbines from 2.5 kW to 1.5 MW.
Source: Data from United States Department of Energy, National Renewable Energy Laboratory, National Wind Technology Center, <http://www.nrel.gov/wind/>.

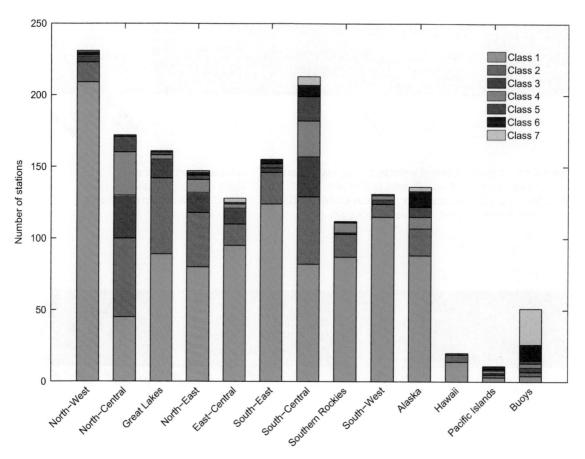

Chart 4.14. Number of United States stations falling into each wind power class at 80 meters height. Speed <5.9 m/s (class 1 at 80 m); 5.9<=speed<6.9 m/s (class 2 at 80 m); 6.9<=speed<7.5 m/s (class 3 at 80 m); 7.5<=speed<8.1 m/s (class 4 at 80 m); 8.1<=speed<8.6 m/s (class 5 at 80 m); 8.6<=speed<9.4 m/s (class 6 at 80 m); speed>=9.4 m/s (class 7 at 80 m).
Source: Data from Archer, Cristina L. and Mark Z. Jacobson, The Spatial and Temporal Distributions of U.S. Winds and Windpower at 80 m Derived from Measurements, <http://www.stanford.edu/group/efmh/winds/us_winds.html>, accessed 23 May 2012; this is an update of Archer, C.L., and M.Z. Jacobson, Journal of Geophysical Research, Vol. 108, No. D9, 4289, May 16, 2003.

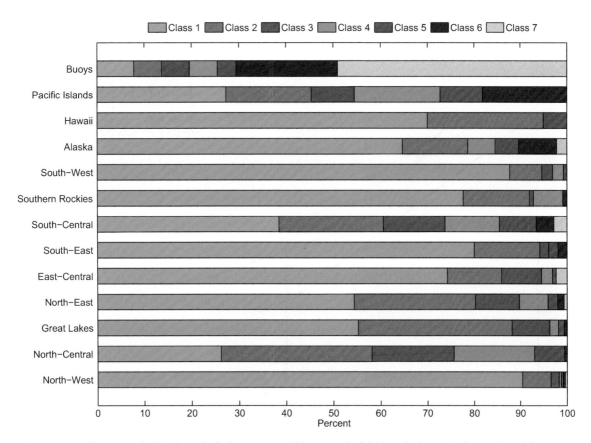

Chart 4.15. Regional distribution of wind resource at 80 meters height by wind speed class in the United States. Speed < 5.9 m/s (class 1 at 80 m); 5.9 <= speed < 6.9 m/s (class 2 at 80 m); 6.9 <= speed < 7.5 m/s (class 3 at 80 m); 7.5 <= speed < 8.1 m/s (class 4 at 80 m); 8.1 <= speed < 8.6 m/s (class 5 at 80 m); 8.6 <= speed < 9.4 m/s (class 6 at 80 m); speed >= 9.4 m/s (class 7 at 80 m).
Source: Data from Archer, Cristina L. and Mark Z. Jacobson, The Spatial and Temporal Distributions of U.S. Winds and Windpower at 80 m Derived from Measurements, <http://www.stanford.edu/group/efmh/winds/us_winds.html>, accessed 23 May 2012; this is an update of Archer, C.L., and M.Z. Jacobson, Journal of Geophysical Research, Vol. 108, No. D9, 4289, May 16, 2003.

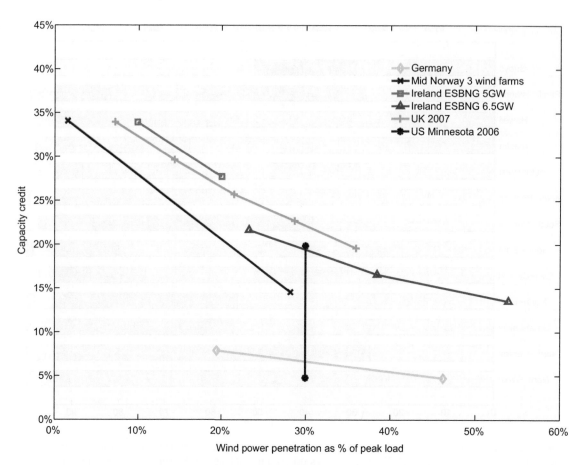

Chart 4.16. Estimates of the capacity credit (capacity value) for wind power. The capacity value of a generator is the amount of additional load that can be served at the target reliability level with the addition of the generator in question.

Source: Data from Holttinen, H. Meibom, P. Orths, A. van Hulle, F. Lange, B. O'Malley, M. Pierik, J. Ummels, B. Tande, J.O. Estanqueiro, A. Matos, M. Gomez, E. Söder, L. Strbac, G. Shakoor, A. Ricardo, J. Smith, J.C. Milligan, M. Ela, E. 2009. Design and Operation of Power Systems with Large Amounts of Wind Power. IEA WIND Task 25, Phase One 2006–2008. (Vuorimiehentie, Finland, VTT).

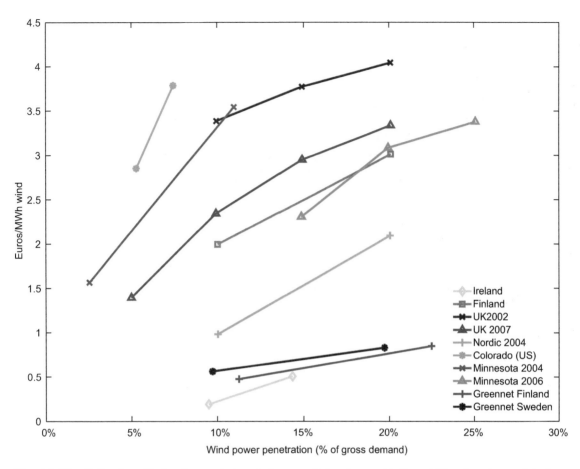

Chart 4.17. Estimates of balancing costs for wind power. Balancing cost is the additional cost associated with adding a generator to the system in order to maintain the target reliability level.

Source: Data from Holttinen, H. Meibom, P. Orths, A. van Hulle, F. Lange, B. O'Malley, M. Pierik, J. Ummels, B. Tande, J.O. Estanqueiro, A. Matos, M. Gomez, E. Söder, L. Strbac, G. Shakoor, A. Ricardo, J. Smith, J.C. Milligan, M. Ela, E. 2009. Design and Operation of Power Systems with Large Amounts of Wind Power. IEA WIND Task 25, Phase One 2006–2008. (Vuorimiehentie, Finland, VTT).

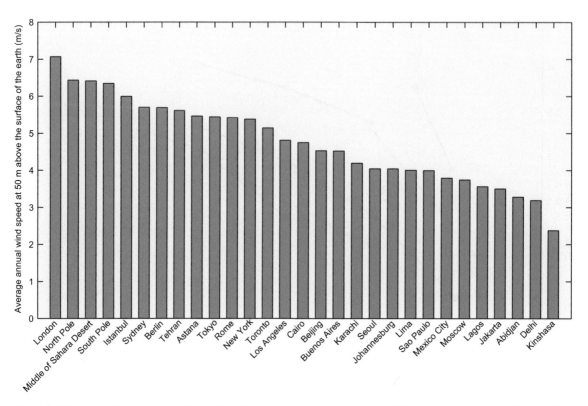

Chart 4.18. Annual average wind speed at 50 meters above the surface of the earth for selected cities and locations (10-year annual average, July 1983 - June 1993).
Source: Data from United States National Aeronautics and Space Administration (NASA), Surface meteorology and Solar Energy database, <http://eosweb.larc.nasa.gov/sse/>.

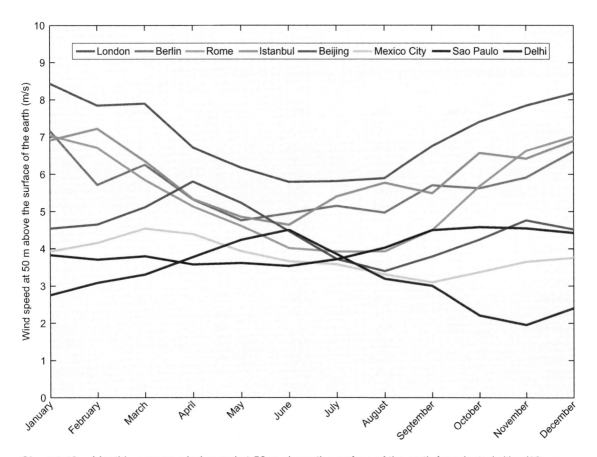

Chart 4.19. Monthly average wind speed at 50 m above the surface of the earth for selected cities (10-year annual average, July 1983 - June 1993).
Source: Data from United States National Aeronautics and Space Administration (NASA), Surface meteorology and Solar Energy database, <http://eosweb.larc.nasa.gov/sse/>.

Tables

Table 4.1. Net generation of electricity by wind, top 20 nations, 1990-2010 (billion kWh)

	1990	2000	2010	Rank 2010	Rank 1990	Change 2000-2010
United States	2.79	5.59	94.65	1	1	1592.2%
China	0	0.58	48.00	2	9	8119.2%
Spain	0.01	4.49	41.50	3	5	824.1%
India	0.03	1.60	20.00	4	4	1150.0%
United Kingdom	0.01	0.90	9.70	5	6	977.8%
France	0	0.07	9.20	6	a	12502.7%
Portugal	0	0.16	8.60	7	a	5275.0%
Italy	0	0.54	8.00	8	11	1395.3%
Denmark	0.58	4.03	7.42	9	2	84.2%
Canada	0	0.25	5.50	10	12	2091.2%
Netherlands	0.05	0.79	3.78	11	3	379.7%
Australia	0	0.06	3.62	12	a	6472.7%
Japan	0	0.10	3.60	13	a	3361.5%
Sweden	0.01	0.43	3.30	14	8	660.4%
Turkey	0	0.03	2.80	15	a	8932.3%
Ireland	0	0.23	2.67	16	a	1050.9%
Brazil	0	0.00	2.07	17	a	206900.0%
Greece	0	0.43	2.03	18	10	374.3%
Austria	0	0.06	1.92	19	a	2900.0%
Poland	0	0.01	1.60	20	a	31900.0%
World	3.5	28.4	327.8			1056.3%

[a]No gneration reported in 1990.
Source: United States Department of Energy, Energy Information Administration, International Energy Statistics,
<http://www.eia.gov/countries/data.cfm>.

Table 4.2. Estimates of global wind power

Authors[a]	Kinetic Energy in Global Wind (TW)	Technical power (TW)[b]	Economic/sustainable power (TW)
Archer and Caldeira (2009)		1500 (jet stream, % feasible?)	
Capps and Zender (2010)		39 (offshore)	
de Castro et a. (2011)	100 (200 m closest to surface)	~1	
DeVries et al. (2007)			4.5 (in 2050)
EEA (2009)		8.6 (Europe)	3.5 (Europe 2030)
Elliott et al. (2004)+Musial (2005)		1 (EEUU)	
Felllows (2000)		5.4	
Greenblatt (2005)		70.4 (global)	
Greenpeace (2008) Greenpeace 2010			1 (in 2050) 1.2 (in 2050)
Grubb and Meyer (1993)		6-57	
Gustavson (1979)	3600		
Hoogwijk et al. (2004)		11	2,4-6 (economic)

Table 4.2.   Continued

Authors[a]	Kinetic Energy in Global Wind (TW)	Technical power (TW)[b]	Economic/sustainable power (TW)
Hoogwijk and Graus (2008)		13	
Keith et al. (2004)	522		
Krewitt et al. (2009)		14	
Lorenz (1967)	1270		
Lu et al. (2009)	340	78 (onshore), >7 (offshore)	
Miller et al., 2010		17–38 (onshore, geographical potential)	
Peixoto and Oort (1992)	768		
Schindler et al. (2007)			6.9 (sustainable)
Skinner (1986)	350		
Sorensen (1979, 2004)	1200		
Wang and Prinn (2010	860		
WBGU (2004)		4-32	
WEC (1994)		55.2 (global)	
Wijk and Coelingh (1993)		2.3 (onshore, OCDE)	
Zerta et al. (2008)			6.9 (sustainable)

[a]Cited work:

[b]Most technical powers are primary, not electrical.

C.L. Archer, K. Caldeira, Global assessment of high-altitude wind power, Energies, 2 (2009), pp. 307–319.

S.B. Capps, C.S. Zender, The estimated global ocean wind power potential from QuickScat observations, accounting for turbine characteristics and sitting, Journal of Geophysical Research, 115 (D01101) (2010), p. 13.

Carlos de Castro, Margarita Mediavilla, Luis Javier Miguel, Fernando Frechoso, Global wind power potential: Physical and technological limits, Energy Policy, Volume 39, Issue 10, October 2011, Pages 6677-6682.

B.J.M. DeVries et al., Renewable energy sources: their global potential for the first-half of the 21st century at a global level: an integrated approach, Energy Policy, 35 (2007), pp. 2590–2610.

EEA, 2009. Europe's Onshore and Offshore Wind Energy Potential. An Assessment of Environmental and Economic Constraints. European Environment Agency. EEA Technical Report No. 6/2009.

D. Elliott et al., Wind Resource Base. Encyclopedia of Energy, Elsevier, NY (2004) pp. 465–479.

Fellows, A. (2000). The Potential of Wind Energy to Reduce Carbon Dioxide Emissions. Garrad Hassan and Partners Ltd, Glasgow, Scotland, 146 pp.

Greenblatt J.B., 2005. Wind as a source of energy, now and in the future. InterAcademy Council, Amsterdam, Netherlands October 1, 2005.

Greenpeace, Global Wind Energy Outlook 2008, GWEC, Greenpeace International, DLR and Ecofys (2008).

Greenpeace, Energy revolution: a sustainable world energy outlook, European Renewable Energy Council, Greenpeace (2010).

Grubb, M.J., and N.I. Meyer (1993). Wind energy: Resources, systems and regional strategies. In: Renewable Energy: Sources for Fuels and Electricity. T.B. Johansson, H. Kelly, A.K. Reddy, and R.H. Williams (eds.), Island Press, Washington, DC, USA, pp. 157-212.

M.R. Gustavson, Limits to wind power utilization, Science, 203 (4388) (1979), pp. 13–17.

M. Hoogwijk et al., Assessment of the global and regional geographical, technical and economic potential of onshore wind energy, Energy Economics, 26 (2004), pp. 889–919X.

Hoogwijk, M., and W. Graus (2008). Global Potential of Renewable Energy Sources: A Literature Assessment. Ecofys, Utrecht, The Netherlands, 45 pp.

D.W. Keith et al., The influence of large-scale wind power on global climate, Proceedings of the National Academy of Sciences USA, 101 (16115–16120) (2004), p. 2004.

Krewitt, W., K. Nienhaus, C. Kleßmann, C. Capone, E. Stricker, W. Graus, M. Hoogwijk, N. Supersberger, U. von Winterfeld, and S. Samadi (2009). Role and Potential of Renewable Energy and Energy Efficiency for Global Energy Supply. Climate Change 18/2009, ISSN 1862-4359, Federal Environment Agency, Dessau-Roßlau, Germany, 336 pp.

E. Lorenz, The Nature and Theory of the General Circulation of the Atmosphere, World Meteorological Organization, Geneva (1967)

Lu et al., Global potential for wind-generated electricity, Proceedings of the National Academy of Sciences, 106 (27) (2009), pp. 10933–10938.

L.M. Miller et al., Estimating maximum global land surface wind power extractability and associated climatic consequences, Earth System Dynam. Discussion, 1 (169–189) (2010), p. 2010.

W. Musial, Offshore wind energy potential for the United States, B. Ruth (Ed.), Wind Powering America: Annual State Summit, Evergreen, CO (2005).

Continued

Table 4.2. Continued

J.P. Peixoto, A.H. Oort, Physics of climate, American Institute of Physics, 1 (379–385) (1992), p. 109 1992

J. Schindler et al., Where will the Energy for Hydrogen Production Come From? Status and Alternatives, European Hydrogen Association (2007).

B. Sorensen, Renewable Energy; Academic Press, London, UK (1979); Renewable Energy: Its Physics, Engineering Use, Environment Impacts, Economy and Planning Aspects, Elsevier Academic Press (2004).

C. Wang, R.G. Prinn, Potential climatic impacts and reliability of very large-scale wind farms, Atmospheric Chemistry and Physics, 10 (2010), pp. 2053–2061.

WBGU, H. Graßl, J. Kokott, M. Kulessa, J. Luther, F. Nuscheler, R. Sauerborn, H.J. Schellnhuber, R. Schubert, E.D. Schulze (2004). World in Transition: Towards Sustainable Energy Systems. German Advisory Council on Global Change (WBGU), Earthscan, London, UK and Sterling, VA, USA, 352 pp.

WEC, New Renewable Energy Resources: A Guide to the Future. World Energy Council, Kogan Page Limited, London, UK (1994) 1994

van Wijk, A.J.M., Coelingh, J.P., 1993. Wind Power Potential in the OECD Countries. December 1993. Report commissioned by the Energy Research Center, The Netherlands (ECN).

M. Zerta et al., Alternative world energy outlook (AWEO) and the role of hydrogen in a changing energy landscape, International Journal of Hydrogen Energy, 33 (12) (2008), pp. 3021–3025.

Source: Adapted from Carlos de Castro, Margarita Mediavilla, Luis Javier Miguel, Fernando Frechoso, Global wind power potential: Physical and technological limits, Energy Policy, Volume 39, Issue 10, October 2011, Pages 6677-6682; Moomaw, W., F. Yamba,

M. Kamimoto, L. Maurice, J. Nyboer, K. Urama, T. Weir, 2011: Introduction. In IPCC Special Report on Renewable Energy Sources and Climate Change Mitigation [O. Edenhofer, R. Pichs-Madruga, Y. Sokona, K. Seyboth, P. Matschoss, S. Kadner, T. Zwickel, P. Eickemeier, G. Hansen, S. Schlömer, C.von Stechow (eds)], Cambridge University Press, Cambridge, United Kingdom and New York, NY, USA.

Table 4.3. Largest wind farms in the world, 2012

Wind farm name	Capacity (MW)	Country
ONSHORE		
Roscoe Wind Farm	781.5	USA
Horse Hollow Wind Energy Center	735.5	USA
Capricorn Ridge Wind Farm	662.5	USA
Fowler Ridge Wind Farm	599.8	USA
Sweetwater Wind Farm	585.3	USA
Buffalo Gap Wind Farm	523.3	USA
Dabancheng Wind Farm	500	China
Meadow Lake Wind Farm	500	USA
Panther Creek Wind Farm	458	USA
Biglow Canyon Wind Farm	450	USA
OFFSHORE		
Wanley	367	UK
Thanet	300	UK
Horns Rev II	209	Denmark
Rœïdsand II	207	Denmark
Lynn and Inner Dowsing	194	UK
Robin Rigg (Solway Firth)	180	UK
Gunfleet Sands	172	UK
Nysted (Rœïdsand I)	166	Denmark
Bligh Bank (Belwind)	165	Belgium
Horns Rev I	16	Denmark

Source: Adapted from Leung, Dennis Y.C., Yuan Yang. 2012. Wind energy development and its environmental impact: A review, Renewable and Sustainable Energy Reviews, Volume 16, Issue 1, Pages 1031-1039.

Table 4.4. Cumulative installed wind turbine capacity (MW)

	2000	2005	2010	Rank 2010	Rank 2000	Change 2000-10
China	352	1,264	44,781	1	9	12622%
US	2,610	9,181	40,274	2	3	1443%
Germany	6,107	18,445	27,364	3	1	348%
Spain	2,836	10,027	20,300	4	2	616%
India	1,220	4,388	12,966	5	5	963%
France	63	775	5,961	6	19	9362%
United Kingdom	425	1,336	5,862	7	7	1279%
Italy	424	1,713	5,793	8	8	1266%
Canada	139	683	4,011	9	13	2786%
Portugal	111	1,087	3,837	10	15	3357%
Denmark	2,341	3,087	3,805	11	4	63%
Japan	142	1,159	2,429	12	12	1611%
Netherlands	473	1,221	2,241	13	6	374%
Sweden	265	554	2,141	14	11	708%
Australia	30	717	2,084	15	24	6847%
World	18,450	59,398	199,523			981%

Source: BP, Statistical Review of World Energy 2011; European Wind Energy Association.

Table 4.5. Annual additions to installed wind turbine capacity (MW)

	2000	2005	2010	Rank 2010	Rank 2000	Change 2000-10
US	165	2,431	5,115	1	5	3000%
India	185	1,388	2,139	2	4	1056%
Germany	1,665	1,796	1,551	3	1	−7%
United Kingdom	63	447	1,522	4	10	2316%
Spain	1,024	1,764	1,516	5	2	48%
France	38	389	1,186	6	16	3021%
Italy	147	452	948	7	6	545%
Canada	13	239	690	8	20	5208%
Sweden	45	76	604	9	13	1242%
Turkey	10	0	528	10	21	5180%
Denmark	603	4	397	11	3	−34%
Poland	1	10	382	12	25	38100%
Portugal	50	502	363	13	11	626%
Belgium	8	71	350	14	22	4275%
Romania	0	0	341	15	30	NA
World	4,518	11,486	39,436	NA	NA	773%

Source: BP, Statistical Review of World Energy 2011; European Wind Energy Association.

Table 4.6. Annual electricity production from a wind turbine based on the Weibull probability distribution

Wind velocity (m s⁻¹)	Weibull distribution (%)	Number of hours per year	Wind turbine power curve (kW)	Electricity production (kWh)
0	0	0	0	0
1	3.2	280	0	0
2	5.3	464	0	0
3	6.8	596	0	0
4	7.7	675	27	18,212
5	8.3	727	70.4	51,186
6	8.4	736	130	95,659
7	8.3	727	211	153,414
8	7.8	683	314	214,550
9	7.2	631	435	274,363
10	6.5	569	562	320,003
11	5.7	499	678	338,539
12	4.9	429	764	327,939
13	4.1	359	814	292,356
14	3.4	298	837	249,292
15	2.8	245	846	207,507
16	2.2	193	849	163,619
17	1.7	149	850	126,582
18	1.3	114	850	96,798
19	1	88	850	74,460
20	0.8	70	850	59,568
21	0.6	53	850	44,676
22	0.4	35	850	29,784
23	0.3	26	850	22,338
24	0.2	18	850	14,892
25	0.1	9	850	7,446
26	0.1	9	0	0
27	0.1	9	0	0
28	0	0	0	0
Total	100	8,760		3,183,184

Source: Adapted from Katsaprakakis, D. Al and D.G. Christakis, 2.07 - Wind Parks Design, Including Representative Case Studies, In: Editor-in-Chief: Ali Sayigh, Editor(s)-in-Chief, Comprehensive Renewable Energy, (Oxford, Elsevier), Pages 169-223.

Table 4.7. Global technical potential for onshore wind energy by region

Grubb and Meyer (1993)		WEC (1994)		Krewitt et al. (2009)		Lu et al. (2009)	
Region	%	Region	%	Region	%	Region	%
Western Europe	9	Western Europe	7	OECD Europe	5	OECD Europe	4
North America	26	North America	26	OECD North America	42	North America	22
Latin America	10	Latin America and Caribbean	11	Latin America	10	Latin America	9
Eastern Europe and Former Soviet Union	20	Eastern Europe and CIS	22	Transition Economies	17	Non-OECD Europe and Former Soviet Union	26
Africa	20	Sub-Saharan Africa	7	Africa and Middle East	9	Africa and Middle East	17
Australia	6	Middle East and North Africa	8	OECD Pacific	14	Oceania	13
Rest of Asia	9	Pacific	14	Rest of Asia	4	Rest of Asia	9
		Rest of Asia	4				

Regions shown in the table are defined by each individual study. Some regions have been combined to improve comparability among the four studies.

[a]Cited work:

Grubb, M.J., and N.I. Meyer (1993). Wind energy: Resources, systems and regional strategies. In: Renewable Energy: Sources for Fuels and Electricity. T.B. Johansson, H. Kelly, A.K. Reddy, and R.H. Williams (eds.), Island Press, Washington, DC, USA, pp. 157-212.

Krewitt, W., K. Nienhaus, C. Kleßmann, C. Capone, E. Stricker, W. Graus, M. Hoogwijk, N. Supersberger, U. von Winterfeld, and S. Samadi (2009). Role and Potential of Renewable Energy and Energy Efficiency for Global Energy Supply. Climate Change 18/2009, ISSN 1862-4359, Federal Environment Agency, Dessau-Roßlau, Germany, 336 pp.

Lu et al., Global potential for wind-generated electricity, Proceedings of the National Academy of Sciences, 106 (27) (2009), pp. 10933–10938.

WEC, New Renewable Energy Resources: A Guide to the Future. World Energy Council, Kogan Page Limited, London, UK (1994) 1994

Source: Adapted from Moomaw, W., F. Yamba, M. Kamimoto, L. Maurice, J. Nyboer, K. Urama, T. Weir, 2011: Introduction. In IPCC Special Report on Renewable Energy Sources and Climate Change Mitigation [O. Edenhofer, R. Pichs-Madruga, Y. Sokona, K. Seyboth, P. Matschoss, S. Kadner, T. Zwickel, P. Eickemeier, G. Hansen, S. Schlömer, C.von Stechow (eds)], Cambridge University Press, Cambridge, United Kingdom and New York, NY, USA.

Table 4.8. Learning curve data for onshore wind energy

Authors	Learning By Doing Rate[d] (%)	Global or National Independent Variable (cumulative capacity)	Dependent Variable	Data Years
Neij (1997)	4	Denmark[c]	Denmark (turbine cost)	1982–1995
Mackay and Probert (1998)	14	USA	USA (turbine cost)	1981–1996
Neij (1999)	8	Denmark[c]	Denmark (turbine cost)	1982–1997
Wene (2000)	32	USA[b]	USA (generation cost)	1985–1994
Wene (2000)	18	EU[b]	EU (generation cost)	1980–1995
Miketa and Schrattenholzer (2004)[a]	10	Global	Global (investment cost)	1971–1997
Junginger et al. (2005)	19	Global	UK (investment cost)	1992–2001
Junginger et al. (2005)	15	Global	Spain (investment cost)	1990–2001

Continued

Table 4.8.   Continued

Authors	Learning By Doing Rate[d] (%)	Global or National Independent Variable (cumulative capacity)	Dependent Variable	Data Years
Klassen et al. (2005)	5	Germany, Denmark, and UK	Germany, Denmark, and UK (investment cost)	1986–2000
Kobos et al. (2006)[a]	14	Global	Global (investment cost)	1981–1997
Jamasb (2007)[a]	13	Global	Global (investment cost)	1980–1998
Söderholm and Sundqvist (2007)	5	Germany, Denmark, and UK	Germany, Denmark, and UK (investment cost)	1986–2000
Söderholm and Sundqvist (2007)[a]	4	Germany, Denmark, and UK	Germany, Denmark, and UK (investment cost)	1986–2000
Neij (2008)	17	Denmark	Denmark (generation cost)	1981–2000
Kahouli-Brahmi (2009)	17	Global	Global (investment cost)	1979–1997
Nemet (2009)	11	Global	California (investment cost)	1981–2004
Ek and Söderholm (2010)[a]	17	Global	Germany, Denmark, Spain, Sweden, and UK (investment cost)	1986–2002
Wiser and Bolinger (2010)	9	Global	USA (investment cost)	1982–2009

[a]Two-factor learning curve that also includes R&D; others are one-factor learning curves.
[b]Independent variable is cumulative production of electricity.
[c]Cumulative turbine production used as independent variable; others use cumulative installations.
[d]The rate at which unit cost declines with each doubling of cumulative output.
Cited work:

Neij, L. (1997). Use of experience curves to analyse the prospects for diffusion and adoption of renewable energy technology. Energy Policy, 25, pp. 1099-1107.
Neij, L. (1999). Cost dynamics of wind power. Energy, 24, pp. 375-389.
Neij, L. (2008). Cost development of future technologies for power generation – A study based on experience curves and complementary bottom-up assessments. Energy Policy, 36, pp. 2200-2211.
Mackay, R.M., and S.D. Probert (1998). Likely market-penetrations of renewable energy technologies. Applied Energy, 59, pp. 1-38.
Wene, C.O. (2000). Experience curves for energy technology policy. International Energy Agency and Organisation for Economic Co-operation and Development, Paris, France, 127 pp.
Miketa, A., and L. Schrattenholzer (2004). Experiments with a methodology to model the role of R&D expenditures in energy technology learning processes; first results. Energy Policy, 32, pp. 1679-1692.
Junginger, M., A. Faaij, and W.C. Turkenburg (2005). Global experience curves for wind farms. Energy Policy, 33, pp. 133-150.
Klaassen, G., A. Miketa, K. Larsen, and T. Sundqvist (2005). The impact of R&D on innovation for wind energy in Denmark, Germany and the United Kingdom. Ecological Economics, 54, pp. 227-240.
Kobos, P.H., J.D. Erickson, and T.E. Drennen (2006). Technological learning and renewable energy costs: implications for US renewable energy policy. Energy Policy, 34, pp. 1645-1658.
Jamasb, T. (2007). Technical change theory and learning curves: patterns of progress in electricity generation technologies. The Energy Journal, 28, pp. 51–71.
Söderholm, P., and T. Sundqvist (2007). Empirical challenges in the use of learning curves for assessing the economic prospects of renewable energy technologies. Renewable Energy, 32, pp. 2559-2578.
Kahouli-Brahmi, S. (2009). Testing for the presence of some features of increasing returns to adoption factors in energy system dynamics: An analysis via the learning curve approach. Ecological Economics, 68, pp. 1195-1212.
Nemet, G.F. (2009). Interim monitoring of cost dynamics for publicly supported energy technologies. Energy Policy, 37, pp. 825-835.
Wiser, R., and M. Bolinger (2010). 2009 Wind Technologies Market Report. US Department of Energy, Washington, DC, USA, 88 pp.
Source: Adapted from Moomaw, W., F. Yamba, M. Kamimoto, L. Maurice, J. Nyboer, K. Urama, T. Weir, 2011: Introduction. In IPCC Special Report on Renewable Energy Sources and Climate Change Mitigation [O. Edenhofer, R. Pichs-Madruga, Y. Sokona, K. Seyboth, P. Matschoss, S. Kadner, T. Zwickel, P. Eickemeier, G. Hansen, S. Schlömer, C.von Stechow (eds)], Cambridge University Press, Cambridge, United Kingdom and New York, NY, USA.

Table 4.9. Twenty five largest offshore wind farms in 2011, rated by nameplate capacity

Wind farm	Total (MW)	Country	Coordinates	Turbines & model	Start Date
Walney (phases 1 & 2)	367.2	United Kingdom	54°02′38″N 3°31′19″W	102 × Siemens SWT-3.6-107	2011 (phase 1) 2012 (phase 2)
Thanet	300	United Kingdom	51°26′N 01°38′E	100 × Vestas	2010
Horns Rev II	209	Denmark	55°36′00″N 7°35′24″E	V90-3MW 91 × Siemens 2.3-93	2009
Rødsand II	207	Denmark	54°33′0″N 11°42′36″E	90 × Siemens 2.3-93	2010
Lynn and Inner Dowsing	194	United Kingdom	53°07′39″N 00°26′10″E	54 × Siemens 3.6-107	2008
Robin Rigg (Solway Firth)	180	United Kingdom	54°45′N 3°43′W	60 × Vestas V90-3MW	2010
Gunfleet Sands	172	United Kingdom	51°43′16″N 1°17′31″E	48 × Siemens 3.6-107	2010
Nysted (Rødsand I)	166	Denmark	54°33′0″N 11°42′36″E	72 × Siemens 2.3	2003
Bligh Bank (Belwind)	165	Belgium	51°39′36″N 2°48′0″E	55 × Vestas V90-3MW	2010
Horns Rev I	160	Denmark	55°31′47″N 7°54′22″E	80 × Vestas V80-2MW	2002
Ormonde	150	United Kingdom	54°6′N 3°24′W	30 × REpower 5M	2012
Longyuan Rudong Intertidal	131.3	China		21 × Siemens 2.3-93; 2 × 3MW 2 × 2.5MW 6 × 2MW 6 × 1.5MW Sinovel	2012
Princess Amalia	120	Netherlands	52°35′24″N 4°13′12″E	60 × Vestas V80-2MW	2008
Lillgrund	110	Sweden	55°31′N 12°47′E	48 × Siemens 2.3-93	2007
Egmond aan Zee	108	Netherlands		36 × Vestas V90-3MW	2006
Donghai Bridge	102	China	30°46′12″N 121°59′38″E	34 × Sinovel SL3000/90	2010
Kentish Flats	90	United Kingdom	51°27′36″N 1°5′24″E	30 × Vestas V90-3MW	2005
Barrow	90	United Kingdom	53°59′N 3°17′W	30 × Vestas V90-3MW	2006
Burbo Bank	90	United Kingdom	53°29′N 03°10′W	25 × Siemens 3.6-107	2007
Rhyl Flats	90	United Kingdom	53°22′N 03°39′W	25 × Siemens 3.6-107	2009
North Hoyle	60	United Kingdom	53°26′N 3°24′W	30 × Vestas V80-2MW	2003
Scroby Sands	60	United Kingdom	52°38′56″N 1°47′25″E	30 × Vestas V80-2MW	2004

Continued

Table 4.9.   Continued

Wind farm	Total (MW)	Country	Coordinates	Turbines & model	Start Date
Alpha Ventus	60	Germany	54°1′N 6°36′E	6 × REpower 5M, 6 × AREVA Wind M5000-5M	2009
Baltic 1	48	Germany	54°36′36″N 12°39′0″E	21 × Siemens 2.3-93	2011
Middelgrunden	40	Denmark	55°41′27″N 12°40′13″E	20 × Bonus (Siemens) 2MW	2001

Source: European Wind Energy Association (EWEA), Wind in power: 2011 European statistics; Wikipedia, List of offshore wind farms, <http://en.wikipedia.org/wiki/List_of_offshore_wind_farms>, accessed 15 May 2012.

Table 4.10.   Characteristics of offshore wind farms (OWF) with a rated power above 25 MW

Name	State[a]	Year[b]	Operator[c]	Turbine	N_{wt}[d]	Power (MW)	D_{sb}[e] (m)	F_o[f]	D_{sh}[g] (km)
Huaneng Rongcheng[h]	CH	2011	Huaneng	SL 3000/90	34	102	0–1	P	0
Walney 1	UK	2011	Dong-SSE	SWT-3.6-107	51	184	28	M	15
Baltic 1	G	2011	EnBW	SWT-2.3-93	21	48	19	M	16
Belwind 1	B	2011	Belwind	V90/3000	55	165	37	M	46
Chenjagang Xiangshui[h]	CH	2011	Yangtze	FD77-1.5	134	201	0–1	P	0
Thanet	UK	2010	Vattenfall	V90/3000	100	300	25	M	12
Robin Rigg	UK	2010	EON	V90/3000	60	180	9	M	10
Longyuan Rudong 1[h]	CH	2010	Guodian	SE 2.0	16	32	0–1	P	3.5
Rodsand 2	D	2010	EON	SWT-2.3-93	90	207	10	G	4
Donghai Bridge	CH	2010	SD	SL 3000/90	34	102	7	M	9
Gunfleet Sands	UK	2010	Dong	SWT-3.6-107	48	173	15	M	7
Rhyl Flats	UK	2009	RWE	SWT-3.6-107	25	90	11	M	8
Gäslingergrund	SW	2009	VV	WWD-3-100	10	30	13	–	0
Alpha Ventus	G	2009	DOTI	M5000-5M	6-6	60	33	T-J4	45
Horns Rev 2	D	2009	Dong	SWT-2.3-93	91	209	17	M	27
Thorntonbank I	B	2009	CP	5M	6	30	19	G	29
Lynn-Inner Dowsing	UK	2009	Centrica	SWT-3.6-107	54	194	11	M	5
Kemi Ajos	F	2008	PV	WWD-3-100	10	30	7	–	0
Prinses Amalia	N	2008	Eneco	V80/2000	60	120	24	M	23
Egmond aan Zee	N	2008	V-NZW	V90/3000	36	108	18	M	10
Lillgrund	SW	2007	Vattenfall	SWT-2.3-93	48	110	9	G	7
Burbo Bank	UK	2007	Dong	SWT-3.6-107	25	90	8	M	7
Barrow	UK	2006	Centrica	V90/3000	30	90	20	M	8
Kentish Flats	UK	2005	Vattenfall	V90/3000	30	90	5	M	10
Scroby Sands	UK	2004	EON	V80/2000	30	60	15	M	3
Arklow Bank 1	IR	2004	SSE	GE 3.6 Off	7	25	25	M	10
Rodsand 1 (Nysted)	D	2004	D-EON	SWT-2.3-82	72	166	10	G	10
North Hoyle	UK	2003	RWE	V80/2000	30	60	12	M	8

Table 4.10.   Continued

Name	State[a]	Year[b]	Operator[c]	Turbine	N_{wt}[d]	Power (MW)	D_{sb}[e] (m)	F_o[f]	D_{sh}[g] (km)
Horns Rev	D	2002	V-D	V80/2000	80	160	14	M	14
Middlegrunden	D	2001	D-MWTC	B76/2000	20	40	6	G	2

[a]B stands for Belgium, CH stands for China, D stands for Denmark, F stands for Finland, G stands for Germany, IR stands for Ireland, N stands for the Netherlands, SW stands for Sweden and UK stands for the United Kingdom.
[b]Year in which the OWF was commissioned and fully connected to the grid.
[c]CP stands for C-Power, DOTI for Deutsche Offshore-Testfeld und Infrastruktur GmbH, D for Dong Energy, EnBW for Energie Baden-Württemberg AG, MWTC for Middlegrunden Wind Turbine Cooperative, NZW for Nordzeewind, PV for Pohjolan Voima, RWE for RWE Npower Renewables, SSE stands for Scottish and Southern Energy, SD stands for Shanghai Donghai Wind Power Generation Company Ltd., V for Vattenfall and VV for Vindpark Vänern.
[d]Number of wind turbines in the OWF.
[e]Maximum water depth to the lowest astronomical tide within the OWF area.
[f]Foundation type, where P stands for high-rise pile cap foundations, G stands for gravity based foundations, M for monopiles, T for tripods and J4 for four legged jacket.
[g]Distance to the nearest shore, which does not necessarily correspond to the length of the submarine cable.
[h]Chinese OWPPs projects do not have dedicated web pages as is usual in Europe. These installations seem to be in intertidal areas, only partially covered at high tides, so it is not clear if they should be more appropriately classified as near-shore wind farms.
Source: Adapted from Madariaga, A., I. Martínez de Alegría, J.L. Martín, P. Eguía, S. Ceballos. 2012. Current facts about offshore wind farms, Renewable and Sustainable Energy Reviews, Volume 16, Issue 5, Pages 3105-3116.

Table 4.11.   Capital costs of several offshore wind farms

Project	Rated capacity (MW)	Average water depth (m)	Average distance from shore (km)	Support structure	Turbine capacity (MW)	Investment costs (million €)	Year online
Donghai Bridge	102	10	10.5	Gravity based	3	258	2010
Alpha Ventus	60	30	53	(6) tripods (6) Jackets	5	250	2009
Horns Rev II	209	13	30	Monopiles	2.3	470	2009
Princess Amalia	120	22	23	Monopiles	2	380	2008
Lynn/Inner Dowsing	194.4	10	5	Monopiles	3.6	349	2009
Rhyl Flats	90	8	8	Monopiles	3.6	216	2009
Robin Rigg	180	5	9.5	Monopiles	3	500	2009
Thanet	300	22.5	11.5	Monopiles	3	912	2010
Middelgrunden	40	8	3	Gravity based	2	50	2000
Horns Rev I	160	10	16	Monopiles	2	270	2002
Nysted	165.2	8	8	Gravity based	2.3	268	2003
Egmond aan Zee	108	20	10	Monopiles	3	200	2006

Source: Adapted from Kapsali, M. and J.K. Kaldellis. 2012. 2.14 - Offshore Wind Power Basics, In: Editor-in-Chief: Ali Sayigh, Editor(s)-in-Chief, Comprehensive Renewable Energy, (Oxford, Elsevier), Pages 431-468.

Table 4.12. Collector and transmission network characteristics of operating offshore wind farms (OWF) with rated power above 25 MW

Name	Collector System				Offshore Substation	Transmission System			
	N_{clu}[a]	V_{cs}[b] (kV)	D_{wt}[c]	L_{cs}[d] (km)		V_{ts}[b] (kV)	N_{cab}[e]	L_{lf}[f] (km)	L_{cpss}[f] (km)
Huaneng Roncheng	–	–	–	–	–	–	–	–	–
Walney 1	5	33	–	–	1	132	1	45	1
Baltic 1	–	33	–	–	1	150	–	61	–
Belwind 1	6	33	5.6–7.2	50	1	150	1	52	3
Chenjagang Xiangshui	–	–	–	–	–	–	–	–	–
Thanet	10	33	5.6–8.8	55	1	132	2	26	3
Robin Rigg	–	–	–	–	2	–	–	–	–
Longyuan Rudong 1	–	–	–	–	–	–	–	–	–
Rodsand 2	–	33	–	–	1	132	–	–	–
Donghai Bridge	–	–	–	–	–	–	–	–	–
Gunfleet Sands	–	33	–	–	1	132	–	8.5	–[g]
Rhyl Flats	3	33	–	–	0[h]	33	3	–	2
Gäslingergrund	1	20	–	–	–	–	–	–	–
Alpha Ventus	2	33	6.6	16	1	110	1	61	5
Horns Rev 2	7	–	–	70	1[i]	–	–	–	–
Thorntonbank I	1	36	7	50.8	1[j]	150	2	36	3.3
Lynn-Inner Dowsing	6	36	–	32	0[j]	36	6	40	–
Kemi Ajos	–	–	–	–	–	110	–	0	–
Prinses Amalia	10	22	–	45	1	150	1	28	7[g]
Egmond aan Zee	3	34	–	–	0	34	3	–	7
Lillgrund	5	33	–	22	1	130	1	7	2
Burbo Bank	3	33	5.0–6.7	5	0	33	3	8	3.5
Barrow	4	33	5.6–8.3	–	1	132	–	27	3
Kentish Flats	3	33	7.8–7.8	18.9	0	33	3	10	2.6
Scroby Sands	–	–	–	–	0	–	–	–	–
Arklow Bank 1	–	–	–	–	–	–	–	–	–
Rodsand 1 (Nysted)	8	33	5.9–10.4	48	1	132	–	11	18
North Hoyle	2	–	–	16	0	–	2	22	–
Horns Rev	5	33	7	–	1	150	1	21	–
Middlegrunden	1	30	2.4	3.5	0	30	2	2	1.5

[a]Number of clusters of turbines according to their electrical arrangement, not their geometrical disposition.

[b]V_{cs} and V_{ts} stand for the voltages of the CS (also known as *internal grid*) and the TS, respectively.

[c]Separation among the turbines factorized by the turbine diameter. When two values are given, they account for the geometrical distances between turbines in a row and between rows, respectively.

[d]Total length of the MVAC collector system of the farm.

[e]Number of cables for the TS. Three phase high voltage alternating current (HVAC) submarine cable with optic fibre for communications is the general case.

[f]The length of the TS is the length of the submarine cable (L_{lf}), in most cases between the OS and the landfall, plus the length of the onshore transmission line (L_{cpss}), between the landfall and the CPS.

[g]A three phase submarine cable is connected in the landfall to three single phase and communication cables.

[h]Data corresponding to ongoing second and third phases of Thorntonbank OWPP.

[i]It is the first OS to offer accommodation facilities for O&M staff and visitors.

[j]When there is no OS, each cluster of OWTs has its own transmission cable, at least as far as the landfall, the voltages of the collection and transmission systems being the same.

Source: Adapted from Madariaga, A., I. Martínez de Alegría, J.L. Martín, P. Eguía, S. Ceballos. 2012. Current facts about offshore wind farms, Renewable and Sustainable Energy Reviews, Volume 16, Issue 5, Pages 3105-3116.

Table 4.13.   Nominal power versus rotor swept area for wind turbines

Wind turbine model	Rated power (kW)	Rated wind speed (m s⁻¹)	Rotor diameter (m)	Rotor swept area (m²)	kW m⁻²
Enercon E-48	800	14	48	1810	0.44
Enercon E-53	800	13	52	2123	0.38
Suzlon S64-1250	1250	14	64	3215	0.39
Suzlon S66-1250	1250	14	66	3420	0.37
GE 1.5se	1500	12	70.5	3902	0.38
GE 1.5sle	1500	12	77	4416	0.34
GE 1.5xle	1500	12	82.5	5343	0.28
Enercon E70	2300	15	71	3959	0.58
Enercon E82	2000	12	82	5281	0.38
Vestas V90-1	1800	12	90	6362	0.28
Vestas V90-2	2000	13	90	6362	0.31
Vestas V90-3	3000	15	90	6362	0.47

Source: Adapted from Wizelius, T. 2012. 2.13 - Design and Implementation of a Wind Power Project, In: Editor-in-Chief: Ali Sayigh, Editor(s)-in-Chief, Comprehensive Renewable Energy, (Oxford, Elsevier), Pages 391-430.

Table 4.14.   International Electrotechnical Commission classes for wind turbines

	IEC classes				
	I	II	III	IV	S
V_{ref} (m s⁻¹)	50	42.5	37.5	30	Values to be specified by the designer
V_{ave} (m s⁻¹)	10	8.5	7.5	6	
AI_{15}	0.18	0.18	0.18	0.18	
a	2	2	2	2	
BI_{15}	0.16	0.16	0.16	0.16	
a	3	3	3	3	

The values apply at hub height. '**A**' designates higher turbulence; '**B**' designates lower turbulence; I_{15} is the turbulence intensity at 15 m s⁻¹; a is the slope parameter in the normal turbulence model equation.
Source: International Electrotechnical Commission.

Table 4.15. Material composition of wind turbines

Component	Item	Wind turbine 1 (850 kW)		Wind turbine 2 (3.0 MW)	
		Weight	Materials	Weight	Materials
Foundation	Reinforced concrete	495 t	480 t concrete	1176 t	1140 t concrete
			15 t steel		36 t steel
Tower	Painted steel	70 t	69.07 t steel	160 t	158.76 t steel
			0.93 t paint		1.24 t paint
Nacelle	Bedplate/frame	3.35 t	3.35 t steel	13 t	13 t steel
	Cover	2.41 t	2.41 t steel	9.33 t	9.33 t steel
	Generator	1.84 t	1.47 t steel	7.14 t	5.71 t steel
			0.37 t copper		1.43 t copper
	Main shaft	4.21 t	4.21 t steel		
	Brake system	0.26 t	0.26 t steel	1.02 t	1.02 t steel
	Hydraulics	0.26 t	0.26 t steel		
	Gearbox	6.2 t	6.08 t steel	24.06 t	23.58 t steel
			0.062 t copper		0.241 t copper
			0.062 t aluminium		0.241 t aluminium
	Cables	0.42 t	0.18 t aluminium	1.63 t	0.69 t aluminium
			0.24 t copper		0.94 t copper
	Revolving system	1 t	1 t steel	3.87 t	3.87 t steel
	Crane	0.26 t	0.26 t steel	1.02 t	1.02 t steel
	Transformer/ sensors	1.79 t	0.894 t steel	6.93 t	3.47 t steel
			0.357 t copper		1.38 t copper
			0.357 t aluminium		1.38 t aluminium
			0.18 t plastic		0.7 t plastic
Rotor	Hub	4.8 t	4.8 t steel	19.2 t	19.2 t steel
	Blades	5.02 t	3.01 t fiber glass	20.07 t	12.04 t fibre glass
			2.01 t epoxy		8.03 t epoxy
	Bolts	0.18 t	0.18 t steel	0.73 t	0.73 t steel

Source: Adapted from Crawford, R.H. 2009. Life cycle energy and greenhouse emissions analysis of wind turbines and the effect of size on energy yield, Renewable and Sustainable Energy Reviews, Volume 13, Issue 9, December 2009, Pages 2653-2660.

Table 4.16. Main categories of generator/converter pairs in wind turbines

Description	Speed control (%)	Efficiency	Advantages	Disadvantages
Squirrel-cage induction generator	100 ± 0.5	0.96	Robust, reliable, low cost	Low efficiency, excitation current
Double-fed asynchronous generator with static inverter, harmonic filter, and reactive power compensation	100 ± 25	0.94	Best compromise of controllability, cost, and performance	Poor reliability
Direct-drive synchronous generator with permanent magnet excitation, static inverter (AC-DC-AC), and harmonic filter	100 ± 50	0.94	Inherently grid compliant, affordable power electronics	Average reliability, large rotor size

Adapted from Stavrakakis,G.S. 2012. 2.10 - Electrical Parts of Wind Turbines, In: Editor-in-Chief: Ali Sayigh, Editor(s)-in-Chief, Comprehensive Renewable Energy, (Oxford, Elsevier), Pages 269-328.

Table 4.17. Technical specifications for the Vestas 7.0 and 2.0 megawatt (MW) wind turbines

	V164-7.0 MW (offshore)	V80-2.0 MW GridStreamer (onshore)
Power Regulation	pitch regulated with variable speed	pitch regulated with variable speed
Operating Data		
Rated power	7.0 MW	2,000 kW
Cut-in wind speed	4 m/s	4.0 m/s
Operational temperature range	−10° C to 25° C	−20° C to 40° C
Rotor		
Rotor diameter	164 m	80m
Swept area	21,124 m^2	5,027 m^2
Electrical		
Frequency	50 Hz	50/60 Hz
Converter type	Full Scale converter	full scale converter
Generator type	Permanent magnet	permanent magnet generator
Tower		
Type	Tubular steel tower	Tubular steel tower
Hub heights	Site specific	65 m and 80 m
Blade Dimensions		
Length	80 m	39 m
Max. Chord	5.4 m	3.4 m
Nacelle Dimensions (incl. hub and coolers)		
Length	24 m	10.4 m
Width	12 m	3.4 m

Source: Vestas Wind Systems A/S.

Table 4.18. **Classes of wind power density at 10 m and 50 m[a] used by the U.S. National Renewable Energy Laboratory**

Wind Power	10 m (33 ft)		50 m (164 ft)	
Class*	Wind Power Density (W/m²)	Speed[b] m/s (mph)	Wind Power Density (W/m²)	Speed[b] m/s (mph)
	0	0	0	0
1	100	4.4 (9.8)	200	5.6 (12.5)
2	150	5.1 (11.5)	300	6.4 (14.3)
3	200	5.6 (12.5)	400	7.0 (15.7)
4	250	6.0 (13.4)	500	7.5 (16.8)
5	300	6.4 (14.3)	600	8.0 (17.9)
6	400	7.0 (15.7)	800	8.8 (19.7)
7	1000	9.4 (21.1)	2000	11.9 (26.6)

[a]Vertical extrapolation of wind speed based on the 1/7 power law.
[b]Mean wind speed is based on Rayleigh speed distribution of equivalent mean wind power density. Wind speed is for standard sea-level conditions. To maintain the same power density, speed increases 3%/1000 m (5%/5000 ft) elevation.

Table 4.19. **The Hellman exponent in wind speed estimation**

Location	α
Unstable air above open water surface	0.06
Neutral air above open water surface	0.1
Unstable air above flat open coast	0.11
Neutral air above flat open coast	0.16
Stable air above open water surface	0.27
Unstable air above human inhabited areas	0.27
Neutral air above human inhabited areas	0.34
Stable air above flat open coast	0.4
Stable air above human inhabited areas	0.6

The Hellman exponent (α) is used to estimate the change in wind speed with altitude according to:

$$v_w(h) = v_{10} \cdot \left(\frac{h}{h_{10}} \right)^{\alpha}$$

where $v_w(h)$=wind velocity at height h (m/s); v_{10} = wind velocity at height 10 meters (m/s). The coefficient depends upon the coastal location and the shape of the terrain on the ground, and the stability of the air.
Source: Adapted from "Renewable energy: technology, economics, and environment" by Martin Kaltschmitt, Wolfgang Streicher, Andreas Wiese, (Springer, 2007).

Table 4.20. **Power per unit area available from steady wind (air density = 1.225 kg/m³)**

Wind Speed (m/s)	Power/area (W/m²)
0	0
5	80
10	610
15	2070
20	4900
25	9560
30	16,550

Source: National Renewable Energy Laboratory.

Table 4.21. Number (and percent with respect to each region) of U.S. weather stations falling into each wind power class at 80 m

Region	Total	Wind Class at 80 m															
		1		2		3		3		5		6		7		>= 3	
		V< 5.9 m/s		5.9 <= V< 6.9 m/s		6.9 <= V< 7.5 m/s		7.5 <= V< 8.1 m/s		8.1 <= V< 8.6 m/s		8.6 <= V< 9.4 m/s		V> 9.4 m/s		V> 6.9 m/s	
	#	#	%	#	%	#	%	#	%	#	%	#	%	#	%	#	%
North-West	231	209	90.5	14	6.1	4	1.7	1	0.4	1	0.4	1	0.4	1	0.4	8	3.5
North-Central	180	45	25	55	30.6	30	16.7	30	16.7	11	6.1	1	0.6	0	0	80	44.4
Great Lakes	161	89	55.3	53	32.9	13	8.1	3	1.9	2	1.2	1	0.6	0	0	19	11.8
North-East	147	80	54.4	38	25.9	14	9.5	9	6.1	3	2	2	1.4	1	0.7	29	19.7
East-Central	128	95	74.2	15	11.7	11	8.6	3	2.3	1	0.8	0	0	3	2.3	18	14.1
South-Central	155	124	80	22	14.2	3	1.9	0	0	3	1.9	3	1.9	0	0	9	5.8
South-East	213	82	38.5	47	22.1	28	13.1	25	11.7	17	8	8	3.8	6	2.8	84	39.4
Southern Rockies	112	87	77.7	16	14.3	1	0.9	7	6.3	0	0	1	0.9	0	0	9	8
South-West	131	115	87.8	9	6.9	3	2.3	3	2.3	1	0.8	0	0	0	0	7	5.3
Alaska	144	88	61.1	19	13.2		5.6	8	5.6	7	4.9	11	7.6	3	2.1	37	25.7
Hawaii	20	14	70	5	25	0	0	0	0	1	5	0	0	0	0	1	5
Pacific Islands	11	3	27.3	2	18.2	1	9.1	2	18.2	1	9.1	2	18.2	0	0	6	54.5
Buoys	51	4	7.8	3	5.9	3	5.9	3	5.9	2	3.9	11	21.6	25	49	44	86.3
Total U.S.	1684	1035	61.5	298	17.7	127	7.5	94	5.6	50	3	41	2.4	39	2.3	351	20.8

V = wind velocity at 80 meters measured in meters per second.

Source: Adapted from Archer, Cristina L. and Mark Z. Jacobson, The Spatial and Temporal Distributions of U.S. Winds and Windpower at 80 m Derived from Measurements, Stanford University, <http://www.stanford.edu/group/efmh/winds/us_winds.html>.

Table 4.22.   Estimates of wind power learning rates[a]

Study	Geographical scope of the cost estimates (geographical domain of learning)	Obs	Learning rates
Andersen and Fuglsang (1996)	Denmark (national)	1	20
Anderson (2010)	USA (national)	4	3.3–13.5
Christiansson (1995)	USA (national)	1	16
Coulomb & Neuhoff (2006)	Germany (global/national)	5	10.9–17.2/7.2
Durstewitz and Hoppe-Kilpper (1999)	Germany (national)	1	8
Ek and Söderholm (2010)	Denmark, Germany, Spain, Sweden, UK (global)	1	17.1
Goff (2006)	Denmark, Germany, Spain, UK, USA (national)	4	5.1–7.3
Hansen et al. (2001)	Denmark (national)	4	6.1–15.3
Hansen et al. (2003)	Denmark (national)	4	7.4–11.2
Ibenholt (2002)	Denmark, Germany, UK (national)	5	− 3.0–25.0
IEA (2000): EU Atlas project	EU (EU)	1	16
IEA (2000): Kline/Gripe	USA (national)	1	32
Isoard and Soria (2001)	EU (global)	3	14.7–17.6
Jamasb (2007)	World (global)	1	13.1
Jensen (2004a)	Denmark (national)	3	9.9–11.7
Jensen (2004b)	Denmark (national)	1	8.6
Junginger et al. (2005)	UK, Spain (global)	3	15.0–19.0
Kahouli-Brahmi (2009)	World (global)	5	17.1–31.2
Klaassen et al. (2005)	Denmark, Germany, UK (national)	1	5.4
Kobos (2002)	World (global)	2	14.0–17.1
Kobos et al. (2006)	World (global)	1	14.2
Kouvaritakis et al. (2000)	OECD (global)	1	15.7
Loiter and Norberg-Bohm (1999)	California (national)	1	18
Mackay and Probert (1998)	USA (national)	1	14.3
Madsen et al. (2002)	Denmark (national)	4	8.6–18.3
Miketa and Schrattenholzer (2004)	World (global)	1	9.7
Neij (1997)	Denmark (national)	1	9
Neij (1999)	Denmark (national)	10	− 1.0–8.0
Neij et al. (2003)	Denmark, Germany, Spain, Sweden (national)	10	4.0–17.0
Neij et al. (2004)	Denmark (national)	3	− 1.0–33.0
Nemet (2009)	World (global)	1	11
Papineau (2006)	Denmark, Germany (national)	12	1.0–13.0
Sato and Nakata (2005)	Japan (national)	2	7.9–10.5

Table 4.22. Continued

Study	Geographical scope of the cost estimates (geographical domain of learning)	Obs	Learning rates
Söderholm and Klaassen (2007)	Denmark, Germany, Spain, UK (national)	1	3.1
Söderholm and Sundqvist (2007)	Denmark, Germany, Spain, UK (national)	12	1.8–8.2
Wiser and Bolinger (2010)	USA (global)	1	9.4

[a]The percentage decrease in wind power cost for each doubling of cumulative capacity or production.

Work cited:

Andersen, P.D., Fuglsang, P., 1996. Vurdering af udviklingsforløb for vindkraftteknologien (Estimation of the Future Advances of the Wind Power Technology). Research Centre Risø, Roskilde, Denmark. Risø-R-829 (D).

Anderson, J., 2010. Learning-by-doing in the US Wind Energy Industry. Paper presented at the Freeman Spogli Institute for International Studies, Stanford University, USA. 24 February.

Christiansson, L., 1995. Diffusion and Learning Curves of Renewable Energy Technologies. IIASA Working Paper WP-95-126, International Institute for Applied Systems Analysis, Laxemburg, Austria.

Coulomb, L., Neuhoff, K., 2006. Learning Curves and Changing Product Attributes: The Case of Wind Turbines. Cambridge Working Papers in Economics 0618, University of Cambridge, UK.

Durstewitz, M., Hoppe-Kilpper, M., 1999. Wind Energy Experience Curve from German "250 MW Wind"-Programme. IEA International Workshop on Experience Curves for Policy Making — The Case of Energy Technologies, Stuttgart, Germany. May 1999.

Ek, K., Söderholm, P., 2010. Technology learning in the presence of public R&D: the case of European wind power. Ecological Economics 69 (12), 2356–2362.

Goff, C. (2006). Wind Energy Cost Reductions: A Learning Curve Analysis with Evidence from the United States. Germany. Denmark. Spain. and the United Kingdom, Master's Thesis, Graduate School of Arts and Sciences of Georgetown University, Washington, DC.

Hansen, J.D., Jensen, C., Madsen, E.S., 2001. Green Subsidies and Learning-by-doing in the Windmill Industry. Discussion Paper 2001–06, University of Copenhagen, Department of Economics, Denmark.

Hansen, J.D., Jensen, C., Madsen, E.S., 2003. The establishment of the Danish windmill industry — was it worthwhile? Review of the World Economics 139 (2), 324–347.

Ibenholt, K., 2002. Explaining learning curves for wind power. Energy Policy 30 (13), 1181–1189.

IEA, 2000. Experience Curves for Energy Technology Policy. International Energy Agency (IEA), Paris.

Isoard, S., Soria, A., 2001. Technical change dynamics: evidence from the emerging renewable energy technologies. Energy Economics 23, 619–636.

Jamasb, T., 2007. Technical change theory and learning curves: patterns of progress in energy technologies. The Energy Journal 28 (3), 51–72.

Jensen, S.G., 2004a. Describing technological development with quantitative models. Energy & Environment 15 (2), 187–200.

Jensen, S.G., 2004b. Quantitative analysis of technological development within renewable energy technologies. International Journal of Energy Technology and Policy 2 (4), 335–353.

Junginger, M. (2005). Learning in Renewable Energy Technology Development, Ph.D. Thesis, Copernicus Institute, Utrecht University, The Netherlands.

Junginger, M., Faaij, A., Turkenburg, W.C., 2005. Global experience curves for wind farms. Energy Policy 33 (2), 133–150.

Junginger, M., van Sark, W., Faaij, A. (Eds.), 2010. Technological Learning in the Energy.

Sector: Lessons for Policy. Industry and Science. Edward Elgar, Cheltenham.

Kahouli-Brahmi, S., 2009. Testing for the presence of some features of increasing returns to adoption factors in energy system dynamics: an analysis via the learning curve approach. Ecological Economics 68 (4), 1195–1212.

Klaassen, G., Miketa, A., Larsen, K., Sundqvist, T., 2005. The impact of R&D on innovation for wind energy in Denmark, Germany and the United Kingdom. Ecological Economics 54 (2–3), 227–240.

Kobos, P.H., 2002. The Empirics and Implications of Technological Learning for Renewable Energy Technology Cost Forecasting. Proceedings of the 22nd Annual.

North American Conference of the USAEE/IAEE, Vancouver, B.C., Canada. October 2002.

Kobos, P.H., Erickson, J.D., Drennen, T.E., 2006. Technological learning and renewable energy costs: implications for US renewable energy Policy. Energy Policy 34 (13), 1645–1658.

Kouvaritakis, N., Soria, A., Isoard, S., 2000. Modelling energy technology dynamics: methodology for adaptive expectations models with learning by doing and learning by searching. International Journal of Global Energy Issues 14 (1–4), 104–115.

Loiter, J.M., Norberg-Bohm, V., 1999. Technology policy and renewable energy: public roles in the development of new energy technologies. Energy Policy 27 (2), 85–97.

Continued

Table 4.22. Continued

Mackay, R.M., Probert, S.D., 1998. Likely market-penetrations of renewable-energy technologies. Applied Energy 59 (1), 1–38.

Madsen, E.S., Jensen, C., Hansen, J.D., 2002. Scale in Technology and Learning-by-doing in the Windmill Industry. Working Papers, Department of Economics, Aarhus, School of Buisness, University of Aarhus, No. 02–2. January 1, 2002. Miketa, A., Schrattenholzer, L., 2004. Experiments with a methodology to model the role of R&D expenditures in energy technology learning processes; first results. Energy Policy 32 (15), 1679–1692.Neij, L., 1997. Use of experience curves to analyse the prospects for diffusion and adoption of renewable energy technology. Energy Policy 23 (13), 1099–1107.

Neij, L., 1999. Cost dynamics of wind power. Energy 24 (5), 375–389.

Neij, L., Andersen, P.D., Dustewitz, M., Helby, P., Hoppe-Kilpper, M., Morthorst, P.E., 2003. Experience Curves: A Tool for Energy Policy Assessment. Research Report funded in part by the European Commission within the Fifth Framework: Energy. Environment and Sustainable Development (Contract ENG1-CT2000-00116).

Neij, L., Andersen, P.D., Durstewitz, M., 2004. Experience curves for wind power. International Journal of Energy Technology and Policy 2 (1–2), 15–32.

Nemet, G.F., 2009. Interim monitoring of cost dynamics for publicly supported energy technologies. Energy Policy 37 (3), 825–835.

Papineau, M., 2006. An economic perspective on experience curves and dynamic economies in renewable energy technologies. Energy Policy 34 (4), 422–432.

Sato, T., Nakata, T., 2005. Learning Curve of Wind Power Generation in Japan.Proceedings of the 28th Annual IAEE International Conference, Taipei, Taiwan. June 2005.

Söderholm, P., Klaassen, G., 2007. Wind power in Europe: a simultaneous innovation diffusion model. Environmental and Resource Economics 36 (2), 163–190.

Söderholm, P., Sundqvist, T., 2007. The empirical challenges of measuring technology learning in the renewable energy sector. Renewable Energy 32 (15), 2559–2578.

Stanley, V.K., 2001. Wheat from chaff: meta-analysis as quantitative literature review.

Wiser, R.H., Bolinger, M., 2010. 2009 Wind Technologies Market Report. US Department of Energy, Washington, DC.

Source: Adapted from Lindman, Åsa, Patrik Söderholm. 2012. Wind power learning rates: A conceptual review and meta-analysis, Energy Economics, Volume 34, Issue 3, Pages 754-761.

Coal

Figures

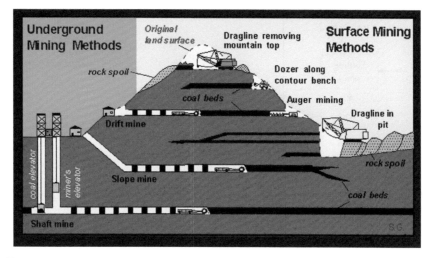

Figure 5.1. Types of coal mining methods.
Source: Kentucky Geological Survey, Methods of Mining, <http://www.uky.edu/KGS/coal/coal_mining.htm>.

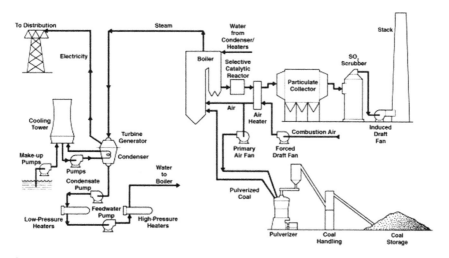

Figure 5.2. Schematic of a typical coal-fired power plant.
Source: Miller, Bruce G. 2011. Clean Coal Engineering Technology, (Boston, Butterworth-Heinemann).

Handbook of Energy. http://dx.doi.org/10.1016/B978-0-08-046405-3.00005-X

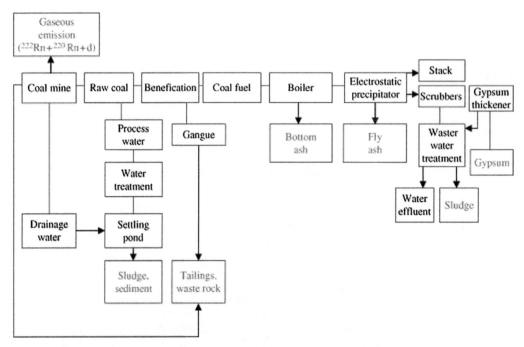

Figure 5.3. The coal fuel cycle from mining to combustion as fuel in a coal-fired power plant. Releases of NORM (naturally occurring radioactive materials) are shown in color.
Source: Paschoa A.S. and F. Steinhäusler, Editors. 2010. Radioactivity in the Environment, Volume 17, Pages 1-18.

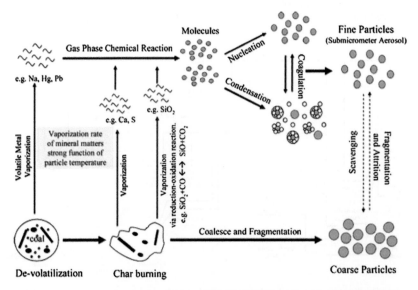

Figure 5.4. Mineral transformation and particle formation pathways during coal combustion.
Source: Chen, Lei, Sze Zheng Yong, Ahmed F. Ghoniem. 2012. Oxy-fuel combustion of pulverized coal: Characterization, fundamentals, stabilization and CFD modeling, Progress in Energy and Combustion Science, Volume 38, Issue 2, Pages 156-214.

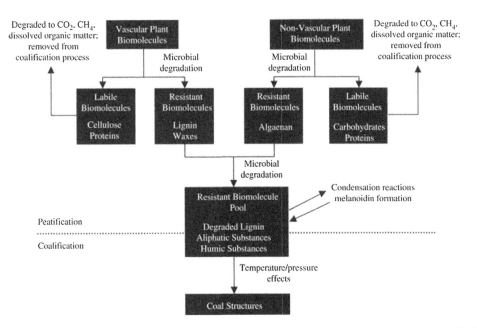

Figure 5.5. Schematic outline showing the major changes occurring in organic matter during peatification and coalification.
Source: Orem, W.H. and R.B. Finkelman. 2003. Coal Formation and Geochemistry, In: Heinrich D. Holland and Karl K. Turekian, Editor-in-Chief, Treatise on Geochemistry, (Oxford, Pergamon), Pages 191-222.

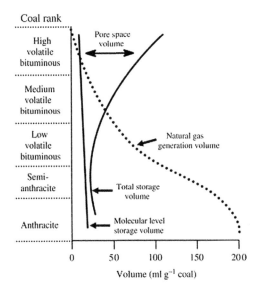

Figure 5.6. Schematic plot showing natural gas yield and changes in storage volume (molecular storage and pore space storage) as a function of coal rank.
Source: Orem, W.H. and R.B. Finkelman. 2003. Coal Formation and Geochemistry, In: Heinrich D. Holland and Karl K. Turekian, Editor-in-Chief, Treatise on Geochemistry, (Oxford, Pergamon), Pages 191-222.

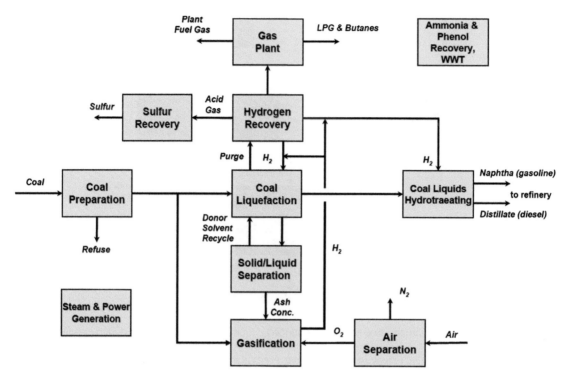

Figure 5.7. Direct liquefaction of coal.
Source: United States Department of Energy, National Energy Technology Laboratory

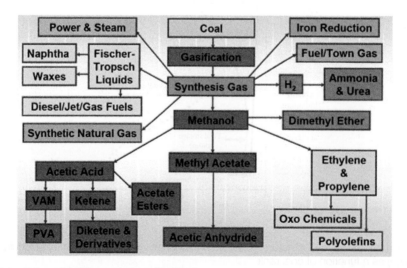

Figure 5.8. Potential products from coal gasification.
Source: Speight, James G. 2011. Handbook of Industrial Hydrocarbon Processes, (Boston, Gulf Professional Publishing).

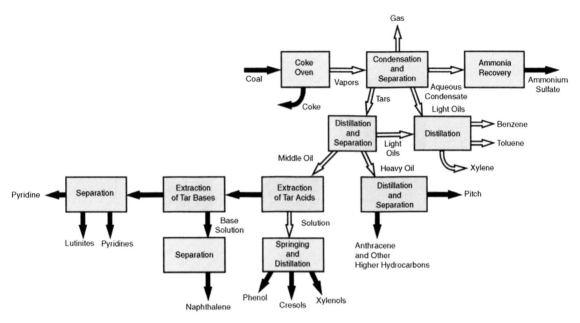

Figure 5.9. A block flow diagram of the recovery of by-products from a coke oven.
Source: Miller, Bruce G. 2011. Clean Coal Engineering Technology, (Boston, Butterworth-Heinemann).

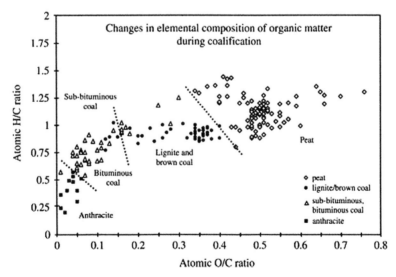

Figure 5.10. An H/C versus O/C (van Krevelen plot) plot of peat and coals of different rank. The plot illustrates that loss of oxygen (change in O/C) is most important during peatification and early stage coalification, with H loss (change in H/C) becoming more important during coalification to higher rank.
Source: Orem, W.H. and R.B. Finkelman. 2003. Coal Formation and Geochemistry, In: Heinrich D. Holland and Karl K. Turekian, Editor-in-Chief, Treatise on Geochemistry, (Oxford, Pergamon), Pages 191-222.

Charts

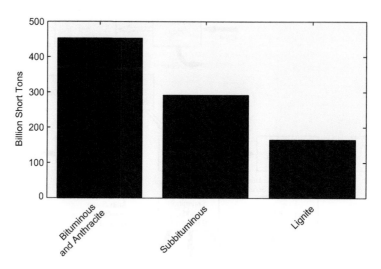

Chart 5.1.   World total recoverable coal reserves by rank in 2008.
Source: Data from United States Department of Energy, Energy Information Administration, International Energy Statistics, <http://www.eia.gov/countries/data.cfm>.

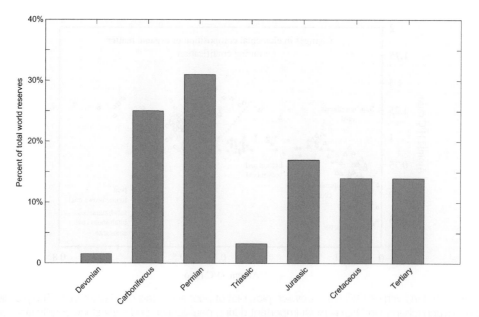

Chart 5.2.   Distribution of world coal resources by geologic age.
Source: Data from Orem, W.H. and R.B. Finkelman. 2003. Coal Formation and Geochemistry, In: Heinrich D. Holland and Karl K. Turekian, Editor(s)-in-Chief, Treatise on Geochemistry, (Oxford, Pergamon), Pages 191-222.

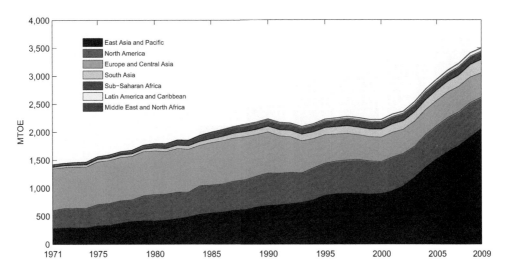

Chart 5.3. Coal production by region, 1971-2009.
Source: Data from International Energy Agency (IEA), Energy statistics database, <http://www.iea.org/stats/index.asp>.

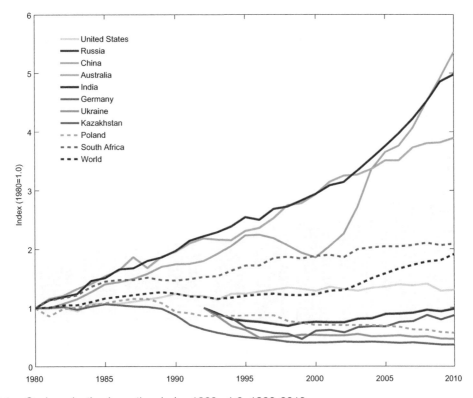

Chart 5.4. Coal production by nation, index 1980 = 1.0, 1980-2010.
Source: Data from United States Department of Energy, Energy Information Administration, International Energy Statistics, <http://www.eia.gov/countries/data.cfm>.

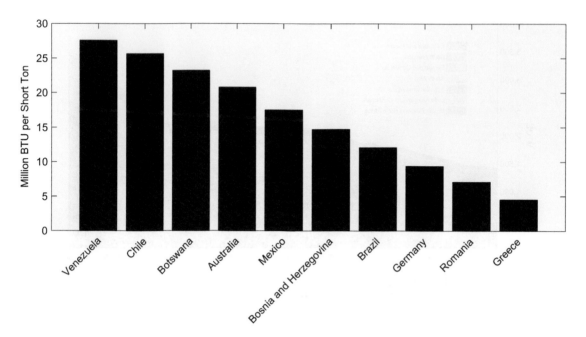

Chart 5.5. Gross heat content of coal production in 2009 for 10 nations spanning highest and lowest heat content.
Source: Data from United States Department of Energy, Energy Information Administration, International Energy Statistics, <http://www.eia.gov/countries/data.cfm>.

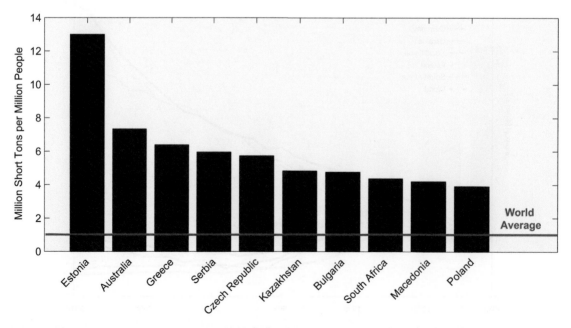

Chart 5.6. Coal consumption per capita for top 10 nations.
Source: United States Department of Energy, Energy Information Administration, International Energy Statistics, <http://www.eia.gov/countries/data.cfm>.

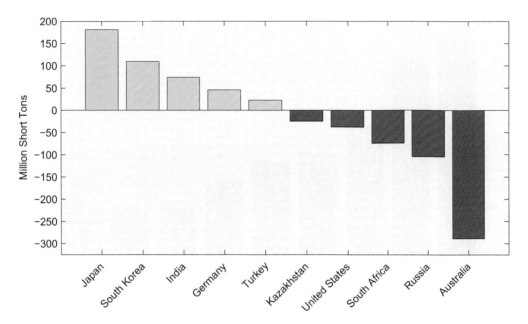

Chart 5.7. Net imports of coal for 10 nations spanning highest and lowest importers in 2009.
Source: Data from United States Department of Energy, Energy Information Administration, International Energy Statistics, <http://www.eia.gov/countries/data.cfm>.

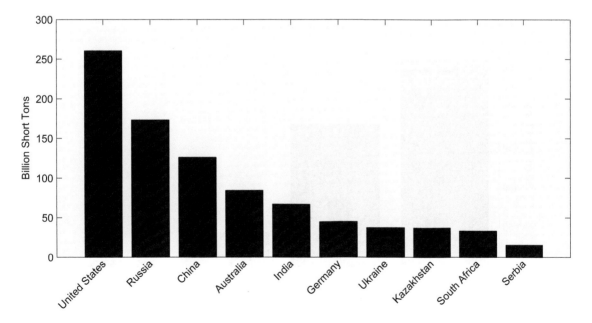

Chart 5.8. Coal reserves for top 10 nations.
Source: Data from United States Department of Energy, Energy Information Administration, International Energy Statistics, <http://www.eia.gov/countries/data.cfm>.

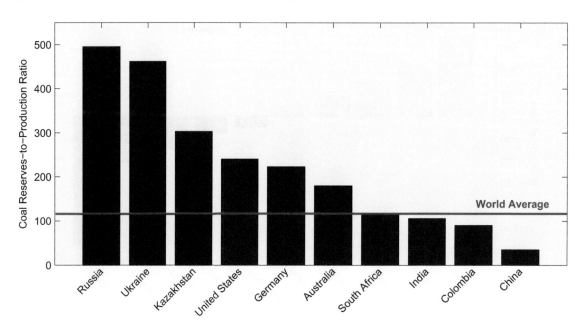

Chart 5.9. Reserve/production ratio (R/P) for the top 10 holders of coal reserves, 2010. The R/P ratio Is the hypothetical number of years that it would take to deplete proved reserves at the current production rate, and is normally calculated as current proved reserves divided by current annual production.
Source: Data from BP, Statistical Review of World Energy 2012,
<http://www.bp.com/sectionbodycopy.do?categoryId=7500&contentId=7068481>

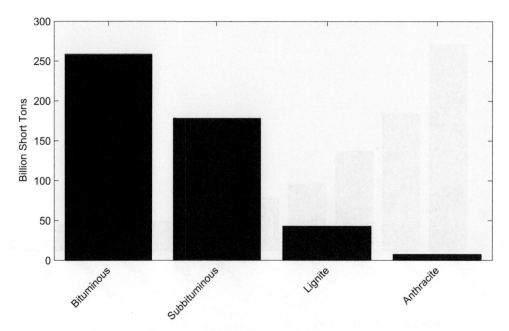

Chart 5.10. Coal reserves in the United States by rank of coal, 2009.
Source: Data from United States Department of Energy, Energy Information Administration, International Energy Statistics, <http://www.eia.gov/countries/data.cfm>.

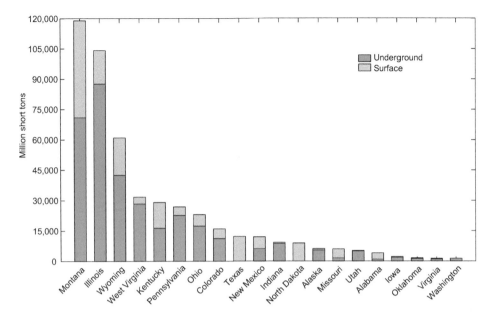

Chart 5.11. Recoverable coal reserves by method of mining, top 20 states, United States, 2010.
Source: Data from United States Department of Energy, Energy Information Administration, International Energy Statistics, <http://www.eia.gov/coal/data.cfm#reserves>.

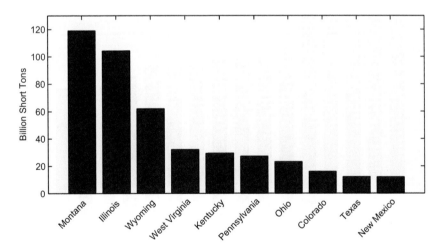

Chart 5.12. Top 10 states holding coal reserves in the United States, 2009.
Source: Data from United States Department of Energy, Energy Information Administration, International Energy Statistics, <http://www.eia.gov/countries/data.cfm>.

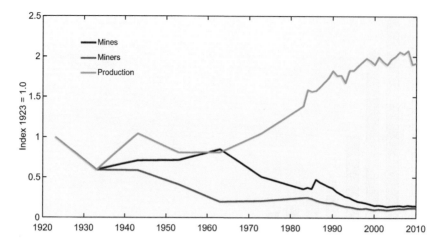

Chart 5.13. Coal production and the productivity of coal mines and coal miners in the United States, 1923-2010. Coal output per worker and per mine have increased sharply due to (i) the regional shift to Western, thick-seam, surface-mined coal; (ii) strong interfuel and intrafuel competition that led to the exit of less efficient (and generally smaller) producers, and (iii) technological change throughout the coal industry, which has more than offset resource depletion effects.
Source: Data from U.S. Department of Energy, Energy Information Administration.

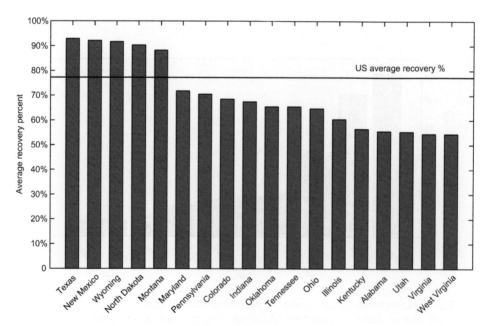

Chart 5.14. Average recovery of coal at producing mines by state in the United States, 2010.
Source: Data from United States Department of Energy, Energy Information Administration, International Energy Statistics, <http://www.eia.gov/coal/data.cfm#reserves>.

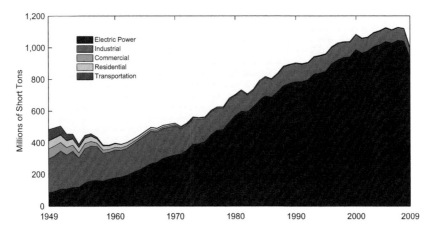

Chart 5.15. Coal consumption by end use sector in the United States, 1949-2009. Electric power generation has steadily grown to be the dominant use for coal, accounting for 94% of coal use in 2009. Most of industrial consumption is for the production of steel. The sharp drop in consumption in 2009 is associated with a severe recession.
Source: Data from United States Department of Energy, Energy Information Administration, International Energy Statistics, <http://www.eia.gov/countries/data.cfm>.

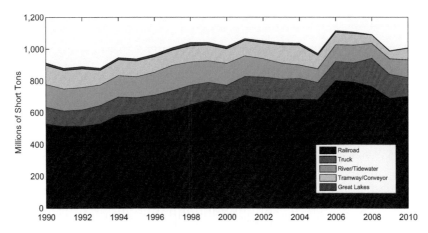

Chart 5.16. Volume of coal transported by mode in the United States, 1990-2010. Much of the transport by rail is done by unit trains that are composed of cars carrying only coal that are all bound for the same destination. Unit trains often connect one or more coal mines to power stations.
Source: Data from United States Department of Energy, Energy Information Administration.

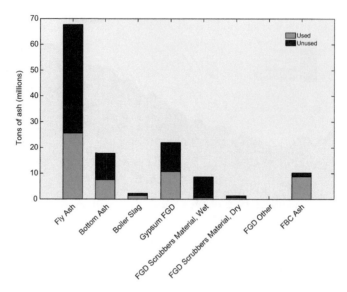

Chart 5.17. Production and beneficial use of coal fly ash in the United States, 2010. FGD = flue gas desulfurization; FBC = fluidized bed combustion.
Source: Data from American Coal Ash Association, Coal Combustion Products Production & Use Statistics, <http://acaa.affiniscape.com/displaycommon.cfm?an=1&subarticlenbr=3>.

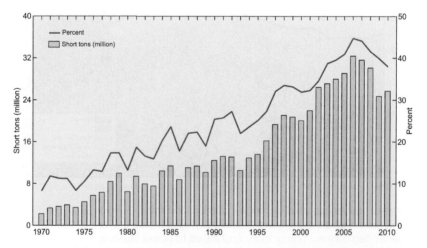

Chart 5.18. Coal fly ash production and productive use in the United States, 1970-2010. The right hand vertical axis represents the percentage of fly ash generated that is captured and put to beneficial use.
Source: Data from American Coal Ash Association, Coal Combustion Products Production & Use Statistics, <http://acaa.affiniscape.com/displaycommon.cfm?an=1&subarticlenbr=3>.

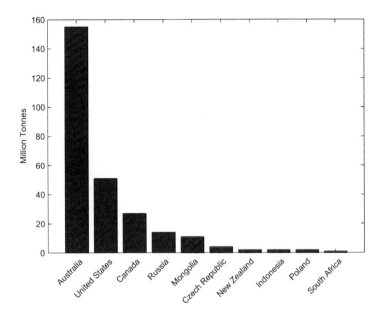

Chart 5.19. Top 10 exporters of coking coal, 2010.
Source: Data from World Coal Association, Coal Statistics,
<http://www.worldcoal.org/resources/coal-statistics/>.

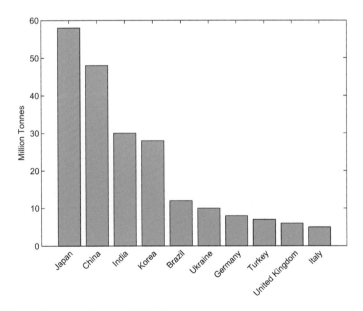

Chart 5.20. Top 10 importers of coking coal, 2010.
Source: Data from World Coal Association, Coal Statistics,
<http://www.worldcoal.org/resources/coal-statistics/>.

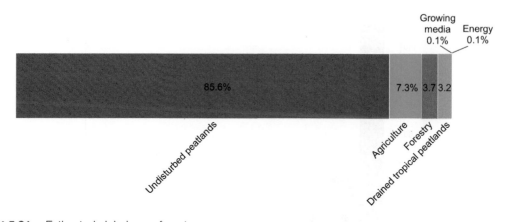

Chart 5.21. Estimated global use of peat.
Source: Data from International Peat Society, <http://www.peatsociety.org/>.

Tables

Table 5.1. Proved Reserves of coal by nation and region, 2010 (Million metric tons)

Million tonnes	Anthracite and bituminus	Sub-bituminous and lignite	Total	Share of Total	R/P ratio
US	108,501	128,794	237,295	27.6%	241
Canada	3,474	3,108	6,582	0.8%	97
Mexico	860	351	1,211	0.1%	130
Total North America	**112,835**	**132,253**	**245,088**	**28.5%**	**231**
Brazil	-	4,559	4,559	0.5%	a
Colombia	6,366	380	6,746	0.8%	91
Venezuela	479	-	479	0.1%	120
Other S. & Cent. America	45	679	724	0.1%	a
Total S. & Cent. America	**6,890**	**5,618**	**12,508**	**1.5%**	**148**
Bulgaria	2	2,364	2,366	0.3%	82
Czech Republic	192	908	1,100	0.1%	22
Germany	99	40,600	40,699	4.7%	223
Greece	-	3,020	3,020	0.4%	44
Hungary	13	1,647	1,660	0.2%	183
Kazakhstan	21,500	12,100	33,600	3.9%	303
Poland	4,338	1,371	5,709	0.7%	43
Romania	10	281	291	b	9
Russian Federation	49,088	107,922	157,010	18.2%	495
Spain	200	330	530	0.1%	73
Turkey	529	1,814	2,343	0.3%	27
Ukraine	15,351	18,522	33,873	3.9%	462
United Kingdom	228	-	228	b	13
Other Europe & Eurasia	1,440	20,735	22,175	2.6%	317
Total Europe & Eurasia	**92,990**	**211,614**	**304,604**	**35.4%**	**257**
South Africa	30,156	-	30,156	3.5%	119
Zimbabwe	502	-	502	0.1%	301
Other Africa	860	174	1,034	0.1%	a
Middle East	1,203	-	1,203	0.1%	a
Total Middle East & Africa	**32,721**	**174**	**32,895**	**3.8%**	**127**
Australia	37,100	39,300	76,400	8.9%	180
China	62,200	52,300	114,500	13.3%	35
India	56,100	4,500	60,600	7.0%	106
Indonesia	1,520	4,009	5,529	0.6%	18
Japan	340	10	350	b	382
New Zealand	33	538	571	0.1%	107
North Korea	300	300	600	0.1%	16
Pakistan	-	2,070	2,070	0.2%	a
South Korea	-	126	126	b	60
Thailand	-	1,239	1,239	0.1%	69
Vietnam	150	-	150	b	3
Other Asia Pacific	1,582	2,125	3,707	0.4%	114
Total Asia Pacific	**159,326**	**106,517**	**265,843**	**30.9%**	**57**

[a] More than 500 years.

[b] Less than 0.05%.

Reserves-to-production (R/P) ratio: If the reserves remaining at the end of the year are divided by the production in that year, the result is the length of time that those remaining reserves would last if production were to continue at that rate.

Proved reserves of coal: Generally taken to be those quantities that geological and engineering information indicates with reasonable certainty can be recovered in the future from known deposits under existing economic and operating conditions.

Source: BP Statistical Review of World Energy, 2011.

Table 5.2.   World recoverable coal reserves as of January 1, 2008 (billion short tons)

Region/Country	Recoverable reserves by coal rank				Reserves-to-production ratio (years)
	Bituminous and anthracite	Subbituminous	Lignite	Total	
World	452.9	291.4	165.1	909.4	118
United States[a]	119.6	108.7	33.3	261.6	241
Russia	54.1	107.4	11.5	173.1	495
China	68.6	37.1	20.5	126.2	35
Other Non-OECD Europe and Eurasia	49.1	19.0	27.3	95.3	290
Australia and New Zealand	40.6	2.5	41.5	84.6	195
India	59.5	0.0	5.1	64.6	106
Africa	35.1	0.2	0.0	35.3	127
OECD Europe	9.3	3.4	19.0	31.7	48
Other Central and South America	7.7	1.1	0.0	8.8	102
Other Non-OECD Asia	2.5	2.8	4.5	9.8	24
Brazil	0.0	7.8	0.0	7.8	more than 500
Canada	3.8	1.0	2.5	7.3	97
Other[b]	3.0	0.3	0.1	3.4	181

[a]Data for the United States represent recoverable coal estimates as of January 1, 2009.
[b]Includes Mexico, Middle East, Japan, and South Korea.
Source: U.S. Department of Energy, Energy Information Administration.

Table 5.3.   Coal reserves for top 15 nations, 2010[a]

	Anthracite and Bituminous	Sub-bituminous and Lignite	Total	Rank	R/P ratio
United States	108,501	128,794	237,295	1	241
Russian Federation	49,088	107,922	157,010	2	495
China	62,200	52,300	114,500	3	35
Australia	37,100	39,300	76,400	4	180
India	56,100	4,500	60,600	5	106
Germany	99	40,600	40,699	6	223
Ukraine	15,351	18,522	33,873	7	462
Kazakhstan	21,500	12,100	33,600	8	303
South Africa	30,156	-	30,156	9	119
Colombia	6,366	380	6,746	10	91
Canada	3,474	3,108	6,582	11	97
Poland	4,338	1,371	5,709	12	43
Indonesia	1,520	4,009	5,529	13	18
Brazil	-	4,559	4,559	14	-
Greece	-	3,020	3,020	15	44
World	404,762	456,176	860,938		118

[a]Million metric tons
Source: BP, Statistical Review of World Energy 2011.

Table 5.4.   Coal Demonstrated Reserve Base in the United States, January 1, 2010 (Billion Short Tons)

| Region and State | Anthracite | | Bituminous Coal | | Subbituminous Coal | | Lignite | Total | | |
	Under-ground	Surface	Under-ground	Surface	Under-ground	Surface	Surface[a]	Under-ground	Surface	Total
Appalachian	**4**	**3.3**	**68.6**	**22**	**0**	**0**	**1.1**	**72.6**	**26.5**	**99**
Alabama	0	0	0.9	2.1	0	0	1.1	0.9	3.2	4.1
Kentucky, Eastern	0	0	0.8	9.1	0	0	0	0.8	9.1	10
Ohio	0	0	17.4	5.7	0	0	0	17.4	5.7	23.1
Pennsylvania	3.8	3.3	19	0.8	0	0	0	22.8	4.2	27
Virginia	0.1	0	0.9	0.5	0	0	0	1	0.5	1.5
West Virginia	0	0	28.5	3.4	0	0	0	28.5	3.4	32
Other[b]	0	0	1.1	0.3	0	0	0	1.1	0.3	1.4
Interior	**0.1**	**(s)**	**116.8**	**27.1**	**0**	**0**	**12.6**	**116.9**	**39.7**	**156.6**
Illinois	0	0	87.7	16.5	0	0	0	87.7	16.5	104.2
Indiana	0	0	8.6	0.6	0	0	0	8.6	0.6	9.3
Iowa	0	0	1.7	0.5	0	0	0	1.7	0.5	2.2
Kentucky, Western	0	0	15.7	3.6	0	0	0	15.7	3.6	19.3
Missouri	0	0	1.5	4.5	0	0	0	1.5	4.5	6
Oklahoma	0	0	1.2	0.3	0	0	0	1.2	0.3	1.5
Texas	0	0	0	0	0	0	12.2	0	12.2	12.2
Other[c]	0.1	(s)	0.3	1.1	0	0	0.4	0.4	1.5	1.9
Western	**(s)**	**0**	**21.2**	**2.3**	**121.2**	**56.5**	**29.2**	**142.5**	**88**	**230.5**
Alaska	0	0	0.6	0.1	4.8	0.6	(s)	5.4	0.7	6.1
Colorado	(s)	0	7.5	0.6	3.7	0	4.2	11.2	4.8	16
Montana	0	0	1.4	0	69.6	32.3	15.8	71	48.1	119
New Mexico	(s)	0	2.7	0.9	3.4	5	0	6.1	5.9	12
North Dakota	0	0	0	0	0	0	8.9	0	8.9	8.9
Utah	0	0	4.9	0.3	(s)	0	0	4.9	0.3	5.2
Washington	0	0	0.3	0	1	0	(s)	1.3	(s)	1.3
Wyoming	0	0	3.8	0.5	38.6	18.6	0	42.5	19.1	61.6
Other[d]	0	0	(s)	0	(s)	(s)	0.4	(s)	0.4	0.4
U.S. Total	**4.1**	**3.4**	**206.6**	**51.4**	**121.2**	**56.5**	**42.9**	**331.9**	**154.2**	**486.1**
States East of the Mississippi River	4	3.3	180.6	42.8	0	0	1.1	184.6	47.2	231.8
States West of the Mississippi River	0.1	(s)	26	8.6	121.2	56.5	41.8	147.3	107	254.3

(s)=Less than 0.05 billion short tons.
[a] Lignite resources are not mined underground in the United States.
[b] Georgia, Maryland, North Carolina, and Tennessee.
[c] Arkansas, Kansas, Louisiana, and Michigan.
[d] Arizona, Idaho, Oregon, and South Dakota.
Source: U.S. Energy Information Administration, <http://www.eia.gov/coal/data.cfm#reserves>.

Table 5.5. Estimated available coal reserves and corresponding gas reserves from underground coal gasification (UCG)

	Estimated available coal reserves for UCG billion metric tons	Potential gas reserves from UCG (as Natural Gas) trillion m3	Current natural gas reserves (end-2005) trillion m3
USA	138.1	41.4	5.9
Europe	130.1	21.8	5.7
Russian Federation	87.9	26.3	47.8
China	64.1	19.2	2.4
India	51.8	15.5	1.1
South Africa	48.7	8.2	NA
Australia	44	13.2	0.8
Total	**564.7**	**145.6**	**63.7**

Source: World Energy Council.

Table 5.6. Coal production in top 15 producing nations, 1990-2010[a]

Country	1990	2000	2010	Rank 2010	Rank 1990	Change 2000-10
China	562.3	762.5	1800.4	1	2	136.1%
US	565.9	570.1	552.2	2	1	−3.1%
Australia	109.0	166.5	235.4	3	5	41.4%
India	91.9	132.2	216.1	4	8	63.4%
Indonesia	6.6	47.4	188.1	5	20	297.0%
Russian Federation	176.2	116.0	148.8	6	3	28.2%
South Africa	100.1	126.6	143.0	7	6	13.0%
Kazakhstan	67.7	38.5	56.2	8	10	45.9%
Poland	94.5	71.3	55.5	9	7	−22.2%
Colombia	13.3	24.9	48.3	10	14	94.4%
Germany	117.3	56.5	43.7	11	4	−22.7%
Ukraine	83.9	42.0	38.1	12	9	−9.2%
Canada	40.0	36.1	34.9	13	12	−3.4%
Vietnam	2.9	6.5	24.7	14	27	279.8%
Czech Republic	36.7	25.0	19.4	15	13	−22.6%
World	2267.1	2352.5	3731.4			58.6%

[a]Million metric tons oil equivalent.
Source: BP, Statistical Review of World Energy 2011.

Table 5.7.   Total coal exports for top 15 exporting nations, 1980-2010[a]

Country	1980	1990	2000	2010	Rank 2010	Rank 1980	Change 2000-10
Australia	47,711	114,607	206,115	328,132	1	2	59.2%
Indonesia	110	5,159	63,332	316,152	2	22	399.2%
United States	93,813	106,376	59,635	83,178	3	1	39.5%
South Africa	32,187	54,633	77,061	76,683	4	4	−0.5%
Colombia	165	16,164	39,522	76,380	5	17	93.3%
Canada	16,932	34,404	35,695	36,921	6	5	3.4%
Vietnam	551	870	3,584	24,676	7	13	588.6%
China	7,265	20,481	77,451	22,658	8	6	−70.7%
Mongolia	0	540	1	18,336	9	n/a	n/a
Poland	37,934	35,186	29,474	18,108	10	3	−38.6%
Venezuela	0	2,022	8,741	6,814	11	39	−22.0%
Netherlands	2,479	3,687	10,972	6,634	12	8	−39.5%
Korea, North	110	551	397	5,089	13	21	1182.5%
New Zealand	77	370	1,685	2,666	14	26	58.2%
India	119	110	1,431	2,395	15	20	67.4%
World	325,760	468,061	723,446	1,212,819			67.6%

[a]Thousand Short Tons.
Source: U.S. Deparment of Energy, Energy Information Administration, <http://www.eia.gov/countries/data.cfm>

Table 5.8.   Total Coal Imports for top 15 importing nations, 1980-2010[a]

Country	1980	1990	2000	2010	Rank 2010	Rank 1980	Change 2000-10
Japan	75,681	115,937	168,586	206,702	1	1	22.6%
China	2,196	2,208	2,402	195,063	2	21	8021.1%
South Korea	5,680	26,581	67,944	125,808	3	13	85.2%
India	365	5,570	25,737	101,564	4	38	294.6%
Taiwan	5,172	20,409	50,410	71,130	5	15	41.1%
Germany	--	--	39,387	55,154	6	n/a	40.0%
Turkey	874	6,142	15,127	30,030	7	30	98.5%
United Kingdom	8,084	16,631	26,377	29,359	8	8	11.3%
Italy	18,971	22,810	21,531	23,735	9	3	10.2%
Netherlands	8,977	19,702	25,461	22,831	10	7	−10.3%
Russia	--	--	28,324	21,848	11	n/a	−22.9%
France	35,861	22,671	22,575	20,785	12	2	−7.9%
United States	1,853	3,464	16,294	20,567	13	22	26.2%
Brazil	5,558	11,817	16,365	19,078	14	14	16.6%
Malaysia	88	2,610	3,060	18,768	15	50	513.3%
World	306,123	463,655	720,431	1,178,153			63.5%

[a]Thousand Short Tons.
Source: U.S. Deparment of Energy, Energy Information Administration, <http://www.eia.gov/countries/data.cfm>, accessed 25 May 2012.

Table 5.9.   Coal consumption in top 15 consuming nations, 1970-2010[a]

	1970	1980	1990	2000	2010	Rank 2010	Rank 1970	Change 2000-10
China	162.9	304.9	525.3	737.1	1,713.5	1	2	132.5%
US	309.1	388.6	483.1	569.0	524.6	2	1	−7.8%
India	37.6	56.7	95.5	144.2	277.6	3	8	92.4%
Japan	60.2	57.6	76.0	98.9	123.7	4	6	25.1%
Russian Federation	n/a	n/a	180.6	105.2	93.8	5	n/a	−10.8%
South Africa	27.4	42.7	66.4	74.6	88.7	6	10	18.9%
Germany	151.7	139.6	129.6	84.9	76.5	7	3	−9.9%
South Korea	5.6	13.2	24.4	43.0	76.0	8	20	76.6%
Poland	70.2	101.6	80.2	57.6	54.0	9	5	−6.3%
Australia	18.0	26.1	36.5	46.7	43.4	10	11	−7.1%
Taiwan	2.7	3.8	11.0	28.7	40.3	11	25	40.4%
Indonesia	0.1	0.3	4.0	13.7	39.4	12	45	186.5%
Ukraine	n/a	n/a	74.8	38.8	36.4	13	n/a	−6.1%
Kazakhstan	n/a	n/a	40.2	23.2	36.1	14	n/a	55.7%
Turkey	4.2	7.5	16.8	25.5	34.4	15	23	34.9%
Total World	1,499.6	1,806.4	2,220.3	2,399.7	3,555.8			48.2%

[a]Million metric tons oil equivalent.
Source: BP, Statistical Review of World Energy 2011.

Table 5.10.   Coking Coal Imports and Exports, 2010 (Mt)

Top Exporters	
Australia	155
USA	51
Canada	27
Russia	14
Mongolia	11
Czech Republic	4
New Zealand	2
Indonesia	2
Poland	2
South Africa	1
Top Importers	
Japan	58
China	48
India	30
Korea	28
Brazil	12
Ukraine	10
Germany	8
Turkey	7
UK	6
Italy	5

Source: World Coal Association.

Table 5.11. **Coal Properties for Coke Production**

Coal Parameter	Typical Values
Carbon content, wt.% (dry, ash free)	≈85
Hydrogen content, wt.% (dry, ash free)	≈5.25
Volatile matter content, wt.% (dry, ash free)	24–28
Ash content, wt.%	<10
Sulfur content, wt.%	≈0.5
H/C ratio	0.725
O/C ratio	0.04
Heating value, Btu/lb (moist, mineral matter free)	≈15,500
Vitrinite reflectance, %	≈1.25
Vitrinite/inertinite/exinite, % (dry basis)	≈55/35/10
Free swelling index	6.5.8
Maximum fluidity, dial divisions per minute	≈1,000
Roga index	≈45
Gray-King assay	≥G4

Source: Miller, Bruce G. 2011. Clean Coal Engineering Technology, (Boston,Butterworth-Heinemann).

Table 5.12. **Composition of various coals**

	Anthracite	Bituminous	Subbituminous	Lignite
Moisture (%)	3–6	2–15	10–25	25–45
Volatile matter (%)	2–12	15–45	28–45	24–32
Fixed carbon (%)	75–85	50–70	30–57	25–30
Ash (%)	4–15	4–15	3–10	3–15
Sulfur (%)	0.5–2.5	0.5–6	0.3–1.5	0.3–2.5
Hydrogen (%)	1.5–3.5	4.5–6	5.5–6.5	6–7.5
Carbon (%)	75–85	65–80	55–70	35–45
Nitrogen (%)	0.5–1	0.5–2.5	0.8–1.5	0.6–1.0
Oxygen (%)	5.5–9	4.5–10	15–30	38–48
Heating value				
Btu/lb	12,000–13,500	12,000–14,500	7500–10,000	6000–7500
kJ/kg	27,893–31,380	27,893–33,704	17,433–23,244	13,947–17,433
Density (g/mL)	1.35–1.70	1.28–1.35	1.35–1.40	1.40–1.45

Source: United States Geological Survey.

Table 5.13.   Scratch hardness of coal[a]

Material	Scratch Hardness Relative to Barnsley Soft Coal[b]
Anthracite, Great Mountain	1.70
Anthracite, Red Vein	1.75
Welsh steam	0.29
Barnsley hards	0.85
Barnsley softs	1.00
Illinois coal	1.10
Cannel	0.92
Carbonaceous shale	0.69
Shale	0.32
Pyrite	5.71
Calcite	1.92

[a]Scratch hardness is the measure of how resistant a sample is to fracture or permanent plastic deformation due to friction from a sharper object.
[b]Barnsley is a town in South Yorkshire, England, a coal ming region.
Source: Adapted from Speight, James G. 2005. Handbook of Coal Analysis (New York, Wiley).

Table 5.14.   Proximate analysis of solid fuels

Fuel	Ash content (%)	Moisture (%)	Volatiles (%)	Heating value (HHV GJ Mg^{-1}) [a]
Anthracite coal	7.83	2.80	1.3	30.90
Bituminous coal	2.72	2.18	33.4	34.50
Sub-bituminous	3.71	18.41	44.3	21.24
Softwood	1.00	20.00	85.0	18.60

[a]On a moisture and ash free basis (maf).
Source: Kreith Frank and D.Yogi Goswami. 2004. The CRC Handbook of Mechanical Engineering, Second Edition, (Boca Raton, CRC Press).

Table 5.15.   Coal Inorganic Impurities

Type	Origin	Examples	Physical separation
Strongly chemically bonded elements	From coal-forming organic tissue material	Organic sulfur, nitrogen	No
Adsorbed and weakly bonded groups	Ash-forming components in pure water, adsorbed on the coal surface	Various salts	Very limited
Mineral matter	Minerals washed or blown into the peat during its formation	Clays, quartz	Partly separable by physical methods
a. Epiclastic			
b. Syngenetic	Incorporated into coal from the very earliest peat-accumulation stage	Pyrite, siderite, some clay minerals	Intimately intergrown with coal macerals
c. Epigenetic	Stage subsequent to syngenetic; migration of the minerals-forming solutions through coal fractures	Carbonates, pyrite, kaolinite	Vein type mineralization; epigenetic minerals concentrated along cleats, preferentially exposed during breakage; separable by physical methods

Source: Adapted from Pisupadti, Sarma V. 2003. Fuel Chemistry, In: Robert A. Meyers, Editor(s)-in-Chief, Encyclopedia of Physical Science and Technology (Third Edition), (New York, Academic Press), Pages 253-274.

Table 5.16. Classification of Coal by Rank

Class	Group	Fixed Carbon Limits (%, Dry, Mineral-Matter-Free Basis)		Volatile Matter Limits (%, Dry, Mineral-Matter-Free Basis)		Calorific Value Limits (Btu/lb, Moist, Mineral-Matter-Free Basis)[b]		Agglomerating Character
		Equal to or Greater Than	Less Than	Greater Than	Equal to or Less Than	Equal to or Greater Than	Less Than	
I. Anthracitic	1. Metaanthracite	98	-	-	2	-	-	Nonagglomerating
	2. Anthracite	92	98	2	8	-	-	
	3. Semianthracite[c]	86	92	8	14	-	-	
II. Bituminous	1. Low-volatile bituminous coal	78	86	14	22	-	-	Commonly agglomerating[c]
	2. Medium-volatile bituminous coal	69	78	22	31	-	-	
	3. High-volatile A bituminous coal	-	69	31	-	14,000[d]	-	
	4. High-volatile B bituminous coal	-	-	-	-	13,000	14,000	
	5. High-volatile C bituminous coal	-	-	-	-	11,500	13,000	
						10,500	11,500	
III. Subbituminous	1. Subbituminous A coal	-	-	-	-	10,500	11,500	
	2. Subbituminous B coal	-	-	-	-	9,500	10,500	
	3. Subbituminous C coal	-	-	-	-	8,300	9,500	Nonagglomerating
IV. Lignite	1. Lignite A	-	-	-	-	6,300	8,300	
	2. Lignite B	-	-	-	-	-	6,300	

[b]Moist refers to coal containing its natural inherent moisture but not including visible water on the surface of the coal.

[c]If agglomerating, classify in low-volatile group in the bituminous class.

[d]Coals having 69% or more fixed carbon on the dry, mineral-matter-free basis shall be classified according to fixed carbon, regardless of calorific value.

[e]There may be nonagglomerating varieties in these groups of the bituminous class, and there are notable exceptions in the high-volatile C bituminous group.

Source: United States Geological Survey.

Table 5.17. Classification of Coal Macerals[a]

Maceral group	Maceral (or submaceral)	Principal mode of origin
Vitrinite	Telinite	Mummified cell walls of woody tissues
	Telocollinite (collotelinite)	Gel-impregnated woody tissues
	Vitrodetrinite	Small vitrinitic fragments
	Desmocollinite (collodetrinite)	Gelified smaller plant tissues
	Gelinite	Organic gels, mainly as cell infillings
Liptinite (or exinite)	Sporinite	Remains of spore coatings
	Cutinite	Remains of leaf cuticles
	Resinite	Resin bodies
	Alginite	Remains of algal masses
	Suberinite	Remains of waxy, cork-like tissues
	Liptodetrinite	Fine fragments of liptinite macerals
	Fluorinite	Lenticles of plant oils or fats
	Bituminite	Bitumen-like residues from algae, etc.
	Exsudatinite	Fluorescent cavity infilling
Inertinite	Fusinite	Oxidized cellular woody tissue
	Semifusinite	Partly oxidized tissue
	Inertodetrinite	Oxidized cell wall fragments
	Micrinite	Fine-grained, probable decay product
	Macrinite	Massive, possibly fusinized gels
	Funginite	Cellular fungal remains
Secretinite	Fusinized	Vesicular resin bodies

[a]any of the numerous microscopically recognizable, individual organic constituents of coal with characteristic physical and chemical properties.
Source: Adapted from Ward, Colin R. 2003. Coal Geology, In: Robert A. Meyers, Editor-in-Chief, Encyclopedia of Physical Science and Technology (Third Edition), (New York, Academic Press), Pages 45-77.

Table 5.18. Physical properties used for determining coal suitability for use

Test/Property	Results/Comments
Physical Properties	
Density	True density as measured by helium displacement
Specific Gravity	Apparent density
Pore structure	Specification of the porosity or ultrafine structure of coals and nature of pore structure betwene macro, micro and transitional pores
Surface area	Determination of total surface area by heat of absorption
Reflectivity	Useful in petrographic analyses
Mechanical Poperties	
Strength	Specification of compressibility/strength
Hardness/abrasiveness	Specification of scratch and indentation hardness; also abrasive action of coal
Friability	Ability to withstand degradation in size on handling, tendency toward breakage

Table 5.18. Continued

Test/Property	Results/Comments
Grindability	Relative amount of work needed to pulverize coal
Dustiness index	Amount of dust produced when coal is handled
Thermal properties	
Calorific value	Indication of energy content
Heat capacity	Measurement of the heat required to raise the temperature of a unit amount of coal 1 degree
Thermal conductivity	Time rate of heat transfer through unit area, unit thickness, unit temperature difference
Plastic/agglutinating	Changes in a coal upon heating; caking properties of coal
Agglomerating index	Grading on nature of residue from 1-g sample when heated at 950 degrees Celsius (1550 degrees Farenheit)
Free swelling index	Measure of the increase in volume when a coal is heated without restriction
Electrical properties	
Electirical resisitivity	Electrical resisitivity of coal measured in ohm-centimeters
Dielectric constant	Measure of electrosattic polarizability

Source: James G. Speight, Handbook of Coal Analysis (New York, John Wiley and Sons, 2005).

Table 5.19. Variation of heat content of coal with temperature

Temperature °C	Heat Content		
	As Tested		Ash-Free Basis
	cal/g	Btu/lb	cal/g
Lignite (Texas)			
32.7	11.8	21.2	13.5
69.3	20.2	36.4	22.5
95.3	25.4	45.7	27.7
34.4	39.2	70.6	42.5
Subbituminous B (Wyoming)			
42.3	14.1	25.4	14.5
65.0	19.4	34.9	19.8
89.7	26.4	47.5	2639
112.6	34.0	61.2	34.6

Source: Adapted from Speight, James G. 2005. Handbook of Coal Analysis (New York, Wiley).

Table 5.20. Average specific gravity and average weight of unbroken coal per unit of volume of different ranks

| Rank | Specific gravity | Weight of unbroken coal per unit volume | | | Short tons per square mile-foot | Metric tons per square mile-foot | Metric tons per square hectometer-meter | Metric tons per square hectometer-meter |
		Short tons per acre-foot	Short tons per acre-inch	Metric tons per acre-foot				
Anthracite and semianthracite	1.47	2,000	166.6	1,814	1,280,000	1,160,960	14,700	1,470,000
Bituminous coal	1.32	1,800	150	1,633	1,152,000	1,045,120	13,200	1,320,000
Subbituminous coal	1.3	1,770	147.5	1,605	1,132,800	1,027,200	13,000	1,300,000
Lignite	1.29	1,750	145.8	1,588	1,120,000	1,016,320	12,900	1,290,000

Source: U.S. Geological Survey, Coal Resource Classification System of the U.S. Geological Survey, GEOLOGICAL SURVEY CIRCULAR 891.

Table 5.21. Arithmetic and geometric means for chemical elements in US coal

| Component | Arithmetic | | Geometric | | Max. |
	Mean	SD	Mean	SD	
Ash (%)	13.1	8.3	10.9	1.9	50
Aluminum (Al) (%)	1.5	1.1	1.1	2.1	10.6
Antimony (Sb) (ppm)	1.2	1.6	0.61	3.6	35
Arsenic (As) (ppm)	24	60	6.5	5.5	2,200
Barium (Ba) (ppm)	170	350	93	3	22,000
Beryllium (Be) (ppm)	2.2	4.1	1.3	3.5	330
Bismuth (Bi) (ppm)	(<1.0)	ND	ND	ND	14
Boron (B) (ppm)	49	54	30	3.1	1,700
Bromine (Br) (ppm)	17	19	9.1	4.1	160
Cadmium (Cd) (ppm)	0.47	4.6	0.02	18	170
Calcium (Ca) (%)	0.46	1	0.23	3.3	72
Carbon (C) (%)	63	15	62	1.3	90
Cerium (Ce) (ppm)	21	28	5.1	7.1	700
Cesium (Cs) (ppm)	1.1	1.1	0.7	3.2	15
Chlorine (Cl) (ppm)	614	670	79	41	8,800
Chromium (Cr) (ppm)	15	15	10	2.7	250
Cobalt (Co) (ppm)	6.1	10	3.7	2.9	500
Copper (Cu) (ppm)	16	15	12	2.1	280
Dysprosium (Dy) (ppm)	1.9	2.7	0.008	35	28
Erbium (Er) (ppm)	1	1.1	0.002	73	11
Europium (Eu) (ppm)	0.4	0.33	0.12	5.8	4.8
Fluorine (F) (ppm)	98	160	35	15	4,000
Gadolinium (Gd) (ppm)	−1.8	ND	ND	ND	39
Gallium (Ga) (ppm)	5.7	4.2	4.5	2.1	45
Germanium (Ge) (ppm)	5.7	14	0.59	16	780
Gold (Au) (ppm)	(<0.05)	ND	ND	ND	ND
Hafnium (Hf) (ppm)	0.73	0.68	0.04	38	18

Table 5.21.   Continued

Component	Arithmetic		Geometric		Max.
	Mean	**SD**	**Mean**	**SD**	
Holmium (Ho) (ppm)	−0.35	ND	ND	ND	4.5
Hydrogen (H) (%)	5.2	0.09	5.2	1.2	9.5
Indium (In) (ppm)	(<0.3)	ND	ND	ND	ND
Iodine (I) (ppm)	(<1.0)	ND	ND	ND	ND
Iridium (Ir) (ppm)	(<0.001)	ND	ND	ND	ND
Iron (Fe) (ppm)	1.3	1.5	0.75	2.9	24
Lanthanum (La) (ppm)	12	16	3.9	6	300
Lead (Pb) (ppm)	11	37	5	3.7	1,900
Lithium (Li) (ppm)	16	20	9.2	3.3	370
Lutetium (Lu) (ppm)	0.14	0.1	0.06	4.7	1.8
Magnesium (Mg) (%)	0.11	0.12	0.07	2.7	1.5
Manganese (Mn) (ppm)	43	84	19	3.9	2,500
Mercury (Hg) (ppm)	0.17	0.24	0.1	3.1	10
Molybdenum (Mo) (ppm)	3.3	5.6	1.2	6.5	280
Neodymium (Nd) (ppm)	−9.5	ND	ND	ND	230
Nickel (Ni) (ppm)	14	15	9	2.8	340
Niobium (Nb) (ppm)	2.9	3.1	1	7.7	70
Nitrogen (N) (%)	1.3	0.4	1.3	1.4	13
Osmium (Os) (ppm)	(<0.001)	ND	ND	ND	ND
Oxygen (O) (%)	16	12	12	2	60
Palladium (Pd) (ppm)	(<0.001)	ND	ND	ND	ND
Phosphorus (P) (ppm)	430	1,500	20	20	58,000
Platinum (Pt) (ppm)	(<0.001)	ND	ND	ND	ND
Potassium (K) (%)	0.18	0.21	0.1	3.5	2
Praseodymium (Pr) (ppm)	−2.4	ND	ND	ND	65
Rhenium (Re) (ppm)	(<0.001)	ND	ND	ND	ND
Rhodium (Rh) (ppm)	(<0.001)	ND	ND	ND	ND
Rubidium (Rb) (ppm)	21	20	0.62	41	140
Ruthenium (Ru) (ppm)	(<0.001)	ND	ND	ND	ND
Samarium (Sm) (ppm)	1.7	1.4	0.35	13	18
Scandium (Sc) (ppm)	4.2	4.4	3	2.3	100
Selenium (Se) (ppm)	2.8	3	1.8	3.1	150
Silicon (Si) (%)	2.7	2.4	1.9	2.4	20
Silver (Ag) (ppm)	(<0.1)	0.35	0.01	9.1	19
Sodium (Na) (%)	0.08	0.12	0.04	3.5	1.4
Strontium (Sr) (ppm)	130	150	90	2.5	2,800
Sulfur (S) (%)	1.8	1.8	1.3	2.4	25
Tantalum (Ta) (ppm)	0.22	0.19	0.02	13	1.7
Tellurium (Te) (ppm)	(<0.1)	ND	ND	ND	ND
Terbium (Tb) (ppm)	0.3	0.23	0.09	7.7	3.9
Thallium (Tl) (ppm)	1.2	3.4	0.00004	205	52
Thorium (Th) (ppm)	3.2	3	1.7	5	79
Thulium (Tm) (ppm)	[0.15]	ND	ND	ND	1.9
Tin (Sn) (ppm)	1.3	4.3	0.001	54	140

Continued

Table 5.21.   Continued

Component	Arithmetic		Geometric		
	Mean	**SD**	**Mean**	**SD**	**Max.**
Titanium (Ti) (%)	0.08	0.07	0.06	2.2	0.74
Tungsten (W) (ppm)	1	7.6	0.1	14	400
Uranium (U) (ppm)	2.1	16	1.1	3.5	1,300
Vanadium (V) (ppm)	22	20	17	2.2	370
Ytterbium (Yb) (ppm)	[0.95]	ND	ND	ND	20
Yttrium (Y) (ppm)	8.5	6.7	6.6	2.2	170
Zinc (Zn) (ppm)	53	440	13	3.4	19,000
Zirconium (Zr) (ppm)	27	32	19	2.4	700

All values are on a coal basis. Data are exclusively from the US Geological Survey (USGS) except for estimated values in parentheses which are based on USGS and literature data. Values in brackets are calculated from cerium and lanthanum data and assuming a chondrite normalized rare-earth-element distribution pattern. (ND=no data; SD=standard deviation; Max.=maximum).

Source: Adapted from Orem, W.H. and R.B. Finkelman. 2003. Coal Formation and Geochemistry, In: Heinrich D. Holland and Karl K. Turekian, Editor(s)-in-Chief, Treatise on Geochemistry, (Oxford, Pergamon), Pages 191-222.

Table 5.22.   Coal flotation reagents

Type	Examples	Remarks
Collectors	Insoluble in water, oily hydrocarbons, kerosene, fuel oil	Used in so-called emulsion flotation of coal, in which collector droplets must attach to coal particles
Frothers	Water-soluble surfactants; aliphatic alcohols; MIBC	To stabilize froth; adsorb at oil/water interface and onto coal; have some collecting abilities
Promoters	Emulsifiers	Facilitate emulsification of oily collector and the attachment of the collector droplets to oxidized and/or low-rank coal particles
Depressants/ dispersants	Organic colloids: dextrin, carboxymethyl cellulose, etc.	Adsorb onto coal and make it hydrophilic
Inorganic salts	NaCl, CaCl$_2$, Na$_2$SO$_4$, etc.	Improve floatability; in so-called salt flotation process may be used to float metallurgical coals, even without any organic reagents; are coagulants in dewatering.

Source: Adapted from Pisupadti, Sarma V. 2003. Fuel Chemistry, In: Robert A. Meyers, Editor-in-Chief, Encyclopedia of Physical Science and Technology (Third Edition), (New York, Academic Press), Pages 253-274.

Table 5.23. Coal Combustion Product (CCP) Production and Use in the United States, 2010

CCP Categories:	Fly Ash	Bottom Ash	Boiler Slag	Gypsum FGD	FGD Scrubbers Material Wet	FGD Scrubbers Material Dry	FGD Other	FBC Ash
2010 Total CCPs Produced by Category	67,700,000	17,800,000	2,332,944	22,000,000	8,670,814	1,405,952	3,740	10,267,914
2010 Total CCPs Used by Category	25,723,217	7,541,732	1,418,996	10,713,138	624,223	584,112	0	8,732,008
Concrete/Concrete Products /Grout	11,016,097	615,332	0	21,045	0	16,847	0	0
Blended Cement/ Raw Feed for Clinker	2,045,797	949,183	3,000	1,135,211	0	0	0	0
Flowable Fill	135,321	52,414	0	0	0	13,998	0	0
Structural Fills/ Embankments	4,675,992	3,124,549	78,647	454,430	424,581	358,019	0	0
Road Base/Sub-base	242,952	715,357	3,128	0	3,018	0	0	0
Soil Modification/ Stabilization	785,552	162,065	0	0	0	19,189	0	0
Snow and Ice Control	0	549,520	41,194	0	0	0	0	0
Blasting Grit/Roofing Granules	86,484	19,914	1,257,571	0	0	0	0	0
Mining Applications	2,399,837	528,881	0	835,536	186,624	112,373	0	8,660,408
Gypsum Panel Products	109	0	0	7,661,527	0	0	0	0
Waste Stabilization/ Solidification	3,258,825	41,233	0	0	0	39,283	0	71,600
Agriculture	22,220	4,674	0	481,827	0	0	0	0
Aggregate	6,726	555,031	27,155	0	0	0	0	0
Miscellaneous/Other	1,047,305	223,579	8,301	123,562	10,000	24,403	0	0

Source: American Coal Ash Association, Coal Combustion Products Production & Use Statistics, <http://acaa.affiniscape.com/displaycommon.cfm?an=1&subarticlenbr=3>, accessed 16 June 2012.

Table 5.24.　Properties of physical solvents used in gas treatment and sulfur recovery in coal combustion

Process		Selexol[a]	Rectisol[b]
Solvent		DMPEG	Methanol
Formula		$CH_3O(C_2H_4O)_xCH_3$	CH_3OH
Molecular weight	lb/lb mol	178 to 442	32
Boiling point at 760 Torr	°F	415 to 870	147
Melting point	°F	−4 to −20	−137
Viscosity	cP	4.7 at 86°F	0.85 at 5°F
	cP	5.8 at 77°F	1.4 at −22°F
	cP	8.3 at 59°F	2.4 at −58°F
Specific weight	kg/m³	1.031	790
Selectivity at working temperature	($H_2S:CO_2$)	1:09	1:9.5

[a]The Selexol process was originally developed by Allied Chemical Corporation and is now owned by UOP. It uses dimethyl ethers of poly-ethylene glycol (DMPEG).
[b]The Rectisol process uses cold methanol as a solvent and was originally developed to provide a treatment for gas from the Lurgi moving-bed gasifier, which in addition to H_2S and CO_2 contains hydrocarbons, ammonia, hydrogen cyanide, and other impurities.
Source: Adapted from Miller, Bruce G. 2011. Clean Coal Engineering Technology, (Boston,Butterworth-Heinemann).

Table 5.25.　CO_2 Emission Factors for U.S. Coals (Emission Factor Rating: C)[a]

Coal Type	Average %C[b]	Conversion Factor[c]	Emission Factor[d] lb/ton Coal
Subbituminous	66.3	72.6	4,810
High-volatile bituminous	75.9	72.6	5,510
Medium-volatile bituminous	83.2	72.6	6,040
Low-volatile bituminous	86.1	72.6	6,250

[a]Tons are short tons. This table should be used only when an ultimate analysis is not available. If the ultimate analysis is available, CO_2 emissions should be calculated by multiplying the % carbon (%C) by 72.6. This resultant factor would receive a quality rating of B. The USPEA AP-42 emission factor rating is an overall assessment of how good a factor is, based on both the quality of the test(s) or information that is the source of the factor and on how well the factor represents the emission source. A = Excellent; B = Above average; C = Average; D = Below average; E = Poor.
[b]Based on average carbon contents for each coal type (dry basis) based on extensive sampling of U.S. coals.
[c]Based on the following equation: $\dfrac{44 \text{ ton } CO_2}{12 \text{ ton C}} \times 0.99 \times 2000 \dfrac{\text{lb } CO_2}{\text{ton } CO_2} \times \dfrac{1}{100\ \%} = 72.6 \dfrac{\text{lb } CO_2}{\text{ton \% C}}$ where 44 = molecular weight of CO_2;

12 = molecular weight of carbon; and 0.99 = fraction of fuel oxidized during combustion.
[d]To convert from lb/ton to kg/Mg, multiply by 0.5.
Source: U.S. Environmental Protection Agency, AP 42, Fifth Edition, Compilation of Air Pollutant Emission Factors, Volume 1: Stationary Point and Area Sources, <http://www.epa.gov/ttn/chief/ap42/index.html#toc>, accessed 25 May 2012; Miller, Bruce G. 2011. Clean Coal Engineering Technology, (Boston, Butterworth-Heinemann).

Table 5.26. Uncontrolled Emission Factors for particulate matter (PM) and PM10 from Bituminous and Subbituminous Coal Combustion[a]

Firing Configuration	Filterable PM[b]		Filterable PM$_{10}$	
	Emission Factor (lb/ton)	Emission Factor Rating[k]	Emission Factor (lb/ton)	Emission Factor Rating[k]
PC-fired, dry bottom, wall-fired	10A	A	2.3A	E
PC-fired, dry bottom, tangentially fired	10A	B	2.3A[c]	E
PC-fired, wet bottom	7A[d]	D	2.6A	E
Cyclone furnace	2A[d]	E	0.26A	E
Spreader stoker	66[e]	B	13.2	E
Spreader stoker, with multiple cyclones, and reinjection	17	B	12.4	E
Spreader stoker, with multiple cyclones, no reinjection	12	A	7.8	E
Overfeed stoker[f]	16[g]	C	6	E
Overfeed stoker, with multiple cyclones[f]	9	C	5	E
Underfeed stoker	15[h]	D	6.2	E
Underfeed stoker, with multiple cyclones	11	D	6.2[h]	E
Hand-fed units	15	E	6.2[i]	E
FBC, bubbling bed	—[j]	E	—[j]	E
FBC, circulating bed	—[j]	E	—[j]	E

[a]Factors represents uncontrolled emissions unless otherwise specified and should be applied to coal feed, as fired. Tons are short tons.
[b]Based on EPA Method 5 (front half catch). Where particulate is expressed in terms of coal ash content, the A factor is determined by multiplying weight % ash content of coal (as fired) by the numerical value preceding the A. For example, if coal with 8% ash is fired in a PC-fired, dry bottom unit, the PM emission factor would be 10 × 8, or 80 lb/ton.
[c]No data found; emission factor for PC-fired dry bottom boilers used.
[d]Uncontrolled particulate emissions, when no fly ash reinjection is employed. When control device is installed, and collected fly ash is reinjected to boiler, particulate from boiler reaching control equipment can increase up to a factor of 2.
[e]Accounts for fly ash settling in an economizer, air heater, or breaching upstream of control device or stack. (Particulate directly at boiler outlet typically will be twice this level.) Factor should be applied even when fly ash is reinjected to boiler form air heater or economizer dust hoppers.
[f]Includes traveling grate, vibrating grate, and chain grate stokers.
[g]Accounts for fly ash settling in breaching or stack base. Particulate loadings directly at boiler outlet typically can be 50% higher.
[h]Accounts for fly ash settling in breaching downstream of boiler outlet.
[i]No data found; emission factor for underfeed stoker used.
[j]No data found; use emission factor for spreader stoker with multiple cyclones and reinjection.
[k]The USPEA AP-42 emission factor rating is an overall assessment of how good a factor is, based on both the quality of the test(s) or information that is the source of the factor and on how well the factor represents the emission source. A = Excellent; B = Above average; C = Average; D = Below average; E = Poor.
Source: U.S. Environmental Protection Agency, AP 42, Fifth Edition, Compilation of Air Pollutant Emission Factors, Volume 1: Stationary Point and Area Sources, <http://www.epa.gov/ttn/chief/ap42/index.html#toc>, accessed 25 May 2012; Miller, Bruce G. 2011. Clean Coal Engineering Technology, (Boston, Butterworth-Heinemann).

Table 5.27. Emission Factors for CH₄, TNMOC, and N₂O from Bituminous and Subbituminous Coal Combustion[a]

Firing Configuration	CH_4[b] Emission Factor (lb/ton)	Emission Factor Rating[d]	$TNMOC$[b,c] Emission Factor (lb/ton)	Emission Factor Rating[d]	N_2O Emission Factor (lb/ton)	Emission Factor Rating[d]
PC-fired, dry bottom, wall fired	0.04	B	0.06	B	0.03	B
PC-fired, dry bottom, tangentially fired	0.04	B	0.06	B	0.08	B
PC-fired, wet bottom	0.05	B	0.04	B	0.08	E
Cyclone furnace	0.01	B	0.11	B	0.09c	E
Spreader stoker	0.06	B	0.05	B	0.04	E
Spreader stoker, with multiple cyclones, and reinjection	0.06	B	0.05	B	0.04	E
Spreader stoker, with multiple cyclones, no reinjection	0.06	B	0.05	B	0.04	E
Overfeed stoker	0.06	B	0.05	B	0.04	E
Overfeed stoker, with multiple cyclones	0.06	B	0.05	B	0.04	E
Underfeed stoker	0.8	B	1.3	B	0.04	E
Underfeed stoker, with multiple cyclones	0.8	B	1.3	B	0.04	E
Hand-fed units	5	E	10	E	0.04	E
FBC, bubbling bed	0.06	E	0.05	E	3.5	B
FBC, circulating bed	0.06	E	0.05	E	3.5	B

[a]Tons are short tons. PC=pulverized coal. FBC=fluidized bed combustion. Factors represent uncontrolled emissions unless otherwise specified and should be applied to coal feed, as fired.

[b]Nominal values achievable under normal operating conditions; values 1 or 2 orders of magnitude higher can occur when combustion is not complete.

[c]Total Non-Methane Organic Carbon (TNMOC) are expressed as C_2 to C_{16} alkane equivalents. Because of limited data, the effects of firing configuration on TNMOC emission factors could not be distinguished. As a result, all data were averaged collectively to develop a single average emission factor for pulverized coal units, cyclones, spreaders, and overfeed stokers.

[d]The USPEA AP-42 emission factor rating is an overall assessment of how good a factor is, based on both the quality of the test(s) or information that is the source of the factor and on how well the factor represents the emission source. A=Excellent; B=Above average; C=Average; D=Below average; E=Poor

Source: U.S. Environmental Protection Agency, AP 42, Fifth Edition, Compilation of Air Pollutant Emission Factors, Volume 1: Stationary Point and Area Sources, <http://www.epa.gov/ttn/chief/ap42/index.html#toc>, accessed 25 May 2012; Miller, Bruce G. 2011. Clean Coal Engineering Technology, (Boston, Butterworth-Heinemann).

Table 5.28. Emission Factors for SOx, NOx, and CO from Bituminous and Subbituminous Coal Combustion[a]

Firing Configuration	SOx[b] Emission Factor (lb/ton)	SOx Emission Factor Rating[j]	NOx[c] Emission Factor (lb/ton)	NOx Emission Factor Rating[j]	CO[d,e] Emission Factor (lb/ton)	CO[d,e] Emission Factor Rating[j]
PC, dry bottom, wall-fired[f], bituminous pre-NSPS[g]	38S	A	22	A	0.5	A
PC, dry bottom, wall-fired[f], bituminous pre-NSPS[g] with low-NOx burner	38S	A	11	A	0.5	A
PC, dry bottom, wall-fired[f], bituminous NSPS[g]	38S	A	12	A	0.5	A
PC, dry bottom, wall-fired[f], subbituminous pre-NSPS[g]	35S	A	12	C	0.5	A
PC, dry bottom, wall-fired[f], subbituminous NSPS[g]	35S	A	7.4	A	0.5	A
PC, dry bottom, cell burner fired, bituminous	38S	A	31	A	0.5	A
PC, dry bottom, cell burner fired, subbituminous	35S	A	14	E	0.5	A
PC, dry bottom, tangentially fired, bituminous, pre-NSPS[g]	38S	A	15	A	0.5	A
PC, dry bottom, tangentially fired, bituminous, pre-NSPS[g] with low-NOx burner	38S	A	9.7	A	0.5	A
PC, dry bottom, tangentially fired, bituminous, NSPS[g]	38S		10	A	0.5	A
PC, dry bottom, tangentially fired, subbituminous, pre-NSPS[g]	35S	A	8.4	A	0.5	A
PC, dry bottom, tangentially fired, subbituminous, pre-NSPS[g]	35S	A	7.2	A	0.5	A
PC, wet bottom, wall-fired[f], bituminous, pre-NSPS[g]	38S	A	31	D	0.5	A
PC, wet bottom, tangentially fired, bituminous, NSPS[g]	38S	A	14	E	0.5	A
PC, wet bottom, wall-fired, subbituminous	35S	A	24	E	0.5	A
Cyclone furnace, bituminous	38S	A	33	A	0.5	A
Cyclone furnace, subbituminous	35S	A	17	C	0.5	A
Spreader stoker, bituminous	38S	B	11	B	5	A
Spreader stoker, subbituminous	35S	B	8.8	B	5	A
Overfeed stoker[h]	38S(35S)	B	7.5	A	6	B

Continued

Table 5.28.　Continued

Firing Configuration	SO_x[b] Emission Factor (lb/ton)	SO_x[b] Emission Factor Rating[j]	NO_x[c] Emission Factor (lb/ton)	NO_x[c] Emission Factor Rating[j]	CO[d,e] Emission Factor (lb/ton)	CO[d,e] Emission Factor Rating[j]
Underfeed stoker	31S	B	9.5	A	11	B
Hand-fed units	31S	D	9.1	E	275	E
FBC, circulating bed	—[i]	E	5	D	18	E
FBC, bubbling bed	—[i]	E	15.2	D	18	D

[a]Factors represent uncontrolled emissions unless otherwise specified and should be applied to coal feed, as fired. Tons are short tons.

[b]Expressed as SO_2, including SO_2, SO_3, and gaseous sulfates. Factors in parentheses should be used to estimate gaseous SO_x emissions for subbituminous coal. In all cases, S is weight % sulfur content of coal as fired. Emission factor would be calculated by multiplying the weight percent sulfur in the coal by the numerical value preceding S. For example, if fuel is 1.2% sulfur, then S = 1.2. On average for bituminous coal, 95% of fuel sulfur is emitted as SO_2, and only about 0.7% of fuel sulfur is emitted as SO_3 and gaseous sulfate. An equally small percent of fuel sulfur is emitted as particulate sulfate. Small quantities of sulfur are also retained in bottom ash. With subbituminous coal, about 10% more fuel sulfur is retained in the bottom ash and particulate because of the more alkaline nature of the coal ash. Conversion to gaseous sulfate appears about the same as for bituminous coal.

[c]Expressed as NO_2. Generally, 95 vol.% or more of NO_x present in combustion exhaust will be in the form of NO, the rest NO_2. To express factors as NO, multiply factors by 0.66. All factors represent emissions at baseline operation (i.e., 60 to 110% load and no NO_x control measures).

[d]Nominal values achievable under normal operating conditions. Values 1 or 2 orders of magnitude higher can occur when combustion is not complete.

[e]Emission factors for CO_2 emissions from coal combustion should be calculated using lb CO_2/ton coal = 72.6C, where C is the weight % carbon content of the coal. For example, if carbon content is 85%, then C equals 85.

[f]Wall-fired includes front and rear wall-fired units, as well as opposed wall-fired units.

[g]Pre-NSPS(New Source Performance Standards) boilers are not subject to any NSPS. NSPS boilers are subject to Subpart D or Subpart Da. Subpart D boilers are boilers constructed after August 17, 1971, and with a heat input rate greater than 250 million Btu per hour (MMBtu/h). Subpart Da boilers are boilers constructed after September 18, 1978, and with a heat input rate greater than 250 MMBtu/h.

[h]Includes traveling grate, vibrating grate, and chain grate stokers.

[i]SO_2 emission factors for fluidized bed combustion are a function of fuel sulfur content and calcium-to-sulfur ratio. For both bubbling bed and circulating bed design, use: lb SO_2/ton coal = $39.6(S)(Ca/S)^{-1.9}$. In this equation, S is the weight percent sulfur in the fuel and Ca/S is the molar calcium-to-sulfur ratio in the bed. This equation may be used when the Ca/S is between 1.5 and 7. When no calcium-based sorbents are used and the bed material is inert with respect to sulfur capture the emission factor for underfeed stokers should be used to estimate the SO_2 emissions. In this case, the emission factor ratings are E for both bubbling and circulating units.

[j]The USPEA AP-42 emission factor rating is an overall assessment of how good a factor is, based on both the quality of the test(s) or information that is the source of the factor and on how well the factor represents the emission source. A = Excellent; B = Above average; C = Average; D = Below average; E = Poor.

Source: U.S. Environmental Protection Agency, AP 42, Fifth Edition, Compilation of Air Pollutant Emission Factors, Volume 1: Stationary Point and Area Sources, <http://www.epa.gov/ttn/chief/ap42/index.html#toc>, accessed 25 May 2012; Miller, Bruce G. 2011. Clean Coal Engineering Technology, (Boston, Butterworth–Heinemann).

Table 5.29. Emission Factors for Trace Elements, POM, and HCOH From Uncontrolled Bituminous and Subbituminous Coal Combustion[a] (Emission Factor Rating: E)

Firing Configuration	Emission Factor, lb/10^12 Btu									
	As	Be	Cd	Cr	Pb[b]	Mn	Hg	Ni	POM	HCOH
Pulverized coal, configuration unknown	ND	ND	ND	1,922	ND	ND	ND	ND	ND	112[c]
Pulverized coal, wet bottom	538	81	44–70	1,020–1,570	507	808–2,980	16	840–1,290	ND	ND
Pulverized coal, dry bottom	684	81	44.4	1,250–1,570	507	228–2,980	16	1,030–1,290	2.08	ND
Pulverized coal, dry bottom, tangential	ND	ND	ND	ND	ND	ND	ND	ND	2.4	ND
Cyclone furnace	115	<81	28	212–1,502	507	228–1,300	16	174–1,290	ND	ND
Stoker, configuration unknown	ND	73	ND	19–300	ND	2,170	16	775–1,290	ND	ND
Spreader stoker	264–542	ND	21–43	942–1,570	507	ND	ND	ND	ND	221[d]
Overfeed stoker, traveling grate	542–1,030	ND	43–82	ND	507	ND	ND	ND	ND	140[c]

[a]The emission factors in this table represent the ranges of factors reported in the literature. If only 1 data point was found, it is still reported in this table. To convert from lb/1012 Btu to pg/J, multiply by 0.43. ND = no data. The USPEA AP-42 emission factor rating is an overall assessment of how good a factor is, based on both the quality of the test(s) or information that is the source of the factor and on how well the factor represents the emission source. A = Excellent; B = Above average; C = Average; D = Below average; E = Poor; POM = polycyclic organic matter HCOH = hydroxymethylene

[b]Lead emission factors were taken directly from an EPA background document for support of the National Ambient Air Quality Standards.

[c]Based on 2 units; 133 × 10^6 Btu/h and 155 × 10^6 Btu/h.

[d]Based on 1 unit; 59 × 10^6 Btu/h.

Source: U.S. Environmental Protection Agency, AP 42, Fifth Edition, Compilation of Air Pollutant Emission Factors, Volume 1: Stationary Point and Area Sources, <http://www.epa.gov/ttn/chief/ap42/index.html#toc>, accessed 25 May 2012; Miller, Bruce G. 2011. Clean Coal Engineering Technology, (Boston, Butterworth-Heinemann).

Table 5.30. Emission factors for trace metals from controlled coal combustion[a]

Pollutant	Emission Factor[b] (lb/ton)	Emission Factor Rating
Antimony	1.80E-05	A
Arsenic	4.10E-04	A
Beryllium	2.10E-05	A
Cadmium	5.10E-05	A
Chromium	2.60E-04	A
Chromium (VI)	7.90E-05	D
Cobalt	1.00E-04	A
Lead	4.20E-04	A
Magnesium	1.10E-02	A
Manganese	4.90E-04	A
Mercury	8.30E-05	A
Nickel	2.80E-04	A
Selenium	1.30E-03	A

[a]Tons are short tons. The emission factors were developed from emissions data at eleven facilities firing bituminous coal, fifteen facilities firing subbituminous coal, and two facilities firing lignite. The factors apply to boilers utilizing either venturi scrubbers, spray dryer absorbers, or wet limestone scrubbers with an electrostatic precipitator (ESP) or fabric filter (FF). In addition, the factors apply to boilers using only an ESP, FF, or venturi scrubber. Firing configurations include pulverized coal-fired, dry bottom boilers; pulverized coal, dry bottom, tangentially fired boilers; cyclone boilers; and atmospheric fluidized bed combustors, circulating bed. The USPEA AP-42 emission factor rating is an overall assessment of how good a factor is, based on both the quality of the test(s) or information that is the source of the factor and on how well the factor represents the emission source. A=Excellent; B=Above average; C=Average; D=Below average; E=Poor; POM=polycyclic organic matter HCOH = hydroxymethylene
[b]Mission factor should be applied to coal feed, as fired. To convert from lb/ton to kg/Mg, multiply by 0.5.
Source: U.S. Environmental Protection Agency, AP 42, Fifth Edition, Compilation of Air Pollutant Emission Factors, Volume 1: Stationary Point and Area Sources, <http://www.epa.gov/ttn/chief/ap42/index.html#toc>, accessed 25 May 2012; Miller, Bruce G. 2011. Clean Coal Engineering Technology, (Boston, Butterworth-Heinemann).

Table 5.31. Mercury species in coal

Mercury compound	Temperature at maximum concentration (°C)	Initial temperature of appearance (°C)	Final temperature of appearance (°C)
$HgCl_2$	120 ± 10	70	220
Hg_2Cl_2	80 ± 5 and 130 ± 10	60	220
$HgBr_2$	110 ± 5	60	220
HgS (metacinnabar, black)	205 ± 5 and 245 ± 5	170	290
HgS (cinnabar, red)	310 ± 10	240	350
$HgSO_4$	540 ± 20	500	600
Hg_2SO_4	280 ± 10	120	480
HgO	505 ± 5	430	560

Source: Adapted from: Monterroso, Rodolfo, Maohong Fan, Morris Argyle. 2011. Mercury Removal, In: David Bell and Brian Towler, Coal Gasification and Its Applications, (Boston, William Andrew Publishing), Pages 247-292.

Table 5.32. **Mercury distribution in coals from different regions in the United States**

Coal-producing region	Mercury content (wppm, mean)	Calorific value (Btu/lb)
Appalachian, northern	0.24	12,440
Appalachian, central	0.15	13,210
Appalachian, southern	0.21	12,760
Eastern interior	0.1	11,450
Fort Union	0.1	6360
Green River	0.09	9560
Gulf Coast	0.16	6470
Pennsylvania anthracite	0.1	12,520
Powder River	0.08	8090
Raton Mesa	0.09	12,300
San Juan River	0.08	9610
Uinta	0.07	10,810
Western interior	0.18	11,420
Wind River	0.15	9560

Source: Adapted from: Monterroso, Rodolfo, Maohong Fan, Morris Argyle. 2011. Mercury Removal, In: David Bell and Brian Towler, Coal Gasification and Its Applications, (Boston, William Andrew Publishing), Pages 247-292.

Table 5.33. **Efficiency of conventional mercury control technologies**

Technology	Hg removal (%)
Coal cleaning	21
ESPs	24
Baghouses	28
FGDs	34
SCR combined with scrubber	61

ESP = electrostatic precipitators; FGD = flue gas desulfurization; SCR = selective catalytic reduction.
Source: Adapted from: Monterroso, Rodolfo, Maohong Fan, Morris Argyle. 2011. Mercury Removal, In: David Bell and Brian Towler, Coal Gasification and Its Applications, (Boston, William Andrew Publishing), Pages 247-292.

Table 5.34.   Summary of removal capacities of different mercury capture techniques

Process	Mercury removal (%): maximum achieved value at the conditions presented in each study.	Amount of Hg adsorbed (µg Hg/g sorbent): maximum achieved value at the conditions presented in each study.
Coal cleaning	21	
ESPs	24	
Baghouses	28	
FGDs	34	
SCR combined with scrubber	61	
Coal blending	80	
Regular activated carbon injection	65 to 94.5	78–200 µg/g (in absence of HCl)
Norit Darco FGD activated carbon	>90	200 µg/g (in absence of HCl) to 1610 µg/g (with HCl present)
6 N nitric acid washed	N/A	141 µg/g
Sulfur mixed and nitrogen flowed	N/A	96 µg/g
5% ZnCl$_2$ activated carbon	N/A	800–900 µg/g (after 8 h of exposure)
Bromine impregnated activated carbon	70–90	N/A
Thief sorbents	81	190–1380 µg/g
Bromination	N/A	N/A
Fenton reactions	40–81	N/A
Sodium polysulfide montmorillonite	17 (93 achieved at 70 °C)	22 µg/g
Nanoscale metal sulfides	N/A	N/A
Iridium	96	2,900 µg/g

Source: Adapted from: Monterroso, Rodolfo, Maohong Fan, Morris Argyle. 2011. Mercury Removal, In: David Bell and Brian Towler, Coal Gasification and Its Applications, (Boston, William Andrew Publishing), Pages 247-292.

Table 5.35.   Combined APCD mercury removal capabilities

Post-combustion control strategy	Post-combustion emission control device configuration	Coal burned in pulverized-coal-fired boiler unit		
		Bituminous coal	Sub-bituminous coal	Lignite
PM control only	CS-ESP	36%	3%	0%
	HS-ESP	9%	6%	not tested
	FF	90%	72%	not tested
	PS	not tested	9%	not tested
PM control and spray dryer adsorber	SDA + CS –ESP	not tested	35%	not tested
	SDA + FF	98%	24%	0%
	SDA + FF + SCR	98%	not tested	not tested
PM control and wet FGD system	PS + FGD	12%	0%	33%
	CS-ESP + FGD	75%	29%	44%
	HS-ESP + FGD	49%	29%	not tested
	FF + FGD	98%	not tested	not tested

APCD = air pollution control devices; CS-ESP = cold-side electrostatic precipitator; HS-ESP = hot-side electrostatic precipitator; FF = fabric filter; PS = particle scrubber; SDA = spray dryer absorber system.
Source: Adapted from: Monterroso, Rodolfo, Maohong Fan, Morris Argyle. 2011. Mercury Removal, In: David Bell and Brian Towler, Coal Gasification and Its Applications, (Boston, William Andrew Publishing), Pages 247-292.

Table 5.36.   Activity concentration range (Bq/kg) in coal power plant ash and slag residues

	^{238}U series	^{232}Th series	^{40}K series
Hungary	200–2000	20–300	300–800
USA	100–600[a]	30–300[a]	100–1200[a]
Egypt	16–41[a]	9–11[a]	
	41–90[b]	24–34[b]	
Germany	6–166[a]	3–120[a]	125–742[a]
	68–245[b]	76–170[b]	337–1240[b]

[a]Fly ash or ash.
[b]Slags.
The becquerel (symbol Bq) is the SI-derived unit of radioactivity. One Bq is defined as the activity of a quantity of radioactive material in which one nucleus decays per second. The Bq unit is thus equivalent to an inverse second, s–1.
Source: Adapted from International Atomic Energy Agency. 2003. Extent of environmental contamination by naturally occurring radioactive material (NORM) and technological options for mitigation. Technical Report Series no. 419.

Oil and Gas

Figures

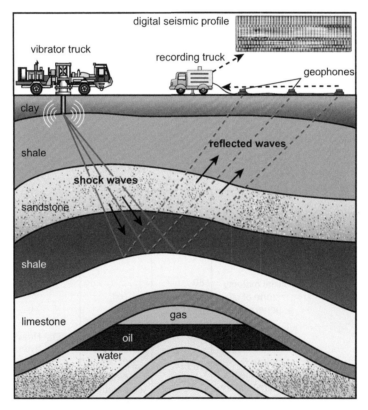

Figure 6.1. Seismic survey used in oil and gas exploration. A seismic survey measures the the physical properties of the Earth using principles such as magnetic, electric, gravitational, thermal, and elastic theories. It is based on the theory of elasticity and tries to deduce elastic properties of materials by measuring their response to elastic disturbances called seismic (or elastic) waves.

Handbook of Energy. http://dx.doi.org/10.1016/B978-0-08-046405-3.00006-1

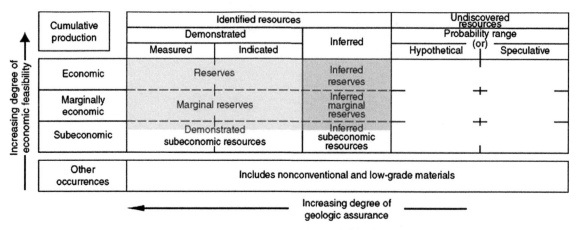

Figure 6.2. The natural resource classification scheme of the U.S. Geological Survey that describes deposits according to their degree of geological assurance of existence (increasing from right to left), and their degree of economic and technologic feasibility of recovery (increasing from bottom to top). Reserves, or proved reserves, of oil and natural gas are those quantities which, by analysis of geological and engineering data, can be estimated with a high degree of confidence to be commercially recoverable from a given date forward, from known reservoirs and under current economic conditions.
Source: United States Geological Survey.

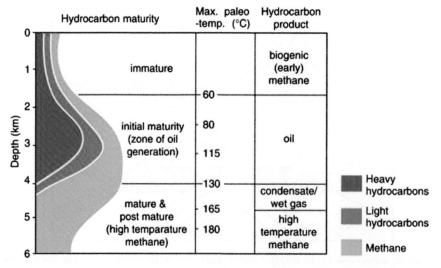

Figure 6.3. Hydrocarbon maturation is the conversion of sedimentary organic matter into petroleum. The resulting products are largely controlled by the composition of the original matter. The peak conversion of kerogen occurs at a temperature of about 100°C. If the temperature is raised above 130°C for even a short period of time, crude oil itself will begin to 'crack' and gas will start to be produced. When kerogens are present in high concentrations in shale, and have not been heated to a sufficient temperature to release their hydrocarbons, they may form oil shale deposits.
Source: Jahn, Frank, Mark Cook and Mark Graham, Editors. 2008. Developments in Petroleum Science, Volume 55, (Amsterdam, Elsevier).

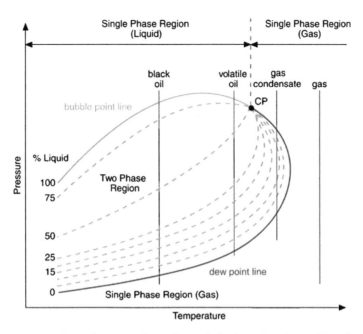

Figure 6.4. Pressure–temperature phase envelopes for main hydrocarbon types. The phase diagrams of the main types of reservoir fluid are used to predict fluid behavior during production, which in turn influences field development planning. The four vertical lines on the diagram show the isothermal depletion loci for the main types of hydrocarbon: gas (incorporating dry gas and wet gas), gas condensate, volatile oil and black oil. *Source: Jahn, Frank, Mark Cook and Mark Graham, Editors. 2008. Developments in Petroleum Science, Volume 55, (Amsterdam, Elsevier).*

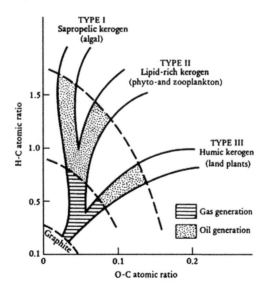

Figure 6.5. Graph of hydrogen/carbon ratio plotted against oxygen/carbon ratio. This shows the composition of the different types of kerogen. Type I kerogen generates principally oil with minor gas. Type II kerogen generates both oil and gas. Type III kerogen generates only gas. Type IV kerogen, not shown, is inert. *Source: Selley, Richard C. 2003. Petroleum Geology, In: Robert A. Meyers, Editor-in-Chief, Encyclopedia of Physical Science and Technology (Third Edition), (New York, Academic Press), Pages 729-740.*

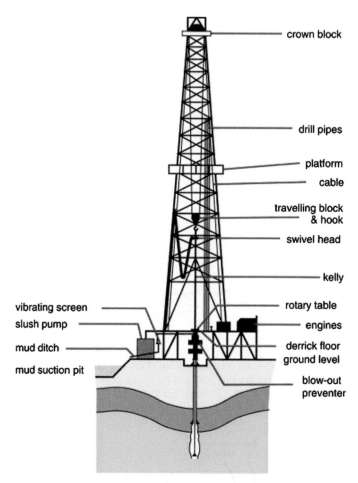

Figure 6.6. The basic rotary rig used in the drilling of oil and gas wells.
Source: Jahn, F., M. Cook, M. Grahm. 2008. Drilling Engineering, In: Frank Jahn, Mark Cook and Mark Graham, Editors, Developments in Petroleum Science, (Amsterdam, Elsevier), Volume 55, Pages 47-81.

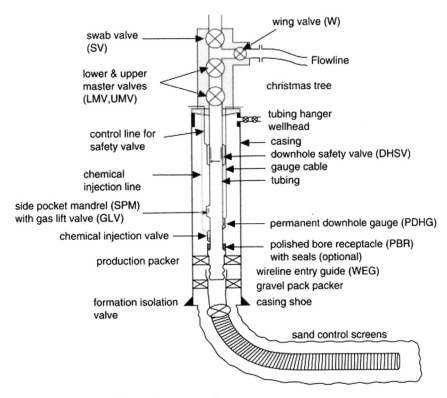

Figure 6.7. A typical completion of an oil or gas well.
Source: Jahn, Frank, Mark Cook and Mark Graham, Editors. 2008. Developments in Petroleum Science, Volume 55, (Amsterdam, Elsevier).

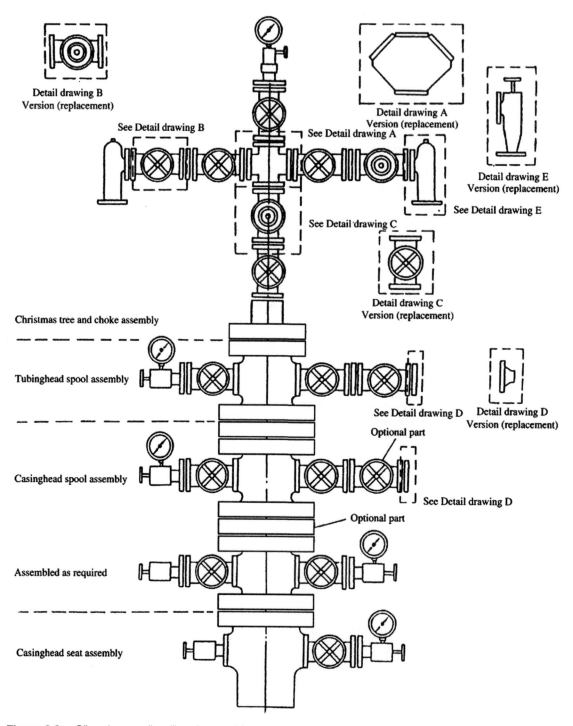

Figure 6.8. Oil and gas well wellhead assembly.
Source: Renpu, Wan. 2011. Advanced Well Completion Engineering (Third Edition), (Boston, Gulf Professional Publishing).

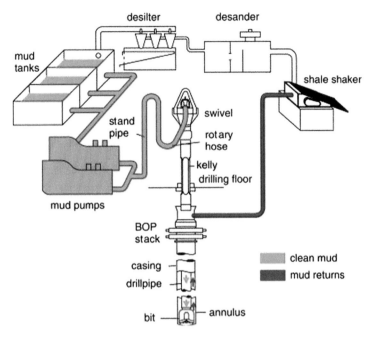

Figure 6.9. A drilling mud circulation system.
Source: Jahn, Frank, Mark Cook and Mark Graham, Editors. 2008. Developments in Petroleum Science, Volume 55, (Amsterdam, Elsevier).

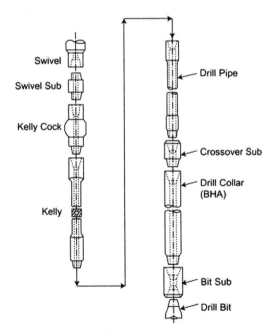

Figure 6.10. A drill string used in oil and gas drilling.
Source: Guo, Boyun and Gefei Liu. 2111. Applied Drilling Circulation Systems, (Boston, Gulf Professional Publishing).

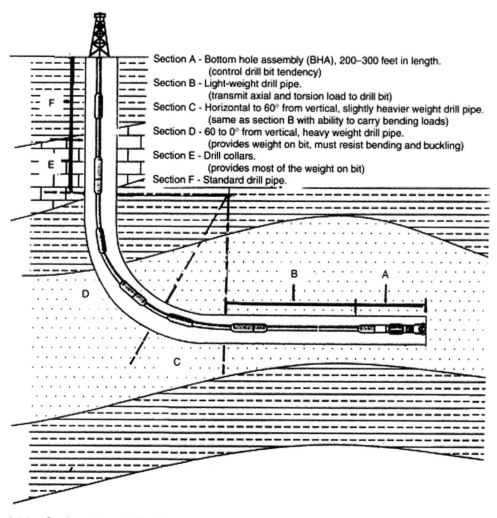

Section A - Bottom hole assembly (BHA), 200–300 feet in length.
      (control drill bit tendency)
Section B - Light-weight drill pipe.
      (transmit axial and torsion load to drill bit)
Section C - Horizontal to 60° from vertical, slightly heavier weight drill pipe.
      (same as section B with ability to carry bending loads)
Section D - 60 to 0° from vertical, heavy weight drill pipe.
      (provides weight on bit, must resist bending and buckling)
Section E - Drill collars.
      (provides most of the weight on bit)
Section F - Standard drill pipe.

Figure 6.11. Configuration of drill string used in horizontal well drilling.
Source: Azar, J.J. 2004. Oil and Natural Gas Drilling, In: Cutler J. Cleveland, Editor-in-Chief, Encyclopedia of Energy, (New York, Elsevier), Pages 521-534.

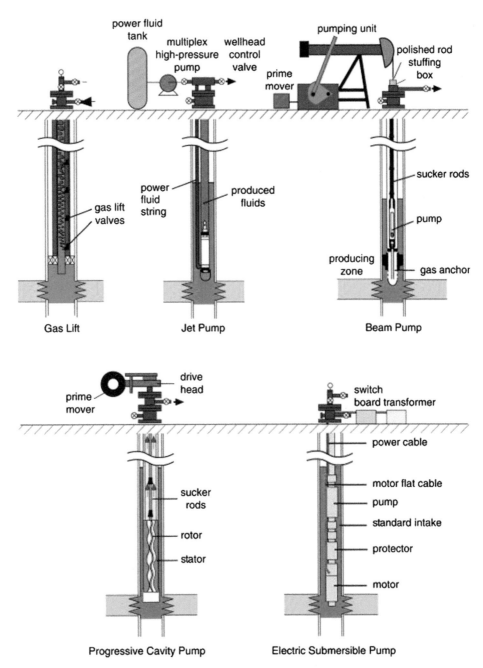

Figure 6.12. Artifical lift systems used in the recovery of oil.
Source: Jahn, Frank, Mark Cook and Mark Graham, Editors. 2008. Developments in Petroleum Science, Volume 55, (Amsterdam, Elsevier).

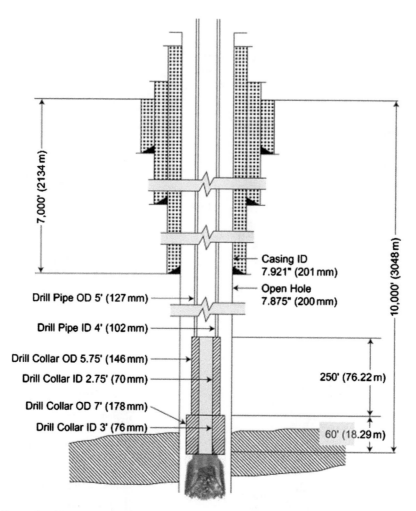

Figure 6.13. Example of borehole configuration at the total depth of an oil or gas well
Source: Guo, Boyun and Gefei Liu. 2111. Applied Drilling Circulation Systems, (Boston, Gulf Professional Publishing).

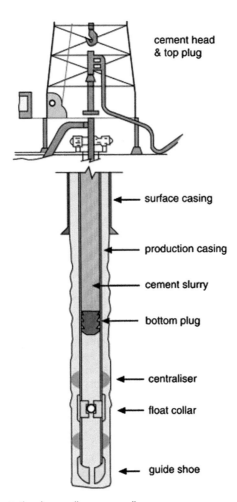

cement head
& top plug

surface casing

production casing

cement slurry

bottom plug

centraliser

float collar

guide shoe

Figure 6.14. Principle of cementation in an oil or gas well.
Source: Jahn, Frank, Mark Cook and Mark Graham, Editors. 2008. Developments in Petroleum Science, Volume 55, (Amsterdam, Elsevier).

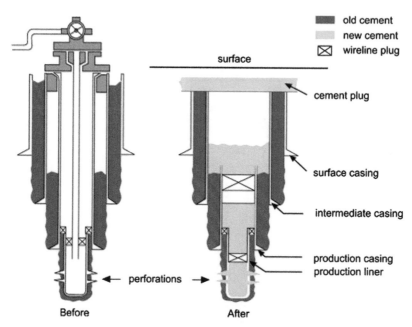

Figure 6.15. An oil well before and after abandonment. Cement is placed across the open perforations and partially squeezed into the formation to seal off all production zones. The production casing is cut and removed above the top of cement, and a cement plug positioned over the casing stub.
Source: Jahn, Frank, Mark Cook and Mark Graham, Editors. 2008. Developments in Petroleum Science, Volume 55, (Amsterdam, Elsevier).

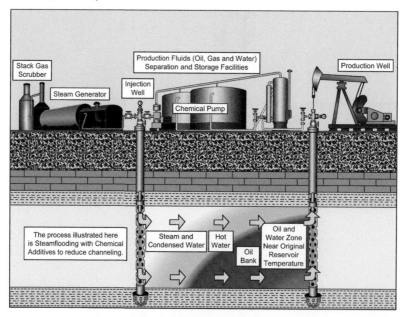

Figure 6.16. Enhanced oil recovery: thermal methods. Thermal recovery comprises the techniques of steamflooding, cyclic steam stimulation, and in situ combustion. The crude oil is heated which improves its mobility.
Source: United States Department of Energy, National Energy Technology Laboratory.

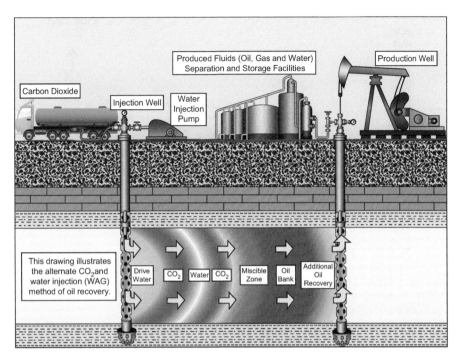

Figure 6.17. Enhanced oil recovery: miscible methods. Techniques for gas miscible recovery include carbon dioxide flooding, cyclic carbon dioxide stimulation, nitrogen flooding, and nitrogen-CO_2 flooding.
Source: United States Department of Energy, National Energy Technology Laboratory.

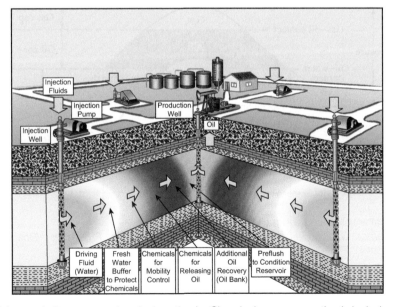

Figure 6.18. Enhanced oil recovery: chemical methods. Chemical recovery methods include polymer, micellar-polymer, and alkaline flooding.
Source: United States Department of Energy, National Energy Technology Laboratory.

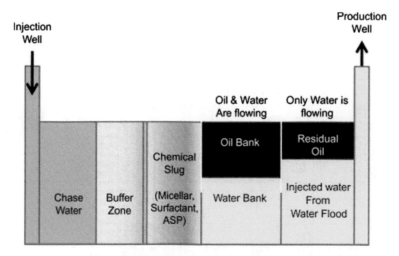

Figure 6.19. Enhanced oil recovery using chemical flooding. Micellar, alkaline and soap-like substances are used to reduce surface tension between oil and water in the reservoir; polymers such as polyacrylamide or polysaccharide improve sweep efficiency. The chemical solutions are pumped through specially distributed injection wells to mobilize oil left behind after primary or secondary recovery.
Source: Ahmed, Tarek, and D. Nathan Meehan. 2012. Advanced Reservoir Management and Engineering (Second Edition), (Boston, Gulf Professional Publishing).

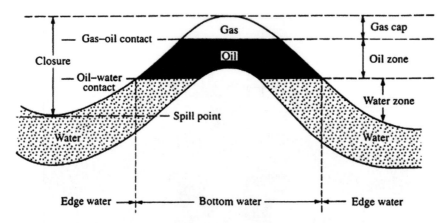

Figure 6.20. Typical anticlinal petroleum trap.
Source: Speight, James G. 2011. Handbook of Industrial Hydrocarbon Processes, (Boston, Gulf Professional Publishing).

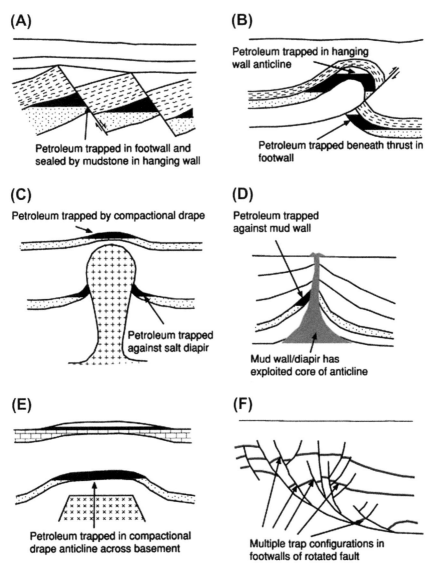

Figure 6.21. Structural traps. (A) Tilted fault blocks in an extensional regime. The seals are overlying mudstones and cross-fault juxtaposition against mudstones. (B) Rollover anticline on thrust. Petroleum accumulations may occur on both the hanging wall and the footwall. The hanging wall accumulation is dependent on a subthrust fault seal, whereas at least part of the hanging wall trap is likely to be a simple, four-way, dip-closed structure. (C) Lateral seal of a trap against a salt diapir and compactional drape trap over the diapir crest. (D) Diapiric mudstone associated trap with lateral seal against mud wall. Diapiric mud associated traps share many common features with those of salt. In this diagram, the diapiric mud wall developed at the core of a compressional fold. (E) Compactional drape over a basement block commonly creates enormous low-relief traps. (F) Gravity-generated trapping commonly occurs in deltaic sequences. Sediment loading causes gravity-driven failure and produces convex-down (listric) faults. The hanging wall of the fault rotates, creating space for sediment accumulation adjacent to the fault planes. The marker beds (grey) illustrate the form of the structure that has many favorable sites for petroleum accumulation.
Source: Gluyas, J. 2005. Petroleum Geology: Overview, In: Richard C. Selley, L. Robin M. Cocks, and Ian R. Plimer, Editors-in-Chief, Encyclopedia of Geology, (Oxford,Elsevier), Pages 229-247.

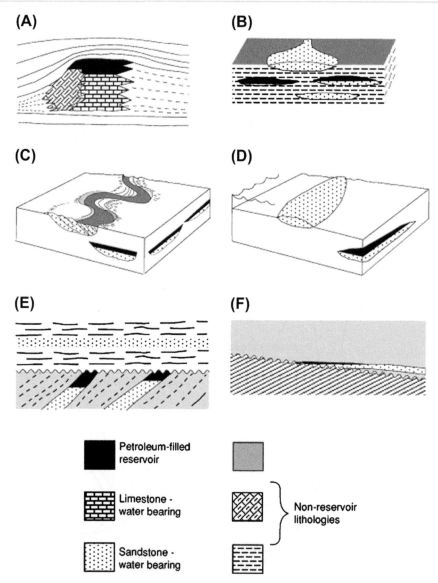

Figure 6.22. Stratigraphical traps. (A) 'Reef' oil is trapped in the core of the reef, with fore-reef talus and back-reef lagoonal muds acting as lateral seals and basinal mudstones as top seals. (B) Pinchout (sandstone) trap within stacked submarine fan sandstones. The upper surface of the diagram shows the plan geometry of a simple fan lobe. Lateral, bottom, and top seals are the surrounding basinal mudstones. (C) Channel-fill sandstone trap. The oil occurs in ribbon-shaped sandstone bodies. The top surface of the diagram shows the depositional geometry of the sandstone. Total seal may be provided by interdistributary mudstones or a combination of these and marine flooding surfaces. (D) Shallow marine sandstone bar completely encased in shallow marine mudstone. The upper surface of the diagram shows the prolate bar. (E) Subunconformity trap. The reservoir horizon is truncated at its up-dip end by an unconformity and the sediments overlying the unconformity provide the top seal. Lateral and bottom seals, like the reservoir interval, pre-date the unconformity. (F) Onlap trap. A basal or near-basal sandstone onlaps a tilted unconformity. The sandstone pinches out on the unconformity and is overstepped by a top seal mudstone.

Source: Gluyas, J. 2005. Petroleum Geology: Overview, In: Richard C. Selley, L. Robin M. Cocks, and Ian R. Plimer, Editors-in-Chief, Encyclopedia of Geology, (Oxford,Elsevier), Pages 229-247.

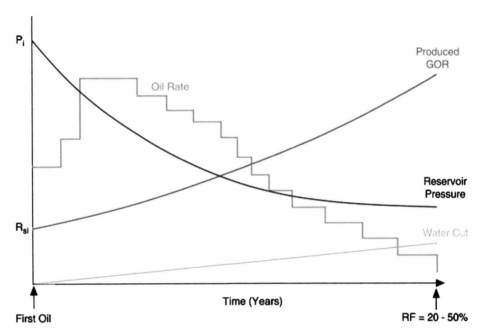

Figure 6.23.

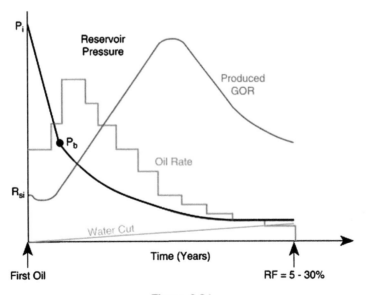

Figure 6.24.

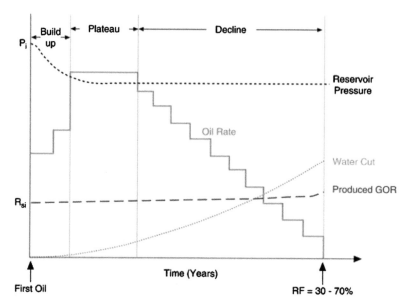

Figures 6.23-6.25. Characteristic oil production profiles for gas cap drive (top), solution gas drive (middle), and water drive (bottom). GOR = ratio of gas produced to oil produced; Water cut = the ratio of water produced to water-plus-oil produced; RF = recovery factor, the ratio of petroleum recovered to petroleum originally in place; P = reservoir pressure; R_{si} = intial solution GOR.
Source: Jahn, Frank, Mark Cook and Mark Graham, Editors. 2008. Developments in Petroleum Science, Volume 55, (Amsterdam, Elsevier).

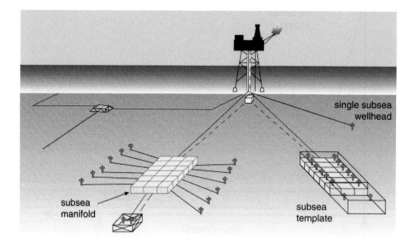

Figure 6.26. Typical subsea field development technology that is tied back to a host facility. A single subsea wellhead with subsea tree is connected to a production facility by a series of pipelines and umbilicals. A control module, usually located on the subsea tree, allows the production platform to remotely operate the subsea facility via its valves and chokes.
Source: Jahn, Frank, Mark Cook and Mark Graham, Editors. 2008. Developments in Petroleum Science, Volume 55, (Amsterdam, Elsevier).

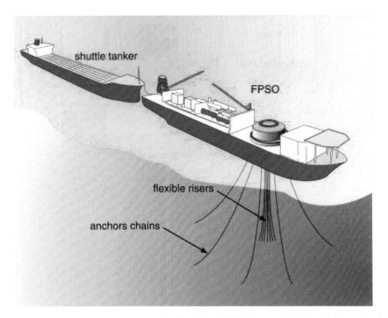

Figure 6.27.   A floating production, storage, and offloading (FPSO) vessel with offshore loading to a shuttle tanker.
Source: Jahn, Frank, Mark Cook and Mark Graham, Editors. 2008. Developments in Petroleum Science, Volume 55, (Amsterdam, Elsevier).

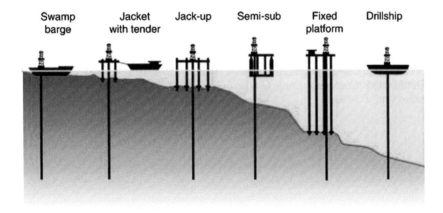

Figure 6.28.   *Swamp barges* operate in very shallow water (less than 20 ft). They can be towed onto location and are then ballasted so that they 'sit on bottom.' *Drilling jackets* are small steel platform structures that are used in areas of shallow and calm water. *Jack-up rigs* are either towed to the drilling location (or alongside a jacket) or are equipped with a propulsion system. The three or four legs of the rig are lowered onto the seabed. After some penetration the rig will lift itself to a determined operating height above the sea level. A *semi-submersible rig* is used in water depths too great for a jack-up; it is movable offshore vessel consisting of a large deck area built on columns of steel. Barge-shaped hulls called pontoons are partially filled with water and submersed to give stability. *Drill ships* are used for deep and very deep water work. *Fixed platforms* can accommodate production and processing facilities as well as living quarters.
Source: Jahn, Frank, Mark Cook and Mark Graham, Editors. 2008. Developments in Petroleum Science, Volume 55, (Amsterdam, Elsevier).

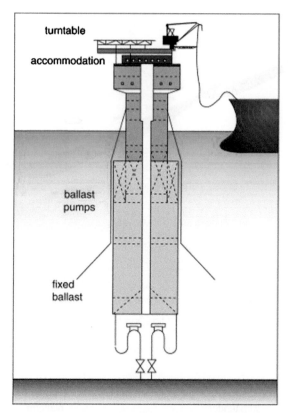

Figure 6.29. SPAR type offshore oil storage terminal. Such systems may receive crude from a number of production centers and serve as a central loading point. The name dervies from the Brent Spar oil storage terminal located in the North Sea and operated by Shell UK from 1976-1991.
Source: Jahn, Frank, Mark Cook and Mark Graham, Editors. 2008. Developments in Petroleum Science, Volume 55, (Amsterdam, Elsevier).

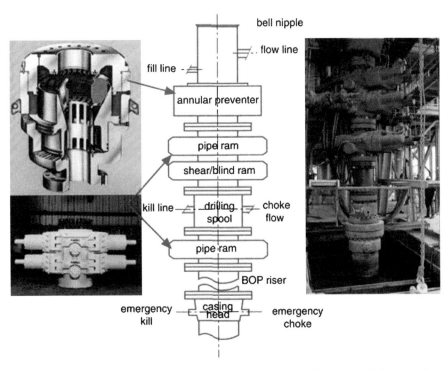

Figure 6.30. A blowout preventer (BOP), a large valve at the top of an oil or gas well that may be closed if the drilling crew loses control of formation fluids. By closing this valve (usually operated remotely via hydraulic actuators), the drilling crew usually regains control of the reservoir, and procedures can then be initiated to increase the mud density until it is possible to open the BOP and retain pressure control of the formation. *Source: Jahn, Frank, Mark Cook and Mark Graham, Editors. 2008. Developments in Petroleum Science, Volume 55, (Amsterdam, Elsevier).*

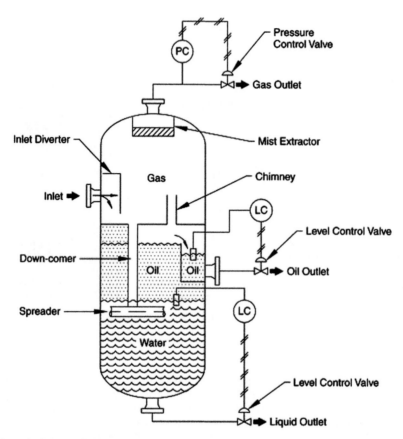

Figure 6.31. A vertical three-phase seperator used to separate the oil, water, and gaseous components of a steam emerging from an oil well.
Source: Stewart, Maurice and Ken Arnold. 2008. Gas-Liquid And Liquid-Liquid Separators, (Burlington, Gulf Professional Publishing).

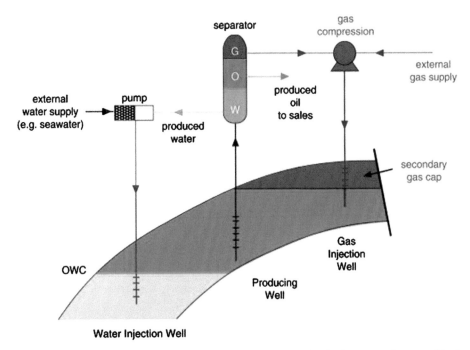

Figure 6.32.   Secondary recovery: gas or water injection schemes. This is the second stage of hydrocarbon production during which an external fluid such as water or gas is injected into the reservoir through injection wells so as to maintain reservoir pressure and to displace hydrocarbons toward the wellbore.
Source: Jahn, Frank, Mark Cook and Mark Graham, Editors. 2008. Developments in Petroleum Science, Volume 55, (Amsterdam, Elsevier).

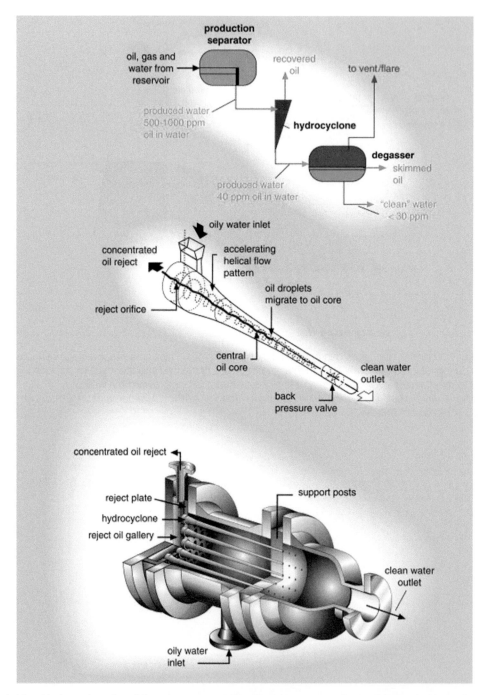

Figure 6.33. Hydrocyclone for oil-in-water removal. The device separates liquids based on their different densities by applying centrifugal force to a liquid mixture so as to promote the separation of heavy and light components.
Source: Jahn, Frank, Mark Cook and Mark Graham, Editors. 2008. Developments in Petroleum Science, Volume 55, (Amsterdam, Elsevier).

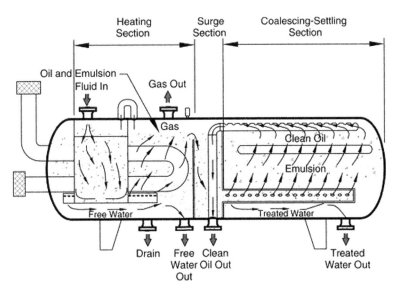

Figure 6.34. A horizontal heater-treater. This unit heats an oil-and-water emulsion and then removes the water and gas. Heaters are especially used when producing natural gas or condensate to avoid the formation of ice and gas hydrates. These solids can plug the wellhead, chokes, and flowlines. Heaters may also be used to heat emulsions before further treating procedures, or when producing crude oil in cold weather to prevent freezing of oil or formation of paraffin accumulations so that the oil can be accepted by the pipeline or other transport.
Source: Arnold, Ken and Maurice Stewart. 2008. Surface Production Operations (Third Edition), (Burlington, Gulf Professional Publishing). F 7-16, p. 374.

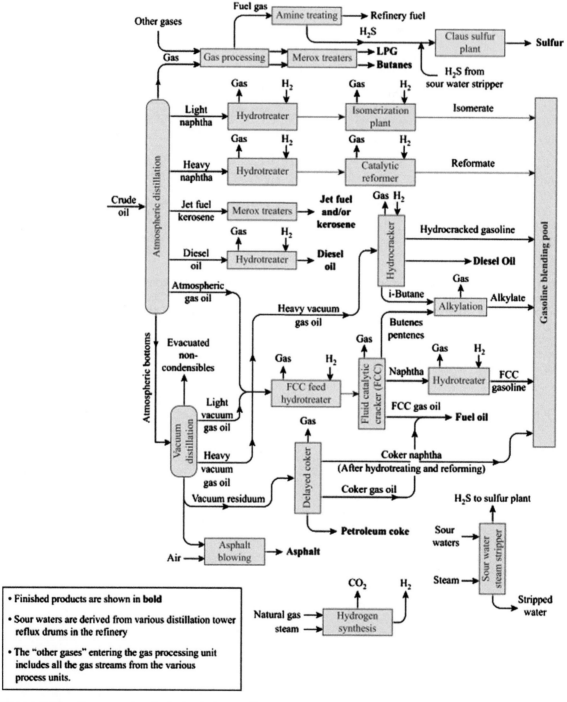

Figure 6.35. Processes in a typical petroleum refinery.
Source: Ross, Julian R.H. 2012. Heterogeneous Catalysis, (Amsterdam, Elsevier).

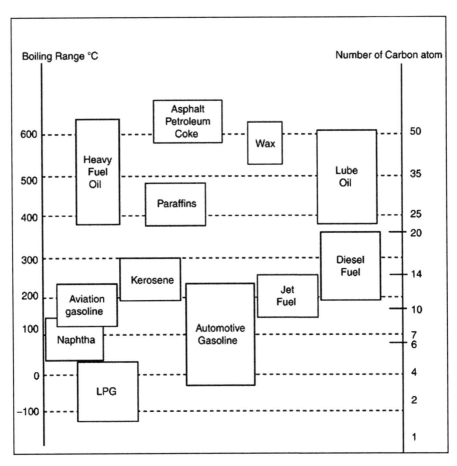

Figure 6.36. Principal petroleum products produced in a refinery with carbon numbers and boiling ranges.
Source: Fahim, Mohamed A., Taher A. Alsahhaf, Amal Elkilani. 2010. Fundamentals of Petroleum Refining, (Amsterdam, Elsevier).

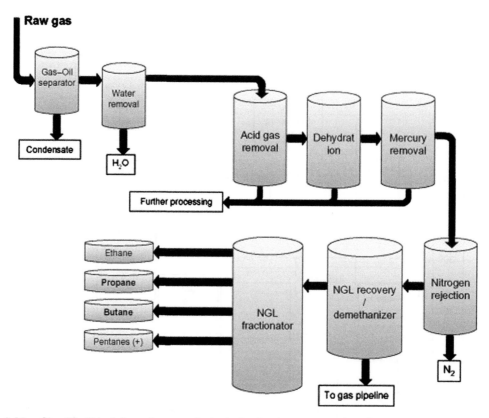

Figure 6.37. Simplified block flow diagram of a typical natural gas processing plant.
Source: Kaltschmitt, Torsten and Olaf Deutschmann. 2012. Fuel Processing for Fuel Cells, In: Kai Sundmacher, Editor, Advances in Chemical Engineering, (New York Academic Press), Pages 1-64.

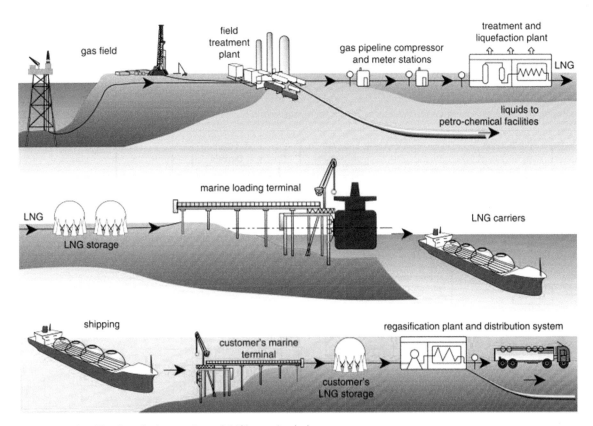

Figure 6.38. The liquefied natural gas (LNG) supply chain.
Source: Jahn, Frank, Mark Cook and Mark Graham, Editors. 2008. Developments in Petroleum Science, Volume 55, (Amsterdam, Elsevier).

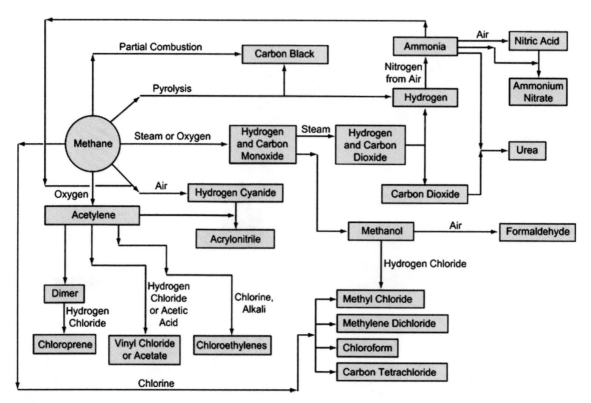

Figure 6.39. Chemicals derived from the processing of methane (natural) gas, CH_4.
Source: Speight, James G. 2011. Handbook of Industrial Hydrocarbon Processes, (Boston, Gulf Professional Publishing).

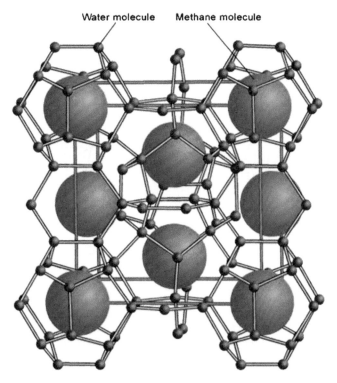

Figure 6.40. A methane or natural gas hydrate is a type of solid clathrate compound (more specifically, a clathrate hydrate) in which a large amount of methane is trapped within a crystal structure of water, forming a solid similar to ice. Large amounts of methane naturally frozen in this form occur both in permafrost formations and under the ocean sea-bed.
Source: Liu, Xiaoyang. 2011. High Pressure Synthesis and Preparation of Inorganic Materials, In: Ruren Xu, Wenqin Pang and Qisheng Huo, Editors, Modern Inorganic Synthetic Chemistry, (Amsterdam, Elsevier), Pages 97-128.

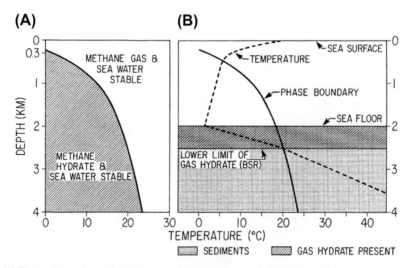

Figure 6.41. (A) Phase boundary of methane hydrate in the ocean (solid line). The pressure axis has been converted to depth into the ocean, so pressure increases downward. (B) The same phase boundary as shown in (A) with a seafloor inserted at 2 km depth and a typical temperature curve (dashed line).Hatched region shows the vertical extent of the gas hydrate stability zone under these assumed conditions.
Source: Dillon, William P. 2001. Gas Hydrate in the Ocean Environment, In: Robert A. Meyers, Editor-in-Chief, Encyclopedia of Physical Science and Technology (Third Edition), (New York, Academic Press), Pages 473-486.

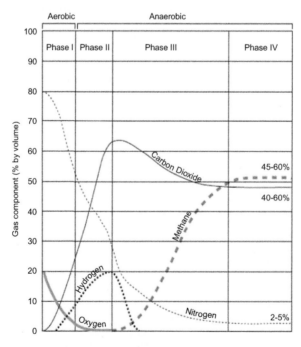

Figure 6.42. Production phases of typical landfill gases.
Source: United States Environmental Protection Agency.

Charts

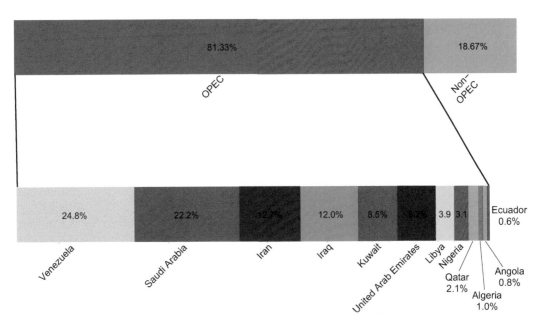

Chart 6.1. Distribution of world oil reserves, 2010.
Source: Data from BP, Statistical Review of World Energy 2012, <http://www.bp.com/ sectionbodycopy.do?categoryId=7500&contentId=7068481>. Organization of Petroleum Exporting Countries, Statistical database, <http://www.opec.org/library/ Annual%20Statistical%20Bulletin/interactive/current/FileZ/Main.htm>.

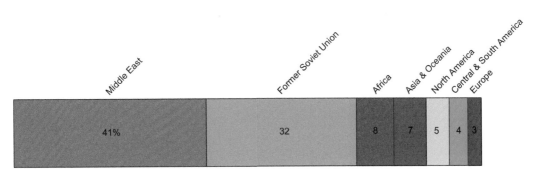

Chart 6.2. Natural gas reserves, by region, 2009.
Source: Data from United States Department of Energy, Energy Information Administration, International Energy Statistics, <http://www.eia.gov/countries/data.cfm>.

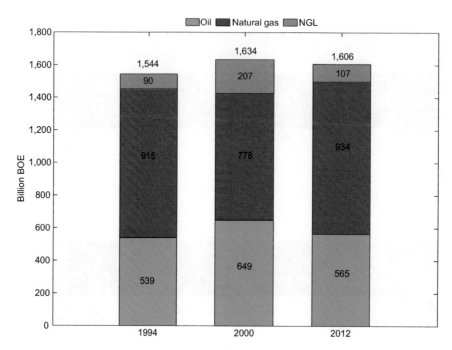

Chart 6.3.   Estimates of undiscovered, technically recoverable, conventional oil, natural gas, and natural gas liquids (NGL) outside of the United States. As of 2010, the total U.S. endowment of technically recoverable oil and natural gas was 163 billion barrels of oil (including NGLs) and 1421 Tcf (237 BBOE) of natural gas. *Source: Data from United States Geological Survey, World Energy Assessment Team. 2000. U.S. Geological Survey world petroleum assessment 2000—Description and results: U.S. Geological Survey Digital Data Series DDS–60, 4 CD-ROMs; Schenk, C.J., 2012, An estimate of undiscovered conventional oil and gas resources of the world, 2012: U.S. Geological Survey Fact Sheet 2012–3042, 6 p; United States Bureau of Ocean Energy Management (BOEM). 2011. Assessment of Undiscovered Technically Recoverable Oil and Gas Resources of the Nation's Outer Continental Shelf, 2011, BOEM Fact Sheet RED-2011-01a; Whitney, Gene, Carl E. Behrens and Carol Glover. 2010. U.S. Fossil Fuel Resources: Terminology, Reporting, and Summary, (Washington, D.C., Congressional Research Service).*

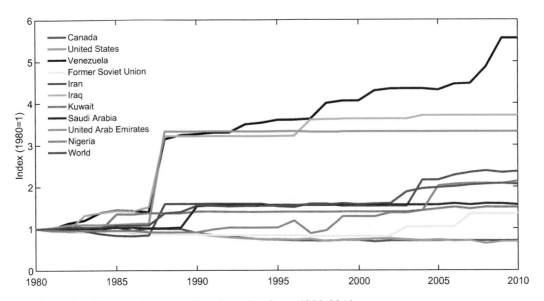

Chart 6.4. Index of crude oil reserves for selected nations, 1980-2010.
Source: Data from BP, Statistical Review of World Energy 2012, <http://www.bp.com/ sectionbodycopy.do?categoryId=7500&contentId=7068481>; United States Department of Energy, Energy Information Administration, International Energy Statistics, <http://www.eia.gov/countries/>.

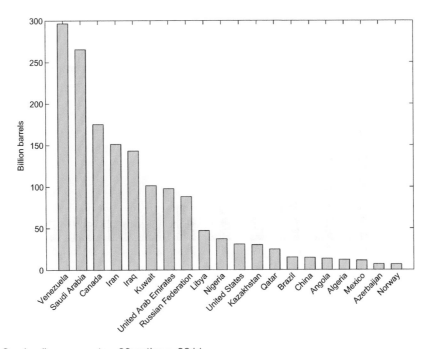

Chart 6.5. Crude oil reserves, top 20 nations, 2011.
Source: Data from BP, Statistical Review of World Energy 2012, <http://www.bp.com/ sectionbodycopy.do?categoryId=7500&contentId=7068481>.

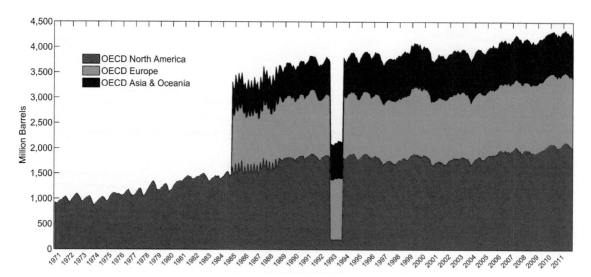

Chart 6.6. Monthly crude oil stocks by region, 1970-2011.
Source: Data from United States Department of Energy, Energy Information Administration; Organization of Petroleum Exporting Countries, Statistical database, <http://www.opec.org/library/ Annual%20Statistical%20Bulletin/interactive/current/FileZ/Main.htm>.

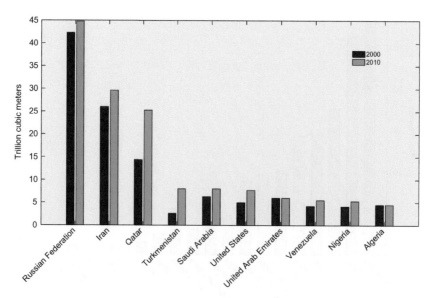

Chart 6.7. Natural gas reserves, top 10 nations, 2000 and 2010.
Source: Data from BP, Statistical Review of World Energy 2012, <http://www.bp.com/ sectionbodycopy.do?categoryId=7500&contentId=7068481>.

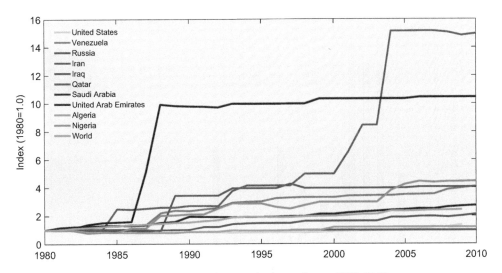

Chart 6.8. Index of natural gas reserves in leading producing nations, 1980-2010.
Source: Data from Organization of Petroleum Exporting Countries, Statistical database,
<http://www.opec.org/library/Annual%20Statistical%20Bulletin/interactive/current/FileZ/Main.htm>.

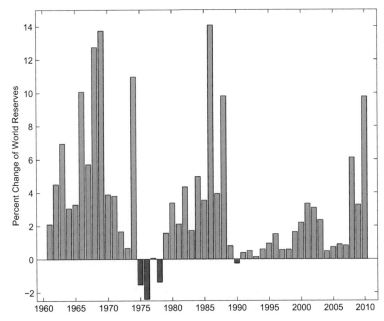

Chart 6.9. Annual percent change in world oil reserves, 1960-2010. The large increase in world reserves in the mid-1980s is due in large part to upward revisions by members of the Organization of Petroleum Exporting Countries (OPEC). The veracity of these estimates is the subject of intense debate. Regardless, revisions in official data in some OPEC nations had little to do with the actual discovery of new reserves.
Source: Data from BP, Statistical Review of World Energy 2012, <http://www.bp.com/
sectionbodycopy.do?categoryId=7500&contentId=7068481>.
Organization of Petroleum Exporting Countries, Statistical database, <http://www.opec.org/
library/Annual%20Statistical%20Bulletin/interactive/current/FileZ/Main.htm>.

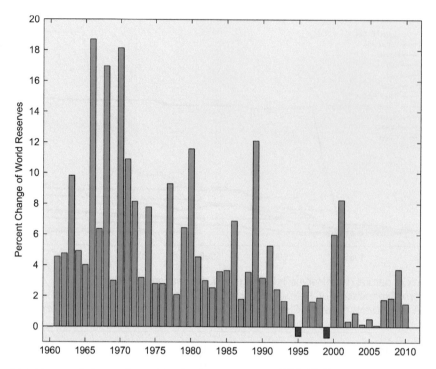

Chart 6.10. Annual percent change in world natural gas reserves, 1960-2010.
Source: Data from Organization of Petroleum Exporting Countries, Statistical database,
<http://www.opec.org/library/Annual%20Statistical%20Bulletin/interactive/current/FileZ/Main.htm>.

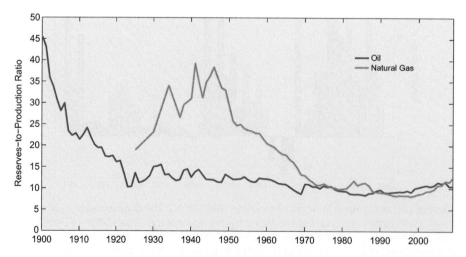

Chart 6.11. The reserve-to-production ratio for crude oil and natural gas in the United States. The R/P ratio
Is the hypothetical number of years that it would take to deplete proved reserves at the current production rate,
and is normally calculated as current proved reserves divided by current annual production.
Source: Data from United States Department of Energy, Energy Information Administration.

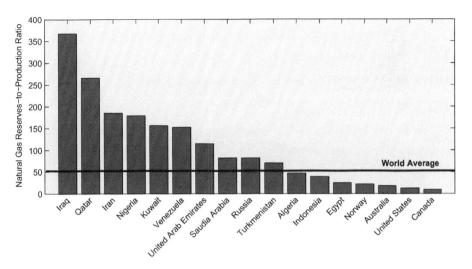

Chart 6.12.   Reserve-to-Production (R/P) ratio for natural gas for major producing nations in 2009. The R/P ratio Is the hypothetical number of years that it would take to deplete proved reserves at the current production rate, and is normally calculated as current proved reserves divided by current annual production. The large ratios for Russian and Middle East nations is due not only to their substantial gas resources, but also because their methods for reporting reserves are less conservative than nations such as the United States and Norway. *Source: Data from United States Department of Energy, Energy Information Administration, International Energy Statistics, <http://www.eia.gov/countries/data.cfm>.*

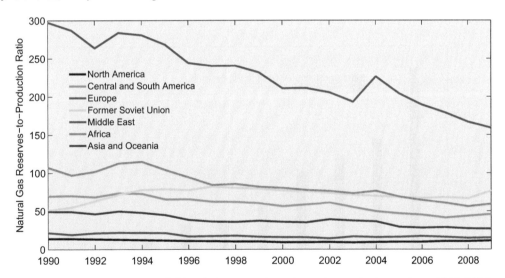

Chart 6.13.   Reserve-to-Production (R/P) ratio for natural gas for major producing nations, 1990-2009. The R/P ratio Is the hypothetical number of years that it would take to deplete proved reserves at the current production rate, and is normally calculated as current proved reserves divided by current annual production. Large discoveries in the Middle East produced a high R/P ratio; the decline thereafter was driven by economic considerations. Incentive to explore for new gas reserves declined because producers had little incentive to develop gas that would not be produced until far into the future. *Source: Data from United States Department of Energy, Energy Information Administration, International Energy Statistics, <http://www.eia.gov/countries/data.cfm>.*

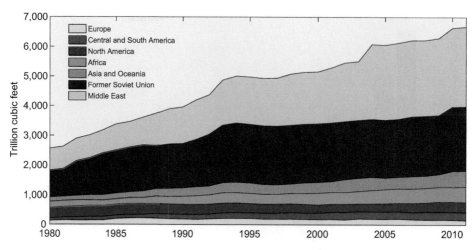

Chart 6.14. Proved reserves of dry natural gas by region, 1980- 2009. Almost three-quarters of the world's natural gas reserves are located in Eurasia and the Middle East. Eurasia includes the nations of the former USSR. Despite high rates of increase in natural gas consumption, gas reserves have increased due to new discoveries, especially in Russia, the Middle East and North Africa, and to improved recovery techniques. *Source: Data from United States Department of Energy, Energy Information Administration, International Energy Statistics, <http://www.eia.gov/countries/data.cfm>.*

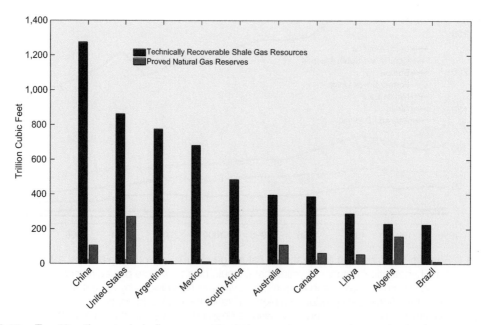

Chart 6.15. Top 10 nations, technically recoverable shale gas resources and proved natural gas reserves. *Source: Data from United States Department of Energy, Energy Information Administration, International Energy Statistics, <http://www.eia.gov/countries/data.cfm>.*

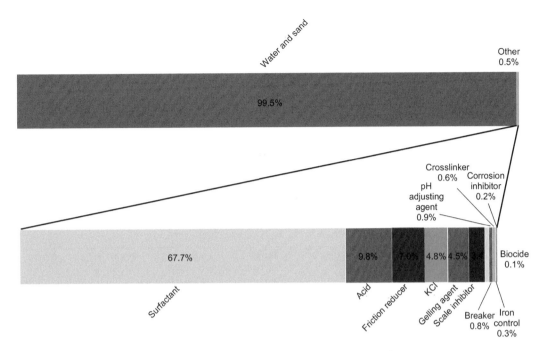

Chart 6.16. Volumetric composition of fracture fluid used in shale gas extraction.
Source: Data from Ground Water Protection Council. 2009. Modern Shale Gas Development in the United States: A Primer, Prepared for U.S. Department of Energy, Office of Fossil Energy and National Energy Technology Laboratory.

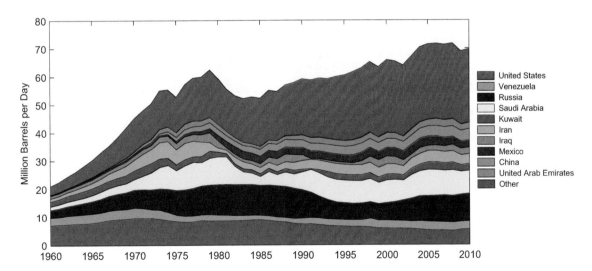

Chart 6.17. World crude oil production by country, 1960-2010.
Source: Data from Organization of Petroleum Exporting Countries, Statistical database, <http://www.opec.org/library/Annual%20Statistical%20Bulletin/interactive/current/FileZ/Main.htm>.

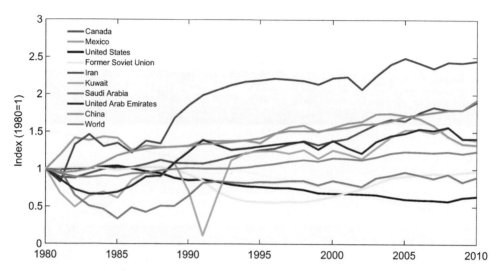

Chart 6.18. Top crude oil producing nations (index), 1980-2010. Excluded is Brazil which experienced an 11-fold increase in production over this period.
Source: Data from Data from United States Department of Energy, Energy Information Administration, International Energy Statistics, <http://www.eia.gov/countries/>.

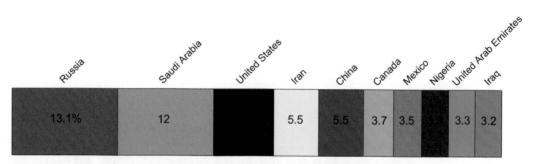

Chart 6.19. Country shares of crude oil production, top 10 nations, 2010.
Source: Data from United States Department of Energy, Energy Information Administration, International Energy Statistics, <http://www.eia.gov/petroleum/>.

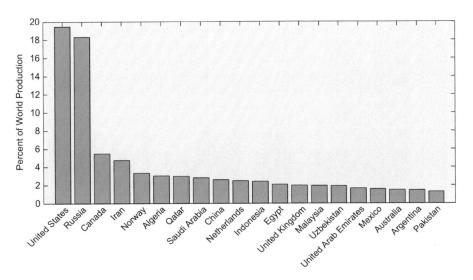

Chart 6.20. Country shares of world natural gas production, 2009.
Source: Data from International Energy Agency (IEA), Energy statistics database, <http://www.iea.org/stats/index.asp>.

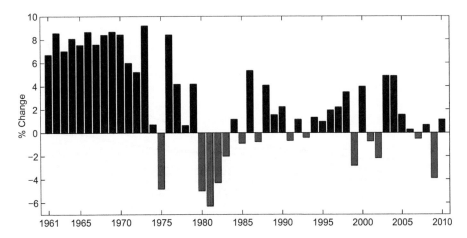

Chart 6.21. Annual percent change in world crude oil production, 1960-2010.
Source: Data from Organization of Petroleum Exporting Countries, Statistical database, <http://www.opec.org/library/Annual%20Statistical%20Bulletin/interactive/current/FileZ/Main.htm>.

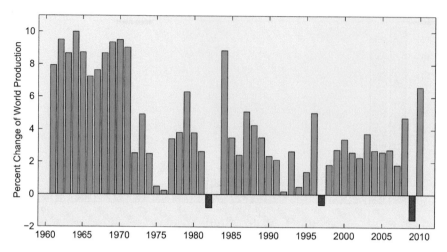

Chart 6.22.   Annual percent change in world natural gas production, 1960-2010.
Source: Data from Organization of Petroleum Exporting Countries, Statistical database,
<http://www.opec.org/library/Annual%20Statistical%20Bulletin/interactive/current/FileZ/Main.htm>.

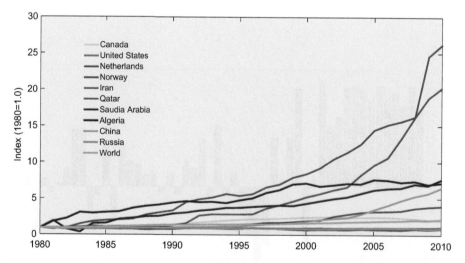

Chart 6.23.   Index of natural gas production in leading producing nations, 1980-2010.
Source: Data from Organization of Petroleum Exporting Countries, Statistical database,
<http://www.opec.org/library/Annual%20Statistical%20Bulletin/interactive/current/FileZ/Main.htm>.

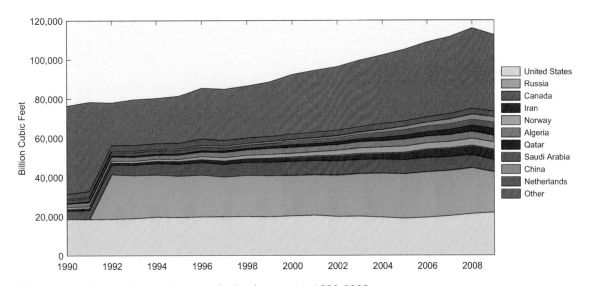

Chart 6.24. Marketed natural gas production by country, 1990-2009.
Source: Data from United States Department of Energy, Energy Information Administration, International Energy Statistics, <http://www.eia.gov/countries/data.cfm>.

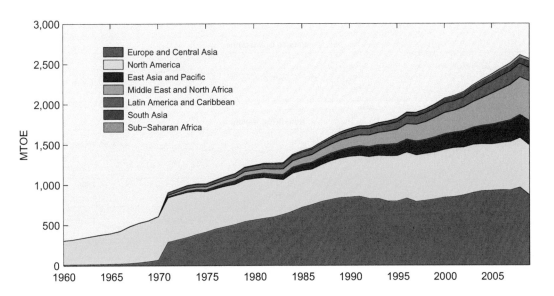

Chart 6.25. Natural gas production, by region, 1960-2009.
Source: Data from International Energy Agency (IEA), Energy statistics database, <http://www.iea.org/stats/index.asp>.

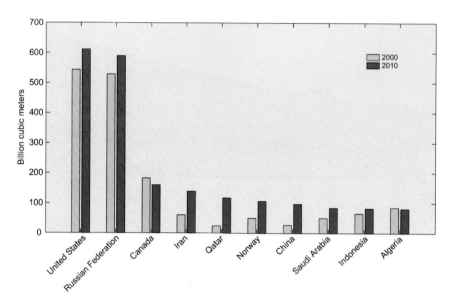

Chart 6.26. Natural gas production, top 10 nations, 2000 and 2010.
Source: Data from BP, Statistical Review of World Energy 2012, <http://www.bp.com/sectionbodycopy.do?categoryId=7500&contentId=7068481>.

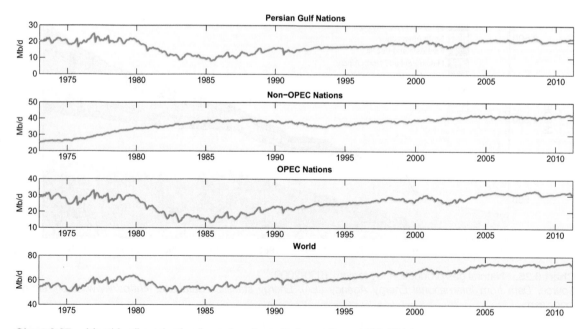

Chart 6.27. Monthly oil production for major oil-producing regions, 1973-2011.
Source: Data from United States Department of Energy, Energy Information Administration, International Energy Statistics, <http://www.eia.gov/countries/>.

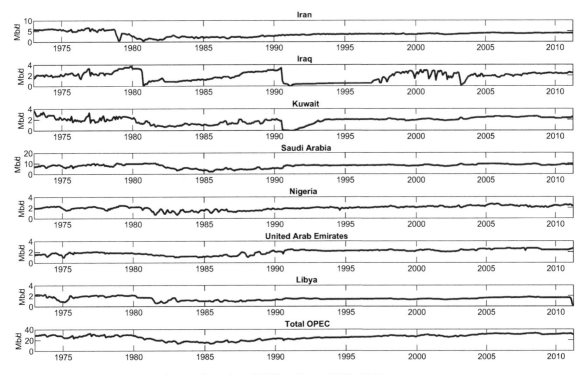

Chart 6.28. Monthly oil production for select OPEC nations, 1973-2011.
Source: Data from United States Department of Energy, Energy Information Administration, International Energy Statistics, <http://www.eia.gov/countries/>.

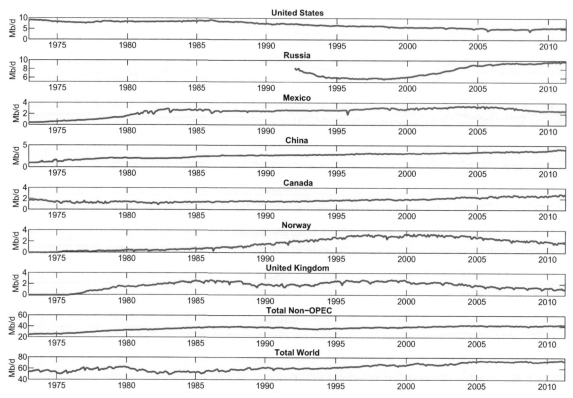

Chart 6.29. Monthly oil production for select non-OPEC nations, 1973-2011.
Source: Data from United States Department of Energy, Energy Information Administration, International Energy Statistics, <http://www.eia.gov/countries/>.

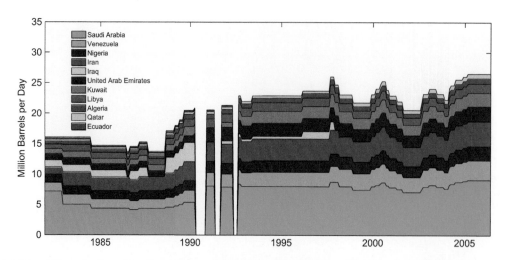

Chart 6.30. Oil production allocations for members of the Organization of Petroleum Exporting Countries (OPEC), 1980-2008.
Source: Data from United States Department of Energy, Energy Information Administration.

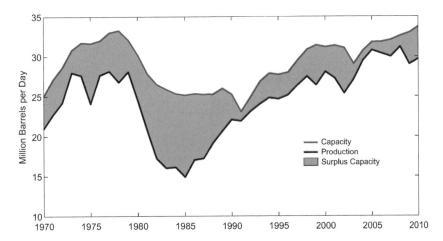

Chart 6.31. Oil production and capacity, Organization of Petroleum Exporting Countries (OPEC) countries, 1970-2010.
Source: Data from United States Department of Energy, Energy Information Administration; Organization of Petroleum Exporting Countries, Statistical database, <http://www.opec.org/ library/Annual%20Statistical%20Bulletin/interactive/current/FileZ/Main.htm>.

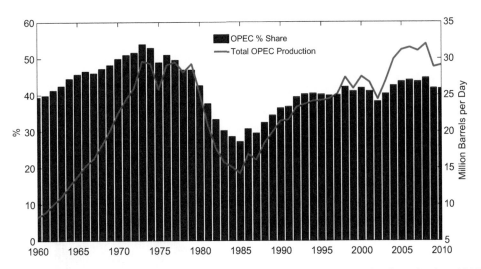

Chart 6.32. Organization of Petroleum Exporting Countries (OPEC) share of world oil production, 1960-2010.
Source: Data from Organization of Petroleum Exporting Countries, Statistical database, <http://www.opec.org/ library/Annual%20Statistical%20Bulletin/interactive/current/FileZ/Main.htm>.

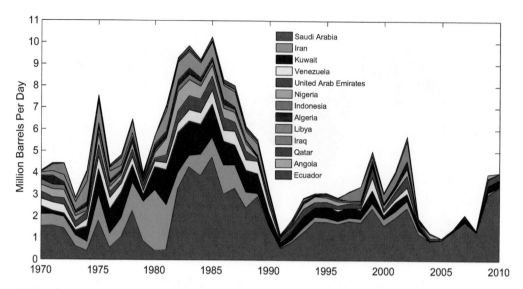

Chart 6.33. Surplus oil production capacity for the Organization of Petroleum Exporting Countries (OPEC), 1970-2010.
Source: Data from United States Department of Energy, Energy Information Administration; Organization of Petroleum Exporting Countries, Statistical database, <http://www.opec.org/ library/Annual%20Statistical%20Bulletin/interactive/current/FileZ/Main.htm>.

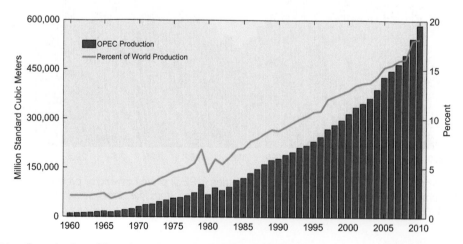

Chart 6.34. Organization of Petroleum Exporting Countries (OPEC) share of world natural gas production, 1960-2010.
Source: Data from Organization of Petroleum Exporting Countries, Statistical database, <http://www.opec.org/library/Annual%20Statistical%20Bulletin/interactive/current/FileZ/Main.htm>.

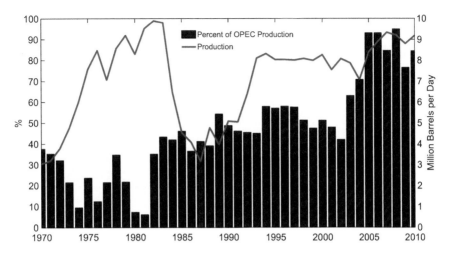

Chart 6.35. Saudi Arabia oil production and capacity,1970-2010.
*Source: Data from United States Department of Energy, Energy Informa-
tion Administration; Organization of Petroleum Exporting Countries, Statistical database,
<http://www.opec.org/library/Annual%20Statistical%20Bulletin/interactive/current/FileZ/Main.htm>.*

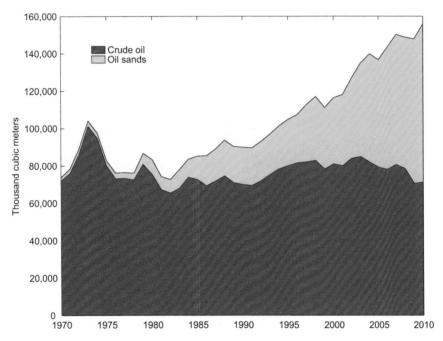

Chart 6.36. Oil production in Canada, 1970-2010.
*Source: Data from Canadian Association of Petroleum Producers (CAPP), Statistical Handbook,
<http://www.capp.ca/library/statistics/handbook/Pages/default.aspx>.*

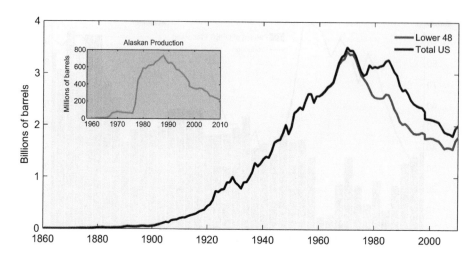

Chart 6.37. Crude oil production in the United States, 1860-2010. Oil production peaked in 1970, and declined substantially thereafter due to the depletion of the domestic resource base. Production stabilized for a period in the 1980s due to production from the Prudhoe Bay field in Alaska that came on line in 1977. Production increased after 2008 due to the development of "tight oil" and offshore resources in the Gulf of Mexico.
Source: Data from United States Department of Energy, Energy Information Administration.

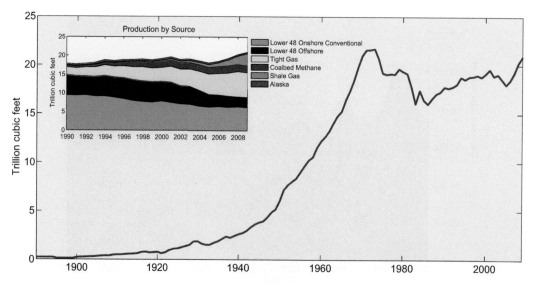

Chart 6.38. United States production of natural gas, 1890-2010.
Source: Data from United States Department of Energy, Energy Information Administration,
<http://www.eia.gov/naturalgas/>.

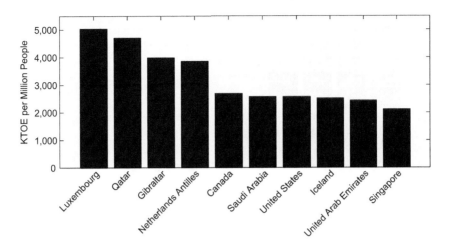

Chart 6.39. Total oil consumption per capita, top 10 nations, 2009.
Source: Data from International Energy Agency (IEA). Statistical database, <http://www.iea.org/stats/index.asp>.

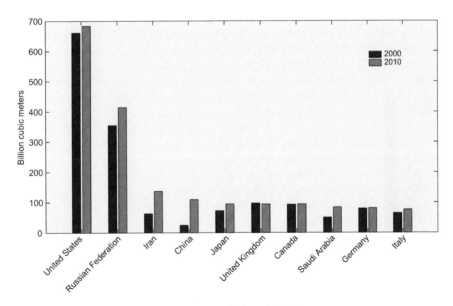

Chart 6.40. Natural gas production, top 10 nations, 2000 and 2010.
Source: Data from BP, Statistical Review of World Energy 2012, <http://www.bp.com/ sectionbodycopy.do?categoryId=7500&contentId=7068481>

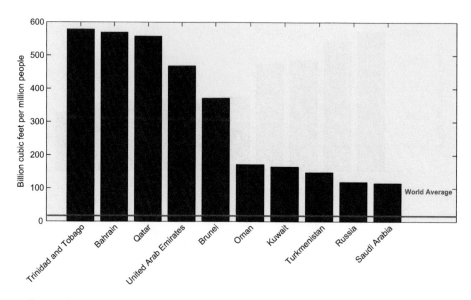

Chart 6.41.   Per capita consumption of natural gas, top 10 nations, 2010. Most of these nations have rich domestic sources of natural gas, as well as government policies such as price subsidies that encourage gas consumption.
Source: Data from United States Department of Energy, Energy Information Administration, International Energy Statistics, <http://www.eia.gov/countries/data.cfm>.

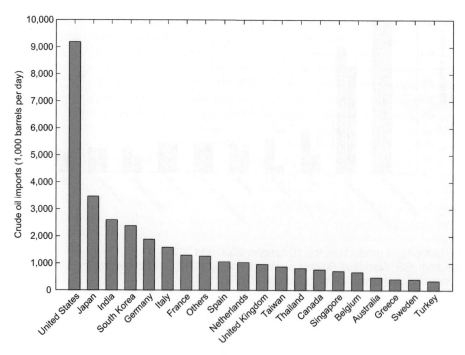

Chart 6.42.   Crude oil imports, top 20 nations, 2010.
Source: Data from Organization of Petroleum Exporting Countries, Statistical database, <http://www.opec.org/library/Annual%20Statistical%20Bulletin/interactive/current/FileZ/Main.htm>.

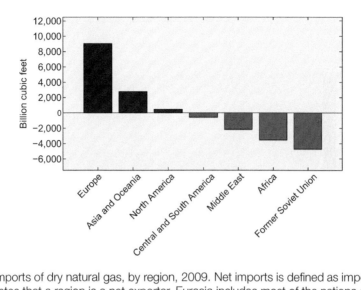

Chart 6.43. Net imports of dry natural gas, by region, 2009. Net imports is defined as imports-exports; a negative value indicates that a region is a net exporter. Eurasia includes most of the nations of the former USSR, and is dominated by exports from Russia and Turkmenistan. Algeria is the leading African exporter, while Qatar is the main Middle East exporter. The large importers in Europe are Germany, Italy, the UK, and Spain; Norway is a large net exporter.
Source: Data from United States Department of Energy, Energy Information Administration, International Energy Statistics, <http://www.eia.gov/countries/data.cfm>.

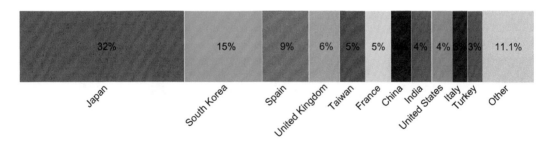

Chart 6.44. Liquefied natural gas imports by nation, 2010.
Source: Data from International Gas Union, World LNG Report 2011, <http://www.igu.org/>.

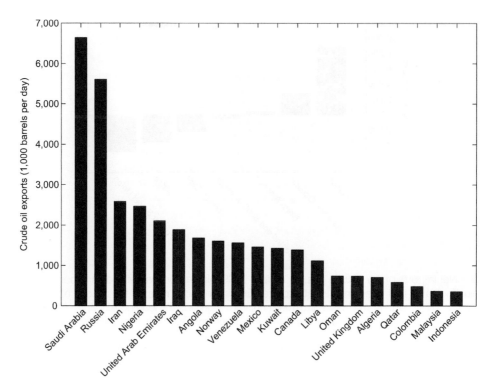

Chart 6.45. Crude oil exports, top 20 nations, 2010.
Source: Data from Organization of Petroleum Exporting Countries, Statistical database, <http://www.opec.org/library/Annual%20Statistical%20Bulletin/interactive/current/FileZ/Main.htm>, accessed 16 July 2012.

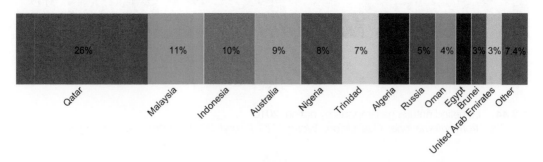

Chart 6.46. Liquefied natural gas exports, by nation, 2010.
Source: Data from International Gas Union, World LNG Report 2011, <http://www.igu.org/>.

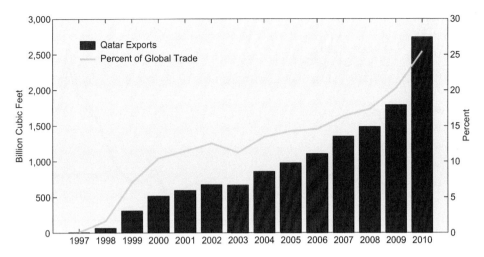

Chart 6.47. Qatar exports of liquefied natural gas (LNG), 1997-2010.
Source: Data from United States Department of Energy, Energy Information Administration, International Energy Statistics; LNG Word News, <http://www.lngworldnews.com/>.

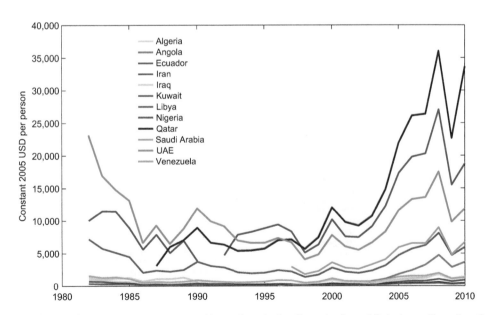

Chart 6.48. Per capita export revenue earned by nations in the Organization of Petroleum Exporting Countries (OPEC) 1982-2011.
Source: Data from United States Department of Energy, Energy Information Administration.

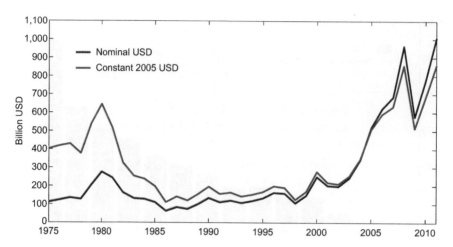

Chart 6.49. Total export revenue earned by the Organization of Petroleum Exporting Countries (OPEC) 1975-2011.
Source: Data from United States Department of Energy, Energy Information Administration.

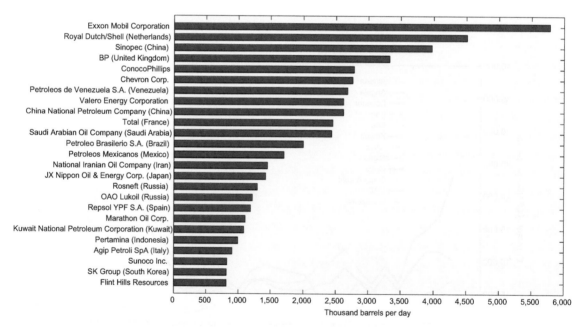

Chart 6.50. The word's largest refiners of crude oil, 2011.
Source: Data from PetroStrategies, Inc., <http://www.petrostrategies.org>.

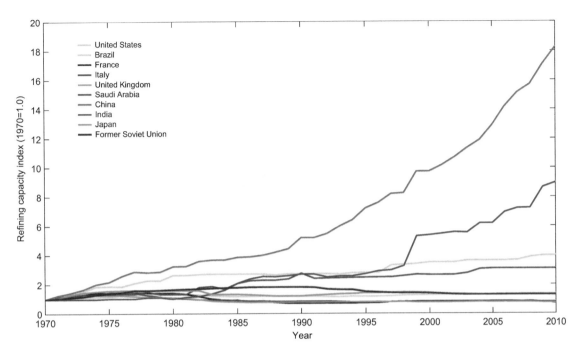

Chart 6.51. Index of oil refining capacity by region, 1970-2010.
Data from BP, Statistical Review of World Energy 2012, <http://www.bp.com/sectionbodycopy.do?categoryId=7500&contentId=7068481>.

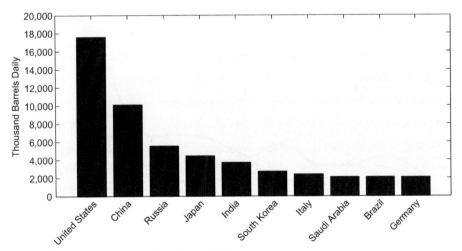

Chart 6.52. Oil refining capacity, top 10 nations, 2010.
Source: Data from BP, Statistical Review of World Energy 2012,
<http://www.bp.com/sectionbodycopy.do?categoryId=7500&contentId=7068481>.

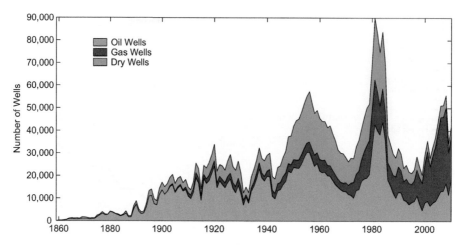

Chart 6.53.   Oil, gas, and dry wells drilled in the United States,1859-2010.
Source: Data complied from various soures by authors.

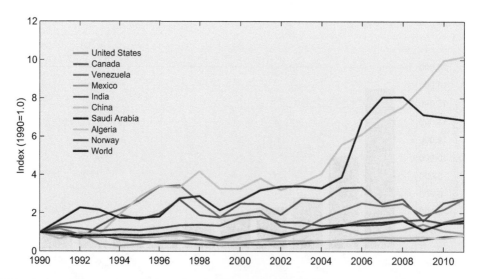

Chart 6.54.   Index of the number of international drilling rigs,1990-2011.
Source: Data from PennEnergy Research, Oil & Gas Statistical Tables, <http://ogjresearch.stores.
yahoo.net/oil-gas-statistical-tables.html>.

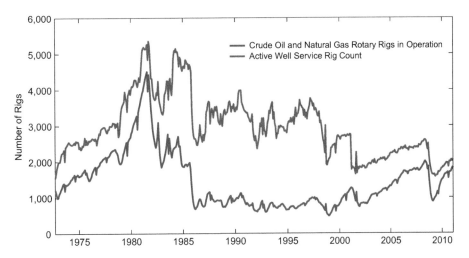

Chart 6.55. Oil and gas drilling rig activity in the United States, 1973-2011.
Source: Data from Baker Hughes Inc., Rotary Rig Count, <http://investor.shareholder.com/bhi/rig_counts/rc_index.cfm>.

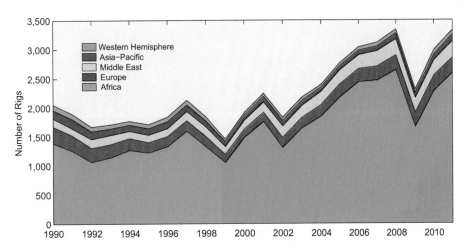

Chart 6.56. Number of active drilling rigs, by region, 1990-2011.
Source: Data from PennEnergy Research, Oil & Gas Statistical Tables, <http://ogjresearch.stores.yahoo.net/oil-gas-statistical-tables.html>.

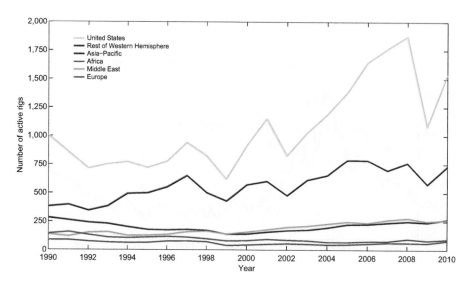

Chart 6.57. Rotary drilling rig count by region, 1990-2010.
*Source: Data from Baker Hughes Inc., Rotary Rig Count, <http://investor.shareholder.com/bhi/rig_counts/
rc_index.cfm>.*

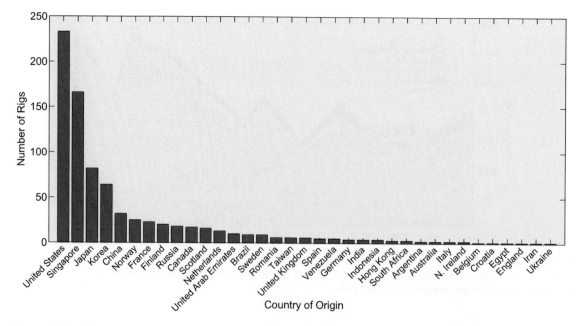

Chart 6.58. Total number of drilling rigs by country of origin, as of November 2011.
*Source: Data from Shipbuilding History, The World Fleet of Offshore Drilling Rigs, <http://www.
shipbuildinghistory.com>.*

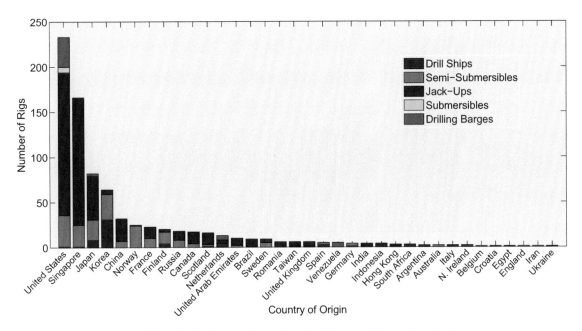

Chart 6.59. Number of drilling rigs by type and country of origin, as of November 2011.
Source: Data from Shipbuilding History, The World Fleet of Offshore Drilling Rigs, <http://www.shipbuildinghistory.com>.

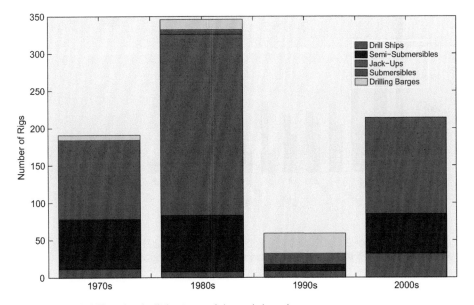

Chart 6.60. Number of drilling rigs built by type of rig and decade.
Source: Data from Shipbuilding History, The World Fleet of Offshore Drilling Rigs, <http://www.shipbuildinghistory.com>.

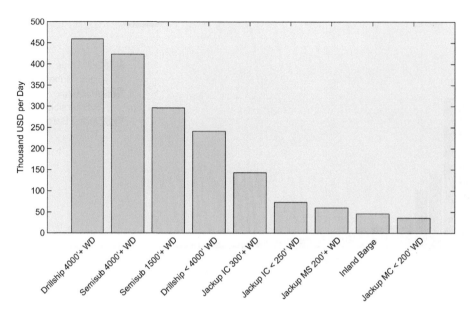

Chart 6.61.   Day rates for offshore drilling rigs by type, August 2011.
Source: Data from Rigzone, Rigzone Rig Data Center, <http://www.rigzone.com/data/>.

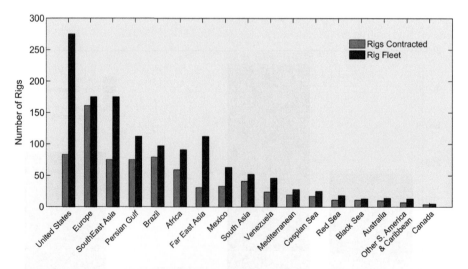

Chart 6.62.   Utilization rates for offshore drilling rigs by region, August 2011.
Source: Data from Rigzone, Rigzone Rig Data Center, <http://www.rigzone.com/data/>.

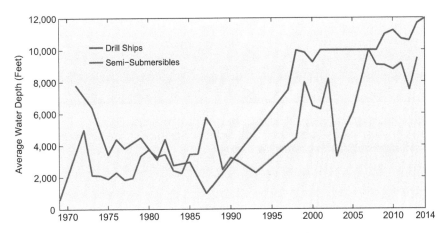

Chart 6.63. Average water depth of drill ships and semi-submersibles by year.
Source: Data from Shipbuilding History, The World Fleet of Offshore Drilling Rigs, <http://www. shipbuildinghistory.com>.

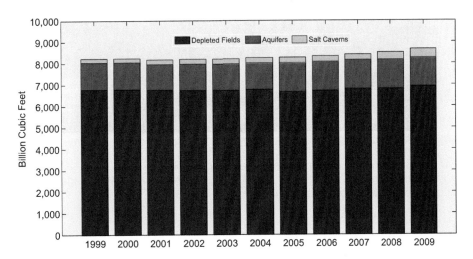

Chart 6.64. United States natural gas storage by mode of storage, 1999-2009.
Source: Data from United States Department of Energy, Energy Information Administration, Weekly Natural Gas Storage Report, <http://ir.eia.gov/ngs/ngs.html>.

Chart 6.65. United States natural gas pipeline mileage by type of material, 2009.
Source: Data from American Gas Association, Gas Industry Statistics, <http://www.aga.org/Kc/analyses-and-statistics/statistics/Pages/default.aspx>.

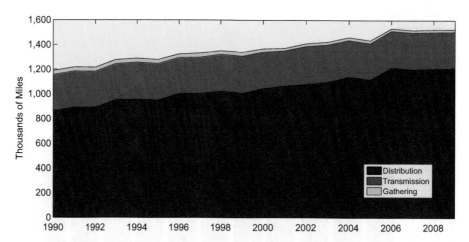

Chart 6.66. United States natural gas pipeline mileage by type of pipeline, 2009.
Source: Data from American Gas Association, Gas Industry Statistics, <http://www.aga.org/Kc/analyses-and-statistics/statistics/Pages/default.aspx>.

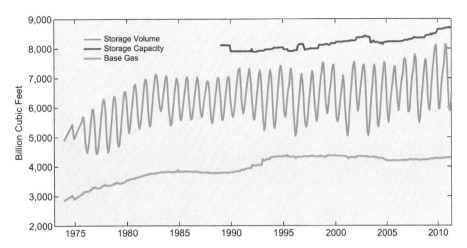

Chart 6.67. United States natural gas storage and storage capacity, 1973-2011. *Base gas* is the volume of gas intended as permanent inventory in a storage reservoir to maintain adequate pressure and deliverability rates throughout the withdrawal season. *Storage capacity* is the maximum volume of gas that can be stored in an underground storage facility in accordance with its design.
Source: Data from United States Department of Energy, Energy Information Administration, Weekly Natural Gas Storage Report, <http://ir.eia.gov/ngs/ngs.html>.

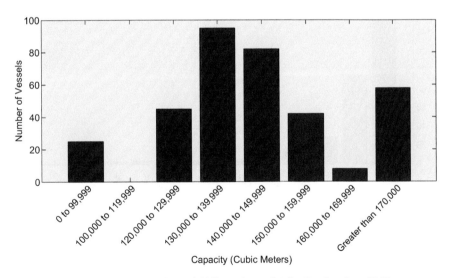

Chart 6.68. World fleet of liquefied natural gas (LNG) carriers, distribution by size, 2010.
Source: Data from International Gas Union, World LNG Report 2011, <http://www.igu.org/>.

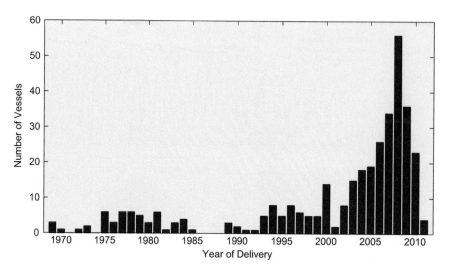

Chart 6.69. World fleet of liquefied natural gas (LNG) carriers, distribution by delivery date, 2010.
Source: Data from International Gas Union, World LNG Report 2011, <http://www.igu.org/>.

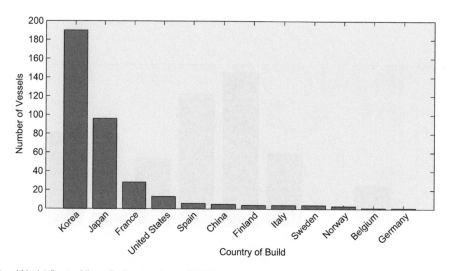

Chart 6.70. World fleet of liquefied natural gas (LNG) carriers, distribution by country of build, 2010.
Source: Data from International Gas Union, World LNG Report 2011, <http://www.igu.org/>.

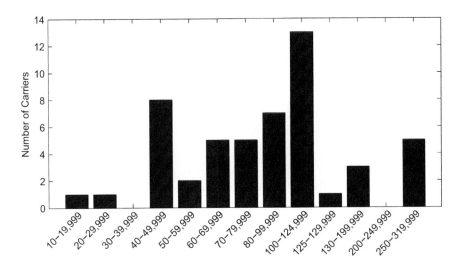

Chart 6.71. Distribution of world combined carrier fleet by size, 2010.
Source: Data from Data from Organization of Petroleum Exporting Countries, Statistical database, <http://www.opec.org/library/Annual%20Statistical%20Bulletin/interactive/current/FileZ/Main.htm>.

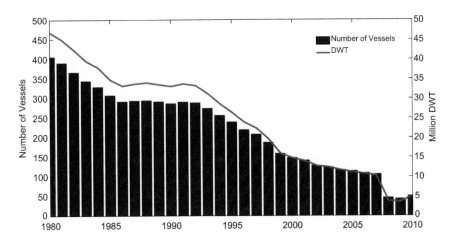

Chart 6.72. World combined carrier fleet by size, 10,000 DWT and over, 1980 - 2010.
Source: Data from Organization of Petroleum Exporting Countries, Statistical database, <http://www.opec.org/library/Annual%20Statistical%20Bulletin/interactive/current/FileZ/Main.htm>.

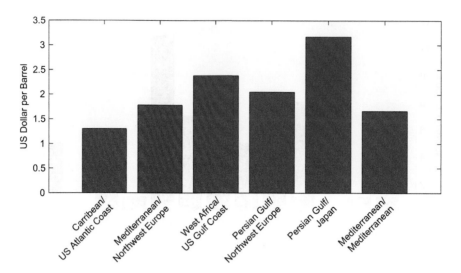

Chart 6.73. Crude oil freight costs in the spot market, 2010.
Source: Data from Organization of Petroleum Exporting Countries, Statistical database,
<http://www.opec.org/library/Annual%20Statistical%20Bulletin/interactive/current/FileZ/Main.htm>.

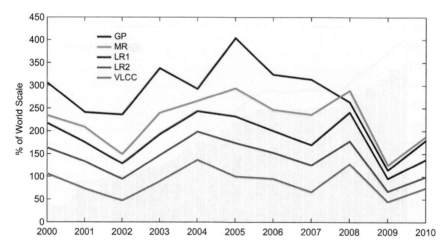

Chart 6.74. Average spot freight rates for all routes, 2000-2010. General Purpose (GP) = 6,5000–24,999
DWT; Medium Range (MR) = 25.000-44,999 DWT; Large Range 1 (LR1) = 45,000-79,999 DWT; Large Range
2 (LR2) = 80,000-159,999 DWT; Very Large Crude Carrier (VLCC) = 160,000-319,999 DWT. Worldscale is
the standard system for assessing freight rates. Once a year, a set of base voyage costs is published for a
theoretical standard vessel plying its trade between each of the world's main loading and discharging ports. Spot
freight rates are commonly expressed as percentages of those theoretical rates. For example, if VLCC rates are
said to be WS67, actual rates are two-thirds of the base or flat rates published at the beginning of the year.
Source: Data from Organization of Petroleum Exporting Countries, Statistical database, <http://www.opec.org/
library/Annual%20Statistical%20Bulletin/interactive/current/FileZ/Main.htm>, accessed 25 August 2011.

Tables

Table 6.1.   Liquid hydrocarbon reserves for top 20 nations, 1980-2010[a]

Country	1980	1990	2000	2010	Rank 2010	Rank 1980	Change 2000–2010
Saudi Arabia	168.0	260.3	262.8	264.5	1	1	0.7%
Venezuela[b]	19.5	60.1	76.8	211.2	2	10	174.8%
Canada (Oil Sands)	n/a	n/a	163.3	143.1	3	n/a	−12.4%
Iran	58.3	92.9	99.5	137.0	4	4	37.7%
Iraq	30.0	100.0	112.5	115.0	5	7	2.2%
Kuwait	67.9	97.0	96.5	101.5	6	3	5.2%
United Arab Emirates	30.4	98.1	97.8	97.8	7	7	0.0%
Russian Federation	n/a	n/a	59.0	77.4	8	n/a	31.1%
Libya	20.3	22.8	36.0	46.4	9	9	29.0%
Kazakhstan	n/a	n/a	25.0	39.8	10	n/a	59.2%
Nigeria	16.7	17.1	29.0	37.2	11	11	28.3%
Canada (Conventional)	8.7	11.2	18.3	32.1	12	14	75.2%
United States	36.5	33.8	30.4	30.9	13	6	1.6%
Qatar	3.6	3.0	16.9	25.9	14	18	53.6%
China	13.3	16.0	15.2	14.8	15	12	−2.7%
Brazil	1.3	4.5	8.5	14.2	16	28	68.3%
Angola	1.4	1.6	6.0	13.5	17	27	126.1%
Algeria	8.2	9.2	11.3	12.2	18	16	7.8%
Mexico	47.2	51.3	20.2	11.4	19	5	−43.5%
India	2.8	5.6	5.3	9.0	20	20	71.0%
Total World	668	1,003	1,105	1,383			25.2%

[a]Reserves include gas condensate and natural gas liquids (NGLs) as well as crude oil.
[b]Venezuelan reserves are taken from the OPEC Annual Statistical Bulletin that noted in 2008 that the figure included "proven reserves of the Magna Reserve Project in the Orinoco Belt, which amounted to 94,168mb"
Source: BP, Statistical Review of World Energy 2011.

Table 6.2.   Liquid hydrocarbon production for top 20 nations, 1970-2010[a]

Country	1970	1980	1990	2000	2010	Rank 2010	Rank 1970	Change 2000–10
Russian Federation	n/a	n/a	10,405	6,536	10,270	1	n/a	57.1%
Saudi Arabia	3,851	10,270	7,105	9,491	10,007	2	3	5.4%
United States	11,297	10,170	8,914	7,733	7,513	3	1	–2.8%
Iran	3,848	1,479	3,270	3,855	4,245	4	4	10.1%
China	615	2,119	2,774	3,252	4,071	5	14	25.2%
Canada	1,473	1,764	1,965	2,721	3,336	6	9	22.6%
Mexico	487	2,129	2,977	3,450	2,958	7	15	–14.3%
United Arab Emirates	762	1,745	2,283	2,620	2,849	8	13	8.7%
Kuwait	3,036	1,757	964	2,206	2,508	9	7	13.7%
Venezuela	3,754	2,228	2,244	3,239	2,471	10	5	-23.7%
Iraq	1,549	2,658	2,149	2,614	2,460	11	8	–5.9%
Nigeria	1,084	2,059	1,870	2,155	2,402	12	10	11.4%
Brazil	167	188	650	1,268	2,137	13	23	68.6%
Norway	-	528	1,716	3,346	2,137	14	n/a	–36.1%
Angola	103	150	475	746	1,851	15	28	148.1%
Algeria	1,052	1,139	1,347	1,578	1,809	16	11	14.6%
Kazakhstan	n/a	n/a	551	744	1,757	17	n/a	136.1%
Libya	3,357	1,862	1,424	1,475	1,659	18	6	12.5%
Qatar	363	476	434	757	1,569	19	17	107.3%
United Kingdom	4	1,663	1,918	2,667	1,339	20	38	-49.8%
Total World	48,064	62,948	65,460	74,893	82,095			9.6%

[a]Thousand barrels per day. Includes crude oil, shale oil, oil sands and NGLs (the liquid content of natural gas where this is recovered separately). Excludes liquid fuels from other sources such as biomass and coal derivatives.
Source: BP, Statistical Review of World Energy 2011.

Table 6.3.   Liquid hydrocarbon consumption for top 20 nations, 1970-2010[a]

Country	1970	1980	1990	2000	2010	Rank 2010	Rank 1970	Change 2000–10
United States	14,710	17,062	16,988	19,701	19,148	1	1	−3%
China	556	1,690	2,320	4,766	9,057	2	11	90%
Japan	3,874	4,900	5,234	5,530	4,451	3	2	−20%
India	391	644	1,213	2,261	3,319	4	18	47%
Russian Federation	n/a	n/a	5,049	2,698	3,199	5	n/a	19%
Saudi Arabia	408	607	1,175	1,578	2,812	6	17	78%
Brazil	522	1,155	1,432	2,018	2,604	7	13	29%
Germany	2,774	3,020	2,689	2,746	2,441	8	3	−11%
South Korea	162	476	1,042	2,252	2,384	9	30	6%
Canada	1,472	1,898	1,747	1,922	2,276	10	7	18%
Mexico	412	1,048	1,580	1,950	1,994	11	16	2%
Iran	330	621	947	1,304	1,799	12	20	38%
France	1,867	2,221	1,895	1,994	1,744	13	5	−13%
United Kingdom	2,030	1,647	1,754	1,704	1,590	14	4	−7%
Italy	1,664	1,930	1,924	1,930	1,532	15	6	−21%
Spain	536	1,044	1,026	1,425	1,505	16	12	6%
Indonesia	138	396	644	1,143	1,304	17	34	14%
Singapore	138	178	444	645	1,185	18	33	84%
Thailand	103	233	422	835	1,128	19	43	35%
Netherlands	702	780	748	879	1,057	20	8	20%
Total World	45,406	61,177	66,503	76,605	87,382			14%

[a]Thousand barrels per day. Inland demand plus international aviation and marine bunkers and refinery fuel and loss. Consumption of fuel ethanol and biodiesel is also included.
Source: BP, Statistical Review of World Energy 2011.

Table 6.4. Natural gas reserves for top 20 nations, 2010 (trillion cubic meters)

Country	1980	1990	2000	2010	Rank 2010	Rank 1980	Change 2000–2010
Russian Federation	n/a	n/a	42.3	44.8	1	n/a	5.9%
Iran	14.1	17.0	26.0	29.6	2	1	13.9%
Qatar	2.8	4.6	14.4	25.3	3	5	75.3%
Turkmenistan	n/a	n/a	2.6	8.0	4	n/a	210.6%
Saudi Arabia	3.2	5.2	6.3	8.0	5	4	27.2%
US	5.6	4.8	5.0	7.7	6	2	53.6%
United Arab Emirates	2.4	5.6	6.0	6.0	7	7	0.6%
Venezuela	1.3	3.4	4.2	5.5	8	10	31.4%
Nigeria	1.2	2.8	4.1	5.3	9	11	28.9%
Algeria	3.7	3.3	4.5	4.5	10	3	−0.4%
Iraq	0.8	3.1	3.1	3.2	11	15	1.9%
Indonesia	0.8	2.9	2.7	3.1	12	14	14.4%
Australia	0.2	0.9	2.2	2.9	13	32	32.5%
China	0.7	1.0	1.4	2.8	14	17	105.4%
Malaysia	0.9	1.6	2.3	2.4	15	13	2.6%
Egypt	0.1	0.4	1.4	2.2	16	40	54.2%
Norway	0.4	1.7	1.3	2.0	17	21	62.2%
Kazakhstan	n/a	n/a	1.8	1.8	18	n/a	4.0%
Kuwait	1.1	1.5	1.6	1.8	19	12	14.6%
Canada	2.5	2.7	1.7	1.7	20	6	2.6%
Total World	81.0	125.7	154.3	187.1			21.3%

Source: BP, Statistical Review of World Energy 2011.

Table 6.5.   Natural gas production for top 20 producing nations, 2010 (billion cubic feet per day)

	1970	1980	1990	2000	2010	Rank 2010	Rank 1970	Change 2000–2010
US	57.6	53.0	48.8	52.4	59.1	1	1	12.8%
Russian Federation	n/a	n/a	57.1	51.0	57.0	2	n/a	11.7%
Canada	5.5	7.2	10.5	17.6	15.5	3	2	−12.1%
Iran	1.2	0.7	2.2	5.8	13.4	4	5	130.5%
Qatar	0.1	0.5	0.6	2.3	11.3	5	22	393.8%
Norway	-	2.4	2.5	4.8	10.3	6	-	114.4%
China	0.3	1.4	1.5	2.6	9.4	7	14	256.7%
Saudi Arabia	0.2	0.9	3.2	4.8	8.1	8	19	69.0%
Indonesia	0.1	1.8	4.2	6.3	7.9	9	21	26.2%
Algeria	0.2	1.4	4.8	8.1	7.8	10	15	−4.5%
Netherlands	2.6	7.4	5.9	5.6	6.8	11	3	21.7%
Malaysia	-	-	1.7	4.4	6.4	12	-	47.3%
Egypt	a	0.2	0.8	2.0	5.9	13	-	192.8%
Uzbekistan	n/a	n/a	3.6	4.9	5.7	14	n/a	16.1%
United Kingdom	1.0	3.4	4.4	10.5	5.5	15	9	−47.1%
Mexico	1.1	2.5	2.6	3.7	5.3	16	7	44.8%
United Arab Emirates	0.1	0.7	1.9	3.7	4.9	17	23	33.3%
India	0.1	0.1	1.2	2.5	4.9	18	25	93.6%
Australia	0.2	1.1	2.0	3.0	4.9	19	18	62.0%
Trinidad & Tobago	0.2	0.3	0.5	1.4	4.1	20	17	192.3%
Total World	96.9	138.4	191.6	232.9	309.0			32.7%

aLess than 0.05
Source: BP, Statistical Review of World Energy 2011.

Table 6.6.   Natural gas consumption for top 20 consuming nations, 2010 (billion cubic feet per day)

	1970	1980	1990	2000	2010	Rank 2010	Rank 1970	Change 2000–2010
US	57.9	54.3	52.5	63.8	**66.1**	1	1	3.7%
Russian Federation	n/a	n/a	39.4	34.2	**40.1**	2	n/a	17.3%
Iran	0.9	0.7	2.2	6.1	**13.2**	3	10	118.3%
China	0.3	1.4	1.5	2.4	**10.5**	4	19	346.2%
Japan	0.3	2.3	4.7	7.0	**9.1**	5	16	31.1%
United Kingdom	1.1	4.3	5.1	9.3	**9.1**	6	7	−2.8%
Canada	3.5	5.0	6.5	8.9	**9.1**	7	2	1.5%
Saudi Arabia	0.2	0.9	3.2	4.8	**8.1**	8	23	69.0%
Germany	1.5	5.5	5.8	7.7	**7.9**	9	5	2.6%
Italy	1.3	2.5	4.2	6.3	**7.4**	10	6	17.6%
Mexico	1.0	2.2	2.7	4.0	**6.7**	11	9	68.4%
India	0.1	0.1	1.2	2.5	**6.0**	12	31	135.6%
United Arab Emirates	0.1	0.5	1.6	3.0	**5.9**	13	29	93.1%
Ukraine	n/a	n/a	12.0	6.8	**5.0**	14	n/a	−26.3%
France	1.0	2.5	2.8	3.8	**4.5**	15	8	19.6%
Uzbekistan	n/a	n/a	3.5	4.4	**4.4**	16	n/a	−0.1%
Egypt	a	0.2	0.8	1.9	**4.4**	17	-	126.2%
Thailand	-	-	0.6	2.1	**4.4**	18	-	106.0%
Netherlands	1.6	3.2	3.3	3.8	**4.2**	19	4	12.3%
Argentina	0.6	1.1	2.0	3.2	**4.2**	20	12	30.8%
Total World	95.5	139	190	233	306.6			31.8%

[a]Less than 0.05
Source: BP, Statistical Review of World Energy 2011.

Table 6.7. Statistical values for estimates of the ultimate recovery (URR)[a] of non-renewable energy resources

Resource	Coal	Conventional oil		Conventional gas	Uranium
Number of studies surveyed	40	201		70	16
Skewness	2.5679	2.248		1.1227	2.9646
	Exajoules	**Billion barrels**	**Exajoules**	**Exajoules**	**Exajoules**
Grand mean	40,815	2,144	10,897	14,072	1643
Standard deviation	44,877	1,111	4404	5660	2291
Standard error	7096		532	399	573
Mode	17,585	2,000	12,240	12,240	
Percentile					223
Minimum	4015		1840	2754	223
5%	9791		4221	6405	272
10%	12,801		5797	9792	360
25%	18,691		7650	11,016	827
Median	24,142	2,050	10,500	12,956	2363
75%	33,398		12,347	16,409	4914
90%	124,850		17,136	20,210	9467
95%	145,960		18,925	24,345	9467
Maximum	222,740		28,764	52,020	9244
Range	218,723	8,500	26,924	49,266	9021

[a]URR is an estimate of the total amount of a resource that will ever be produced.
Source: Adapted from Michael Dale, Meta-analysis of non-renewable energy resource estimates, Energy Policy, Volume 43, April 2012, Pages 102-122.

Table 6.8. Selected forecasts of global oil production, made between 1956 and 2005, which gave a date for the peak

Date	Author/group	Hydrocarbon	Ultimate (Gb)	Date of peak
1956	Hubbert	Cv oil	1250	'About the year 2000' (at 35 Mb/day)
1969	Hubbert	Cv oil	1350; 2100	1990 (at 65 Mb/day); 2000 (at 100 Mb/day)
1972	ESSO	Pr Cv oil	2100	'Increasingly scarce from ~2000'
1972	Report: UN Conference	Pr Cv oil	2500	'Likely peak by 2000'
1974	SPRU, UK	Pr Cv oil	1800–2480	No prediction
1976	UK DoE	Pr Cv oil	na	'About 2000'
1977	Hubbert	Cv oil	2000	1996 If unconstrained logistic; plateau to 2035 if production flat
1977	Ehrlich et al.	Cv oil	1900	2000
1978	WEC/IFP	Pr Cv oil	1803	No prediction
1979	Shell	Pr Cv oil	na	'Plateau within the next 25 years'
1979	BP	Pr Cv oil	na	Peak (non-communist world): 1985
1981	World Bank	Pr Cv oil	1900	'Plateau — turn of the century'
1992	Meadows et al.	Pr Cv oil	1800–2500	No prediction
1995	Petroconsultants	Cv Oil (xN)	1800	About 2005
1996	Ivanhoe	Cv oil	~2000	About 2010 (production mirrors discovery)
1997	Edwards	Pr Cv oil	2836	2020
1997	Laherrère	All liquids	2700	No prediction
1998	IEA	Cv oil	2300	Ref. case 2014
1999	USGS	Pr Cv oil	~2000	Peak ~2010
2000	Bartlett	Pr Cv oil	2000 and 3000	2004 and 2019, respectively
2002	BGR	Cv and Ncv oil	Cv: 2670	Combined peak in 2017
2003	Deffeyes	Cv oil		~2005 (Hubbert linearization)
2003	Bauquis	All liquids	3000	Combined peak in 2020
2003	Campbell — Uppsala	All hydrocarbons		Combined peak ~2015 (includes gas infrastructure constraints)
2003	Laherrère	All liquids	3000	
2003	Energyfiles Ltd.	All liquids	Cv: 2338	2016 (if 1% demand growth)
2003	Energyfiles Ltd.	All hydrocarbons		Peak ~2020 (includes gas infrastructure constraints)
2003	Bahktiari	Pr Cv oil		2006–2007
2004	Miller, BP: own model	Cv and Ncv oil		2025 all possible OPEC production used
2004	PFC Energy	Cv and Ncv oil		2018 (base case)
2005	Deffeyes	Cv oil		2005 (Hubbert linearization)

Pr: probable; Cv: conventional; Ncv: non-conventional; xN: excluding natural gas liquids.
Source: Adapted from Hughes, Larry, Jacinda Rudolph. 2011. Future world oil production: growth, plateau, or peak?, Current Opinion in Environmental Sustainability, Volume 3, Issue 4, Pages 225–234.

Table 6.9.   Selected forecasts of global oil production that forecast no peak before 2030

Date	Author/group	Hydrocarbon	Ultimate (Gb)	Forecast date of peak (by study end-date)	World production	
					2020	**2030**
1998	WEC/IIASA-A2	Cv oil		No peak	90	100
2000	IEA: WEO 2000	Cv oil (+N)	3345	No peak	103	n/a
2001	US DoE EIA	Cv oil	3303	2016/2037	Various	Various
2002	US DoE	Ditto		No peak	109	n/a
2002	Shell Scenario	Cv and Ncv oil	~4000	Plateau: 2025–2040	100	105
2003	WETO study	Cv oil (+N)	4500	No peak	102	120
2004	Exxon-Mobil	Cv and Ncv oil		No peak	114	118
2005	IEA: WEO 2005					
	Reference Sc.	Ditto		No peak	105	115
	Deferred Invest.	Ditto		No peak	100	105
2007	IEA: WEO 2007					
	Reference Sc.	Ditto		No peak	–	116

Cv: conventional; Ncv: non-conventional; +N: plus natural gas liquids.
Source: Adapted from Hughes, Larry, Jacinda Rudolph. 2011. Future world oil production: growth, plateau, or peak?, Current Opinion in Environmental Sustainability, Volume 3, Issue 4, Pages 225–234.

Table 6.10.   Comparison of global oil production decline rate studies

	IEA	Hook *et al.*	CERA
No. of fields in sample	651 (54 supergiant, 263 giant, 334 large)	331 (all giant)	811 (400 large and above)
No. post-peak fields	580[a,b]	261[c]	
% of total production of crude oil in sample	~58%	~50%	~66%
Cumulative discoveries of crude oil in sample	1241 Gb	1130 Gb	1155 Gb
Definition of plateau	Production >85% of peak	Production >96% of peak	Production >80% of peak
Definition of onset of decline	After year of peak production	After last year of plateau	After last year of plateau
Production weighting	Cumulative production[d]	Annual production	Annual production

Studies Cited:
CERA. (2008). "Finding the Critical Numbers." Cambridge Energy Research Associates: London.
Höök, M. (2009). "Depletion and decline curve analysis in crude oil production." Licentiate thesis, Uppsala University.
Höök, M. and K. Aleklett. (2008). "A decline rate study of Norwegian oil production." Energy Policy, 36:11, pp. 4262-71.
Höök, M., R. L. Hirsch, and K. Aleklett. (2009a). "Giant oil field decline rates and the influence on world oil production." Energy Policy, 37:6, pp. 2262-72.
IEA. (2005). "Resources to Reserves - Oil and Gas Technologies for the Energy Markets of the Future." International Energy Agency: OECD, Paris.
[a]101 fields in plateau (production >85% of peak), 117 fields in 'phase 1 decline' (production >50% of peak), 362 fields in 'phase 3' decline (production <50% of peak).
[b]387 onshore, 264 offshore, 185 OPEC and 466 non-OPEC.
[c]261 onshore, 214 offshore, 143 OPEC and 188 non-OPEC.
[d]IEA weights by annual production when estimating historical trends in decline rates.

Table 6.11.   Oil refining capacity, leading nations (thousand barrels per day)

	1970	1980	1990	2000	2010	Rank 2010	Rank 1980	Change 2000–10
United States	12,860	18,620	15,680	16,595	17,594	1	1	6.0%
China	554	1,805	2,892	5,407	10,121	2	10	87.2%
Russian Federation	n/a	7,012	7,294	5,655	5,555	3	2	−1.8%
Japan	3,504	5,643	4,324	5,010	4,463	4	3	−10.9%
India	412	557	1,122	2,219	3,703	5	24	66.9%
South Korea	209	608	798	2,598	2,712	6	22	4.4%
Italy	2,939	3,136	2,520	2,485	2,396	7	6	−3.6%
Saudi Arabia	676	700	1,885	1,806	2,100	8	20	16.3%
Brazil	525	1,393	1,440	1,849	2,095	9	12	13.3%
Germany	2,643	3,422	2,024	2,262	2,091	10	4	−7.6%
Canada	1,396	2,155	1,920	1,861	1,914	11	8	2.9%
Iran	594	1,085	865	1,597	1,860	12	16	16.5%
United Kingdom	2,305	2,614	1,850	1,778	1,757	13	7	−1.2%
France	2,339	3,326	1,699	1,984	1,703	14	5	−14.2%
Mexico	562	1,207	1,595	1,481	1,463	15	14	−1.2%
Total World	51,344	79,363	74,647	82,473	91,791			11.3%

Source: Statistical Review of World Energy 2012.

Table 6.12.   Overview of petroleum refinng processes

Process name	Action	Method	Purpose	Feedstock(s)	Product(s)
FRACTIONATION PROCESSES					
Atmospheric distillation	Separation	Thermal	Separate fractions	Desalted crude oil	Gas, gas oil, distillate, residual
Vacuum distillation	Separation	Thermal	Separate w/o cracking	Atmospheric tower residual	Gas oil, lube stock, residual
CONVERSION PROCESSED–DECOMPOSITION					
Catalytic cracking	Alteration	Catalytic	Upgrade gasoline	Gas oil, coke distillate	Gasoline, petrochemical feedstock
Coking	Polymerize	Thermal	Convert vacuum residuals	Gas oil, coke distillate	Gasoline, petrochemical feedstock
Hydro-cracking	Hydrogenate	Catalytic	Convert to lighter HC's	Gas oil, cracked oil, residual	Lighter, higher-quality products
Hydrogen steam reforming	Decompose	Thermal/ catalytic	Produce hydrogen	Desulfurized gas, O_2, steam	Hydrogen, CO, CO_2
Steam cracking	Decompose	Thermal	Crack large molecules	Atm tower hvy fuel/ distillate	Cracked naphtha, coke, residual
Visbreaking	Decompose	Thermal	reduce viscosity	Atmospheric tower residual	Distillate, tar

Table 6.12. Continued

Process name	Action	Method	Purpose	Feedstock(s)	Product(s)
CONVERSION PROCESSES--UNIFICATION					
Alkylation	Combining	Catalytic	Unite olefins & isoparaffins	Tower isobutane/ cracker olefin	Iso-octane (alkylate)
Grease compounding	Combining	Thermal	Combine soaps & oils	Lube oil, fatty acid, alky metal	Lubricating grease
Polymerizing	Polymerize	Catalytic	Unite 2 or more olefins	Cracker olefins	High-octane naphtha, petrochemical stocks
CONVERSION PROCESSES--ALTERATION OR REARRANGEMENT					
Catalytic reforming	Alteration/ dehydration	Catalytic	Upgrade low-octane naphtha	Coker/ hydro-cracker naphtha	High oct. Reformate/ aromatic
Isomerization	Rearrange	Catalytic	Convert straight chain to branch	Butane, pentane, hexane	Isobutane/ pentane/ hexane
TREATMENT PROCESSES					
Amine treating	Treatment	Absorption	Remove acidic contaminants	Sour gas, HCs w/CO_2 & H_2S	Acid free gases & liquid HCs
Desalting	Dehydration	Absorption	Remove contaminants	Crude oil	Desalted crude oil
Drying & sweetening	Treatment	Abspt/ therm	Remove H_2O & sulfur cmpds	Liq Hcs, LPG, alky feedstk	Sweet & dry hydrocarbons
Furfural extraction	Solvent extr.	Absorption	Upgrade mid distillate & lubes	Cycle oils & lube feed-stocks	High quality diesel & lube oil
Hydrodesulfurization	Treatment	Catalytic	Remove sulfur, contaminants	High-sulfur residual/ gas oil	Desulfurized olefins
Hydrotreating	Hydrogenation	Catalytic	Remove impurities, saturate HC's	Residuals, cracked HC's	Cracker feed, distillate, lube
Phenol extraction	Solvent extr.	Abspt/ therm	Improve visc. index, color	Lube oil base stocks	High quality lube oils
Solvent deasphalting	Treatment	Absorption	Remove asphalt	Vac. tower residual, propane	Heavy lube oil, asphalt
Solvent dewaxing	Treatment	Cool/ filter	Remove wax from lube stocks	Vac. tower lube oils	Dewaxed lube basestock
Solvent extraction	Solvent extr.	Abspt/ precip.	Separate unsat. oils	Gas oil, reformate, distillate	High-octane gasoline
Sweetening	Treatment	Catalytic	Remv H_2S, convert mercaptan	Untreated distillate/ gasoline	High-quality distillate/gasoline

Source: U.S. Occupational Safety and Health Administration, Technical Manual, <http://www.osha.gov/dts/osta/otm/otm_toc.html>.

Table 6.13. Refinery emission factors

Refinery process	Air pollutant	Emission factor lb/1000 bbl feed
Catalytic cracking (fluid bed)	Particulates	242
	Carbon monoxide	13,700
	Sulphur dioxide	493
	Nitrogen oxides	71
	Hydrocarbons	220
	Aldehydes	19
	Ammonia	54
Catalytic cracking (moving bed)	Particulates	17
	Carbon monoxide	3800
	Sulphur dioxide	60
	Nitrogen oxides	5
	Hydrocarbons	87
	Aldehydes	12
	Ammonia	6
Catalytic reforming	Hydrocarbons	25
	Inorganic chlorine	4450
Sulphur recovery plant	Sulphur dioxide	359 lb/ton sulphur recovered
	Reduced Sulphur	0.65 lb/ton sulphur recovered
Storage vessels	Hydrocarbons	No single emission factor
Fluid coking	Particulates	523
Wastewater streams	Hydrocarbons	0.097
Cooling towers	Hydrocarbons	0.0048
Equipment leaks	Hydrocarbons	0.034
Blowdown system	Hydrocarbons	580
Vacuum distillation	Hydrocarbons	50
Steam boiler, furnace or process heater (below 100 MBtu/h capacity)	Particulates	2 lb/1000 gallon distillate oil
	Nitrogen oxides	20 lb/1000 gallon distillate oil
	Carbon monoxide	5 lb/1000 gallon distillate oil
	Sulphur oxides	142 × sulphur percentage in fuel/1000 gallon distillate oil
Steam boiler, furnace or process heater (above 100 MBtu/h capacity)	Particulates	2 lb/1000 gallon fuel oil
	Nitrogen oxides	24 lb/1000 gallon fuel oil
	Carbon monoxide	5 lb/1000 gallon fuel oil
	Sulphur oxides	157 wt% S in fuel/1000 gallon fuel
Compressor engine (reciprocating)	Hydrocarbons	1.4 lb/1000 cubic feet gas fuel
	Carbon monoxide	0.43 lb/1000 cubic feet gas fuel
	Nitrogen oxides	3.4 lb/1000 cubic feet gas fuel
	Sulphur oxides	2 wt% S in fuel/1000 cubic feet gas
Compressor engine (gas turbine)	Hydrocarbons	0.02 lb/1000 cubic feet gas fuel
	Carbon monoxide	0.12 lb/1000 cubic feet gas fuel
	Nitrogen oxides	0.3 lb/1000 cubic feet gas fuel
	Sulphur oxides	2 wt% S in fuel /1000 cubic feet gas
Vessel loading (barge)	Hydrocarbons	3.4 lb/1000 gallons transferred
Vessel loading (ship)	Hydrocarbons	1.8 lb/1000 gallons transferred

Source: U.S Environmental Protection Agency, AP 42, Fifth Edition, Volume I Chapter 5: Petroleum Industry.

Table 6.14. World's largest peroleum provinces[a]

	Province name	Region(s)	Oil BB	Gas TCF	NGL BB	Total BBOE	Percent of world total BBOE
	West Siberian Basin	Russia, Kazakhstan	140.4	1271.8	3.1	355.6	12.5%
1	Mesopotamian Foredeep Basin	Saudi Arabia, Iran, Iraq, Kuwait	292.4	298.3	1.8	344	12.1%
2	Greater Ghawar Uplift	Saudi Arabia, Qatar, Bahrain	141.7	248.6	8.6	191.7	6.7%
3	Zagros Fold Belt	Iran, Iraq	121.6	399.4	1.4	189.5	6.7%
4	Rub Al Khali Basin	Saudi Arabia, Oman, Yemen, United Arab Emirates	89.9	182.3	2.6	122.8	4.3%
5	Qatar Arch	Saudi Arabia, Qatar, United Arab Emirates, Iran	1.2	465.6	13.8	92.5	3.3%
6	Volga-Ural Region	Russia	64	99.2	1.1	81.6	2.9%
7	North Sea Graben	Denmark, Germany, the Netherlands, Norway, and the United Kingdom	44.1	160.6	6	76.9	2.7%
8	Western Gulf	USA (Texas, Louisiana)	26.9	251.6	7.5	76.2	2.7%
9	Permian Basin	USA (Texas, New Mexico)	32.7	94	6.7	55	1.9%
10	**Total: Top ten provinces**						**55.7%**

BB = billion barrels of oil, TCF = trillion cubic feet, BBOE = billion barrels of oil equivalent, NGL = natural gas liquids.
[a]A peroleum province is a region in which a number of oil and gas pools occur in a simlar geologic environment.
Source: T. R. Klett, T. S. Ahlbrandt, J. W. Schmoker, and G. L. Dolton. 1997. RANKING OF THE WORLD'S OIL AND GAS PROVINCES BY KNOWN PETROLEUM VOLUMES, U.S. Geological Survey, OPEN FILE REPORT 97-463.

Table 6.15. Characteristics of selected petroleum-producing geologic environments

Depositional Environment	Reservoir Distribution	Production Characteristic
Deltaic (distributary channel)	Isolated or stacked channels usually with fine-grained sands. May or may not be in communication	Good producers; permeabilities of 500–5000 mD. Insufficient communication between channels may require infill wells in late stage of development
Shallow marine/coastal (clastic)	Sand bars, tidal channels. Generally coarsening upwards. High subsidence rate results in stacked reservoirs. Reservoir distribution dependent on wave and tide action	Prolific producers as a result of clean and continuous sand bodies. Shale layers may cause vertical barriers to fluid flow
Shallow water carbonate (reefs and carbonate muds)	Reservoir quality governed by diagenetic processes and structural history (fracturing)	Prolific production from karstified carbonates. High and early water production possible. Dual porosity systems in fractured carbonates. Dolomites may produce H_2S
Shelf (clastics)	Sheet-like sand bodies resulting from storms or transgression. Usually thin but very continuous sands, well sorted and coarse between marine clays	Very high productivity but high-quality sands may act as 'thief zones' during water or gas injection. Action of sediment burrowing organisms may impact on reservoir quality

Source: Adapted form Jahn, Frank, Mark Cook and Mark Graham, Editors. 2008. Developments in Petroleum Science, Volume 55, (Amsterdam, Elsevier).

Table 6.16. Simple hydrocarbons

Number of carbon atoms	Alkane	Alkene	Alkyne	Cycloalkane
1	Methane		–	–
2	Ethane	Ethylene (ethene)	Acetylene (ethyne)	–
3	Propane	Propylene (propene)	Methylacetylene (propyne)	Cyclopropane
4	Butane	Butylene (butene)	Butyne	Cyclobutane
5	Pentane	Pentylene (pentene)	Pentyne	Cyclopentane
6	Hexane	Hexene	Hexyne	Cyclohexane
7	Heptane	Heptene	Heptyne	Cycloheptane
8	Octane	Octene	Octyne	Cyclooctane
9	Nonane	Nonene	Nonyne	Cyclononane
10	Decane	Decene	Decyne	Cyclodecane

Source: Speight, James G. 2011. Handbook of Industrial Hydrocarbon Processes, (Boston, Gulf Professional Publishing).

Table 6.17. Flash points of hydrocarbon fuels

Fuel	Flash point	Auto-ignition temperature
Gasoline	<−40°C (−40°F)	246°C (475°F)
Diesel fuel	>62°C (143°F)	210°C (410°F)
Jet fuel	>60°C (140°F)	210°C (410°F)
Kerosene	>38–72°C (100–162°F)	220°C (428°F)

Source: Adapted from Speight, James G. 2011. Handbook of Industrial Hydrocarbon Processes, (Boston, Gulf Professional Publishing).

Table 6.18. Flash points, auto-ignition temperatures, and flammability limits for various hydrocarbons

Hydrocarbon	Flash point (°C)	Auto-ignition temperature (°C)	Flammable limits	
			upper (vol % at 25°C)	lower
Methane	−188	630	5	15
Ethane	−135	515	3	12.4
Propane	−104	450	2.1	9.5
n-Butane	−74	370	1.8	8.4
n-Pentane	−49	260	1.4	7.8
n-Hexane	−23	225	1.2	7.4
n-Heptane	−3	225	1.1	6.7
n-Octane	14	220	0.95	6.5
n-Nonane	31	205	0.85	−
n-Decane	46	210	0.75	5.6
n-Dodecane	74	204	0.6	−
n-Tetradecane	99	200	0.5	−

Source: Adapted from Speight, James G. 2011. Handbook of Industrial Hydrocarbon Processes, (Boston, Gulf Professional Publishing).

Table 6.19. Hydrocarbon number range for petroleum products

Product	Lower carbon limit	Upper carbon limit	Lower boiling point °C	Upper boiling point °C
Refinery gas	C_1	C_4	−161	−1
Liquefied petroleum gas	C_3	C_4	−42	−1
Naphtha	C_5	C_{17}	36	302
Gasoline	C_4	C_{12}	−1	216
Kerosene/diesel fuel	C_8	C_{18}	126	258
Aviation turbine fuel	C_8	C_{16}	126	287
Fuel oil	C_{12}	$>C_{20}$	216	421
Lubricating oil	$>C_{20}$		>343	
Wax	C_{17}	$>C_{20}$	302	>343
Asphalt	$>C_{20}$		>343	
Coke	$>C50$[a]		>1000[a]	

[a]Carbon number and boiling point difficult to precisely assess.
Source: Adapted from Speight, James G. 2011. Handbook of Industrial Hydrocarbon Processes, (Boston, Gulf Professional Publishing).

Table 6.20. Heat content of petroleum products

Energy source	Heat content (million Btu/barrel)	Heat content (MJ/barrel)
Asphalt	6.64	7,001.4
Aviation gasoline	5.05	5,325.9
Butane	4.33	4,564.2
Distillate fuel oil	5.83	6,145.7
Ethane	3.08	3,251.7
Isobutane	3.97	4,192.8
Jet fuel, kerosene-type	5.67	5,982.2
Jet fuel, naphtha-type	5.36	5,649.8
Kerosene	5.67	5,982.2
Motor gasoline		
Conventional	5.25	5,542.2
Oxygenated	5.15	5,433.5
Reformulated	5.15	5,433.5
Fuel ethanola	3.54	3,733.8
Natural gasoline	4.62	4,874.4
Pentanes plus	4.62	4,874.4
Petrochemical Feedstocks		
Naphtha less than 401 °F	5.25	5,536.9
Still gas	6	6,330.3
Petroleum coke	6.02	6,355.7
Plant condensate	5.42	5,716.3
Propane	3.84	4,047.2
Residual fuel oil	6.29	6,633.1
Special naphthas	5.25	5,536.9
Still gas	6	6,330.3

Source: U.S. Department of Energy.

Table 6.21. Types of oil and gas reservoirs and classification of storage and flow space system

Type	Growth degree of storage space (%)			Relative porosity of fractures (%)			Fracture-matrix permeability ratio
	Pore	Vug	Fracture	Macro-fracture	Medium fracture	Crackle	
Pore	>75	<25	<5	0	0	100	≤1
Fracture–pore	>50	<25	>5	25	25	50	≥10
Microfracture–pore	50~95	<25	<25	0	0	90	≥3
Fracture	<25	<25	>50	30	30	40	∞
Pore–fracture	25~75	<25	25~75	40	40	20	≥10
Vug–fracture	<25	25~75	25~75	50	40	10	∞
Vug, fracture–vug	<25	>75	5~25	50	40	10	∞
Fracture–vug–pore	25~75	25~75	5~50	33	30	40	≥10
Para pore	>50	<1	≈50	0	0	100	1

Pore: The storage space and percolation channels are the intergranular pores. Therefore, the flow is called flow in porous media such as sandstone, conglomerate, bioclastic limestone, and oolitic limestone reservoirs.
Fractured reservoir: Natural fractures are not only the main storage space, but also the flow channels. There may be no primary pores or be disconnected pores.
Fracture porosity reservoir: Intergranular pores are the main storage space, whereas fractures are the main flow channels. The flow is called flow in dual-porosity single-permeability media.
Porous fractured reservoir: Both intergranular pores and fractures are the storage space and the flow channels. The flow is called flow in dual-porosity dual-permeability media.
Combined fracture–vug–pore reservoir. Fractures, vugs, and pores are both the storage space and the flow channels. All reservoirs of this type are soluble salinastone.
Source: Adapted from Renpu, Wan. 2011. Advanced Well Completion Engineering (Third Edition), (Boston, Gulf Professional Publishing).

Table 6.22. Gas reservoir classification in accordance with nonhydrocarbon gas content

Type	Subtype/classification index						
	Subtype	Slight-sulfur gas reservoir	Low-sulfur gas reservoir	Mid-sulfur gas reservoir	High-sulfur gas reservoir	Extra-sulfur gas reservoir	Hydrogen sulfide gas reservoir
H_2S-bearing gas reservoir	H_2S, g/cm^3	<0.02	0.02~5.0	5.0~30.0	30.0~150.0	150.0~770.0	≥770.0
	H_2S, %	<0.0013	0.0013~0.3	0.3~2.0	2.0~10.0	10.0~50.0	≥50.0
CO_2-bearing gas reservoir	Subtype	Slight-CO_2 gas reservoir	Low-CO_2 gas reservoir	Mid-CO_2 gas reservoir	High-CO_2 gas reservoir	Extra-CO_2 gas reservoir	CO_2 gas reservoir
	CO_2,%	<0.01	0.01~2.0	2.0~10.0	10.0~50.0	50.0~70.0	≥70.0
N_2-bearing gas reservoir	Subtype	Slight-N_2 gas reservoir	Low-N_2 gas reservoir	Mid-N_2 gas reservoir	High-N_2 gas reservoir	Extra-N_2 gas reservoir	N_2 gas reservoir
	NO_2, %	<2.0	2.0~5.0	5.0~10.0	10.0~50.0	50.0~70.0	≥70.0

Source: Adapted from Renpu, Wan. 2011. Advanced Well Completion Engineering (Third Edition), (Boston, Gulf Professional Publishing).

Table 6.23.   Formation damage mechanism of tight sand gas reservoir

Type of damage	Reason for and process of damage	Operation	Precautions and treatment
Liquid phase trap (liquid blocking)	Water saturation is increased.	Drilling by using water-based drilling and completion fluid.	Avoiding using water-based working fluid
	Capillary percolation (overbalance pressure)	Cementing	Reducing pressure difference
	Imbibition	Perforating	Forming cake with zero permeability
	Displacement leakage	Underbalanced drilling by using water-based working fluid	Reducing capillary pressure
		Fracturing, acidizing, workover	Gas injection
			Heat treatment
			Fracturing (N_2, CO_2)
	Oil saturation is increased.	Drilling by using oil-based drilling fluid	Avoiding using oil-based working fluid
	Capillary percolation	Too low flowing pressure during production	Controlling pressure difference
	Condensation		Gas injection
			Heat treatment
			Fracturing
Solids invasion	Particles are deposited in throat of matrix.	Drilling (overbalance pressure, pulsing underbalance pressure)	Forming cake with zero permeability
		Cementing	Underbalanced drilling
		Well completion (openhole completion)	Fluid to be used in well should be strictly filtrated
		Fracturing	Deep penetrating perforating
		Acidizing	Fracturing, acidizing
		Well killing	Heat treatment
	Particles are deposited in fracture.	Workover, well cleanout	Ultrasonic treatment
	Packing in fracture		
	Cake in fracture face		
Formation damage of clay minerals	Action of foreign fluid on rock.	Drilling	Forming cake with zero permeability
	Alkali sensitivity (pH=9)	Cementing	Underbalanced drilling
	Water sensitivity	Well completion	Potential damage evaluation
	Salt sensitivity	Acidizing	Adding clay stabilizing agent
	Rate sensitivity	Fracturing	Reducing pH value
	Action of formation fluid on rock.	Well killing	Reducing flow rate (using horizontal well)
	Rate sensitivity	Workover, well cleanout	Stable pressure difference, reducing surge
		Production	Production under flow rate lower than critical flow rate

Table 6.23.   Continued

Type of damage	Reason for and process of damage	Operation	Precautions and treatment
Chemical adsorption	High-molecular polymer adsorption and retained oil wetting surfactant adsorption	Drilling	Laboratary evaluation
	Throat surface	Well completion	Gas-based fracturing fluid
	Fracture face (wall)	Well killing	Using oxidizing agent (enzyme)
	Emulsification	Fracturing	Microbial degradation polymer
		Workover, well cleanout	
Stress sensitivity	Change in throat size of matrix	Overbalanced drilling (lost circulation)	Forming cake with zero permeability
	Change in fracture width (open, close)	Underbalnced drilling	Controlling rational pressure difference
		Long cemented section cementing	Horizontal well
		Perforating	Inhibiting hydration swell of mud shale of adjacent bed
		Production	Optimizing perforating parameters
Inorganic scale deposition	Action of foreign fluid on formation fluid.	Drilling	Fluid compatibility test
	Inorganic salt and water $BaSO_4$, $CaCO_3 \cdot 2H_2O$, $CaCO_3$	Cementing	Controlling rational pressure difference
	Environmental change of formation fluid	Well killing	Prolonging water breakthrough time
	Inorganic salt, $CaCO_3$	Workover, well cleanout	Using anti-scaling agent
		Production	Mechanical and chemical scale removed
			Fracturing, acidizing
Rock face glazing	Tight porcelain layer is formed by rock powder and water under the action of drill bit and drill pipe (high temperature).	Gas drilling	Increasing lubricity and heat transfer capacity
	Plugging throat	Openhole completion	Mist drilling fluid
	Plugging fracture	Horizontal well	Perforating
			Cave-in stress completion

Source: Renpu, Wan. 2011. Advanced Well Completion Engineering (Third Edition), (Boston, Gulf Professional Publishing).

Table 6.24.   Types of petroleum reservoir fluids

Type	Dry Gas	Wet Gas	Gas Condensate	Volatile Oil	Black Oil
Appearance at surface	Colourless gas	Colourless gas+some clear liquid	Colourless+significant clear/straw liquid	Brown liquid some red/ green colour	Black viscous liquid
Initial GOR (scf/ stb)	No liquids	>15000	3000–15000	2500–3000	100–2500
Degrees API	–	60–70	50–70	40–50	<40
Gas specific gravity (air = 1)	0.60–0.65	0.65–0.85	0.65–0.85	0.65–0.85	0.65–0.8
Composition (mol%)					
C_1	96.3	88.7	72.7	66.7	52.6
C_2	3	6	10	9	5
C_3	0.4	3	6	6	3.5
C_4	0.17	1.3	2.5	3.3	1.8
C_5	0.04	0.6	1.8	2	0.8
C_6	0.02	0.2	2	2	0.9
C_{7+}	0	0.2	5	11	27.9

Source: Adapted from Jahn, Frank, Mark Cook and Mark Graham, Editors. 2008. Developments in Petroleum Science, Volume 55, (Amsterdam, Elsevier).

Table 6.25.   Classification of petroleum traps

Structural traps
Anticlines
Fault traps
Diapiric traps
Salt diapirs
Mud diapirs
Stratigraphic traps
Unconformity related (truncation and onlap)
Unassociated with unconformities (channels, bars, and reefs)
Diagenetic traps (due to solution or cementation)
Hydrodynamic traps (due to water flow)
Combination traps (due to a combination of any 2 or more of the above)

Source: Selley, Richard C. 2003. Petroleum Geology, In: Robert A. Meyers, Editor-in-Chief, Encyclopedia of Physical Science and Technology (Third Edition), (New York, Academic Press), Pages 729-740.

Table 6.26. Types of fracturing fluids and their applications

Type of fracturing fluid	Advantage	Disadvantage	Applicable range
Water-based crosslinked vegetable gum fracturing fluid	Low cost, high safety, strong operability, high overall performance, and wide applicable range	High residue and high damage to fracture propped and reservoir	Normal reservoirs except those with extra-low pressure, oil wettability and strong water sensitivity
Oil-based crosslinked gel fracturing fluid	Good compatibility with reservoir, ease of flowing back	Low safety, high cost, poor temperature tolerance, and high filtration loss	Strong water-sensitive and low-pressure reservoirs
Emulsified fracturing fluid	Low filtration loss, low residue, high viscosity, and low damage	Low overall performance	Water-sensitive and low-pressure reservoirs, and low- and medium-temperature wells
Foamed fracturing fluid	Ease of flowing back, low damage, high proppant-carrying capacity, low friction resistance, and low filtration loss	High operating pressure, requiring special operational equipment and N_2 or CO_2 gas source	Low-pressure water-sensitive reservoirs (especially gas reservoir)
Acid-based fracturing fluid	Good dissolution behavior	High filtration loss and high corrosivity	Carbonate reservoir and sandstone reservoir with more lime
Alcohol-based fracturing fluid	Low surface tension, enabling to eliminate water blocking	High cost, low safety, and low viscosity	Oil reservoirs with water sensitivity, low pressure, and low permeability

Source: Adapted from Renpu, Wan. 2011. Advanced Well Completion Engineering (Third Edition), (Boston, Gulf Professional Publishing).

Table 6.27. Components in fracturing fluids

Component/category	Function/remark
Water-based polymers	Thickener, to transport proppant, reduces leak-off in formation
Friction reducers	Reduce drag in tubing
Fluid loss additives	Form filter cake, reduce leak-off in formation if thickener is not sufficient
Breakers	Degrade thickener after job or disable crosslinker (wide variety of different chemical mechanisms)
Emulsifiers	For diesel premixed gels
Clay stabilizers	For clay-bearing formations
Surfactants	Prevent water-wetting of formation
Nonemulsifiers	
pH-Control additives	Increase the stability of fluid (e.g., for elevated temperature applications)
Crosslinkers	Increase the viscosity of the thickener
Foamers	For foam-based fracturing fluids
Gel stabilizers	Keep gels active longer
Defoamers	
Oil-gelling additives	Same as crosslinkers for oil-based fracturing fluids
Biocides	Prevent microbial degradation
Water-based gel systems	Common
Crosslinked gel systems	Increase viscosity
Alcohol-water systems	
Oil-based systems	Used in water sensitive formation
Polymer plugs	Used also for other operations
Gel concentrates	Premixed gel on diesel base
Resin coated proppants	Proppant material
Ceramics	Proppant material

Source: Fink, Johannes Karl. 2012. Petroleum Engineer's Guide to Oil Field Chemicals and Fluids, (Boston, Gulf Professional Publishing).

Table 6.28. Fracturing fluids used in shale gas extraction, their main compounds, and common uses

Additive type	Main compound(s)	Purpose	Common use of main compound
Diluted Acid (15%)	Hydrochloric acid or muriatic acid	Help dissolve minerals and initiate cracks in the rock	Swimming pool chemical and cleaner
Biocide	Glutaraldehyde	Eliminates bacteria in the water that produce corrosive byproducts	Disinfectant; sterilize medical and dental equipment
Breaker	Ammonium persulfate	Allows a delayed break down of the gel polymer chains	Bleaching agent in detergent and hair cosmetics, manufacture of household plastics
Corrosion Inhibitor	N,n-dimethyl formamide	Prevents the corrosion of the pipe	Used in pharmaceuticals, acrylic fibers, plastics
Crosslinker	Borate salts	Maintains fluid viscosity as temperature increases	Laundry detergents, hand soaps, and cosmetics
Friction Reducer	Polyacrylamide	Minimizes friction between the fluid and the pipe	Water treatment, soil conditioner
	Mineral oil		Make-up remover, laxatives, and candy
Gel	Guar gum or hydroxyethyl cellulose	Thickens the water in order to suspend the sand	Cosmetics, toothpaste, sauces, baked goods, ice cream
Iron Control	Citric acid	Prevents precipitation of metal oxides	Food additive, flavoring in food and beverages; Lemon Juice ~7% Citric Acid
KCl	Potassium chloride	Creates a brine carrier fluid	Low sodium table salt substitute
Oxygen Scavenger	Ammonium bisulfite	Removes oxygen from the water to protect the pipe from corrosion	Cosmetics, food and beverage processing, water treatment
pH Adjusting Agent	Sodium or potassium carbonate	Maintains the effectiveness of other components, such as crosslinkers	Washing soda, detergents, soap, water softener, glass and ceramics
Proppant	Silica, quartz sand	Allows the fractures to remain open so the gas can escape	Drinking water filtration, play sand, concrete, brick mortar
Scale Inhibitor	Ethylene glycol	Prevents scale deposits in the pipe	Automotive antifreeze, household cleansers, and deicing agent
Surfactant	Isopropanol	Used to increase the viscosity of the fracture fluid	Glass cleaner, antiperspirant, and hair color

The specific compounds used in a given fracturing operation will vary depending on company preference, source water quality and site-specific characteristics of the target formation. The compounds shown above are representative of the major compounds used in hydraulic fracturing of gas shales.
Source: Ground Water Protection Council. 2009. Modern Shale Gas Development in the United States: A Primer, Prepared for U.S. Department of Energy, Office of Fossil Energy and National Energy Technology Laboratory.

Table 6.29.   Example of a single stage of a sequenced hydraulic fracture treatment in shale gas extraction

Hydraulic Fracture Treatment Sub-Stage	Volume (gallons)	Rate (gal/min)
Diluted Acid (15%)	5,000	500
Pad	100,000	3,000
Prop 1	50,000	3,000
Prop 2	50,000	3,000
Prop 3	40,000	3,000
Prop 4	40,000	3,000
Prop 5	40,000	3,000
Prop 6	30,000	3,000
Prop 7	30,000	3,000
Prop 8	20,000	3,000
Prop 9	20,000	3,000
Prop 10	20,000	3,000
Prop 11	20,000	3,000
Prop 12	20,000	3,000
Prop 13	20,000	3,000
Prop 14	10,000	3,000
Prop 15	10,000	3,000
Flush	13,000	3,000

Volumes are presented in gallons (42 gals = one barrel, 5,000 gals = ~120 bbls).

Rates are expressed in gals/minute, 42 gals/minute = 1 bbl/min, 500 gal/min = ~12 bbls/min.

Flush volumes are based on the total volume of open borehole, therefore as each stage is completed the volume of flush decreases as the volume of borehole is decreased.

Total amount of proppant used is approximately 450,000 pounds.

Source: Ground Water Protection Council. 2009. Modern Shale Gas Development in the United States: A Primer, Prepared for U.S. Department of Energy, Office of Fossil Energy and National Energy Technology Laboratory.

Table 6.30. Types and degrees of formation damage during well construction and oil and gas production

	Well construction phase		Oil and gas production phase				
	Drilling and cementing	Well completion	Downhole operation	Stimulation	Well testing	Natural recovery	Supplementary recovery
Drilling fluid solids invading	****	**	***	—	*	—	—
Fine particle migration	***	****	***	****	****	***	****
Clay mineral swelling	****	**	***	—	—	—	**
Emulsification plugging/ water phase trap damage	***	****	**	****	*	****	****
Wettability reversal	**	***	***	****	—	—	****
Relative permeability reduction	***	***	****	***	—	**	—
Organic scale	*	*	***	****	—	****	—
Inorganic scale	**	***	****	*	—	***	***
Foreign fluid solids invading	—	****	***	***	—	—	****
Secondary mineral precipitation	—	—	—	****	—	—	***
Bacteria invasion damage	**	**	**	—	—	**	****
Sand production	—	***	*	****	—	***	**

"*" represents the degree of severity whereas "—" represents "no damage."
Source: Renpu, Wan. 2011. Advanced Well Completion Engineering (Third Edition), (Boston, Gulf Professional Publishing).

Table 6.31.   Classification of drilling muds

Class	Description
Fresh water muds[a]	pH from 7–9.5, include spud muds, bentonite-containing muds, phosphate-containing muds, organic thinned muds (red muds, lignite muds, lignosulfonate muds), organic colloid muds
Inhibited muds[a]	Water-based drilling muds that repress hydration of clays (lime muds, gypsum muds, sea water muds, saturated salt water muds)
Low-solids muds[b]	Contain less than 3–6% of solids. Most contain an organic polymer
Emulsions	Oil in water and water in oil (reversed phase, with more than 5% water)
OBMs	Contain less than 5% water; mixture of diesel fuel and asphalt

[a]Dispersed systems.
[b]Nondispersed systems.
Source: Fink, Johannes Karl. 2012. Petroleum Engineer's Guide to Oil Field Chemicals and Fluids, (Boston, Gulf Professional Publishing).

Table 6.32.   Thermal conductivities of various dry and saturated rocks

Rock	Density, lb/ft^3	Specific heat, BTU/lb- F	Thermal conductivity, BTU/hour-ft- F	Thermal diffusivity, ft^2/hour
(a) Dry rocks				
Sandstone	130	0.183	0.507	0.0213
Silty sand	119	0.202	0.4	0.0167
Siltstone	120	0.204	0.396	0.0162
Shale	145	0.194	0.603	0.0216
Limestone	137	0.202	0.983	0.0355
Sand (fine)	102	0.183	0.362	0.0194
Sand (coarse)	109	0.183	0.322	0.0161
(b) Water-saturated				
Sandstone	142	0.252	1.592	0.0445
Silty sand	132	0.288	1.5	0.0394
Siltstone	132	0.276	1.51	0.0414
Shale	149	0.213	0.975	0.0307
Limestone	149	0.266	2.05	0.0517
Sand (fine)	126	0.339	1.59	0.0372
Sand (coarse)	130	0.315	1.775	0.0433

Source; Adapted from Ahmed, Tarek, and D. Nathan Meehan. 2012. Advanced Reservoir Management and Engineering (Second Edition), (Boston, Gulf Professional Publishing).

Table 6.33. Thermal cracking process

Visbreaking
Mild heating 471–493 °C (880–920 °F) at 50–200°psig
Reduce viscosity of fuel oil
Low conversion (10%) at 221 °C (430 °F)
Heated coil or soaking drum
Delayed coking
Moderate heating 482–516 °C0020(900–960 °F) at 90°psig
Soak drums 452–482 °C (845–900 °F)
Residence time: until they are full of coke
Coke is removed hydraulically
Coke yield ~ 30 wt%
Fluid coking and flexicoking
Severe heating 482–566 °C (900–1050 °F) at 10 psig
Fluidized bed with steam
Higher yields of light ends
Less coke yield (20% for fluid coking and 2% for flexicoking)

Source: Fahim, Mohamed A., Taher A. Alsahhaf, Amal Elkilani. 2010. Fundamentals of Petroleum Refining, (Amsterdam, Elsevier).

Table 6.34. Factors invoved in the appraisal of an oil or gas field

Input Parameter	Controlling Factors
Gross rock volume	Shape of structure; dip of flanks; position of bounding faults; position of internal faults; depth of fluid contacts (e.g. OWC)
Net:gross ratio	Depositional environment; diagenesis
Porosity	Depositional environment; diagenesis
Hydrocarbon saturation	Reservoir quality; capillary pressures
Formation volume factor	Fluid type; reservoir pressure and temperature
Recovery factor (initial conditions only)	Physical properties of the fluids; formation dip angle; aquifer volume; gas cap volume

Source: Adapted from Jahn, Frank, Mark Cook and Mark Graham, Editors. 2008. Developments in Petroleum Science, Volume 55, (Amsterdam, Elsevier).

Table 6.35.   Information and application of conventional physical property and petrographic analyses

Main analysis item	Main information acquired	Main applications
1. Conventional physical property analysis (porosity, permeability, and saturation measurement, grain size distribution)	Porosity, permeability, grain size distribution of clastic rock	1. Reservoir evaluation, reserves calculation, development program design 2. Well completion method selection, completion fluid design, perforated completion optimization, optimal gravel size selection of gravel packing
2. Pore throat configuration analysis (cast thin section and pore cast analysis, capillary pressure curve measurement, nuclear magnetic resonance analysis)	Types of pores, geometric parameters and distribution of pore configuration; types of throats, geometric parameters and distribution of throats, etc.	1. Reservoir evaluation, reserves calculation, development performance analysis 2. Potential damage evaluation, reservoir protection program formulation 3. Designing drilling and completion, perforating, killing, and workover fluids of temporary plugging
3. Thin section analysis (ordinary polaroid, cast thin section, fluorescence thin section, and cathodeluminescence thin section)	Rock texture and frame; grain composition; matrix composition and distribution; type and distribution of cement; pore characteristics; types, concentrations, and occurrences of sensitive minerals	1. Reservoir evaluation, petrological properties, potential damage evaluation 2. Reservoir protection program formulation 3. Completion fluid design 4. Appraisal of degree of consolidation and strength of rock
4. Scanning electron microscope analysis (including environment scanning electron microscope analysis)	Pore throat characteristics analysis; rock texture analysis; occurrence and type analyses of clay minerals	1. Reservoir evaluation, petrological properties, potential damage evaluation 2. Completion fluid and reservoir protection program designs
5. X-ray diffraction analysis	Absolute concentrations, types, and relative concentrations of clay minerals, interlayer ratio, degree of order	1. Potential damage evaluation 2. Completion fluid and reservoir protection program designs
6. Electron probe analysis (electron probe wave and energy spectra)	Mineral composition identification, crystal structure analysis	1. Potential damage evaluation 2. Completion fluid and reservoir protection program designs

Source: Adapted from Renpu, Wan. 2011. Advanced Well Completion Engineering (Third Edition), (Boston, Gulf Professional Publishing).

Table 6.36. Acting mechanisms and applicable conditions of various measures for putting an oil or gas well into production

Measures for putting a well into production		Acting mechanism	Applicable condition and range
Chemical blocking removal	Chemical solvent	Dissolving blocking matter due to the strong solvability of solvent	Blocking of organic matter precipitated due to change of temperature and pressure, sulfur precipitation blocking for high-sulfur oil and gas wells
	Anti-swelling and deswelling agent	Inhibiting the hydration, swelling, dispersion, and migration of smectite fine, so that crystal sheet may be compressed	Underproduction oil wells and underinjection water injection wells with low acid treatment effectiveness due to clay particle migration and clay mineral swelling
	Viscosity-reducing agent	Wetting and dispersing effect, forming O/W (oil-in-water) emulsion with heavy oil for visbreaking	Oil wells with high flow resistance due to the deposition of heavy components (gum, asphaltenes, paraffin) in the vicinity of the wellbore
	Heat chemical	Generating a great amount of heat energy and gas by chemical reaction, softening and dissolving blocking matter, reducing heavy oil viscosity	Poor physical properties, complicated relation between oil and water, viscous oil with low flowability, organic matter precipitation, contamination and blocking due to emulsified oil caused by invasion of killing fluid filtrate in the vicinity of the wellbore
	Oxidant	Generating oxidizing and decomposing effects on organic scale blocking matter due to strong oxidizing ability	Oil well and water injection well blocking caused by bacterial slime and the organic matter and organic gel in drilling fluid system
	Surface tension reducer	Reducing the surface tension of water, reducing the starting pressure difference necessary for water phase flow	Water block and emulsion block caused by mud filtrate, water-based well-flushing fluid and killing fluid in reservoir throats and the block caused by oil sullage
	Active enzyme	Releasing reservoir oil adsorbed on rock surface, changing rock surface wettability	When water content in reservoir is less than 50% and formation damage is caused by reservoir rock wettability reversal, active enzyme can be effectively used for removing blocking
Physical blocking removal	Ultrasonic wave	Using the vibration and cavitation effects of ultrasonic wave	Appropriate for oil reservoirs sensitive to water and acid and inappropriate for reservoirs treated by conventional hydraulic fracturing; the lower the reservoir permeability and the higher the oil viscosity, the poorer the effectiveness of ultrasonic treatment

Continued

Table 6.36. Continued

Measures for putting a well into production		Acting mechanism	Applicable condition and range
	Blocking removal by microwave	Using the heating effect, fracture-creating effect, and unheating effect of microwave	Appropriate mainly for heavy oil reservoirs, high pour-point oil reservoirs, and low-permeability oil reservoirs
	Magnetic treatment	Using magnetocolloid effect, hydrogen bond variation, and inner crystal nucleus effect to change the microstructures of paraffin crystal and the crystal with dissolved salt	Appropriate mainly for reservoirs with high viscosity and reservoirs that are easy to paraffin
	Hydraulic vibration	Generating fluid pressure pulse for removing blocking at bottomhole	Appropriate for wells for which conventional stimulation is inefficient and of which the reservoir has a slight sand production
	Mechanical vibration	Generating mechanical wave field for removing blocking by using mechanical device	Appropriate to oil reservoirs that have simple structure, integral tract, good connectedness, medium-viscosity oil, and no or less sand production
	Low-frequency electric pulse	Generating vibration effect, cavitation effect, relative movement effect, twice shock wave effect, and heat effect for removing reservoir blocking	Appropriate for oil and gas reservoirs that have rapid production rate decline and low degree of production and are sensitive to water and acid
Acidizing	Acid cleaning	Generating decomposition and dissolution effect of acid on blocking matter for removing the blocking matter in pores and fracture	Appropriate for skin blocking removal of sandstone and carbonatite oil and gas reservoirs, and interconnection of perforation
	Matrix acidizing	Decomposing and dissolving blocking matter, interconnecting and enlarging original pores and fractures in the reservoir, increasing reservoir permeability	Appropriate for wells that have serious blocking in the vicinity of the wellbore and treatment range of about 1 m
	Acid fracturing	Forming etched channel and etched fracture with a certain geometric size and flow conductivity	The most commonly used treatment technology in carbonatite reservoir stimulation
Fracturing	Hydraulic fracturing	Forming the propped fracture with a certain geometric shape in the reservoir	

Source: Adapted from Renpu, Wan. 2011. Advanced Well Completion Engineering (Third Edition), (Boston, Gulf Professional Publishing).

Table 6.37. Average spot freight rates for all routes, 2000-2010 (percentage of Worldscale)

Date	General purpose (GP) 16,5000–24,999 DWT	Medium range (MR) 25.000-44,999 DWT	Large range 1 (LR1) 45,000-79,999 DWT	Large range 2 (LR2) 80,000-159,999 DWT	Very large crude carrier (VLCC) 160,000-319,999 DWT
2000	306.6	234.8	217.7	164.0	106.9
2001	241.6	209.2	175.9	133.9	74.4
2002	236.5	150.0	129.2	95.6	48.1
2003	338.2	240.3	193.3	147.7	90.0
2004	292.9	266.4	243.9	199.5	137.6
2005	404.4	294.4	232.8	174.7	100.8
2006	324.2	246.7	200.8	153.0	95.4
2007	313.2	235.6	169.3	125.2	66.2
2008	263.8	289.6	241.8	178.7	128.4
2009	114.4	125.3	95.8	67.9	45.2
2010	180.0	188.4	137.6	99.9	75.1

DWT = Deadweight tonnage is a measure of how much weight a ship is carrying or can safely carry. It is the sum of the weights of cargo, fuel, fresh water, ballast water, provisions, passengers, and crew.
Worldscale is the standard system for assessing freight rates. Once a year, a set of base voyage costs is published for a theoretical standard vessel plying its trade between each of the world's main loading and discharging ports. Spot freights rates are commonly expressed as percentages of those theoretical rates. For example, if VLCC rates are said to be WS67, actual rates are two-thirds of the base or flat rates published at the beginning of the year.
Source: Organization of Petroleum Exporting Countires, International Energy Agency.

Table 6.38. Freight costs in the spot market ($/b)

Date effective	Crude oil						Petroleum products	
	Caribs/ USAC	Med/ NWE	WAF/ USGC	Gulf/ NWE	Gulf/ Japan	Med/ Med	Caribs/USAC	Spore/Japan
2000	1.49	1.44	2.01	2.27	3.03	0.84	1.35	1.84
2001	1.29	1.35	1.62	1.91	3.55	0.99	1.43	2.12
2002	0.93	1.53	1.12	1.18	2.54	0.69	0.98	1.42
2003	1.37	1.51	1.88	2.16	3.24	0.96	1.22	1.83
2004	1.89	1.79	2.79	3.66	3.65	1.21	1.69	2.20
2005	2.07	1.76	2.27	2.79	4.99	1.31	1.69	2.86
2006	1.85	2.15	3.05	3.30	6.22	1.36	1.96	3.62
2007	1.60	1.51	2.74	2.68	4.44	1.76	2.49	2.82
2008	2.15	3.14	4.17	4.06	6.45	2.56	2.81	3.22
2009	1.03	1.58	2.09	1.70	2.98	1.29	1.35	1.38
2010	1.30	1.78	2.38	2.05	3.17	1.66	1.60	1.63

Notes: Caribs = Carribean; USAC = US Atlantic Coast; Med = Mediterranean; NWE = Northwest Europe; USGC = US Gulf Coast; WAF = West Africa; Gulf = Persian Gul; Spore = Singapore.
Source: Organization of Petroleum Exporting Countries.

Table 6.39.   Categories of crude oil

Category	Name	Description
Class A	Light, volatile oils	Highly fluid
		Strong odor
		Spreads rapidly
		High volatiles
		Penetrates soil
		Flammable
Class B	Nonsticky oils	Adheres to surface
		Waxy feel
		Can be washed away
		Mild volatiles
		Nonpenetrating
Class C	Heavy, sticky oils	Tarry/sticky
		Adheres to surfaces
		Cannot be washed away
		Nonpenetrating
		Low volatiles
Class D	Nonfluid oils	Black/brown solid
		Nonpenetrating
		Cannot be washed away
		Melts upon heating
		Nonvolatile

Source: U.S. Environmental Protection Agency.

Table 6.40.   Typical average characteristics of various crude oils

Crude source	Paraffins (% vol)	Aromatics (% vol)	Naphthenes (% vol)	Sulfur (% wt)	API gravity (approx.)	Napht. yield (% vol)	Octane no. (typical)
Nigerian -Light	37	9	54	0.2	36	28	60
Saudi -Light	63	19	18	2	34	22	40
Saudi -Heavy	60	15	25	2.1	28	23	35
Venezuela -Heavy	35	12	53	2.3	30	2	60
Venezuela -Light	52	14	34	1.5	24	18	50
USA -Midcont. Sweet	-	-	-	0.4	40	-	-
USA -W. Texas Sour	46	22	32	1.9	32	33	55
North Sea -Brent	50	16	34	0.4	37	31	50

Source :U.S. Occupational Safety and Health Admininstration, Technical Manual, <http://www.osha.gov/dts/osta/otm/otm_toc.html>, accessed 28 May 2012.

Table 6.41.   Oil refinery throughputs by region, 1980–2010 (thousand barrels per day)[a]

Country	1980	1990	2000	2010	Rank 2010	Rank 1980	Change 2000–10
Europe & Eurasia	24,398	22,682	19,240	19,664	1	1	2.2%
US	13,481	13,409	15,067	14,722	2	2	−2.3%
China	1,510	2,153	4,218	8,492	3	8	101.3%
Other Asia Pacific	2,591	3,842	6,831	7,062	4	5	3.4%
Middle East	2,427	4,470	5,430	6,513	5	6	19.9%
S. & Cent. America	5,249	4,315	5,337	4,621	6	3	−13.4%
India	502	1,038	2,039	3,903	7	12	91.4%
Japan	4,015	3,437	4,145	3,619	8	4	−12.7%
Africa	1,439	2,171	2,188	2,454	9	9	12.2%
Canada	1,893	1,584	1,765	1,827	10	7	3.5%
Mexico	1,129	1,490	1,363	1,184	11	10	−13.1%
Australasia	609	716	851	756	12	11	−11.2%
Total World	59,243	61,306	68,475	74,816			9.3%

[a]Input to primary distallation units only.
Source: BP Statistical Review of Energy, 2011

Table 6.42.   Rotary drilling rig count by region, 1990–2010.

	1990	1995	2000	2005	2010
United States	1008	724	916	1381	1540
Other W. Hemisphere	384	501	572	789	734
Asia-Pacific	285	179	140	225	269
Africa	91	67	46	50	83
Middle East	139	128	156	248	265
Europe	146	113	83	70	94
World	2,053	1,712	1,913	2,763	2,985

Source: Baker Hughes International, Oil & Gas Journal Energy Database.

Table 6.43.   Common wireline tool types and their applications

Generic device	Tool examples	Measurement type	Application
Gamma	GR, NGT, Spectralog	Natural gamma radiation	Lithology, correlation
SP	SP	Spontaneous potential	Lithology, permeability (indicator)
Density	LDL, ZDL, SDL	Bulk density	Porosity, lithology
Neutron	CN, CNL, DSN	Hydrogen index	Lithology, porosity, gas indicator
Acoustic	BHC, XMAC, DSI	Travel time, acoustic waveform	Porosity, seismic calibration
Resistivity	DLL, HRLA, HDLL	Electrical resistance of formation	Saturation, permeability indicator
Induction	ILD, AIT, HILT, HDIL, HRAI	Induced electrical current	Saturation (OBMs)
Image	FMI, STAR, CBIL, EI, OBMI, CAST	Resistivity or acoustic pixellated image	Sedimentology, fracture/fault analysis
NMR	MRIL, MREX, CMR	Nuclear magnetic resonance	Porosity, permeability, saturation
Formation tester	RFT, MDT, RCI pressure	Pore pressure	Fluid types, pressures and contacts

Source: Adapted from Jahn, Frank, Mark Cook and Mark Graham, Editors. 2008. Developments in Petroleum Science, Volume 55, (Amsterdam, Elsevier).

Table 6.44.   Criteria for enhanced oil recovery methods

Process	Crude oil	Reservoir
N_2 and flue gas	>35° API	S_o>40%
	<1.0 cp	Formation: SS or carbonate with few fractures
	High percentage of light hydrocarbons	Thickness: relatively thin unless formation is dipping
		Permeability: not critical
		Depth >6000 ft
		Temperature: not critical
Chemical	>20° API	S_o>35%
	<35 cp	Formation: SS preferred
	ASP: organic acid groups in the oi are needed	Thickness: not critical
		Permeability >10 md
		Depth <9000 ft (function of temperature)
		Temperature <200° F
Polymer	>15° API	S_o>50%
	<100 cp	Formation: SS but can be used in carbonates
		Thickness: not critical
		Permeability >10 md
		Depth <9000 ft
		Temperature <200° F
Miscible CO_2	>22° API	S_o>20%
	<10 cp	Formation: SS or carbonate
	High percentage of intermediate components (C5–C12)	Thickness: relatively thin unless dipping
		Permeability: not critical
		Depth: depends on the required minimum miscibility pressure "MMP"
First-contact miscible flood	>23° API	S_o>30%
	<3 cp	Formation: SS or carbonate with min fractures
	High Cm	Thickness: relatively thin unless formation is dipping
		Permeability: not critical
		Depth >4000 ft
		Temperature: can have a significant effect on MMP
Steam flooding	10–25° API	S_o>40%
	<10,000 cp	Formation: SS with high permeability
		Thickness >20 ft
		Permeability >200 md
		Depth <5000 ft
		Temperature: not critical

Table 6.44.   Continued

Process	Crude oil	Reservoir
In situ combustion	10–27°API	S_o>50%
	<5000 cp	Formation: SS with high porosity
		Thickness>10 ft
		Permeability>50 md
		Depth<12,000 ft
		Temperature>100°F

Source; Adapted from Ahmed, Tarek, and D. Nathan Meehan. 2012. Advanced Reservoir Management and Engineering (Second Edition), (Boston, Gulf Professional Publishing).

Table 6.45.   Effects of different fracture network models on oil well productivity ratio

Type	Perforation density (Shots/m)			Perforation penetration depth (mm)			Fracture density (Fractures/m)		
	40	24	16	450	300	100	10	5	1
One group of vertical fractures	0.79	0.76	0.74	0.77	0.74	0.52	0.78	0.76	0.07
Two groups of vertical fractures	1.05	1.03	1.02	1.04	0.98	0.78	1.19	1.07	0.79
One group of horizontal fractures	0.89	0.88	0.77	0.88	0.85	0.66	0.94	0.89	0.8
Three groups of orthogonal fractures	1.22	1.21	1.19	1.2	1.16	0.82	1.43	1.26	0.87

The oil productivity ratio is defined as the oil productivity after treatment divided by the oil productivity before treatment. The basic parameters of perforation density, penetration depth, and fracture density are, respectively, 24,400 and 4.
Source: Adapted from Renpu, Wan. 2011. Advanced Well Completion Engineering (Third Edition), (Boston, Gulf Professional Publishing).

Table 6.46.   Crude distillation unit products

Fraction	Typical boiling ranges		Disposition
	°C	°F	
Gases	−160 to 0	−260 to 32	To gas plant
LSR naphtha	25 to 85	80 to 180	To isomerizer or gasoline blending
HSR naphtha	90 to 180	200 to 350	To catalytic reformer
Kerosine	160 to 260	320 to 500	To jet, diesel, or #1 or #2 fuel oil blending
Atmospheric gas oil	200 to 350	400 to 650	To diesel or #2 fuel oil blending
Light vacuum gas oil	310 to 400	600 to 750	To FCC or HC feed
Heavy vacuum gas oil	370 to 565	700 to 1050	To FCC or HC feed
Residuum	565+	1050+	To asphalt, #6 fuel oil, or coker feed

FCC = fluid catalytic cracking; HC = bydrocracker.
Source: Adapted from Gary, James H. 2003. Petroleum Refining, In: Robert A. Meyers, Editor-in-Chief, Encyclopedia of Physical Science and Technology (Third Edition), (New York, Academic Press), Pages 741-761.

Table 6.47. Crude petroleum fractions in a typical refinery

Fraction	Boiling range	
	°C	°F
Light naphtha	−1 to 150	30–300
Gasoline	−1 to 180	30–355
Heavy naphtha	150–205	300–400
Kerosene	205–260	400–500
Light gas oil	260–315	400–600
Heavy gas oil	315–425	600–800
Lubricating oil	>400	>750
Vacuum gas oil	425–600	800–1100
Residuum	>510	>950

Source: Speight, James G. 2011. Handbook of Industrial Hydrocarbon Processes, (Boston, Gulf Professional Publishing).

Table 6.48. Properties of hydrocarbon products from petroleum

	Molecular weight	Specific gravity	Boiling point °F	Ignition temperature °F	Flash point °F	Flammability limits in air % v/v
Benzene	78.1	0.88	176.2	1040	12	1.35–6.65
n-Butane	58.1	0.6	31.1	761	−76	1.86–8.41
iso-Butane	58.1		10.9	864	−117	1.80–8.44
n-Butene	56.1	0.6	21.2	829	Gas	1.98–9.65
iso-Butene	56.1		19.6	869	Gas	1.8–9.0
Diesel fuel	170–198	0.88			100–130	
Ethane	30.1	0.57	−127.5	959	Gas	3.0–12.5
Ethylene	28		−154.7	914	Gas	2.8–28.6
Fuel oil No. 1		0.88	304–574	410	100–162	0.7–5.0
Fuel oil No. 2		0.92		494	126–204	
Fuel oil No. 4	198	0.96		505	142–240	
Fuel oil No. 5		0.96			156–336	
Fuel oil No. 6		0.96			150	
Gasoline	113	0.72	100–400	536	−45	1.4–7.6
n-Hexane	86.2	0.66	155.7	437	−7	1.25–7.0
n-Heptane	100.2	0.67	419	419	25	1.00–6.00
Kerosene	154	0.8	304–574	410	100–162	0.7–5.0
Methane	16	0.55	−258.7	900–1170	Gas	5.0–15.0
Naphthalene	128.2		424.4	959	174	0.90–5.90
Neohexane	86.2	0.65	121.5	797	−54	1.19–7.58
Neopentane	72.1		49.1	841	Gas	1.38–7.11
n-Octane	114.2	0.71	258.3	428	56	0.95–3.2
iso-Octane	114.2	0.7	243.9	837	10	0.79–5.94
n-Pentane	72.1	0.63	97	500	−40	1.40–7.80
iso-Pentane	72.1	0.62	82.2	788	−60	1.31–9.16
n-Pentene	70.1	0.64	86	569	−	1.65–7.70
Propane	44.1		−43.8	842	Gas	2.1–10.1
Propylene	42.1		−53.9	856	Gas	2.00–11.1
Toluene	92.1	0.87	321.1	992	40	1.27–6.75
Xylene	106.2	0.86	281.1	867	63	1.00–6.00

Source: Adapted from Speight, James G. 2011. Handbook of Industrial Hydrocarbon Processes, (Boston, Gulf Professional Publishing).

Table 6.49. General properties of liquid fuels from petroleum

	Chemical formula	Mol. wt	Composition (wt%) C	Composition (wt%) H	Specific gravity	Boiling point (°C)	Autoignition temperature (°C)	Flash point (°C)	Heating values (kJ/kg) LHV	Heating values (kJ/kg) HHV	Flammability limits (vol%)
Gasoline	C4–C12	113	85–88	12–15	0.72	35–225	257	–43	43,500	46,500	1.4–7.6
Kerosene	C10–C16	170	85	15	0.8	150–300	210	37–72	43,100	46,200	0.7–5.0
JP-4	C6–C11	119	86	14	0.751–0.802	45–280	246	–23 – 1	42,800	45,800	1.3–8.0
JP-7	C10–C16	166	87	13	0.779–0.806	60–300	241	43–66	43,500	46,800	0.6–4.6
Diesel fuel	C8–C25	200	87	13	0.85	180–340	315	60–80	42,800	45,800	1.0–6.0
Fuel oil #1	C9–C16	200	87	13	0.88	150–300	210	37–72			0.7–5.0
Fuel oil #2	C11–C20				0.92		256	52–96			
Fuel oil #4		198			0.96		263	61–115			
Fuel oil #5					0.96			69–169			
Fuel oil #6					0.96	175–345		66			

Source: Adapted from Speight, James G. 2011. Chapter 3 - Fuels for Fuel Cells, In: Dushyant Shekhawat, J.J. Spivey And David A. Berry, Editors, Fuel Cells: Technologies for Fuel Processing, (Amsterdam, Elsevier), 2011, Pages 29-48.

Table 6.50. Typical primary products from light and heavy feedstocks

	Product (wt%)			
Light feedstock	**Ethane**	**Propane**	**n-Butane**	**i-Butane**
H_2	3.7	1.6	1.5	1.1
CH_4	3.5	23.7	19.3	16.6
C_2H_2	0.4	0.8	1.1	0.7
C_2H_4	48.8	41.4	40.6	5.6
C_2H_6	40	3.5	3.8	0.9
C_3H_6	1	12.9	13.6	26.4
C_3H_8	0.03	7	0.5	0.4
$i\text{-}C_4H_8$				19.6
$i\text{-}C_4H_{10}$	{<0.2	{<0.8	{<1.9	20

	Product (wt%)		
Heavy feedstock	**Naphtha**	**Gas oil**	**Vacuum distillate**
CH_4	10.3	8	6.6
C_2H_4	25.8	19.5	19.4
C_2H_6	3.3	3.3	2.8
C_3H_6	16	14	13.9
C_4H_6	4.5	4.5	5
C_4H_8	7.9	6.4	7
BTX[a]	10	10.7	18.9
C_5 to 200 °C (not BTX)	17	10	—
Fuel Oil	3	21.8	25
$H_2+C_2H_2+C_3H_4+C_3H_8$	2.2	1.8	1.4

[a]BTX refers to mixtures of benzene, toluene, and the three xylene isomers, all of which are aromatic hydrocarbons.
Source: Crynes, B.L,. Lyle F. Albright, Loo-Fung Tan, Thermal Cracking, In: Robert A. Meyers, Editor-in-Chief, Encyclopedia of Physical Science and Technology (Third Edition), (New York, Academic Press), Pages 613-626.

Table 6.51. Typical properties for gasoline blending components

Component	RVP[a] (psi)	MON[b]	RON[b]	API
iC_4	71	92	93	
nC_4	52	92	93	
iC_5	19.4	90.8	93.2	
nC_5	14.7	72.4	71.5	
iC_6	6.4	78.4	79.2	
LSR gasoline	11.1	61.6	66.4	78.6
HSR gasoline	1	58.7	62.3	48.2
Light hydrocracker gasoline	12.9	82.4	82.8	79
Heavy hydrocracker gasoline	1.1	67.3	67.6	49
Coker gasoline	3.6	60.2	67.2	57.2

Table 6.51.   Continued

Component	RVP[a] (psi)	MON[b]	RON[b]	API
FCC Light gasoline	1.4	77.1	92.1	49.5
FCC gasoline	13.9	80.9	83.2	51.5
Reformate 94 RON	2.8	84.4	94	45.8
Reformate 98 RON	2.2	86.5	98	43.1
Alkylate	5.7	87.3	90.8	
Alkylate	4.6	95.9	97.3	70.3
Alkylate	1	88.8	89.7	

[a]RVP =Reid Vapor Pressure = the vapor pressure at 37.8°C (100°F) of petroleum products and crude oils with initial boiling point above 0°C (32°F).

[b]The Research Octane Number (RON) simulates fuel performance under low severity engine operation. The Motor Octane Number (MON) simulates more severe operation that might be incurred at high speed or high load. In practice the octane of a gasoline is reported as the average of RON and MON or R + M/2.

[c]API A specific gravity scale developed by the American Petroleum Institute (API) for measuring the relative density of various petroleum liquids, expressed in degrees.

Table 6.52.   Hydrocarbon component streams for gasoline

Stream	Producing process	Boiling range	
		°C	°F
Paraffinic butane	Distillation	0	32
Iso-pentane	Distillation	27	81
Alkylate	Alkylation	40–150	105–300
Isomerate	Isomerization	40–70	105–160
Naphtha	Distillation	30–100	85–212
Hydrocrackate olefinic	Hydrocracking	40–200	105–390
Catalytic naphtha	Catalytic cracking	40–200	105–390
Cracked naphtha	Steam cracking	40–200	105–390
Polymer aromatic	Polymerization	60–200	140–390
Catalytic reformate	Catalytic reforming	40–200	105–390

Source: Adapted from Speight, James G. 2011. Handbook of Industrial Hydrocarbon Processes, (Boston, Gulf Professional Publishing).

Table 6.53.   Research octane numbers of pure hydrocarbons

Paraffins		Naphthenes	
n-Butane	94	Cyclopentane	>100
Isobutane	>100	Cyclohexane	83
n-Pentane	61.8	Methylcyclopentane	91.3
2-Methyl-1-butane	92.3	Methylcyclohexane	74.8
n-Hexane	24.8	1,3-Dimethylcyclopentane	80.6
2-Methyl-1-pentane	73.4	1,1,3-Trimethylcyclopentane	87.7
2,2-Dimethyl-1-butane	91.8	Ethylcyclohexane	45.6
n-Heptane	0	Isobutylcyclohexane	33.7
2-Methylhexane	52	**Aromatic**	
2,3-Dimethylpentane	91.1	Benzene	-
2,2,3-Trimethylbutane	>100	Toluene	>100
n-Octane	<0	o-Xylene	-

Continued

Table 6.53. Continued

Paraffins		Naphthenes	
3,3-Dimethylhexane	75.5	*m*-Xylene	>100
2,2,4-Trimethylpentane	100	p-Xylene	>100
n-Nonane	<0	Ethylbenzene	>100
2,2,3,3-Tetramethylpentane	>100	*n*-Propylbenzene	>100
n-decane	<0	Isopropylbenzene	>100
Olefins		1-Methyl-3-ethylbenzene	>100
1-Hexene	76.4	*n*-butylbenzene	>100
1-Heptene	54.5	1-Methyl-3-isopropylbenzene	–
2-Methyl-2-hexene	90.4	1,2,3,4-Tetramethylbenzene	>100
2,3-Dimethyl-1-pentene	99.3		

Source: Fahim, Mohamed A., Taher A. Alsahhaf, Amal Elkilani. 2010. Fundamentals of Petroleum Refining, (Amsterdam, Elsevier).

Table 6.54. Naphtha production processes

Process	Primary product	Secondary process	Secondary product
Atmospheric distillation	Naphtha		Light naphtha
			Heavy naphtha
	Gas oil	Catalytic cracking	Naphtha
	Gas oil	Hydrocracking	Naphtha
Vacuum distillation	Gas oil	Catalytic cracking	Naphtha
		Hydrocracking	Naphtha
	Residuum	Coking	Naphtha
		Hydrocracking	Naphtha

Source: Adapted from Speight, James G. 2011. Handbook of Industrial Hydrocarbon Processes, (Boston, Gulf Professional Publishing).

Table 6.55. Selected hydrocarbon addition polymers

Name(s)	Formula	Monomer	Properties	Uses
Polyethylene - low density (LDPE)	$-(CH_2\text{-}CH_2)_n-$	Ethylene $CH_2 = CH_2$	Soft, waxy solid	Film wrap, plastic bags
Polyethylene - high density (HDPE)	$-(CH_2\text{-}CH_2)_n-$	Ethylene $CH_2 = CH_2$	Rigid, translucent solid	Electrical insulation, bottles, toys
Polypropylene (PP) different grades	$-[CH_2\text{-}CH(CH_3)]_n-$	Propylene	Atactic: soft, elastic solid	Similar to LDPE
		$CH_2 = CHCH_3$	Isotactic: hard, strong solid	Carpet, upholstery
Polystyrene (PS)	$-[CH_2\text{-}CH(C_6H_5)]_n-$	Styrene	Hard, rigid, clear solid	Toys, cabinets, packaging (foamed)
		$CH_2 = CHC_6H_5$	soluble in organic solvents	
cis-**Polyisoprene** - natural rubber	$-[CH_2\text{-}CH = C(CH_3)\text{-}CH_2]_n-$	Isoprene $CH_2 = CH\text{-}C(CH_3)$ CH_2	Soft, sticky solid	Requires vulcanization for practical use

Source: Adapted from Speight, James G. 2011. Handbook of Industrial Hydrocarbon Processes, (Boston, Gulf Professional Publishing).

Table 6.56. Hydrocarbon intermediates used in the petrochemical industry

Carbon number	Hydrocarbon type		
	Saturated	Unsaturated	Aromatic
1	Methane		
2	Ethane	Ethylene	
		Acetylene	
3	Propane	Propylene	
4	Butanes	n-Butenes	
		Isobutene	
		Butadiene	
5	Pentanes	Isopentenes (Isoamylenes)	
6	Hexanes	Methylpentenes	Benzene
	Cyclohexane		
7		Mixed heptenes	Toluene
8		di-Isobutylene	Xylenes
			Ethylbenzene
			Styrene
9			Cumene
12		Propylene tetramer	
		tri-Isobutylene	
18			Dodecylbenzene
6–18		n-Olefins	
11–18	n-Paraffins		

Source: Adapted from Speight, James G. 2011. Handbook of Industrial Hydrocarbon Processes, (Boston, Gulf Professional Publishing).

Table 6.57. Sources of petrochemical intermediates

Hydrocarbon	Source
Methane	Natural gas
Ethane	Natural gas
Ethylene	Cracking processes
Propane	Natural gas, catalytic reforming, cracking processes
Propylene	Cracking processes
Butane	Natural gas, reforming and cracking processes
Butene(s)	Cracking processes
Cyclohexane	Distillation
Benzene	Catalytic reforming
Toluene	Catalytic reforming
Xylene(s)	Catalytic reforming
Ethylbenzene	Catalytic reforming
Alkylbenzenes	Alkylation
>C_9	Polymerization

Source: Adapted from Speight, James G. 2011. Handbook of Industrial Hydrocarbon Processes, (Boston, Gulf Professional Publishing).

Table 6.58. Hydroconversion techniques[a]

Feedstock	Products	Type of process	Impurities removal
Naphtha	Reformer feed	Hydrotreating	S
	LPG	Hydrocracking	
Atmospheric gas oil	Diesel	Hydrotreating	S, aromatics
	Jet fuel	Mild hydrocracking	S, aromatics
	Naphtha	Hydrocracking	
Vacuum gas oil	Low sulfur fuel oil (LSFO)	Hydrotreating	S
	FCC feed	Hydrotreating	S, N, metals
	Diesel	Hydrotreating, Hydrocracking	S, aromatics
	Kerosene/Jet	Hydrotreating	S, aromatics
	Naphtha	Hydrotreating	S, aromatics
	Lube oil	Hydrotreating	S, N, aromatics
Residuum	LSFO	Hydrotreating	S
	FCC feedstock	Hydrotreating	S, N, CCR and metals
	Coker feedstock	Hydrotreating	S, CCR, metals
	Diesel	Hydrocracking	

[a]Hydroconversion is a term used to describe the processes in which hydrocarbon reacts with hydrogen. It includes hydrotreating, hydrocracking and hydrogenation.
Source: Adapted from Fahim, Mohamed A., Taher A. Alsahhaf, Amal Elkilani. 2010. Fundamentals of Petroleum Refining, (Amsterdam, Elsevier).

Table 6.59. Process parameters for hydrotreating different feedstocks

Feedstock	Naphtha	Kerosene	Gas oil	Vacuum gas oil	Residue
Boiling range, °C	70–180	160–240	230–350	350–550	>550
Operating temperature, °C	260–300	300–340	320–350	360–380	360–380
Hydrogen pressure, bar	5–10	15–30	15–40	40–70	120–160
Hydrogen consumption, wt%	0.05–0.1	0.1–0.2	0.3–0.5	0.4–0.7	1.5–2.0
LHSV, hr^{-1} [a]	4–10	2–4	1–3	1–2	0.15–0.3
H_2/HC ratio, std m^3/m^3	36–48	36–48	36–48	36–48	12–24

[a]LHSV = Liquid volumetric flow rate at 15°C (ft3/h)/Volume of catalyst (ft3)
Source: Adapted from Fahim, Mohamed A., Taher A. Alsahhaf, Amal Elkilani. 2010. Fundamentals of Petroleum Refining, (Amsterdam, Elsevier).

Table 6.60. Steam reformer outlet composition for different feeds at 850 °C, 24 bar and steam to carbon ratio of 4

Feedstock	Methane	Natural gas	LPG	Naphtha
Composition (vol%)				
CH_4	3.06	2.91	2.39	2.12
CO	12.16	12.62	13.62	14.17
CO_2	9.66	10.4	12.73	14.19
H_2	75.12	73.98	71.86	69.52
N_2	–	0.38	–	–

Source: Fahim, Mohamed A., Taher A. Alsahhaf, Amal Elkilani. 2010. Fundamentals of Petroleum Refining, (Amsterdam, Elsevier).

Table 6.61.   Coke characterization and uses from delayed coker

Type of coke	Operating condition	Feed characterization	Coke property	End use as calcinated coke
Sponge	Reflux ratio >35%	Low metal	$M < 200$	Anodes for aluminium industry
	Operating pressure 2–4 bar	Low S	$S < 2.5\%$	
		Tar residue FCC heavy dist	High density >780	
		Low to moderate asphaltene	HGI[a] ~ 100	
Shot	Low pressure	High S	High S and metal	Fuel (green)
	Low reflux ratio	High metal	Low HGI<50	
	Large drums	Low asphaltene	Low surface area	
Needle	Pressure > 4 bar	High aromatic content	Crystalline structure	Graphite electrodes
	Reflux ratio = 60–100% to maximize coke yield	Tars, FCC decant	Small needles of high conductivity	
	High temperature to reduce volatile material	Low S <0.5 wt%		
		Low ash <0.1 wt%		
		No asphaltene		

[a]HGI = Hard grove grindability index.
Source: Fahim, Mohamed A., Taher A. Alsahhaf, Amal Elkilani. 2010. Fundamentals of Petroleum Refining, (Amsterdam, Elsevier).

Table 6.62.   General properties of natural gas

Relative molar mass	16–20
Carbon content (wt%)	73–75
Hydrogen content (wt%)	27–25
Oxygen content (wt%)	0–0.4
Hydrogen-to-carbon atomic ratio	3.5–4.0
Density relative to air @15 °C	0.6–1.5
Boiling temperature (°C)	−162
Autoignition temperature (°C)	540–560
Vapor flammability limits (vol%)	5–15
Flammability limits	0.7–2.1
Lower heating/calorific value (kJ/mole)	950
Methane concentration (vol%)	80–100
Ethane concentration (vol%)	0–5
Nitrogen concentration (vol%)	0–15
Carbon dioxide concentration (vol%)	0–5
Sulfur concentration (ppmw)	0–5

Source: Adapted from Speight, James G. 2011. Chapter 3 - Fuels for Fuel Cells, In: Dushyant Shekhawat, J.J. Spivey And David A. Berry, Editor(s), Fuel Cells: Technologies for Fuel Processing, (Amsterdam, Elsevier), 2011, Pages 29-48.

Table 6.63.   General properties of the principal constituents of natural gas

	Molecular weight	Specific gravity	Vapor density air = 1	Boiling point °C	Ignition temperature °C	Flash point °C
Methane	16	0.553	0.56	−160	537	−221
Ethane	30	0.572	1.04	−89	515	−135
Propane	44	0.504	1.5	−42	468	−104
Butane	58	0.601	2.11	−1	405	−60
Pentane	72	0.626	2.48	36	260	−40
Hexane	86	0.659	3	69	225	−23
Benzene	78	0.879	2.8	80	560	−11
Heptane	100	0.668	3.5	98	215	−4
Octane	114	0.707	3.9	126	220	13
Toluene	92	0.867	3.2	161	533	4
Ethyl benzene	106	0.867	3.7	136	432	15
Xylene	106	0.861	3.7	138	464	17

Source: Speight, James G. 2011. Handbook of Industrial Hydrocarbon Processes, (Boston, Gulf Professional Publishing).

Table 6.64.   Physical properties and hydrate formation of some common natural gas components

	Hydrate structure	Molar mass (g/mol)	Hydrate press. at 0°C (MPa)	Normal boil. point (K)	Density (kg/m³)	Solubility (mol frac × 10^4)
Methane	Type I	16.043	2.603	111.6	19.62	0.46
Ethane	Type I	30.07	0.491	184.6	6.85	0.8
Propane	Type II	44.094	0.173	231.1	3.49	0.74
Isobutane	Type II	58.124	0.113	261.4	3.01	0.31
Acetylene	Type I	26.038	0.557	188.4	6.7	14.1
Ethylene	Type I	28.054	0.551	169.3	7.11	1.68
Propylene	Type II	42.081	0.48	225.5	9.86	3.52
c-propane	Type II	42.081	0.0626	240.3	1.175	2.81
CO_2	Type I	44.01	1.208	194.72	25.56	13.8
N_2	Type II	28.013	16.22	77.4	196.6	0.19
H_2S	Type I	34.08	0.099	213.5	1.5	38.1

Source: Carroll, John J. 2009. Natural Gas Hydrates (Second Edition), (Burlington, Gulf Professional Publishing).

Table 6.65.   Composition (% v/v) of natural gas

Methane	CH_4	70–90%
Ethane	C_2H_6	0–20%
Propane	C_3H_8	
Butane	C_4H_{10}	
Pentane and higher boiling hydrocarbons	$C_5H_{12}^+$	0–10%
Carbon dioxide	CO_2	0–8%
Nitrogen	N_2	0–5%
Hydrogen sulfide, carbonyl sulfide	H_2S, COS	0–5%
Oxygen	O_2	0–0.2%
Rare gases: argon, helium, neon, xenon	A, He, Ne, Xe	Trace

Source: Natural Gas Supply Association, <http://www.naturaglas.org/overview/background.asp>, Accessed 27 May 2012.

Table 6.66.   **Properties of propane and butane**

	Propane	Butane
Formula	C_3H_8	C_4H_{10}
Boiling point, °F	−44°	32°
Specific gravity – gas (air = 1.00)	1.53	2
Specific gravity – liquid (water = 1.00)	0.51	0.58
lb/gallon – liquid at 60°F	4.24	4.81
Btu/gallon – gas at 60°F	91,690	102,032
Btu/lb – gas	21,591	21,221
Btu/ft³ – gas at 60°F	2,516	3,280
Flash point, °F	−156	−96
Ignition temperature in air, °F	920–1,020	900–1,000
Maximum flame temperature in air, °F	3,595	3,615
Octane number (iso-octane = 100)	100+	92

Source: Adapted from Speight, James G. 2011. Handbook of Industrial Hydrocarbon Processes, (Boston, Gulf Professional Publishing).

Table 6.67.   **Heating value of gaseous fuels**

Type of gas	Btu/ft³ Gross	Btu/ft³ Net	kcal/m³ Gross	kcal/m³ Net	MJ/m³ Gross	MJ/m³ Net
Acetylene, commercial	1,410	1,360	12,548	12,105	52.5	50.7
Blast furnace	92	91	819	819	3.4	3.4
Blue (water), bituminous	260	239	2,314	2,127	9.7	8.9
Butane, commercial, natural gas	3,210	2,961	28,566	26,350	119.6	110.3
Butane, commercial, refinery gas	3,184	2,935	28,334	26,119	118.6	109.3
Carbureted blue, low gravity	536	461	4,770	4,102	20.0	17.2
Carbureted blue, heavy oil	530	451	4,716	4,013	19.7	16.8
Coke oven, by-product	569	509	5,064	4,530	21.2	19.0
Mapp	2,406	2,282	21,411	20,308	89.6	85.0
Natural, Alaska	998	906	8,879	8,063	37.2	33.7
Natural, Algerian LNG, Canvey	1,122	1,014	9,985	9,024	41.8	37.8
Natural, Gaz de Lacq	1,011	911	8,997	8,107	37.7	33.9
Natural, Groningen, Netherlands	875	789	7,787	7,021	32.6	29.4
Natural, Libyan LNG	1,345	1,223	11,969	10,883	50.1	45.6
Natural, North Sea, Bacton	1,023	922	9,104	8,205	38.1	34.3
Natural, Birmingham, AL	1,002	904	8,917	8,045	37.3	33.7
Natural, Cleveland, OH	1,059	959	9,424	8,534	39.4	35.7

Continued

Table 6.67   Continued

Type of gas	Btu/ft³ Gross	Btu/ft³ Net	kcal/m³ Gross	kcal/m³ Net	MJ/m³ Gross	MJ/m³ Net
Natural, Kansas City, MO	974	879	8,668	7,822	36.3	32.7
Natural, Pittsburgh, PA	1,129	1,021	10,047	9,086	42.1	38.0
Producer, Koppers-Totzek[a]	288	271	2,563	2,412	10.7	10.1
Producer, Lurgi[b]	183	167	1,629	1,486	6.8	6.2
Producer, W-G, bituminous[b]	168	158	1,495	1,406	6.3	5.9
Producer, Winkler[b]	117	111	1,041	988	4.4	4.1
Propane, commercial, natural gas	2,558	2,358	22,764	20,984	95.3	87.8
Propane, commercial, refinery gas	2,504	2,316	22,283	20,610	93.3	86.3
Sasol, South Africa	500	448	4,450	3,986	18.6	16.7
Sewage, Decatur	690	621	6,140	5,526	25.7	23.1
SNG, no methanation	853	765	7,591	6,808	31.8	28.5

[a]O_2-blown.
[b]Air-blown
Source: Adapted from Kutz, M. 1998. Mechanical Engineers Handbook, (New York, Wiley-Interscience).

Table 6.68.   Heat capacities for some gas hydrates

	Hydrate type	Heat capacity (J/g•°C)	Heat capacity (J/mol•°C)	Heat capacity (Btu/lb•°F)
Methane	I	2.25	40	0.54
Ethane	I	2.2	43	0.53
Propane	II	2.2	43	0.53
Isobutane	II	2.2	45	0.53
Ice	–	2.06	37.1	0.492

Source: Carroll, John J. 2009. Natural Gas Hydrates (Second Edition), (Burlington, Gulf Professional Publishing).

Table 6.69.   Densities of some gas hydrates at 0°C

	Hydrate type	Density (g/cm³)	Density (lb/ft³)
Methane	I	0.913	57
Ethane	I	0.967	60.3
Propane	II	0.899	56.1
Isobutane	II	0.934	58.3
CO_2	I	1.107	69.1
H_2S	I	1.046	65.3
Ice	–	0.917	57.2
Water	–	1	62.4

Source: Carroll, John J. 2009. Natural Gas Hydrates (Second Edition), (Burlington, Gulf Professional Publishing).

Table 6.70.   **Enthalpies of fusion for some gas hydrates**

	Hydrate type	Enthalpy of fusion (kJ/g)	Enthalpy of fusion (kJ/mol)	Enthalpy of fusion (Btu/lb)
Methane	I	3.06	54.2	1320
Ethane	I	3.7	71.8	1590
Propane	II	6.64	129.2	2850
Isobutane	II	6.58	133.2	2830
Ice	–	0.333	6.01	143

Source: Carroll, John J. 2009. Natural Gas Hydrates (Second Edition), (Burlington, Gulf Professional Publishing).

Table 6.71.   **Hydrate forming conditions for methane**

Temp. (°C)	Press. (MPa)	Phases	Vapor composition		
			ppm	mg/Sm^3	lb/MMCF
0	2.6	L$_A$-H-V	270	20.6	12.6
2.5	3.31	L$_A$-H-V	260	20	12.3
5	4.26	L$_A$-H-V	260	19.4	12
7.5	5.53	L$_A$-H-V	250	18.9	11.6
10	7.25	L$_A$-H-V	240	18.6	11.4
12.5	9.59	L$_A$-H-V	240	18.5	11.4
15	12.79	L$_A$-H-V	250	18.7	11.5
17.5	17.22	L$_A$-H-V	250	19.4	11.9
20	23.4	L$_A$-H-V	270	20.2	12.4
22.5	32	L$_A$-H-V	280	21.1	13
25	44.1	L$_A$-H-V	290	21.7	13.4
27.5	61.3	L$_A$-H-V	290	22	13.5
30	85.9	L$_A$-H-V	290	21.7	13.3

Source: Carroll, John J. 2009. Natural Gas Hydrates (Second Edition), (Burlington, Gulf Professional Publishing).

Table 6.72.   **Amount of methane per cubic meter for hydrate, pipeline gas, and liquid**

	Pressure (MPa)	Temperature (°C)	Amount (kg)	Comment
Hydrate	2.6	0	115	1 m^3 hydrate contains 170 Sm^3 of gas
Pipeline Gas	7	27	50	Typical pipeline conditions
Liquid	0.101	−161	422	At the normal boiling pt
	Pressure (psia)	Temperature (°F)	Amount (lb)	Comment
Hydrate	377	32	254	1 ft^3 hydrate contains 170 SCF of gas
Pipeline Gas	1015	81	110	Typical pipeline conditions
Liquid	14.7	−258	930	At the normal boiling pt

Source: Carroll, John J. 2009. Natural Gas Hydrates (Second Edition), (Burlington, Gulf Professional Publishing).

Table 6.73.   Summary of methane inventories in marine gas hydrates

Study	CH$_4$ Mass/10^6 Tg
McIver, 1981	2.3
Dobrynin et al., 1981	5,500
MacDoland, 1990	10
Gornitz and Fung, 1994	18
Harvey and Huang, 1995	32
Kvenvolden and Lorenson, 2001	10
Buffet and Archer, 2004	6.7
Milkov, 2004	2[a]
Klauda and Sandler, 2005	74

References

Buffett, B. and D. Archer. 2004. Global inventory of methane clathrate: sensitivity to changes in the deep ocean. Earth and Planetary Science Letters 227(3-4): 185-199.

Dobrynin, V. M., Y. P. Korotajev and D. V. Plyuschev. 1981. Gas Hydrates: A possible energy resource. Long-Term Energy Resources. R. F. Meyer and J. C. Olson. Boston, MA., Pitman. 1: 727-729.

Gornitz, V. and I. Fung. 1994. Potential Distribution Of Methane Hydrates In The Worlds Oceans. Global Biogeochemical Cycles 8(3): 335-347.

Harvey, L. D. D. and Z. Huang. 1995. Evaluation Of The Potential Impact Of Methane Clathrate Destabilization On Future Global Warming. Journal Of Geophysical Research-Atmospheres 100(D2): 2905-2926.

MacDonald, G. J. 1990. Role Of Methane Clathrates In Past And Future Climates. Climatic Change 16(3): 247-281.

McIver, R. D. 1981. Gas Hydrates. Long-Term Energy Resources. R. F. Meyer and J. C. Olson. Boston, MA., Pitman. 1: 713-726.

Milkov, A. V. 2000. Worldwide distribution of submarine mud volcanoes and associated gas hydrates. Marine Geology 167(1-2): 29-42.

Klauda, J. B. and S. I. Sandler. 2005. Global distribution of methane hydrate in ocean sediment. Energy & Fuels 19(2): 459-470.

Kvenvolden, K. A. and T. D. Lorenson. 2001. The Global Occurrence of Natural Gas Hydrates. Natural Gas Hydrates: Occurrence, Distribution, and Detection. C. K. Paull and W. P. Dillon. Washington DC, American Geophysical Union. 124: 3-18.

[a]midpoint value.

Adapted from U.S. Environmental Protection Agency,
<http://www.epa.gov/outreach/pdfs/Methane-and-Nitrous-Oxide-Emissions-From-Natural-Sources.pdf>.

Table 6.74.   Typical sulfur compounds and odorants in natural gas and LPG

Name	Formula	Structure	Boiling point (°C)
Hydrogen sulfide	H_2S	H——S——H	−60
Carbonyl sulfide	COS	O══C══S	−50
Methyl mercaptan	CH_3SH	H_3C——S——H	6
Ethyl mercaptan	C_2H_5SH	H_3C——H_2C——S——H	36
Dimethyl sulfide	C_2H_6S	H_3C——S——CH_3	38
t-Butyl mercaptan	$C_4H_{10}SH$	CH_3 $\|$ H_3C——C——S——H $\|$ CH_3	64
Tetrahydrothiophene	C_4H_8S		120

Source: Adapted from Gangwal, Santosh K. 2011. Chapter 11 - Desulfurization for Fuel Cells, In: Dushyant Shekhawat, J.J. Spivey And David A. Berry, Editor(s), Fuel Cells: Technologies for Fuel Processing, (Amsterdam, Elsevier), Pages 317-360.

Table 6.75. Liquified natural gas (LNG) trade volumes, 2010, (Mmtpa)[a]

Importer / Exporter	Argentina	Belgium	Brazil	Canada	Chile	China	Dom Rep	France	Greece	India	Italy	Japan	Korea	Kuwait	Mexico	Portugal	Spain	Tawain	Turkey	UAE	UK	US	Total
Algeria	-	-	-	-	0.18	-	-	4.77	0.71	-	1.23	0.06	-	-	-	-	3.54	-	2.83	-	0.95	-	14.3
Australia	-	-	-	-	-	3.90	-	-	-	0.06	-	13.28	0.91	0.06	-	-	-	0.83	-	-	-	-	19.0
Belgium	-	-	0.06	-	-	-	-	-	-	-	-	0.06	0.07	0.07	-	-	0.06	-	0.07	-	-	-	0.4
Brunei	-	-	-	-	-	-	-	-	-	-	-	5.93	0.73	-	-	-	-	-	-	-	-	-	6.7
Egypt	-	0.13	-	-	0.36	-	-	0.53	0.06	0.06	0.44	0.43	0.81	0.21	0.12	-	2.11	0.06	0.19	-	0.12	1.49	7.1
Eq. Guinea	-	-	0.02	-	1.17	0.07	-	-	-	0.12	0.06	0.54	1.45	0.19	-	-	-	0.45	-	-	-	-	4.1
Indonesia	-	-	-	-	-	1.88	-	-	-	-	-	12.75	5.54	-	1.38	-	-	1.98	-	-	-	-	23.5
Libya	-	-	-	-	-	-	-	-	-	-	-	-	-	-	-	-	0.25	-	-	-	-	-	0.2
Malaysia	-	-	-	-	-	1.19	-	-	-	-	-	13.89	4.96	0.13	-	-	-	2.96	-	-	-	-	23.1
Nigeria	-	0.06	0.68	-	-	0.20	-	2.81	-	0.25	-	0.58	0.87	0.06	1.73	2.06	5.71	0.81	1.08	-	0.31	0.88	18.1
Norway	-	0.06	-	0.06	-	-	-	0.33	-	-	0.13	-	0.13	-	-	-	1.33	0.06	0.12	-	0.70	0.55	3.5
Oman	-	0.08	-	-	-	-	-	-	-	-	-	2.86	4.64	0.71	-	-	0.12	0.39	-	-	-	-	8.7
Peru	0.18	-	0.06	0.12	-	-	-	-	-	-	-	-	0.07	-	0.18	-	0.49	-	-	-	-	0.34	1.3
Qatar	-	4.51	0.38	0.18	0.12	1.26	-	1.77	0.03	8.05	4.56	7.91	7.50	-	0.81	0.06	4.19	2.89	1.46	0.12	10.46	0.96	57.4
Russia	-	-	-	-	-	0.38	-	-	-	-	-	6.23	3.39	-	-	-	-	0.51	-	-	-	-	10.6
Trinidad	1.10	0.06	0.70	1.18	0.37	0.05	0.59	0.24	0.06	0.48	0.24	0.11	0.66	0.29	-	0.13	2.51	0.37	0.19	-	1.29	4.52	15.1
UAE	-	-	0.04	-	-	-	-	-	-	-	-	5.10	0.19	0.18	-	-	-	0.33	-	-	-	-	5.8
US	-	-	0.06	-	-	-	-	-	-	-	-	0.63	0.26	-	-	-	0.09	-	-	-	0.14	-	1.2
Yemen	-	-	-	-	0.06	0.47	-	0.07	-	0.28	-	0.12	1.88	0.14	0.13	-	0.13	-	-	-	0.20	0.82	4.3
Re-exports	-	-0.33	-	-	-	-	-	-	-	-	-	-	-	-	-	-	-	-	-	-	-	-0.74	-1.1
Total	1.3	4.6	2.0	1.5	2.3	9.4	0.6	10.5	0.9	9.3	6.7	70.5	34.1	2.1	4.3	2.2	20.5	11.6	5.9	0.1	14.2	8.8	224

[a]Million metric tons per annum.

Source: International Gas Union, World LNG Report 2010.

Table 6.76. Liquified natural gas (LNG) trade volumes between regions, 2010, (Mmtpa)[a]

Importing Region	Europe	Asia-Pacific	Middle East	N. America	S. America	Total
Exporting Region						
Africa	12.2	5.3	0.3	2.6	1.9	22.2
Asia-Pacific	-	81.4	0.3	1.6	-	83.0
Europe	2.4	0.3	0.1	0.6	0.1	3.5
MENA	45.4	45.4	1.3	4.5	1.3	98.0
N. America	0.2	0.9	-	(0.5)	0.1	0.6
S. America	5.3	1.7	0.3	6.2	2.8	16.5
Total	66.5	135.1	2.2	14.9	6.1	223.8

[a]Million metric tons per annum.
Source: International Gas Union, World LNG Report 2010.

Table 6.77. Exports of liquefied natural gas (LNG) from Qatar, 1997–2010

Year	Qatar Exports	Global Exports	Percent of Global Trade
1997	6	3,950	0.10%
1998	68	3,992	1.70%
1999	311	4,415	7.10%
2000	517	4,933	10.50%
2001	596	5,194	11.50%
2002	676	5,370	12.60%
2003	669	5,912	11.30%
2004	860	6,375	13.50%
2005	979	6,843	14.30%
2006	1,110	7,627	14.60%
2007	1,353	8,260	16.40%
2008	1,488	8,550	17.40%
2009	1,792	8,810	20.30%
2010	2,742	10,752	25.50%

Table 6.78. Estimated shale gas technically recoverable resources in 32 countries, compared to conventional gas resources

	2009 Natural Gas Market (trillion cubic feet, dry basis)			Proved natural gas reserves (trillion cubic feet)	Technically recoverable shale gas resources (trillion cubic feet)
	Production	Consumption	Imports (Exports)		
France	0.03	1.73	98%	0.2	180
Germany	0.51	3.27	84%	6.2	8
Netherlands	2.79	1.72	−62%	49	17

Table 6.78. Continued

	2009 Natural Gas Market (trillion cubic feet, dry basis)			Proved natural gas reserves (trillion cubic feet)	Technically recoverable shale gas resources (trillion cubic feet)
	Production	Consumption	Imports (Exports)		
Norway	3.65	0.16	−2156%	72	83
U.K.	2.09	3.11	33%	9	20
Denmark	0.3	0.16	−91%	2.1	23
Sweden	-	0.04	100%		41
Poland	0.21	0.58	64%	5.8	187
Turkey	0.03	1.24	98%	0.2	15
Ukraine	0.72	1.56	54%	39	42
Lithuania	-	0.1	100%		4
Other Europe	0.48	0.95	50%	2.71	19
United States	20.6	22.8	10%	272.5	862
Canada	5.63	3.01	−87%	62	388
Mexico	1.77	2.15	18%	12	681
China	2.93	3.08	5%	107	1,275
India	1.43	1.87	24%	37.9	63
Pakistan	1.36	1.36	-	29.7	51
Australia	1.67	1.09	−52%	110	396
South Africa	0.07	0.19	63%	-	485
Libya	0.56	0.21	−165%	54.7	290
Tunisia	0.13	0.17	26%	2.3	18
Algeria	2.88	1.02	−183%	159	231
Morocco	0	0.02	90%	0.1	11
Western Sahara	-	-		-	7
Mauritania	-			1	0
Venezuela	0.65	0.71	9%	178.9	11
Colombia	0.37	0.31	−21%	4	19
Argentina	1,46	1.52	4%	13.4	774
Brazil	0.36	0.66	45%	12.9	226
Chile	0.05	0.1	52%	3.5	64
Uruguay	-	0	100%		21
Paraguay	-	-			62
Bolivia	0.45	0.1	−346%	26.5	48
Total of above areas	53.1	55	−3%	1,274	6,622
Total world	106.5	106.7	0%	6,609	

Source: U.S. Department of Energy, Energy Information Administration, World Shale Gas Resources: An Initial Assessment of 14 Regions Outside the United States, April 5, 2011, <http://www.eia.gov/analysis/studies/worldshalegas/>.

Table 6.79.   Comparison of data for gas shales in the United States

Gas Shale Basin	Barnett	Fayetteville	Haynesville	Marcellus	Woodford	Antrim	New Albany
Estimated Basin Area, square miles	5000	9000	9000	95000	11000	12000	43500
Depth, ft	6, 500 - 8,500	1,000 - 7,000	10,500 - 13,500	4,000 - 8,500	6,000 - 11,000	600 - 2,200	500 - 2,000
Net Thickness, ft	100-600	20 - 200	200-300	50-200	120-220	70-120	50-100
Depth to Base of Treatable Water, ft	~1200	~500	~400	~850	~400	~300	~400
Rock Column Thickness between Top of Pay and Bottom of Treatable Water, ft	5,300 - 7,300	500 - 6,500	10,100 - 13,100	2,125 - 7650	5,600 - 10,600	300 - 1,900	100 - 1,600
Total Organic Carbon, %	4.5	4.0-9.8	0.5-4.0	3-12	1-14	1-20	1-25
Total Porosity, %	4-5	2-8	8-9	10	3-9	9	10-14
Gas Content, scf/ton	300-350	60-220	100-330	60-100	200-300	40-100	40-80
Water Production, Barrels water/day	N/A	N/A	N/A	N/A	N/A	5-500	5-500
Well spacing, acres	60-160	80-160	40-560	40-160	640	40-160	80
Original Gas-In-Place, tcf	327	52	717	1500	23	76	160
Technically Recoverable Resources, tcf	44	41.6	251	262	11.4	20	19.2

Mcf = thousands of cubic feet of gas; scf = standard cubic feet of gas; tcf = trillions of cubic feet of gas; N/A = Data not available
Source: Ground Water Protection Council. 2009. Modern Shale Gas Development in the United States: A Primer, Prepared for U.S. Department of Energy, Office of Fossil Energy and National Energy Technology Laboratory.

Table 6.80. Estimates of the ultimately recoverable resources (URR)[b] of unconventional gas[a]

Resource	Year	Author	URR (billion barrels) unless stated	Mean URR (Exajoules)	Error (Exajoules)
Tight gas (coal seam)	1985	Edmonds and Reilly	370-860	3760	1500
Tight gas (shale gas)	1986	Edmonds and Reilly	170-i260	1320	280
Tight gas	2002	Bentley, (2002a) and Bentley, (2002b)	180	1100	0
Unconventional gas	2001	Laherrere (2001)	2500 (trillion cubic feet)	2650	0

Authors of studies cited: Edmonds, J., Reilly, J.M., 1985. Global Energy: Assessing the Future. Oxford University Press; R.W. Bentley Global oil & gas depletion: an overview, Energy Policy, 30 (3) (2002), pp. 189–205; R.W. Bentley, Oil forecasts, past and present, Energy Exploration & Exploitation, 20 (2002), pp. 481–491; Laherrere, J., 2001. Estimates of oil reserves. EMF/IEA/IEW meeting IIASA. Laxenburg, Austria, June 19, 2001.
[a]Adapted from Michael Dale, Meta-analysis of non-renewable energy resource estimates, Energy Policy, Volume 43, April 2012, Pages 102-122.
[b]URR is an estimate of the total amount of a resource that will ever be produced.

Table 6.81. Miles of natural gas pipeline in the United States, 1990-2009 (thousands)

	Total	Gathering	Transmission	Distribution
1990	1,189.20	32.4	292.2	864.6
1991	1,208.20	32.7	294.1	891.4
1992	1,216.10	32.6	291.5	892
1993	1,277.20	32.1	293.3	951.8
1994	1,288.40	31.3	301.5	955.6
1995	1,277.60	30.9	296.9	949.8
1996	1,323.60	29.6	292.2	1,001.80
1997	1,331.80	34.7	294	1,003.10
1998	1,351.20	29	300.1	1,022.10
1999	1,340.30	31.8	301	1,007.50
2000	1,369.30	27.1	296.6	1,045.60
2001	1,373.50	20.1	287.1	1,066.30
2002	1,411.40	22.3	309.5	1,079.60
2003	1,424.20	22.3	304	1,097.90
2004	1,462.30	23.7	298.9	1,139.80
2005	1,437.50	23.3	296.4	1,117.80
2006	1,534.30	19.9	300.4	1,214.00
2007	1,520.20	19.1	300	1,201.10
2008	1,525.00	20	299	1,206.00
2009	1,526.40	19.7	297.2	1,209.50

Source: American Gas Association

Electricity

Figures

Figure 7.1. Electromagnetic induction. Electromagnetic induction is the production of an electric current across a conductor moving through a magnetic field. Voltage is induced in a conductor when it moves at a right angle to the magnetic field. Induction underlies the operation of generators, transformers, induction motors, electric motors, synchronous motors, and solenoids.
Source: Cadena, Richard. 2009. Electricity for the Entertainment Electrician and Technician, (Boston, Focal Press).

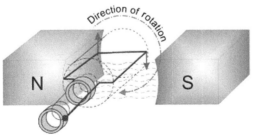

Figure 7.2. Basic concept of the alternating current (AC) generator. As the loop of wire spins, it cuts through the magnetic lines of flux.
Source: Cadena, Richard. 2009. Electricity for the Entertainment Electrician and Technician, (Boston, Focal Press).

Handbook of Energy. http://dx.doi.org/10.1016/B978-0-08-046405-3.00007-3

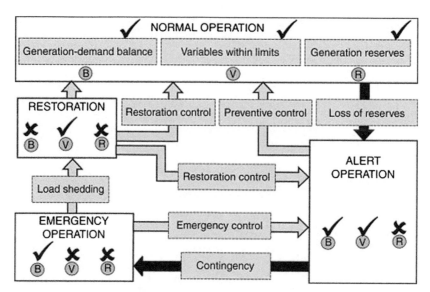

Figure 7.3. Operating states of an electrical power system.
Source: Carta, J.A. 2012. Wind Power Integration, In: Ali Sayigh, Editor-in-Chief, Comprehensive Renewable Energy, (Oxford, Elsevier), Pages 569-622.

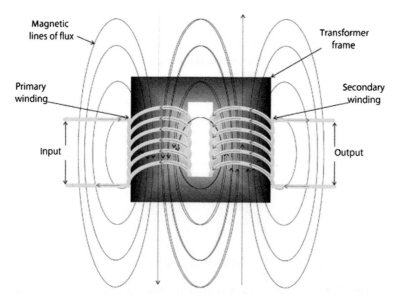

Figure 7.4. A simple transformer, a device used to transfer electric energy from one circuit to another.
Source: Cadena, Richard. 2009. Electricity for the Entertainment Electrician and Technician, (Boston, Focal Press).

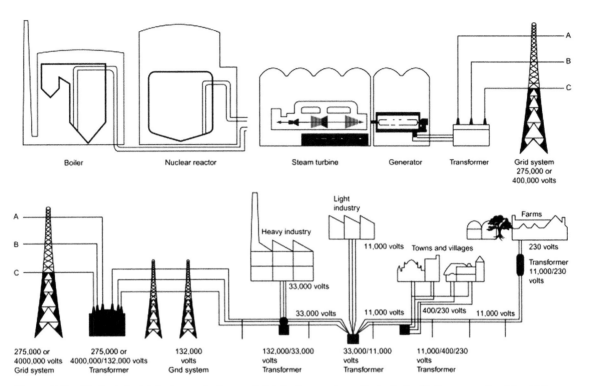

Figure 7.5. Typical electricity transmission and distribution system, illustrated with the example of the UK public electricity supply.
Source: McEvoy, Augustin, Tom Markvart and Luis Castaner. 2012. Practical Handbook of Photovoltaics (Second Edition), (Boston, Academic Press).

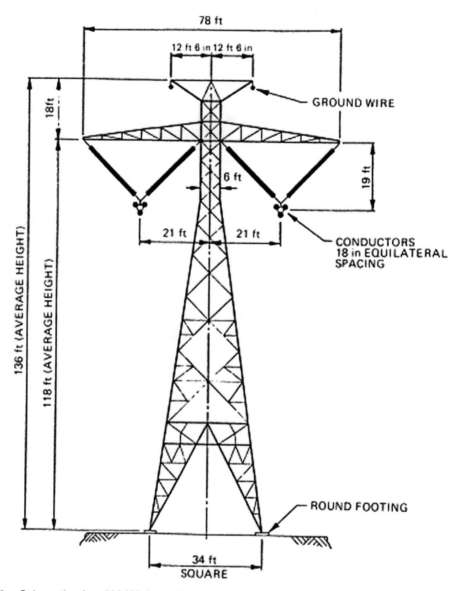

Figure 7.6. Schematic of a ±500 kV electricity transmission tower.
Source: Hingorani, Narain G. and A. Figueroa. 2003. Direct Current Power Transmission, High Voltage, In: Robert A. Meyers, Editor-in-Chief, Encyclopedia of Physical Science and Technology (Third Edition), (New York, Academic Press), Pages 501-522.

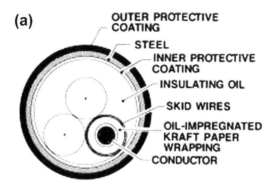

(a)

OUTER PROTECTIVE COATING

STEEL

INNER PROTECTIVE COATING

INSULATING OIL

SKID WIRES

OIL-IMPREGNATED KRAFT PAPER WRAPPING

CONDUCTOR

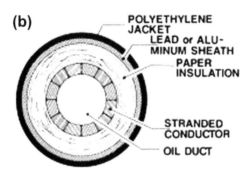

(b)

POLYETHYLENE JACKET

LEAD or ALU-MINUM SHEATH

PAPER INSULATION

STRANDED CONDUCTOR

OIL DUCT

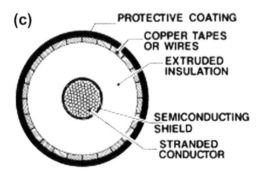

(c)

PROTECTIVE COATING

COPPER TAPES OR WIRES

EXTRUDED INSULATION

SEMICONDUCTING SHIELD

STRANDED CONDUCTOR

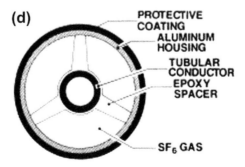

(d)

PROTECTIVE COATING

ALUMINUM HOUSING

TUBULAR CONDUCTOR

EPOXY SPACER

SF$_6$ GAS

Figure 7.7. Cross-sections of common power transmission cables.
Source: Annestrand, S.A. 2003. Power Transmission, High-Voltage, In: Robert A. Meyers, Editor-in-Chief, Encyclopedia of Physical Science and Technology (Third Edition), (New York, Academic Press), Pages 35-55.

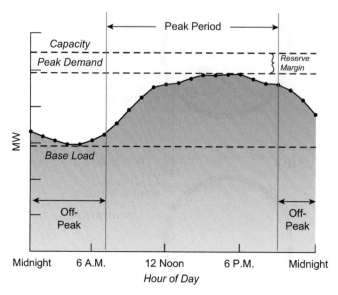

Figure 7.8. Typical load curve for electricity demand.

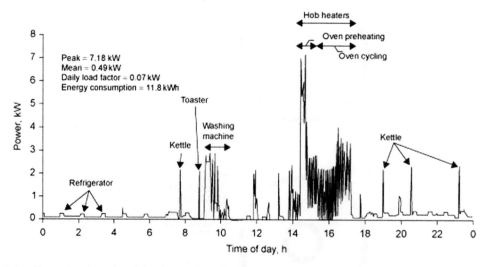

Figure 7.9. Example of an electricity demand profile from an individual household recorded on a 1-minute time basis.
Source: Flick, Tony and Justin Morehouse. 2011. Securing the Smart Grid, (Boston, Syngress).

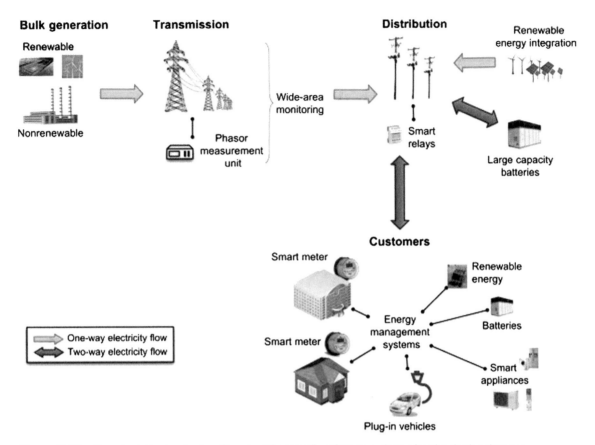

Figure 7.10. Smart grid technologies. Smart grid generally refers to a class of technologies that use computer-based remote control and automation to increase the efficiency, reliability, and security of electricity delivery systems.

Source: Alvaro A. Cárdenas, Reihaneh Safavi-Naini. 2012. Security and Privacy in the Smart Grid, In: Sajal K. Das, Krishna Kant and Nan Zhang, Security and Privacy in the Smart Grid, Handbook on Securing Cyber-Physical Critical Infrastructure, (Boston, Morgan Kaufmann), Pages 637-654.

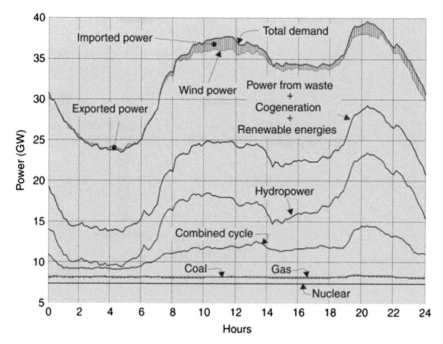

Figure 7.11. Evolution of the generation structure of the Spanish mainland electricity system, 24 February 2010.
Source: Carta, J.A. 2012. Wind Power Integration, In: Ali Sayigh, Editor-in-Chief, Comprehensive Renewable Energy, (Oxford, Elsevier), Pages 569-622.

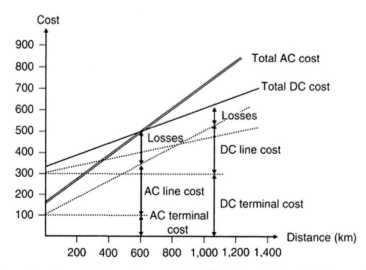

Figure 7.12. Cost as a function of distance for equivalent high voltage AC and DC transmission links.
Source: Bayliss, C.R. and B.J. Hardy. 2012. Transmission and Distribution Electrical Engineering (Fourth Edition), (Oxford, Newnes).

Charts

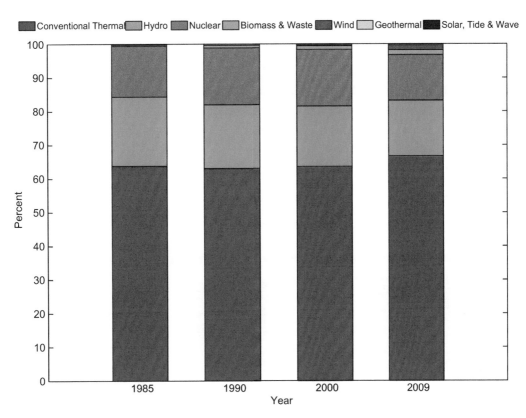

Chart 7.1. World net electricity generation by fuel, 1985, 1990, 2000, 2009.
Source: Data from United States Department of Energy, Energy Information Administration, International Energy Statistics, <http://www.eia.gov/countries/data.cfm>.

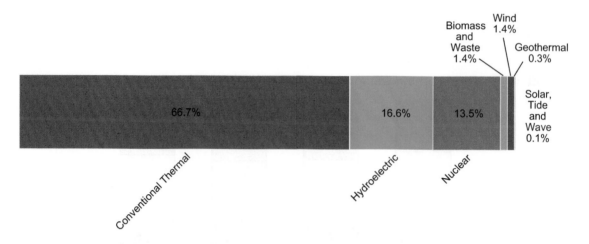

Chart 7.2. Composition of global net electricity generation by fuel, 2009.
Source: Data from United States Department of Energy, Energy Information Administration, International Energy Statistics, <http://www.eia.gov/countries/data.cfm>.

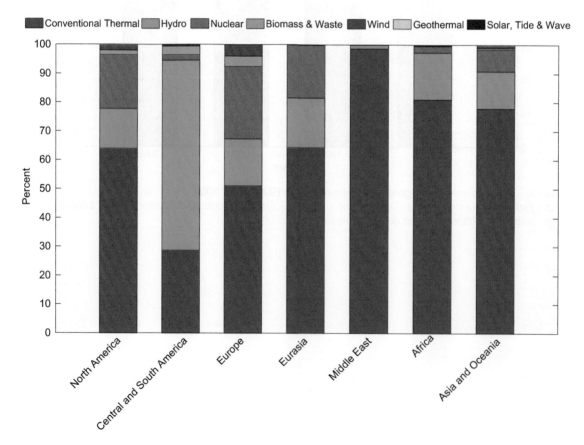

Chart 7.3. Composition of global net electricity generation by fuel and region, 2009.
Source: Data from United States Department of Energy, Energy Information Administration, International Energy Statistics, <http://www.eia.gov/countries/data.cfm>.

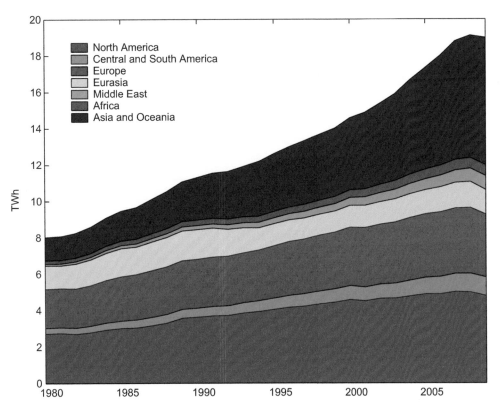

Chart 7.4. Net generation of electricity by region, 1980-2010.
Source: Data from United States Department of Energy, Energy Information Administration, International Energy Statistics, <http://www.eia.gov/countries/data.cfm>.

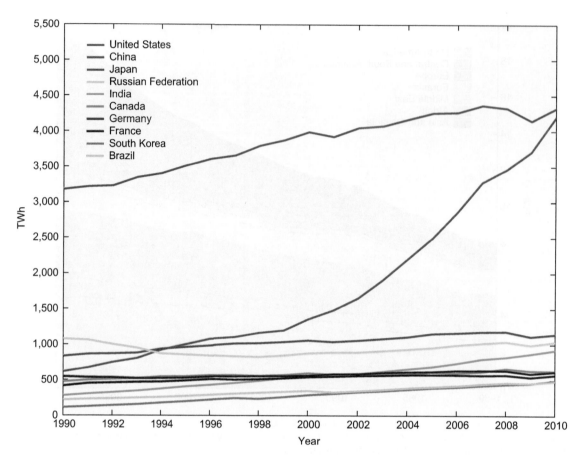

Chart 7.5. Net generation of electricity, top 10 nations, 1990-2010.
Source: Data from United States Department of Energy, Energy Information Administration, International Energy Statistics, <http://www.eia.gov/countries/data.cfm>. BP, Statistical Review of World Energy 2012, <http://www.bp.com/sectionbodycopy.do?categoryId=7500&contentId=7068481>

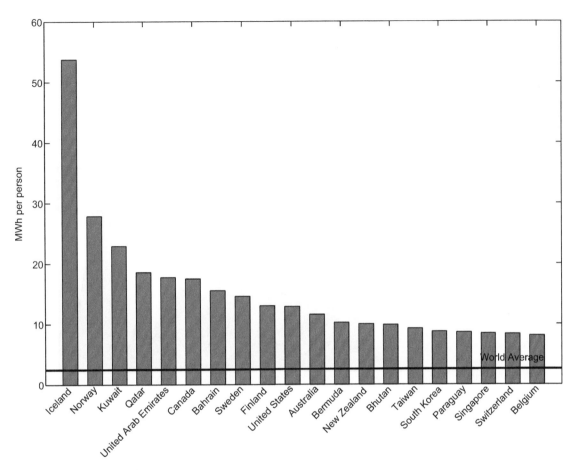

Chart 7.6. Electricity generation per capita, top 20 nations, 2009.
Source: Data from United States Department of Energy, Energy Information Administration, International Energy Statistics, <http://www.eia.gov/countries/data.cfm>. BP, Statistical Review of World Energy 2012, <http://www.bp.com/sectionbodycopy.do?categoryId=7500&contentId=7068481>

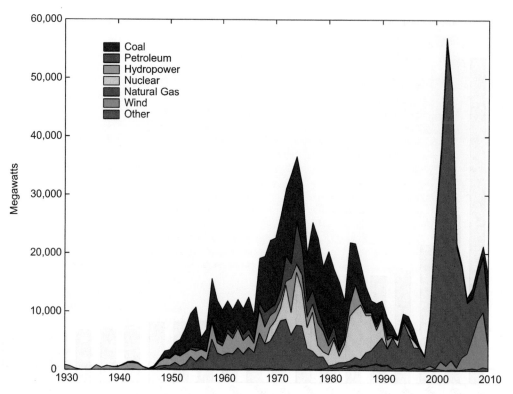

Chart 7.7. Electric power generation in the United States by initial year of operation and fuel type. The 'other' category includes solar, biomass, and geothermal generators, as well as landfill gas, municipal solid waste, and a variety of small-magnitude fuels such as byproducts from industrial processes (e.g., black liquor, blast furnace gas).
Source: Data from United States Department of Energy, Energy Information Administration, <http://www.eia.gov/todayinenergy/detail.cfm?id=1830>.

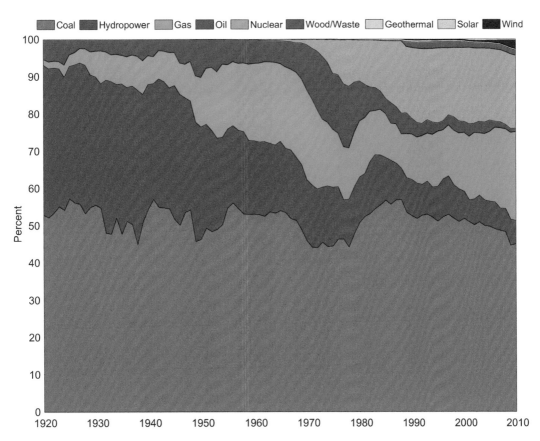

Chart 7.8. Generation of electricity in the United States, by source, 1920-2010.
Source: Data from United States Department of Energy, Energy Information Administration.

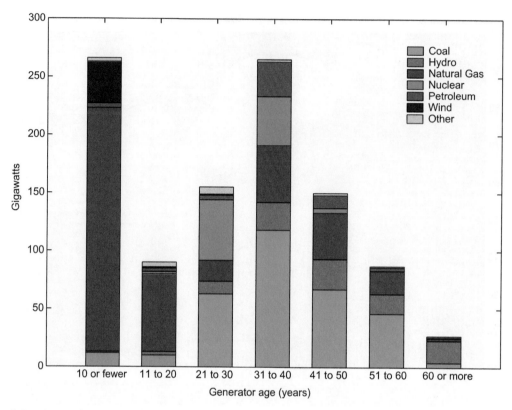

Chart 7.9. Age and capacity of existing electric generators by fuel type in the United States, 2010. *Source: Data from United States Department of Energy, Energy Information Administration, <http://www.eia.gov/todayinenergy/detail.cfm?id=1830>.*

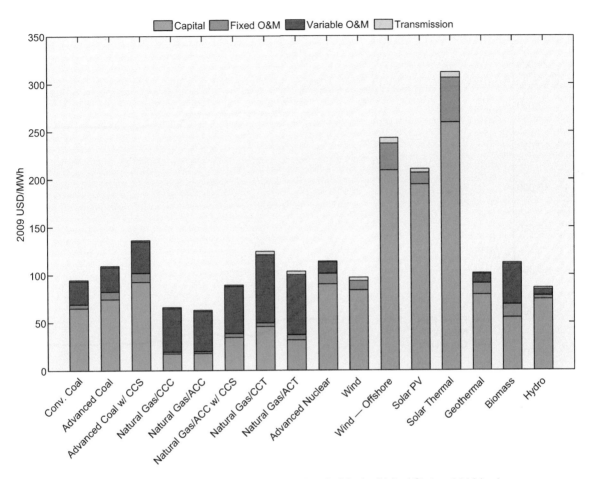

Chart 7.10. Levelized cost of new electricity generation in 2016 in the United States. 2016 is chosen as the base year because the long lead times needed for some technologies means that they could not be brought on line prior to 2016 unless they were already under construction. CCS = carbon capture and storage; ACT = advanced combustion turbine; ACC = advanced combined cycle; CCT = conventional combustion turbine. *Source: Data from United States Department of Energy, Energy Information Administration, 2016 Levelized Cost of New Generation Resources from the Annual Energy Outlook 2010, <www.eia.gov/oiaf/aeo/pdf/2016levelized_costs_aeo2010.pdf>.*

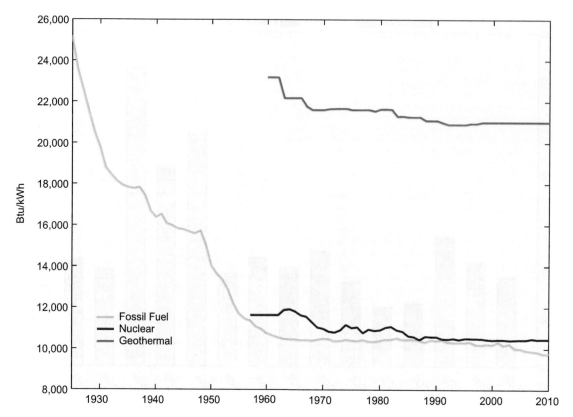

Chart 7.11. The heat rate of thermal power generation in the United states. Heat rate is the amount of fuel energy required by a power plant to produce one kilowatt-hour of electrical output.
Source: Data from United States Department of Energy, Energy Information Administration.

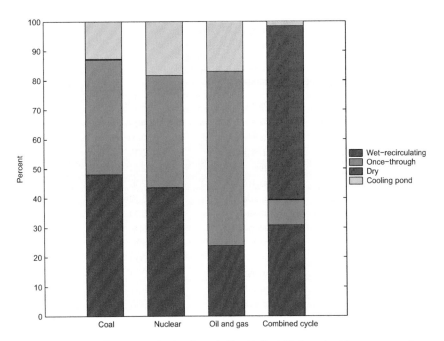

Chart 7.12. Distribution of cooling tower technology in the United States electric power sector.
Source: Data from Fthenakis, Vasilis, Hyung Chul Kim. 2010. Life-cycle uses of water in U.S. electricity generation, Renewable and Sustainable Energy Reviews, Volume 14, Issue 7, Pages 2039-2048.

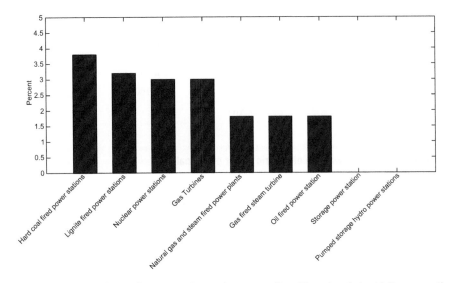

Chart 7.13. Frequency of unplanned power outages for conventional baseload electricity generation. Storage power stations and pumped storage are assumed to be zero.
Source: Data from Holttinen, H. Meibom, P. Orths, A. van Hulle, F. Lange, B. O'Malley, M. Pierik, J. Ummels, B. Tande, J.O. Estanqueiro, A. Matos, M. Gomez, E. Söder, L. Strbac, G. Shakoor, A. Ricardo, J. Smith, J.C. Milligan, M. Ela, E. 2009. Design and Operation of Power Systems with Large Amounts of Wind Power. IEA WIND Task 25, Phase One 2006–2008. (Vuorimiehentie, Finland, VTT).

Tables

Table 7.1. World's largest electricity generating facilities, 2011

Rank	Facility	Country	Capacity (MW)	Fuel type
1	Three Gorges	China	18,460	Hydroelectricity
2	Itaipu	Brazil/Paraguay	14,750	Hydroelectricity
3	Simon Bolivar (Guri)	Venezuela	10,055	Hydroelectricity
4	Tucurui	Brazil	8,370	Hydroelectricity
5	Kashiwazaki-Kariw	Japan	8,206	Nuclear
6	Bruce	Canada	6,830	Nuclear
7	Sayanao-Shushenskaya	Russia	6,500	Hydroelectricity
8	Grand Coulee	United States	6,809	Hydroelectricity
9	Longtan	China	6,426	Hydroelectricity
10	Krasnoyarsk	Russia	6,000	Hydroelectricity
11	Zaporizhzhya	Ukraine	6,000	Nuclear
12	Poryong	South Korea	5,954	Coal
13	Ulchin Nuclear	South Korea	5,900	Nuclear
14	Yonggwang	South Korea	5,901	Nuclear
15	Taichung	Taiwan	5,834	Coal
16	Gravelines	France	5,706	Nuclear
17	Futtsu	Japan	5,598	Natural gas
18	Surgut-2	Russia	5,597	Oil
19	Al-Shuaibah	Saudi Arabia	5,600	Oil
20	Paluel	France	5,528	Nuclear
21	Cattenom	France	5,448	Nuclear
22	Churchill Falls	Canada	5,429	Hydroelectricity
23	Bourassa (La Grande-II)	Canada	5,328	Hydroelectricity
24	Belchatow	Poland	5,298	Lignite
25	Waigaoqiao	China	5,160	Coal

Source: Platts World Electric Power Plants Database, December 2011.

Table 7.2. Order of magnitude comparison of electric power plant scales

Type	kW
Nuclear Plant	1,300,000
Coal Plant	500,000
Gas Turbine Combine cycle	250,000
Gas Turbine Sigel Cycle	100,000
Industrial Cogeneration Plan	50,000
Wind Turbine	1,000
Micro-turbine	50
Residential Fuel Cell	7
Household Solar Panel	3
Stirling Engine	1

Source: Adapted from Union of the Electricity Industry (EURELECTRIC). 2003. Efficiency in Electricity Generation (Essen, VGB PowerTech e.V.).

Table 7.3. **Net generation of electricity, top 20 nations, 1990-2010 (Terrawatt-hours)**

Country	1990	2000	2010	Rank 2010	Rank 1990	Change 2000-10
US	3,185.4	3,990.5	4,325.9	1	1	8.4%
China	621.2	1,355.6	4,206.5	2	4	210.3%
Japan	841.1	1,057.9	1,145.3	3	3	8.3%
Russian Federation	1,082.2	877.8	1,036.8	4	2	18.1%
India	284.2	554.7	922.2	5	10	66.2%
Canada	478.2	599.2	629.9	6	6	5.1%
Germany	549.9	564.5	621.0	7	5	10.0%
France	420.2	540.8	573.2	8	7	6.0%
South Korea	118.5	290.4	497.2	9	20	71.2%
Brazil	222.8	348.9	484.8	10	11	38.9%
United Kingdom	319.7	377.1	381.2	11	8	1.1%
Spain	164.6	232.0	300.4	12	14	29.5%
Italy	216.9	276.6	298.2	13	12	7.8%
Mexico	122.4	204.4	270.0	14	18	32.1%
South Africa	165.4	210.7	268.1	15	13	27.3%
Taiwan	90.2	184.9	247.0	16	21	33.6%
Australia	156.0	212.3	245.3	17	15	15.5%
Iran	57.7	119.3	226.1	18	29	89.4%
Saudi Arabia	70.1	126.2	214.0	19	25	69.5%
Turkey	57.5	124.9	210.2	20	30	68.3%
Total World	11,860.6	15,394.2	21,325.1			38.5%

Source: BP, Statistical Review of World Energy 2011.

Table 7.4.   Typical efficiency of electricity generating technologies

Technology	Low	High	Greater than
Steam turbine fuel-oil	38	44	
Steam turbine coal-fired	39	47	
Pulv. coal boilers w/ ultra-critical steam		47	
Atmospheric Circulating Fluidized Bed Combustion			40
Pressurised Fluidized Bed Combustion (PFBC)			40
Coal fired IGCC			43
Large gas turbine (MW range)		39	
Large gas fired CCGT		58	
Biomass and biogas	30	40	
Biomass gasification combined cycle		40	
Waste-to-electricity	22	38	
Nuclear	33	36	
Geothermal		15	
Parabolic trough	14	18	
Power tower	14	19	
Dish stirling	18	23	
Photovoltaic cells		15	
Wind turbine		35	
Large hydro		95	
Small hydro		90	
Tidal		90	
Proton Exchange Membrane Fuel Cell (PEMFC)		40	
Phosphoric Acid Fuel Cell (PAFC)		40	
Solid Oxide Fuel Cell (SOFC)	46	>60	
Molten Carbonate Fuel Cell (MCFC)	52	65	
Small and Microturbines	17	22	
Diesel engine as decentr. CHP unit	22	40	

Source: Adapted from Union of the Electricity Industry (EURELECTRIC). 2003. Efficiency in Electricity Generation (Essen, VGB PowerTech e.V.).

Table 7.5.   Current (2010) state of development of electricity-generating technologies (representative characteristics)

Technology	Annual generation(TWh$_{el}$/y)	Capacity factor (%)	Mitigation potential [a] (GtCO$_2$)	Energy Requirments[d] (kWh$_{th}$/kWh$_{el}$)	CO$_2$ emissions (g/kWh$_{el}$)	Generating cost (UScents/ kWh)	Barriers/Challenges
Coal	7755	70–90		2.6–3.5	900	3–6	Greenhouse gas emissions
Oil	1096	60–90		2.6–3.5	700	3–6	Resource constraints
Gas	3807	~60		2–3	450	4–6	Fuel price
Carbon capture and storage	-	n.a.	150–250	2–2.5 + 0.3–1	170–280	3–6 + 0–4	Energy penalty, large-scale storage, late deployment
Nuclear fission[q]	2793	86	>180	0.12	65	3–7	Waste disposal, profileration, public acceptance
Large hydro	3121	41	200–300	0.1	45–200	4–10	Resource potential, social and environmental impact
Small hydro	~250	~50	~100?	n.a.	45	4–20	Resource potential
Wind	260y	24.5	~450–500	0.05	~65	3–7	Variability and grid integration
Solar-photovoltaic	12	15	25–200?	0.4/1–0.8/1	40–150–100–200	10–20	Generating cost
Concentrating Solar	~1	20–40	25–200?	0.3	50–90	15–25	Generating cost
Geothermal	60	70–90	25–500?	n.a.	20–140	6–8	Uncertain field capacity

[a]Difference between the IPCC SRES A2 and a modified SRES B1 scenario where the technology assumes its economic or resource potential.

[b]kWh$_{th}$ are counted as fossil energy only, hydropotential, nuclear energy and ambient energy are excluded.

Source: Adapted from Lenzen, Manfred. 2010. Current State of Development of Electricity Generating Technologies: A Literature Review, Energies, 3, no. 3: 462–591.

Table 7.6.   Historical trends in high-voltage power transmission

System voltage (kV)				
Nominal	Maximum	Year introduced	Typical transmission capacity (MW)	Typical right-of-way width (m)
Alternating current				
115	121	1915	50–200	15–25
230	242	1921	200–500	30–40
345	362	1952	400–1500	35–40
500	550	1964	1000–2500	35–45
765	800	1965	2000–5000	40–55
1100	1200	Tested 1970s	3000–10000	50–75
Direct current				
50		1954	50–100	25–30
200	(± 100)	1961	200–500	30–35
500	(± 250)	1965	750–1500	30–35
800	(± 400)	1970	1500–2000	35–40
1000	(± 500)	1984	2000–3000	35–40
1200	(± 600)	1985	3000–6000	40–55

Source: Adapted from Annestrand, S.A. 2003. Power Transmission, High-Voltage, In: Robert A. Meyers, Editor-in-Chief, Encyclopedia of Physical Science and Technology (Third Edition), (New York, Academic Press), Pages 35–55.

Table 7.7.   Notable high-voltage direct current (HVDC) transmission systems

HVDC link	Supplier[a]	Year	Power (MW)	DC voltage (kV)	Length (km)	Location
Itaipu	A	1986	3150	± 600	785	Brazil
Itaipu II	A	1987	3150	± 600	805	Brazil
Three Gorges-Changzhou	AB	2003	3000	± 500	890	China
Three Gorges-Quangdong	AB	2004	3000	± 500	940	China
Guizhou-Guangdong	G	2004	3000	± 500	940	China
Three Gorges-Shanghai	AB	2007	3000	± 500	1059	China
Itaipu	A	1985	2383		785	Brazil
Quebec-New England	A	1990	2000[b]	± 450	1500	Canada-U.S.A.
Pacific Intertie upgrade	A	1984	2000	± 500	1362	U.S.A.
(Sellindge)	I	1986	2000	2 × ± 270		
East-South Intercon.	G	2003	2000	± 500	1450	India
Celilo Conv. station	AB	2004	2000	± 500	upgrade	U.S.A. Mozambique-S.
Cahora Bassa	J	1978	1920	± 533	1360	Africa
Intermountain power project	A	1986	1920	± 500	784	U.S.A. Mozambique-S.
Cahora Bassa	G	1998	1920	± 533	1456	Africa
Nelson River II	J	1985	1800	± 500	930	Canada
Nicolet Tap	A	1992	1800			Canada
TSQ-Beijao	G	2000	1800	± 500	903	China
Nelson River I[c]	I	1972	1620	± 450	892	Canada
Pacific intertie	JV	1982	1600	± 400	1362	U.S.A
Leyte-Luzun	AB	1997	1600	400	440	Philippines
Itaipu	A	1984	1575	± 300	785	Brazil

[a]A - ASEA; B - Brown Boveri; C - General Electric; D - Toshiba; E - Hitachi; F - Russian; G - Siemens; H - CGEE Alsthom; I - GEC (Formerly English Electric); J - HVDC Working Group. (AEG, BBC, Gmens); K - (Independent); AB - ABB (ASEA Brown Boveri); JV - Joint Venture (GE and ASEA).

[b]Multiterminal system. Largest terminal is rated 2250 MW.

[c]two valve groups in Pole 1 replaced with thyristors by GEC in 1991.

Source: Adapted from Sood, Vijay K. 2011. HVDC Transmission, In: Muhammad H. Rashid, Editor-in-Chief, Power Electronics Handbook (Third Edition), (Boston, Butterworth-Heinemann), Pages 823-849.

Table 7.8. Leading generation of electricity by renewable sources, 2009 (billion kilowatt-hours)

Country	Hydroelectric	Geothermal	Wind	Solar, Tide and Wave	Biomass and Waste	Total Renewables	Total Net Generation	% Renewables	% Non-hydropwer renewables
Denmark	0.0	0.0	6.4	0.0	3.8	10.2	34.3	29.8%	29.8%
El Salvador	1.5	1.4	0.0	0.0	0.2	3.1	5.7	54.2%	27.9%
Iceland	12.2	4.3	0.0	0.0	0.0	16.5	16.5	100.0%	26.2%
Kenya	2.1	1.3	0.0	0.0	0.3	3.7	6.6	56.7%	24.0%
Portugal	8.2	0.2	7.2	0.2	2.3	18.0	46.8	38.4%	20.9%
Guatemala	2.1	0.4	0.0	0.0	1.1	3.6	7.2	49.5%	20.5%
Nicaragua	0.3	0.3	0.1	0.0	0.2	0.9	3.4	26.0%	17.3%
Costa Rica	7.2	1.1	0.3	0.0	0.1	8.7	9.1	95.3%	16.9%
Philippines	9.7	9.8	0.1	0.0	NA	19.6	58.8	33.3%	16.8%
Spain	26.1	0.0	35.9	5.7	4.0	71.7	274.7	26.1%	16.6%
Belize	0.2	0.0	a	a	0.1	0.2	0.3	77.5%	16.3%
Mauritius	0.1	0.0	0.0	0.0	0.5	0.6	2.9	20.1%	16.0%
New Zealand	24.0	4.6	1.4	0.0	0.6	30.6	42.1	72.6%	15.6%
Germany	18.5	0.0	36.7	6.3	33.8	95.2	546.8	17.4%	14.0%
Finland	12.6	0.0	0.3	0.0	8.5	21.3	68.3	31.2%	12.8%
Papua New Guinea	0.8	0.4	0.0	0.0	0.0	1.2	3.3	37.4%	12.0%
Austria	39.9	0.0	1.9	0.0	5.5	47.3	64.0	73.9%	11.6%
Ireland	0.9	0.0	2.8	a	0.2	3.9	26.1	14.9%	11.4%
Netherlands	0.1	0.0	4.4	0.0	7.4	11.9	106.9	11.1%	11.0%
Sweden	65.2	0.0	2.4	0.0	11.6	79.1	132.3	59.8%	10.5%
World	3,145.2	63.0	262.6	18.8	271.1	3,760.6	18,979.9	19.8%	3.2%

a Value is too small for the number of decimal places shown.

Source: United States Department of Energy, Energy Information Administration, International Energy Statistics, <http://www.eia.gov/countries/data.cfm>.

Table 7.9. **Comparison between today's grid and the smart grid**

Goal	Current Grid	Smart Grid
Self heals.	Responds to prevent further damage. Focus is on protection of assets following system faults.	Automatically detects and responds to actual and emerging transmission and distribution problems. Focus is on prevention. Minimizes consumer impact.
Motivates and includes the consumer.	Consumers are uninformed and nonparticipative with the power system.	Informed, involved and active consumers. Broad penetration of demand response.
Resists attacks.	Vulnerable to malicious acts of terror and natural disasters.	Resilient to attack and natural disasters with rapid restoration capabilities.
Provides power quality (PQ) for twenty-first century needs.	Focused on outages rather than power quality problems. Slow response in resolving PQ issues.	Quality of power meets industry standards and consumer needs. PQ issues identified and resolved prior to manifestation. Various levels of PQ at various prices.
Accommodates all generation and storage options.	Relatively small number of large generating plants. Numerous obstacles exist for interconnecting distributed energy resources.	Very large numbers of diverse distributed generation and storage devices deployed to complement the large generating plants. *Plug-and-play* convenience. Significantly more focus on and access to renewables.
Enables markets.	Limited wholesale markets still working to find the best operating models. Not well integrated with each other. Transmission congestion separates buyers and sellers.	Mature wholesale market operations in place; well integrated nationwide and integrated with reliability coordinators. Retail markets flourishing where appropriate. Minimal transmission congestion and constraints.
Optimizes assets and operates efficiently.	Minimal integration of limited operational data with asset management processes and technologies. Siloed business processes. Time-based maintenance.	Greatly expanded sensing and measurement of grid conditions. Grid technologies deeply integrated with asset management processes to most effectively manage assets and costs. Condition based maintenance.

Source: U.S. Department of Energy, National Energy Technology Laboratory.

Table 7.10. **Power outage categories**

Category	Description
Dropout	A loss of power that has a short duration, on a timescale of seconds, and is usually fixed quickly.
Brownout	The electrical power supply encounters a partial drop in voltage, or temporary reduction in electric power. In the case of a three-phase electric power supply, when a phase is absent, at reduced voltage, or incorrectly phased.
Blackout	An affected area experiences a complete loss of electrical power, ranging from several hours to several weeks.
Load shedding	An electric company either reduces or completely shuts off the available power to sections of the grid. Sometimes referred to as rolling brownouts and rolling blackouts.

Source: Flick, Tony and Justin Morehouse. 2011. Securing the Smart Grid, (Boston, Syngress).

Table 7.11.   Main phenomena causing electromagnetic and power quality disturbances

Conducted low-frequency phenomena
Harmonics, interharmonics
Signaling voltage
Voltage fluctuations
Voltage dips
Voltage imbalance
Power frequency variations
Induced low-frequency voltages
DC components in AC networks
Radiated low-frequency phenomena
Magnetic fields
Electric fields
Conducted high-frequency phenomena
Induced continuous wave (CW) voltages or currents
Unidirectional transients
Oscillatory transients
Radiated high-frequency phenomena
Magnetic fields
Electric fields
Electromagnetic field
Steady-state waves
Transients
Electrostatic discharge phenomena (ESD)
Nuclear electromagnetic pulse (NEMP)

Source: Fuchs, Ewald F. and Mohammad A.S. Masoum, Editors. 2008. Power Quality in Power Systems and Electrical Machines, (Burlington, Academic Press).

Table 7.12.   Characteristics of subcritical, supercritical, and ultra-supercritical coal power plants

Type of Unit	Main Steam Pressure (psia/MPa)	Main Steam Temperature (°F/°C)	Reheat Steam Temperature (°F/°C)	Efficiency (%)[a]
Subcritical	<3,201/<22.1	Up to 1,050/565	Up to 1,050/565	33–39
Supercritical	3,207–3,628/ 22.1–25	1,000–1,075/ 540–580	1,000–1,075/ 540–580	38–42
Ultra-supercritical	>3,628/>25	>1,075/>580	>1,075/>580	>42

[a]based on coal's higher heating value and bituminous coal).
Source: Adapted from Miller, Bruce G. 2011. Clean Coal Engineering Technology, (Boston,Butterworth-Heinemann).

Table 7.13. General comparison of IGCC, PC, and FBC power plants

	IGCC Plant	PC Plant	FBC Plant
Operating Principal	Feedstock is only partially oxidized. The high-pressure synthesis gas produced is combusted and expanded in a combustion turbine to produce power. Heat is recoevered from the turbine exhaust gas to produce steam for expansion in a steam turbine to produce added power.	Pulverized coal is combined in a boiler where the heat is directly transferred to produce high-pressure steam that is expanded in a steam turbine to produce power.	Air- suspended coal is combusted together with sorbents for sulfur control. Heat is directly transferred to produce high-pressure steam. Boiler operates at either atmospheric pressure or may be pressurized. Key designs are bubbling bed and circulating bed boilers.
Oxidant	Air or oxygen in the gasifier. Air in the combustion turbine.	Air in the boiler...	Air in the boiler.
Operating Pressure	25 to 40 atmospheres	1 atmosphere	1 to 100+ atmospheres
Coal Sulfur Conversion	Sulfur is primarily cinverted to H_2S and COS in the synfuel.	Sulfur is converted to SO_2 in the combustion process and exits boiler with flue gas.	Sulfur is converted to SO_2 in the combustion process and is mostly captured by an in-bed sorbent such as limestone. Residual SO_2 exits the boiler with flue gas.
Coal Nitrogen Conversion	Converted to ammonia and nitrogen in the gasifier. Ammonia is removed from the syngas prior to combustion in the combustion turbine. NOx is formed in the combustion turbine. Exits turbine as constituent of flue gas.	Converted to Nox. Low-NOx burners are used to minimize conversion to NOx	Converted to NOx. FBC is an inherently low NOx producer due to its low combustion temperature. NOx exits boiler as consituent of flue gas.
Process Solids	Most of the coal ash if recovered as inert slag or bottom ash from the gasifier. Only a small portion of the ash is entrained with the synfuel.	Approximately 80% of the coal ash is entrained in the flue gas as fly ash. The remaining ash is recovered as bottom ash or inert slag.	Ash and spent sorbent limestone are entrained in the flue gas collected in a conrol device such as cyclone and returned to the boiler. Most solids collect as bottom ash.
Thermal Efficiency, % (HHV Basis)	38–50	34–43	36–45

IGCC = integrated gasification combined cycle; PC = pulverized coal; FBC = fluidized bed combustion.
Source: Adapted from Ratafia-Brown, J., L. Manfredo, J. Hoffmann, and M. Ramezan. 2002. Major Environmental Aspects of Gasification-Based Power Generation Technologies," prepared for U.S. Department of Energy, Office of Fossil Energy, National Energy Technology Laboratory.

Table 7.14. Relative permittivity of solids[a]

Material	Temperature (°C)	Frequency	Permittivity
Vacuum			1 (by definition)
Cellulose (see also paper)			
Cellophane	20	50 Hz/1 MHz	7.6/6.7
	−30/70	50 Hz	7.2/8.0
Paper fibers	20	50 Hz	6.5
Ceramics			
Alumina	20/100	50 Hz/1 MHz	8.5
	20	1 MHz	10.8
Calcium titanate	20	1 MHz	150
Lead zirconate	20	1 MHz	110
Magnesium titanate	20/150	50 Hz/1 MHz	14
Porcelain	20/100	50 Hz/1 MHz	5.5
Rutile	20	1 MHz/1 GHz	80
	20	1 MHz/1 GHz	40
	20	1 MHz/100 MHz	12
	20	1 MHz/100 MHz	15
Steatite	20	1 MHz/1 GHz	6
(low loss)	20	1 MHz/1 GHz	6
Strontium titanate	20	1 MHz	200
Strontium zirconate	20	1 MHz	38
Crystals (single, inorganic)			
Alkali halides			
LiF	20/25	1 kHz/10 GHz	8.9/9.1
LiCl	20	1 kHz/1 MHz	11.8/11.0
LiBr	20	1 kHz/1 MHz	13.2/12.1
LiI	20	1 kHz/1 MHz	16.8/11.0
NaF	20	1 kHz/1 MHz	5.1/6.0
NaCl	20/25	1 kHz/10 GHz	6.1/5.9
NaBr	20	1 kHz/1 MHz	6.5/6.0
NaI	20	1 kHz/1MHz	7.3/6.6
KF	20	1 kHz/1 MHz	5.3/6.0
KCl	20	1 kHz/10 GHz	4.9/4.8
KBr	20/25	1 kHz/10 GHz	5.0/4.9
KI	20	1 kHz/1 MHz	5.1/5.0
RbF	20	1 kHz	6.5
RbCl	20	1 kHz	4.9
RbBr	20	1 kHz	4.9
RbI	20	1 kHz	4.9
Calcite	20	1 kHz/10 kHz	8.5
	20	1 kHz/10 kHz	8
Diamond	20	500 Hz/100 MHz	5.7/5.5
Fluorite	20	10 kHz/2 MHz	7.4/6.8
Gallium Arsenide	20	1 kHz	12
Germanium	20	1 kHz	16.3
Iodine	17/22	100 MHz	4

Table 7.14. Continued

Material	Temperature (°C)	Frequency	Permittivity
Mica, muscovite (best)	20/100	50 Hz/100 MHz	7
Periclase	25	100 Hz/100 MHz	9.7
Quartz	20/25	1 kHz/35 MHz	4.43/4.43
	20/25	1 kHz/35 MHz	4.63/4.63
Ruby	17/22	10 kHz	13.3
	17/22	10 kHz	11.3
Rutile	20	50 Hz/100 MHz	86
	17/22	100 MHz	170
Sapphire	20	50 Hz/1 GHz	9.4
	20	50 Hz/1 GHz	11.6
Selenium	17/22	100 MHz	6.6
Silicon	20	1 kHz	11.7
Sulphur	25	1 kHz	3.8
	25	1 kHz	4
	25	1 kHz	4.4
Urea	17/22	400 MHz	3.5
Zircon	17/22	100 MHz	12
Glasses			
Borosilicate	20	1 kHz/1 MHz	5.3
	20	1 MHz	5
	20	50 Hz/100 MHz	4
Fused quartz	20/150	50 Hz/100 MHz	3.8
Lead	20	1 kHz/1 MHz	6.9
Soda	20	1 MHz/100 MHz	7.5
Minerals			
Amber	20	1 MHz/3 GHz	2.8/2.6
Asbestos (chrysotile)	25	50 Hz/1 MHz	5.8/3.1
	20	1 MHz	3
Bitumen	25	50 Hz/100 MHz	2.7/2.55
	20	1 kHz	3.5
Granite	20	1 MHz	8
Gypsum	20	10 kHz	5.7
Marble	20	1 MHz	8
Sand	20	1 MHz	2.5
	20	1 MHz	9
Sandstone	20	1 MHz	10
Soil	20	1 MHz	3
	20	1 MHz	10
Sulphur	20	3 GHz/10 GHz	3.4
Paper and Pressboard			
(see also cellulose)			
Unimpregnated, dry			
Kraft (tissue)	20/90	1 kHz	1.8
	20/90	1 kHz	3
Rag (cotton)	20/90	50 Hz/50 kHz	1.7

Continued

Table 7.14.　Continued

Material	Temperature (°C)	Frequency	Permittivity
Impregnated, mineraloil ($\varepsilon_r{'} = 2.2$)			
Kraft (tissue)	20	50 Hz	3.6
	20	50 Hz	4.3
Rag (cotton)	20	50 Hz	3.5
	20	50 Hz	4.2
Impregnated (Pentachlordiphenyl)			
Kraft (tissue)	20	50 Hz	5.7
	20	50 Hz	6
Fibre	20	1 MHz	4.5
Pressboard	20	50 Hz	3.2
Plastics (non-polar, synthetic)			
Poly-ethylene	20	50 Hz/1 GHz	2.3
isobutylene	20	50 Hz/3 GHz	2.2
4-methylpentene (TPX) (dimethyl)	20	100 Hz/10 kHz	2.1
phenyloxide (PPO)	25	100 Hz/1 MHz	2.6
propylene	20	50 Hz/1 MHz	2.2
styrene	20	50 Hz/1 GHz	2.6
tetrafluoroethylene (PTFE)	20	50 Hz/3 GHz	2.1
Plastics (polar, synthetic)			
Poly-amides	20	50 Hz/100 MHz	3-Apr
carbonates	20	50 Hz/1 MHz	3.2/3.0
ethyleneterephthalate	20	50 Hz/100 MHz	3.2/2.9
imides	20	1 MHz	3.4
methylmethacrylate	20	50 Hz/100 MHz	3.4/2.6
vinylcarbazole	20	50 Hz/100 MHz	2.8
vinylchloride	20	50 Hz/100 MHz	3.2/2.8
Plastics (miscellaneous)			
Aniline resin	20	3 GHz	3.5
	20	1 MHz/1 GHz	4-May
	100	1 MHz	6
Cellulose acetate	20	1 MHz/1 GHz	3.5
Cellulose triacetate	20	50 Hz/100 MHz	3.8/3.2
Ebonite	20	1 kHz/1 GHz	3/2.7
	20	50 kHz/1 GHz	4.1/3.8
Epoxy resin	25	1 kHz/100 MHz	3.6/3.5
Melamine resin	20	3 GHz	4.7
Phenolic resin	20	1 MHz	5.5
	20	1 MHz/1 GHz	5
	140	1 MHz/10 MHz	6
	20	1 MHz	5
Urea resin	20	1 MHz	6
Vinyl acetate (poly-)	20	1 MHz/10 MHz	4
Vinyl chloride (poly-) (PVC)	20	1 MHz/10 MHz	4

Table 7.14. Continued

Material	Temperature (°C)	Frequency	Permittivity
Rubbers			
Natural	20/80	1 MHz/10 MHz	2.4
	20	1 MHz/10 MHz	3.2
Butadiene/styrene	20/80	50 Hz/100 MHz	2.5
(GR-S)	20/80	50 Hz/100 MHz	2.5
Butyl	20	50 Hz/100 MHz	2.4
Chloroprene	20	1 kHz/1 MHz	6.5/5.7
Silicone	20	50 Hz/100 MHz	8.6/8.5
Silicone	25	1 kHz/100 MHz	3.2/3.1
Waxes, etc.			
Chlornaphthalene			
(tri and tetrachlor-)	20	50 Hz/100 MHz	5.4/4.2
Ozokerite	20	50 Hz/100 MHz	2.3
Paraffin wax	20	1 MHz/1 GHz	2.2
Petroleum jelly	20/60	50 Hz	2.1/1.9
Rosin	20	3G Hz	2.4
Wood (% water)			
Balsa 0%	20	50 Hz/3 GHz	1.4/1.2
Beech 16%	20	1 MHz/100 MHz	9.4/8.5
Birch 10%	20	1 MHz/100 MHz	3.1
Douglas fir 11%	15	1 MHz/10 MHz	3.2
compressed	15	1 MHz/10 MHz	4.3
Scots pine 15%	20	1 MHz/100 MHz	8.2/7.3
Walnut 0%	20	10 MHz	2
Walnut 17%	20	10 MHz	5
Whitewood 10%	20	1 MHz/100 MHz	3

[a]The relative permittivity of a material under given conditions reflects the extent to which it concentrates electrostatic lines of flux. In technical terms, it is the ratio of the amount of electrical energy stored in a material by an applied voltage, relative to that stored in a vacuum. Likewise, it is also the ratio of the capacitance of a capacitor using that material as a dielectric, compared to a similar capacitor that has a vacuum as its dielectric. The relative permittivity of a material for a frequency of zero is known as its static relative permittivity or as its dielectric constant. A dielectric is an electrical insulator that can be polarized by an applied electric field.

Source: Adapted from Kaye and Laby Table of Physical and Chemical Constants,
<http://www.kayelaby.npl.co.uk/general_physics/2_6/2_6_5.html>.

Table 7.15.   Relative permittivity of liquids[a]

Material	Temperature (°C)	Permittivity[b]
Vacuum		
Alcohols (primary)		
Methanol	25	32.65 P
Ethanol	25	24.51 P
Propanol	25	20.51 P
Butanol	25	17.59 P
Pentanol	25	15.09 P
Hexanol	25	13.3 P
Hydrocarbons		
n-Pentane	20	1.84
n-Hexane	20	1.89
n-Heptane	20	1.92
n-Octane	20	1.95
n-Nonane	20	1.97
n-Decane	20	1.99
n-Undecane	20	2
n-Dodecane	20	2.01
Benzene	20	2.284
Cyclopentane	20	1.96
Cyclohexane	20	2.025
Toulene	20	2.39
(Chloro/Fluoro)-hydrocarbons		
CCl_4	20	2.24
CCl_3F	29	2.28
CCl_2F_2	29	2.13
$CClF_3$	−30	2.3
$CHCl_3$	20	4.80 P
$CHCl_2F$	28	5.34 P
$CHClF_2$	24	6.11 P
$(-CCl_2F)_2$	25	2.52
$(-CClF_2)_2$	25	2.26
$(-CH_2Cl)_2$	20	10.66 P
$(CCl_2)_2$	25	2.30
$CCl_2\,CHCl$	20	3.4 P
F-pentane	20	4.24 P
F-benzene	25	5.42 P
Cl-benzene	20	5.70 P
Miscellaneous		
Aniline	20	6.89
Acetone	25	20.7 P
Diethylketone	20	17.0 P
Diethylether	20	4.34 P
Cyclohexanone	20	18.3 P
Nitrobenzene	25	34.8 P

Table 7.15. Continued

Material	Temperature (°C)	Permittivity[b]
CS_2	20	2.64
Liquid gases	T/K	
Argon	82	1.53
Helium	4.19	1.048
Helium	2.06	1.055
Hydrogen	20.4	1.22
Nitrogen	70	1.45
Oxygen	80	1.5

[a]The relative permittivity of a material under given conditions reflects the extent to which it concentrates electrostatic lines of flux. In technical terms, it is the ratio of the amount of electrical energy stored in a material by an applied voltage, relative to that stored in a vacuum. Likewise, it is also the ratio of the capacitance of a capacitor using that material as a dielectric, compared to a similar capacitor that has a vacuum as its dielectric. The relative permittivity of a material for a frequency of zero is known as its static relative permittivity or as its dielectric constant. A dielectric is an electrical insulator that can be polarized by an applied electric field.

[b]P = polarliquid

Source: Adapted from Kaye and Laby Table of Physical and Chemical Constants, <http://www.kayelaby.npl.co.uk/general_physics/2_6/2_6_5.html>.

Table 7.16. Resistivities of some metallic elements[a]

Metal	ρ (10⁻⁸ Ω m)			
	78.2 K	273.2 K	573.2 K	973.2 K
Aluminium (1.17K)	0.229	2.42	5.83	25.5
Arsenic	5.5	26	—	—
Barium	6.65	30.2	90.8	—
Beryllium	0.72	3.02	12.2	26.5
Bismuth	35	107	129	155
Cadmium (0.54K)	1.6	6.8	—	—
Calcium	0.63	3.11	7.8	20
Caesium	4.06	18.8	67.3	128
Chromium	0.5	12.7	25.2	47.2
Cobalt	0.9	5.6	19.7	48
Copper	0.204	1.54	3.59	6.55
Gallium (1.1K)	2.75	13.6	31	—
Gold	0.468	2.05	4.61	8.59
Indium (3.35K)	1.8	8.0	36.7	47
Iridium (0.14K)	0.9	4.7	10.8	22
Iron	0.64	8.57	30.2	85.7
Lithium	0.93	8.53	28.5	39.0
Lutetium	16	54	—	—
Magnesium	0.53	4.05	9.0	15.4
Manganese	132	143	152	—
Mercury (4.12K)	5.8	94.1	128	214
Molybdenum (0.92K)	0.454	4.85	12.5	23.3
Nickel	0.51	6.16	23.1	40.7
Platinum	1.96	9.81	21.0	34.3
Plutonium	~150	146	109	—
Polonium	—	~40	—	—
Potassium	1.30	6.49	27.7	64.7
Sodium	0.76	4.33	17.4	38.9
Strontium	3.55	12.3	25.5	
Thorium (1.37K)	3.9	14.7	32.5	53.6
Thulium	31	67	—	—
Tin (3.69K)	2.1	11.5	50	60
Titanium (0.39K)	4.6	39	90	142
Tungsten (0.01K)	0.573	4.82	12.3	23.7
Uranium (0.68K)	11	28	47	—
Vanadium (5.03K)	2.30	18.1	39.5	63.1
Yttrium	15.5	55	—	—
Zinc (0.85K)	1.04	5.48	13.3	
Zirconium (0.55K)	6.36	38.8	87.7	128

[a]The electrical resistivity or specific resistance ρ is the resistance between the opposite faces of a metre cube of a material. The values given below are in ohm meter units (Ω m). The reciprocal of ρ is the electrical conductivity.
Source: Adapted from Kaye and Laby Table of Physical and Chemical Constants,
<http://www.kayelaby.npl.co.uk/general_physics/2_6/2_6_1.html>.

Table 7.17. Resistivities of semi-conducting elements at normal temperature[a]

Material	$\rho/(\Omega\ m)$
Carbon	
Amorphous	$\sim 6 \times 10^{-5}$
Graphite	$(3-60) \times 10^{-6}$
Pyrolytic graphite, along planes	$\sim 5 \times 10^{-6}$
Pyrolytic graphite, normal to planes	$\sim 5 \times 10^{-3}$
Germanium	$(1-500)10^{-3}$
Selenium	~ 0.1
Silicon	$(1-600) \times 10^{-1}$
Tellurium	$\sim 3 \times 10^{-3}$

[a]The electrical resistivity or specific resistance ρ is the resistance between the opposite faces of a metre cube of a material. The values given below are in ohm meterunits (Ω m). The reciprocal of ρ is the electrical conductivity.
Source: Adapted from Kaye and Laby Table of Physical and Chemical Constants,
<http://www.kayelaby.npl.co.uk/general_physics/2_6/2_6_1.html>.

Table 7.18. Resistivities of insulators[a]

Material	Volume resistivity Ω m	Surface resistivity Ω (per square)
Beeswax (fresh surface)	$10^{12}-10^{13}$	$\sim 10^{14}$
Boron	10^{10}	
Ceramics		
Alumina	$10^{9}-10^{12}$	
Porcelain	$10^{10}-10^{12}$	10^{11} glazed
		10^{9} unglazed
Diamond	$10^{10}-10^{11}$	
Glass		
Soda–lime	$10^{9}-10^{11}$	$10^{10}-10^{12}$
Borosilicate (Pyrex)	10^{12}	
Plate	2×10^{11}	5×10^{10}
Hard rubber (Ebonite, etc.)	$10^{13}-10^{15}$	$10^{10}-10^{18}$
Iodine	10^{13}	
Ivory	10^{6}	10^{9}
Marble	$10^{7}-10^{9}$	10^{9}
Mica, sheet	$10^{11}-10^{15}$	$10^{10}-10^{13}$
Mica, sheet moulded	10^{13}	5×10^{13}
Mineral oil	$>10^{10}$	
Paper (dry)	$\sim 10^{10}$	$10^{9}-10^{10}$
Paraffin (kerosene, coloured)	$10^{11}-10^{12}$	
Paraffin wax	$10^{13}-10^{17}$	10^{15}
Plastics		
Acrylic (Perspex, etc.)	$>10^{13}$	$>10^{14}$
Alkydes or polyester (no filler)	$10^{12}-10^{13}$	$10^{13}-10^{14}$
Aminos, melamines (cellulose)	$>10^{9}$	$10^{12}-10^{14}$
Aminos, melamines (mineral)	10^{9}	$10^{12}-10^{14}$

Continued

Table 7.18.　Continued

Material	Volume resistivity Ω m	Surface resistivity Ω (per square)
Cellulose acetate	10^8–10^{11}	10^{11}–10^{12}
Epoxy cast resin (no filler)	10^{12}–10^{13}	3×10^7–$>10^{14}$
Phenolics	10^6–10^{12}	10^8–10^{14}
Polyamides (nylon)	10^8–10^{13}	10^{11}–10^{15}
P.C.T.F.E.[†]	10^{16}	10^{12}–10^{13}
P.E.T.[‡]	10^{15}–10^{17}	
Polyformaldehyde	$\sim6 \times 10^{12}$	$>2\times10^{13}$
Polypropylene	10^{13}–10^{15}	$>10^{15}$
Polystyrene (general purpose)	10^{15}–10^{19}	$>10^{14}$
Polystyrene (toughened)	10^{10}–10^{15}	$>10^{14}$
P.T.F.E.[§]	10^{15}–10^{19}	4×10^{12}–10^{17}
Polythene (high density)	10^{14}–10^{15}	10^{12}–10^{17}
Polythene (low densiy)	10^{14}–10^{18}	$>10^{14}$
Polyurethanes	10^9–10^{12}	10^{13}–10^{15}
P.V.C.[*] (rigid)	5×10^{12}–10^{13}	10^{12}–10^{15}
P.V.C. (flexible)	5×10^6–5×10^{12}	$>10^{14}$
Silicone (glass)	10^8–10^{12}	$>10^{11}$
Quartz, parallel optic axis	10^{12}	
Quartz, perpindicular optic axis	10^{14}	
Quartz, fused optic axis	$>10^{16}$	3×10^{12}
Silicone, oils	10^{12}	
Silicone, rubber	10^9	
Slate	10^5–10^6	10^7
Soil	10^2–10^4	
Wood (paraffined)	10^8–10^{11}	10^{12}
Water (distilled)	10^2–10^5	

[a]The SI unit of electrical resistance is the ohm (Ω).
[†]Polychlorotrifluoroethylene.
[‡]Polyethyleneterephthalate.
[§]Polytetrafluorethylene.
[*]Polyvinylchloride.
Source: Adapted from Kaye and Laby Table of Physical and Chemical Constants,
<http://www.kayelaby.npl.co.uk/general_physics/2_6/2_6_3.html>.

Table 7.19.   Properties of resistance wires[a]

S.W.G[b]	Diameter		Copper	Copper–Manganese alloys	Copper–nickel alloys	Nickel–chromium
	mm	in	$\Omega\ m^{-1}$ at 20°C	$\Omega\ m^{-1}$	$\Omega\ m^{-1}$	$\Omega\ m^{-1}$
12	2.642	0.104	0.003 12	0.076	0.09	0.197
14	2.032	0.080	0.005 32	0.128	0.151	0.333
16	1.626	0.064	0.008 31	0.200	0.235	0.520
18	1.219	0.048	0.014 8	0.355	0.42	0.92
20	0.914	0.036	0.026 3	0.630	0.745	1.65
22	0.711	0.028	0.043 4	1.05	1.23	2.72
24	0.559	0.022	0.070 3	1.69	2.00	4.40
26	0.457	0.018	0.105	2.53	3.00	6.60
28	0.376	0.0148	0.155	3.75	4.40	9.7
30	0.315	0.0124	0.221	5.3	6.30	13.9
32	0.274	0.0108	0.292	7	8.30	18.3
34	0.234	0.0092	0.402	9.7	11.4	25.2
36	0.193	0.0076	0.589	14.2	16.7	37
38	0.152	0.006	0.945	22.7	27.0	59
40	0.122	0.0048	1.48	35.5	42.0	92
42	0.102	0.004	2.13	51.0	60.5	133
44	0.0813	0.0032	3.32	80	94	208
46	0.0610	0.0024	5.91	142	168	370
48	0.0406	0.0016	13.3	320	380	835
50	0.0254	0.001	34.0	820	970	2130

[a]The SI unit of electrical resistance is the ohm (Ω).
[b]British Standard Wire Gauge.
Source: Adapted from Kaye and Laby Table of Physical and Chemical Constants,
<http://www.kayelaby.npl.co.uk/general_physics/2_6/2_6_2.html>.

Nuclear

Figures

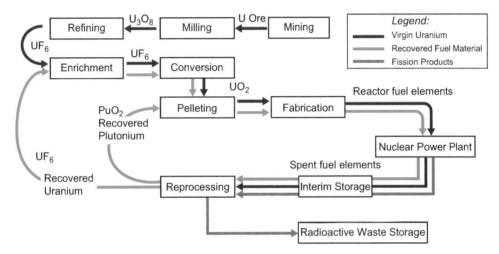

Figure 8.1. Nuclear fuel cycle.

Handbook of Energy. http://dx.doi.org/10.1016/B978-0-08-046405-3.00008-5

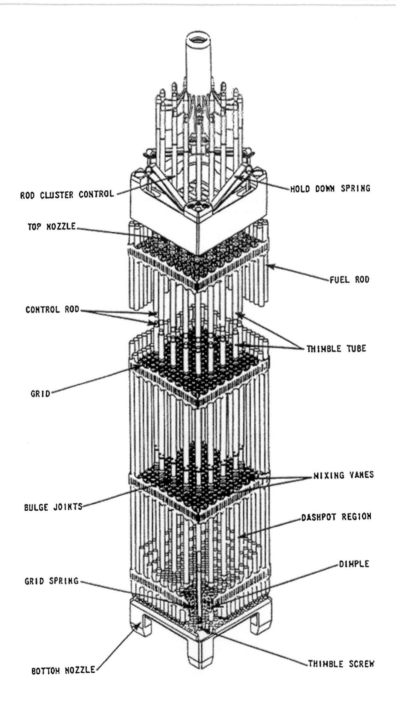

ROD CLUSTER CONTROL

HOLD DOWN SPRING

TOP NOZZLE

FUEL ROD

CONTROL ROD

THIMBLE TUBE

GRID

MIXING VANES

BULGE JOINTS

DASHPOT REGION

DIMPLE

GRID SPRING

BOTTOM NOZZLE

THIMBLE SCREW

Reactor Fuel Assembly

Figure 8.2. Nuclear reactor fuel assembly.
Source: United States Nuclear Regulatory Commission

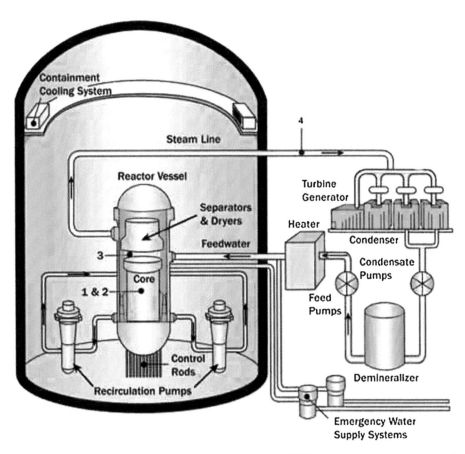

Figure 8.3. Schematic diagram of boiling water nuclear reactor (BWR). In a typical commercial boiling-water reactor, (1) the core inside the reactor vessel creates heat, (2) a steam-water mixture is produced when very pure water (reactor coolant) moves upward through the core, absorbing heat, (3) the steam-water mixture leaves the top of the core and enters the two stages of moisture separation where water droplets are removed before the steam is allowed to enter the steam line, and (4) the steam line directs the steam to the main turbine, causing it to turn the turbine generator, which produces electricity.
Source: United States Nuclear Regulatory Commission, Power Reactors, <http://www.nrc.gov/reactors/ power.html>.

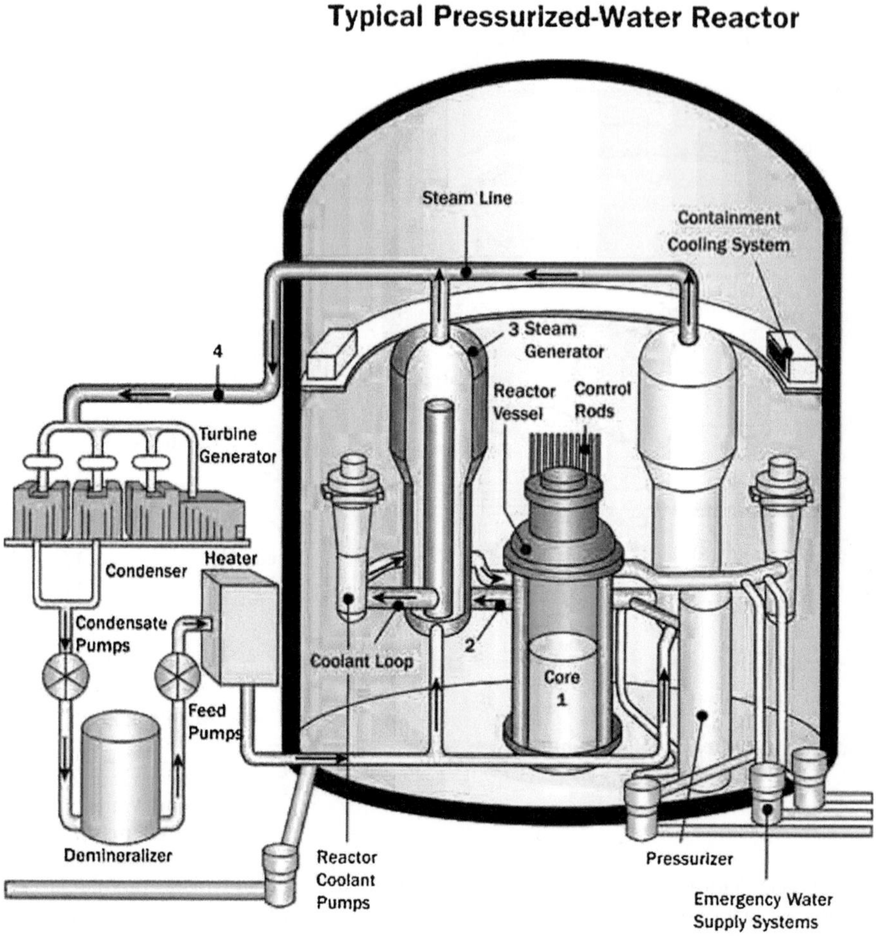

Typical Pressurized-Water Reactor

Figure 8.4. Schematic diagram of pressurized water nuclear reactor (PWR). In a typical commercial pressurized light-water reactor(1) the core inside the reactor vessel creates heat, (2) pressurized water in the primary coolant loop carries the heat to the steam generator, (3) inside the steam generator, heat from the steam, and (4) the steam line directs the steam to the main turbine, causing it to turn the turbine generator, which produces electricity.

Source: Source: U.S. Nuclear Regulatory Commission Source: United States Nuclear Regulatory Commission, Power Reactors, <http://www.nrc.gov/reactors/power.html>.

Typical Pressurized Water Reactor

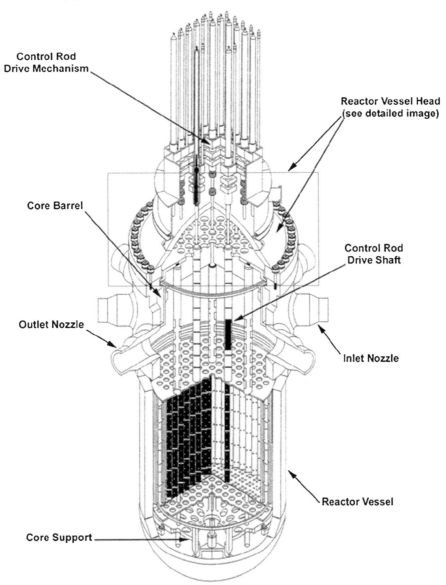

Control Rod Drive Mechanism

Reactor Vessel Head (see detailed image)

Core Barrel

Control Rod Drive Shaft

Outlet Nozzle

Inlet Nozzle

Reactor Vessel

Core Support

Figure 8.5. Schematic diagram of the reactor vessel in a typical pressurized water nuclear reactor (PWR). The reactor vessel is a protective containment vessel surrounding the nuclear fission core in a nuclear reactor. *Source: United States Nuclear Regulatory Commission, Power Reactors, <http://www.nrc.gov/reactors/power.html>.*

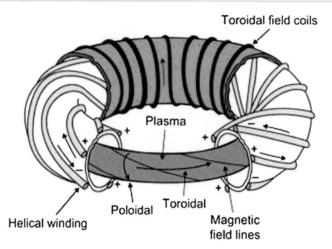

Figure 8.6.   Schematic of a stellarator, a device used to confine a hot plasma with magnetic fields in order to sustain a controlled nuclear fusion reaction. The outer winding provides the toroidal field, and the inner helical winding provides the poloidal field that gives the field lines a twist, causing the magnetic field lines to spiral around inside the chamber. The toroidal field is much stronger than the poloidal field.
Source: McCracken, Garry, Peter Stott. 2013. Fusion in the Sun and Stars, Fusion (Second Edition), (Boston, Academic Press).

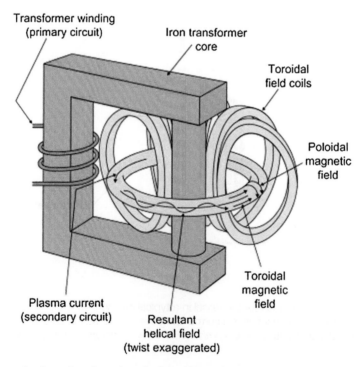

Figure 8.7.   A schematic view of a tokamak, a device using a magnetic field to confine a plasma in the shape of a torus. The current is induced in the plasma by a primary transformer winding. The toroidal magnetic field due to the external coils and the poloidal field due to the current flowing in the plasma combine to produce a helical magnetic field.
Source: McCracken, Garry, Peter Stott. 2013. Fusion in the Sun and Stars, Fusion (Second Edition), (Boston, Academic Press).

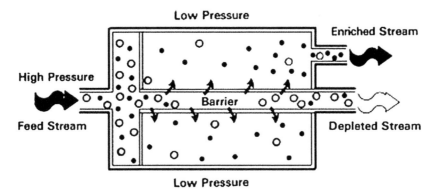

Figure 8.8. Schematic view of the gaseous diffusion uranium enrichment stage. The solid circles represent U-235; the open circles represent U-238.
Source: Knief, Ronald A. 2003. Nuclear Fuel Cycles, In: Robert A. Meyers, Editor-in-Chief, Encyclopedia of Physical Science and Technology (Third Edition), (New York, Academic Pres), Pages 655–670.

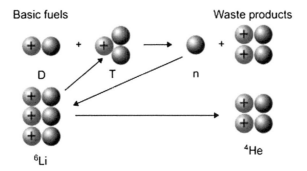

Figure 8.9. The overall fusion reaction. D = deuterium; T = tritium.

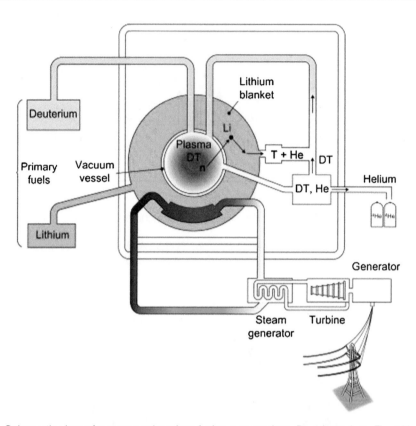

Figure 8.10. Schematic view of a proposed nuclear fusion power plant. D = deuterium; T = tritium. The deuterium and tritium fuel burns at a very high temperature in the central reaction chamber. The energy is released as charged particles, neutrons, X-rays, and ultraviolet radiation and it is absorbed in a lithium blanket surrounding the reaction chamber. The neutrons convert the lithium into tritium fuel. A conventional steam-generating plant is used to convert the nuclear energy to electricity. The waste product from the nuclear reaction is helium.
Source: McCracken, Garry, Peter Stott. 2013. Fusion in the Sun and Stars, Fusion (Second Edition), (Boston, Academic Press).

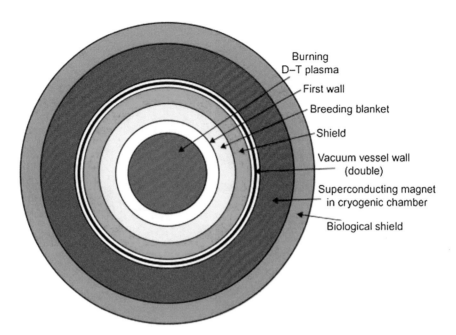

Figure 8.11.   Cross-section of a conceptual design for a fusion power plant. The power is extracted from the blanket and is used to drive a turbine and generator. For illustration, the cross-section has been shown circular, which would be the case for inertial confinement, but in a tokamak it would probably be D-shaped. No magnets are required for inertial confinement.
Source: McCracken, Garry, Peter Stott. 2013. Fusion in the Sun and Stars, Fusion (Second Edition), (Boston, Academic Press).

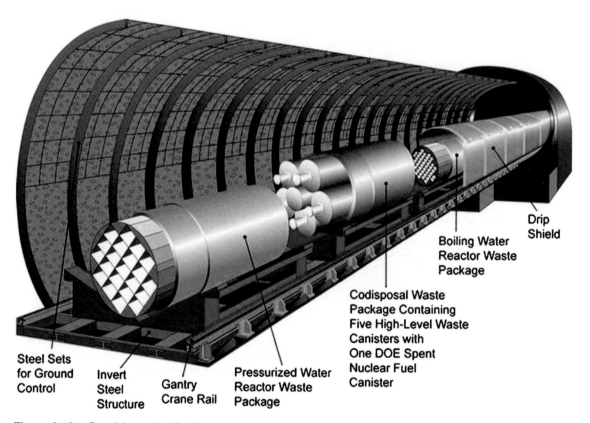

Figure 8.12. Possible method for the underground, long-term storage of nuclear waste.
Source: Research.gov.

Charts

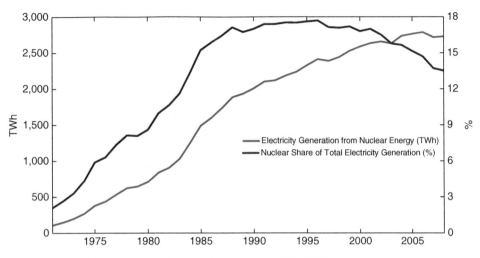

Chart 8.1. World electricity generation from nuclear energy, 1970-2010.
Source: Data from International Atomic Energy Agency (IAEA), Database on Nuclear Power Reactors,
<http://www.iaea.org/pris/>.

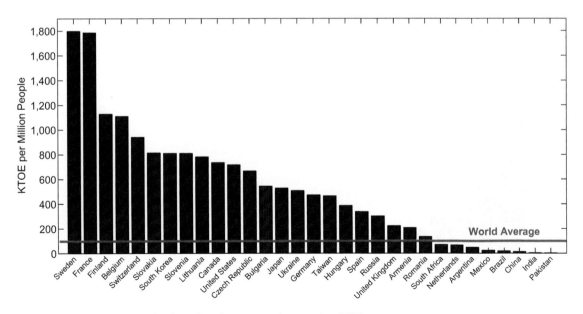

Chart 8.2. Per capita production of nuclear energy by country, 2009.
Source: Data from International Energy Agency (IEA), Energy statistics database, <http://www.iea.org/
stats/index.asp>

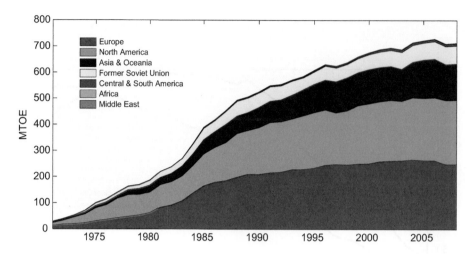

Chart 8.3. Generation of nuclear energy by region, 1970-2009.
Source: Data from International Energy Agency (IEA), Energy statistics database, <http://www.iea.org/stats/index.asp>.

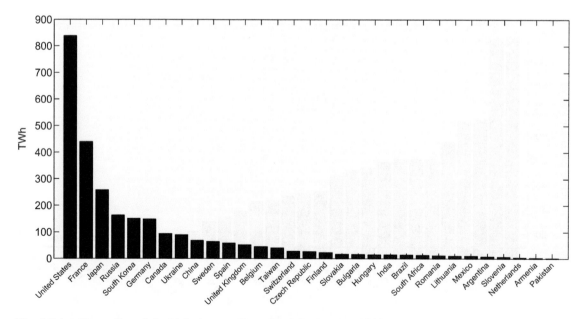

Chart 8.4. Generation of electricity from nuclear energy by country, 2010.
Source: Data from International Atomic Energy Agency (IAEA), Database on Nuclear Power Reactors, <http://www.iaea.org/pris/>.

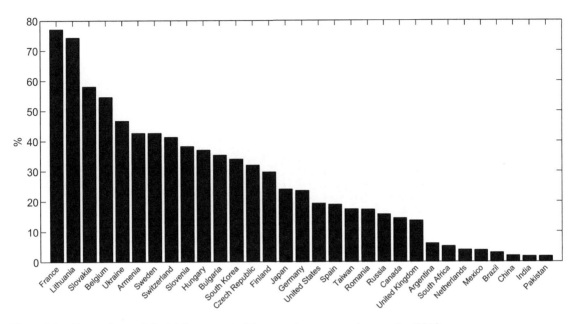

Chart 8.5. Percentage of electricity generated from nuclear energy, by country, 2010.
Source: Data from International Atomic Energy Agency (IAEA), Database on Nuclear Power Reactors, <http://www.iaea.org/pris/>.

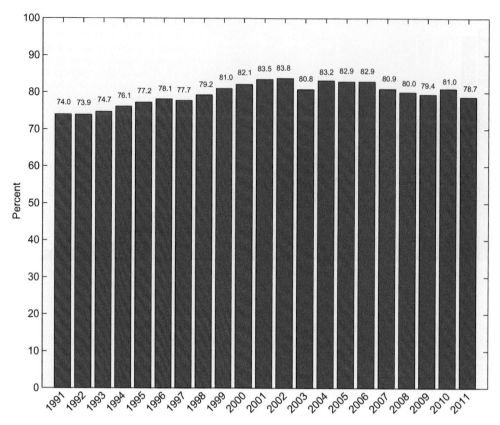

Chart 8.6. Worldwide weighted average energy availability factor for nuclear power, 1991-2011. Energy availability is the ratio of available energy to theoretically possible energy in a given period; it characterizes the reliability of a plant in general considering all complete and partial outages.
Source: Data from International Atomic Energy Agency (IAEA), Database on Nuclear Power Reactors, <http://www.iaea.org/pris/>.

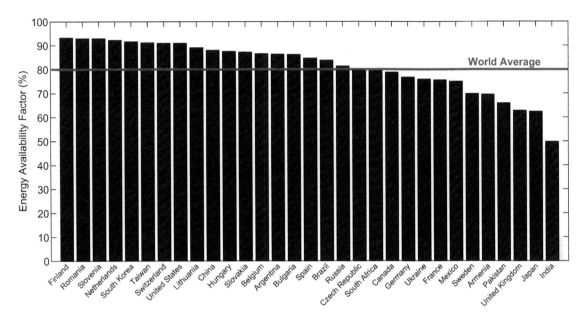

Chart 8.7. Average energy availability factor (2008-2010) for nuclear power, by country. Energy availability is the ratio of available energy to theoretically possible energy in a given period; it characterizes the reliability of a plant in general considering all complete and partial outages.
Source: Data from International Atomic Energy Agency (IAEA), Database on Nuclear Power Reactors, <http://www.iaea.org/pris/>.

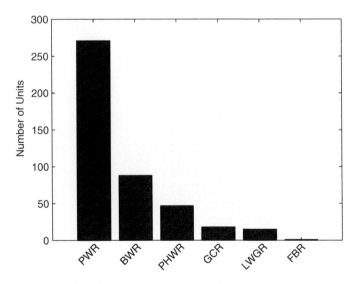

Chart 8.8. Number of operational nuclear reactors by type, 2010.
Source: Data from International Atomic Energy Agency (IAEA), Database on Nuclear Power Reactors, <http://www.iaea.org/pris/>.

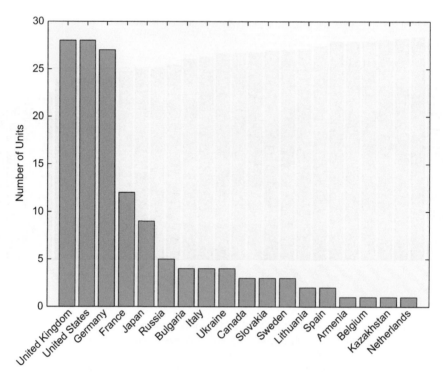

Chart 8.9. Number of shutdown nuclear reactors by country, 2011.
Source: Data from International Atomic Energy Agency (IAEA), Database on Nuclear Power Reactors, <http://www.iaea.org/pris/>.

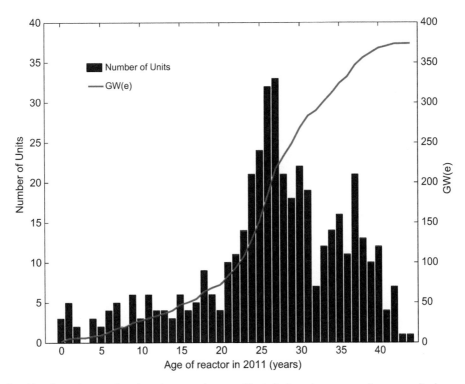

Chart 8.10. Number of operational nuclear reactors and installed nuclear generation capacity, by age of reactor in 2011.
Source: Data from International Atomic Energy Agency (IAEA), Net-Enabled Radioactive Waste Management Database (NEWMDB), <http://newmdb.iaea.org>.

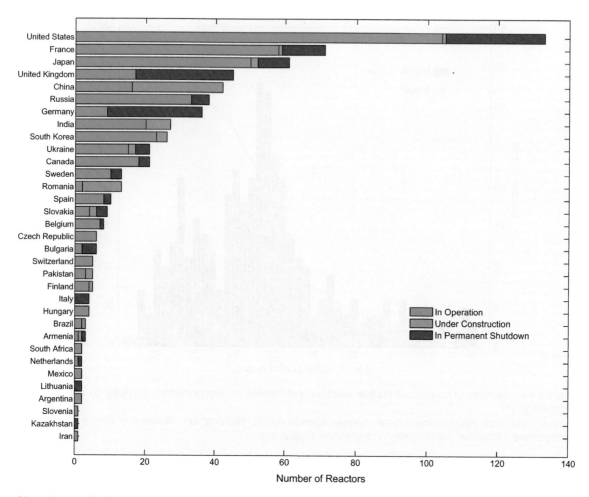

Chart 8.11. Operating, under construction, and shutdown nuclear reactors by country, 2011.
Source: Data from International Atomic Energy Agency (IAEA), Database on Nuclear Power Reactors, <http://www.iaea.org/pris/>.

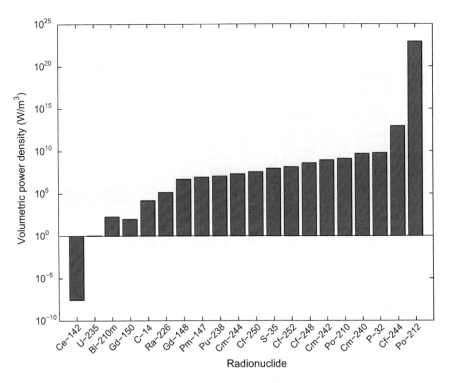

Chart 8.12. Volumetric radioactive power density of selected radionuclides.
Source: Data from Freitas, Robert A. 1999. Nanomedicine, Volume I: Basic Capabilties (Austin, TX, Landes Bioscience).

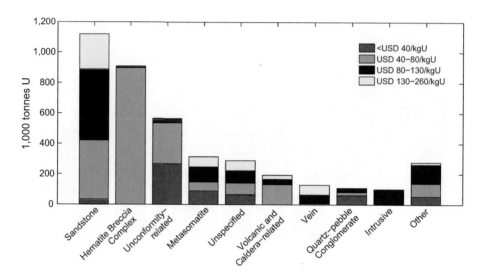

Chart 8.13. Global reasonably assured resources by deposit type, 2010.
Source: Data from Organisation for Economic Cooperation and Development (OECD), Nuclear Energy Agency, <http://www.oecd-nea.org>.

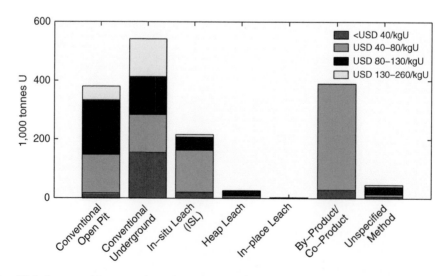

Chart 8.14. Global reasonably assured uranium resources by production method, 2010.
Source: Data from Organisation for Economic Cooperation and Development (OECD), Nuclear Energy Agency, <http://www.oecd-nea.org>.

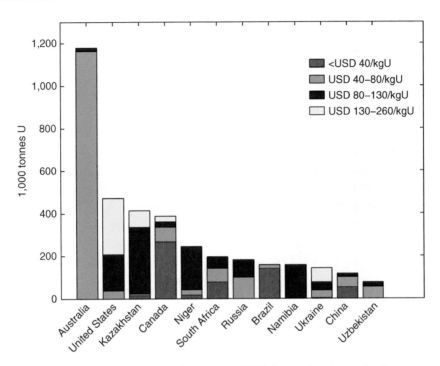

Chart 8.15. Distribution of reasonably assured resources (RAR) for uranium by cost category among countries with major resources. RAR includes uranium that occurs in known mineral deposits of delineated size, grade, and configuration such that the quantities which could be recovered within the given production cost ranges with currently proven mining and processing technology, can be specified.
Source: Data from Organisation for Economic Cooperation and Development (OECD), Nuclear Energy Agency, <http://www.oecd-nea.org>.

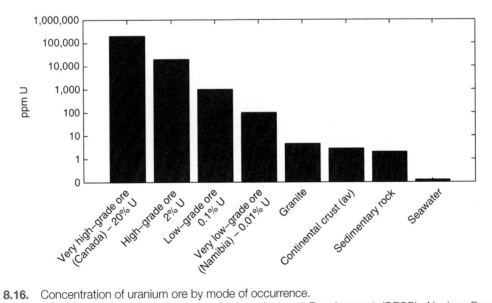

Chart 8.16. Concentration of uranium ore by mode of occurrence.
Source: Data from Organisation for Economic Cooperation and Development (OECD), Nuclear Energy Agency, <http://www.oecd-nea.org>.

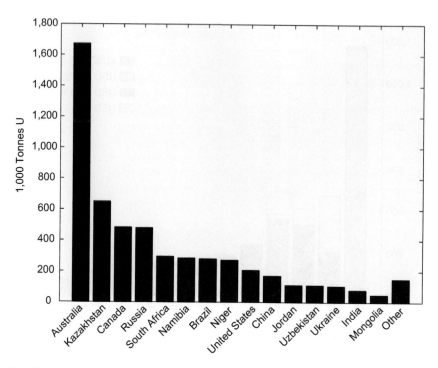

Chart 8.17.   Uranium reserves by nation, 2009.
Source: Data from Organisation for Economic Cooperation and Development (OECD), Nuclear Energy Agency, <http://www.oecd-nea.org>.

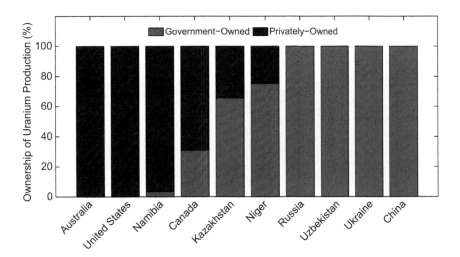

Chart 8.18. Ownership of uranium production in top 10 producing nations, 2010.
Source: Data from Organisation for Economic Cooperation and Development (OECD), Nuclear Energy Agency, <http://www.oecd-nea.org>.

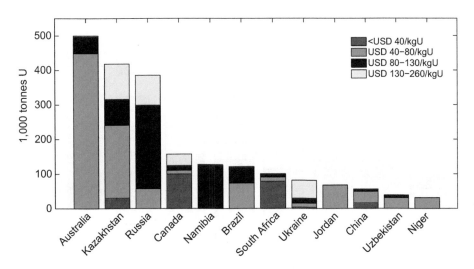

Chart 8.19. Distribution of inferred uranium resources by cost category among countries with major resources. Uranium in addition to reasonably assured resources (RAR) that is inferred to occur based on direct geological evidence, in extensions of well-explored deposits, or in deposits in which geological continuity has been established but where specific data, including measurements of the deposits, and knowledge of the deposit's characteristics are considered to be adequate to classify the resource as RAR.
Source: Data from Organisation for Economic Cooperation and Development (OECD), Nuclear Energy Agency, <http://www.oecd-nea.org>.

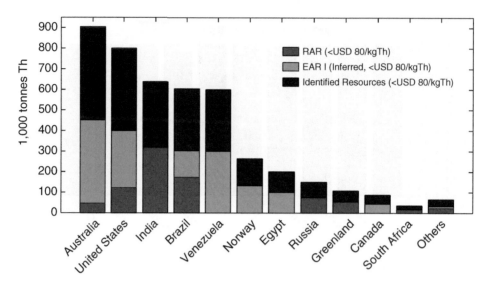

Chart 8.20. World thorium resources by country and category, 2010. RAR=reasonably assured reserves; EAR=estimated additional reserves.
Source: Data from Organisation for Economic Cooperation and Development (OECD), Nuclear Energy Agency, <http://www.oecd-nea.org>.

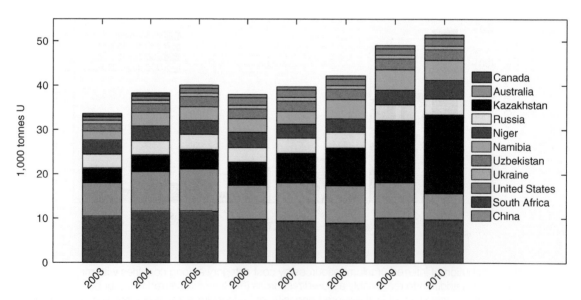

Chart 8.21. Uranium production by top 10 producing nations, 2003-2010.
Source: Data from World Nuclear Association, Public Information Service, <http://www.world-nuclear.org/infomap.aspx>.

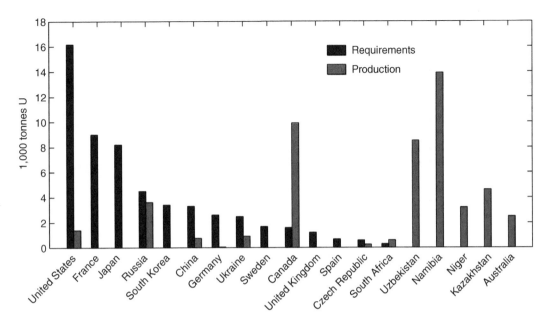

Chart 8.22. Estimated uranium production and reactor-related requirements for major producing and consuming countries, 2009.
Source: Data from Organisation for Economic Cooperation and Development (OECD), Nuclear Energy Agency, <http://www.oecd-nea.org>.

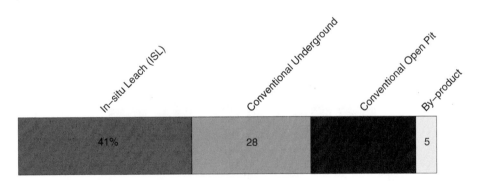

Chart 8.23. Share of world uranium production by mining method, 2010.
Source: Data from World Nuclear Association, Public Information Service, <http://www.world-nuclear.org/infomap.aspx>.

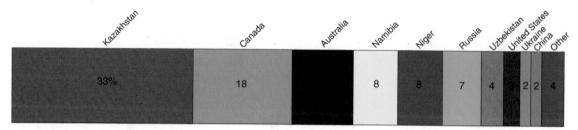

Chart 8.24. Share of uranium production from mines for top 10 producing nations.
Source: Data from World Nuclear Association, Public Information Service, <http://www.world-nuclear.org/infomap.aspx>.

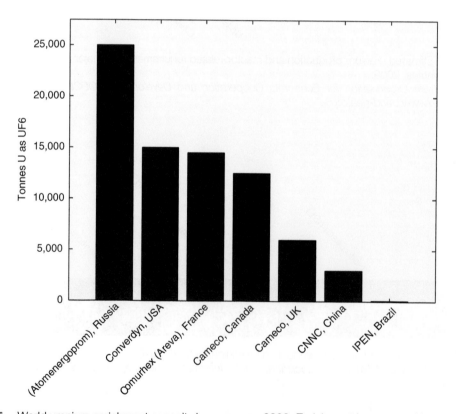

Chart 8.25. World uranium enrichment capacity by company, 2009. Enrichment is measured in separative work units (SWU). A SWU is a unit that expresses the energy required to separate U-235 and U-238.
Source: Data from World Nuclear Association, Public Information Service, <http://www.world-nuclear.org/infomap.aspx>.

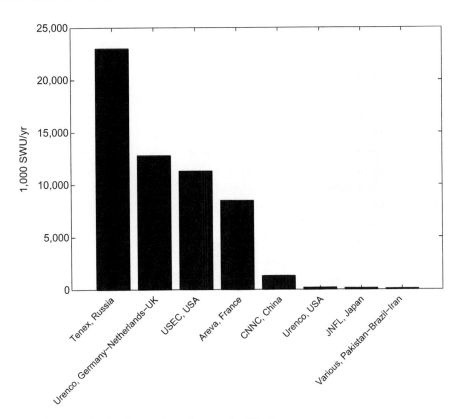

Chart 8.26. Uranium production from mines, by country, 2010.
Source: Data from World Nuclear Association, Public Information Service, <http://www.world-nuclear.org/infomap.aspx>.

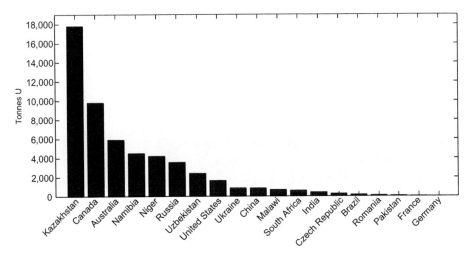

Chart 8.27. Uranium mill tailings for top 10 nations.
Source: Data from WISE Uranium Project, <http://www.wise-uranium.org/>.

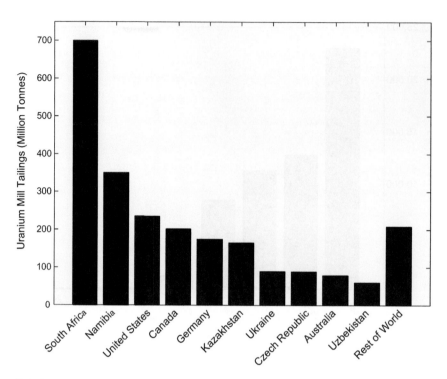

Chart 8.28.   Relative proportions of the major types of fission products and transuranium elements that are found in spent fuel of moderate burnup.
Source: Data from Buck, E.C, Hanson, B.D., McNamara, B.K. 2004. The geochemical behaviour of Tc, Np and Pu in spent nuclear fuel in an oxidizing environment. In: Gieré, R. and P. Stille, Editors, Energy, Waste, and the Environment: a Geochemical Perspective. The Geological Society of London Special Publication 236: pp 65-88.

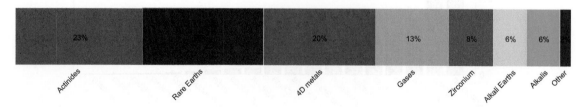

Chart 8.29.   World nuclear waste by waste class, 2010.
Source: Data from International Atomic Energy Agency (IAEA), Net-Enabled Radioactive Waste Management Database (NEWMDB), <http://newmdb.iaea.org>, accessed 10 August 2011.

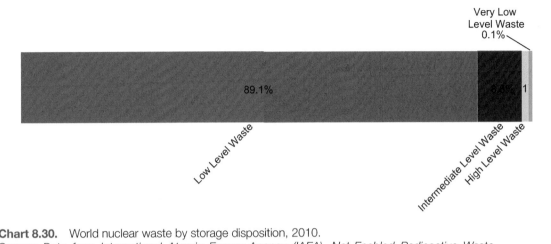

Chart 8.30. World nuclear waste by storage disposition, 2010.
Source: Data from International Atomic Energy Agency (IAEA), Net-Enabled Radioactive Waste Management Database (NEWMDB), <http://newmdb.iaea.org>.

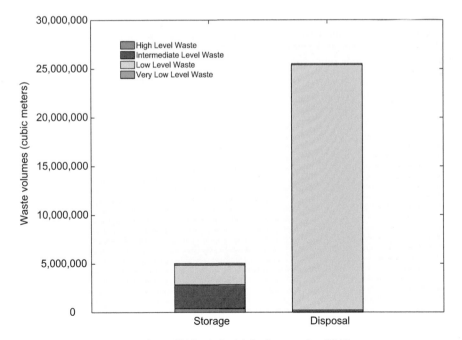

Chart 8.31. Nuclear waste generated per GWh of electricity, by country, 2010.
Source: Data from International Atomic Energy Agency (IAEA), Net-Enabled Radioactive Waste Management Database (NEWMDB), <http://newmdb.iaea.org>.

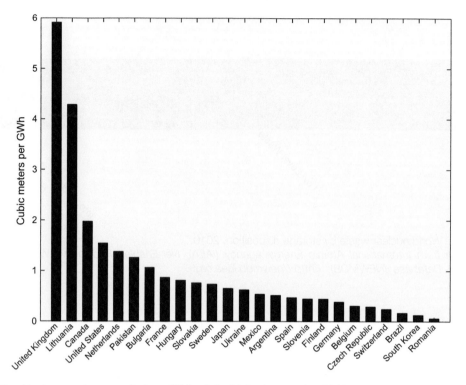

Chart 8.32. Nuclear waste generated per GWh of electricity, by country, 2010.
Source: International Atomic Energy Agency (IAEA), Net-Enabled Radioactive Waste Management Database (NEWMDB), <http://newmdb.iaea.org>.

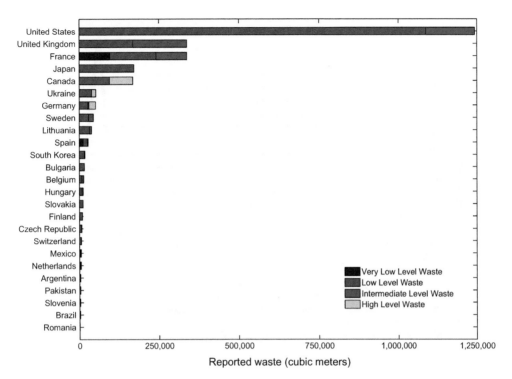

Chart 8.33. Waste generated by nuclear power production, by waste class and country, 2010. *Source: Data from International Atomic Energy Agency (IAEA), Net-Enabled Radioactive Waste Management Database (NEWMDB), <http://newmdb.iaea.org>.*

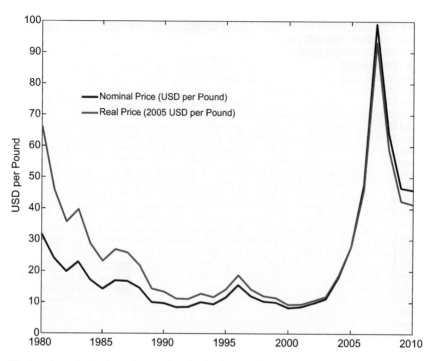

Chart 8.34. Real and nominal prices for uranium in the United States, 1980-2010.
Source: Data from United States Department of Energy, Energy Information Administration,
<http://www.eia.gov/nuclear/data.cfm>.

Tables

Table 8.1. Typical attributes of nuclear power reactors and their fuels

Reactors	PWR	BWR	ALWR		HTGR	FR (CR=0.5)
Thermal Power (MWth)	2647	2647		2647	600	843
Electric Power (MWe)	900	900		900	284	320
Thermal Efficiency (%)	34	34		34	47	38
Capacity Factor (%)	90	90		90	90	85
Technical Lifetime (yr)	50	50		50	50	50
Fuels	UOX	UOX	UOX		MOX	Particle
Average Burnup (GWd/tHM)	50	40	50	50	120	Metal
# fuel batches	5	5	5	3	7	120
Cycle Length (mo)	12	12	12		12	12
Initial U (t/tIHM)	1	1	1		0	1
Initial Enrichment (%)	4.2	3.7	4.2	0.25	15.5	0
Initial DU (t/tIHM)	0	0	0	0.91903	0	
Initial REPU (t/tIHM)	0	0	0		0	0.061
Initial Pu (t/tIHM)	0	0	0	0.08097	0	0.5936
Initial MA (t/tIHM)	0	0	0		0	0.2919
Spent U (t/tIHM)	0.93545	0.94576	0.93545	0.88753	0.85917	0.0535
Spent Enrichment (%)	0.82	0.8	0.82	0.15	4.8	0.5936
Spent Pu (t/tIHM)	0.012	0.1085	0.012	0.05512	0.01883	
Spent MA (t/tIHM)	0.00125	0.00114	0.00125	0.0074	0.002	0.2365
Spent FP (t/tIHM)	0.0513	0.04225	0.0513	0.04996	0.12	0.0452

BWR = boiling water reactor; PWR = pressurized water reactor; ALWR = advanced light water reactor; FR = fast reactor.
CR = conversion ratio (the ratio of the rate of production of fissile materials to the rate of destruction of the existing fissile materials).
UOX = urnaium oxide; MOX = mixed oxide fuel.
Source: Adapted from OECD Nuclear Energy Agency. 2011. A Preliminary assessment of raw material inputs that would be required for rapid growth in nuclear generating capacity, (Paris, France).

Table 8.2.   Characteristics of nuclear reactors

Reactor Type	Function	Coolant	Moderator	Chemical Form of Fuel	Fuel Enrichment Level[a]
Thermal Neutron Reactors					
Boiling Water	electricity	light water	light water	uranium dioxide	low enriched uranium
Pressurized Water	electricity, nautical power	light water	light water	uranium dioxide	low enriched uranium
Heavy Water	electricity, plutonium production	heavy water	heavy water	uranium dioxide or uranium metal	natural, unenriched uranium
Gas Cooled Graphite Moderated	electricity, plutonium production	carbon dioxide or helium	graphite	uranium dicarbide or uranium metal	slightly enriched or natural uranium
Water Cooled Graphite Moderated	electricity, plutonium production	light water	graphite	uranium dicarbide or uranium metal	slightly enriched uranium
Pebble Bed Gas Cooled Graphite Moderated[b]	electricity	pressurized helium	graphite and silicon carbide	uranium dioxide or thorium dioxide	low enriched uranium
Fast Neutron Reactors					
Fast Neutron Breeder	electricity, plutonium production	molten sodium or lead	none required	various mixtures of plutonium dioxide and uranium dioxide	various mixtures of plutonium dioxide and uranium dioxide

Notes:

[a]Percentage of U-235 isotope in the fuel compared to U-238 isotope. Natural uranium contains 0.7% U-235, slightly enriched uranium from 0.8 to 3.0% U-235, and low enriched uranium from 3.0 to 5.0% U-235.

[b]Currently under development.

Source: Settle, Frank, Nuclear Chemistry, Nuclear Reactors, Chemcases.com, <http://www.chemcases.com/nuclear/nc-10.html>.

Table 8.3.   Performance factors for world nuclear power plants, 2000-2011

Year	Energy Availability Factor (%)	Unplanned Capability Loss Factor (%)	Unit Capability Factor (%)
2000	82	4.1	83.5
2001	83.5	3.6	85
2002	83.7	3.7	84.6
2003	80.8	5.8	81.7
2004	83.2	4.6	84
2005	82.8	3.9	84
2006	82.9	4.3	83.9
2007	80.9	5	82.6
2008	80	5.3	80.8
2009	79.4	5.5	80.9
2010	81	5.7	82
2011	78.7	3.8	80.2

The Energy Availability Factor over a specified period is the ratio of the energy that the available capacity could have produced during this period, to the energy that the reference unit power could have produced during the same period.

Unplanned Capability Loss Factor is the ratio of the unplanned energy losses during a given period of time, to the reference energy generation, expressed as a percentage.

Unit Capability Factor is defined as the ratio of the available energy generation over a given time period to the reference energy generation over the same time period, expressed as a percentage.

Source: International Atomic Energy Agency, Power Reactor Information System, <http://pris.iaea.org/Public/home.aspx>.

Table 8.4.   Nuclear reactors and nuclear-generated electricity by country, 2011

Country	Operational		Under Construction		Nuclear Electricity Supplied (GWh)	Nuclear Share (%)
	Number of reactors	Total Net Electrical Capacity (MW)	Number of reactors	Total Net Electrical Capacity (MW)		
ARGENTINA	2	935			5,894	5
ARMENIA	1	375	1	692	2,357	33.2
BELGIUM	7	5,927			45,942	54
BRAZIL	2	1,884	1	1,245	14,795	3.2
BULGARIA	2	1,906			15,264	32.6
CANADA	18	12,604			88,318	15.3
CHINA	16	11,816	26	26,620	82,569	1.8
CZECH REPUBLIC	6	3,766			26,696	33
FINLAND	4	2,736	1	1,600	22,266	31.6
FRANCE	58	63,130	1	1,600	423,509	77.7
GERMANY	9	12,068			102,311	17.8
HUNGARY	4	1,889			14,707	43.2
INDIA	20	4,391	7	4,824	28,948	3.7
IRAN, ISLAMIC REPUBLIC OF	1	915			98	0
JAPAN	50	44,215	2	2,650	156,182	18.1

Continued

Table 8.4.   Continued

Country	Operational		Under Construction		Nuclear Electricity Supplied (GWh)	Nuclear Share (%)
	Number of reactors	Total Net Electrical Capacity (MW)	Number of reactors	Total Net Electrical Capacity (MW)		
KOREA, REPUBLIC OF	23	20,671	3	3,640	147,763	34.6
MEXICO	2	1,300			9,313	3.6
NETHERLANDS	1	482			3,917	3.6
PAKISTAN	3	725	2	630	3,843	3.8
ROMANIA	2	1,300	11	9,270	10,811	19
RUSSIA	33	23,643			162,018	17.6
SLOVAKIA	4	1,816	2	782	14,342	54
SLOVENIA	1	688			5,902	41.7
SOUTH AFRICA	2	1,830			12,939	5.2
SPAIN	8	7,567			55,121	19.5
SWEDEN	10	9,331			58,098	39.6
SWITZERLAND	5	3,263			25,694	40.8
UKRAINE	15	13,107	2	1,900	84,894	47.2
UNITED KINGDOM	17	9,736			62,658	17.8
UNITED STATES OF AMERICA	104	101,465	1	1,165	790,439	19.2
Total	436	370,499	62	59,218	2,517,980	

Source: International Atomic Energy Agency, Power Reactor Information System, <http://pris.iaea.org/public/>.

Table 8.5. Summary of Generation IV nuclear power systems

System	Abbreviation	Neutron spectrum	Coolant	Maximum temperature (°C)	Pressure	Fuel[a]	Fuel cycle	Output (MW$_e$)	Output
Gas-cooled fast reactor	GFR	Fast	Helium	850	High	U-238, MOX	In situ closed	288	Electricity, hydrogen production
Liquid metal (e.g., Pb) cooled fast reactor	LMFR	Fast	Pb-Bi	550-800	Low	U-238, MOX	Closed, regional	50-150, 300-400, 1200	Electricity, hydrogen production
Molten-salt reactor	MSR	Epithermal	Fluoride salts	700-800	Low	UF$_6$ in salt	In situ closed	1000	Electricity, hydrogen production
Sodium-cooled fast reactor	SFR	Fast	Sodium	550	Low	U-238, MOX	Closed	300-1500	Electricity
Supercritical, water-cooled reactor	SCWR	Thermal/ fast	Water	510-550	Very high	UO2	Open (th), closed (f)	1500	Electricity
Very high-temperature gas-cooled reactor (as in the GA system)	VHTR	Thermal	Helium	1000	High	UO2	Open	250	Electricity, hydrogen production

[a]MOX = Mixed oxide fuel, a nuclear fuel that contains more than one oxide of fissile material.

Source: Adapted from Penner, S.S., R. Seisera, and K.R. Schultz, Steps toward passively safe, proliferation-resistant nuclear power, Progress in Energy and Combustion Science. Volume 34, Issue 3, June 2008, Pages 275–287.

Table 8.6.   World light water reactor nuclear fuel fabrication capacity, metric tons per year

	Fabricator	Location	Conversion	Pelletizing	Rod/ assembly
Belgium	AREVA NP-FBFC	Dessel	0	700	700
Brazil	INB	Resende	160	160	280
China	CNNC	Yibin Batou	400	400	450
France	AREVA NP-FBFC	Romans	1,800	1,400	1,400
Germany	AREVA NP-ANF	Lingen	800	650	650
India	DAE Nuclear Fuel Complex	Hyderabad	48	48	48
Japan	NFI (BWR)	Kumatori	0	360	284
	NFI (PWR)	Tokai-Mura	0	250	250
	Mitsubishi Nuclear Fuel	Tokai-Mura	475	440	440
	GNF-J	Kurihama	0	750	750
Kazakhstan	Ulba	Ust Kamenogorsk	2,000	2,000	0
Korea	KNFC	Daejon	600	600	600
Russia	TVEL-MSZ[a]	Elektrostal	1,450	1,200	120
	TVEL-NCCP	Novosibirsk	250	200	400
Spain	ENUSA	Juzbado	0	300	300
Sweden	Westinghouse AB	Vasteras	600	600	600
UK	Westinghouse[b]	Springfields	950	600	860
USA	AREVA Inc	Richland	1,200	1,200	1,200
	Global NF	Wilmington	1,200	1,200	750
	Westinghouse	Columbia	1,500	1,500	1,500
Total			**13,433**	**14,558**	**12,662**

[a]Includes approx. 220 tHM for RBMK reactors.
[b]Includes approx. 200 tHM for AGR reactors.
Source: Adapted from World Nuclear Association.

Table 8.7.   Composition of fresh and spent nuclear reactor fuel from a typical light water reactor

Material	Fresh Fuel %	Spent Fuel %	Type of Waste
Transuranic elements	0.00	0.68	Transuranic
U-236	0.00	0.46	Depleted uranium
Pu isotopes	0.00	0.89	Transuranic
Fission products	0.00	0.35	High level
U-235	3.30	0.08	Depleted uranium
U-238	96.70	94.30	Depleted uranium

After 3 years in a reactor, a fuel rod assembly containing 264 rods weighing 1,450 lbs would contain 12.9 pounds of plutonium, 5.07 pounds of fission products, and 1,367 pounds of U-238.
Source: U.S. Department of Energy.

Table 8.8.   World nuclear enrichment capacity, 2010, and planned, 2020 (thousand SWU/yr)

| Country | Company and plant | Separative Work Units (SWU) | |
		2010	2020
France	Areva, Georges Besse I & II	8500[a]	7050
Germany-Netherlands-UK	Urenco: Gronau, Germanu; Almelo, Netherlands; Capenhurst, UK.	12,800	12300
Japan	JNFL, Rokkaasho	150	1500
USA	USEC, Paducah & Piketon	11,300[a]	3800
USA	Urenco, New Mexico	200	5900
USA	Areva, Idaho Falls	0	3300
USA	Global Laser Enrichment	0	3500
Russia	Tenex: Angarsk, Novouralsk, Zelenogorsk, Seversk	23,000	30-35,000
China	CNNC, Hanzhun & Lanzhou	1300	6-8,000
Pakistan, Brazil, Iran	various	100	300
	Total SWU approx	57,350	74-81,000

[a]Diffusion.
SWU = Separative work units, a measure of the work expended during an enrichment process.
Source: Adapted from World Nuclear Association, http://www.world-nuclear.org.

Table 8.9.   World commercial nuclear fuel reprocessing capacity (metric tons per year)

LWR fuel	France, La Hague	1700	
	UK, Sellafield (THORP)	900	
	Russia, Ozersk (Mayak)	400	
	Japan (Rokkasho)	800[a]	
	Total LWR (approx)		3800
Other nuclear fuels	UK, Sellafield (Magnox)	1500	
	India (PHWR, 4 plants)	330	
	Total other (approx)		1830
Total civilian capacity			5630

[a]Expected to start operation in October 2012.
Source: Adapted from World Nuclear Association, http://www.world-nuclear.org.

Table 8.10.   Worldwide nuclear waste volumes, 2010

Waste Class/Origin	Unprocessed Storage(m³)	Processed Storage (m³)	Unprocessed Disposal (m³)	Processed Disposal (m³)
High Level Waste	370,532	5,678	3,960	0
Decommissioning/Remediation	558	3	3,960	0
Defense	354,998	2,000	0	0
Fuel Fabrication/Enrichment	17	56	0	0
Nuclear Application	66	364	0	0
Reactor Operation	11,355	34	0	0
Reprocessing	3,539	3,222	0	0
Intermediate Level Waste	2,313,321	152,760	124,053	82,791
Decommissioning/Remediation	1,859,466	1,823	117,264	1,070
Defense	82,110	888	6,479	66,918
Fuel Fabrication/Enrichment	59,779	12,848	0	108
Not Determened	149	729	164	369
Nuclear Application	12,200	37,323	79	2,647
Reactor Operation	185,157	21,848	66	11,562
Reprocessing*	114,459	77,301	0	118
Low Level Waste	1,547,157	486,065	20,334,522	4,872,503
Decommissioning/Remediation	1,102,570	38,035	17,263,835	591,622
Defense*	5,308	42,499	1,264,351	1,866,356
Fuel Fabrication/Enrichment	28,152	23,966	104,000	354,380
Not Determened	6,404	5,807	65,345	26,403
Nuclear Application	120,716	67,917	634,234	531,345
Reactor Operation	234,407	266,475	586,758	1,235,053
Reprocessing	49,600	41,367	416,000	267,345
Very Low Level Waste	304	152,846	304	112,413
Decommissioning/Remediation	33	2,349	59	22,528
Defense	0	28,423	5	4,467
Fuel Fabrication/Enrichment	0	33,091	0	17,866
Not Determened	186	0	133	341
Nuclear Application	0	35,666	53	37,519
Reactor Operation	85	24,457	53	26,120
Reprocessing*	0	28,861	0	3,573

Source: International Atomic Energy Agency, Net-Enabled Waste Management Database, <http://newmdb.iaea.org>.

Table 8.11. Waste generated by nuclear power production, by country

Year	Country	Total reported Nuclear Production (GWh)	Total reported Waste (m³)	VLLW (%)	LLW (%)	ILW (%)	HLW (%)	Reported Waste per generated GWh (m³)	Types of Reactors in use
2008	USA	806208	1243300	0	87.5	12.5	0	1.54	35 BWR, 69 PWR
2007	United Kingdom	57249	338353	0	50	50	0	5.91	16 GCR, 1 PWR
2009	France	390000	338292	28.57	42.86	28.57	0	0.87	58 PWR
2009	Japan	263071	171790	0	100	0	0	0.65	26 BWR, 24 PWR
2009	Canada	85315	168318	0	0	56.52	43.48	1.97	18 PHWR
2010	Ukraine	83800	52536	0	74.25	0.75	25	0.63	15 PWR
2007	Germany	133209	51692	0	54	6	40	0.39	2 BWR, 7 PWR
2008	Sweden	61336	45037	0	66.67	33.33	0	0.73	7 BWR, 3 PWR
2008	Lithuania	9140	39141	0	83.33	16.67	0	4.28	
2010	Spain	59256	28196	43.75	47.5	2.5	6.25	0.48	2 BWR, 6 PWR
2010	Korea, Republic of	141894	17835	0	90	10	0	0.13	4 PHWR, 19 PWR
2010	Bulgaria	15249	16206	0	100	0	0	1.06	2 PWR
2007	Belgium	45853	14066	0	60	40	0	0.31	7 PWR
2009	Hungary	14571	11821	5.38	63.08	30	1.54	0.81	4 PWR
2008	Slovakia	15453	11734	0	100	0	0	0.76	4 PWR
2010	Finland	21884	9781	0	99.07	0.93	0	0.45	2 BWR, 2 PWR
2010	Czech Republic	26441	7676	0	90	10	0	0.29	6 PWR
2010	Switzerland	25200	6103	7.15	91.52	1.33	0	0.24	2 BWR, 3 PWR
2009	Mexico	10108	5461	33.33	0	33.33	33.33	0.54	2 BWR
2010	Netherlands	3755	5170	0	30	36.67	33.33	1.38	1 PWR
2010	Argentina	6692	3464	0	87.5	12.5	0	0.52	2 PHWR
2010	Pakistan	2560	3223	0	100	0	0	1.26	1 PHWR, 2 PWR
2010	Slovenia	5381	2420	0	57.14	28.57	14.29	0.45	1 PWR
2008	Brazil	14004	2335	0	66.67	33.33	0	0.17	2 PWR
2008	Romania	10334	613	0	40	43.33	16.67	0.06	2 PHWR

VLLW Very low level waste
LLW Low level waste
ILW Intermediate level waste
HLW High level waste
BWR Boiling Light-Water-Cooled and Moderated Reactor
FBR Fast Breeder Reactor
GCR Gas-Cooled, Graphite-Moderated Reactor
HTGR High-Temperature Gas-Cooled, Graphite-Moderated Reactor
HWGCR Heavy-Water-Moderated, Gas-Cooled Reactor
HWLWR Heavy-Water-Moderated, Boiling Light-Water-Cooled Reactor
LWGR Light-Water-Cooled, Graphite-Moderated Reactor
PHWR Pressurized Heavy-Water-Moderated and Cooled Reactor
PWR Pressurized Light-Water-Moderated and Cooled Reactor
SGHWR Steam-Generating Heavy-Water Reactor
X Others
Source: International Atomic Energy Agency, Net-Enabled Radioactive Waste Management Database, <http://newmdb.iaea.org/>.

Table 8.12.   Typical Characteristics of Nuclear Waste Classes

Waste Classes	Typical Characterisics	Disposal Options
1. Exempt Waste (EW):	Activity levels at or below clearance levels, which are based on an annual dose to members of the public of less than 0.01 mSv	No radiological restrictions
2. Low and Intermediate Level Waste (LILW):	Activity levels above clearance levels and thermal power below about 2kW/m^3	
2.1. Short Lived Waste (LILW-SL):	Restricted long-lived radionuclide concentrations (limitation of long lived alpha emitting radionuclides to 4000 Bq/g in individual waste packages and to an overall average of 400 Bq/g per waste package).	Near surface or geological disposal facility
2.2. Long Lived Waste (LILW-LL):	Long lived radionuclide concentrations exceeding limitations for short lived waste	Geological disposal facility
3. High Level Waste (HLW):	Thermal power above about 2kW/m^3 and long lived radionuclide concentrations exceeding limitations for short lived waste	Geological disposal facility

Source: International Atomic Energy Agency.

Table 8.13.   Options for long-term nuclear waste management

Option	Examples
Near-surface disposal at ground level, or in caverns below ground level (at depths of tens of meters)	• Implemented for LLW in many countries, including Czech Republic, Finland, France, Japan, Netherlands, Spain, Sweden, UK and USA • Implemented in Finland and Sweden for LLW and short-lived ILW
Deep geological disposal (at depths between 250m and 1000 m)	• Most countries with high-level and long-lived radioactive waste have investigated deep geological disposal and it is official policy in various countries (variations also include multinational facilities) • Implemented in USA for defense-related ILW • Preferred sites for HLW/spent fuel selected in France, Sweden, Finland and USA • Geological repository site selection process commenced in UK and Canada
Ideas	**Examples**
Long-term above ground storage	• Investigated in France, Netherlands, and Switzerland, UK and USA • Not currently planned to be implemented anywhere
Disposal in outer space (proposed for wastes that are highly concentrated)	• Investigated by USA • Investigations now abandoned due to cost and potential risks of launch
Deep boreholes (at depths of a few kilometers)	• Investigated by Australia, Denmark, Italy, Russia, Sweden, Switzerland, and USA • Not implemented anywhere
Rock-melting (proposed for wastes that are heat-generating)	• Investigated by Russia, UK and USA • Not implemented anywhere • Laboratory studies performed in the UK

Table 8.13.   Continued

Ideas	Examples
Disposal at subduction zones	• Investigated by USA • Not implemented anywhere • Not permitted by international agreements
Sea disposal	• Implemented by Belgium, France, Germany, Italy, Japan, Netherlands, Russia, South Korea, Switzerland, UK and USA • Not permitted by international agreements
Sub seabed disposal	• Investigated by Sweden and UK (and organizations such as the OECD Nuclear Energy Agency) • Not implemented anywhere • Not permitted by international agreements
Disposal in ice sheets (proposed for wastes that are heat-generating)	• Investigated by USA • Rejected by countries that have signed the Antarctic Treaty or committed to providing solutions within national boundaries
Direct injection (only suitable for liquid wastes)	• Investigated by Russia and USA • Implemented in Russia for 40 years and in USA (grouts) • Investigations abandoned in USA in favor of deep geological disposal of solid wastes

Source: Adapted from World Nuclear Organization, Storage and Disposal Options, Radioactive Waste Management Appendix 2, http://www.world-nuclear.org/info/inf04ap2.html.

Table 8.14.   Inventory of separated recyclable nuclear materials as of August, 2011

	Quantity (metric tons)	Natural U Equivalent (metric tons)
Plutonium from reprocessed fuel	320	60,000
Uranium from reprocessed fuel	45,000	50,000
Ex-military plutonium	70	15,000
Ex-military high-enriched uranium	230	70,000

There is an additional 1.6 million metric tons of enrichment tails with recoverable fissile uranium.
Source: Adapted from World Nuclear Association, <http://www.world-nuclear.org/info/inf29.html#1>.

Table 8.15. Concentration of fission products and those that result from neutron capture and decay reactions in a low burn-up (30 MWD/kg U) spent nuclear fuel

Element	Concentration (ppm)	Notable long-lived radionuclides (isotope (half-live))
Xe	5657	
I	259	^{129}I (16 million years)
Cs	2605	^{135}Cs (2.3 million years)
Sr	794	
Ba	1750	
Se	58	^{79}Se (1 million years)
Te	529	
Zr	3639	
Mo	3497	
Tc	799	^{99}Tc (200,000 years)
Ru	2404	
Rh	484	
Pd	1684	
Ag	92	
La	1269	
Ce	2469	
Pr	1161	
Nd	4190	
Sm	815	
Eu	155	
Gd	142	
Np-237	468	^{237}Np (2.1 million years)
Pu (total)	9459	^{239}Pu (24,100 years)
Am (total)	484	^{241}Am (432 years)
Cm (total)	39	

Source: Adapted from Bruno, Jordi and Rodney C. Ewing. 2006. Spent Nuclear Fuel. *Elements,* 2:343-349.

Table 8.16. Neutrons from fission of various elements

Fuel	Neutrons per Fission	Neutrons per absorption
Uranium-233	2.51	2.28
Uranium-235	2.43	2.07
Natural uranium	2.43	1.34
Plutonium-239	2.9	2.1
Plutonium-241	3.06	2.24

Source: Adapted from Babcock and Wilcox. 1978. Steam: Its Generation and Use (Charlotte, Babcock and Wilcox).

Table 8.17. Composition of plutonium produced from different sources

Reactor Type	Mean fuel burn-up (MW d/t)	Percentage of Pu Isotopes at Discharge					Fissile Content %
		Pu-238	Pu-239	Pu-240	Pu-241	Pu-242	
Magnox	3,000	0.1	80	16.9	2.7	0.3	82.7
	5,000	N/A	68.5	25	5.3	1.2	73.8
CANDU	7,500	N/A	66.6	26.6	5.3	1.5	71.9
AGR	18,000	0.6	53.7	30.8	9.9	5.0	63.6
BWR	27,500	2.6	59.8	23.7	10.6	3.3	70.4
	30,400	N/A	56.8	23.8	14.3	5.1	71.1
PWR	33,000	1.3	56.6	23.2	13.9	4.7	70.5
	43,000	2.0	52.5	24.1	14.7	6.2	67.2
	53,000	2.7	50.4	24.1	15.2	7.1	65.6

PWR = pressurized water reactor; BWR = boiling water reactor; magox = pressurized, carbon dioxide cooled, graphite moderated reactors using natural uranium as fuel and magnox alloy as fuel cladding; CANDU = CANada Deuterium Uranium, a pressurized heavy water reactor. The acronym refers to its deuterium-oxide (heavy water) moderator and its use of uranium fuel. AGR = advanced gas-cooled reactor. Source: Adapted from World Nuclear Association.

Table 8.18. Uranium-238 decay series

Symbol	Element	Radiation	Half-Life	Decay Product
U-238	Uranium-238	alpha	4,460,000,000 years	Th-234
Th-234	Thorium-234	beta	24.1 days	Pa-234
Pa-234	Protactinium-234	beta	1.17 minutes	U-234
U-234	Uranium-234	alpha	247,000 years	Th-230
Th-230	Thorium-230	alpha	80,000 years	Ra-226
Ra-226	Radium-226	alpha	1,602 years	Rn-222
Rn-222	Radon-222	alpha	3.82 days	Po-218
Po-218	Polonium-218	alpha	3.05 minutes	Pb-214
Pb-214	Lead-214	beta	27 minutes	Bi-214
Bi-214	Bismuth-214	beta	19.7 minutes	Po-214
Po-214	Polonium-214	alpha	1 microsecond	Pb-210
Pb-210	Lead-210	beta	22.3 years	Bi-210
Bi-210	Bismuth-210	beta	5.01 days	Po-210
Po-210	Polonium-210	alpha	138.4 days	Pb-206
Pb-206	Lead-206	none	stable	(none)

Source: New York State Department of Health, <http://www.health.ny.gov/environmental/radiological/radon/chain.htm>.

Table 8.19.　Activity releases of globally dispersed radionuclides from reactors and reprocessing plants

Years	Electrical energy generated (GW a)	Fuel reprocessed (GW a)	Release (TBq)				
			3H	3H (to sea)	14C	85Kr	129I
Pre-1970	28.8	2.3	2 146	919	38	32 060	0.11
1970-1974	87.7	7.04	6 543	2 809	116	97 970	0.32
1975-1979	277	22.2	24 200	8 858	364	308 900	1.01
1980-1984	514	36.3	44 330	13 640	523	424 400	1.53
1985-1989	937	62.5	77 960	23 660	672	454 000	1.79
1990-1994	1147	130	98 900	35 390	650	823 700	3.87
1995-1997	767	160	42 830	40 770	442	1 102 000	6.14
Total	3757	420	296 900	126 000	2 805	3 243 000	14.8

Source: United Nations Scientific Committee on the Effects of Atomic Radiation (UNSCEAR), UNSCEAR 2000 Report Vol. 1: Sources and Effects of Ionizing Radiation. United Nations Scientific Committee on the Effects of Atomic Radiation.

Table 18.20.　Example ranges and/or averages of radionuclide activity concentrations in coal (Bq/kg)

Country	^{238}U	^{230}Th	^{226}Ra	^{210}Pb	^{210}Po	^{232}Th	^{228}Ra	^{40}K
Australia	8.5–47	21–68	19–24	20–33	16–28	11–69	11–64	23–140
Brazil[a]	72		72	72		62	62	
Egypt	59		26			8	8	
Germany			10–145 32[a] <158		10–63 21[a] <1-58		10–700 225[a] <4-220	
Lignite								
Former East Germany			10[a]			8[a]	22[a]	
Greece[b]	117–390		44–206	59–205			9–41	
Hungary	20–480					12–97		30–384
Italy[c]	23+-3					18+-4		218+-15
Poland	<159					<123		<785
	18[d]					11[d]		
Romania	<415		<557	<510	<580	<170		
	80[a]		126[a]	210[a]	262[a]	62[a]		
UK	7–19	8.5–25.5	7.8–21.8		7–19			
USA	6.3–73		8.9–59.2	12.2–77.7	3.3–51.8	3.7–21.1		

Notes:
[a]Average
[b]Lignite
[c]Lignite, average
[d]Average for all coal seams

Source: International Atomic Energy Agency. 2003. Extent of environmental contamination by naturally occurring radioactive material (NORM) and technological options for mitigation. Technical Reports Series No. 419.

Table 8.21. **Radionuclides in oil and gas residues**

Material	Example activity concentrations (Bq/kg)
Scale in downhole tubing, pipes, and other equipment for handling oil/gas and formation waters	^{226}Ra: Background to 15 000 000 (average 1000 to hundreds of thousands)
Sludges in separations and production equipment	^{226}Ra: 10 000 to 1 000 000
Sludges, films in natural gas supply equipment	^{210}Pb: Background to about 40 000
Sludges from soils beneath ponds of produced water	^{226}Ra: 10 000 to 40 000

The becquerel (symbol Bq) is the SI-derived unit of radioactivity. One Bq is defined as the activity of a quantity of radioactive material in which one nucleus decays per second. The Bq unit is thus equivalent to an inverse second, s^{-1}.
Source: International Atomic Energy Agency. 2003. Extent of environmental contamination by naturally occurring radioactive material (NORM) and technological options for mitigation. Technical Reports Series No. 419.

Table 8.22. **Human Produced Nuclides**

Nuclide	Symbol	Half-life	Source
Tritium	^{3}H	12.3 yr	Produced from weapons testing and fission reactors; reprocessing facilities, nuclear weapons manufacturing
Iodine-131	^{131}I	8.04 days	Fission product produced from weapons testing and fission reactors, used in medical treatment of thyroid problems
Iodine-129	^{129}I	1.57×10^7 yr	Fission product produced from weapons testing and fission reactors
Cesium-137	^{137}Cs	30.17 yr	Fission product produced from weapons testing and fission reactors
Strontium-90	^{90}Sr	28.78 yr	Fission product produced from weapons testing and fission reactors
Technetium-99	^{99}Tc	2.11×10^5 yr	Decay product of ^{99}Mo, used in medical diagnosis
Plutonium-239	^{239}Pu	2.41×10^4 yr	Produced by neutron bombardment of ^{238}U (^{238}U + n--> ^{239}U--> ^{239}Np +β--> ^{239}Pu+β)

Source: Adapted from The Radiation Information Network, Idaho State University, Radioactivity in Nature, <http://www.physics.isu.edu/radinf/natural.htm>.

Renewables

Charts

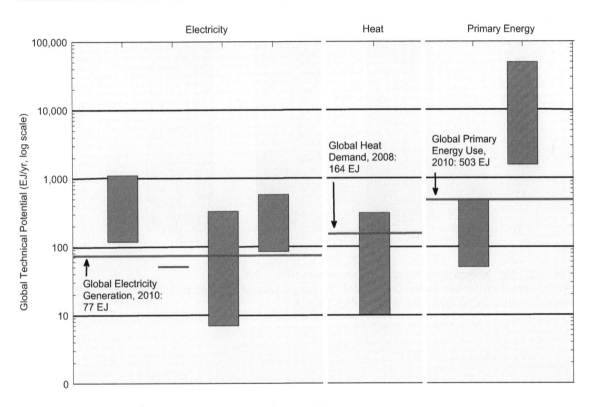

Chart 9.1. Range of global technical potentials of renewable sources.
Source: Data from Moomaw, W., F. Yamba, M. Kamimoto, L. Maurice, J. Nyboer, K. Urama, T. Weir, 2011: Introduction. In IPCC Special Report on Renewable Energy Sources and Climate Change Mitigation [O. Edenhofer, R. Pichs-Madruga, Y. Sokona, K. Seyboth, P. Matschoss, S. Kadner, T. Zwickel, P. Eickemeier, G. Hansen, S. Schlömer, C.von Stechow (eds)], Cambridge University Press, Cambridge, United Kingdom and New York, NY, USA.

Handbook of Energy. http://dx.doi.org/10.1016/B978-0-08-046405-3.00009-7

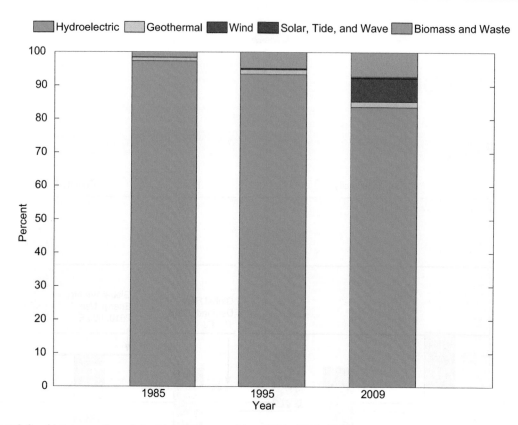

Chart 9.2. Net generation of electricity by renewables 1985, 1995, 2009.
Source: Data from United States Department of Energy, Energy Information Administration, International Energy Statistics, <http://www.eia.gov/countries/data.cfm>

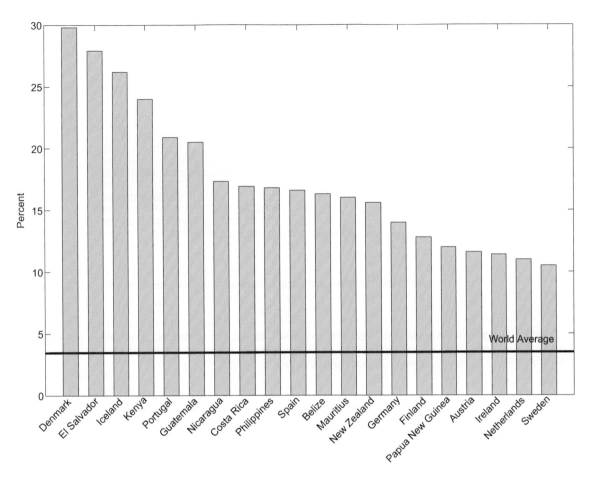

Chart 9.3.   Net generation of electricity from non-hydro renewables, top 20 nations, 2009.
Source: Data from United States Department of Energy, Energy Information Administration, International Energy Statistics, <http://www.eia.gov/countries/data.cfm>

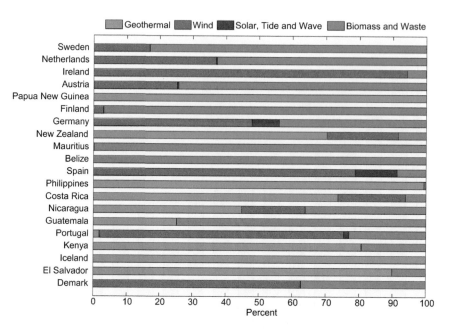

Chart 9.4. Net generation of electricity from non-hydro renewables, top 20 nations, by source of fuel, 2009. *Source: Data from United States Department of Energy, Energy Information Administration, International Energy Statistics, <http://www.eia.gov/countries/data.cfm>*

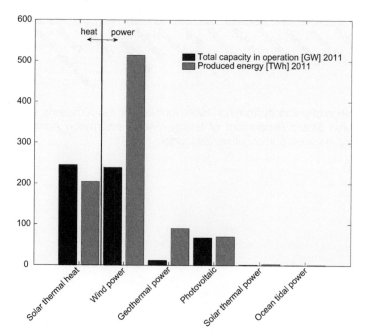

Chart 9.5. Thermal energy and electricity produced from renewable sources in the world, 2011. *Source: Data from Weiss, Werner and Franz Mauthner. 2012. Solar Heat Worldwide, Markets and Contribution to the Energy Supply 2010, Edition 2012, (Gleisdorf, AEE - Institute for Sustainable Technologies; Paris, Solar Heating and Cooling Programme, International Energy Agency).*

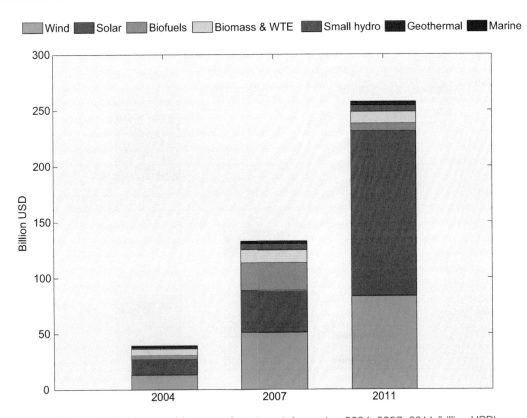

Chart 9.6.   Global trends in renewable energy investment, by sector, 2004, 2007, 2011 (billion USD). WTE = waste-to-energy

Source: Data from Frankfurt School of Finance and Management gGmbH. 2012. Global trends in renewable energy investment 2012, (Frankfurt, Frankfurt School of Finance and Management gGmbH).

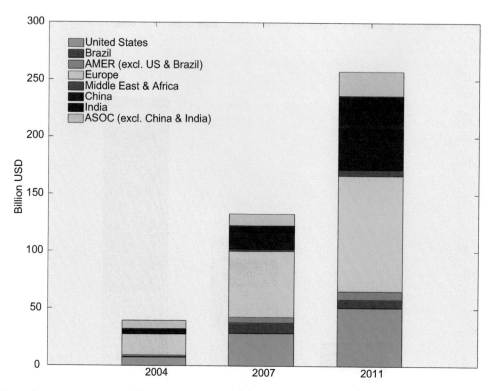

Chart 9.7.   Global trends in renewable energy investment, by region 2004/2007/2011.
Source: Data from Frankfurt School of Finance and Management gGmbH. 2012. Global trends in renewable energy investment 2012, (Frankfurt, Frankfurt School of Finance and Management gGmbH)

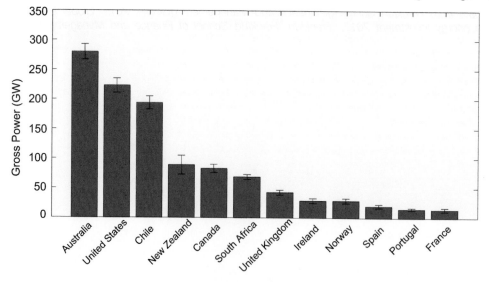

Chart 9.8.   Annual mean wave power resource potential for countries, with 95% confidence intervals.
Source: Data from Gunn, Kester, Clym Stock-Williams, Quantifying the global wave power resource, Renewable Energy, Volume 44, August 2012, Pages 296-304.

Tables

Table 9.1.   Estimates for renewable energy global technical potential (EJ/year)

Study and year of estimate	Solar	Wind	Ocean	Hydro	Biomass	Geothermal Electricity	Heat
Hafele (1981) ("realizable" potential)	NA	95 (32)	33 (16)	95 (47)	189 (161)	3.2 (3.2)	47 (16)
Rogner et al. (2000)	>1,175[e]	640	NA	50	>276	5000[f]	NA
Lightfoot/Green (2002) (range of values)	163 (118–206)	72 (48–72)	0 (1.8–3.6)	19 (1–19)]	539 (373–772)	1.5 (1.5)	NA (NA)
Gross et al. (2003)	43–114	72–144[a]	7–14[b]	NA	29‚Äì90	NA	14–144[c]
Johansson et al. (2004)	>1600	600	NA	50	>250		5000[c]
Sims et al. (2007)	1650	600	7	62	250	NA	5000[c]
Field et al. (2008)	NA	NA	NA	NA	27	NA	NA
Resch et al. (2008)	1600	600	NA	50	250	NA	5000[c]
Krewitt et al. (2009)	7538[g]	398	166	49	129	18[g]	429
Klimenko et al. (2009) ("economic" potential)	2592 (19)	191 (8.6)	22 (2.2)	54 (29)	NA (NA)	22 (3.6)	NA (NA)
Cho (2010)	>1577	631	NA	50	284	NA	120
Tomabechi (2010)[d]	1600	700	11	59	200	NA	310,000[c]
WEC (2010)	NA	NA	7.6[b]	57.4	50–1500	1.1–4.4	140
Arvizu et al. (2011)	1,575-49,837[e]	85-850	7-331	50-52	50-500[e]	118-1,109[g]	10-312[h]
All studies range	118–49,837	48–850	1.8–331	50–95	27–1500	1.1–5,000	14–310,000

[a]Onshore only.
[b]Wave only.
[c]Includes both electricity and direct heat.
[d]'Usable maximum'
[e]Primary energy, i.e., energy before conversion to secondary or final energy
[f]Geothermal is estimated technically recoverable primary heat flow
[g]Electricity only
[h]Heat from geothermal only

References:

- R.E.H. Sims, R.N. Schock, A. Adegbululgbe, J. Fenhann, I. Konstantinaviciute, W. Moomaw et al., Energy supply, B. Metz, O.R. Davidson, P.R. Bosch, R. Dave, L.A. Meyer (Eds.), Climate change 2007: mitigation, CUP, Cambridge, UK (2007), pp. 251–322.

- W. Hafele, Energy in a finite world: a global systems analysis, Ballinger, Cambridge, MA (1981).

- Lightfoot HD, Green C. An assessment of IPCC Working Group 111 findings in Climate Change 2001: mitigation of the potential contribution of renewable energies to atmospheric carbon dioxide stabilization. Centre for Climate and Global Change Research (C²GCR) C²GCR Report No. 2002–5, 2002.

- R. Gross, M. Leach, A. Bauen, Progress in renewable energy, Environ Int, 29 (2003), pp. 105–122.

- Johansson, Thomas B. J., Kes McCormick, Lena Neij, Wim Turkenburg. 2004.The Potentials of Renewable Energy: Thematic Background Paper for the International Conference for Renewable Energies, Bonn 2004.

- C.B. Field, J.E. Campbell, D.B. Lobell, Biomass energy: the scale of the potential resource, Trends Ecol Evol, 23 (2) (2008), pp. 65–72.

- G. Resch, A. Held, T. Faber, C. Panzer, F. Toro, R. Haas, Potentials and prospects for renewable energies at global scale, Energy Policy, 36 (2008), pp. 4048–4056.

- V.V. Klimenko, A.G. Tereshin, O.V. Mikushina, Global energy and climate of the planet in the XXI century in the context of historical trends, Russ J Gen Chem, 79 (11) (2009), pp. 2469–2476.
- A. Cho, Energy's tricky tradeoffs, Science, 329 (2010), pp. 786–787.
- K. Tomabechi, Energy resources in the future, Energies, 3 (2010), pp. 686–695 doi:10.3390/en3040686
- World Energy Council (WEC), 2010 survey of energy resources, WEC, London (2010).
- Rogner, H.-H., F. Barthel, M. Cabrera, A. Faaij, M. Giroux, D. Hall, V. Kagramanian, S. Kononov, T. Lefevre, R. Moreira, R. Nötstaller, P. Odell, and M. Taylor (2000). Energy resources. In: World Energy Assessment. Energy and the Challenge of Sustainability. United Nations Development Programme, United Nations Department of Economic and Social Affairs, World Energy Council, New York, USA, pp. 508.
- Krewitt, W., K. Nienhaus, C. Kleßmann, C. Capone, E. Stricker, W. Graus, M. Hoogwijk, N. Supersberger, U. von Winterfeld, and S. Samadi (2009). Role and Potential of Renewable Energy and Energy Effi ciency for Global Energy Supply. Climate Change 18/2009, ISSN 1862-4359, Federal Environment Agency, Dessau-Roßlau, Germany, pp. 336.
- Arvizu, D., T. Bruckner, H. Chum, O. Edenhofer, S. Estefen, A. Faaij, M. Fischedick, G. Hansen, G. Hiriart, O. Hohmeyer, K. G. T. Hollands, J. Huckerby, S. Kadner, Å. Killingtveit, A. Kumar, A. Lewis, O. Lucon, P. Matschoss, L. Maurice, M. Mirza, C. Mitchell, W. Moomaw, J. Moreira, L. J. Nilsson, J. Nyboer, R. Pichs-Madruga, J. Sathaye, J. Sawin, R. Schaeffer, T. Schei, S. Schlömer, K. Seyboth, R. Sims, G. Sinden, Y. Sokona, C. von Stechow, J. Steckel, A. Verbruggen, R. Wiser, F. Yamba, T. Zwickel, 2011: Technical Summary. In IPCC Special Report on Renewable Energy Sources and Climate Change Mitigation [O. Edenhofer, R. Pichs-Madruga, Y. Sokona, K. Seyboth, P. Matschoss, S. Kadner, T. Zwickel, P. Eickemeier, G. Hansen, S. Schlömer, C. von Stechow (eds)], Cambridge University Press, Cambridge, United Kingdom and New York, NY, USA.

Source: Adapted from Moriarty, Patrick, Damon Honnery. 2012. What is the global potential for renewable energy?, Renewable and Sustainable Energy Reviews, Volume 16, Issue 1, Pages 244–252.

Solar

Figures

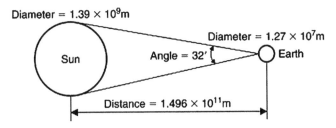

Figure 10.1. Sun-Earth relationship.
Source: Kalogirou, S.A. 2012. 3.01 - Solar Thermal Systems: Components and Applications – Introduction, In: Ali Sayigh, Editor-in-Chief, Comprehensive Renewable Energy, (Oxford, Elsevier), Pages 1-25.

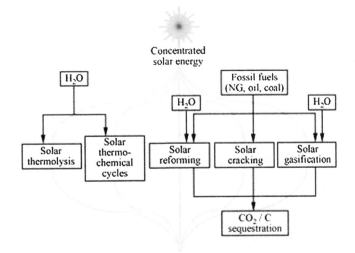

Figure 10.2. Routes for solar hydrogen.
Source: Chen, Haisheng, Thang Ngoc Cong, Wei Yang, Chunqing Tan, Yongliang Li, Yulong Ding. 2009. Progress in electrical energy storage system: A critical review, Progress in Natural Science, Volume 19, Issue 3, Pages 291-312.

Handbook of Energy. http://dx.doi.org/10.1016/B978-0-08-046405-3.00010-3

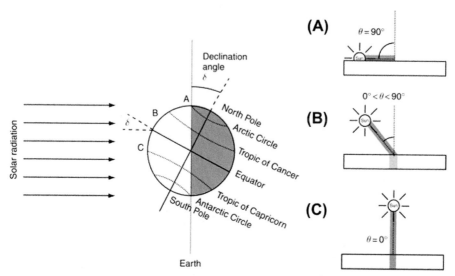

Figure 10.3. Effects of the orientation of the Earth's surfaces toward the Sun on the amount of incident solar radiation at different locations of the Earth's surface. Left: The amount of solar radiation that reaches the surface at a given latitude depends on the declination angle δ. The declination angle measures the angle between the Earth's axis of rotation and the vertical plane of the orbit, or, alternatively, the angle between the direction of solar radiation and the Earth's equator. The declination angle defines Earth's major regions: the tropics (latitudes −δ to + δ) and the polar regions (90° − δ to pole). Earth's declination angle is currently at 23.45°. Right: At a given location on Earth, the zenith angle measures the angle between the vertical and the Sun. It depends on hour, latitude, and time of year. In the situation shown on the left (Northern Hemisphere winter solstice), the zenith angle at location A at noon is 90°, that is, the Sun does not rise above the horizon and no solar radiation is incident at the surface. At location B, the zenith angle is in between 0° and 90°. At location C at the Tropic of Capricorn, the zenith angle is 0° at noon, and the incoming solar radiation is vertical to the surface. In sloped terrain, a correction needs to be applied for the calculation of incident radiation to correct for the slope.
Source: Kleidon, A. 2008. Energy Balance, In: Sven Erik Jorgensen and Brian Fath, Editors-in-Chief, Encyclopedia of Ecology, (Oxford, Academic Press), Pages 1276-1289.

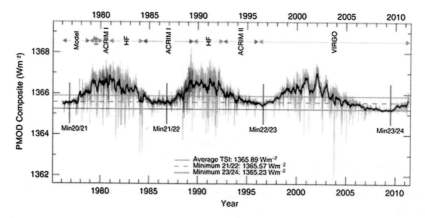

Figure 10.4. Solar irradiance measurements at top of the atmosphere from airborne sensors onboard satellites in the period 1976–2008. The fluctuation of the measurements is due to the 22-year Sun spot cycle. The various sensors are shown at the top of the figure.
Source: Kambezidis, H.D. 2012. The Solar Resource, In: Ali Sayigh, Editor-in-Chief, Comprehensive Renewable Energy, (Oxford, Elsevier), 2012, Pages 27-84.

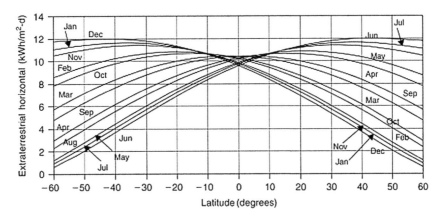

Figure 10.5. Monthly average daily extraterrestrial insolation on horizontal surface.
Source: Kalogirou, S.A. 2012. Solar Thermal Systems: Components and Applications – Introduction, In: Ali Sayigh, Editor-in-Chief, Comprehensive Renewable Energy, (Oxford, Elsevier), Pages 1-25.

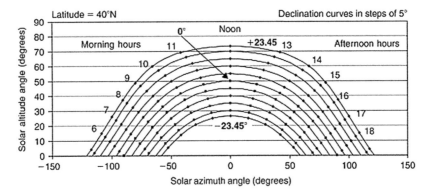

Figure 10.6. Sun path diagram for 40°N latitude.
Source: Kalogirou, S.A. 2012. Solar Thermal Systems: Components and Applications – Introduction, In: Ali Sayigh, Editor-in-Chief, Comprehensive Renewable Energy, (Oxford, Elsevier), Pages 1-25.

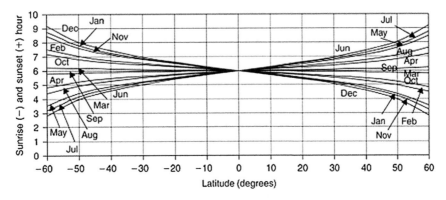

Figure 10.7. Sunset and sunrise angles in various months for a number of latitudes.
Source: Kalogirou, S.A. 2012. Solar Thermal Systems: Components and Applications – Introduction, In: Ali Sayigh, Editor-in-Chief, Comprehensive Renewable Energy, (Oxford, Elsevier), Pages 1-25.

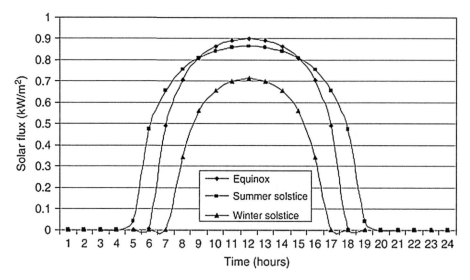

Figure 10.8. Daily variation of solar flux: polar N-S axis with E-W tracking.
Source: Kalogirou, Soteris A. 2009. Solar Energy Engineering, (Boston, Academic Press).

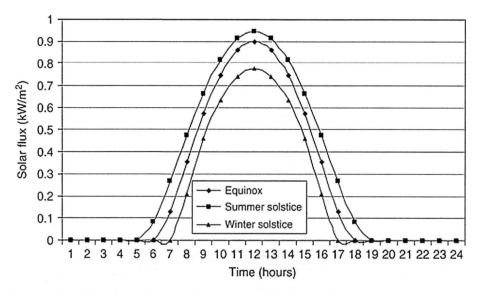

Figure 10.9. Daily variation of solar flux- tilted N-S axis with tilt adjusted daily.
Source: Kalogirou, Soteris A. 2009. Solar Energy Engineering, (Boston, Academic Press).

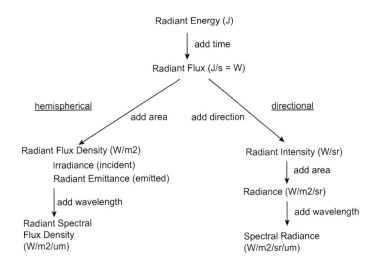

Figure 10.10. Relationships among radiant energy terms.
Source: Adapted from Campbell, Gaylon S. and John M. Norman. 1998. An Introduction to Environmental Biophysics, (New York, Springer-Verlag).

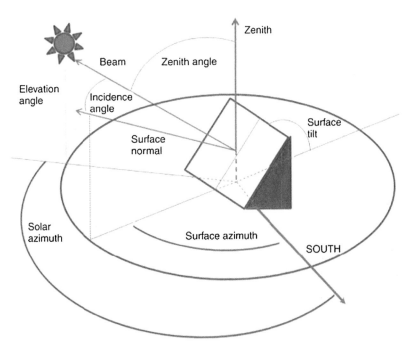

Figure 10.11. Solar geometry terms for a general case.
Source: Myers, D.R. 2012. Solar Radiation Resource Assessment for Renewable Energy Conversion, In: Ali Sayigh, Editor-in-Chief, Comprehensive Renewable Energy, (Oxford, Elsevier), Pages 213-237.

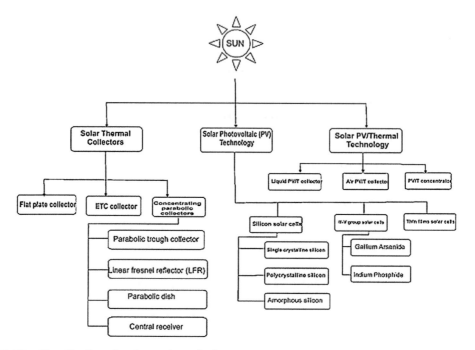

Figure 10.12. Classification of various solar collectors.
Source: Tyagi, V.V., S.C. Kaushik, S.K. Tyag. 2012 Advancement in solar photovoltaic/thermal (PV/T) hybrid collector technology, Renewable and Sustainable Energy Reviews, Volume 16, Issue 3, Pages 1383-1398.

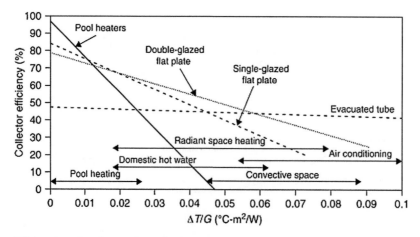

Figure 10.13. Efficiencies of various solar collectors.
Source: Kalogirou, Soteris A. 2009. Solar Energy Engineering, (Boston, Academic Press).

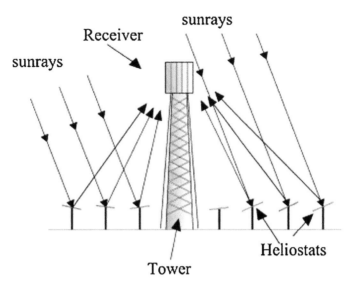

Figure 10.14. A central receiver system.
Source: Tyagi, V.V., S.C. Kaushik, S.K. Tyag. 2012 Advancement in solar photovoltaic/thermal (PV/T) hybrid collector technology, Renewable and Sustainable Energy Reviews, Volume 16, Issue 3, Pages 1383-1398.

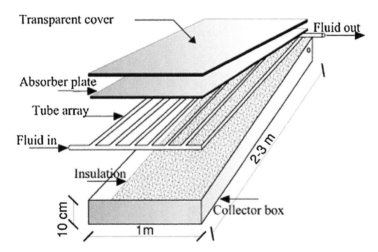

Figure 10.15. Cross-section and isometric view of flat-plate solar collector.
Source: Tyagi, V.V., S.C. Kaushik, S.K. Tyag. 2012 Advancement in solar photovoltaic/thermal (PV/T) hybrid collector technology, Renewable and Sustainable Energy Reviews, Volume 16, Issue 3, Pages 1383-1398.

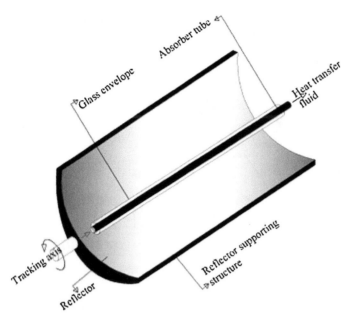

Figure 10.16.   Solar parabolic trough collector.
Source: Tyagi, V.V., S.C. Kaushik, S.K. Tyag. 2012 Advancement in solar photovoltaic/thermal (PV/T) hybrid collector technology, Renewable and Sustainable Energy Reviews, Volume 16, Issue 3, Pages 1383-1398.

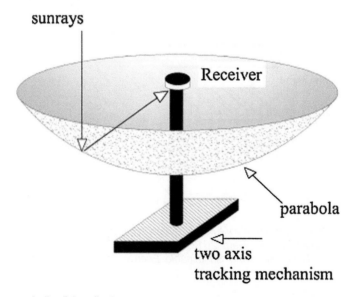

Figure 10.17.   Solar parabolic dish collector.
Source: Tyagi, V.V., S.C. Kaushik, S.K. Tyag. 2012 Advancement in solar photovoltaic/thermal (PV/T) hybrid collector technology, Renewable and Sustainable Energy Reviews, Volume 16, Issue 3, Pages 1383-1398.

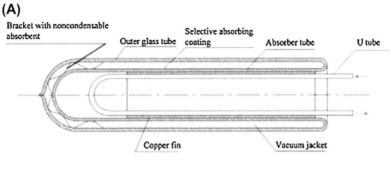

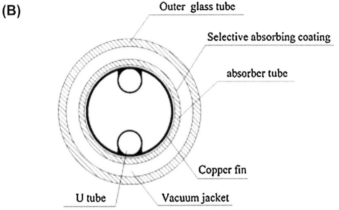

Figure 10.18. Glass evacuated tube solar collector with U-tube. (a) Illustration of the glass evacuated tube and (b) cross section.
Source: Tyagi, V.V., S.C. Kaushik, S.K. Tyag. 2012 Advancement in solar photovoltaic/thermal (PV/T) hybrid collector technology, Renewable and Sustainable Energy Reviews, Volume 16, Issue 3, Pages 1383-1398.

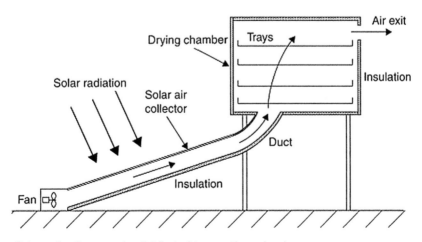

Figure 10.19. Schematic diagram of a distributed-type active solar dryer.
Source: Kalogirou, Soteris A. 2009. Solar Energy Engineering, (Boston, Academic Press).

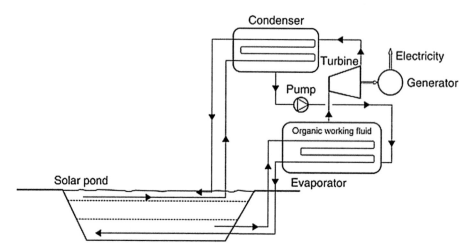

Figure 10.20. Schematic of a solar pond power generation system.
Source: Kalogirou, Soteris A. 2009. Solar Energy Engineering, (Boston, Academic Press).

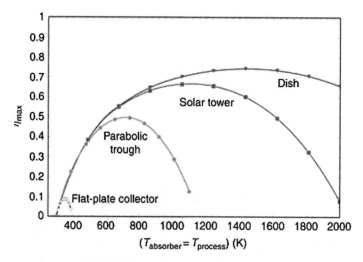

Figure 10.21. Upper boundary of solar thermal power plant efficiencies.
Source: Hoffschmidt, B., S. Alexopoulos, C. Rau, J. Sattler, A. Anthrakidis, C. Boura, B. O'Connor, P. Hilger. 2012. Concentrating Solar Power, In: Ali Sayigh, Editor-in-Chief, Comprehensive Renewable Energy, (Oxford, Elsevier), Pages 595-636.

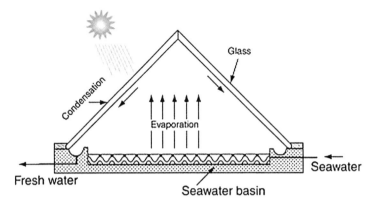

Figure 10.22. Simplified view of a solar still.
Source: Tzen, E., G. Zaragoza, D.-C. Alarcón Padilla. 2012. Solar Desalination, In:Ali Sayigh, Editor-in-Chief, Comprehensive Renewable Energy, (Oxford,Elsevier), Pages 529-565.

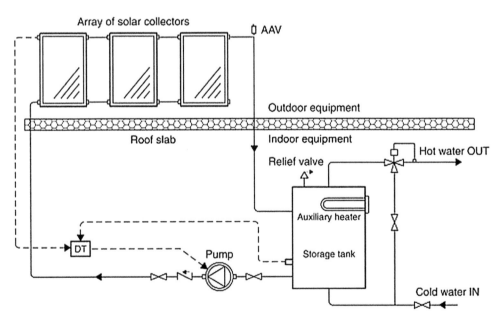

Figure 10.23. Direct circulation active solar hot water system.
Source: Sayigh, Ali. 2012. Editor-in-Chief, Comprehensive Renewable Energy, (Oxford,Elsevier).

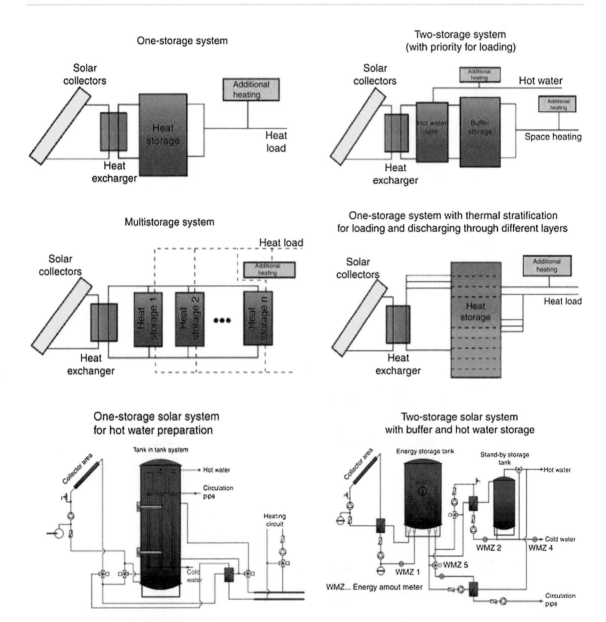

Figure 10.24. Hydraulic principle schematics for solar heating systems with thermal energy storage. *Source: Faninger, G. 2012. Solar Hot Water Heating Systems, In: Ali Sayigh, Editor-in-Chief, Comprehensive Renewable Energy, (Oxford, Elsevier), Pages 419-447.*

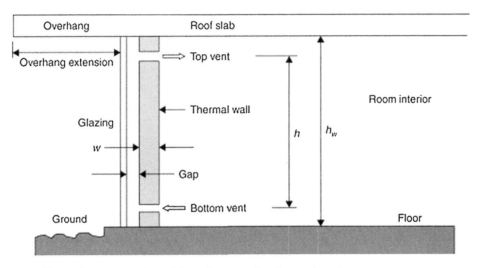

Figure 10.25. Thermal storage wall used for solar space heating and cooling.
Source: Kalogirou, S.A. 2012. Solar Thermal Systems: Components and Applications – Introduction, In: Ali Sayigh, Editor-in-Chief, Comprehensive Renewable Energy, (Oxford, Elsevier), Pages 1-25.

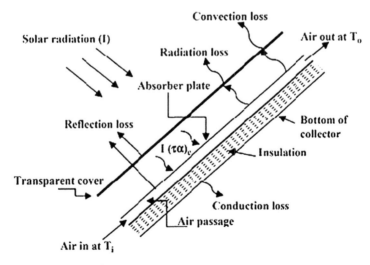

Figure 10.26. Conventional solar air heater.
Source: Tyagi, V.V., N.L. Panwar, N.A. Rahim, Richa Kothari. 2012. Review on solar air heating system with and without thermal energy storage system, Renewable and Sustainable Energy Reviews, Volume 16, Issue 4, Pages 2289-2303.

Charts

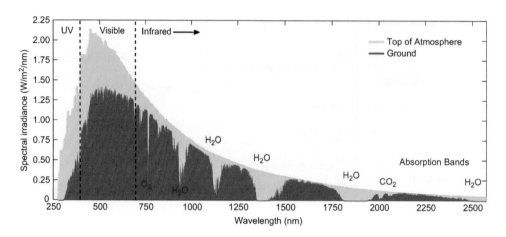

Chart 10.1.   Solar spectra as a function of wavelength.

Source: Data from United States Department of Energy, National Renewable Energy Laboratory, Reference Solar Spectral Irradiance: ASTM G-173, <http://rredc.nrel.gov/solar/spectra/am1.5/ASTMG173/ASTMG173.html>

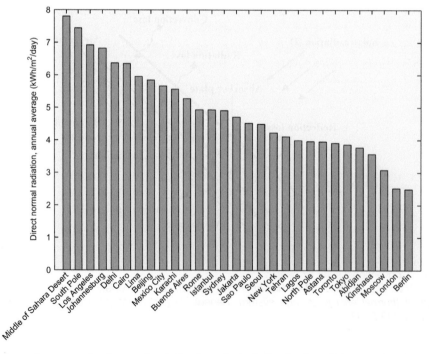

Chart 10.2.   Annual average direct normal radiation for selected cities/regions (22-year annual average, July 1983 - June 2005).
Source: Data from United States National Aeronautics and Space Administration (NASA), Surface meteorology and Solar Energy database, <http://eosweb.larc.nasa.gov/sse/>.

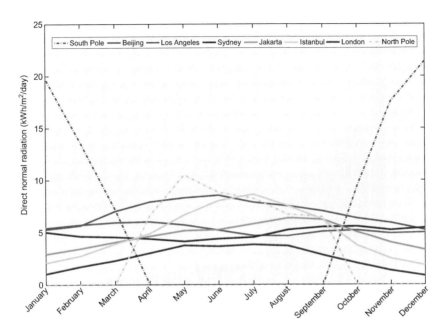

Chart 10.3. Monthly average direct normal radiation for selected cities/regions (22-year annual average, July 1983 - June 2005).
Source: Data from United States National Aeronautics and Space Administration (NASA), Surface meteorology and Solar Energy database, <http://eosweb.larc.nasa.gov/sse/>.

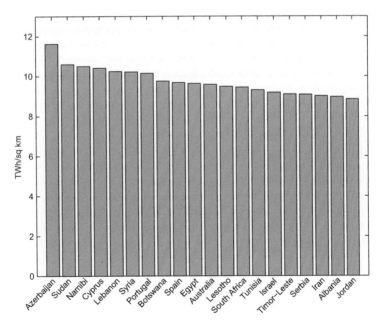

Chart 10.4. Total potential solar energy per year per unit land area, top 20 nations, as measured by direct normal irradiance modeled between 1961 and 2008.
Source: Data from United States Department of Energy, National Renewable Energy Laboratory, Solar Resources By Class Per Country, <http://en.openei.org/datasets/node/498>.

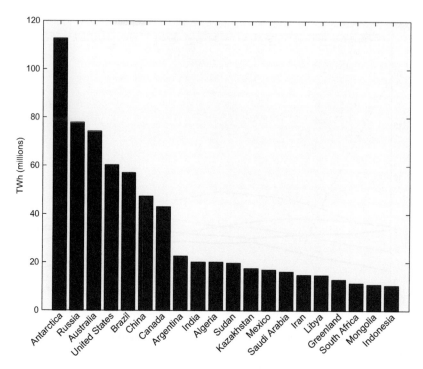

Chart 10.5. Total potential solar energy per year, top 20 nations and Antarctica, as measured by direct normal irradiance modeled between 1961 and 2008.
Source: Data from United States Department of Energy, National Renewable Energy Laboratory, Solar Resources By Class Per Country, <http://en.openei.org/datasets/node/498>, accessed 23 February.

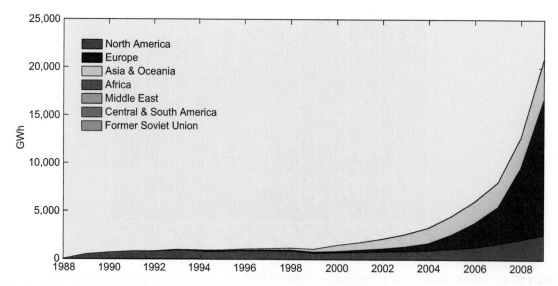

Chart 10.6. Electricity production from solar energy by region, 1988-2009.
Source: Data from International Energy Agency (IEA). Statistical database, <http://www.iea.org/stats/index. asp>.

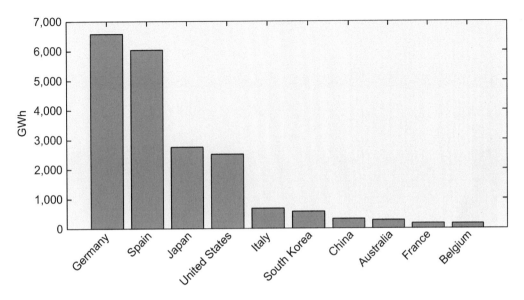

Chart 10.7. Generation of electricity from solar energy, top 10 nations, 2009.
Source: Data from International Energy Agency (IEA). Statistical database, <http://www.iea.org/stats/index. asp>.

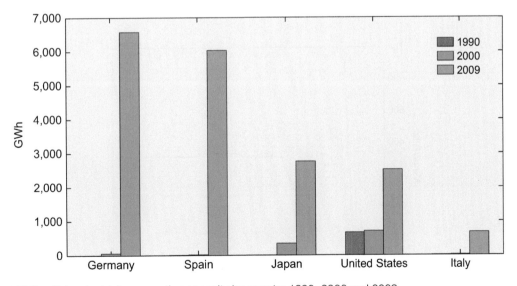

Chart 10.8. Solar electricity generation capacity by country 1990, 2000 and 2009.
Source: Data from International Energy Agency (IEA). Statistical database, <http://www.iea.org/stats/index. asp>.

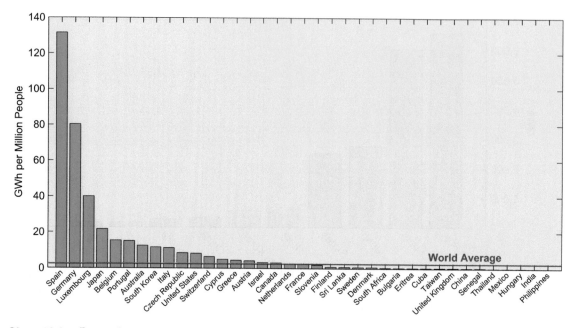

Chart 10.9. Per capita generation of electricity from solar energy by country, 2009.
Source: Data from International Energy Agency (IEA). Statistical database, <http://www.iea.org/stats/index.asp>.

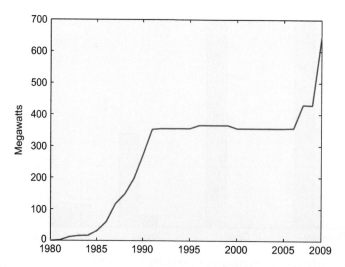

Chart 10.10. World installed concentrating solar thermal energy capacity, 1980-2009.
Source: Data from Earth Policy Institute, Data Center <http://www.earth-policy.org/data_center/>.

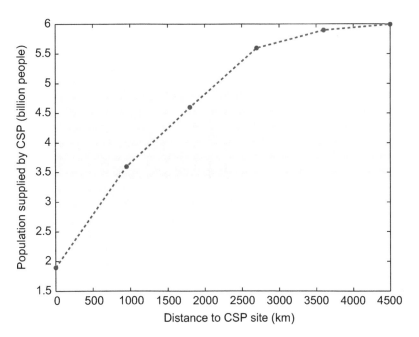

Chart 10.11. Energy supply potential of concentrating solar power (CSP) in the world versus distance to CSP sites in 2000. These estimates are based on assumptions about land use restrictions, current and potential access to transmission networks, and other factors. Based on these assumptions, the energy supply potential of CSP technology for the world population living within 3,000 km distance to potential CSP areas exceeds 90% of world population.
Source: Data from Breyer, Christian and Gerhard Knies. 2009. Global energy supply potential of concentrating solar power. Paper presented at SolarPaces Conference, Berlin, September 2009.

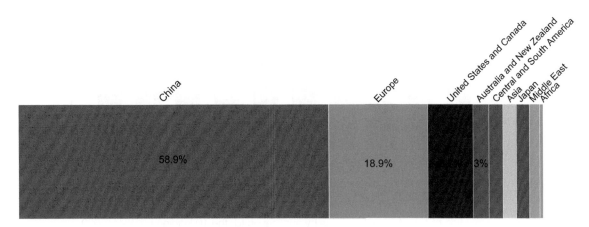

Chart 10.12. Share of installed solar thermal heat capacity by major nation/region, 2009.
Source: Data from Weiss, Werner and Franz Mauthner. 2012. Solar Heat Worldwide, Markets and Contribution to the Energy Supply 2010, Edition 2012, (Gleisdorf, AEE - Institute for Sustainable Technologies; Paris, Solar Heating and Cooling Programme, International Energy Agency).

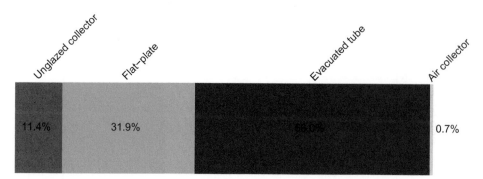

Chart 10.13. Share of installed solar thermal capacity by collector type, worldwide, 2009.
Source: Data from Weiss, Werner and Franz Mauthner. 2012. Solar Heat Worldwide, Markets and Contribution to the Energy Supply 2010, Edition 2012, (Gleisdorf, AEE - Institute for Sustainable Technologies; Paris, Solar Heating and Cooling Programme, International Energy Agency).

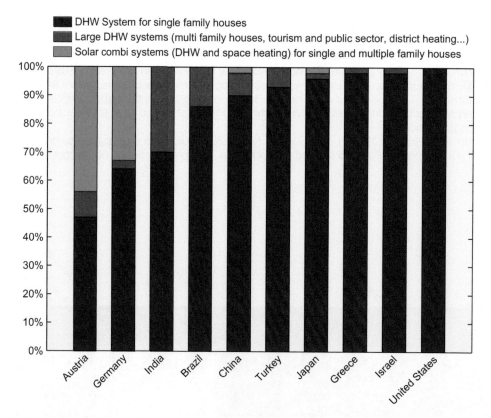

Chart 10.14. Distribution of solar thermal systems by application for the 10 leading markets, 2009. DHW = domestic hot water.
Source: Data from Weiss, Werner and Franz Mauthner. 2012. Solar Heat Worldwide, Markets and Contribution to the Energy Supply 2010, Edition 2012, (Gleisdorf, AEE - Institute for Sustainable Technologies; Paris, Solar Heating and Cooling Programme, International Energy Agency).

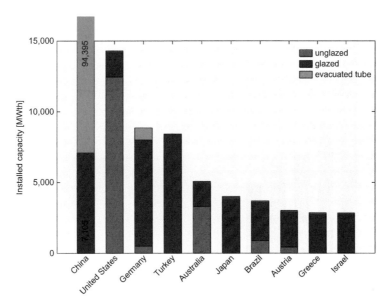

Chart 10.15. Total installed capacity of solar thermal water collectors in operation, top 10 nations, 2009. *Source: Data from Weiss, Werner and Franz Mauthner. 2012. Solar Heat Worldwide, Markets and Contribution to the Energy Supply 2010, Edition 2012, (Gleisdorf, AEE - Institute for Sustainable Technologies; Paris, Solar Heating and Cooling Programme, International Energy Agency).*

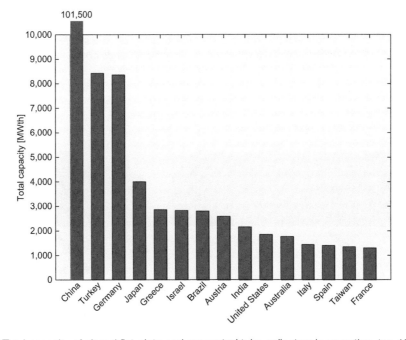

Chart 10.16. Total capacity of glazed flat-plate and evacuated tube collectors in operation, top 15 nations, 2009. *Source: Data from Weiss, Werner and Franz Mauthner. 2012. Solar Heat Worldwide, Markets and Contribution to the Energy Supply 2010, Edition 2012, (Gleisdorf, AEE - Institute for Sustainable Technologies; Paris, Solar Heating and Cooling Programme, International Energy Agency).*

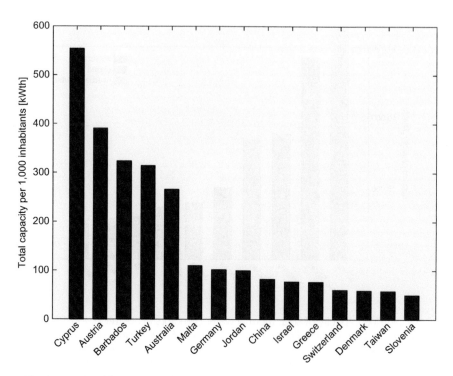

Chart 10.17. Total capacity of flat-plate and evacuated tube collectors in operation per 1000 inhabitants, top 15 nations, 2009.
Source: Data from Weiss, Werner and Franz Mauthner. 2012. Solar Heat Worldwide, Markets and Contribution to the Energy Supply 2010, Edition 2012, (Gleisdorf, AEE - Institute for Sustainable Technologies; Paris, Solar Heating and Cooling Programme, International Energy Agency).

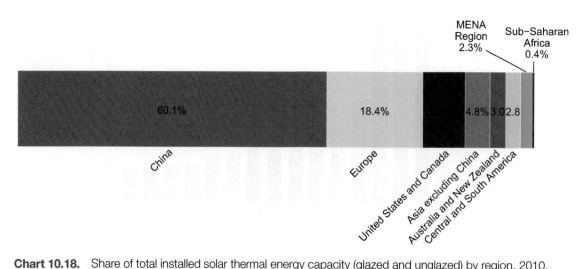

Chart 10.18. Share of total installed solar thermal energy capacity (glazed and unglazed) by region, 2010.
Source: Data from Weiss, Werner and Franz Mauthner. 2012. Solar Heat Worldwide, Markets and Contribution to the Energy Supply 2010, Edition 2012, (Gleisdorf, AEE - Institute for Sustainable Technologies; Paris, Solar Heating and Cooling Programme, International Energy Agency).

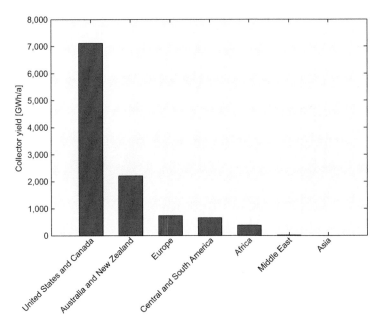

Chart 10.19. Annual yield of unglazed solar thermal water collectors by region, 2009.
Source: Data from Weiss, Werner and Franz Mauthner. 2012. Solar Heat Worldwide, Markets and Contribution to the Energy Supply 2010, Edition 2012, (Gleisdorf, AEE - Institute for Sustainable Technologies; Paris, Solar Heating and Cooling Programme, International Energy Agency).

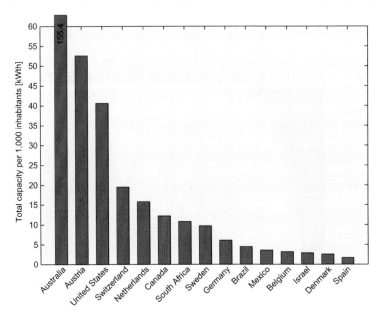

Chart 10.20. Capacity of unglazed solar thermal water collectors per 1,000 inhabitants, leading nations.
Source: Data from Weiss, Werner and Franz Mauthner. 2012. Solar Heat Worldwide, Markets and Contribution to the Energy Supply 2010, Edition 2012, (Gleisdorf, AEE - Institute for Sustainable Technologies; Paris, Solar Heating and Cooling Programme, International Energy Agency).

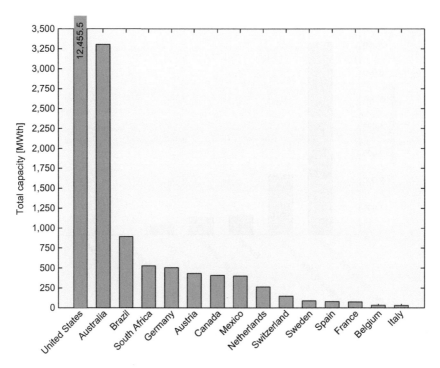

Chart 10.21. Capacity of unglazed solar thermal water collectors, leading nations.
Source: Data from Weiss, Werner and Franz Mauthner. 2012. Solar Heat Worldwide, Markets and Contribution to the Energy Supply 2010, Edition 2012, (Gleisdorf, AEE - Institute for Sustainable Technologies; Paris, Solar Heating and Cooling Programme, International Energy Agency).

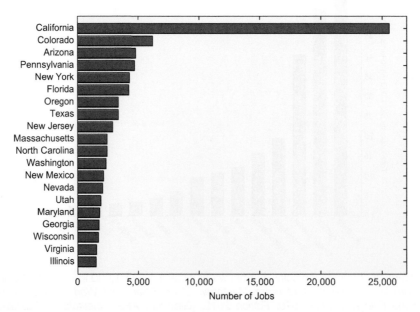

Chart 10.22. Number of solar jobs by state, 2010
Source: Data from The Solar Foundation, <http://thesolarfoundation.org/>.

Tables

Table 10.1.   **Monthly average daily extraterrestrial insolation on horizontal surface (MJ/m^2)**

Latitude	17-Jan	16-Feb	16-Mar	15-Apr	15-May	11-Jun	17-Jul	16-Aug	15-Sep	15-Oct	14-Nov	10-Dec
60°S	41.1	31.9	21.2	10.9	4.4	2.1	3.1	7.8	16.7	28.1	38.4	43.6
55°S	41.7	33.7	23.8	13.8	7.1	4.5	5.6	10.7	19.5	30.2	39.4	43.9
50°S	42.4	35.3	26.3	16.8	10	7.2	8.4	13.6	22.2	32.1	40.3	44.2
45°S	42.9	36.8	28.6	19.6	12.9	10	11.2	16.5	24.7	33.8	41.1	44.4
40°S	43.1	37.9	30.7	22.3	15.8	12.9	14.1	19.3	27.1	35.3	41.6	44.4
35°S	43.2	38.8	32.5	24.8	18.6	15.8	17	22	29.2	36.5	41.9	44.2
30°S	43	39.5	34.1	27.2	21.4	18.7	19.8	24.5	31.1	37.5	41.9	43.7
25°S	42.5	39.9	35.4	29.4	24.1	21.5	22.5	26.9	32.8	38.1	41.6	43
20°S	41.5	39.9	36.5	31.3	26.6	24.2	25.1	29.1	34.2	38.5	41.1	42
15°S	40.8	39.7	37.2	33.1	28.9	26.8	27.6	31.1	35.4	38.7	40.3	40.8
10°S	39.5	39.3	37.7	34.6	31.1	29.2	29.9	32.8	36.3	38.5	39.3	39.3
5°S	38	38.5	38	35.8	33	31.4	32	34.4	36.9	38.1	37.9	37.6
0	36.2	37.4	37.9	36.8	34.8	33.5	33.9	35.7	37.2	37.3	36.4	35.6
5°N	34.2	36.1	37.5	37.5	36.3	35.3	35.6	36.7	37.3	36.3	34.5	33.5
10°N	32	34.6	36.9	37.9	37.5	37	37.1	37.5	37	35.1	32.5	31.1
15°N	29.5	32.7	35.9	38	38.5	38.4	38.3	38	36.5	33.5	30.2	28.5
20°N	26.9	30.7	34.7	37.9	39.3	39.5	39.3	38.2	35.7	31.8	27.7	25.7
25°N	24.1	28.4	33.3	37.5	39.8	40.4	40	38.2	34.7	29.8	25.1	22.9
30°N	21.3	26	31.6	36.8	40	41.1	40.4	37.9	33.4	27.5	22.3	19.9
35°N	18.3	23.3	29.6	35.8	39.9	41.5	40.6	37.3	31.8	25.1	19.4	16.8
40°N	15.2	20.5	27.4	34.6	39.7	41.7	40.6	36.5	30	22.5	16.4	13.7
45°N	12.1	17.6	25	33.1	39.2	41.7	40.4	35.4	27.9	19.8	13.4	10.7
50°N	9.1	14.6	22.5	31.4	38.4	41.5	40	34.1	25.7	16.9	10.4	7.7
55°N	6.1	11.6	19.7	29.5	37.6	41.3	39.4	32.7	23.2	13.9	7.4	4.8
60°N	3.4	8.5	16.8	27.4	36.6	41	38.8	31	20.6	10.9	4.5	2.3

Source: Kalogirou, Soteris A. 2009. Solar Energy Engineering, (Boston, Academic Press).

Table 10.2. Direct Normal Irradiation (DNI) for the world's 15 largest regions and countries[a]

	Area km²	Total DNI TWh/year	Average DNI TWh/km²	Rank Size	Rank Total DNI	Rank Average DNI[b]
Russia	17,098,242	77,861,103	4.6	1	2	152
Antarctica	14,000,000	112,710,104	8.1	2	1	37
Canada	9,984,670	43,000,204	4.3	3	7	156
United States	9,826,675	60,309,982	6.1	4	4	94
China	9,596,961	47,417,343	4.9	5	6	137
Brazil	8,514,877	57,026,648	6.7	6	5	80
Australia	7,741,220	74,233,585	9.6	7	3	11
India	3,287,263	20,163,652	6.1	8	9	95
Argentina	2,780,400	22,586,341	8.1	9	8	34
Kazakhstan	2,724,900	17,557,838	6.4	10	12	86
Algeria	2,381,741	20,129,952	8.5	11	10	29
Congo	2,267,048	9,873,172	4.4	12	22	154
Greenland	2,166,086	12,908,956	6.0	13	17	102
Saudi Arabia	2,149,690	16,192,394	7.5	14	14	53
Mexico	1,964,375	16,991,179	8.6	15	13	24

[a]Direct Normal Irradiation (DNI) = the amount of solar radiation received per unit area by a surface that is always held perpendicular (or normal) to the rays that come in a straight line from the direction of the Sun at its current position in the sky.
[b]The 15 highest ranking nations in terms of averge DNI are: Azerbaijan, Sudan, Namibia, Cyprus, Lebanon, Syria, Portugal, Botswana, Spain, Egypt, Australia, Lesotho, South Africa, Tunisia, Israel.
Source: National Renewable Energy Laboratory, Solar Resources By Class Per Country, downloaded from OpenEI, <http://en.openei.org/datasets/node/498>.

Table 10.3. Direct Normal Irradiation (DNI) by nation

Country	TWh/year	Area	TWh/km2
Afghanistan	4,875,623	652230	7.47531288
Albania	257,640	28748	8.9620076
Algeria	20,129,952	2381741	8.45178035
American Samoa	665	199	3.34170854
Andorra	3,358	468	7.17521368
Angola	7,597,112	1246700	6.09377678
Anguilla	589	91	6.47252747
Antarctica	112,710,104	14000000	8.05072171
Antigua & Barbuda	1,472	443	3.32284533
Argentina	22,586,341	2780400	8.12341441
Armenia	211,720	29743	7.11830396
Aruba	972	180	5.40088081
Australia	74,233,585	7741220	9.58939095
Austria	559,996	83871	6.67687152
Azerbaijan	1,006,624	86600	11.6238308
Bahrain	2,352	760	3.09473684
Bangladesh	824,517	143998	5.72589102
Barbados	1,449	430	3.36871061

Table 10.3. Continued

Country	TWh/year	Area	TWh/km2
Belarus	1,300,715	207600	6.26548478
Belgium	150,147	30528	4.91834852
Belize	110,503	22966	4.81160019
Benin	491,062	112622	4.36026321
Bermuda	114	54	2.11111111
Bhutan	184,730	38394	4.81143216
Bolivia	7,949,442	1098581	7.23610004
Bosnia & Herzegovina	415,658	51197	8.11879952
Botswana	5,672,081	581730	9.75036776
Brazil	57,026,648	8514877	6.69729562
Brunei	39,447	5765	6.84252793
Bulgaria	856,865	110879	7.72793253
Burkina Faso	1,496,595	274200	5.45804238
Burundi	108,863	27830	3.9116993
Cambodia	1,064,304	181035	5.87899355
Cameroon	1,811,941	475440	3.81108237
Canada	43,000,204	9984670	4.30662245
Cape Verde	21,538	4033	5.34044136
Cayman Is.	1,023	264	3.87637522
Central African Republic	2,859,652	622984	4.59025028
Chad	10,295,685	1284000	8.01844637
Chile	5,665,421	756102	7.49293153
China	47,417,343	9596961	4.94087065
Colombia	5,592,978	1138910	4.91081674
Comoros	7,299	2235	3.26564761
Congo	9,873,172	2,267,048	4.35507867
Costa Rica	315,994	51100	6.18383737
Cote d'Ivoire	1,157,021	322463	3.58807396
Croatia	400,831	56594	7.08256364
Cuba	787,373	110860	7.10240627
Cyprus	96,286	9251	10.4081449
Czech Republic	414,223	78867	5.25217632
Denmark	229,667	43094	5.32944695
Djibouti	155,391	23200	6.69788545
Dominica	2,399	751	3.19382483
Dominican Republic	377,324	48670	7.75269809
Ecuador	1,148,064	283561	4.04873748
Egypt	9,647,512	1001450	9.63354352
El Salvador	158,331	21041	7.52490088
Equatorial Guinea	86,821	28051	3.09511247
Eritrea	998,183	117600	8.4879505
Estonia	264,357	45228	5.8449947
Ethiopia	6,163,076	1104300	5.58097944
Fiji	126,445	18274	6.91937241
Finland	1,864,927	338145	5.51516814

Continued

Table 10.3. Continued

Country	TWh/year	Area	TWh/km2
France	4,012,495	643801	6.23250782
French Guiana	501,417	83,534	6.00255357
French Polynesia	7,191	4167	1.72570194
Gabon	855,340	267667	3.19553774
Georgia	494,587	69700	7.09593922
Germany	1,880,139	357022	5.26616992
Ghana	1,284,053	238533	5.38312636
Greece	1,019,766	131957	7.72801936
Greenland	12,908,956	2166086	5.9595767
Guadeloupe	6,774	1,630	4.15590359
Guam	3,269	544	6.00919118
Guatemala	721,879	108889	6.62949601
Guinea	1,038,656	245857	4.22463387
Guinea-Bissau	128,996	36125	3.57082491
Guyana	1,241,212	214969	5.77391047
Haiti	239,698	27750	8.63776557
Honduras	684,756	112090	6.10898028
Hungary	648,109	93028	6.96681197
Iceland	478,366	103000	4.6443314
India	20,163,652	3287263	6.13387256
Indonesia	10,508,763	1904569	5.51765963
Iran	14,854,573	1648195	9.01263089
Iraq	3,867,114	438317	8.82264295
Ireland	332,092	70273	4.72574715
Israel	190,775	20770	9.18512998
Italy	2,573,615	301340	8.54056784
Jamaica	74,918	10991	6.81633376
Japan	1,789,750	377915	4.73585404
Jordan	790,765	89342	8.85098594
Kazakhstan	17,557,838	2724900	6.44347976
Kenya	3,360,202	580367	5.78978809
Kuwait	126,096	17818	7.07686569
Kyrgyzstan	1,517,827	199951	7.59099608
Laos	1,139,319	236800	4.81131436
Latvia	401,473	64589	6.21581617
Lebanon	106,530	10400	10.2432855
Lesotho	287,757	30355	9.47972006
Liberia	311,138	111369	2.79375769
Libya	14,736,682	1759540	8.37530366
Lithuania	382,510	65300	5.8577267
Luxembourg	13,289	2586	5.13869411
Macedonia	202,092	25713	7.85951526
Madagascar	4,724,037	587041	8.04720116
Malawi	688,727	118484	5.81282742
Malaysia	1,816,199	329847	5.50618761

Table 10.3. Continued

Country	TWh/year	Area	TWh/km2
Mali	7,873,713	1240192	6.34878581
Malta	2,184	316	6.91089766
Martinique	4,496	1,128	3.9858156
Mauritania	6,501,969	1030700	6.30830435
Mauritius	7,971	2040	3.90725341
Mayotte	1,595	374	4.26461424
Mexico	16,991,179	1964375	8.64966179
Micronesia	1,233	702	1.75641026
Moldova	266,772	33851	7.88077162
Mongolia	10,942,120	1564116	6.9957215
Montenegro	119,745	13812	8.66966435
Montserrat	513	102	5.02941176
Morocco	3,687,682	446550	8.25816167
Mozambique	5,261,507	799380	6.58198463
Myanmar	3,700,442	676578	5.46935046
Namibia	8,650,543	824292	10.4945123
Nepal	1,030,667	147181	7.00271449
Netherlands	169,265	41543	4.07445711
New Zealand	1,404,171	267710	5.24511878
Nicaragua	693,995	130370	5.32327019
Niger	8,861,214	1267000	6.99385458
Nigeria	4,237,654	923768	4.58735781
North Korea	480,985	120538	3.99031961
Norway	1,482,609	323802	4.57875213
Oman	2,644,529	309500	8.54451905
Pakistan	6,043,107	796095	7.59093739
Panama	359,322	75420	4.76427552
Papua New Guinea	2,664,446	462840	5.7567322
Paraguay	2,812,720	406752	6.91507271
Peru	7,975,495	1285216	6.20556773
Philippines	1,424,125	300000	4.74708494
Poland	1,720,599	312685	5.50265779
Portugal	934,267	92090	10.1451487
Puerto Rico	68,525	13790	4.96919827
Qatar	93,947	11586	8.10862907
Romania	1,881,750	238391	7.89354645
Russia	77,861,103	17098242	4.55374903
Rwanda	83,848	26338	3.18352461
Saudi Arabia	16,192,394	2149690	7.53243228
Senegal	977,626	196722	4.96958374
Serbia	703,560	77474	9.08124195
Seychelles	1,087	455	2.3879824
Sierra Leone	238,070	71740	3.3185145
Singapore	1,926	697	2.76327116
Slovakia	296,386	49035	6.0443822

Continued

Table 10.3.   Continued

Country	TWh/year	Area	TWh/km2
Slovenia	138,469	20273	6.83019474
Solomon Is.	150,631	28896	5.21286412
Somalia	5,196,207	637657	8.14890529
South Africa	11,507,051	1219090	9.43904937
South Korea	409,198	99720	4.10346495
Spain	4,892,431	505370	9.68088845
Sri Lanka	262,397	65610	3.99934326
Sudan	19,715,969	1861484	10.5915328
Suriname	911,199	163820	5.56219816
Swaziland	118,163	17364	6.80505913
Sweden	2,473,994	450295	5.49416346
Switzerland	298,383	41277	7.22879441
Syria	1,891,515	185180	10.2144646
Tajikistan	1,181,395	143100	8.25573357
Tanzania	5,731,577	947300	6.0504346
Thailand	2,946,562	513120	5.74244309
The Bahamas	73,197	13880	5.27352979
The Gambia	45,695	11295	4.04555354
Timor-Leste	135,351	14874	9.0998553
Togo	218,220	56785	3.84290785
Tonga	2,021	747	2.70548862
Trinidad & Tobago	22,281	5128	4.34497932
Tunisia	1,522,039	163610	9.30285089
Turkey	5,379,332	783562	6.86522809
Turkmenistan	4,256,017	488100	8.7195603
Uganda	1,374,522	241038	5.70251093
Ukraine	4,141,146	603550	6.86131459
United Arab Emirates	481,451	83600	5.75898888
United Kingdom	1,164,989	243610	4.78218671
United States	60,309,982	9826675	6.1373742
Uruguay	1,348,871	176215	7.65469122
Uzbekistan	3,526,586	447400	7.88240165
Venezuela	5,949,576	912050	6.52330026
Vietnam	1,292,868	331210	3.90346762
Western Sahara	2,038,442	266000	7.66331702
Yemen	4,044,923	527968	7.66130347
Zambia	5,403,289	752618	7.1793253
Zimbabwe	3,108,107	390757	7.95406512

Direct Normal Irradiation (DNI) is the amount of solar radiation received per unit area by a surface that is always held perpendicular (or normal) to the rays that come in a straight line from the direction of the sun at its current position in the sky.
Source: National Renewable Energy Laboratory, Solar Resources By Class Per Country, downloaded from OpenEI, <http://en.openei.org/datasets/node/498>.

Table 10.4.   Comparison of solar radiation received for various modes of tracking

Tracking mode	Solar energy received (kWh m^{-2})			Percentage to full tracking		
	E	SS	WS	E	SS	WS
Full tracking	8.43	10.6	5.7	100	100	100
E–W polar	8.43	9.73	5.23	100	91.7	91.7
N–S horizontal	7.51	10.36	4.47	89.1	97.7	60.9
E–W horizontal	6.22	7.85	4.91	73.8	74.0	86.2

Notes: E: equinoxes; SS: summer solstice; WS: winter solstice.
Source: Adapted from Kalogirou, S.A. 2012. 3.01 - Solar Thermal Systems: Components and Applications – Introduction,
In: Ali Sayigh, Editor-in-Chief, Comprehensive Renewable Energy, (Oxford, Elsevier), Pages 1–25.

Table 10.5.   Comparison of energy received for various modes of solar tracking

Tracking mode	Solar energy received (kWh/m^2)			Percentage to full tracking		
	E	SS	WS	E	SS	WS
Full tracking	8.43	10.6	5.7	100	100	100
E-W polar	8.43	9.73	5.23	100	91.7	91.7
N-S horizontal	7.51	10.36	4.47	89.1	97.7	60.9
E-W horizontal	6.22	7.85	4.91	73.8	74	86.2

E = equinoxes, SS = summer solstice, WS = winter solstice.
Source: Kalogirou, Soteris A. 2009. Solar Energy Engineering, (Boston, Academic Press).

Table 10.6.   Globally aggretated potential of concentrating solar power (CSP)[a]

	Potential CSP area million km^2	Solar radiation potential 1000 TWh$_{rad}$/y	Average direct normal irradiance (DNI) quality kWh/ m^2/y	Electric power potential[b] 1000 TWh$_e$/y
North America	4.9	11,500	2410	1,150
South America	5.9	13,500	2330	1,350
Africa/Europe/Asia	32.3	73,500	2600	7,350
Pacific	6.8	23,000	2950	2,300
Total	49.9	121,500		12,150

[a]Average Direct Normal Irradiation (DNI) quality and power potential limited to sites of at least 2000 kWh/m^2/y DNI. Seventy percent of all identified areas are classified as potential CSP areas. Land use efficiency of 10% is assumed.
[b]Power potential for electricity supply is calculated assuming an exclusion of 30% of the area that fulfills the solar quality conditions but not land availability constraints.
Source: Adapted from Breyer, Christian and Gerhard Knies. 2009. Global energy supply potential of concentrating solar power. Paper presented at SolarPaces Conference, Berlin, September 2009.

Table 10.7. Annual direct normal irradiation (DNI) by region for potential use via concentrating solar power (CSP)[a]

DNI Class kWh/m²/y	Africa km²	Australia km²	Central Asia, Caucasus km²	Canada km²	China km²	Central South America km²	India km²	Japan km²
2000–2099	1,082,050	70,164	151,109		88,171	334,096	83,522	
2100–2199	1,395,900	187,746	3,025		184,605	207,927	11,510	
2200–2299	1,351,050	355,188	3,594		415,720	232,678	5,310	
2300–2399	1,306,170	812,512	1,642		263,104	191,767	7,169	
2400–2499	1,862,850	1,315,560	569		99,528	57,041	3,783	
2500–2599	1,743,270	1,775,670			96,836	31,434	107	
2600–2699	1,468,970	1,172,760			17,939	42,139	976	
2700–2800+	2,746,100	393,850			24,435	93,865	120	
Total km²	12,956,360	6,083,450	159,939	0	1,190,338	1,190,948	112,497	0

DNI Class kWh/m²/y	Middle East km²	Mexico km²	Other Developing Asia km²	Other East Europe km²	Russia km²	South Korea km²	EU27+ km²	USA km²
2000–2099	36,315	16,999	47,520	59			9,163	149,166
2100–2199	125,682	34,123	52,262	129			5,016	172,865
2200–2299	378,654	35,263	105,768	23			6,381	210,128
2300–2399	557,299	53,765	284,963				1,498	151,870
2400–2499	633,994	139,455	172,043				800	212,467
2500–2599	298,755	60,972	37,855				591	69,364
2600–2699	265,541	12,628	2,084				257	191,144
2700–2800+	292,408	14,903	1,082				270	
Total km²	2,588,648	368,108	703,577	211	0	0	23,975	985,005

[a]Direct Normal Irradiation (DNI) the amount of solar radiation received per unit area by a surface that is always held perpendicular (or normal) to the rays that come in a straight line from the direction of the sun at its current position in the sky. Unsuitability for CSP is defined by slope > 2.1 %, land cover such as permanent or non-permanent water, forests, swamps, agricultural areas, shifting sands including a security margin of 10 km, salt pans, glaciers, airports, settlements, airports, oil or gas fields, mines, quarries, desalination plants, protected areas and restricted areas.

Source: Adapted from Trieb, Franz, Christoph Schillings, Marlene O'Sullivan, Thomas Pregger, Carsten Hoyer-Klick. 2009. Global Potential of Concentrating Solar Power, paper presented at SolarPaces Conference, Berlin, September 2009.

Table 10.8. Solar-electric efficiency, land use factor and land use efficiency of different concentrating solar power technologies[a]

Collector & Power Cycle Technology	Solar-Electric Aperture Related Efficiency	Land Use Factor	Land Use Efficiency
Parabolic Trough Steam Cycle	11–16%	25–40%	3.5–5.6%
Central Receiver Steam Cycle	12–16%	20–25%	2.5–4.0%
Linear Fresnel Steam Cycle	8–12%	60–80%	4.8–9.6%
Central Receiver Combined Cycle[b]	20–25%	20–25%	4.0–6.3%
Multi-Tower Solar Array Steam or Combined Cycle[b]	15–25%	60–80%	9.0–20.0%

[a]A parabolic trough system with 12% annual solar-electric efficiency, 37% land use factor and 4.5% land use efficiency is used as reference. Solar electric efficiency = annual net power generation/annual direct irradiance on aperture; land use factor = aperture area of reflectors/total land area required; land use efficiency = (solar electric efficiency) x (land use factor).
[b]Proposed technologies.
Source: Adapted from Trieb, Franz, Christoph Schillings, Marlene O'Sullivan, Thomas Pregger, Carsten Hoyer-Klick. 2009. Global Potential of Concentrating Solar Power, paper presented at SolarPaces Conference, Berlin, September 2009.

Table 10.9. Technically feasible electricity generation by concentrating solar power classified by annual direct normal irradiation (DNI) by region

DNI Class kWh/m²/y	Africa TWh/y	Australia TWh/y	Central Asia, Caucasus TWh/y	Canada TWh/y	China TWh/y	Central South America TWh/y	India TWh/y	Japan TWh/y
2000–2099	102,254	6,631	14,280	0	8,332	31,572	7,893	0
2100–2199	138,194	18,587	300	0	18,276	20,585	1,140	0
2200–2299	139,834	36,762	372	0	43,027	24,082	550	0
2300–2399	141,066	87,751	177	0	28,415	20,711	774	0
2400–2499	209,571	148,001	64	0	11,197	6,417	426	0
2500–2599	203,963	207,753	0	0	11,330	3,678	13	0
2600–2699	178,480	142,490	0	0	2,180	5,120	119	0
2700–2800+	346,009	49,625	0	0	3,079	11,827	15	0
Total TWh/y	1,459,370	697,600	15,193	0	125,835	123,992	10,928	0

DNI Class kWh/m²/y	Middle East TWh/y	Mexico TWh/y	Other Developing Asia TWh/y	Other East Europe TWh/y	Russia TWh/y	South Korea TWh/y	EU27+ TWh/y	USA TWh/y
2000–2099	3,432	1,606	4,491	6	0	0	866	14,096
2100–2199	12,443	3,378	5,174	13	0	0	497	17,114
2200–2299	39,191	3,650	10,947	2	0	0	660	21,748
2300–2399	60,188	5,807	30,776	0	0	0	162	16,402
2400–2499	71,324	15,689	19,355	0	0	0	90	23,903
2500–2599	34,954	7,134	4,429	0	0	0	69	8,116
2600–2699	32,263	1,534	253	0	0	0	31	2,326
2700–2800+	36,843	1,878	136	0	0	0	34	0
Total TWh/y	290,639	40,675	75,561	21	0	0	2,409	103,704

Source: Adapted from Trieb, Franz, Christoph Schillings, Marlene O'Sullivan, Thomas Pregger, Carsten Hoyer-Klick. 2009. Global Potential of Concentrating Solar Power, paper presented at SolarPaces Conference, Berlin, September 2009.

Table 10.10. Segmentation of world population into world regions and their distance to potential concentrating solar power (CSP) sites[a]

	Potential CSP area million km²	Electric power potential[b] 1000 TWh$_e$/y	Site million people	900 km million people	1,800 km million people	2,700 km million people	3,600 km million people	4,500 km million people
North America	4.9	1,150	160	400	460	470	470	470
South America	5.9	1,350	160	310	360	370	370	370
Africa/Europe/ Asia	32.3	7,350	1,610	2,730	3,510	4,400	4,840	5,120
Pacific	6.8	2,300	20	190	260	430	740	1,800
Insersection								
Asia/Pacific			0	0	0	130	520	1,750
Total	49.9	12,150	1,950	3,630	4,590	5,540	5,900	6,010

[a]Average Direct Normal Irradiation (DNI) quality and power potential limited to sites of at least 2000 kWh/m²/y DNI. Seventy percent of all identified areas are classified as potential CSP areas. Land use efficiency of 10% is assumed.

[b]Power potential for electricity supply is calculated assuming an exclusion of 30% of the area that fulfills the solar quality conditions but not land availability constraints.

Source: Adapted from Breyer, Christian and Gerhard Knies. 2009. Global energy supply potential of concentrating solar power. Paper presented at SolarPaces Conference, Berlin, September 2009.

Table 10.11. U.S. Parabolic trough power plant data

Plant Name[a]	Location	First Year of Operation	Net Output (MW$_e$)	Solar Field Outlet (°C)	Solar Field Area (m²)	Solar Turbine Effic. (%)	Power Cycle	Dispatchability Provided By
Nevada Solar One	Boulder City, NV	2007	64	390	357,200	37.6	100 bar, reheat	None
APS Saguaro	Tucson, AZ	2006	1	300	10,340	20.7	Organic Rankine Cycle	None
SEGS IX	Harper Lake, CA	1991	80	390	483,960	37.6	100 bar, reheat	HTF heater[b]
SEGS VIII	Harper Lake, CA	1990	80	390	464,340	37.6	100 bar, reheat	HTF heater
SEGS VI	Kramer Junction, CA	1989	30	390	188,000	37.5	100 bar, reheat	Gas boiler
SEGS VII	Kramer Junction, CA	1989	30	390	194,280	37.5	100 bar, reheat	Gas boiler
SEGS V	Kramer Junction, CA	1988	30	349	250,500	30.6	40 bar, steam	Gas boiler
SEGS III	Kramer Junction, CA	1987	30	349	230,300	30.6	40 bar, steam	Gas boiler
SEGS IV	Kramer Junction, CA	1987	30	349	230,300	30.6	40 bar, steam	Gas boiler
SEGS II	Daggett, CA	1986	30	316	190,338	29.4	40 bar, steam	Gas boiler
SEGS I	Daggett, CA	1985	13.8	307	82,960	31.5	40 bar, steam	3-hrs TES[c]

[a]SEGS = Solar Energy Generating System.
[b]HTF = Heat Transfer Fluid.
[c]TES = Thermal Energy Storage.
Source: National Renewable Energy Laboratory.

Table 10.12.   Advantages and disadvantages of two types of solar chimney

Type	Advantages	Disadvantages
Vertical solar chimney	• The external glass gain sun radiation, solar collector is not needed	• Insulation is needed to prevent direct heat transfer between chimney and interior room because of high temperature and high contact area
	• The air flow in chimney could go upward directly without bends	• Barriers are strictly prevented because the solar gained wall is lower than roof solar collector
	• Easier to be control with inlet and outlet for different climatic condition	
	• Stack height is not restricted by roof height	
Roof solar chimney	• Very large collector areas easily achieved	• Stack height is restricted by roof height
	• May be more aesthetically pleasing than a tower	• Heat transfer between heated air and glass is higher than for a vertical surface
	• No additional towers needed	• Additional bends create greater pressure-losses
	• Likely to be cheaper than a tower design	• Incorporation of thermal mass may be more difficult
	• Easier to retrofit	

Source: Adapted from Shi, Long, Michael Yit Lin Chew. 2012. A review on sustainable design of renewable energy systems, Renewable and Sustainable Energy Reviews, Volume 16, Issue 1, Pages 192–207.

Table 10.13.   Types of solar collectors

Motion	Collector type	Absorber type	Concentration ratio	Indicative temperature range (°C)
Stationary	Flat-plate collector (FPC)	Flat	1	30–80
	Evacuated tube collector (ETC)	Flat	1	50–500
	Compound parabolic collector (CPC)	Tubular	1–5	60–240
Single-axis tracking	Linear Fresnel reflector (LFR)	Tubular	10–40	60–250
	Cylindrical trough collector (CTC)	Tubular	15–50	60–300
	Parabolic trough collector (PTC)	Tubular	10–85	60–400
Two-axis tracking	Parabolic dish reflector (PDR)	Point	600–2000	100–1500
	Heliostat field collector (HFC)	Point	300–1500	150–2000

Concentration ratio is defined as the aperture area divided by the receiver/absorber area of the collector.
Source: Adapted from Kalogirou, Soteris A. 2009. Solar Energy Engineering, (Boston, Academic Press).

Table 10.14. Thermal and radiative properties of solar collector cover materials

Material name	Index of Refraction	Transmittance (τ) (solar)[a] (%)	Transmittance (τ) (solar)[b] (%)	Transmittance (τ) (infrared)[c] (%)	Expansion coefficient (in/in · °F)	Temperature limits (°F)
Temper glass (glass)	1.518 (D 542)	125 mil 84.3 (+0.1)	125 mil 78.6 (+0.2)	125 mil 2.0 (est)	$4.8\ (10^{-6})$ (D 696)	450–500 continous use; 500–550 short-term use
Clear lime sheet glass (low iron oxide glass)	1.51 (D 542)	Insufficient data	125 mil 87.5	125 mil 2.0 (+0.5)	$5.0\ (10^{-6})$ (D 696) (est)	400 for continous operation
Clear lime temper glass (low iron oxide glass)	1.51 (D 542)	Insufficient data	125 mil 87.5 (+0.5)	125 mil 2.0	$5.0\ (10^{-6})$ (D 696)	400 for continous operation
Sunadex white crystal glass (0.01% iron oxide glass)	1.5 (D 542)	Insufficient data	125 mil 91.5 (+0.2)	125 mil 2.0	$4.7\ (10^{-6})$ (D 696)	400 for continous operation
Lexan (polycarbonate)	1.586 (D 542)	125 mil 64.1 (±0.8)	125 mil 72.6 (±0.1)	125 mil 2.0 (est)	$3.75\ (10^{-5})$ (H 696)	250–270 service temperature
Plexiglas (acrylic)	1.49 (D 542)	125 mil 89.6 (±03)	125 mil 79.6 (±0.8)	125 mil 2.0 (est)	$3.9\ (10^{-9})$ at 60°F; $4.6\ (10^{-6})$ at 100°F	180–200 service temperature
Teflon F.E.P (fluorocarbon)	1.343 (D 542)	5 mil 92.3 (±02)	5 mil 89.8 (±0.4)	5 mil 25.6 (±0.5)	$5.9\ (10^{-5})$ at 160°F; $9.0\ (10^{-5})$ at 212°F	400 continuous use; 475 short-term use
Tedlar P.V.F. (fluorocarbon)	1.46 (D 542)	4 mil 92.2 (±0.1)	4 mil 88.3 (±0.9)	4 mil 20.7 (±0.2)	$2.8\ (10^{-5})$ (D 696)	225 continuous use; 350 short-term use
Mylar (polyester)	1.64–1.67 (D 542)	5 mil 86.9 (±03)	5 mil 80.1 (±0.1)	5 mil 17.8 (±0.5)	$0.94\ (10^{-5})$ (D 696–44)	300 continuous use; 400 short-term use
Sunlite (fiberglass)	1.54 (D 542)	25 mil (P) 86.5 (0.2) 25 mil (R) 87.5 (±0.2)	25 mil (P) 75.4 (±0.1) 25 mil (R) 77.1 (±0.7)	25 mil (P) 7.6 (±0.1) 25 mil (R) 3.3 (±0.1)	$1.4\ (10^{-5})$ (D 696)	200 continuous use causes 5% loss in T
Float glass (glass)	1.518 (D 542)	125 mil 84.3 (±0.2)	125 mil 78.6 (±0.2)	125 mil 2.0 (est)	$4.8\ (10^{-5})$ (D 696)	1350 softening point; 100 thermal shock

[a] Compiled data based on ASTM Code E 424 Method B.

[b] Numerical integration for $\left(\sum \tau_{avg} F_{\lambda_1 T - \lambda_2 T}\right)$ for $\lambda = 0.2 - 0.4\ \mu M$.

[c] Numerical integration for $\left(\sum \tau_{avg} F_{\lambda_1 T - \lambda_2 T}\right)$ for $\lambda = 3.0 - 50.0\ \mu M$.

Source: Adapted from Kreith, Frank and D. Yogi. Goswami, Eds. 2005. CRC Handbook of Mechanical Engineering, Second Edition (Boca Raton, CRC Press).

Table 10.15. Cumulative solar water and space heating installations in top ten countries and the world, 2009

	Cumulative Installed Capacity
Country	**Thousand Square Meters**
China	145,000
Turkey	12,035
Germany	11,933
Japan	5,720
Greece	4,077
Israel	4,039
Brazil	4,000
Austria	3,689
India	3,081
United States	2,642
World Total	**229,000**

Source: Compiled by Earth Policy Institute with world total from Renewable Energy Policy Network for the 21st Century (REN21), *Renewables 2011 Global Status Report* (Paris: REN21 Secretariat, 2011), p. 74; country data from Werner Weiss and Franz Mauthner, *Solar Heat Worldwide: Markets and Contribution to the Energy Supply 2009* (Gleisdorf, Austria: International Energy Agency, Solar Heating & Cooling Programme, May 2011), p. 27.

Table 10.16. Solar domestic hot water systems for single-family houses, 2010

Country	Reference climate	Collector area (gross area) for single sys. [m²]	Total collector area-SFH [m²]	Total number of systems SFH	Type of System
Albania	Tirana	2.5	22,543	9,017	TS
Australia	Sydney	6.0	2,857,251	476,209	PS
Austria	Graz	6.0	1,858,739	309,790	PS
Barbados	Grantley Adams	4.0	131,690	32,923	TS
Belgium	Brussels	4.0	316,634	79,159	PDS
Brazil	Brasília	4.0	4,157,540	1,039,385	TS
Bulgaria	Sofia	4.0	33,306	8,327	PS
Canada	Montreal	6.0	33,107	5,518	PS
Chile	Santiago de Chile	4.0	12,953	3,238	PS
China	Shanghai	4.0	151,200,000	37,800,000	TS
Cyprus	Nicosia	4.0	786,811	196,703	TS
Czech Republic	Prague	6.0	178,605	29,767	PS
Denmark	Copenhagen	4.0	455,708	113,927	PS
Estonia	Tallinn	4.0	2,841	710	PS
Finland	Helsinki	4.0	32,738	8,184	PS
France incl. DOM	Paris	4.0	1,645,272	411,318	PS
Germany	Wurzburg	6.0	5,887,419	981,237	PS
Greece	Athens	2.5	4,005,260	1,602,104	TS
Hungary	Budapest	6.0	74,907	12,485	PS
India	Delhi	4.0	3,176,000	794	TS
Ireland	Dublin	4.0	136,059	34,015	PS
Israel	Jerusalem	4.0	827,579	206,895	TS

Table 10.16. Continued

Country	Reference climate	Collector area (gross area) for single sys. [m²]	Total collector area-SFH [m²]	Total number of systems SFH	Type of System
Italy	Bologna	4.0	2,547,578	636,895	PS
Japan	Tokyo	4.0	5,175,606	1,293,901	TS
Jordan	Amman	4.0	790,050	197,512	TS
Korea, South	Seoul	4.0	845,812	211,453	PS
Latvia	Riga	4.0	7,244	1,811	PS
Lebanon	Beirut	4.0	348,312	87,078	TS
Lithuania	Vilnius	4.0	4,518	1,130	PS
Luxembourg	Luxembourg	4.0	30,800	7,700	PS
Macedonia	Skopje	4.0	12,100	3,025	PS
Malta	Luqa	4.0	43,469	10,867	PS
Mexico	Mexico City	4.0	243,561	60,890	PS
Morocco	Rabat	4.0	341,260	85,315	TS
Namibia	Windhoek	4.0	9,903	2,476	TS
Netherlands	Amsterdam	3.0	331,857	110,619	PDS
New Zealand	Wellington	4.0	144,989	36,247	PS
Norway	Oslo	6.0	15,151	2,525	PS
Poland	Warsaw	6.0	459,060	76,510	PS
Portugal	Lisbon	4.0	525,951	131,488	PS
Romania	Bucharest	4.0	109,996	27,499	PS
Slovakia	Bratislava	6.0	135,746	22,624	PS
Slovenia	Ljubljana	6.0	135,869	22,645	PS
South Africa	Johannesburg	4.0	359,682	89,920	TS
Spain	Madrid	4.0	818,300	204,575	PS
Sweden	Gothenburg	6.0	30,200	5,033	PS
Switzerland	Zürich	6.0	532,824	88,804	PS
Taiwan	Taipei	4.0	1,933,245	483,311	TS
Thailand	Bangkok	4.0	91,392	22,848	TS
Tunisia	Tunis	4.0	475,009	118,752	TS
Turkey	Ankara	4.0	12,253,166	3,063,292	TS
United Kingdom	London	4.0	564,783	141,196	PS
United States	LA, Indianapolis	6.0	2,446,342	407,724	PS
Uruguay	Montevideo	4.0	12,096	3,024	PS
Zimbabwe	Harare	4.0	18,196	4,549	PS
Total			209,627,029	51,818,147	

DHW-SFH: domestic hot water systems PS: pumped system for single-family houses; TS: thermosiphon system PDS: pumped drain back system.

Source; Adapted from Weiss, Werner and Franz Mauthner. 2012. Solar Heat Worldwide, Markets and Contribution to the Energy Supply 2010, Edition 2012, (Gleisdorf, AEE - Institute for Sustainable Technologies; Paris, Solar Heating and Cooling Programme, International Energy Agency).

Table 10.17.   Types of solar water heating systems

Passive systems	Active systems
Thermosiphon (direct and indirect)	Direct circulation (or open loop active) systems
Integrated collector storage	Indirect circulation (or closed loop active) systems, internal and external heat exchanger
	Air systems
	Heat pump systems
	Pool heating systems

Source: Kalogirou, Soteris A. 2009. Solar Energy Engineering, (Boston, Academic Press).

Table 10.18.   Characteristics of potential phase change materials for solar air heating system

Materials	Melting Point (°C)	Latent heat (kJ/kg)
Organic Materials		
Capric acid	36	152
Eladic acid	47	218
Lauric acid	49	178
Pentadeconoic acid	52.5	178
Tristearin	56	191
Myristic acid	58	199
Palmatic acid	55	163
Stearic acid	69.4	199
Acetamide	81	241
Inorganic materials		
$Zn(NO_3)_2 \cdot 6H_2O$	36.1	134
$FeCl_3 \cdot 6H_2O$	37.0	223
$Mn(NO_3)_2 \cdot 4H_2O$	37.1	115
$Na_2HPO_4 \cdot 12H_2O$	40.0	279
$CoSO_4 \cdot 7H_2O$	40.7	170
$KF \cdot 2H_2O$	42	162
$MgI_2 \cdot 8H_2O$	42	133
$CaI_2 \cdot 6H_2O$	42	162
$K_2HPO_4 \cdot 7H_2O$	45.0	145
$Zn(NO_3)_2 \cdot 4H_2O$	45	110
$Mg(NO_3) \cdot 4H_2O$	47.0	142
$Ca(NO_3) \cdot 4H_2O$	47.0	153
$Fe(NO_3)_3 \cdot 9H_2O$	47	155
$Na_2SiO_2 \cdot 4H_2O$	48	168
$K_2HPO_4 \cdot 3H_2O$	48	99
$Na_2S_2O_3 \cdot 5H_2O$	48.5	210
$MgSO_4 \cdot 7H_2O$	48.5	202
$CA(NO_3)_2 \cdot 3H_2O$	51	104
$Zn(NO_3)_2 \cdot 2H_2O$	55	68
$FeCl_3 \cdot 2H_2O$	56	90
$Ni(NO_3)_2 \cdot 6H_2O$	57.0	169
$MnCl_2 \cdot 4H_2O$	58.0	151
$MgCl_2 \cdot 4H_2O$	58.0	178

Table 10.18. **Continued**

Materials	Melting Point (°C)	Latent heat (kJ/kg)
$CH_3COONa \cdot 3H_2O$	58.0	265
$Fe(NO_3)_2 \cdot 6H_2O$	60.5	126
$NaAl(SO_4)_2 \cdot 10H_2O$	61.0	181
$NaOH \cdot H_2O$	64.3	273
$Na_3PO_4 \cdot 12H_2O$	65.0	190
$LiCH_3COO \cdot 2H_2O$	70	150
$Al(NO_3)_2 \cdot 9H_2O$	72	155
$Ba(OH)_2 \cdot 8H_2O$	78	265

Source: Adapted from Tyagi, V.V., N.L. Panwar, N.A. Rahim, Richa Kothari. 2012. Review on solar air heating system with and without thermal energy storage system, Renewable and Sustainable Energy Reviews, Volume 16, Issue 4, Pages 2289–2303.

Table 10.19. **Absorptance and emittance of selective surfaces commonly used in solar thermal energy**

Coating/substrate	Absorptance	Emittance
Copper, aluminum, or nickel plate with CuO coating	0.8–0.93	0.09–0.21
Black nickel on Zn/Fe substrate	0.94	0.09
Black copper (BlCu-Cu$_2$O:Cu) on Cu substrate	0.97–0.98	0.02
Metal, plated black chrome	0.87	0.09
Metal, plated nickel oxide	0.92	0.08

Source: Adapted from Caouris,Y.G. 2012. 3.04 - Low Temperature Stationary Collectors, In: Ali Sayigh, Editor-in-Chief, Comprehensive Renewable Energy, (Oxford, Elsevier), 2012, Pages 103–147.

Table 10.20. **Total emissivity and solar absorptivity of selected surfaces**

	Temperature (°C)	Total normal emissivity	Extraterrestrial solar absorptivity
Alumina, flame-sprayed	−25	0.8	0.28
Aluminum foil	20	0.04	
Aluminum, vacuum-deposited	20	0.025	0.1
Hard-anodized	−25	0.84	0.92
Highly polished plate, 98.3%	225–575	0.039–0.057	
Commercial sheet	100	0.09	
Rough polish	100	0.18	
Rough plate	40	0.055–0.07	
Oxidized at 600°C	200–600	0.11–0.19	
Antimony, polished	35–260	0.28–0.31	
Asbestos	35–370	0.93–0.94	
Beryllium	150	0.18	0.77
	370	0.21	

Continued

Table 10.20.   Continued

	Temperature (°C)	Total normal emissivity	Extraterrestrial solar absorptivity
	600	0.3	
Bismuth, bright	75	0.34	
Black paint			
Parson's optical black	−25	0.95	0.975
Black silicone	−25–750	0.93	0.94
Black epoxy paint	−25	0.89	0.95
Black enamel paint	95–425	0.81–0.80	
Brass, polished	40–315	0.1	
Carbon, graphitized	100–320	0.76–0.75	
	320–500	0.75–0.71	
Graphite, pressed, filed surface	250–510	0.98	
Chromium, polished	40–1100	0.08–0.36	
Copper, electroplated	20	0.03	0.47
Carefully polished electrolytic copper	80	0.018	
Polished	115	0.023	
Plate heated at 600°C	200–600	0.57	
Cuprous oxide	800–1100	0.66–0.54	
Molten copper	1075–1275	0.16–0.13	
Glass, Pyrex, lead, and soda	260–540	0.95–0.85	
Gypsum	20	0.903	
Gold, pure, highly polished	225–625	0.018–0.035	
Inconel X, oxidized	−25	0.71	0.9
Lead, pure (99.96%), unoxidized	125–225	0.057–0.075	
Oxidized at 150°C	200	0.63	
Magnesium oxide	275–825	0.55–0.20	
	900–1705	0.2	
Magnesium, polished	35–260	0.07–0.13	
Mercury	0–100	0.09–0.12	
Molybdenum, polished	35–260	0.05–0.08	
	540–1370	0.10–0.18	
Nickel, elecroplated	20	0.03	22
Polished	100	0.072	
Platinum, pure, polished	225-625	0.054–0.104	
Silica, sintered, powdered fused	35	0.84	0.08
Silicon Carbide	150–650	0.83–0.96	
Silver, polished, pure	40–625	0.020–0.032	
Stainless steel			
Type 312, heated 300 hour at 260 °C	95–425	0.27	
Type 301 with Armco black oxide	−25	0.75	0.89
Type 410, heated to 700 °C in air	35	0.13	0.76
Type 303, sandblasted	95	0.42	0.68
Titanium, 75A	95–425	0.10–0.19	
75A, oxidized at 450 °C	35–425	0.21–0.25	0.8

Table 10.20. Continued

	Temperature (°C)	Total normal emissivity	Extraterrestrial solar absorptivity
Anodized	−25	0.73	0.51
Tungsten, filament,aged	27–3300	0.032–0.35	
Zinc, pure, polished	225–325	0.045–0.053	
Galvanized sheet	100	0.21	

Emissivity is the relative ability of its surface to emit energy by radiation, measured by the ratio of energy radiated by a particular material to energy radiated by a black body at the same temperature. Solar absorptivity is the property of a body that determines the fraction of the incident solar radiation absorbable by the body.
Source: Adapted from Kreith, F., Boehm, R. F., Raithby, G. D., Hollands, K. G. T., Suryanarayana N.V., et al.."Heat and Mass Transfer, "In Kreith, Frank, Ed. 2000. CRC Handbook of Thermal Engineering (Boca Raton, CRC Press).

Table 10.21. Total emittance of selected surfaces

	Temperature °C	Total Normal Emittance[a]
Alumina, Flame Sprayed	−25	0.8
Aluminum foil		
As received	20	0.04
Bright dipped	20	0.025
Aluminum, vacuum deposited	20	0.025
Hard-anodized	−25	0.84
Highly polished plate, 98.3% pure	225–575	0.039–0.057
Commercial sheet	100	0.09
Rough polish	100	0.18
Heavily oxidized	95–500	0.20–0.31
Antimony, polished	35–260	0.28–0.31
Asbestos	35–370	0.93–0.94
Beryllium	150	0.18
	370	0.21
	600	0.3
Beryllium, anodized	150	0.9
	370	0.88
	600	0.82
Bismuth, bright	75	0.34
Black paint		
Parson's optical black	−25	0.95
Black silicone	−25–750	0.93
Black epoxy paint	−25	0.89
Black enamel paint	95–425	0.81–0.80
Brass, polished	40–315	0.1
Rolled plate, natural surface	22	0.06
Dull plate	50–350	0.22
Oxidized by heating at 600 °C	200–600	0.61–0.59
Carbon, graphitized	100–320	0.76–0.75

Continued

Table 10.21.   Continued

	Temperature °C	Total Normal Emittance[a]
	320–500	0.75–0.71
Candle soot	95–270	0.952
Graphite, pressed, filed surface	250–510	0.98
Chromium, polished	40–1100	0.08–0.36
Copper, electroplated	20	0.03
Carefully polished electrolytic	80	0.018
Copper		
Polished	115	0.023
Plate heated at 600 °C	200–600	0.57
Molten copper	1075–1275	0.16–0.13
Glass, Pyrex, lead, and soda	260–540	0.95–0.85
Gypsum	20	0.903
Gold, pure, highly polished	225–625	0.018–0.035
Inconel X, oxidized	–25	0.71
Lead, pure (99.96%), unoxidized	125–225	0.057–0.075
Gray oxidized Oxidized at 150 °C	25 200	0.28 0.63
Magnesium oxide	275–825	0.55–0.20
	900–1705	0.2
Magnesium, polished	35–260	0.07–0.13
Mercury	0–100	0.09–0.12
Molybdenum, polished	35–260	0.05–0.08
	540–1370	0.10–0.18
	2750	0.29
Nickel, electroplated	20	0.03
Polished	100	0.072
Platinum, pure, polished	225–625	0.054–0.104
Silica, sintered, powdered, fused silica	35	0.84
Silicon carbide	150–650	0.83–0.96
Silver, polished, pure	40–625	0.020–0.032
Stainless steel		
Type 312, heated 300 h at 260 °C	95–425	0.27–0.32
Type 301 with Armco black oxide	–25	0.75
Type 410, heated to 700 °C in air	35	0.13
Type 303, sandblasted	95	0.42
Titanium, 75A	95–425	0.10–0.19
75A, oxidized 300 h at 450 °C	35–425	0.21–0.25
Anodized	–25	0.73
Tungsten, filament, aged	27–3300	0.032–0.35
Zinc, pure, polished	225–325	0.045–0.053
Galvanized sheet	100	0.21

a(energy emitted from a surface/energy emitted by a black surface at same temperature), with direction perpendicular to the surface.
Source: Adapted from Modest, Michael F., Radiation, in Kreith, Frank and D. Yogi. Goswami, Eds. 2005. CRC Handbook of Mechanical Engineering, Second Edition (Boca Raton, CRC Press).

Table 10.22.   **Shortwave absorptivities (%) of leaves and animals**

Leaves		Mammals	
Silver maple	0.48–0.56	bison	0.78
American beach	0.47–0.52	wolf	0.80
Sunflower	0.52–0.57	cat (white)	0.445
Cottonwood	0.50	bobcat	0.70
Cottonwood (yellow)	0.39	Reptiles	
Birds		alligator	0.90
Stellar's jay	0.88	lizard	0.90
Sparrow (dorsal)	0.75	Humans	
Quail (dorsal)	0.72	Eurasian	0.68–0.72
Quail egg	0.18	Negroid	0.79
White swan	0.37		

Source: Adapted from Campbell, Gaylon S. and John M. Norman. 1998. An Introduction to Environmental Biophysics, (New York, Springer-Verlag) and Gates, D. M. 1980. Biophysical Ecology. (New York, Springer-Verlag).

Table 10.23.   **Radiometric terminology and units**

Term	Description	Units
Absorptance	The ratio of the radiation absorbed by the surface of the Earth to the radiation incident upon it	Dimensionless
Albedo	The ratio of the radiation reflected from the surface of the Earth to the radiation incident upon it	Dimensionless
Diffuse solar radiation	The downward scattered and reflected short-wave radiation coming from the whole sky vault with the exception of the solid angle subtended by the Sun's disk	$W\ m^{-2}$ (instantaneous value)
		$Wh\ m^{-2}$ (integrated value over 1 h)
Direct solar radiation	The short-wave radiation emitted from the solid angle of the Sun's disk, comprising mainly unscattered and unreflected solar radiation	$W\ m^{-2}$ (instantaneous value)
		$Wh\ m^{-2}$ (integrated value over 1 h)
Global (or total) solar radiation	The sum of diffuse and direct short-wave radiation components	$W\ m^{-2}$ (instantaneous value)
		$Wh\ m^{-2}$ (integrated value over 1 h)
IR (or terrestrial or long-wave or thermal) radiation	The radiation coming from the sky at wavelengths longer than about 4 μm	$W\ m^{-2}$ (instantaneous value)
		$Wh\ m^{-2}$ (integrated value over 1 h)
Irradiance	The radiant flux incident on a surface from all directions per unit area of this surface	$W\ m^{-2}$ (instantaneous value)
		$Wh\ m^{-2}$ (integrated value over 1 h)
Radiance	The radiant flux emitted by a unit solid angle of a source or scatterer incident on a unit area of a surface	$W\ m^{-2}$ (instantaneous value)
		$Wh\ m^{-2}$ (integrated value over 1 h)

Continued

Table 10.23.   Continued

Term	Description	Units
Radiant flux	The amount of radiation coming from a source per unit time	W
Radiant intensity	The radiant flux leaving a source point per unit solid angle of space surrounding the point	W sr^{-1}
Reflectance/ transmittance	The fraction of radiant flux reflected by a surface or transmitted by a semitransparent medium	Dimensionless
Spectroradiometry	The radiant flux per unit wavelength	W m^{-2} nm^{-1} or W m^{-2} μm^{-1} (instantaneous value)
		Wh m^{-2} nm^{-1} or Wh m^{-2} μm^{-1} (integrated value over 1 h)

Source: Kambezidis, H.D. 2012. 3.02 - The Solar Resource, In: Editor-in-Chief: Ali Sayigh, Comprehensive Renewable Energy, (Oxford, Elsevier), 2012, Pages 27-84.

Photovoltaic

Figures

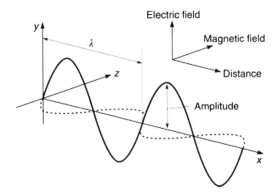

Figure 11.1. An electromagnetic wave at a fixed moment of time.
Source: Allen, Elizabeth, Sophie Triantaphillidou. 2011. The Manual of Photography (Tenth Edition), (Oxford, Focal Press).

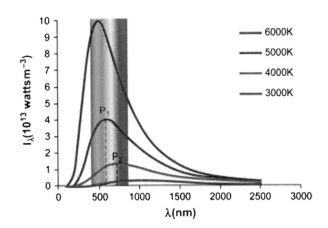

Figure 11.2. Black-body radiation curves (Iλ) at different temperatures.
Source: Allen, Elizabeth, Sophie Triantaphillidou. 2011. The Manual of Photography (Tenth Edition), (Oxford, Focal Press).

Handbook of Energy. http://dx.doi.org/10.1016/B978-0-08-046405-3.00011-5

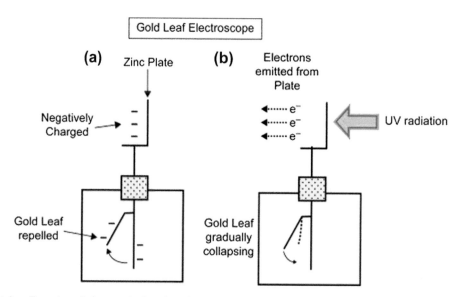

Figure 11.3. Experiment demonstrating the photoelectric effect. A zinc plate is placed on a gold leaf electroscope and negatively charged. In (a), the gold leaf is repelled as a result of the negative charge. (b), When the zinc plate is bathed in UV radiation the gold leaf gradually collapses due to the emission of electrons.
Source: Allen, Elizabeth, Sophie Triantaphillidou. 2011. The Manual of Photography (Tenth Edition), (Oxford, Focal Press).

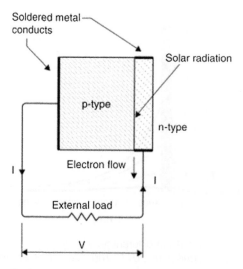

Figure 11.4. The photovoltaic effect.
Source: Kalogirou, Soteris A. 2009. Solar Energy Engineering, (Boston, Academic Press).

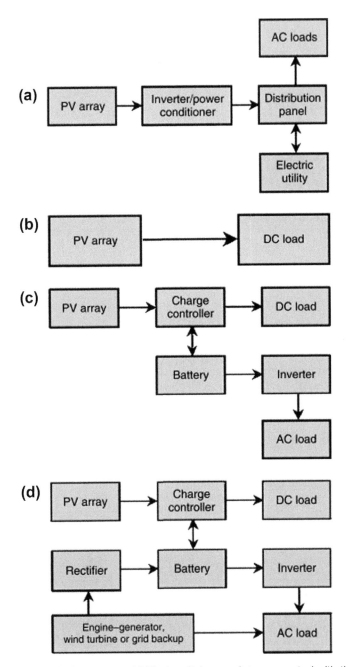

Figure 11.5. Types of photovoltaic systems. (a) Photovoltaic array interconnected with the electric grid; (b) Photovoltaic array directly connected to a load; (c) stand-alone photovoltaic array with storage battery; (d) hybrid system having an auxiliary power source.
Source: Pistoia, Gianfranco. 2009. Battery Operated Devices and Systems, (Amsterdam, Elsevier).

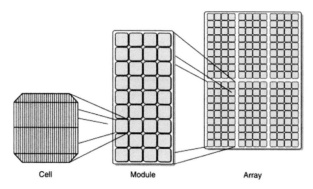

Figure 11.6. Photovoltaic cell, module, panel, and array.
Source: Pistoia, Gianfranco. 2009. Battery Operated Devices and Systems, (Amsterdam, Elsevier).

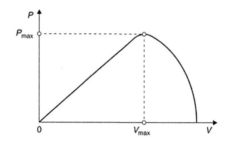

Figure 11.7. Representative power-voltage curve for photovoltaic cells.
Source: Kalogirou, Soteris A. 2009. Solar Energy Engineering, (Boston, Academic Press).

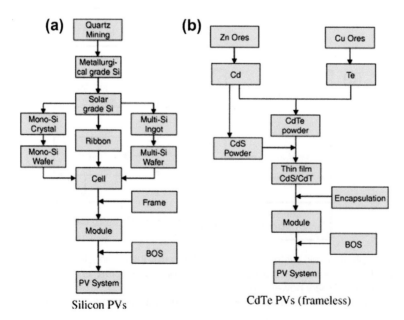

Figure 11.8. Flow diagram from raw material acquisition to manufacturing of PVs.
Source: Fthenakis, V.M. and H.C. Kim. 2012. Environmental Impacts of Photovoltaic Life Cycles, In: Ali Sayigh, Editor-in-Chief, Comprehensive Renewable Energy, (Oxford, Elsevier), Pages 47-72.

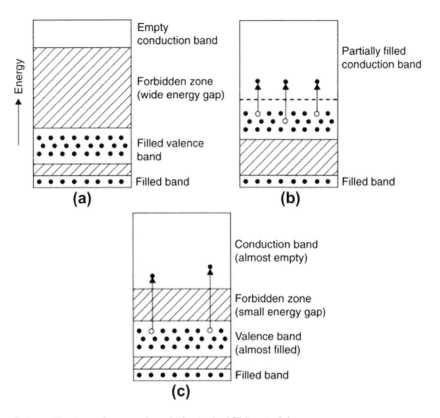

Figure 11.9. Schematic view of energy bands for typical PV materials.
Source: Kalogirou, Soteris A. 2009. Solar Energy Engineering, (Boston, Academic Press).

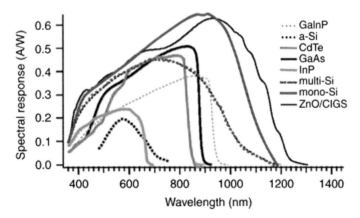

Figure 11.10. Photovoltaic material system spectral responses. GaInP = gallium indium phosphide; a-Si = amorphous silicon thin film; CdTe = cadmium telluride; GaAs = gallium arsenide; InP = indium phosphide; multi-Si = multicrystalline silicon; mono-Si = monocrystalline silicon; ZnO/CIGS = zinc oxide-coated copper indium gallium diselenide film.
Source: Myers, D.R. 2012. Solar Radiation Resource Assessment for Renewable Energy Conversion, In: Ali Sayigh, Editor-in-Chief, Comprehensive Renewable Energy, (Oxford, Elsevier), Pages 213-237.

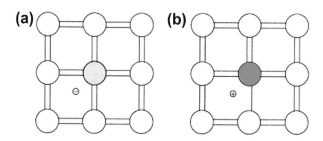

Figure 11.11. The principle of doping in semiconductor materials. Doping is the process of replacing atoms in a semiconductor lattice with impurity atoms with a different number of valence electrons. (a) A donor atom (gray) in a semiconductor lattice. The additional electron makes the material n-type. (b) An acceptor atom (black) in a semiconductor lattice. The electron vacancy makes the material p-type.
Source: Myers, D.R. 2012. Solar Radiation Resource Assessment for Renewable Energy Conversion, In: Ali Sayigh, Editor-in-Chief, Comprehensive Renewable Energy, (Oxford, Elsevier), Pages 213-237.

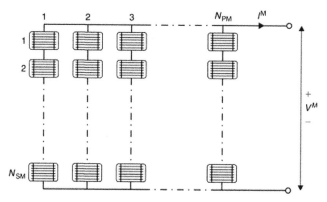

Figure 11.12. Schematic diagram of a photovoltaic module consisting of NPM parallel branches, each with NSM cells in series.
Source: Kalogirou, Soteris A. 2009. Solar Energy Engineering, (Boston, Academic Press).

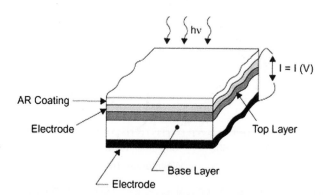

Figure 11.13. Cross-section of a typical solar cell. The area of photon impingement and the area of current production are the same. The anti-reflection (AR) coating has the function of reducing reflection losses. The collecting electrodes (cathode and anode) are shown with the top electrode being transparent.
Source: Fonash, Stephen J. 2010. Solar Cell Device Physics (Second Edition), (Boston, Academic Press).

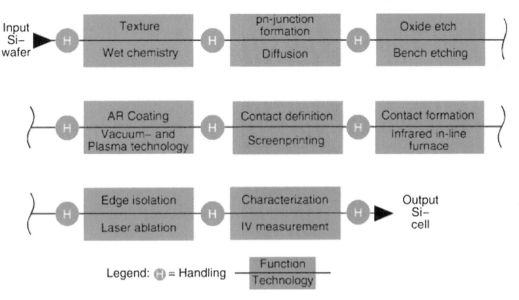

Figure 11.14. Schematic process flow for an industrial crystalline silicon solar cell line.
Source: Glunz, S.W., R. Preu, D. Biro. 2012. Crystalline Silicon Solar Cells: State-of-the-Art and Future Developments, In: Ali Sayigh, Editor-in-Chief, Comprehensive Renewable Energy, (Oxford, Elsevier), Pages 353-387.

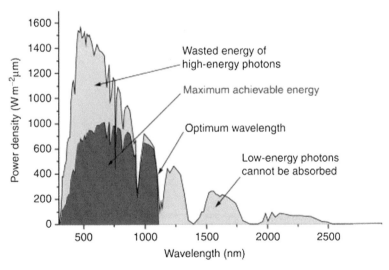

Figure 11.15. Spectral losses in a solar cell. The figure shows the maximum achievable energy of a silicon solar cell in relation to the sun spectrum (AM1.5).
Source: Glunz, S.W., R. Preu, D. Biro. 2012. Crystalline Silicon Solar Cells: State-of-the-Art and Future Developments, In: Ali Sayigh, Editor-in-Chief, Comprehensive Renewable Energy, (Oxford, Elsevier), Pages 353-387.

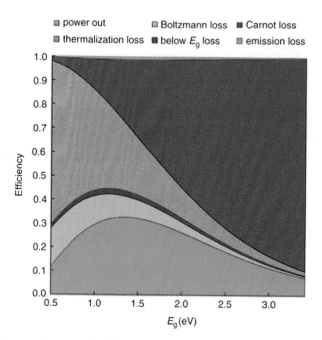

Figure 11.16. Energy loss mechanisms in a Shockley–Queisser solar cell. Intrinsic loss mechanisms and extracted electrical work cumulatively account for all incident solar radiation, demonstrating that intrinsic loss mechanisms lead to fundamental limiting efficiency.
Source: Myers, D.R. 2012. Solar Radiation Resource Assessment for Renewable Energy Conversion, In: Ali Sayigh, Editor-in-Chief, Comprehensive Renewable Energy, (Oxford, Elsevier), Pages 213-237.

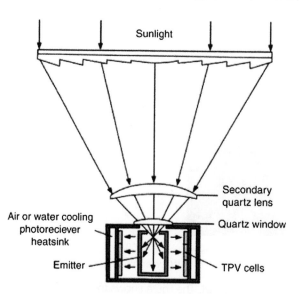

Figure 11.17. Concept of an solar thermophotovoltaic (STPV) system with a Fresnel lens as primary concentrator.
Source: van der Heide, J. 2012. Thermophotovoltaics, In: Ali Sayigh, Editor-in-Chief, Comprehensive Renewable Energy, (Oxford, Elsevier), Pages 603-618.

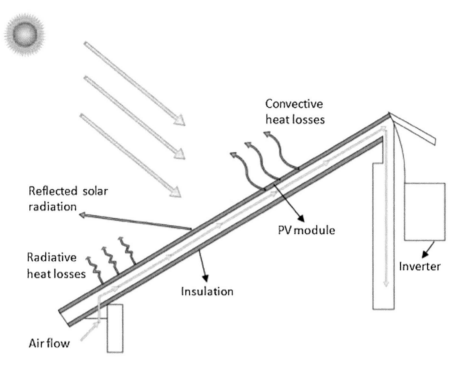

Figure 11.18. Schematic of a typical air-based open-loop, building-integrated photovoltaic system. Shi, Long, Michael Yit Lin Chew. 2012. A review on sustainable design of renewable energy systems, Renewable and Sustainable Energy Reviews, Volume 16, Issue 1, Pages 192-207.

Charts

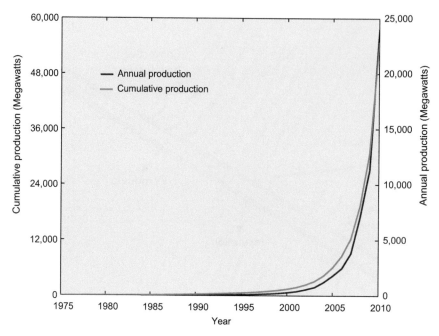

Chart 11.1. Global annual and cumulative photovoltaic production, 1975-2010.
Source: Data from Earth Policy Institute Data Center, http://www.earth-policy.org/data_center/.

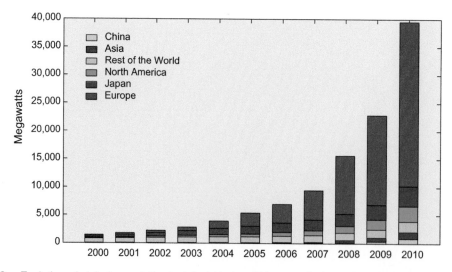

Chart 11.2. Evolution of global cumulative installed photovoltaic capacity by region, 2000-2010.
Source: Data from European Photovoltaic Industry Association. 2012. Global Market Outlook for Photovoltaics until 2016.

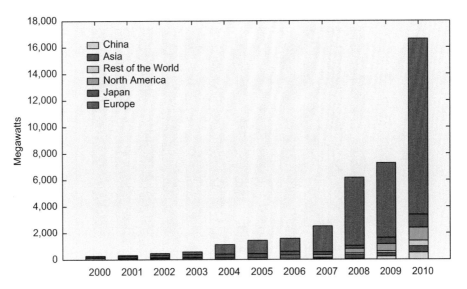

Chart 11.3. Evolution of the global annual photovoltaic market by region, 2000-2010.
Source: Data from European Photovoltaic Industry Association. 2012. Global Market Outlook for Photovoltaics until 2016.

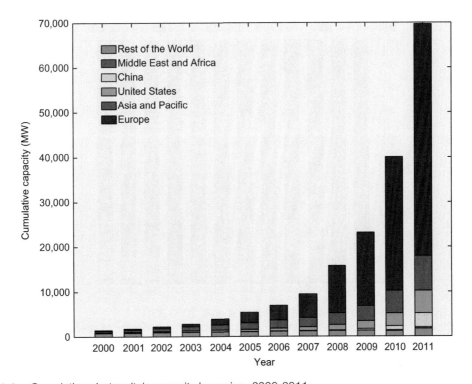

Chart 11.4. Cumulative photovoltaic capacity by region, 2000-2011.
Source: Data from European Photovoltaic Industry Association. 2012. Global Market Outlook for Photovoltaics until 2016.

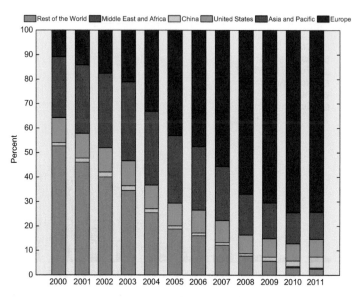

Chart 11.5. Shares of cumulative photovoltaic capacity by region, 2000-2011.
Source: Data from European Photovoltaic Industry Association. 2012. Global Market Outlook for Photovoltaics until 2016.

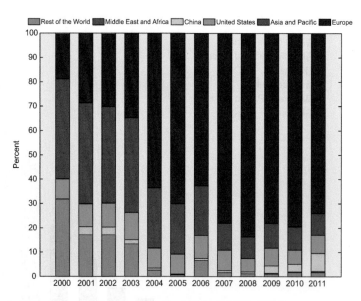

Chart 11.6. Annual photovoltaic installations by region-shares, 2000-2011.
Source: Data from European Photovoltaic Industry Association. 2012. Global Market Outlook for Photovoltaics until 2016.

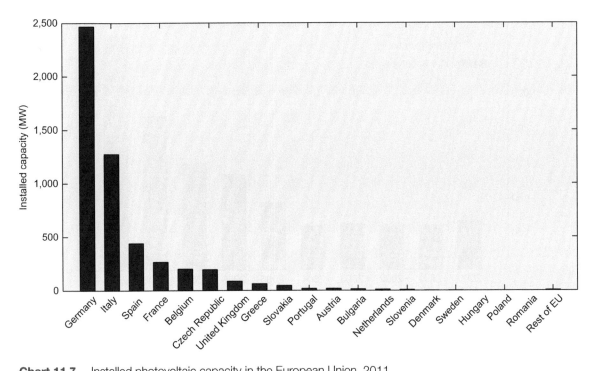

Chart 11.7. Installed photovoltaic capacity in the European Union, 2011.
Source: Data from European Photovoltaic Industry Association. 2012. Global Market Outlook for Photovoltaics until 2016.

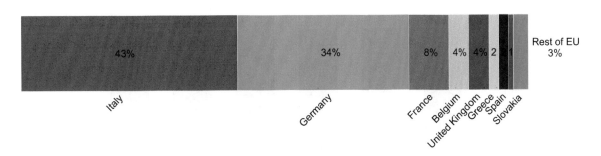

Chart 11.8. Country shares of installed photovoltaic capacity in the European Union, 2011.
Source: Data from European Photovoltaic Industry Association. 2012. Global Market Outlook for Photovoltaics until 2016.

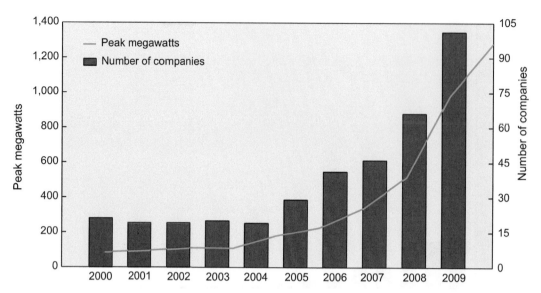

Chart 11.9.   Number of companies and annual shipments of photovoltaic cells and modules in the United States, 2000 - 2009.
Source: Data from United States Department of Energy, Energy Information Administration, Renewable amd Alternative Fuels, <http://www.eia.gov/renewable/>.

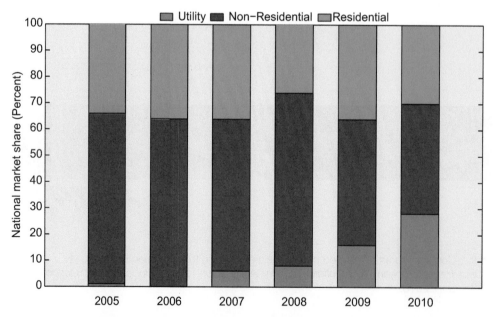

Chart 11.10.   United States photovoltaic market by segment, 2005-2011.
Source: Data from Solar Energy Industries Association, <http://www.seia.org/>.

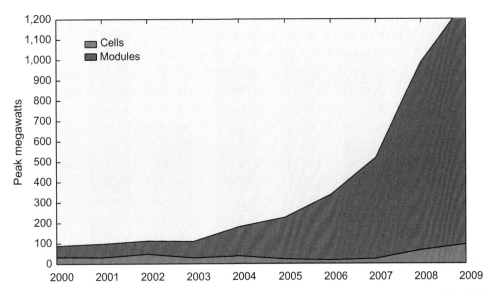

Chart 11.11.   Annual shipments of photovoltaics in the United States by cells and modules, 2000-2009. *Source: Data from United States Department of Energy, Energy Information Administration, Renewable and Alternative Fuels, <http://www.eia.gov/renewable/>.*

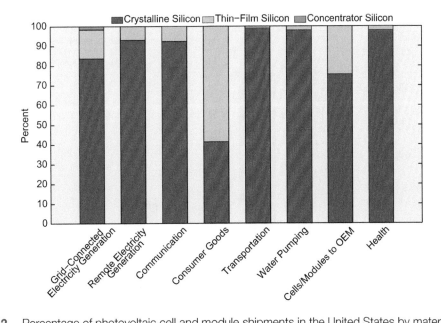

Chart 11.12.   Percentage of photovoltaic cell and module shipments in the United States by material type and end use, 2008-2009. *Source: Data from United States Department of Energy, Energy Information Administration, Renewable and Alternative Fuels, <http://www.eia.gov/renewable/>.*

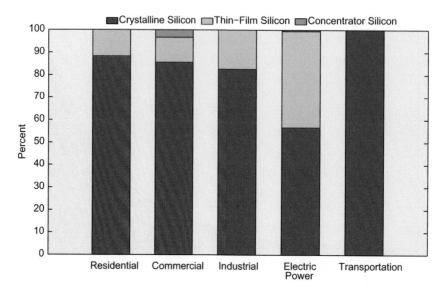

Chart 11.13.   Percentage of photovoltaic cell and module shipments in the United States by material type and market sector, 2009.
Source: Data from United States Department of Energy, Energy Information Administration, Renewable and Alternative Fuels, <http://www.eia.gov/renewable/>.

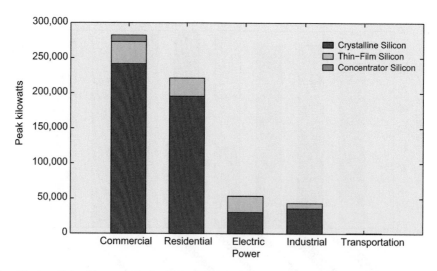

Chart 11.14.   Photovoltaic shipments in the United States by market sector and type, 2009.
Source: Data from United States Department of Energy, Energy Information Administration, Renewable and Alternative Fuels, <http://www.eia.gov/renewable/>.

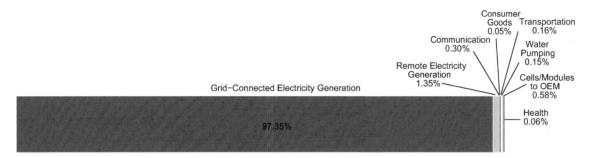

Chart 11.15. Photovoltaic shipments in the United States by end use market sector, 2009.
Source: Data from United States Department of Energy, Energy Information Administration, Renewable and Alternative Fuels, <http://www.eia.gov/renewable/>.

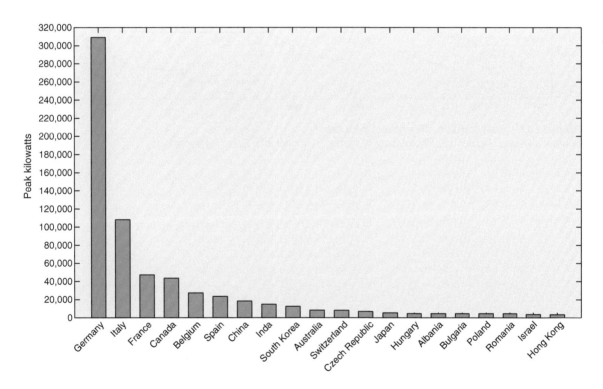

Chart 11.16. Destination of U.S. PV exports, leading nations.
Source: U.S. Energy Information Administration.

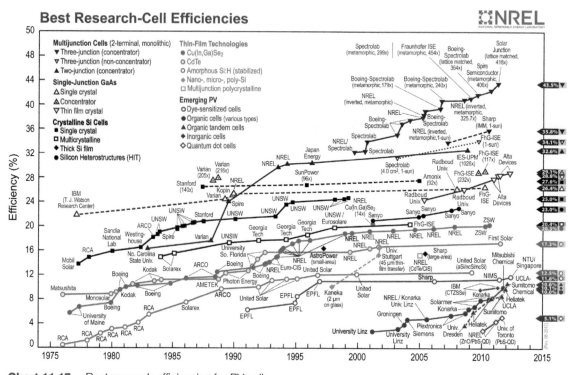

Chart 11.17.   Best research efficiencies for PV cells.
Source: National Center for Photovoltaics, National Renewable Energy Laboratory.

Tables

Table 11.1. Global annual photovoltaic installations by region 2000-2011 (MW)

	2000	2005	2011		
ROW	88	10	508	−88.6%	4980.0%
MEA	N/A	N/A	131	N/A	N/A
China	0	4	2,200	-	54900.0%
America	23	117	2,234	408.7%	1809.4%
APAC	114	296	2,653	159.6%	796.3%
Europe	53	1,002	21,939	1790.6%	2089.5%
Total	277	1,429	29,665	415.9%	1975.9%

ROW: Rest of the World; MEA: Middle East and Africa; APAC: Asia and Pacific.
Source: European Photovoltaic Industry Association. 2012. Global Market Outlook for Photovoltaics until 2016.

Table 11.2. Global cumulative installed photovoltiac capacity by region, 2000-2011 (MW)

	2000	2005	2011	Change 2000-2005	Change 2005-2011
ROW	751	1,003	1,717	33.6%	71.2%
MEA	N/A	N/A	336	N/A	N/A
China	19	68	3,093	257.9%	4448.5%
America	146	496	5,053	239.7%	918.8%
APAC	355	1,475	7,769	315.5%	426.7%
Europe	154	2,299	51,716	1392.9%	2149.5%
Total	1,425	5,340	69,684	274.7%	1204.9%

ROW: Rest of the World; MEA: Middle East and Africa; APAC: Asia and Pacific
Source: European Photovoltaic Industry Association. 2012. Global Market Outlook for Photovoltaics until 2016

Table 11.3. Cumulative and newly-Installed solar photovoltaics capacity in ten leading countries and the world, 2010

Country	Cumulative Installed Capacity	Country	Newly-Installed Capacity
	Megawatts		Megawatts
Germany	17,193	Germany	7,408
Spain	3,784	Italy	2,321
Japan	3,622	Czech Republic	1,490
Italy	3,494	Japan	990
United States	2,528	United States	878
Czech Republic	1,953	France	719
France	1,025	China	520
China	893	Belgium	424
Belgium	803	Spain	369
South Korea	655	Australia	320
World Total	39,529	World Total	16,629

Values include both grid-connected and off-grid PV systems.
Source: Earth Policy Institute, Data Center <http://www.earth-policy.org/data_center/>.

Table 11.4.   Annual Installed solar photovoltaics capacity in selected countries and the world, 1998-2010

Year	Germany	Italy	Czech Republic	Japan	United States	France	China	Spain	Others	World
	Megawatts									
1998	10	n.a.	n.a.	69	n.a.	n.a.	n.a.	0	76	155
1999	12	n.a.	n.a.	72	17	n.a.	n.a.	1	112	197
2000	40	n.a.	n.a.	112	22	n.a.	0	n.a.	106	280
2001	78	n.a.	n.a.	135	29	n.a.	11	2	76	331
2002	80	n.a.	n.a.	185	44	n.a.	15	9	138	471
2003	150	n.a.	n.a.	223	63	n.a.	10	10	125	581
2004	600	n.a.	n.a.	272	90	n.a.	9	6	142	1,119
2005	850	n.a.	n.a.	290	114	n.a.	4	26	155	1,439
2006	843	10	0	287	145	8	12	102	174	1,581
2007	1,271	70	3	210	207	11	20	542	179	2,513
2008	1,809	338	61	230	342	46	45	2,708	589	6,168
2009	3,806	717	398	483	477	219	228	17	912	7,257
2010	7,408	2,321	1,490	990	878	719	520	369	1,934	16,629

n.a. = data not available. Values include both grid-connected and off-grid PV systems.
Source: Compiled by Earth Policy Institute with 1998-1999 data for the world, Japan, and China, and with 1998-2005 data for all other countries, from European Photovoltaic Industry Association (EPIA), *Global Market Outlook for Photovoltaics Until 2013* (Brussels: April 2009), p. 4; 2000-2010 data for the world, Japan, and China, and 2006-2010 data for all other countries, from EPIA, *Global Market Outlook for Photovoltaics Until 2015* (Brussels: May 2011), pp. 9, 14-29.

Table 11.5.   Annual solar photovoltaics production by country, 1995-2010

Year	1995	2000	2005	2010	Change 2000-2005	Change 2005-2010
China	NA	2.5	128.3	10852.4542	50.3	83.6
Taiwan	NA	NA	88	3639.33886	-	40.4
Japan	16.4	128.6	833	2169.3	5.5	1.6
Germany	NA	22.5	339	2021.77261	14.1	5.0
United States	34.75	75	153.1	1115.43473	1.0	6.3
Others	NA	48.2	241	4248.35519	4.0	16.6
World	77.6	276.8	1782.4	24046.6555	5.4	12.5

Source: Earth Policy Institute, Data Center <http://www.earth-policy.org/data_center/>.

Table 11.6.   Top photovoltaic cell manufacturers by capacity, 2011

Rank	Company	Country	MW
1	Suntech	China	2,400
2	JA Solar	China	2,100
3	Trina	China	1,900
4	Yingli	China	1,700
5	Motech	Taiwan	1,500
5	Gintech	Taiwan	1,500
7	Canadian Solar	China (Canada headquarters)	1,300
7	Neo Solar Power	Taiwan	1,300
9	Hanwha Solar One	China	1,100
9	Jinko Solar	China	1,100

Source: Strube, Oliver. 2010. Top 10: Ten Largest Solar PV Companies.
<http://www.renewableenergyworld.com/rea/blog/post/2010/06/top-10-ten-largest-solar-pv-companies>.

Table 11.7.   Top polysilicon manufacturers by capacity, 2011

Rank	Company	Country	Capacity (short tons)
1	GCL	China	65,000
1	OCI	Korea	65,000
3	Hemlock	US	43,000
4	Wacker	Germany	33,000
5	LDK	China	25,000
6	REC	Norway	19,000
7	MEMC	US	1,000
8	Tokuyama	Japan	9,200
9	LCY	Taiwan	8,000
10	Woongjin	Korea	5,000

Source: Strube, Oliver. 2010. Top 10: Ten Largest Solar PV Companies.
<http://www.renewableenergyworld.com/rea/blog/post/2010/06/top-10-ten-largest-solar-pv-companies>.

Table 11.8.   Design aspects of different photovoltaic technologies

Type of PV cell	Maturity	Color/surface/other	Typical area(mm)	Typical thickness(µm)	Flexibility of cell	Operations on cells during design and manufacturing
c-Si	Highly commercially available	Blue, dark-gray, or black/smooth surface with silver grid patterns on top/cells can be colored (gold, orange, pink, red, green, silver) by variable Si_3N_4 layer/decorative grid patterns possible	156 × 156	>180 to 220	Low	Bending only to a limited extent; laser cutting; heating; injection transfer molding in plastics; lamination in plastics
m-Si	Highly commercially available	Shiny blue, dark blue/shiny grains, smooth surface with silver grid patterns on top/cells can be colored (gold, orange, pink, red, green, silver) by variable Si_3N_4 layer/decorative grid patterns possible	156 × 156	>180 to 220	Low	Bending only to a limited extent; laser cutting; heating; injection transfer molding in plastics; lamination in plastics
a-Si	Commercially available	Dark brown or black/smooth surface with light lines/cell interconnects/patterned deposition is possible	Customizable from 10 × 10 to 1000 × 2000	< 1	High	Bending; lamination in plastics; deposition on curved surfaces; cutting not possible
CIGS	Commercially available	Gray or black/smooth surface with light lines/cell interconnects/patterned screen printing is possible	Customizable from 10 × 10 to 1000 × 2000	1–3	High	Bending; lamination in plastics; deposition on curved surfaces; cutting not possible
CdTe	Highly commercially available	Brownish/smooth	Customizable from 10 × 10 to 1000 × 2000	1–3	Low	Heating
III–V three junction	Mainly available for space applications, concentrators	Black/smooth	40 × 80 80 × 80	140–200	Low	Connection to ceramics
III–V single junction, thin film	Commercially available for terrestrial applications	Black/smooth	40 × 80 80 × 80	5	High	Bending; heating; injection transfer molding in plastics; lamination in plastics
DSC	Available	Red or brown/transparent and smooth/cells can be colored by dye molecules	Customizable	1–10	High	Bending; lamination in plastics
Polymer	Limitedly available	Orange, red, or brown/smooth	Long strips, customizable	< 1	High	Bending; lamination in plastics

c-Si: single-crystal silicon wafer-based solar cell; m-Si:multicrystalline silicon solar cell; DSC: dye-sensitized cells; CIGS: copper indium gallium diselenide; CdTe: cadmium-telluride; a-Si:H: hydrogenated amorphous silicon.

Source: Adapted from Reinders, A.H.M.E., W.G.J.H.M. van Sark. 2012. 1.34 - Product-Integrated Photovoltaics, In: Ali Sayigh, Editor-in-Chief, Comprehensive Renewable Energy, (Oxford, Elsevier), Pages 709–732.

Table 11.9. Confirmed terrestrial solar cell and submodule efficiencies measured under the global AM1.5 spectrum (1000W/m^2) at 25 degrees C (IEC 60904-3: 2008, ASTM G-173-03 global)

Classification[a]	Efficiency[b] (%)
Silicon	
Si (crystalline)	25.0 ± 0.5
Si(multicrystalline)	20.4 ± 0.5
Si (thin film transfer)	**19.1 ± 0.4**
Si (thin film submodule)	10.5 ± 0.3
III-V cells	
GaAs (thin film)	**28.1 ± 0.8**
GaAs (multicrystalline)	18.4 ± 0.5
InP (crystalline	22.1 ± 0.7
Thin film chalcogenide	
CIGS (cell)	19.6 ± 0.6[j]
CIGS (submodule)	16.7 ± 0.4
CdTe (cell)	16.7 ± 0.5[j]
Amorphous/nanocrystalline Si	
Si (amorphous)	10.1 ± 0.3[l]
Si (nanocrystalline)	10.1 ± 0.2[m]
Photochemical	
Dye-sensitized	**10.9 ± 0.3[n]**
Dye-sensitized (submodule)	9.9 ± 0.4[n]
Organic	
Organic polymer	8.3 ± 0.3[n]
Organic (submodule)	3.5 ± 0.3[n]
Multijunction devices	
GaInP/GaAs/Ge	32.0 ± 1.5[m]
GaAs/CIS (thin film)	25.8 ± 1.3[m]
a-Si/nc-Si/nc-Si (thin film)	**12.4 ± 0.7[o]**
a-Si/nc-Si (thin film cell)	11.9 ± 0.8
a-Si/nc-Si (thin film submodule)	11.7 ± 0.4[m,q]
Organic (two cell tandem)	8.3 ± 0.3[n]

[a]CIGS, CuInGaSe2; a-Si, amorphous silicon/hydrogen alloy.
[b]Effic., efficiency.
[c](ap), aperture area; (t), total area; (da), designated illumination area.
[d]FF, fill factor.
[e]FhG-ISE, Fraunhofer Institut für Solare Energiesysteme; JQA, Japan Quality Assurance; AIST, Japanese National Institute of Advanced Industrial Science and Technology.
[f]Spectral response reported in Version 36 of these Tables.
[g]Recalibrated from original measurement.
[h]Spectral response and current-voltage curve reported in present version of these Tables.
[i]Reported on a "per cell" basis.
[j]Not measured at an external laboratory.
[k]Spectral response reported in Version 37 of these Tables.
[l]Light soaked at Oerlikon prior to testing at NREL (1000 h, 1 sun, 508C).
[m]Measured under IEC 60904-3 Ed. 1: 1989 reference spectrum.
[n]Stability not investigated. Refs. [30,31] review the stability of similar devices.
[o]Light soaked under 100 mW/cm2 white light at 508C for over 1000 h.
[q]Stabilized by 174 h, 1 sun illumination after 20 h, 5 sun illumination at a sample temperature of 508C.
Source: Martin A. Green, Keith Emery, Yoshihiro Hishikawa, Wilhelm Warta, Ewan D. Dunlop, Solar cell efficiency tables (Version 38), Progress in Photovoltaics: Research and Applications, Volume 19, Issue 5, pages 565–572, August 2011.

Table 11.10.   Important technical advances in photovoltaic technology and their associated U.S. patents

Breakthrough	Patent (year)	Breakthrough description
Boron diffusion	2794846 (1957)	P–N junctions created out of boron rather than lithium increase cell efficiency from 4% to 6%.
	3015590 (1962)	
Contacts (grid and fingers)	2862160 (1958)	Many thin contact fingers reduce the losses experienced by the current as it travels through the cell.
	2919299 (1959)	
	3040416 (1962)	
	3046324 (1962)	
	3450568 (1969)	
	3493437 (1970)	
Antireflective coatings	3533850 (1970)	Transparent layer of antireflective material on top of the cell reduces reflection and increases the amount of sunlight that is absorbed by the cell.
Better, thinner top junctions	3811954 (1974)	Shallower P–N junctions increase the cell response to shorter wavelengths and thus the amount of sunlight the cell could convert.
Tunneling metal–insulator–semiconductor contact	3928073 (1975)	A thin insulator placed between the semiconductor layer and the metal contacts reduces recombination losses and allows more electrons to reach the contacts.
	4104084 (1978)	
Oxide surface passivation	3990100 (1976)	Silicon dioxide attached to broken silicon bonds on the surface of the cell increases cell efficiency.
	4086102 (1978)	
	4171997 (1979)	
Laminating cells to glass with polyvinyl butyral (PVB)	4009054 (1977)	Replacing the layer of exposed silicone with glass allows PV modules to weather the elements better.
Unexposed silicone rubber	4057439 (1977)	Replacing the layer of exposed silicone with glass allows PV modules to weather the elements better.
Aluminum-based pastes	4086102 (1978)	Al-based pastes used at the rear of the cell helped remove impurities and give the modules better weatherization.
Screen printing	4105471 (1978)	By printing contacts on the top and bottom of the cells, PV modules can be made more easily, reducing costs.
Hydrogen plasma passivation	4113514 (1978)	Hydrogen atoms attach to broken silicon bonds throughout the cell and increase cell efficiency.
	4321420 (1982)	
	4322253 (1982)	
	4557037 (1985)	
Pulse annealing	4154625 (1979)	Pulse annealing creates larger silicon crystals, improving the quality of the silicon at a lower cost.
Metallization	4235644 (1980)	Metallization uses thin films of metal to create contacts, reducing the amount of metal used and thus the cost.
	4348546 (1982)	
	4361718 (1982)	
Quasi-square wafers	4356141 (1982)	Square wafers use space and silicon more efficiently than circular wafers.
Reduced metallization resistance	4395583 (1983)	Reduced resistance between the metal contacts and silicon decreases losses and allows more electrons to reach the contacts.
	4694115 (1987)	

Table 11.10. Continued

Breakthrough	Patent (year)	Breakthrough description
Metal–insulator N–P (MINP) cell	4404422 (1983)	Combining metal–insulator–semiconductor contacts with shallow N–P junctions, increasing efficiency.
Ribbon on sacrificial growth plate	4478880 (1984)	Thin ribbons of silicon are continuously produced with less waste silicon, reducing costs.
Ethylene–vinyl acetate (EVA) laminate	4499658 (1985)	EVA doesn't yellow with exposure to sunlight as quickly as PVB, allowing the PV module to last longer.
Plasma deposition of SiN passivation	4640001 (1987)	Thin films of silicon can be made by heating silicon nitride until the nitrogen escapes, reducing the cost and amount of silicon used.
Low-contact resistance with anti-reflective films	4643913 (1987)	Silver pastes lowered the contact resistance, reducing recombination losses and increasing efficiency.
Reactive-ion etching	4664748 (1987)	Highly reactive ions texture the surface of the cell, reducing reflective losses.
	4667058 (1987)	
Passivated emitter solar cell (PESC)	4589191 (1986)	Contacts are made through slits in the top oxide layer, increasing contact passivation and thus efficiency.
Microgrooving	4626613 (1986)	Selective surface etching of microgrooves reduces reflectivity, reduces resistance losses, and is easier to work with than etched pyramids.

Source: Adapted from Nemet, G.F., D. Husmann. 2012. 1.05 - Historical and Future Cost Dynamics of Photovoltaic Technology, In: Ali Sayigh, Editor-in-Chief, Comprehensive Renewable Energy, (Oxford, Elsevier) Pages 47-72.

Table 11.11.   Characteristics of different heat extraction methods for combined photovoltaic/thermal (PV/T) systems

PV/T models	Average efficiency (%)	Advantage	Disadvantages
IS model for air based PV/T type[a]	24–47	Low cost	Low thermal mass
		Simple structure	Poor thermal removal effectiveness
			High heat loss
			Large air volume
IPVTS model for water based PV/T type[b]	33–59	Low cost	Still-high PV temperature
		Direct contribution	Unstable heat removal effectiveness
		High thermal mass	Complex structure
		Low flow volume	Possible piping freezing
PV-SAHP model for refrigerant based PV/T type[c]	56–74	Low PV temperature	Risk of leakage
		Stable performance	Unbalanced liquid distribution
		High efficiency	High cost
		Effective heat removal	Difficult to operate
PV/FPHP model for heat pipe based PV/T type[d]	42–68	Low PV temperature	High cost
		Stable performance	Risk of damage
		High solar efficiency	Complex structure
		Effective heat removal	
		Reduce power input	

[a]Indoor-simulator (IS) model for air-based PV/T.
[b]Integrated PV/T system model for water-based PV/T.
[c]PV solar assisted heat pump model for refrigerant-based PV/T.
[d]PV/flat-plate heat pipe model for heat-pipe-based PV/T.
Source: Adapted from Zhang, Xingxing, Xudong Zhao, Stefan Smith, Jihuan Xu, Xiaotong Yu. 2012. Review of progress and practical application of the solar photovoltaic/thermal (PV/T) technologies, Renewable and Sustainable Energy Reviews, Volume 16, Issue 1, Pages 599-617.

Table 11.12.   Direct, atmospheric cadmium (Cd) emissions during the life cycle of the cadmium-telluride PV module

	Air emissions (g Cd tonne^{-1} Cd[a])	Allocation (%)[c]	Air emissions (g Cd tonne^{-1} Cd[a])	mg Cd GWh^{-1}[b]
Mining of Zn ores	2.7	0.58	0.016	0.02
Zn smelting/refining	40	0.58	0.23	0.3
Cd purification	6	100	6	9.1
CdTe production	6	100	6	9.1
PV manufacturing	3	100	3	4.5
Operation	0.3	100	0.3	0.3
Disposal/recycling	0	100	0	0
Total			15.55	23.3

[a]metric ton of Cd produced.
[b]Energy produced assuming average Southern European insolation (i.e., 1700 kWh m^{-2} yr^{-1}), 9% electrical conversion efficiency, and a 30-year life for the modules.
[c]allocation of emissions to coproduction of Zn, Cd, Ge, and In.
Source: Adapted from Fthenakis, V.M. and H.C. Kim. 2012. 1.08 - Environmental Impacts of Photovoltaic Life Cycles, In: Ali Sayigh, Editor-in-Chief, Comprehensive Renewable Energy, (Oxford, Elsevier) Pages 47-72.

Table 11.13. Summary of test levels for crystalline silicon terrestrial photovoltaic (PV) modules

Name of test	Test conditions
Outdoor exposure test	Exposure to solar irradiation of total of 60 kWh m^{-2}
UV test	Exposure to UV irradiation of total of 15 kWh m^{-2}. The UV wavelength ranges from 280 to 385 nm, with 5 kWh m^{-2} in the wavelength range from 280 to 320 nm
Hot spot test	5 h exposure to 1000 W m^{-2} irradiance in the worst-case hot-spot conditions
Thermal cycling test	50 and 200 thermal cycles from −40 °C to +85 °C, with STC peak power current during 200 cycles
Humidity freeze test	10 cycles from +85 °C, 85% RH to −40 °C
Damp heat test	1000 h at +85 °C, 85% RH
Robustness of termination test	Determination of terminations' capability to withstand appropriate mechanical stress
Mechanical load test	2.4 kPa uniform load applied for 1 h to front and back surface in turn
Hail test	25 mm diameter ice ball at 23 m s^{-1}, directed at 11 impact locations
Bypass diode thermal test	1 h at I_{sc} and 75 °C, 1 h at 1.25 times I_{sc} and 75 °C

Source: International Electrotechnical Commission. 2005. Crystalline silicon terrestrial photovoltaic (PV) modules –Design qualification and type approval, IEC 61215, (Geneva, IEC).

Geothermal

Figures

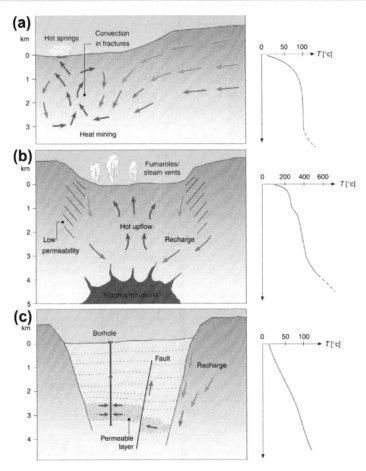

Figure 12.1. Main types of geothermal systems. (a) Volcanic systems: heat sources are hot intrusions or magma; (b) Convective systems: heat source is the hot crust at depth in tectonically active areas, with above average heat flow. (c) Sedimentary systems; heat source derives from permeable sedimentary layers at great depths (> 1 km) and above average geothermal gradients (> 30 °C km⁻¹). These systems are conductive in nature rather than convective. Some convective systems (b) may, however, be embedded in sedimentary rocks. *Source: Axelsson, G. 2012. The Physics of Geothermal Energy, In: Ali Sayigh, Editor-in-Chief, Comprehensive Renewable Energy, (Oxford, Elsevier), Pages 3-50.*

Handbook of Energy. http://dx.doi.org/10.1016/B978-0-08-046405-3.00012-7

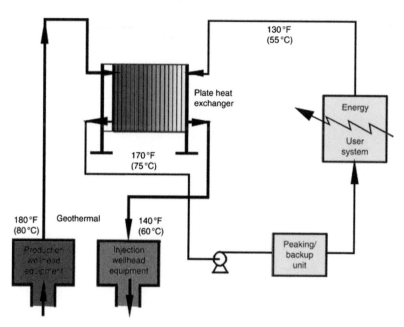

Figure 12.2. Typical configuration for a direct-use geothermal heating system.
Source: Lund, J.W. 2012. Direct Heat Utilization of Geothermal Energy, In: Ali Sayigh, Editor-in-Chief, Comprehensive Renewable Energy, (Oxford, Elsevier), Pages 169-186.

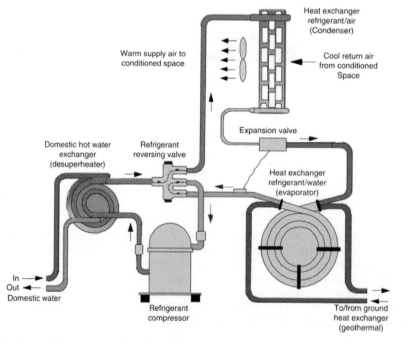

Figure 12.3. Geothermal heat pump in heating mode.
Source: Rybach, L. 2012. Shallow Systems: Geothermal Heat Pumps, In: Ali Sayigh, Editor-in-Chief, Comprehensive Renewable Energy, (Oxford, Elsevier), Pages 187-205.

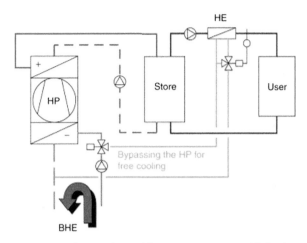

Figure 12.4. The main components of a geothermal heat pump system: (1) the heat source (in this case, a borehole heat exchanger, BHE); (2) the heat pump; (3) the heating/cooling system (in this case, floor panel heating).
Source: Rybach, L. 2012. Shallow Systems: Geothermal Heat Pumps, In: Ali Sayigh, Editor-in-Chief, Comprehensive Renewable Energy, (Oxford, Elsevier), Pages 187-205.

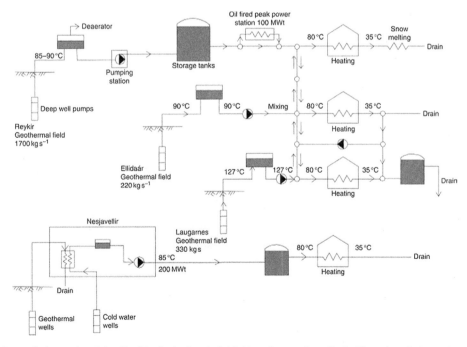

Figure 12.5. Schematic of the Reykjavik, Iceland district heating system that utilizes low-temperature geothermal areas within and in the vicinity of Reykjavik as well as the high-temperature field at Nesjavellir, about 27 km away. The water is transported through 3,846 km of pipeline. About 200,000 people are served in buildings totaling 58 million m^3. The installed capacity of the system is 1,264 MW_t with a peak load (2006) of 924 MW_t.
Source: Lund, J.W. 2012. Direct Heat Utilization of Geothermal Energy, In: Ali Sayigh, Editor-in-Chief, Comprehensive Renewable Energy, (Oxford, Elsevier), Pages 169-186.

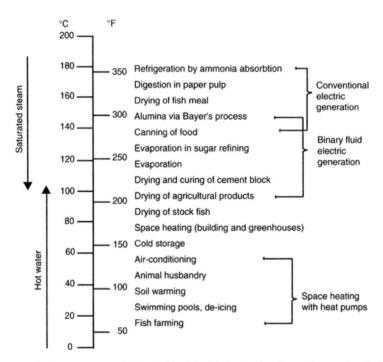

Figure 12.6. The Lindal diagram, named after Baldur Lindal, the Icelandic engineer who first proposed it, indicates the temperature range suitable for various direct uses of geothermal energy.
Source: Lund, J.W. 2012. Direct Heat Utilization of Geothermal Energy, In: Ali Sayigh, Editor-in-Chief, Comprehensive Renewable Energy, (Oxford, Elsevier), Pages 169-186.

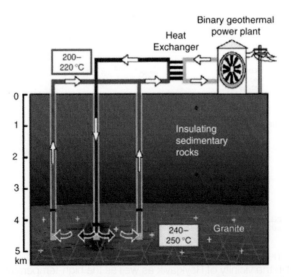

Figure 12.7. Enhanced geothermal system (EGS). These are reservoirs that have been created to extract economical amounts of heat from low permeability and/or porosity geothermal resources.
Source: DiPippo, R. 2012. Geothermal Power Plants, In: Ali Sayigh, Editor-in-Chief, Comprehensive Renewable Energy, (Oxford, Elsevier), Pages 207-237.

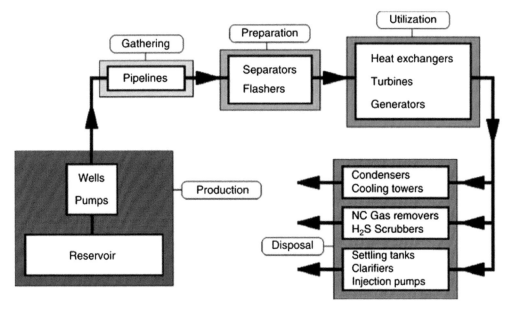

Figure 12.8. General sequence of processes for a geothermal power plant.
Source: DiPippo, R. 2012. Geothermal Power Plants, In: Ali Sayigh, Editor-in-Chief, Comprehensive Renewable Energy, (Oxford, Elsevier), Pages 207-237.

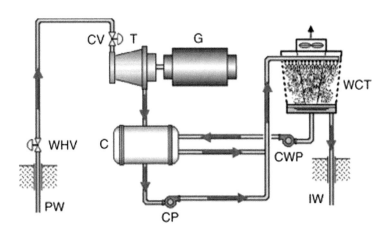

Figure 12.9. Dry-steam geothermal power plant. C, Condenser; CP, Condensate pump; CV, Control valve; CWP, Cooling water pump; G, Generator; IW, Injection well; PW, Production well; T, Turbine; WCT, Water-cooling tower; WHV, Wellhead valve.
Source: DiPippo, R. 2012. Geothermal Power Plants, In: Ali Sayigh, Editor-in-Chief, Comprehensive Renewable Energy, (Oxford, Elsevier), Pages 207-237.

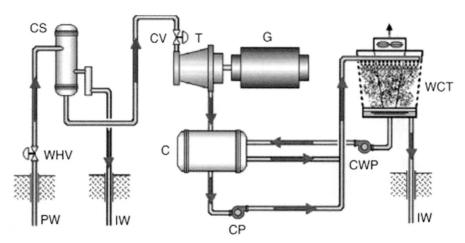

Figure 12.10. Single-flash power geothermal plant. C, Condenser; CP, Condensate pump; CS, Cyclone separator; CV, Control valve; CWP, Cooling water pump; G, Generator; IW, Injection well; PW, Production well; T, Turbine; WCT, Water-cooling tower; WHV, Wellhead valve.
Source: DiPippo, R. 2012. Geothermal Power Plants, In: Ali Sayigh, Editor-in-Chief, Comprehensive Renewable Energy, (Oxford, Elsevier), Pages 207-237.

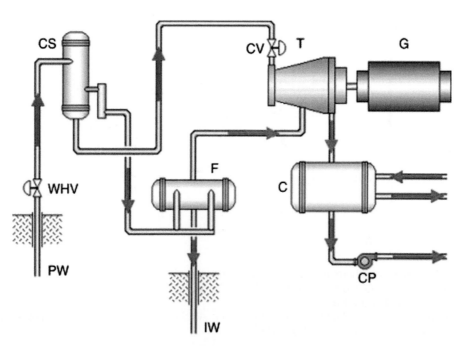

Figure 12.11. Double-flash geothermal power plant. C, Condenser; CP, Condensate pump; CS, Cyclone Separator; CV, Control valve; F, Flash vessel; G, Generator; IW, Injection well; PW, Production well; WHV, Wellhead valve. Note: water-cooling tower omitted for clarity.
Source: DiPippo, R. 2012. Geothermal Power Plants, In: Ali Sayigh, Editor-in-Chief, Comprehensive Renewable Energy, (Oxford, Elsevier), Pages 207-237.

Charts

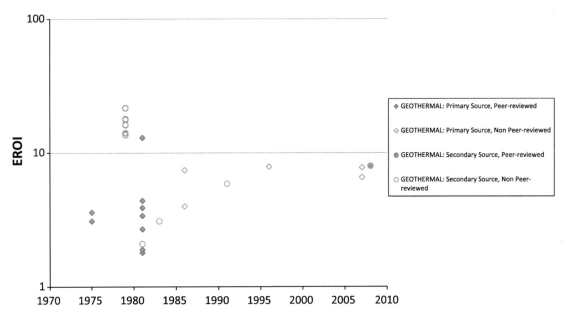

Chart 12.1. Estimates of the energy return on investment (EROI) for geothermal power. EROI is the ratio of energy produced to the direct plus indirect enrgy used in the production process.
Source: Data from Dale, M., S. Krumdieck, P. Bodger. 2012. Global energy modelling — A biophysical approach (GEMBA) Part 2: Methodology, Ecological Economics, Volume 73, Pages 158-167.

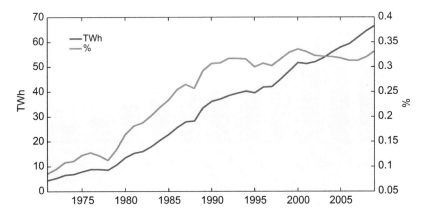

Chart 12.2. Global electricity generation from geothermal energy, and its share of total annual generation.
Source: Data from International Energy Agency (IEA), Energy statistics database, <http://www.iea.org/stats/index.asp>.

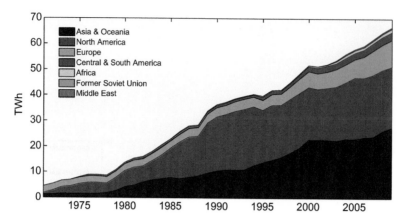

Chart 12.3. Electricity production from geothermal energy by region, 1970-2009.
Source: Data from International Energy Agency (IEA), Energy statistics database, <http://www.iea.org/stats/index.asp>.

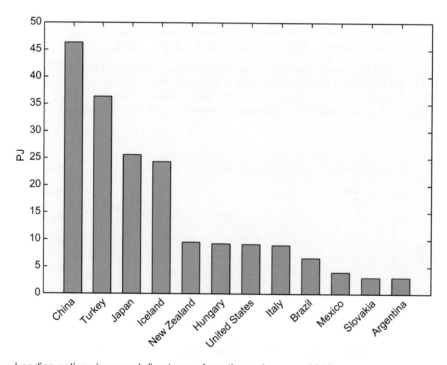

Chart 12.4. Leading nations in annual direct use of geothermal energy, 2010.
Source: Data from Lund, J.W. 2012. Direct Heat Utilization of Geothermal Energy, In: Ali Sayigh, Editor-in-Chief, Comprehensive Renewable Energy, (Oxford, Elsevier), Pages 169-186.

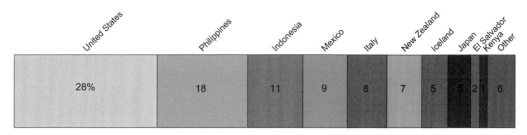

Chart 12.5. Top 10 nations share of installed geothermal capacity in 2010.
Source: Data from International Energy Agency (IEA), Energy statistics database, <http://www.iea.org/stats/index.asp>.

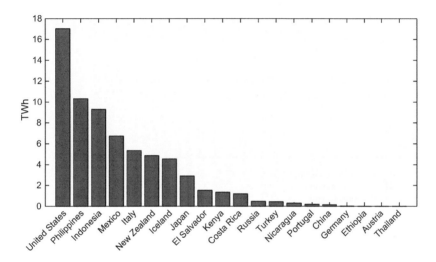

Chart 12.6. Generation of electricity from geothermal energy by country, 2009.
Source: Data from International Energy Agency (IEA), Energy statistics database, <http://www.iea.org/stats/index.asp>.

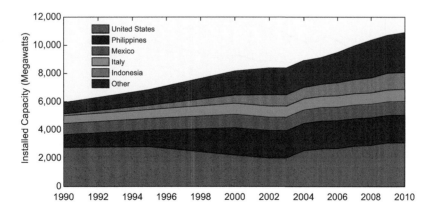

Chart 12.7. Installed geothermal electricity capacity, 1990-2010 for top 5 nations.
Source: Data from BP, Statistical Review of World Energy 2012, <http://www.bp.com/sectionbodycopy.do?categoryId=7500&contentId=7068481>.

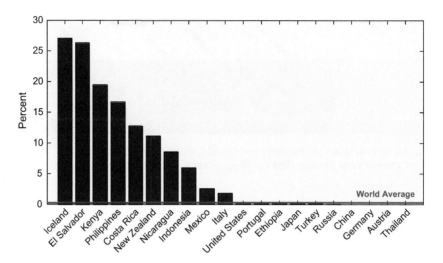

Chart 12.8. Percentage of electricity from geothermal energy for leading producing nations, 2009.
Souece: Data from International Energy Agency (IEA), Energy statistics database, <http://www.iea.org/stats/index.asp>.

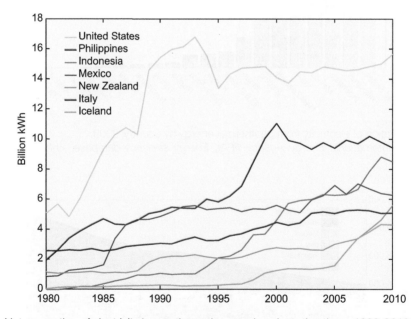

Chart 12.9. Net generation of electricity by geothermal energy in selected nations, 1980-2010.
Source: Data from United States Department of Energy, Energy Information Administration, International Energy Statistics, <http://www.eia.gov/countries/data.cfm>.

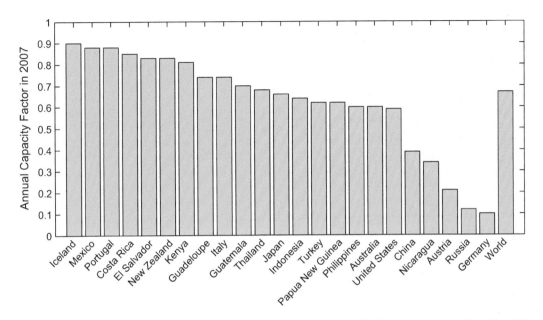

Chart 12.10. Geothermal capacity factors, 2007 for all nations and world. Capacity factor = the ratio of the energy actually produced to the amount that could have been produced by continuous full-power operation
Source: Data from World Energy Council. 2010. Survey of World Energy Resources, (London, United Kingdom).

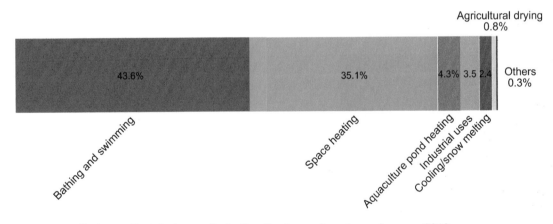

Chart 12.11. End use of installed capacity for the direct use of geothermal energy, 2010.
Source: Data from Lund, J.W. 2012. Direct Heat Utilization of Geothermal Energy, In: Ali Sayigh, Editor-in-Chief, Comprehensive Renewable Energy, (Oxford, Elsevier), Pages 169-186.

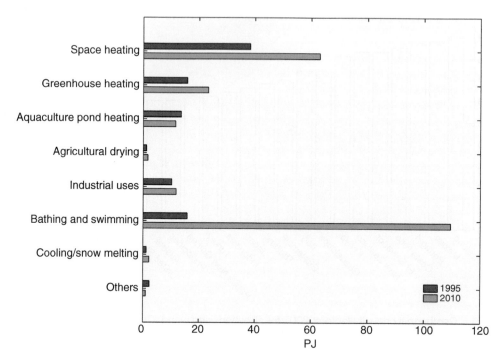

Chart 12.12. Utilization of direct geothermal energy by end use category, 1995 and 2010.
Source: Data from Lund, J.W. 2012. Direct Heat Utilization of Geothermal Energy, In: Ali Sayigh, Editor-in-Chief, Comprehensive Renewable Energy, (Oxford, Elsevier), Pages 169-186.

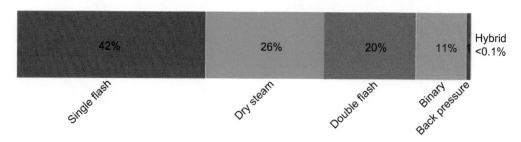

Chart 12.13. Geothermal electric plant types, 2010.
Source: Data from Bertani, Ruggero. 2012. Geothermal power generation in the world 2005–2010 update report, Geothermics, Volume 41, Pages 1-29.

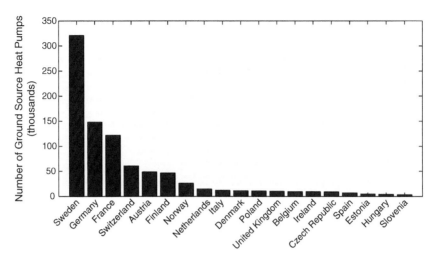

Chat 12.14. Number of ground source heat pumps in Europe, 2008.
Source: Data from Bayer, Peter, Dominik Saner, Stephan Bolay, Ladislaus Rybach, Philipp Blum. 2012. Greenhouse gas emission savings of ground source heat pump systems in Europe: A review, Renewable and Sustainable Energy Reviews, Volume 16, Issue 2, Pages 1256-1267.

Tables

Table 12.1. Classification of geothermal resources by temperature

	Source of estimate			
	(a)	(b)	(c)	(d)
Low enthalpy	<90°C	<125°C	<100°C	<150°C
Intermediate	90–150	125–225	100–200	–
High enthalpy	>150	>225	>200	>150

(a) P. Muffler, R. Cataldi, Methods for regional assessment of geothermal resources, Geothermics, 7 (1978), pp. 53–89.
(b) M.P. Hochstein, Classification and assessment of geothermal resources, M.H. Dickson, M. Fanelli (Eds.), Small geothermal resources, UNITAR/UNDP Centre for Small Energy Resources, Rome Italy (1990), pp. 31–59.
(c) Benderitter, G. Cormy, Possible approach to geothermal research and relative cost estimate, M.H. Dickson, M. Fanelli (Eds.), Small Geothermal Resources, UNITAR/UNDP Centre for Small Energy Resources, Rome, Italy (1990), pp. 61–71.
(d) R. Haenel, L. Rybach, L. Stegena, Fundamentals of geothermics, R. Haenel, L. Rybach, L.A. Stegena (Eds.), Handbook of Terrestrial Heat-Flow Density Determination, Kluwer Academic, Dordrecht, Netherlands (1988), pp. 9–57.

Table 12.2. Properties of geothermal fluids and concentration of key corrosive species in the fluid at various high-temperature geothermal fields

Location	Temperature (°C) (location)	Fluid description[a]	pH	Cl⁻	Concentration of key species in the fluid (ppm)				
					Total CO_2	Total H_2S	Total NH_3	SO_4^{2-}	CH_4
Salton Sea, California, USA	250 (borehole)	Unflashed wellhead fluid	5.2	115000	1000	10–30	300	20	
Baca (Valles Caldera), New Mexico, USA	171 (wellhead)	Flashed fluid	6.8	3770	128	6		59	
Bjarnarflag, North Iceland, Iceland	171 (wellhead)	Flashed fluid	7.9	283	529	1333	1.7		7.6
Reykjanes, Southwest Iceland, Iceland	295 (borehole)	Unflashed wellhead fluid	4.7	19319	1779	53	2	12.2	0.2

[a]Measurements were made at different points of production, before or after flashing; thus, the source fluids cannot be directly compared because often during flashing Cl concentration increases while the concentration of CO_2 and H_2S decreases and increase in pH will occur.
Source: Adapted from Karlsdóttir, N. 2012. Corrosion, Scaling and Material Selection in Geothermal Power Production, IIn: Ali Sayigh, Editor-in-Chief, Comprehensive Renewable Energy, (Oxford, Elsevier), Pages 239–257.

Table 12.3. Global technical and economic geothermal potential by region[a]

Region	Theoretical potential (10^6 EJ)	Technical potential		Economic potential		
		Heat for direct utilization (EJ/year)	Heat for electricity (EJ/year)	Heat for direct utilization (EJ/year)	Heat for electricity (EJ/year)	Produced electricity (TWh/year)
USA	4.738	7	75	1.215	34.9	508
Canada	3.287	4.8	52	0.099	0.307	8.3
Western Europe	2.019	3	32	4.311	6.216	125
Central and Eastern Europe	0.323	0.5	5.1	0.852	1.243	25
Former Soviet Union	6.607	9.9	104	0.508	3.097	67
Northern Africa	1.845	2.8	29	0.103	0	0
Eastern Africa	0.902	1.3	14	0.004	0.918	25
Western and Central Africa	2.103	3.2	33	0	0	0
Southern Africa	1.233	1.8	19	0	0	0
Middle East	1.355	2	21	0.175	0.612	17
China	3.288	4.7	52	1.764	1.856	42
Other East Asia	0.216	0.3	3.4	0.018	0	0
India	0.938	1.4	15	0.062	0.613	17
Other South Asia	2.424	3.7	38	0.002	0	0
Japan	0.182	0.2	2.9	0.201	0.612	17
Other Pacific Asia	1.092	1.4	17	0.004	7.424	166
Oceania	2.304	3.5	36	0.391	1.568	25
Latin America	6.886	9.9	109	0.383	6.216	125
World	41.743	61.4	657	10.092	65.582	1167
Equivalent capacity		5000 GW$_{th}$	1200 GW$_{el}$	800 GW$_{th}$	140 GW$_{el}$	

[a]The values refer to the underground heat available for direct utilization or electricity, with the exception of the expected electricity production, calculated using a weighted average conversion efficiency (about 17 J of heat for 1 J of electricity) and 95% capacity factor. For the direct utilization, the thermal capacity is calculated with an average 40% capacity factor.
Source: Adapted from Bertani, Ruggero. 2012. Geothermal power generation in the world 2005–2010 update report, Geothermics, Volume 41, Pages 1–29.

Table 12.4.   Range of technically recoverable heat energy from accessible geothermal resources

Probability	99%	90%	50%	Log-normal mean	10%	1%	
Recovery Factor	0.05%	1.34%	4.47%	7.00%	14.95%	40.00%	
Accessible Stored Thermal Energy Estimates	$EJ \times 10^6$	$EJ \times 10^6$	$EJ \times 10^6$	$EJ \times 10^6$	$EJ \times 10^6$	$EJ \times 10^6$	$EJ \times 10^6$
< 10 km under continents[a]	403	0.2	5.4	18	28.2	60.2	161.2
< 10 km under continents[c]	110.4	0.1	1.5	4.9	7.7	16.5	44.2
<5 km under continents[d,e]	139.5	0.1	1.9	6.2	9.8	20.9	55.8
<5 km under continents[d,f]	55.9	0.03	0.7	2.5	3.9	8.4	22.4
15 degrees C to 3 km under continents[b]	42.7	0.02	0.6	1.9	3	6.4	17.1

[a]Rowley, J., C., 1982. Worldwide geothermal results In: Handbook of Geothermal Energy, Edwards et al., eds. Chapter 2, pp 44–176, Gulf Publishing Houston, Texas.
[b]Electric Power Research Institute (EPRI), 1978. Geothermal energy prospects for the next 50 years. ER-611-SR, Special Report for the World Energy Conference 1978.
[c]Tester, J.W., E.M. Drake, M.W. Golay, M.J. Driscoll, and W.A. Peters, 2005. Sustainable Energy – Choosing Among Options, MIT Press, Cambridge, MA, 850 pp.
[d]Goldstein, B.A., Hiriart, G., Tester, J.,. B., Bertani, R., Bromley, Gutierrez-Negrin, L.,C.J., Huenges, E., H, Ragnarsson, A., Mongillo, M.A. Muraoka, and V. I. Zui, Great expectations for geothermal energy to 2100, PROCEEDINGS, Thirty-Sixth Workshop on Geothermal Reservoir Engineering Stanford University, Stanford, California, January 31 - February 2, 2011 SGP-TR-191.
[e]Based on interpolation between Rowley, *supra*, to 10 km and EPRI, *supra*, to 3 km.
[f]Based on interpolation between Tester, *supra*, to 10 km and EPRI, *supra*, to 3 km.
Source: Adapted from Goldstein et. al., *supra*.

Table 12.5.   Types of geothermal resources, temperatures, and applications

Type	In-situ fluids	Sub-type	Temperature Range[a]	Utilization Current	Utilization Future
Convective systems (hydrothermal)	Yes	Continental	H, I &L	Power, direct use	
		Submarine	H	None	power
Conductive systems	No	Shallow (<400 m)	L	Direct use (GHP)[c]	
		Hot rock (EGS)[b]	H, I	Prototypes	Power, direct use
		Magma bodies	H	None	Power, direct use
Deep aquifer systems	Yes	Hydrostatic aquifers	H, I &L	Direct use	Power, direct use
		Geo-pressured		Direct use	

[a]Temperature range: H: High (>180 °C), I: Intermediate (100–180 °C), L: Low (ambient to 100 °C).
[b]EGS: Enhanced (or engineered) geothermal systems.
[c]GHP: Geothermal heat pumps.
Source: Adapted from Goldstein, B., G. Hiriart, R. Bertani, C. Bromley, L. Gutiérrez-Negrín, E. Huenges, H. Muraoka, A. Ragnarsson, J. Tester, V. Zui, 2011: Geothermal Energy. In IPCC Special Report on Renewable Energy Sources and Climate Change Mitigation [O. Edenhofer, R. Pichs-Madruga, Y. Sokona, K. Seyboth, P. Matschoss, S. Kadner, T. Zwickel, P. Eickemeier, G. Hansen, S. Schlömer, C. von Stechow (eds)], Cambridge University Press, Cambridge, United Kingdom and New York, NY, USA.

Table 12.6.   Estimated U.S. geothermal resource base to 10 km depth by category

Category of Resource	Thermal Energy EJ
Conduction-dominated EGS[a]	
Sedimentary rock formations	100,000
Crystalline basement rock formations	13,300,000
Supercritical Volcanic EGS[b]	74,100
Hydrothermal	2,400–9,600
Coproduced fluids	0.0944–0.4510
Geopressured systems	71,000–170,000[c]

[a]Enhanced (or engineered) Geothermal Systems (EGS) are engineered reservoirs that have been created to extract economical amounts of heat from low permeability and/or porosity geothermal resources.
[b]Excludes Yellowstone National Park and Hawaii.
[c]Includes methane content.
Source: Adapted from Tester, J.W., B.J. Anderson, A.S. Batchelor, D.D. Blackwell, R. DiPippo, E.M. Drake, J. Garnish, B. Livesay, M.C. Moore, K. Nichols, S. Petty, M.N, Toksöks, and R.W. Veatch Jr. (2006). The Future of Geothermal Energy: Impact of Enhanced Geothermal Systems on the United States in the 21st Century. Prepared by the Massachusetts Institute of Technology, under Idaho National Laboratory Subcontract No. 63 00019 for the U.S. Department of Energy, Assistant Secretary for Energy Efficiency and Renewable Energy, Office of Geothermal Technologies, Washington, DC, USA.

Table 12.7.   World's 20 largest geothermal fields based on installed capacity

Geothermal field name	Nation	Installed capacity (MW)	Produced energy (MWh/year)
CA – The Geysers	USA	1584	7062
Cerro Prieto	Mexico	720	5176
Tongonan/Leyte	Philippines	716	4746
Larderello	Italy	595	3666
Java – Gunung Salak	Indonesia	377	3024
CA – Salton Sea	USA	329	2634
CA – Coso	USA	270	2381
Mak-Ban/Laguna	Philippines	458	2144
Java – Darajat	Indonesia	260	2085
Java – Wayang Windu	Indonesia	227	1821
Hellisheidi	Iceland	213	1704
Wairakei	New Zealand	233	1693
Java – Kamojang	Indonesia	200	1604
Los Azufres	Mexico	188	1517
Olkaria	Kenya	202	1430
Palinpinon/Negros Oriental	Philippines	193	1257
Travale–Radicondoli	Italy	160	1209
Miravalles	Costa Rica	166	1131
Oita	Japan	152	1106
Tiwi/Albay	Philippines	234	1007

Source: Adapted from Bertani, Ruggero. 2012. Geothermal power generation in the world 2005–2010 update report, Geothermics, Volume 41, Pages 1–29.

Table 12.8.   World's largest geothermal turbine manufacturers based on installed capacity

Company	Country	Total number of units	Total MW	MW decommissioned	MW in operation
Mitsubishi	Japan	100	2,882	150	2,628
Toshiba	Japan	44	2,746	222	2,524
Fuji	Japan	60	2,387	58	2,279
Ansaldo/Tosi	Italy	72	1,556	398	1,158
Ormat	Israel	174	1,234	55	1,152
General Electric/Nuovo Pignone	USA	23	533	0	533
Alstom	France	11	155	0	155
Associated Electrical Industries	New Zealand	3	90	0	90
Kaluga Turbine Works	Russia	11	82	0	82
British Thompson Houston	UK	8	82	0	82
Mafi Trench	USA	6	72	0	72
Qingdao Jieneng	China	9	62	1	21
Kawasaki	Japan	3	16	0	16
Westinghouse	USA	1	14	0	14
UTC/Turboden	USA	57	14	0	13
Elliot	New Zealand	3	13	0	10
Enex	Iceland	2	11	0	11
Harbin	China	2	11	0	11
Makrotek	Mexico	1	5	0	5
Parsons	New Zealand	1	5	0	5

Source: Adapted from Bertani, Ruggero. 2012. Geothermal power generation in the world 2005–2010 update report, Geothermics, Volume 41, Pages 1–29.

Table 12.9. World's 20 largest geothermal power plants ranked by installed capacity

Country	Geothermal field	Plant name	Unit	COD[a]	Type	Manufacturer	Plant owner	Installed capacity (MW)
New Zealand	Rotokawa	Nga Awa Purua	1	2010	Single flash	Fuji	Mighty River Power	132
Indonesia	Java – Wayang Windu	Wayang Windu	2	2009	Single flash	Fuji	Star Energy Ltd	117
USA	CA – The Geysers	Grant	1	1985	Dry steam	Toshiba	Calpine	113
USA	CA – The Geysers	Lake View	1	1985	Dry steam	Toshiba	Calpine	113
USA	CA – The Geysers	Quicksilver	1	1985	Dry steam	Toshiba	Calpine	113
USA	CA – The Geysers	Socrates	1	1983	Dry steam	Toshiba	Calpine	113
USA	CA – The Geysers	Cobb Creek	1	1979	Dry steam	Toshiba	Calpine	110
USA	CA – The Geysers	Eagle Rock	1	1975	Dry steam	Toshiba	Calpine	110
Mexico	Cerro Prieto	Cerro Prieto II	1	1986	Double flash	Toshiba	Comision Federal de Electricidad	110
Mexico	Cerro Prieto	Cerro Prieto II	2	1987	Double flash	Toshiba	Comision Federal de Electricidad	110
Mexico	Cerro Prieto	Cerro Prieto III	1	1986	Double flash	Toshiba	Comision Federal de Electricidad	110
Mexico	Cerro Prieto	Cerro Prieto III	2	1986	Double flash	Toshiba	Comision Federal de Electricidad	110
Indonesia	Java – Darajat	Darajat	3	2008	Dry steam	Mitsubishi	PLN	110
Indonesia	Java – Wayang Windu	Wayang Windu	1	2000	Single flash	Fuji	Star Energy Ltd	110
USA	CA – The Geysers	Sulfur Spring	1	1980	Dry steam	Toshiba	Calpine	109
New Zealand	Kawerau	Kawerau	1	2008	Double flash	Fuji	Mighty River Power	100
USA	CA – The Geysers	Big Geyser	1	1980	Dry steam	General Electric/ Nuovo Pignone	Calpine	97
Indonesia	Java – Darajat	Darajat	2	1999	Dry steam	Mitsubishi	PLN	90
USA	CA – The Geysers	Calistoga	1	1984	Dry steam	Toshiba	Calpine	80
Philippines	Tongonan/Leyte	Malitbog	1	1997	Single flash	Fuji	Energy Development Corporation	77.9

[a]Date of commission.

Source: Adapted from Bertani, Ruggero. 2012. Geothermal power generation in the world 2005–2010 update report, Geothermics, Volume 41, Pages 1–29.

Table 12.10.   Geothermal electricity net generation for top 20 nations (billion kilowatthours)

Country	1980	1990	2000	2010	Rank 2010	Rank 1980	Change 2000–10
United States	5.07	15.43	14.09	15.67	1	1	11%
Philippines	1.97	5.19	11.05	9.43	2	3	−15%
Indonesia	0	1.07	4.63	8.50	3	10	84%
Mexico	0.87	4.87	5.61	6.29	4	6	12%
New Zealand	1.15	2.10	2.78	5.59	5	4	101%
Italy	2.57	3.06	4.47	5.09	6	2	14%
Iceland	0.05	0.29	1.26	4.30	7	8	242%
Japan	1.09	1.65	3.18	2.52	8	5	−21%
Kenya	0	0.32	0.41	1.50	9	12	268%
El Salvador	0.47	0.38	0.74	1.43	10	7	94%
Costa Rica	0	0	0.93	1.10	11	13	19%
Turkey	0	0.08	0.07	0.60	12	14	731%
Russia	--	--	0.06	0.44	13	n/a	700%
Papua New Guinea	0	0	0	0.40	14	15	n/a
Nicaragua	0	0.40	0.12	0.27	15	16	125%
Guatemala	0	0	0.20	0.26	16	18	30%
Portugal	0	0	0.08	0.19	17	9	150%
China	0	0	0	0.15	18	18	n/a
Guadeloupe	0	0.03	0.03	0.10	19	n/a	233%
Ethiopia	0	0	0.01	0.02	20	n/a	380%
World	13.24	34.91	49.69	63.88			29%

Source: BP, Statistical Review of World Energy 2011.

Table 12.11.   Geothermal plant types by country for top 10 nations (MW installed)

	Hybrid	Back pressure	Binary	Single flash	Double flash	Dry steam	Total
United States	2	0	656	60	796	1,584	3,098
Philippines	0	0	209	1,330	365	0	1,904
Indonesia	0	2	0	735	0	460	1,197
Mexico	0	75	3	410	470	0	958
Italy	0	0	0	88	0	755	843
New Zealand	0	47	138	387	190	0	762
Iceland	0	0	10	474	90	0	575
Japan	0	0	2	350	160	24	535
El Salvador	0	0	9	160	35	0	204
Kenya	0	2	14	186	0	0	202
World	2	147	1,193	4,552	2,183	2,822	10,898

Source: Adapted from Bertani, Ruggero. 2012. Geothermal power generation in the world 2005–2010 update report, Geothermics, Volume 41, Pages 1–29.

Table 12.12. Typical efficiencies for various geothermal plant types

Plant type	Utilization efficiency (%)	Thermal efficiency (%)
Dry-steam	45–55	NA
Single-flash	25–35	NA
Double-flash	35–45	NA
Basic subcritical binary	15–45	5–15
Supercritical binary	16–50	5–15
Binary with recuperator	18–55	14–18
Binary with mixture WF	20–55	15–20

NA = not applicable.
Source: Adapted from DiPippo, R. 2012. Geothermal Power Plants, In: Ali Sayigh, Editor-in-Chief, Comprehensive Renewable Energy, (Oxford, Elsevier), Pages 207–237.

Table 12.13. Average capacity and energy produced for types of geothermal power plants

Type	Average energy (GWh/unit)	Average capacity (MW/unit)
Binary	27	5
Back pressure	96	6
Single flash	199	31
Double flash	236	34
Dry steam	260	46

Source: Adapted from Bertani, Ruggero. 2012. Geothermal power generation in the world 2005–2010 update report, Geothermics, Volume 41, Pages 1–29.

Table 12.14. The age of selected geothermal systems

Area	Age (years)	Remarks
Broadlands (New Zealand)	370,000	Sphalerite deposits >10,000 1 million
Wairakei (New Zealand)	500,000	Palynological examination of ignimbrites in conglomerates < intersecting faults
Nesjavellir (Iceland)	80,000	Vug fillings in cutting samples
The Geysers (United States)	>57,000	Stored heat calculation
Larderello (Italy)	3 million	Rb/Sr, K/Ar dating

Source: Adapted from Axelsson, G. 2012. The Physics of Geothermal Energy, In: Ali Sayigh, Editor-in-Chief, Comprehensive Renewable Energy, (Oxford, Elsevier), Pages 3–50.

Table 12.15. The impact of large-scale production on selected geothermal systems

System (location)	Production initiated	Number of production wells	Average production (kg s⁻¹)	Reservoir temperature (°C)	Draw down	Temperature decline (°C)
Svartsengi (SW-Iceland)	1976	10	380	240	275 m	0
Laugarnes (SW-Iceland)	1930	10	160	127	110 m	0
Reykir (SW-Iceland)	1944	34	850	70–97	100 m	0–13[a]
Nesjavellir (SW-Iceland)	1975	11	390	280–340	7 bar	0
Hamar (N-Iceland)	1970	2	30	64	30 m	0
Laugaland (N-Iceland)	1976	3	40	95	370 m	0
Krafla (N-Iceland)	1978	21	300	210–340	10–15 bar	0
Urridavatn (E-Iceland)	1980	3	25	75	40 m	2–15[b]
Gata (S-Iceland)	1980	2	17	100	250 m	1–2
Urban Area (China)	late 1970s	90–100	~100	~40–90	45 m	0
Palinpinion-1 (the Philipinnes)	1983	23	710	240	55 bar	-
Ahuachapan (El Salvador)	1976	~16	~700	240–260	14 bar	-

Data are approximate, but representative, values based on information from 2000 to 2006.
[a]Only 3 of the 34 production wells have experienced some cooling.
[b]Two older production wells, not used after 1983, experienced up to 15 °C cooling.
Source: Adapted from Axelsson, G. 2012. The Physics of Geothermal Energy, In: Ali Sayigh, Editor-in-Chief, Comprehensive Renewable Energy, (Oxford, Elsevier), Pages 3–50.

Table 12.16. **Land requirements for typical geothermal power generation systems**

Type of Power Plant[a]	Land use	
	m^2/MW_e	$m^2/GWh/yr$
110-MW_e geothermal flash plants (excluding wells)	1260	160
56-MW_e geothermal flash plant (including wells, pipes, etc.)	7460	900
49-MW_e geothermal FC-RC plant (excluding wells)	2290	290
20-MW_e geothermal binary plant (excluding wells)	1415	170

[a]FC: Flash cycle. RC: Rankine cycle.
Source: Goldstein, B., G. Hiriart, R. Bertani, C. Bromley, L. Gutiérrez-Negrín, E. Huenges, H. Muraoka, A. Ragnarsson, J. Tester, V. Zui, 2011: Geothermal Energy. In IPCC Special Report on Renewable Energy Sources and Climate Change Mitigation [O. Edenhofer, R. Pichs-Madruga, Y. Sokona, K. Seyboth, P. Matschoss, S. Kadner, T. Zwickel, P. Eickemeier, G. Hansen, S. Schlömer, C. von Stechow (eds)], Cambridge University Press, Cambridge, United Kingdom and New York, NY, USA.

Table 12.17. **Summary of the applications for direct use of global geothermal energy, 1995 and 2010**

	1995		2010		Change in Capacity
	Capacity[a]	Capacity factor	Capacity[a]	Capacity factor	1995-2010
Space heating	2,579	0	5,394	0.37	109.2%
Greenhouse heating	1,085	0	1,544	0.48	42.3%
Aquaculture pond heating	1,097	0	653	0.56	−40.5%
Agricultural drying	67	1	125	0.41	86.6%
Industrial uses	544	1	533	0.7	−2.0%
Bathing and swimming	1,085	0	6,701	0.52	517.6%
Cooling/snow melting	115	0	368	0.18	220.0%
Others	238	0	42	0.72	−82.4%
Total	6,810		15,360		125.6%

[a]Units are MW_t.
Source: Adapted from Lund, J.W. 2012. Direct Heat Utilization of Geothermal Energy, In: Ali Sayigh, Editor-in-Chief, Comprehensive Renewable Energy, (Oxford, Elsevier), Pages 169–186.

Table 12.18.   Laws concerning the use of shallow (< 400 meters) geothermal energy

Country	Laws (publishing date)
Australia	Energy Resources Act (1967); The Petroleum and Geothermal Rights in Water and Irrigation Act (1914); Water Act (2000)
Austria	Österreichischer Wasser- und Abfallzweckverband (ÖWAV) Regelblatt 207: Thermal Use of Groundwater and Subsurface – Heating and Cooling (2009)
Belgium	Decree on Environmental Permits (28/06/1985)
Bulgaria	Constitution (1991); Law on the Renewable and Alternative Sources of Energy and the Biofuels (2007); Water Act (1999)
Canada	Water Act (1985)
China	Renewable Energy Law (2006)
Czech Republic	Building and Planning Act (No 183/2006)
Denmark	Order on Heat Abstraction and Groundwater Cooling Plants (BEK-1206, 24/11/2006); Order on Groundwater Heating BEK-1203, 20/11/2006)
Ecuador	Water Act (n.a.)
Finland	Environmental Protection Act (2000); Water Act (1961)
France	Decree 74-498 (24/03/1978); Decree 77-620 (16/06/1977); Decree 78-498 (28/03/1978); Mining Law (16/08/1956)
Germany	Mining Law (13/08/1980); Federal Water Act (27/07/1957)
Great Britain	Water Environment Regulations (2005)
Greece	Decision of Minister of Development No. Δ9B, Δ/Φ166/OiK 18508/5552/207 on Installation Permits for Ground Source Heat Pumps (n.a.)

Country	Laws (publishing date)
Indonesia	Geothermal Law (23/10/2003)
Lithuania	Underground Law (I-1034, 05/01/1996)
Mexico	Water Act (1992)
Netherlands	Groundwater Law (1981); Mining Law (01/01/2003)
Norway	Neighbor Law (n.a.)
Philippines	Economic Activity Law (n.a.); Geological and Mining Law (19/11/1999); Renewable Energy Act (2008)
Poland	Water Act (1974)
Portugal	Decree-Law 87/90 (16/03/1990)
Romania	Environment Protection Law (No. 265/2006); Mining Law (No. 61/1998); Water Law (No. 310/2004)
Slovakia	Water Law (2004) (Law No. 364/2004)
Slovenia	Mining Law 2004 (Official Gazette, 98/2004); Water Law 2002 (Official Gazette, 67/2002)
Sweden	Normbrunn 97 (2002)/Normbrunn 07 (2008)
Switzerland	Water Protection Order (28/10/1998)

Source: Adapted from Haehnlein, Stefanie, Peter Bayer, Philipp Blum. 2010. International legal status of the use of shallow geothermal energy, Renewable and Sustainable Energy Reviews, Volume 14, Issue 9, Pages 2611–2625.

Table 12.19. Porosity values, heat capacities and bulk specific weights for water-saturated rock types[a]

	ϕ	Specific Weight (lb/ft³)	C (Btu/lb-°F)	T (°F)	p (psia)	Q (cal)
Sandstone	0.196	142	0.252	90	14.7	5.44×10^{22}
Shale	0.071	149	0.213	90	14.7	4.83×10^{22}
Limestone	0.186	149	0.266	90	14.7	6.04×10^{22}

[a]Mean surface temperature = 55°F.
ϕ = fractional porosity.
C = thermal capacity.
Q = heat flow.
p = pressure.
Source: Adapted from Edwards, L. M., H.H. Rieke, G. V. Chilingar. 1982. Handbook of Geothermal Energy, (Houston, Gulf Publishing Company).

Table 12.20. Characteristic performance of a borehole heat exchanger (BHE) in different rock types[a]

Rock type	Thermal conductivity (W m⁻¹ K⁻¹)	Specific extraction rate (W per m)	Energy yield (kWh m⁻¹ yr⁻¹)
Hard rock	3	Max. 70	100–120
Unconsolidated rock, saturated	2	45–50	90
Unconsolidated rock, dry	1.5	Max. 25	50

[a]single BHE, depth ~ 150 m.
Source: Adapted from Rybach, L. 2012. Shallow Systems: Geothermal Heat Pumps, In: Ali Sayigh, Editor-in-Chief, Comprehensive Renewable Energy, (Oxford, Elsevier), Pages 187–205.

Table 12.21. The heat budget of the solid Earth (all values in units of 10^{12} W)

Income	
Crustal radioactivity	8.2
Mantle radioactivity	20.0
Core radioactivity	0.2
Latent heat and gravitational energy released by core evolution	1.0
Gravitational energy of mantle differentiation	0.1
Gravitational energy released by thermal contraction	3.1
Tidal dissipation	0.1
TOTAL	32.7
Expenditure	
Crustal heat loss	8.2
Mantle heat loss	32.5
Core heat loss	3.5
TOTAL	44.2
Net Loss of Heat	11.5

Source: Adapted from Stacey, F.D. and P.M. Davis. 2008. Physics of the Earth, 4th ed., (Cambridge University Press, Cambridge, UK).

Table 12.22.   Thermally important radioactive elements in the Earth

Isotope	Energy/atom[a] (MeV)	μ W/kg of isotope	W/kg of element	Estimated total Earth content (kg)	Total heat (10^{12} W)	Total heat 4.5×10^9 y ago (10^{12} W)
^{238}U	47.7	95.0	94.35	$12.86 \times | 10^{16}$	12.21	24.5
^{235}U	43.9	562.0	4.05	0.0940×10^{16}	0.53	44.4
^{232}Th	40.5	26.6	26.6	47.9×10^{16}	12.74	15.9
^{40}K	0.71	30.0	0.003 50	7.77×10^{20}	2.72	33.0
				(Total K)	28.2	117.8

[a]These energies include all series decays to final daughter products. Average locally absorbed energies are considered; neutrino energies are ignored.

Source: Adapted from Stacey, F.D. and P.M. Davis. 2008. Physics of the Earth, 4th ed., (Cambridge University Press, Cambridge, UK).

Table 12.23.   Thermal properties of soil materials

Material	Density (Mg m^{-3})	Specific Heat (J g^{-1} K^{-1})	Thermal Conductivity (W m^{-1} K^{-1})	Volumetric heat capacity (MJ m^{-3} K^{-1})
Soil minerals	2.65	0.87	2.5	2.31
Granite	2.64	0.82	3.0	2.16
Quartz	2.66	0.80	8.8	2.13
Glass	2.71	0.84	0.8	2.28
Organic matter	1.30	1.92	0.25	2.50
Water	1.00	4.18	0.56 + 0.0018T	4.18
Ice	0.92	2.1 + 0.0073T	2.22 - 0.011T	1.93 + .0067T
Air (101 kPa)	(1.29 - 0.0041T) × 10^{-3}	1.01	0.024 + 0.00007T	(1.3 - 0.0041T) × 10^{-3}

Source: Adapted from Campbell, Gaylon S. and John M. Norman. 1998. An Introduction to Environmental Biophysics, (New York, Springer-Verlag).

Table 12.24. **Thermal properties of major heat reservoirs on the Earth**

Substance	Condition	Density ρ (10^3 kg m^{-3})	Specific Heat c (10^3 J kg^{-1} K^{-1})	Heat Capacity ρc (10^6 J m^{-3} K^{-1})	Thermal Conductivity k (W m^{-1} K^{-1})	Thermal Diffusivity k^* (10^{-6} m^2 s^{-1})[a]	Conductive Capacity c[a] (10^3 J m^{-2} K^{-1} s$^{-1/2}$)
Air	20 °C, still	0.0012	1	0.0012	0.026	21.5	0.006
	stirred					4×10^6 [b]	2.4
Water	20 °C, still	1	4.19	4.19	0.58	0.14	1.57
	stirred					130 [b]	48
Ice	0 °C, pure	0.92	2.1	1.93	2.24	1.16	2.08
Snow	Fresh	0.1	2.09	0.21	0.08	0.38	0.13
Sandy soil	Dry	1.6	0.8	1.28	0.3	0.24	0.63
(40% pore space)	Saturated	2	1.48	2.98	2.2	0.74	2.56
Clay soil	Dry	1.6	0.89	1.42	0.25	0.18	0.6
(40% pore space)	Saturated	2	1.55	3.1	1.58	0.51	2.21
Peat soil	Dry	0.3	1.92	0.58	0.06	0.1	0.18
(80% pore space)	Saturated	1.1	3.65	4.02	0.5	0.12	1.39

[a]$k^* = k/\rho c$, $c^* = \rho c \sqrt{k^*}$, $d = (Pk^*/\pi)^{1/2}$.

[b]The values of k* for stirred water and air are much greater than those for still conditions because turbulent eddy mixing is a more efficient process to transport heat vertically than molecular conduction.

Source: Adapted from Shaw, Henry, Environmental Heat Transfer, in Kreith, Frank, Ed. 2000. CRC Handbook of Thermal Engineering (Boca Raton, CRC Press).

Hydrogen

Figures

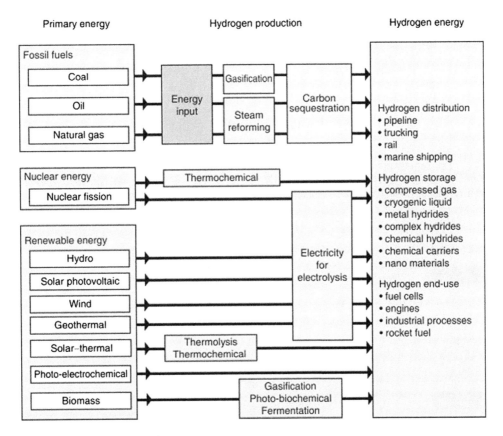

Figure 13.1. Hydrogen energy pathways.
Source: Rand, D.A.J. and R.M. Del. 2009. Fuels - Hydrogen Production: Coal Gasification, In: Jürgen Garche, Editor-in-Chief, Encyclopedia of Electrochemical Power Sources, (Amsterdam, Elsevier), Pages 276-292.

Handbook of Energy. http://dx.doi.org/10.1016/B978-0-08-046405-3.00013-9

H metal
Liquid metal
H_2
gas
Critical point
H_2
liquid
H_2
solid
Triple point
H
gas
Pressure (bar)
10^8
10^6
10^4
10^2
10^0
10^{-2}
10^0
10^1
10^2
10^3
10^4
10^5
10^6
Temperature (K)

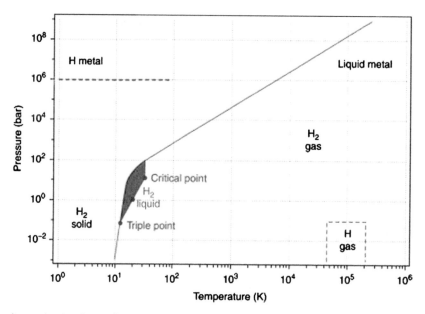

Figure 13.2. Approximate phase diagram for hydrogen.
Source: Züttel, A. 2009. FUELS – HYDROGEN STORAGE: Hydrides, In: Jürgen Garche, Editor-in-Chief, Encyclopedia of Electrochemical Power Sources, (Amsterdam, Elsevier), Pages 440-458.

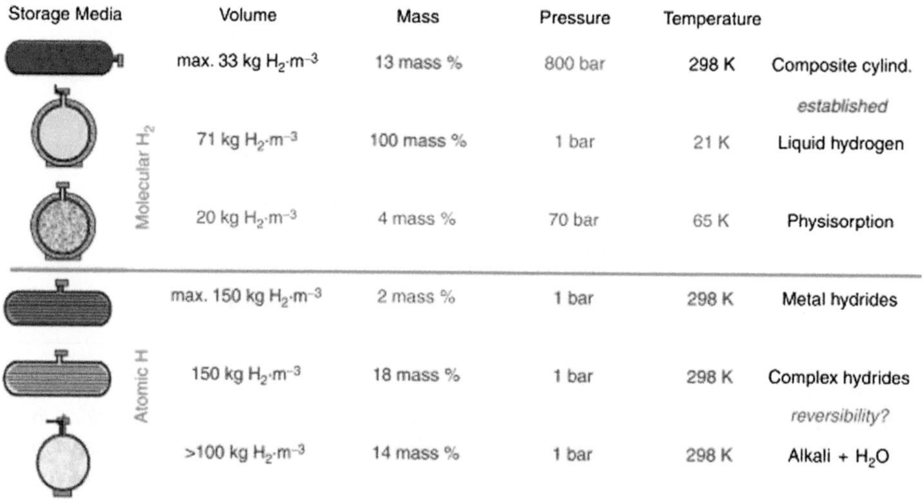

Storage Media
Volume
Mass
Pressure
Temperature
max. 33 kg $H_2 \cdot m^{-3}$
13 mass %
800 bar
298 K
Composite cylind.
established
71 kg $H_2 \cdot m^{-3}$
100 mass %
1 bar
21 K
Liquid hydrogen
20 kg $H_2 \cdot m^{-3}$
4 mass %
70 bar
65 K
Physisorption
Molecular H_2
max. 150 kg $H_2 \cdot m^{-3}$
2 mass %
1 bar
298 K
Metal hydrides
150 kg $H_2 \cdot m^{-3}$
18 mass %
1 bar
298 K
Complex hydrides
reversibility?
>100 kg $H_2 \cdot m^{-3}$
14 mass %
1 bar
298 K
Alkali + H_2O
Atomic H

Figure 13.3. Hydrogen energy storage.
Source: Adapted from Schüth, Ferdi. 2005. Technology: Hydrogen and hydrates, Nature 434, 712-713.

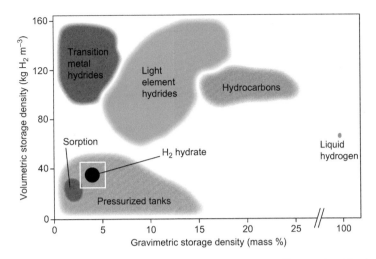

Figure 13.4. The six basic hydrogen storage methods and phenomena. The gravimetric density ϱm, the volumetric density ϱv, the working temperature T, and pressure p are listed. RT stands for room temperature (25 °C). From top to bottom: compressed gas (molecular H2); liquid hydrogen (molecular H2); physisorption (molecular H2) on materials, e.g., carbon, with a very large specific surface area; hydrogen (atomic H) intercalation in host metals, metallic hydrides working at RT are fully reversible; complex compounds ($[AlH4]-$ or $[BH4]-$), desorption at elevated temperature, adsorption at high pressures; chemical oxidation of metals with water and liberation of hydrogen.
Source: Züttel, A. 2009. FUELS – HYDROGEN STORAGE: Hydrides, In: Jürgen Garche, Editor-in-Chief, Encyclopedia of Electrochemical Power Sources, (Amsterdam, Elsevier), Pages 440-458.

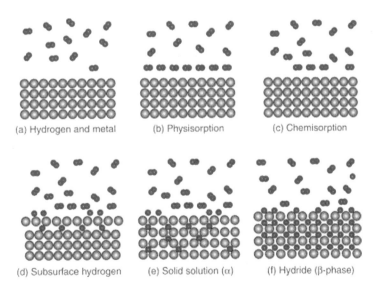

Figure 13.5. Schematic representation of the six distinct hydrogen absorption steps: (a) The metal lattice is exposed to hydrogen gas in the nonequilibrium state; (b) at low temperature, hydrogen molecules adsorb at the surface of the metal; (c) dissociation of hydrogen molecules and chemisorption of hydrogen atoms; (d) subsurface hydrogen; (e) solid solution phase; and (f) hydride phase.
Source: Züttel, A. 2009. FUELS – HYDROGEN STORAGE: Hydrides, In: Jürgen Garche, Editor-in-Chief, Encyclopedia of Electrochemical Power Sources, (Amsterdam, Elsevier), Pages 440-458.

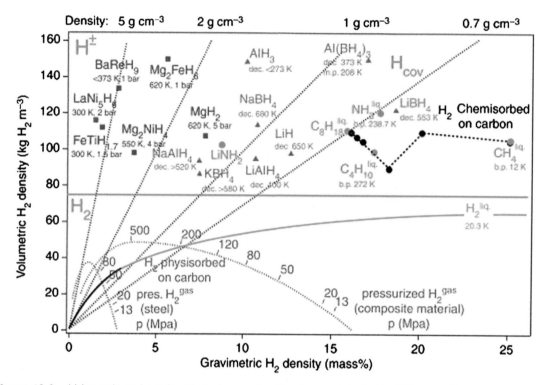

Figure 13.6. Volumetric and gravimetric hydrogen density of some selected hydrides used in hydrogen storage. Mg2FeH6 shows the highest known volumetric hydrogen density of 150 kg m^{-3}, which is more than double that of liquid hydrogen. BaReH9 has the largest H/M ratio of 4.5, i.e., 4.5 hydrogen atoms per metal atom. LiBH4 exhibits the highest gravimetric hydrogen density of 18 mass%. Pressurized gas storage is shown for steel (tensile strength $\sigma v = 460$ MPa, density 6500 kg m^{-3}) and a hypothetical composite material ($\sigma v = 1500$ MPa, density 3000 kg m^{-3}).

Source: Züttel, A. 2009. FUELS – HYDROGEN STORAGE: Hydrides, In: Jürgen Garche, Editor-in-Chief, Encyclopedia of Electrochemical Power Sources, (Amsterdam, Elsevier), Pages 440-458.

Centralized production

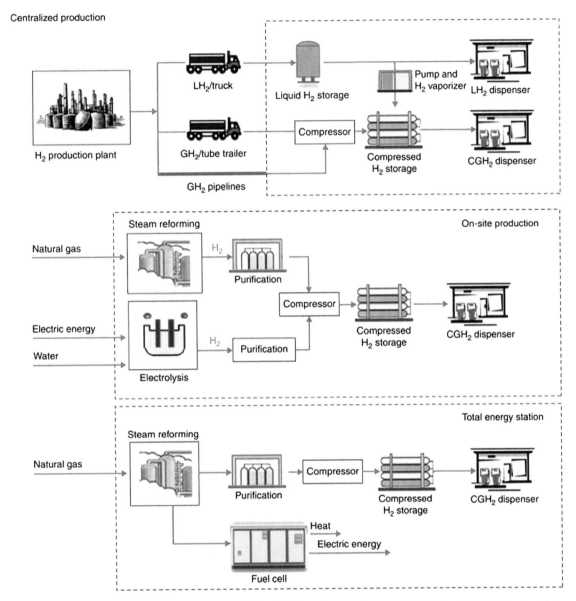

Figure 13.7. Possible hydrogen transport and distribution schemes from centralized to distributed networks. *Source: Conte, M. 2009. Hydrogen Economy, In: Jürgen Garche, Editor-in-Chief, Encyclopedia of Electrochemical Power Sources, (Amsterdam, Elsevier), Pages 232-254.*

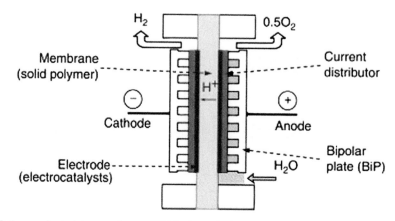

Figure 13.8. Polymer electrolyte membrane (PEM) electrolysis cell used to produce hydrogen.
Source: Smolinka, T. 2009. Water Electrolysis, In: Jürgen Garche, Editor-in-Chief, Encyclopedia of Electrochemical Power Sources, (Amsterdam, Elsevier), Pages 394-413.

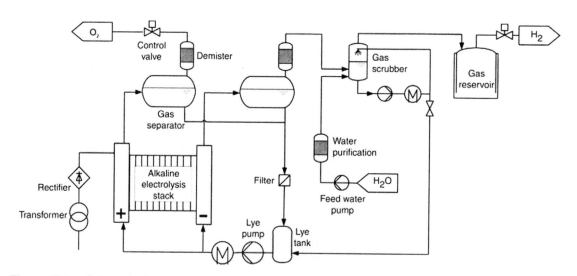

Figure 13.9. Schematic diagram of an alkaline electrolysis system used to produce hydrogen. Balance-of-plant does not include gas purification and compression.
Source: Smolinka, T. 2009. Water Electrolysis, In: Jürgen Garche, Editor-in-Chief, Encyclopedia of Electrochemical Power Sources, (Amsterdam, Elsevier), Pages 394-413.

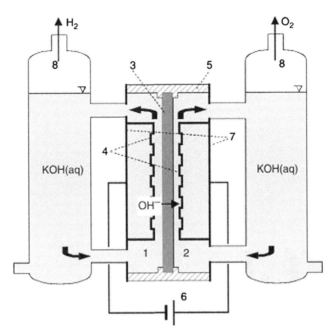

Figure 13.10. Simplified schematic diagram of a conventional alkaline electrolysis cell used to produce hydrogen: anode compartment (1), cathode compartment (2), diaphragm (3), electrodes (4), cell frame (5), direct current (dc) power supply (6), end-plates (7), and gas separators (8).
Source: Smolinka, T. 2009. Water Electrolysis, In: Jürgen Garche, Editor-in-Chief, Encyclopedia of Electrochemical Power Sources, (Amsterdam, Elsevier), Pages 394-413.

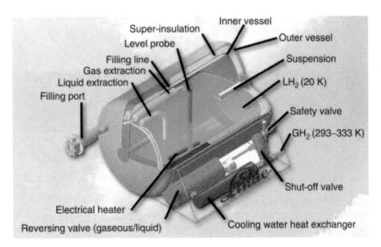

Figure 13.11. Schematic illustration of a liquid hydrogen dewar for automotive applications. Courtesy of Linde Gas.
Source: Bonadio, L. 2009. HYDROGEN STORAGE: Liquid, In: Jürgen Garche, Editor-in-Chief, Encyclopedia of Electrochemical Power Sources, (Amsterdam, Elsevier), Pages 421-439.

Charts

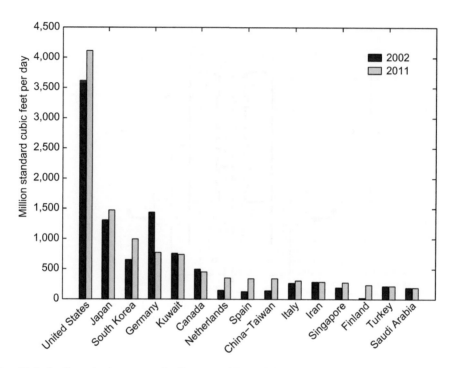

Chart 13.1.   Global refinery hydrogen production capacities, 2002 and 2011.
Source: Data from United States Department of Energy, Hydrogen Analysis Resource Center,
<http://hydrogen.pnl.gov/cocoon/morf/hydrogen/article/706>.

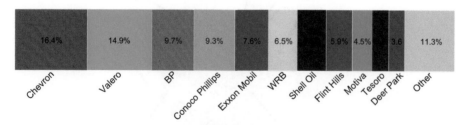

Chart 13.2.   Top refinery hydrogen producers in the United States, 2010.
Source: Data from United States Department of Energy, Hydrogen Analysis Resource Center,
<http://hydrogen.pnl.gov/cocoon/morf/hydrogen/article/706>.

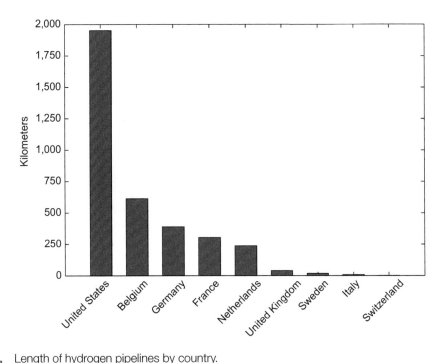

Chart 13.3. Length of hydrogen pipelines by country.
Source: Data from United States Department of Energy, Hydrogen Analysis Resource Center, <http://hydrogen.pnl.gov/cocoon/morf/hydrogen/article/706>.

Tables

Table 13.1. Basic properties of hydrogen

Property	Values	Units	Notes
Autoignition temperature	585	°C	[3, a]
	1085	°F	
Boiling point (1 atm)	−252.9	°C	[b]
	−423.2	°F	
Density (NTP)	0.08375	kg/m³	[1, b]
	0.005229	lb/ft³	
Diffusion coefficient in air (NTP)	0.610	cm²/s	[1, a]
	6.57×10^{-4}	ft²/s	
Enthalpy (NTP)	3858.1	kJ/kg	[1, b]
	1659.8	Btu/lb	
Entropy	53.14	J/g-K	[1, b]
	12.70	Btu/lb-°R	
Flame temperature in air	2045	°C	[a]
	3713	°F	
Flammable range in air	4.0–75.0	vol%	[1, a]
Ignition energy in air	2×10^{-5}	J	[c]
	1.9×10^{-8}	Btu	
Internal Energy (NTP)	2648.3	kJ/kg	[1, 2, b]
	1139.3	Btu/lb	
Molecular weight	2.02		[b]
Specific gravity (air = 1) (NTP)	0.0696		[1, c]
Specific volume (NTP)	11.94	m³/kg	[1, b]
	191.3	ft³/lb	
Specific heat at constant pressure, C_p (NTP)	14.29	J/g-K	[1, b]
	3.415	Btu/lb-°R	
Specific heat at constant volume, C_v (NTP)	10.16	J/g-K	[1, b]
	2.428	Btu/lb-°R	
Thermal conductivity (NTP)	0.1825	W/m-K	[1, b]
	0.1054	Btu/ft-h-°R	
Viscosity (NTP)	8.813×10^{-5}	g/cm-sec	[1, b]
	5.922×10^{-6}	lb/ft-sec	

Notes:
[1] NTP (normal temperature and pressure) = 20 °C (68 °F) and 1 atm.
[2] Reference state: Internal Energy U=0 at 273.16 K for saturated liquid; Entropy S=0 at 273.16 K for saturated liquid.
[3] The autoignition temperature depends on hydrogen concentration (minimum at stoichiometric combustion conditions), pressure, and even the surface characteristics of the vessel. Reported figures range from 932-1085 °F.
Source: U.S. Department of Energy, Hydrogen Resource Center, <http://hydrogen.pnl.gov/cocoon/morf/hydrogen/article/401>, accessed 23 May 2012.

Table 13.2. Hydrogen properties at normal boiling point

Property	Units	Liquid	Vapor	Units	Liquid	Vapor
Temperature	°F	−423.17	−423.17	°C	−252.87	−252.87
Pressure	psi	14.696	14.696	pascals (Pa)	1.013×10^5	1.013×10^5
Density	lb/ft^3	4.4197	0.08362	kg/m^3	70.797	1.3394
Specific volume	ft^3/lb	0.2263	11.959	m^3/kg	0.01413	0.7466
Compressibility factor	Z = PV/RT	0.01698	0.9051			
Heats of fusion and vaporization	Btu/lb	191.9	---	J/g	446.3	---

Source: U.S. Department of Energy, Hydrogen Resource Center, <http://hydrogen.pnl.gov/cocoon/morf/hydrogen/article/401>.

Table 13.3. Comparison of hydrogen properties with other fuels

Property	Hydrogen	Methane	Propane	Gasoline	Diesel	Methanol
Higher Heating Value (at 25 °C and 1 atm)	61,000 Btu/lb (141.86 kJ/g)	24,000 Btu/lb (55.53 kJ/g)	21,650 Btu/lb (50.36 kJ/g)	20,360 Btu/lb (47.5 kJ/g)	19,240 Btu/lb (44.8 kJ/g)	8,580 Btu/lb (19.96 kJ/g)
Energy Density (LHV)	270 Btu/ft^3 (10,050 kJ/m3); gas at 1 atm and 60 °F (15 °C) 48,900 Btu/ft^3 (1,825,000 kJ/m^3); gas at 3,000 psig (200 barg) and 60 °F (15 °C) 121,000 Btu/ft^3 (4,500,000 kJ/m^3); gas at 10,000 psig (690 barg) and 60 °F (15 °C) 227,850 Btu/ft^3 (8,491,000 kJ/m^3); liquid	875 Btu/ft^3 (32,560 kJ/m3); gas at 1 atm and 60 °F (15 °C) 184,100 Btu/ft^3 (6,860,300 kJ/m^3); gas at 3,000 psig (200 barg) and 60 °F (15 °C) 561,500 Btu/ft^3 (20,920,400 kJ/m^3); liquid	2,325 Btu/ft^3 (86,670 kJ/m^3); gas at 1 atm and 60 °F (15 °C) 630,400 Btu/ft^3 (23,488,800 kJ/m^3); liquid	836,000 Btu/ft^3 (31,150,000 kJ/m^3); liquid	843,700 Btu/ft^3 (31,435,800 kJ/m^3) minimum; liquid	424,100 Btu/ft^3 (15,800,100 kJ/m^3); liquid
Flashpoint	< −423 °F (< −253 °C; 20 K)	−306 °F (−188 °C; 85 K)	−306 °F (−188 °C; 85 K)	Approximately −45 °F (−43 °C; 230 K)	52 °F (11 °C; 284 K)	
Autoignition Temperature	1085 °F (585 °C)	1003 °F (540 °C)	914 °F (490 °C)	450 to 900 °F (230 to 480 °C)		725 °F (385 °C)
Octane number	130+ (lean burn)	125	100	80-90	30	89

Source: Adapted from U.S. Department of Energy, Office of Energy Efficiency and Renewable Energy, Hydrogen Fuel Cell Engines and Related Technologies Course Manual (2001), <http://www1.eere.energy.gov/hydrogenandfuelcells/tech_validation/h2_manual.html>.

Table 13.4.   The properties of hydrogen and nitrogen as adsorbates

Property	N_2	H_2
M [g mol^{-1}]	28.014	2.0159
ρ [g cm^{-3}]	0.807	0.0708
V_M [nm^3]	0.0426	0.035
d [nm]	0.4335	0.4059
S_M [nm^2]	0.1627	0.1427
$1/S$ [mol m^{-2}]	1.02×10^{-5}	1.16×10^{-5}
$1/S$ [g m^{-2}]	2.86×10^{-4}	2.35×10^{-5}

M the molecular mass, ρ the density of the liquid at the boiling point, V_M the volume of the molecule, d the diameter of the molecule, S_M the surface area occupied by the molecule, $1/S$ the amount of adsorbate per surface area unit in a monolayer.
Source: Adapted from Hu, Xin, Maohong Fan, Brian Francis Towler, Maciej Radosz, David A. Bell, Ovid Augustus Plumb. 2011. Hydrogen Adsorption and Storage, In: David Bell and Brian Towler, Coal Gasification and its Applications, (Boston, William Andrew Publishing), Pages 157–245.

Table 13.5.   Hydrogen conversions

	Weight		Gas		Liquid	
	Pounds (lb)	Kilograms (kg)	Standard cubic feet (SCF)	Normal cubic meter (Nm^3)	Gallons (gal)	Liters (L)
1 lb	1.0	0.4536	192.00	5.047	1.6928	6.408
1 kg	2.205	1.0	423.3	11.126	3.733	14.128
1 SCF gas	0.005209	0.002363	1.0	0.02628	0.00882	0.0339
1 Nm^3 gas	0.19815	0.08988	38.04	1.0	0.3355	1.2699
1 gal liquid	0.5906	0.2679	113.41	2.981	1.0	3.785
1 L liquid	0.15604	0.07078	29.99	0.77881	0.2642	1.0

Source: Davis, Stacy C., Susan W. Diegel, and Robert G. Boundy. 2009. Transportation Energy Data Book: Edition 28, Oak Ridge National Laboratory.

Table 13.6. Conversion efficiencies for the production of hydrogen[a]

Production processes	Central biomass gasification	Central coal gasification	Central coal gasification with CO$_2$ capture	Central natural gas reforming	Central natural gas reforming with CO$_2$ capture	Central water electrolysis	Forecourt water electrolysis 1500 kg/day	Forecourt ethanol reforming, 1500 kg/day	Forecourt natural gas reforming, 1500 kg/day
Energy Inputs, Raw Units									
Farmed Trees, kg	12.839	-	-	-	-	-	-	-	-
Natural Gas, Nm^3	0.170	-	-	4.501	4.489	-	-	-	4.488
Ethanol, gallons	-	-	-	-	-	-	-	2.191	-
Electricity, kWh	1.600	-	1.720	0.569	1.406	53.440	55.178	2.457	3.077
Pittsburgh #8 Coal, kg	-	8.508	7.849	-	-	-	-	-	-
Energy Inputs, Common Units									
Farmed Trees, Btu	237,919	-	-	-	-	-	-	-	-
Natural Gas, Btu	5,901	-	-	156,249	155,833	-	-	-	155,798
Ethanol, Btu	-	-	-	-	-	-	-	167,239	-
Electricity, Btu	5,459	-	5,867	1,942	4,796	182,345	188,277	8,385	10,501
As Received Bituminous Coal, Btu	-	223,253	205,960	-	-	-	-	-	-
Energy Outputs, Raw Units									
Hydrogen, kg	1.000	1.000	1.000	1.000	1.000	1.000	1.000	1.000	1.000
Electricity, kWh	-	3.175	-	-	-	-	-	-	-
Energy Outputs, Common Units									
Hydrogen, Btu	113,940	113,940	113,940	113,940	113,940	113,940	113,940	113,940	113,940
Electricity, Btu	-	10,834	-	-	-	-	-	-	-
Conversion Efficiencies	45.7%	55.9%	53.8%	72.0%	70.9%	62.5%	60.5%	64.9%	68.5%

Lower Heating Value Assumptions

Farmed Trees, Btu/short ton (LHV)	16,811,000
Pittsburgh #8 Coal, Btu/short ton (LHV)	23,804,819
Natural Gas, Btu/ft^3 @ 1 atm, 32 F (LHV)	983
Hydrogen, Btu/ft^3 @ 1 atm, 32 F (LHV)	290
Ethanol, Btu/gallon (LHV)	76,330

Fuel Density Assumption:

Hydrogen, g/ft^3 @ 1 atm, 32 F	2.55

Other Conversion Assumptions:

Electricity, Btu/kWh	3412
lb/short ton	2000
lb/kg	2.2046
ft^3/m^3	35.315
g/kg	1000
kJ/Btu	1.0551

[a]Energy Inputs and Outputs per kg of Hydrogen Produced; Fuel Energy Inputs and Outputs on LHV Basis.
Source: Adapted from U.S. Department of Energy, Hydrogen Resource Center, <http://hydrogen.pnl.gov/cocoon/morf/hydrogen>, accessed 23 May 2012 ; based on data in GREET version 1.8b, released May 8, 2008. http://www.transportation.anl.gov/modeling_simulation/GREET/index.html; Central Coal Gasification: Current Central Hydrogen Production from Coal without CO2 Sequestration version 2.0.1; Hydrogen Program Production Case Studies. http://www.hydrogen.energy.gov/h2a_prod_studies.html.

Table 13.7. Global and U.S hydrogen production, 2005-2010[a]

	2005	2006	2007	2008	2009	2010
Worldwide Total Production [b]	9.610	10.000	12.000	13.000	15.000	13.000
Worldwide Merchant Production[c]	1.037	1.200	1.700	1.900	1.900	2.000
U.S. Total Production	6.410	6.700	7.900	8.200	8.200	8.350
U.S. Merchant Production	0.663	0.740	1.052	1.185	1.200	1.250
U.S. Large Merchant Production	0.628	0.700	1.000	1.127	1.170	1.215
U.S. Small Merchant Production	0.035	0.040	0.052	0.058	0.035	0.035

U.S. Small Merchant Delivery Modes	
Liquid Tanker	90%
Compressed Gas Tube Trailer	7%
Compressed Gas Cylinder	3%

[a]Trillion standard cubic feet/year.
[b]Excludes hydrogen production from syngas, byproduct gases, and on-site plants not owned and operated by the end-user.
[c]"Merchant" hydrogen production is defined here to mean any hydrogen produced by one company for consumption by another company. Large Merchant production is usually delivered to the customer via pipeline as a compressed gas. The Merchant plant may be on the customer's property, adjacent to the customer's property (commonly referred to as "over the fence") or a few hundred miles away in regions served by hydrogen pipeline networks. Merchant production is a subset of total production.
Source: U.S. Department of Energy, Hydrogen Resource Center, <http://hydrogen.pnl.gov/cocoon/morf/hydrogen>.

Table 13.8. Global refinery hydrogen production capacities, 2002-2011[a]

	2002	2011	Change 2002-11
United States	3617	4114.3	14%
Japan	1306.1	1472.8	13%
South Korea	653.4	995.5	52%
Germany	1437.8	772	−46%
Kuwait	758.5	741.6	−2%
Canada	495	448.8	−9%
Netherlands	147.9	351.8	138%
Spain	125.5	341.1	172%
China-Taiwan	142	341	140%
Italy	268	305.4	14%
Iran	286	286	0%
Singapore	193.8	275	42%
Finland	20	235	1075%
Turkey	217.5	217.5	0%
Saudia Arabia	189.7	190.7	1%
Total	**11,701.6**	**13,898.6**	**19%**

[a]million standard cubic feet per day.
Source: U.S. Department of Energy, Hydrogen Resource Center, <http://hydrogen.pnl.gov/cocoon/morf/hydrogen>.

Table 13.9. **Hydrogen production options and costs**

Process	Inputs	Efficiency (%)	Gate cost of H_2 ($ gge^{-1})[a]	Comments
Electrolysis	Water, electricity	65–75	2–6.5	Other uses compete for electricity
Bioelectrolysis	Biological substrate: acetic acid, sewage sludge, etc.	60–80 overall		Limited by availability of biological substrate
Steam methane reforming	Methane	50–75	1–3.5	No CCS at small scale[b]
Gasification	Coal	75	0.96–1.03	Could be integrated with precombustion CCS[b]
Gasification	Biomass	50–64	1.1–4.6	Limited by availability of biomass
Biological	Biomass/biological substrate			Limited by availability of biomass
High-temperature water splitting	Water and heat			Not yet demonstrated. Requires high-temperature nuclear plant

[a]gge = gallon of gasoline equivalent.
[b]CCS = carbon capture and storage.
Source: Adapted from Hughes, N. and P. Agnolucci. 2012. 4.03 - Hydrogen Economics and Policy, In: Ali Sayigh, Editor-in-Chief, Comprehensive Renewable Energy, (Oxford, Elsevier), Pages 45–75.

Table 13.10. **Summary of characteristics of biological hydrogen production processes**

Bio-hydrogen system	H_2 production rate (mmol H_2)	Required inputs and broad classification of microorganism used	By-products	General reactions and advantages
Direct biophotolysis	0.07	Light energy, H_2O, CO_2, trace minerals; Micro-algae	O_2	$2H_2O + light = 2H_2 + O_2$
				It can directly produce H_2 directly from water and sunlight.
Indirect biophotolysis	0.355	Light energy, H_2O, CO_2, trace minerals; Micro-algae, cyanobacteria	O_2	$12H_2O + light = 12H_2 + O_2$
				It has the ability to fix N_2 from atmosphere.
Photo-fermentation	153	Light energy, H_2O, glucose; Purple-bacteria, microalgae	CO_2	$CH_3COOH + 2H_2O + light = 4H_2 + 2CO_2$.
				A wide spectral light energy can be used by these bacteria

Continued

Table 13.10.   Continued

Bio-hydrogen system	H_2 production rate (mmol H_2)	Required inputs and broad classification of microorganism used	By-products	General reactions and advantages
Dark fermentation		H_2O, carbohydrates, heat (ii and iii only); Fermentative bacteria	CO_2, CH_4, CO, H_2S, acetate or other end products	$C_6H_{12}O_6 + 6H_2O = 12H_2 + 6CO_2$
(i) Mesophilic	121			It can produce H_2 all day long without light. A variety of carbon sources can be used as a substrate. It produces valuable metabolites such as butyric, lactic and acetic acid as by-products. There is no O_2 limitation problem.
(ii) Thermophilic	8.2			
(iii) Extreme thermophilic	8.4			
Hybrid reactor system		Light energy, H_2O, carbohydrates; Fermentative bacteria followed by anoxygenic phototrophic bacteria	CO_2	Stage-I: $C_6H_{12}O_6 + 2H_2O = 4H_2 + 2CH_3COOH + 2CO_2$
				Stage-II: $CH_3COOH + 2H_2O + light = 4H_2 + 2CO_2$
				Two stage fermentation can improve the overall yield of hydrogen

Source: Adapted from Kothari, Richa, D.P. Singh, V.V. Tyagi, S.K. Tyagi. 2012. Fermentative hydrogen production – An alternative clean energy source, Renewable and Sustainable Energy Reviews, Volume 16, Issue 4, Pages 2337–2346.

Table 13.11.   Potential biomass materials for hydrogen production

Biomass species	Main conversion process
Bio-nut shell	Steam gasification
Olive husk	Pyrolysis
Tea waste	Pyrolysis
Crop straw	Pyrolysis
Black liquor	Steam gasification
Municipal solid waste	Supercritical water extraction
Crop grain residues	Supercritical fluid extraction
Pulp grain residue	Microbial fermentation
Petroleum basis plastic waste	Supercritical fluid extraction
Manure slurry	Microbial fermentation

Source: Adapted from Panwar, N.L., Richa Kothari, V.V. Tyagi. 2012. Thermo chemical conversion of biomass – Eco friendly energy routes, Renewable and Sustainable Energy Reviews, Volume 16, Issue 4, Pages 1801–1816.

Table 13.12.   Waste biomass feedstock used for hydrogen production

Biomass feedstock	Major conversion technology
Almond shell	Steam gasification
Pine sawdust	Steam reforming
Crumb rubber	Supercritical conversion
Rice straw/Danish wheat water	Pyrolysis
Microalgae	Gasification
Tea waste	Pyrolysis
Peanut shell	Pyrolysis
Maple sawdust slurry	Supercritical conversion
Starch biomass slurry	Supercritical conversion
Composed municipal refuse	Supercritical conversion
Kraft lignin	Steam gasification
Municipal solid waste	Supercritical conversion
Paper and pulp waste	Microbial conversion

Source; Adapted from Kothari, Richa, V.V. Tyagi, Ashish Pathak. 2010. Waste-to-energy: A way from renewable energy sources to sustainable development, Renewable and Sustainable Energy Reviews, Volume 14, Issue 9, Pages 3164–3170.

Table 13.13.   Various hydrogen producing microbial strains reported and their H2 yield

Organism	Substrate	Process	Maximum yield of H_2 (mol H_2/mol substrate)
Enterobacter aerogenes HU-101 (mutantAY-2)	Glucose	Batch (blocking metabolites formation)	1.17
Enterobacter aerogens	Molasses	Ar sparging, batch	1.58
Enterobacter aerogens	Molasses	Batch	0.52
Clostridium butyricum	Glucose	N2 sparging continuous	1.4–2.3
Enterobacter cloacae IIT BT 08	Glucose	Continuous (immobilized bioreactor)	2.3
Citrobacter sp. Y19	Glucose	Batch Ar sparging	2.49
Rhodopseudomonas palustris P4	Glucose	Batch, with intermittent purging of Ar	2.76
Clostridium butyricum EB6	POME	Batch	3.2 (L/L med)
Clostridium butyricum ATCC19398	Glucose	Batch	1.8
Clostridium acetobutyricum M121	Glucose	Batch	2.29
Clostridium tyrobutyricum FYa102	Glucose	Batch	1.47
Clostridium beijerinckii L9	Glucose	Batch	2.81
C. thermolacticum	Lactose	Continuous	3
Clostridium thermocellum 27405	Delignified wood fiber	Batch	1.6
Klebsiella oxytocoa HP1	Glucose	Batch	1
T. thermosaccharolyticum PSU-2	Sucrose	Batch	2.53
T. saccharolyticum JW/ SL-YS485	Xylose	Batch	0.88

Continued

Table 13.13. Continued

Organism	Substrate	Process	Maximum yield of H_2 (mol H_2/mol substrate)
Caldicellulosiruptor	Sucrose	Batch	5.9
Mixed culture (predominantly Clostridium sp.)	Glucose	N2 sparging, continuous HRT: 8.5 h	1.43
Mixed microflora	Wheat starch co-product	N2 sparging continuous	1.9
Mixed microflora	0.75% soluble starch	Chemostat HRT: 17 h	2.14
Mixed microflora	Sewage-sludge	Anaerobic and acidogenic digestion	1.7

Source: Adapted from Kothari, Richa, D.P. Singh, V.V. Tyagi, S.K. Tyagi. 2012. Fermentative hydrogen production – An alternative clean energy source, Renewable and Sustainable Energy Reviews, Volume 16, Issue 4, Pages 2337–2346.

Table 13.14. Typical data for gas-phase oxidation to produce hydrogen

Characteristics	Natural gas	Heavy fuel oil
Feedstock kg	1	1
Operating pressure kPa	31,500	31,500
Oxygen in m^3	1.02	0.79
Steam in kg	0.05	0.4
Crude gas product m^3	3.97	3.11
CO_x vol %	37	51
Hydrogen vol %	61	46

Trimm, D.L. 2009. Natural Gas: Conventional Steam-Reforming, In: Jürgen Garche, Editor-in-Chief, Encyclopedia of Electrochemical Power Sources, (Amsterdam, Elsevier), Pages 293–299.

Table 13.15. Applications for hydrogen storage and associated performance characteristics

	Capacity	Thermodynamics	Kinetics	Reversibility
CCS hydrogen storage	Not space or weight constrained	Could use heat from power plant	Fast	Important
Stand-alone grid intermittency management	Not space or weight constrained	Input heat not available but potential to reuse desorption heat in district heating	Fast	Important
'Island' renewables storage	Not space or weight constrained	Input heat not available but potential to reuse desorption heat in district heating	Fast	Important
Shipping	Some space and weight constraints but less than cars	Excessive desorption heat undesirable, though moderate heat potentially manageable on large craft	Medium-fast	'Spent' fuel can be off-loaded at port
Trailer distribution	Both volumetric and gravimetric density important	High or low temperatures for uptake or desorption could be managed at loading or unloading depots	Less important	Less important

Source: Adapted form Hughes, N. and P. Agnolucci. 2012. 4.03 - Hydrogen Economics and Policy, In: Ali Sayigh, Editor-in-Chief, Comprehensive Renewable Energy, (Oxford, Elsevier), Pages 45–75.

Table 13.16.   Characteristics of compressed and liquid H_2 storage

Storage mode	Storing energy (kJ kg^{-1})	Spent energy/stored energy	Gravimetric energy content (MJ kg^{-1})	Volumetric energy content (MJ m^{-3})
Compressed H_2 (350 bar)	12,264	0.1	8.04	2492
Compressed H_2 (700 bar)	14,883	0.12	7.2	3599
Liquid H_2	42,600	0.36	16.81	3999

Source: U.S. DOE Hydrogen and Fuel Cells Program, <http://www.hydrogen.energy.gov/annual_progress11.html>.

Table 13.17.   Energy density by weight and volume for various hydrogen storage forms, and mass density

Storage form	kJ kg^{-1}	Energy density MJ m^{-3}	Density kg m^{-3}
Hydrogen, gas (ambient 0.1 MPa)	120,000	10	0.09
Hydrogen, gas at 20 MPa	120,000	1900	15.9
Hydrogen, gas at 30 MPa	120,000	2700	22.5
Hydrogen, liquid	120,000	8700	71.9
Hydrogen in metal hydrides	2000–9000	5,000–15,000	
Hydrogen in metal hydride, typical	2100	11,450	5480
Methane (natural gas) at 0.1 Mpa	56,000	37.4	0.668
Methanol	21,000	17,000	0.79
Ethanol	28,000	22,000	0.79

Sørensen, Bent. 2012. Hydrogen and Fuel Cells-Emerging Technologies and Applications (Second Edition), (Boston, Academic Press).

Table 13.18.   Some of the important families of hydride-forming intermetallic compounds being considered for hydrogen storage

Intermetallic compound	Prototype	Hydrides	Structure
AB_5	LaN_{i5}	$LaNi_5H_6$	Haucke phases, hexagonal
AB_2	ZrV_2, $ZrMn_2$, $TiMn_2$	$ZrV_2H_{5.5}$	Laves phase, hexagonal or cubic
AB_3	$CeNi_3$, YFe_3	$CeNi_4H_4$	Hexagonal, $PuNi_3$-type
A_2B_7	Y_2Ni_7, Th_2Fe_7	$Y_2Ni_7H_3$	Hexagonal, Ce_2Ni_7-type
A_6B_{23}	Y_6Fe_{23}	$Ho_6Fe_{23}H_{12}$	Cubic, Th_6Mn_{23}-type
AB	$TiFe$	$TiFeH_2$	Cubic, CsCl- or Ti_2Ni-type
A_2B	Mg_2Ni, Ti_2Ni	Mg_2NiH_4	Cubic, $MoSi_2$- or Ti_2Ni-type

Source: Adapted from Hu, Xin, Maohong Fan, Brian Francis Towler, Maciej Radosz, David A. Bell, Ovid Augustus Plumb. 2011. Hydrogen Adsorption and Storage, In: David Bell and Brian Towler, Coal Gasification and Its Applications, (Boston, William Andrew Publishing), Pages 157–245.

Fuel Cells

Figures

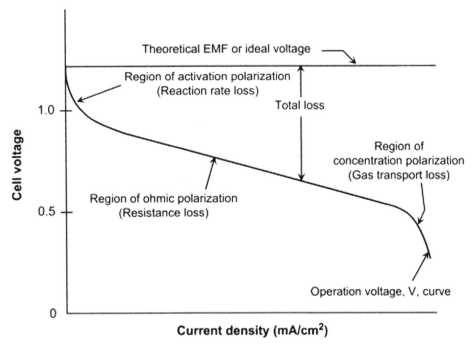

Figure 14.1. Typical current voltage performance in a fuel cell.
Source: Williams, Mark C. 2011. Chapter 2 - Fuel Cells, In: Dushyant Shekhawat, J.J. Spivey And David A. Berry, Editor(s), Fuel Cells: Technologies for Fuel Processing, (Amsterdam, Elsevier), Pages 11-27.

Handbook of Energy. http://dx.doi.org/10.1016/B978-0-08-046405-3.00014-0

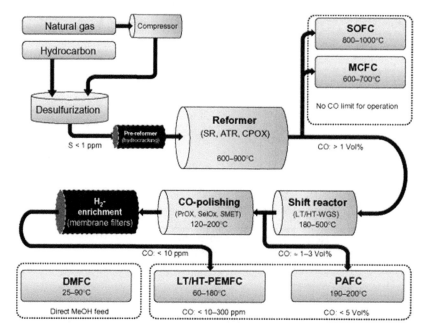

Figure 14.2. Scheme of the requirements of fuel processing in order to meet the specific requirements of various fuel cells. Blackened parts are not necessary in general; their usage depends on the process demands due to fuel source and hydrogen quality. PEMFC = Proton exchange membrane fuel cell; SOFC = solid oxide fuel cell; MCFC = Molten-carbonate fuel cell; DMFC = Direct-methanol fuel cell; HT/LT = high temperature/low temperature.
Source: Kaltschmitt, Torsten and Olaf Deutschmann. 2012. Fuel Processing for Fuel Cells, In: Kai Sundmacher, Editor, Advances in Chemical Engineering, (New York Academic Press), Pages 1-64.

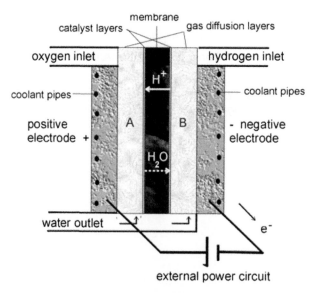

Figure 14.3. Proton exchange membrane fuel cell (PEMFC).
Source: Sørensen, Bent. 2012. Hydrogen and Fuel Cells-Emerging Technologies and Applications (Second Edition), (Boston, Academic Press).

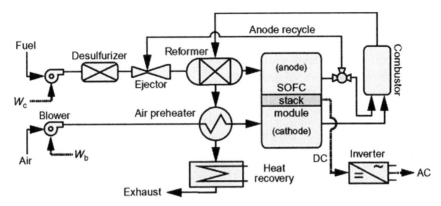

Figure 14.4. Schematic diagram of a solid-oxide fuel cell (SOFC) system.
Source: Kee, Robert J., Huayang Zhu, Robert J. Braun, Tyrone L. Vincent. 2012. Modeling the Steady-State and Dynamic Characteristics of Solid-Oxide Fuel Cells, In: Kai Sundmacher, Editor, Advances in Chemical Engineering, (New York, Academic Press), Volume 41, Pages 331-381.

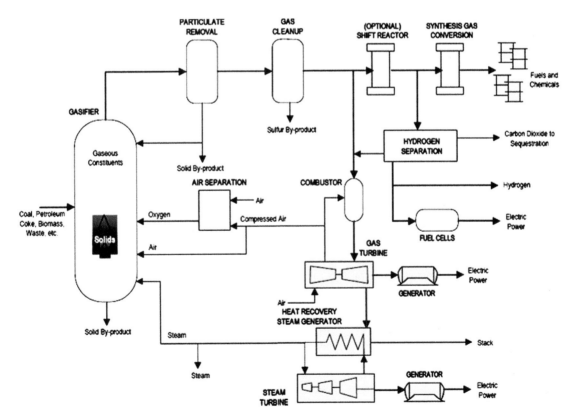

Figure 14.5. Phosphoric acid fuel cell (PAFC) and solid oxide fuel cell (SOFC).
Source: Ashby, Michael F. 2013. Materials and the Environment (Second Edition), (Boston, Butterworth-Heinemann).

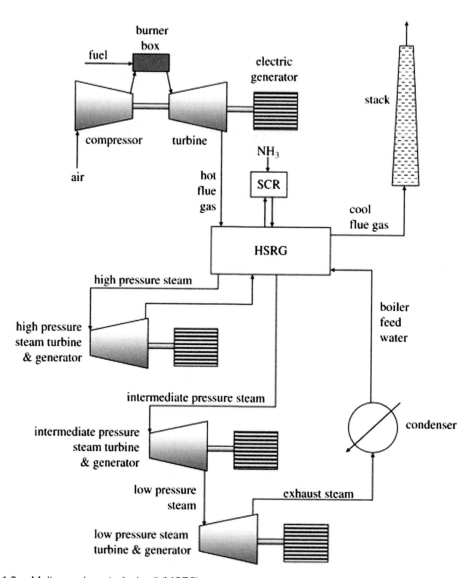

Figure 14.6. Molten carbonate fuel cell (MCFC).
Source: Miller, Bruce G. 2011. Clean Coal Engineering Technology, (Boston, Butterworth-Heinemann).

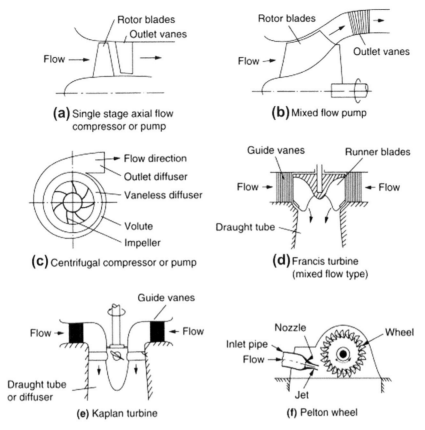

(a) Single stage axial flow compressor or pump

(b) Mixed flow pump

(c) Centrifugal compressor or pump

(d) Francis turbine (mixed flow type)

(e) Kaplan turbine

(f) Pelton wheel

Figure 14.7.   Direct methanol fuel cell (DMFC).
Source: Jörissen, L., and V. Gogel. 2009. FUEL CELLS – DIRECT ALCOHOL FUEL CELLS | Direct Methanol: Overview, In: Jürgen Garche, Editor-in-Chief, Encyclopedia of Electrochemical Power Sources, (Amsterdam, Elsevier), Pages 370-380.

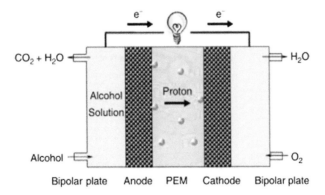

Figure 14.8.   Direct alcohol fuel cell. (DAFC). PEM, proton-exchange membrane.
Source: Zhao, T.S., Z.X. Liang, J.B. Xu. 2009. DIRECT ALCOHOL FUEL CELLS | Overview, In: Jürgen Garche, Editor-in-Chief, Encyclopedia of Electrochemical Power Sources, (Amsterdam, Elsevier), 2009, Pages 362-369.

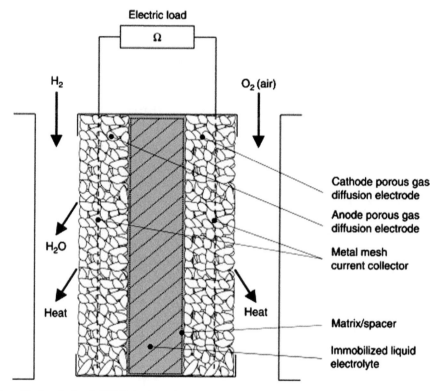

Figure 14.9. Alkaline fuel cell (AFC) single-cell with immobilized liquid electrolyte.
Source: Fraser, S.D., V. Hacker, K. Kordesch. 2009. ALKALINE FUEL CELLS | Cells and Stacks, In: Jürgen Garche, Editor-in-Chief, Encyclopedia of Electrochemical Power Sources, (Amsterdam, Elsevier), Pages 344-352.

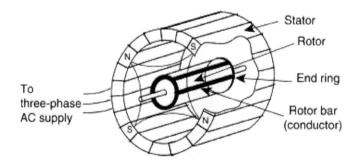

Figure 14.10. Microbial fuel cell.
Source: Scott, K., E.H. Yu, M.M. Ghangrekar, B. Erable, N.M. Duteanu. 2012. Biological and Microbial Fuel Cells, In: Ali Sayigh, Editor-in-Chief, Comprehensive Renewable Energy, (Oxford, Elsevier), Pages 257-280.

Tables

Table 14.1.   Application areas of fuel cell systems and their characteristics

Application	Purpose	Power output	Fuel cell type	Primary fuels	Fuel processor	Technology status
Power plant	Grid supply	100 kW to 500 MW	SOFC MCFC	Natural gas, coal	SR, internal reforming	Commercially available (e.g., DFC®, PureCell®)
Stationary	Electrical power	10–50 kW	SOFC MCFC	Hydrogen, natural gas, biofuels, MeOH	(H_2 tank) SR, ATR, internal reforming	Commercially available (e.g., DFC®, PureCell®)
Residential	Electrical power and heat (CHP systems)	500 W to 5 kW	SOFC PEMFC	Natural gas, biogas, LPG	Mostly ATR	Commercially available (e.g., BlueGen®)
Mobile— drive train	Electrical power	10–500 kW	MCFC PEMFC	Hydrogen, diesel	(H_2 tank) mainly ATR	Under development
Mobile— APU	Electrical power, heat supply	1–5 kW	SOFC PEMFC	Hydrogen, diesel, kerosene	SR, ATR	Under construction/ development
Portable	Electrical power	1–150 W	DMFC PEMFC	MeOH, hydrogen	Internal reforming (H_2 storage)	Under development

Source: Kaltschmitt, Torsten, Olaf Deutschmann. 2012. Chapter 1 - Fuel Processing for Fuel Cells, In: Kai Sundmacher, Editor, Advances in Chemical Engineering, Academic Press, 2012, Volume 41, Pages 1-64.

Table 14.2. Summary of operational specifications of fuel cell technologies[a]

Fuel cell type	AFC	PEMFC	PAFC	MCFC	SOFC	DCFC	DMFC
Common Electrolyte	Aqueous solution of potassium hydroxide soaked in a matrix	Solid organic polymer poly-perfluorosulfonic acid	Liquid phosphoric acid soaked in a matrix	Liquid solution of lithium, sodium, and/or potassium carbonates, soaked in a matrix	Yttria stabilized zirconie		Solid polymer membrane
Anode reaction	$H_2 + 4OH^- \rightarrow 4H_2O + 4e^-$	$H_2 (g) \rightarrow 2H^+ + 2e^-$	$2H_2 \rightarrow 4H^+ + 4e^-$	$H_2O + CO_3^{2-} \rightarrow H_2O + CO_2 + 2e^-$	$O^{2-} (S) + H_2 (g) \rightarrow H_2O (g) + 2e^-$		$CH_3OH + H_2O \rightarrow CO_2 + 6H^+ + 6e^-$
Cathode reaction	$O_2 + 2H_2O + 4e^- \rightarrow 4OH^-$	$1/2 O_2 (g) + 2H^+ + 2e^- \rightarrow H_2O$	$O2 + 4H^+ + 4e^- \rightarrow H_2O$	$1/2 O_2 + CO_2 + 2e^- \rightarrow CO_3^{2-}$	$1/2 O_2 (g) + 2e^- \rightarrow O^{2-}(s)$		$3/2 O_2 + 6e^- + 6H^+ \rightarrow 3H_2O$
Charge carrier	OH^-	H^+	H^+	O^-	O^-		H^+
Fuel	Pure H_2	Pure H_2	Pure H_2	H_2, CO, CH_4, other	H_2, CO, CH_4, other		CH_3OH
Oxidant	O_2 in air	O_2 in air	O_2 in air	O_2 in air	O_2 in air		O_2 in air
Cogeneration	No	No	Yes	Yes	Yes		No
Reformer is required	Yes	Yes	Yes				-
Cell voltage	1	1.1	1.1	0.7–1.0	0.8–1.0		0.2–0.4
T_{op}, °C	<100	<100	200–215	650	700–1000	650–900	
Heat output		60–70	150–200	500–600	600–900	500–800	
Electrical efficiency (%)	45–60	40–45	40–45	45–55	40–50 Planar, 50–65 Tu/MT[b]	>80	
CHP efficiency (%)	>80	>80	>80	>80	>80	>90	
Thermal insulation	Low	Low	Medium	High	High	High	
Internal reforming	Not possible	Not possible	Not possible	Only with steam	Only with steam	Not applicable	
Electrolyte	KOH	PEM	H_3PO_4	Molten carbonate	O^{2-} conducting ceramic	Molten carbonate/molten hydroxide/O^{2-} conducting ceramic	
Impurity sensitivity	CO_2, CO, S	S, CO ≈ 20ppm	Sulfur	Sulfur	Sulfur	Unknown	
Power at cold start	>50%	>50%	Nil	Nil	Nil	Nil	
Thermal cycling	Unlimited	Unlimited	Good	Restricted	Restricted	Restricted	
Start-up/shut-down	Very fast (sec)	Very fast (sec)	Slow (hrs)	Several hrs	Several hrs[c]	Several hours	
Load following	Excellent	Excellent	Limited	Limited	Limited	Limited	
BOP	Simple	Simple	Medium	Complex	Complex	Complex	

[a]Alkaline fuel cell (AFC); Phosphoric acid fuel cell (PAFC); Solid oxide fuel cell (SOFC); Molten carbonate fuel cell (MCFC); Proton exchange membrane fuel cell (PEMFC); Direct methanol fuel cell (DMFC); Direct carbon fuel cell (DCFC).

[b]Tu = Tubular, MT = Microturbine.

[c]Some small tubular design units may heat up to the operating temperature in less than 1 h.

Source: Adapted from Mekhlief, S., R. Saidur, A. Safari. 2012. Comparative study of different fuel cell technologies, Renewable and Sustainable Energy Reviews, Volume 16, Issue 1, Pages 981-989; Giddey, S., S.P.S. Badwal, A. Kulkarni, C. Munnings. 2012. A comprehensive review of direct carbon fuel cell technology, Progress in Energy and Combustion Science, Volume 38, Issue 3, Pages 360–399.

Table 14.3.   Basic fuel cell materials

	Solid oxide	Molten carbonate	Phosphoric acid	Alkaline	Polymer membrane
Electrolyte	Y_2O_3 - stabilized ZrO_2 (YSZ)	Li_2CO_3 K_2CO_3	H_3PO_4	KOH	Perfluorosulfonic acid
Cathode	Sr-doped $LaMnO_3$	Li-doped NiO	Pt on C	Pt-Au	Pt on C
Anode	Ni/YSZ	Ni	Pt on C	Pt-Pd	Pt on C
Temperature	800–1000 °C	650 °C	200 °C	100 °C	90 °C
Fuel	H_2, CO	H_2, CO	H_2	H_2	H_2

Source: Adapted from Williams, Mark C. 2011. Chapter 2 - Fuel Cells, In: Dushyant Shekhawat, J.J. Spivey And David A. Berry, Editor(s), Fuel Cells: Technologies for Fuel Processing, (Amsterdam, Elsevier), Pages 11–27.

Table 14.4.   Physical properties of primary fuels used for fuel processing for fuel cells

Fuel	Molecular weight [g/mol]	Density (15 °C) [kg/m³]	Boiling point [°C]	Autoignition temperature [°C]	Flash point [°C]	Heat of vaporization [kJ/mol]	Heat capacity (C_p) [J/(mol K)]	LHV [kJ/mol]	HHV [kJ/mol]	Sulfur content [ppmw]
Hydrogen	2.02	0.084	−252.76	560	<−253	0.91	29	242.3	286.5	0
Methane	16.04	0.671	−161.5	595	−130	8.17	34.92	802.3	889.3	0
Propane	44.1	1.91	−42.1	540	−104	19.04	73.6	2011	2220.8	0
Methanol (liquid)	32.04	790	64.6	470	11	35.21	79.5	638.1	726.1	0
Ethanol (liquid)	46.07	789.4	78	425	12	38.56	112.4	1329.8	1368	0
								[kJ/kg]	[kJ/kg]	
Natural gas	16.0–20.0	0.66–0.81	−162	540–580	−188			47,141	52,225	0.25
LPG		510–580	−42 to 0	494	−104			46,607	50,152	–
Gasoline	113	720–775	50–200	280	−43			43,500	46,500	10
Kerosene	170	750–845	150–325	210	38			43,100	46,200	<3000
Diesel	202	820–845	160–371	260	>52			42,612	45,575	10–50
Biodiesel (FAME)	292	860–900	315–350	177	>130			37,520	40,160	<10

Source: Kaltschmitt, Torsten, Olaf Deutschmann. 2012. Chapter 1 - Fuel Processing for Fuel Cells, In: Kai Sundmacher, Editor, Advances in Chemical Engineering, Academic Press, 2012, Volume 41, Pages 1–64.

Table 14.5.   Summary of major fuel constituents impact on fuel cells

Gas Species	Proton Exchange fuel cell	Alkaline Fuel Cells	Phosphoric acid fuel cells	Molten carbonate fuel cells	Solid oxide fuel cells
H_2	Fuel	Fuel	Fuel	Fuel	Fuel
CO	Poison (reversible) (50 ppm per stack)	Poison	Poison (<0.5%)	Fuel[a]	Fuel
CH_4	Diluent	Poison	Diluent	Diluent[b]	Fuel[a]
CO_2 & H_2O	Diluent	Poison	Diluent	Diluent	Diluent
S as (H_2S & COS)	No Studies to date	Poison	Poison (<50 ppm)	Poison (<0.5 ppm)	Poison (<1.0 ppm)

[a]In reality, CO, with H_2O, shifts to H_2 and CO_2 and CH_4, with H_2O, reforms to H_2 and CO faster than reacting as a fuel at the electrode.
[b]A fuel in the internal reforming molten carbonate fuel cells.
Source: U.S. Department of Energy, Hydrogen Analysis Resource Center, Hydrogen Data Book,
<http://hydrogen.pnl.gov/cocoon/morf/hydrogen/article/103>.

Table 14.6.   Low and intermediate temperature fuel cells

Fuel cell type	Electrolyte charge carrier	Principal catalyst	Typical operating temperature	Fuel compatibility	Primary contaminant
PEMFC (Proton exchange membrane)	Solid polymer membrane H^+	Platinum	60–80 °C	H_2, methanol	CO, sulfur and NH_3
AAEMFC (Alkaline anion exchange membrane)	Solid polymer membrane OH^-	Platinum SilverNickel	40–60 °C	H_2, methanol …	CO, CO_2 and sulfur
PAFC (Phosphoric acid)	H_3PO_4 solution H^+	Platinum	150–220 °C	H_2	CO < 1%, sulfur
AFC (Alkaline)	KOH solution OH^-	Platinum Silver Nickel	70–250 °C	H_2	CO, CO_2 and sulfur

Source: Adapted from Bidault, F. and P.H. Middleton. 2012. 4.07 - Alkaline Fuel Cells: Theory and Application, In: Ali Sayigh, Editor-in-Chief, Comprehensive Renewable Energy, (Oxford, Elsevier), Pages 159–182.

Table 14.7.   Comparison of fuel cell with other power generating systems

	Reciprocating engine: diesel	Turbine generator	Photovoltaic	Wind turbine	Fuel cells
Capacity range	500 kW–50 MW	500 kW–5 MW	1 kW–1 MW	10 kW–1 MW	200 kW–2 MW
Efficiency	35%	29–42%	6–19%	25%	40–85%
Capital cost ($/kW)	200–350	450–870	6600	1000	1500–3000
O & M cost ($/kW)	0.005–0.015	0.005–0.0065	0.001–0.004	0.01	0.0019–0.0153

Source: Adapted from Mekhilef, S., R. Saidur, A. Safari. 2012. Comparative study of different fuel cell technologies, Renewable and Sustainable Energy Reviews, Volume 16, Issue 1, Pages 981–989.

Table 14.8. Comparison of fuel cell and battery energy densities

Technology	Energy density (kW$_{eh}$/l)	Energy density (kW$_{eh}$/kg)
Secondary cells		
Lead acid	0.07	0.035
Ni–Cd	0.170	0.055
Ni–metal hydride	0.250	0.070
Li–ion	0.350	0.120
Li–polymer	0.350	0.200
Fuel cell different fuel options		
Methanol	4.384	5.600
Butane	7.290	12.600
Iso-octane	8.680	12.340
Ethanol	5.900	7.500
Diesel	8.700	12.400
Hydrogen (gas, 24.8 MPa)	0.64	33.3
Hydrogen (Liquid, -253°C)	2.36	33.3

Source: Adapted from Kundu, Arunabha, Yong Gun Shul, Dong Hyun Kim, Methanol Reforming Processes, In: T.S. Zhao, K.-D. Kreuer and Trung Van Nguyen, Editor(s), Advances in Fuel Cells, (Amsterdam, Elsevier Science), Volume 1, Pages 419–472.

Table 14.9. Efficiencies for proton exchange membrane (PEM) and alkaline electrolyzers

	PEM electrolyzer		Alkaline electrolyzer	
Efficiency	LHV	HHV	LHV	HHV
Stack efficiency				
Low current	80% (5A)	95% (5A)	78% (30A)	92% (30A)
Rated current	63% (5A)	75% (135A)	59% (220A)	70% (220A)
System efficiency				
Low current	0% (15A)	0% (15A)	0% (35A)	0% (35A)
Rated current	49% (135A)	57% (135A)	35% (220A)	41% (220A)

Source: Adapted from Harrison, K.W., W.E. Kramer, G.D. Martin, T.G. Ramsden. 2009. The wind-to-hydrogen project: Operational experience, performance testing, and systems integration, National Renewable Energy Laboratory, Golden, CO, <http://www.nrel.gov/hydrogen/pdfs/44082.pdf>, accessed 23 May 2012.

Table 14.10. Gibbs free energy change, maximum EMF (or reversible open circuit voltage), and efficiency limit (HHV basis) for hydrogen fuel cells

Form of water product	Temperature (°C)	Gibbs free energy change (kJ mol^{-1})	Maximum EMF (V)	Efficiency limit (%)
Liquid	25	−237.2	1.23	83
Liquid	80	−228.2	1.18	80
Gas	100	−225.2	1.17	79
Gas	200	−220.4	1.14	77
Gas	400	−210.3	1.09	74
Gas	600	−199.6	1.04	70
Gas	800	−188.6	0.98	66
Gas	1000	−177.4	0.92	62

Source: Adapted from Tesfai, A. and J.T.S. Irvine. 2012. 4.10 - Solid Oxide Fuel Cells: Theory and Materials, In: Ali Sayigh, Editor-in-Chief, Comprehensive Renewable Energy, (Oxford, Elsevier), Pages 241–256.

Table 14.11. Standard enthalpy and Gibbs function of reaction for candidate fuels and oxidants for fuel cells

Fuel		N	$-\Delta h$ (J/mol)	$-\Delta g$ (J/mol)	$E°_r$ (V)	η (%)[a]
Hydrogen	$H_2 + 0.5O_2 \rightarrow H_2O(l)$	2	286.0	237.3	1.229	82.97
	$H_2 + Cl_2 \rightarrow 2HCl(aq)$	2	335.5	262.5	1.359	78.33
	$H_2 + Br_2 \rightarrow 2HBr(aq)$	2	242.0	205.7	1.066	85.01
Methane	$CH_4 + 2O_2 \rightarrow CO_2 +$ $2H_2O(l)$	8	890.8	818.4	1.060	91.87
Propane	$C_3H_8 + 5O_2 \rightarrow 3CO_2 +$ $4H_2O(l)$	20	2221.1	2109.3	1.093	94.96
Decane	$C_{10}H_{22} + 15.5O_2 \rightarrow 10CO_2$ $+ 11H_2O(l)$	66	6832.9	6590.5	1.102	96.45
Carbon monoxide	$CO + 0.5O_2 \rightarrow CO_2$	2	283.1	257.2	1.333	90.86
Carbon	$C(s) + 0.5O_2 \rightarrow CO$	2	110.6	137.3	0.712	124.18[b]
	$C(s) + O_2 \rightarrow CO_2$	4	393.7	394.6	1.020	100.22[b]
Methanol	$CH_3OH(l) + 1.5O_2 \rightarrow CO_2$ $+ 2H_2O(l)$	6	726.6	702.5	1.214	96.68
Formaldehyde	$CH_2O(g) + O_2 \rightarrow CO_2 +$ $H_2O(l)$	4	561.3	522.0	1.350	93.00
Formic acid	$HCOOH + 0.5O_2 \rightarrow CO_2$ $+ H_2O(l)$	2	270.3	285.5	1.480	105.62[b]
Ammonia	$NH_3 + 0.75O_2 \rightarrow$ $1.5 H_2O(l) + 0.5N_2$	3	382.8	338.2	1.170	88.36
Hydrazine	$N_2H_4 + O_2 \rightarrow 2H_2O(l) + N_2$	4	622.4	602.4	1.560	96.77
Zinc	$Zn + 0.5O_2 \rightarrow ZnO$	2	348.1	318.3	1.650	91.43
Sodium	$Na + 0.25H_2O + 0.25O_2$ $\rightarrow NaOH(aq)$	1	326.8	300.7	3.120	92.00

[a]Energy Conversion Efficiency.

[b]At atmospheric temperature for fuel cell operations, the energy from the thermal bath (or the atmosphere) as heat may be free. But at elevated temperatures, external means must be employed to keep the thermal bath at temperatures above the ambient atmospheric temperature, which constitutes an expense. Therefore, the heat from the thermal bath to the fuel cell system is no long a free energy input; rather it is part of the energy input that has to be paid for. This means that that the ideal reversible efficiency will no longer over 100% for fuel cells.

Source: Adapted from Li, Xianguo. 2007. Thermodynamic Performance of Fuel Cells and Comparison with Heat Engines, In: T.S. Zhao, K.-D. Kreuer and Trung Van Nguyen, Editor(s), Advances in Fuel Cells, (Amsterdam, Elsevier Science), Volume 1, Pages 1–46.

Table 14.12.   Side reactions[a] in the hydrocarbon partial oxidation system to produce H2 and CO from hydrocarbon fuels for fuel cell applications

$CH_4 + 2O_2 \rightarrow CO_2 + 2H_2O$	Combustion
$CH_4 + O_2 \rightarrow CO_2 + 2H_2$	
$CO + H_2O \leftrightarrow CO_2 + H_2$	Water gas shift
$CH_4 + H_2O \leftrightarrow CO + 3H_2$	Steam reforming
$CH_4 + CO_2 \leftrightarrow 2CO + 2H_2$	CO_2 reforming
$CO + H_2 \leftrightarrow C + H_2O$	
$CH_4 \leftrightarrow C + 2H_2$	Methane decomposition
$2CO \leftrightarrow CO_2 + C$	Boudouard
$CO + 0.5O_2 \rightarrow CO_2$	
$H_2 + 0.5O_2 \rightarrow H_2O$	

[a]All reactions involving O_2 are thermodynamically irreversible.
Source: Adapted from Smith, Mark W. and Dushyant Shekhawat. 2011. Chapter 5 - Catalytic Partial Oxidation, In: Dushyant Shekhawat, J.J. Spivey And David A. Berry, Editor(s), Fuel Cells: Technologies for Fuel Processing, (Amsterdam, Elsevier), Pages 73–128.

Table 14.13.   Comparison of direct fuel cell (DFC) power plant emissions to conventional power generation sources

	NO_x (lb MWh^{-1})	SO_x (lb MWh^{-1})	PM-10 (lb MWh^{-1})	CO_2 (lb MWh^{-1})
Average US fossil fuel plant	5.06	11.6	0.27	2031
Microturbine (60 kW)	0.44	0.008	0.09	1596
Small gas turbine (250 kW)	1.15	0.008	0.08	1494
DFC fuel cell on natural gas	0.01	0.0001	0.000 02	940
DFC fuel cell on natural gas with CHP	0.006	0.000 06	0.000 01	550
DFC fuel cell on biogas	0.006	0.000 06	0.000 01	0

Source: Adapted from Leo, T. 2012. 4.09 - Molten Carbonate Fuel Cells: Theory and Application, In: Ali Sayigh, Editor-in-Chief, Comprehensive Renewable Energy, (Oxford, Elsevier), Pages 227–239.

Table 14.14.   Classification of poisons on nickel-based anodes in solid oxide fuel cells

Compound	Gas phase species	Bulk compounds	Chemisorption	Bulk Transformation	Reversible	Impact of Current Density
Sulfur	H_2S	Ni_3S_2	Yes	(No)	Yes	Yes
	COS					
Chlorine	HCl		Yes	No	Yes	No
	CH_3Cl					
Phosphorus	PH_3	$Ni_3P, Ni_5P_2,$ $Ni_{12}P_5, Ni_2P$	Yes	Yes	No	No
Arsenic	AsH_3	Ni_5As_2 $Ni_{11}As_8$	No	Yes	No	No
	As_2					
Antimony	SbO	Ni_3Sb	?	?	?	Yes
Selenium	H_2Se	NiSe	Yes	Yes	(Yes)	Yes
	AsSe, PbSe					

Source: Adapted from Hansen, John Bøgild. 2011. Chapter 13 - Direct Reforming Fuel Cells, In: Dushyant Shekhawat, J.J. Spivey And David A. Berry, Editor(s), Fuel Cells: Technologies for Fuel Processing, (Amsterdam, Elsevier), Pages 409–450.

Table 14.15.   Estimates of the progress ratio[a] for fuel cells

Fuel cell type[b]	Development start	Period investigated	Progress ratio
Manufacturer			
AFC	1952	1964–1970	82 ± 9%
PAFC	1965	1993–2000	75 ± 3%
PEMFC	1959	2002–2005	70 ± 9%
Global			
PEMFC	1959	1995–2006	79 ± 4%

[a]Progress ratio: The rate at which the cost declines each time the cumulative production doubles.
[b]AFC = Alkaline fuel cell; PAFC = Phosphoric acid fuel cell; PEMFC = Proton exchange membrane fuel cell
Source: Adapted from Schoots, K., Kramer, G.J., and van der Zwaan, B.C.C., 2010. Technology learning for fuel cells:
An assessment of past and potential cost reductions, Energy Policy, vol. 38(6), pages 2887–2897.

Table 14.16.   Learning rates for different phases of solid oxide fuel cells (%)

	Pure learning phenomena	Learning + Economies-of scale for only materials required for fuel cell manufacturing	Learning + Economies-of scale for only equipments required for fuel cell manufacturing	All reduction cost phenomena
R&D stage	16	16	16	16
Pilot stage	27	44	28	44
Early commercial stage	1	5	12	12
All stages included	20	27	22	35

Source: Adapted from Rivera-Tinoco, Rodrigo, Koen Schoots, Bob van der Zwaan. 2012. Learning curves for solid oxide fuel cells, Energy
Conversion and Management, Volume 57, Pages 86–96.

Ocean Energy

Figures

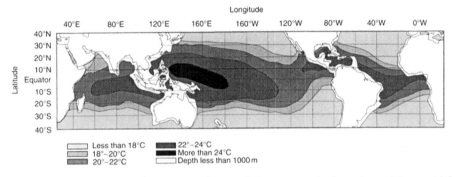

Figure 15.1. Temperature difference between surface and deep sea water in regions of the world. The darkest areas have the greatest temperature difference and therefore are the best locations for Ocean Thermal Energy (OTEC) systems.
Source: Masutani, S.M., P.K. Takahashi, Ocean Thermal Energy Conversion (OTEC). 2001. In: John H. Steele, Karl K. Turekian, and Steve A. Thorpe, Editor-in-Chief, Encyclopedia of Ocean Sciences (Second Edition), (Oxford, Academic Pres), Pages 167-173.

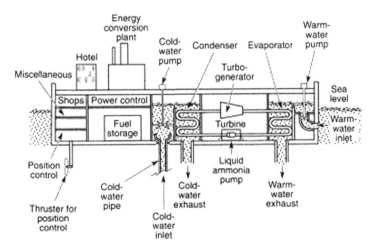

Figure 15.2. Diagram of closed-cycle Ocean Thermal Energy (OTEC) plantship.
Source: Avery, William H. 2003. Ocean Thermal Energy Conversion (OTEC), In: Robert A. Meyers, Editor-in-Chief, Encyclopedia of Physical Science and Technology (Third Edition), (New York, Academic Press), Pages 123-160.

Handbook of Energy. http://dx.doi.org/10.1016/B978-0-08-046405-3.00015-2

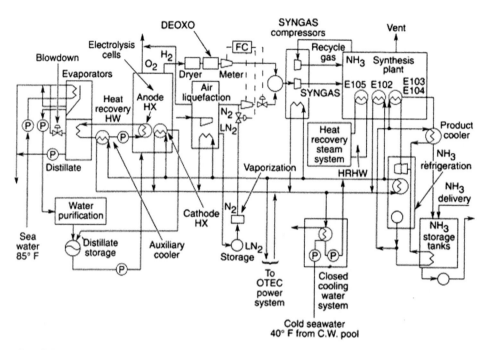

Figure 15.3. Subsystems layout of baseline 40-MWe Ocean Thermal Energy (OTEC) barge.
Source: Avery, William H. 2003. Ocean Thermal Energy Conversion (OTEC), In: Robert A. Meyers, Editor-in-Chief, Encyclopedia of Physical Science and Technology (Third Edition), (New York, Academic Press), Pages 123-160.

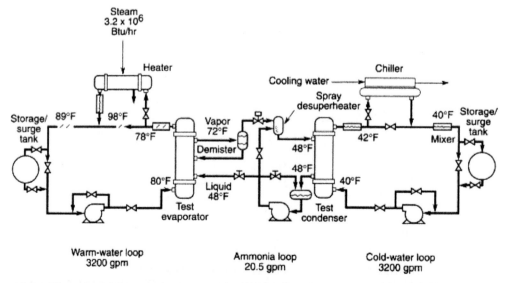

Figure 15.4. Diagram of Claude open-cycle Ocean Thermal Energy (OTEC) power system.
Source: Avery, William H. 2003. Ocean Thermal Energy Conversion (OTEC), In: Robert A. Meyers, Editor-in-Chief, Encyclopedia of Physical Science and Technology (Third Edition), (New York, Academic Press), Pages 123-160.

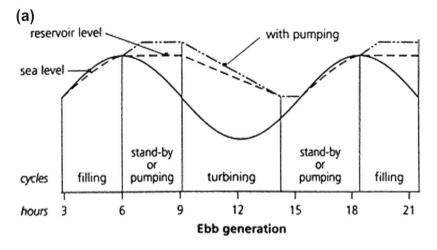

Ebb generation

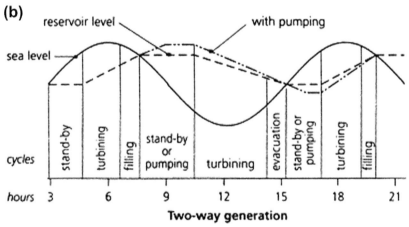

Two-way generation

Figure 15.5. Alternative operational modes at the tidal power plant at La Rance, France.
Source: Charlier, Roger H. 2003. Sustainable co-generation from the tides: A review, Renewable and Sustainable Energy Reviews, Volume 7, Issue 3, Pages 187-213.

Tables

Table 15.1. Annual mean wave power resource and yield estimates for continents and countries (±95% confidence intervals)

Continent	Gross Power (GW)	Extractable Power[a] (GW)	Efficiency of Extraction (%)
North America	427 ± 18	20.5 ± 0.5	4.8
Oceania	400 ± 15	14.9 ± 0.4	3.7
South America	374 ± 16	13.5 ± 0.3	3.6
Africa	324± 12	12.9 ± 0.3	4
Asia	318 ± 14	20.3 ± 0.7	6.4
Europe	270 ± 20	14.6 ± 0.5	5.4
World	**2110 ± 50**	**96.6 ± 1.3**	**4.6**
Country			
Australia	280 ± 13	8.3 ± 0.2	3
United States	223 ± 12	3.59 ± 0.17	1.6
Chile	194 ± 11	4.62 ± 0.17	2.4
New Zealand	89 ± 16	3.51 ± 0.16	4
Canada	83 ± 7	5.13 ± 0.19	6.2
South Africa	69 ± 4	2.17 ± 0.08	3.1
United Kingdom	43 ± 4	2.44 ± 0.14	5.7
Ireland	29 ± 4	1.13 ± 0.09	3.8
Norway	29 ± 4	1.67 ± 0.12	5.7
Spain	20 ± 3	0.65 ± 0.05	3.3
Portugal	15 ± 2	0.49 ± 0.04	3.2
France	14 ± 3	0.57 ± 0.06	3.9

[a]The amount of energy that could be extracted from the gross power using a Pelamis P2 wave machine.
Source: Adapted from Gunn, Kester , Clym Stock-Williams, Quantifying the global wave power resource, Renewable Energy, Volume 44, August 2012, Pages 296-304.

Table 15.2. Regional theoretical potential of wave energy

	Wave Energy TWh/yr (EJ/yr)
Western and Northern Europe	2,800 (10.1)
Mediterranean Sea and Atlantic Archipelagos (Azores, Cape Verde, Canaries)	1,300 (4.7)
North America and Greenland	4,000 (14.4)
Central America	1,500 (5.4)
South America	4,600 (16.6)
Africa	3,500 (12.6)
Asia	6,200 (22.3)
Australia, New Zealand, and Pacific Islands 5,600	5,600 (20.2)
World	**29,500 (106.2)**

Source: Mørk, G., S. Barstow, M.T. Pontes, and A. Kabuth. 2010. Assessing the global wave energy potential. In: Proceedings of OMAE2010 (ASME), 29th International Conference on Ocean, Offshore Mechanics and Arctic Engineering, Shanghai, China, 6-11 June 2010.

PART II

FOUNDATIONS

Physics and Thermodynamics

Tables

Table 16.1. Four fundamental forces

Interaction	Relative Strength	Range (m)	Mediating Particle
Strong	1	10^{-15}	Gluon
Electromagnetic	1/137	Infinite	Photon
Weak	10^{-6}	10^{-18}	W,Z (intermediate vector bosons)
Gravitational	10^{-38}	Infinite	Graviton

The strong force holds a nucleus together against the enormous forces of repulsion of the protons. The electromagnetic force is an infinite-range attractive or repulsive force that acts between charged particles. The weak force transforms an up quark to a down quark (or vice versa), resulting in beta decay. Gravitational force is the force of attraction between all masses in the universe, especially the attraction of the Earth's mass for bodies near its surface.

Source: University Corporation for Atmospheric Research (UCAR), Windows to the Universe, < http://www.windows.ucar.edu>, Accessed 20 November 2009; Nave, Carl R., Hyperphysics, Department of Physics and Astronomy, Georgia State University, <http://hyperphysics.phy-astr.gsu.edu/>.

Table 16.2. Elementary particle masses and charges

Particle	Charge	Mass (kg)	Mass (amu)	Energy (MeV)
Proton	1	1.66×10^{-27}	~1	938.272
Neutron	0	1.67×10^{-27}	~1	939.565
Electron	−1	9.11×10^{-31}	~0	0.511

Handbook of Energy. http://dx.doi.org/10.1016/B978-0-08-046405-3.00016-4

Table 16.3.   Temperature of selected events, locations, processes, and phenomona

Fahrenheit (°F)	Celsius (°C)	Kelvin (K)	Event, location, phenomenon process
		~10^{16}	hottest laboratory experiment (Fermilab)
		~10^{9}	core of hottest stars
		~10^{7}	core of the Sun
		~10^{7}	nuclear explosion
		~10^{6}	solar corona (the Sun's atmosphere)
		25,000	surface of blue stars
		24,000	lightning bolt
		16,000	core of the Earth
		6500	D_{65} standard white hot (effective)
		5933	tungsten boils
		5800	surface of the Sun
		6000	center of earth
		3683	tungsten melts
3034	1668	1941	titanium melts
4900	2700	3000	incandescent light bulb
3100	1700	2000	typical flames
2200	1200	1500	fresh lava
1984.32	1084.62	1357.77	copper freezes
1076	580	853.15	auto-ignition of methane (natural gas)
840	450	720	daytime surface temperature on Mercury
674	357	630	mercury boils
621	327	600	lead melts
574.588	301.438	574.588	fahrenheit and kelvin scales coincide
572	300	573.15	auto-ignition of bituminous coal
530	280	550	very hot home oven
536	280	553.15	auto-ignition of motor gasoline
451	233	506	paper burns
212	100	373.15	water boils
136	58	331	hottest surface temperature on Earth (Libya 1922)
108	42	315	New York City record high
98.6	37	310.2	human body (traditional)
98.2	36.8	309.9	human body (revised)
96	-	-	human body (according to Fahrenheit)
80.6	27	300.15	fastest temperature rise (2 minutes; Spearfish, South Dakota, 1943-01-22)
78.8	26	299.15	fastest temperature decrease (15 minutes; Rapid City, South Dakota, 1911-01-10)
59	15	288	mean surface temperature on Earth
32.018	0.01	273.16	water triple point
32	0	273.15	water freezes
0	−18	255	ice-water-salt mixture (according to Fahrenheit)
−13	−25	248	New York City record low
−38	−39	234	mercury freezes
−40	−40	233	fahrenheit and celsius scales coincide
−56	−49	220	mean surface temperature on Mars
−108	−78	195	sublimation point of dry ice

Table 16.3. Continued

−128	−89	184	coldest surface temperature on Earth (Vostok, Antarctica, 1983)
−300	−180	90	nighttime surface temperature on Mercury
−279	−183	90	oxygen liquefies
−320	−196	77	nitrogen liquefies
		63	nitrogen freezes
		50	mean surface temperature on Pluto
		20.3	hydrogen liquefies
		4.22	helium liquefies
		2.73	mean temperature of the universe
		~1	coldest point in space (Boomerang Nebula)
		0.95	helium freezes (26 atm)
		10^{-10}	coldest laboratory experiment (Helsinki University of Technology)
−459.67	−273.15	0	absolute zero

Source: Adapted from: Elert, Glen, The Physics Hypertextbook, <http://physics.info/>.

Table 16.4. Temperature (or effective temperature) of selected radiant sources

Temperature kelvin	Radiant energy source
306	cosmic background radiation
500	human skin
660	household oven at its hottest
770	minimum temperature for incandescence
1400	dull red heat
1900	glowing coals, electric stove, electric toaster
2000	candle flame
2800	kerosene lamp
2900	incandescent light bulb, 75 W
3000	incandescent light bulb, 100 W
3100	incandescent light bulb, 200 W
3200	sunrise or sunset (effective)
3600	professional studio lights
4000	one hour after sunrise or one hour before sunset (effective)
5500	two hours after sunrise or two hours before sunset (effective)
6500	direct midday sunlight
7000	daylight (effective)
20-30,000	overcast sky (effective)
20-30,000	lightning bolt

Source: Elert, Glen, The Physics Hypertextbook, <http://physics.info/>, accessed 22 June 2012.

Table 16.5.   Regions of the electromagnetic spectrum

	Wavelength (m)	Frequency (Hz)	Energy (J)	Blackbody Temperature (K)
Radio	$> 1 \times 10^{-1}$	$< 3 \times 10^{9}$	$< 2 \times 10^{-24}$	< 0.03
Microwave	$1 \times 10^{-3} - 1 \times 10^{-1}$	$3 \times 10^{9} - 3 \times 10^{11}$	$2 \times 10^{-24} - 2 \times 10^{-22}$	$0.03 - 30$
Infrared	$7 \times 10^{-7} - 1 \times 10^{-3}$	$3 \times 10^{11} - 4 \times 10^{14}$	$2 \times 10^{-22} - 3 \times 10^{-19}$	$30 - 4100$
Optical	$4 \times 10^{-7} - 7 \times 10^{-7}$	$4 \times 10^{14} - 7.5 \times 10^{14}$	$3 \times 10^{-19} - 5 \times 10^{-19}$	$4100 - 7300$
UV	$1 \times 10^{-8} - 4 \times 10^{-7}$	$7.5 \times 10^{14} - 3 \times 10^{16}$	$5 \times 10^{-19} - 2 \times 10^{-17}$	$7300 - 3 \times 10^{6}$
X-ray	$1 \times 10^{-11} - 1 \times 10^{-8}$	$3 \times 10^{16} - 3 \times 10^{19}$	$2 \times 10^{-17} - 2 \times 10^{-14}$	$3 \times 10^{6} - 3 \times 10^{8}$
Gamma-ray	$< 1 \times 10^{-11}$	$> 3 \times 10^{19}$	$> 2 \times 10^{-14}$	$> 3 \times 10^{8}$

Table 16.6.   Energy produced in fission (MeV/fission)

	^{235}U	^{239}Pu	^{241}Pu
Instantaneous			
Kinetic energy of fission fragments	169	175	177
Gamma ray energy	8	8	8
Kinetic energy of fission neutrons	5	6	6
Subtotal	182	189	191
Delayed			
β particles from fission products	8	8	9
γ rays from fission products	7	6	7
Neutron-capture gammas	7	10	10
Subtotal	22	24	26
Total	204	213	217

Source: Adapted from Babcock and Wilcox. 1978. Steam: Its Generation and Use, (Charlotte, Babcock and Wilcox).

Table 16.7. Thermodynamic properties for selected gases

T (K)	Molar Enthalpies (kJ/kmol)					
	N_2	O_2	H_2O	CO	CO_2	H_2
0	0	0	0	0	0	0
298	8,669	8,682	9,904	8,669	9,364	8,468
500	14,581	14,770	16,828	14,600	17,678	14,350
1,000	30,129	31,389	35,882	30,355	42,769	29,154
2,000	64,810	67,881	82,593	65,408	100,804	61,400
3,000	101,407	106,780	136,264	102,210	162,226	97,211

Substance	Formula	$\Delta H_f°$ (kJ/mol)
Carbon	C(s)	0
Hydrogen	H_2	0
Nitrogen	N_2	0
Oxygen	O_2	0
Carbon monoxide	CO	−110,530
Carbon dioxide	CO_2	−393,520
Water liquid	H_2O (1)	−285,830
Water vapor	H_2O(g)	−241,820
Methane	CH_4	−74,850
Ethyl alcohol liquid	C_2H_5OH (1)	−277,690

Source: Adapted from Goswami, Yogi D., Frank Kreith, and Jan F. Kreider. 2000. Principles of Solar Engineering, 2nd edition, (Philadelphia, Taylor & Francis).

Table 16.8. Global entropy budget

	Heat flux (W m⁻²)	T_{cold} (K)	T_{warm} (K)	σ (mW m⁻² K⁻¹)
Scattering of solar radiation	103	n/a[a]	n/a[a]	26
Atmospheric absorption of solar radiation	68	252	5760	258
Surface absorption of solar radiation	170	288	5760	561
Atmospheric absorption of terrestrial radiation	28	252	288	14
Moist convection (evaporation–precipitation)	79	266	288	23
Dry convection (sensible heat into boundary layer)	24	280	288	2
Frictional dissipation of large-scale circulation	10	255	300	6
Biotic activity	8	288	5760	5[b]
Planetary	235	255	5760	881[c]

The global entropy budget characterizes the irreversibility of various Earth system processes. The columns give typical values of the heat flux Q and the temperatures T_{cold} and T_{warm} at which the energy is being transformed. The entropy production σ is then estimated by $\sigma = Q (1/T_{cold} - 1/T_{warm})$ using the steady-state assumption.
[a]Entropy produced by scattering originates from broadening of the solid angle, not from temperature differences.
[b]Term included in surface absorption of solar radiation.
[c]Total does not balance individual contributions due to estimated nature of the budget.
Source: Adapted from Kleidon, A. 2008. Energy Balance, In: Sven Erik Jorgensen and Brian Fath, Editor(s)-in-Chief, Encyclopedia of Ecology, (Oxford, Academic Press), Pages 1276–1289.

Table 16.9. Standard molar chemical exergy e^{CH} (kJ/kmol) of various substances at 298 K and p_0, according to two alternative reference states

Substance	Formula	Model I[a]	Model II[b]
Nitrogen	N_2 (g)	640	720
Oxygen	O_2 (g)	3,950	3,970
Carbon Dioxide	CO_2 (g)	14,175	19,870
Water	H_2O (g)	8,635	9,500
	H_2O (l)	45	900
Carbon (graphite)	C (s)	404,590	410,260
Hydrogen	H_2 (g)	235,250	236,100
Sulfur	S (s)	598,160	609,600
Carbon monoxide	CO (g)	269,410	275,100
Sulfur dioxide	SO_2 (g)	301,940	313,400
Nitrogen monoxide	NO (g)	88,850	88,900
Nitrogen dioxide	NO_2 (g)	55,565	55,600
Hydrogen sulfide	H_2S (g)	799,890	812,000
Ammonia	NH_3 (g)	336,685	337,900
Methane	CH_4 (g)	824,350	831,650
Ethane	C_2H_6 (g)	1,482,035	1,495,840
Methanol	CH_3OH (g)	715,070	722,300
	CH_3OH (l)	710,745	718,000
Ethyl alcohol	C_2H_5OH (g)	1,348,330	1,363,900
	C_2H_5OH (l)	1,342,085	1,357,700

[a]Ahrendts, J. 1977. Die Exergie Chemisch Reaktionsfähiger Systeme, VDI-Forschungsheft. VDI-Verlag, Dusseldorf, 579. Also see Reference States, Energy – The International Journal, 5: 667–677, 1980. In Model I, $p0 = 1.019$ atm.

[b]Szargut, J., Morris, D. R., and Steward, F. R. 1988. Energy Analysis of Thermal, Chemical, and Metallurgical Processes. Hemisphere, New York. In Model II, $p0 = 1.0$ atm.

Source: Adapted from Moran, M. J., Tsatsaronis, G. "Engineering Thermodynamics," In Kreith, Frank, Ed.. 2000. CRC Handbook of Thermal Engineering (Boca Raton, CRC Press).

Table 16.10. Representative steam table data

		(a) Properties of Saturated Water (Liquid-Vapor): Temperature Table						
		Specific Volume (m³/kg)		Internal Energy (kJ/kg)		Enthalpy (kJ/kg)		
Temp (°C)	Pressure (bar)	Saturated liquid ($v_f \times 10^3$)	Saturated vapor (u_f)	Saturated liquid (u_f)	Saturated vapor (u_g)	Saturated liquid (h_f)	Evap. (h_g)	Saturated vapor (h_g)
0.01	0.00611	1.0002	206.136	0	237.5	0.01	2501.3	2501.4
4	0.00813	1.0001	157.232	16.77	2380.9	16.76	2491.9	2508.7
5	0.00872	1.0001	147.12	20.97	2382.3	20.98	2489.6	2510.6
6	0.00935	1.0001	137.734	25.2	2383.6	25.2	2487.2	2512.4
8	0.01072	1.0002	120.917	35.59	2386.4	33.6	2482.5	2516.1
		(b) Properties of Saturated Water (Liquid-Vapor): Pressure Table						
		Specific Volume (m³/kg)		Internal Energy (kJ/kg)		Enthalpy (kJ/kg)		
Pressure (bar)	Temp (°C)	Saturated liquid ($v_f \times 10^3$)	Saturated vapor (u_f)	Saturated liquid (u_f)	Saturated vapor (u_g)	Saturated liquid (h_f)	Evap. (h_g)	Saturated vapor (h_g)
0.04	28.96	1	34.8	121.45	2415.2	121.46	2432.9	2554.4
0.06	36.16	1.01	27.74	151.53	2425	151.53	2415.9	2567.4
0.08	41.51	1.01	18.1	173.87	2432.2	173.88	2403.1	2577
0.1	45.81	1.01	14.67	191.82	2437.9	191.83	2392.8	2584.7
0.2	60.06	1.02	7.65	251.38	2456.7	251.4	2358.3	2609.7

Source: Adapted from Moran, Michael J., Engineering Thermodynamics, in Kreith, Frank and D. Yogi. Goswami, Eds. 2005. CRC Handbook of Mechanical Engineering, Second Edition (Boca Raton, CRC Press).

Table 16.11. Physical and thermodynamic properties of common Rankine cycle working fluids

Property	Water	Methanol	2- Methyl Pyridine, H_2O	Fluorinol 85	Toluene	R-113	Ammonia	Isobutane
Molecular weight	18	32	33	88	92	187	17	58
Boiling point (1 atm) (°C)	100	64	93	75	110	48	−330	−12
Liquid density (kg/m³)	1,000	750	934	1,370	857	2,565	682	594
Specific volume (saturated vapor at boiling point) (m³/kg)	2	1	1	0	0	0	1.12	0.35
Maximum stability temperature (°C)	--	175–230	370–400	290–330	400–425	175–230	300[a]	>200
Wetting-drying	W	W	W	W	D	D	D	D
Heat of vaporization at 1 atm (kJ/kg)	2,256	1,098	879	442	365	1,370	1370	367
Isentropic enthalpy drop across turbine (kJ/kg)	348–1160	162–302	186–354	70–186	116–232	23–46	200–600	120–380

[a]Anhydrous ammonia in the prescence of iron. Small trace of water increases this limit.
Source: Adapted from Goswami, D., Frank Kreith, and Jan F. Kreider. 2000. Principles of Solar Engineering, 2nd edition, (Philadelphia, Taylor & Francis).

Chemistry

Tables

Table 17.1. Physical properties of gases (at 14.7 psia)

Gas	Temp °F	Density lb/cu ft	Specific volume cu ft/lb	Instantaneous specific heat c_p	Instantaneous specific heat c_v	k, c_p/c_v
Air	70	0.0749	13.36	0.241	0.172	1.4
	200	0.0601	16.63	0.242	0.173	1.4
	500	0.0413	24.19	0.248	0.18	1.38
	1000	0.0272	36.79	0.265	0.197	1.34
CO_2	70	0.1148	8.71	0.202	0.155	1.3
	200	0.0922	10.85	0.216	0.17	1.27
	500	0.0634	15.77	0.247	0.202	1.22
	1000	0.0417	23.98	0.28	0.235	1.19
H_2	70	0.0052	191.2	3.44	2.44	1.41
	200	0.0042	238	3.48	2.49	1.4
	500	0.0029	345	3.5	2.515	1.39
	1000	0.0019	526.3	3.54	2.56	1.38
Flue gas[a]	70	0.0776	12.88	0.253	0.187	1.35
	200	0.0623	16.04	0.255	0.189	1.35
	500	0.0429	23.33	0.265	0.199	1.33
	1000	0.0282	35.48	0.283	0.217	1.3
CH_4	70	0.0416	24.05	0.53	0.406	1.3
	200	0.0334	29.95	0.575	0.451	1.27
	500	0.023	43.5	0.72	0.596	1.21
	1000	0.0151	66.22	0.96	0.836	1.15

[a]From coal; 120% total air; flue gas molecular weight 30.
Source: Adapted from Babcock and Wilcox. 1978. Steam: Its Generation and Use, (Charlotte, Babcock and Wilcox).

Handbook of Energy. http://dx.doi.org/10.1016/B978-0-08-046405-3.00017-6

Table 17.2. Specific heat of gases

Gas	Molecular weight	Specific heat at temperature, T (K) (kJ/kmol.K)	Range of validity (K)
H_2S	34	$30.139 + 0.015*T$	300–600
H_2O_{steam}	18	$34.4 + 0.000628*T + 0.0000052T^2$	300–2500
H_2	2	$27.71 + 0.0034*T$	273–2500
CH_4	16	$22.35 + 0.048*T$	273–1200
CO	28	$27.62 + 0.005T$	273–2500
CO_2	44	$43.28 + 0.0114*T-818363/T^2$	273–1200
O_2	32	$34.62 + 0.00108T-785712/T^2$	300–5000
N_2	28	$27.21 + 0.0042T$	300–5000

Source: Adapted from Basu, Prabir. 2010. Biomass Gasification and Pyrolysis, (Boston, Academic Press).

Table 17.3. Thermal conductivity, k, of gases (Btu/ sq ft, hr, °F/in thickness)

Gases at Atmospheric Pressure					
Temp, °F	Air, k	CO_2, k	O_2, k	N_2, k	H_2, k
0	0.168	0.096	0.156	0.168	1.08
500	0.3	0.252	0.312	0.264	2.004
1000	0.408	0.384	0.444	0.336	2.724
1500	0.48	0.516	0.564	0.408	3.36
2000	0.564	0.624	0.672	0.468	3.924
2500	0.636	0.72	0.792	0.528	4.464
3000	0.696	0.804	0.912	–	5.004

Flue Gases from Various Fuels - Atmospheric Pressure			
Temp, °F	Nat. Gas, k[a]	Fuel Oil, k[a]	Coal, k[b]
0	–	–	–
500	0.264	0.264	0.264
1000	0.36	0.348	0.348
1500	0.444	0.432	0.432
2000	0.528	0.516	0.516
2500	0.612	0.588	0.6

[a]For 115% total air.
[b]For 120% total air.
Source: Adapted from Babcock and Wilcox. 1978. Steam: Its Generation and Use, (Charlotte, Babcock and Wilcox).

Table 17.4. Heat of formation of various elements and compounds at standard condition, 25 °C, and 1 bar pressure[a]

Substance	$\Delta H_f°$ (kJ/mol)	$S°$ (J/K.mol)	$\Delta G_f°$ (kJ/mol)	Substance	$\Delta H_f°$ (kJ/mol)	$S°$ (J/K.mol)	$\Delta G_f°$ (kJ/mol)
$Al_{(s)}$	0	28.3	0	$NH_{3 (g)}$	−46.1	192.5	−16.5
$Al_2O_{3 (s)}$	−1675.7	50.9	−1582.3	$N_2H_{4 (l)}$	50.6	121.2	149.3
$Br_{2 (l)}$	0	151.6	0	$NH_4Cl_{(s)}$	−314.4	94.6	−202.9
$HBr_{(g)}$	−36.4	198.7	−53.5	$NH_4NO_{3 (s)}$	−365.6	151.1	−183.9
$Ca_{(s)}$	0	41.4	0	$NO_{(g)}$	90.3	210.8	86.6
$CaCO_{3 (s)}$ (calcite)	−1206.9	92.9	−1128.8	$NO_{2 (g)}$	33.2	240.1	51.3
$CaCl_{2 (s)}$	−795.8	104.6	−748.1	$N_2O_{(g)}$	82.1	219.9	104.2
$C_{(s)}$ (graphite)	0	5.7	0	$N_2O_{4 (g)}$	9.2	304.3	97.9
				$HNO_{3 (l)}$	−174.1	155.6	−80.7
$CCl_{4 (l)}$	−135.4	216.4	−65.2	$O_{(g)}$	249.2	161.1	231.7
$CCl_{4 (g)}$	−96.0	309.9	−60.6	$O_{2 (g)}$	0	205.1	0
$CHCl_{3 (l)}$	−134.5	201.7	−73.7	$O_{3 (g)}$	142.7	238.9	163.2
$CH_{4 (g)}$	−74.8	186.3	−50.7				
$C_2H_{2 (g)}$	226.7	200.9	209.2				
$C_2H_{4 (g)}$	52.3	219.6	68.2				
$C_2H_{6 (g)}$	−84.7	229.6	−32.8				
$C_3H_{8 (g)}$	−103.8	269.9	−23.5				
$C_6H_{6 (l)}$	49	172.8	124.5				
$CH_3OH_{(l)}$	−238.7	126.8	−166.3				
$C_2H_5OH_{(l)}$	−277.7	160.7	−178.8	$K_{(s)}$	0	64.2	0
$CH_3CO_2H_{(l)}$	−484.5	159.8	−389.9	$KCl_{(s)}$	−436.7	82.6	−409.1
$CO_{(g)}$	−110.5	197.7	−137.2	$KClO_{3 (s)}$	−397.7	143.1	−296.3
$CO_{2 (g)}$	−393.5	213.7	−394.4	$KOH_{(s)}$	−428.8	78.9	−379.1
$COCl_{2 (g)}$	−218.8	283.5	−204.6				
$CS_{2 (g)}$	117.4	237.8	67.1				
$Cl_{2 (g)}$	0	223.1	0				
$HCl_{(g)}$	−92.3	186.9	−95.3	$Na_{(s)}$	0	51.2	0
$CrCl_{3 (s)}$	−556.5	123	−486.1	$NaCl_{(s)}$	−411.2	72.1	−384.1
$Cu_{(s)}$	0	33.2	0	$NaOH_{(s)}$	−425.6	64.5	−379.5
$CuO_{(s)}$	−157.3	42.6	−129.7	$Na_2CO_{3 (s)}$	−1130.7	135	−1044.0
$CuCl$	−137.2	86.2	−119.9				
$CuCl_{2 (s)}$	−220.1	108.1	−175.7	$S_{(g)}$	278.8	167.8	238.3
$F_{2 (g)}$	0	202.8	0	$SF_{6 (g)}$	−1209.0	291.8	−1105.3
$HF_{(g)}$	−271.1	173.8	−273.2	$H_2S_{(g)}$	−20.6	205.8	−33.6
$He_{(g)}$	0	126	0	$SO_{2 (g)}$	−296.8	248.2	−300.2
$H_{2 (g)}$	0	130.7	0	$SO_{3 (g)}$	−395.7	256.8	−371.1
$H_2O_{(l)}$	−285.8	69.9	−237.1	$H_2SO_{4 (l)}$	−814.0	156.9	−690.0
$H_2O_{(g)}$	−241.8	188.8	−228.6				
$H_2O_{2 (l)}$	−187.8	109.6	−120.4				

Continued

Table 17.4.   Continued

Substance	$\Delta H_f°$ (kJ/mol)	$S°$ (J/K.mol)	$\Delta G_f°$ (kJ/mol)	Substance	$\Delta H_f°$ (kJ/mol)	$S°$ (J/K.mol)	$\Delta G_f°$ (kJ/mol)
Fe $_{(s)}$	0	27.8	0				
FeO $_{(s)}$	−272.0	57.6	245.1				
Fe$_2$O$_3$ $_{(s)}$	−824.2	87.4	−742.2				
Fe$_3$O$_4$ $_{(s)}$	−1118.4	146.4	−1015.4				
FeCl$_2$ $_{(s)}$	−341.8	118	−302.3				
FeCl$_3$ $_{(s)}$	−399.5	142.3	−344.0				
FeS$_2$ $_{(s)}$	−178.2	52.9	−166.9				
Pb $_{(s)}$	0	64.8	0				
Ne $_{(g)}$	0	146.2	0				
N$_2$ $_{(g)}$	0	191.6	0				

[a]Subscripts in parentheses indicate the state: solid (s), liquid (l), or gaseous (g).
Source: Adapted from Basu, Prabir. 2010. Biomass Gasification and Pyrolysis, (Boston, Academic Press).

Table 17.5.   Physical properties of air at atmospheric pressure

T (K)	ρ (kg/m³)	c_p (kJ/kg-°C)	μ (kg/ms) × 10^{-5}	ν (m²/s) × 10^{-6}	k (W/m-°C)	α (m²/s) × 10^{-4}	Pr
100	3.601	1.0266	0.692	1.923	0.00925	0.025	0.77
150	2.3675	1.0099	1.028	4.343	0.01374	0.0575	0.753
200	1.7684	1.0061	1.329	7.49	0.01809	0.1017	0.739
250	1.4128	1.0053	1.488	9.49	0.02227	0.1316	0.722
300	1.1774	1.0057	1.983	16.84	0.02624	0.2216	0.708
350	0.998	1.009	2.075	20.76	0.03003	0.2983	0.697
400	0.8826	1.014	2.286	25.9	0.03365	0.376	0.689
450	0.7833	1.0207	2.484	31.71	0.03707	0.4222	0.683
500	0.7048	1.0295	2.671	37.9	0.04038	0.5564	0.68
550	0.6423	1.0392	2.848	44.34	0.0436	0.6532	0.68
600	0.5879	1.0551	3.018	51.34	0.04659	0.7512	0.68
650	0.543	1.0635	3.177	58.51	0.04953	0.8578	0.682
700	0.503	1.0752	3.332	66.25	0.0523	0.9672	0.684
750	0.4709	1.0856	3.481	73.91	0.05509	1.0774	0.686
800	0.4405	1.0978	3.625	82.29	0.05779	1.1951	0.689
850	0.4149	1.1095	3.765	90.75	0.06028	1.3097	0.692
900	0.3925	1.1212	3.899	99.3	0.06279	1.4271	0.696
950	0.3716	1.1321	4.023	108.2	0.06225	1.551	0.699
1000	0.3524	1.1417	4.152	117.8	0.06752	1.6779	0.702

T = temperature, ρ = density, c_p = specific heat capacity, μ = viscosity, $\nu = \mu/\rho$ = kinetic viscosity, k = thermal conductivity, $\alpha = c_p\rho/k$ = heat (thermal) diffusivity, Pr = Prandtl number.
Source: Kalogirou, Soteris A. 2009. Solar Energy Engineering, (Boston, Academic Press).

Table 17.6. Physical properties of saturated liquid water

T (°C)	ρ (kg/m³)	c_p (kJ/kg-°C)	v (m²/s) × 10⁻⁶	k (W/m-°C)	α (m²/s) × 10⁻⁷	Pr	β (K⁻¹) × 10⁻³
0	1002.28	4.2178	1.788	0.552	1.308	13.6	
20	1000.52	4.1818	1.006	0.597	1.43	7.02	0.18
40	994.59	4.1784	0.658	0.628	1.512	4.34	
60	985.46	4.1843	0.478	0.651	1.554	3.02	
80	974.08	4.1964	0.364	0.668	1.636	2.22	
100	960.63	4.2161	0.294	0.68	1.68	1.74	
120	945.25	4.25	0.247	0.685	1.708	1.446	
140	928.27	4.283	0.214	0.684	1.724	1.241	
160	909.69	4.342	0.19	0.68	1.729	1.099	
180	889.03	4.417	0.173	0.675	1.724	1.004	
200	866.76	4.505	0.16	0.665	1.706	0.937	
220	842.41	4.61	0.15	0.652	1.68	0.891	
240	815.66	4.756	0.143	0.635	1.639	0.871	
260	785.87	4.949	0.137	0.611	1.577	0.874	
280	752.55	5.208	0.135	0.58	1.481	0.91	
300	714.26	5.728	0.135	0.54	1.324	1.019	

Notes: T = temperature, ρ = density, c_p = specific heat capacity, v = μ/ρ = kinetic viscosity, k = thermal conductivity, α = $c_p\rho$/k = heat (thermal) diffusivity, Pr = Prandtl number.
Source: Kalogirou, Soteris A. 2009. Solar Energy Engineering, (Boston, Academic Press).

Table 17.7. Content of different elements in inanimate and living matter

	Content (% per weight)						
Chemical elements	Solar matter	Atmosphere	Ocean	Earth's crust	Soil	Plants	Animals
Hydrogen (H)	72		10.7	1.6	3.1	10	11
Helium (He)	27						
Oxygen (O)	0.28	20.97	85.8	56.2	66.8	70	65
Carbon (C)	0.12	0.01			1.2	18	19
Nitrogen (N)	0.05	78.08		0.26	0.06	0.9	3
Magnesium (Mg)	0.01		0.13	1.46	0.37	0.08	0.05
Silicon (Si)	0.01			23.05	19.41	0.35	0.24
Sulfur (S)	0.01		0.09	0.24	0.05	0.14	0.18
Iron (Fe)	0.01			3.63	2.24	0.02	0.02
Aluminum (Al)				6.3	4.18	0.01	
Natrium (Na)			1.03	1.95	0.37	0.03	0.05
Potassium (K)			0.04	1.95	0.8	0.03	0.02
Calcium (Ca)			0.04	2.58	0.81	0.03	0.03
Chlorine (Cl)			1.93	0.02	0.01	0.01	0.02

Source: Adapted from Chernyshenko, S.V. 2008. Structure and History of Life, In: Sven Erik Jorgensen and Brian Fath, Editor(s)-in-Chief, Encyclopedia of Ecology,(Oxford, Academic Press), Pages 3403–3416.

Table 17.8. General properties of gaseous hydrocarbons

	Molecular weight	Boiling point (°C)	Ignition temperature (°C)	LHV (kJ/mole)	HHV (kJ/mole)	Flammability limits in air (% v/v)
Methane	16	−161.5	482–632	799.7	889.3	5.0–15.0
Ethane	30.1	−88.6	515	1429.4	1561.6	3.0–12.5
Ethylene	28	−103.7	490	1320.3	1408.2	2.8–28.6
Propane	44.1	−42.1	450	2044.4	2219.7	2.1–10.1
Propylene	42.1	−47.7	458	1927	2060.9	2.00–11.1
n-Butane	58.1	0.5	405	2658.8	2875.8	1.86–8.41
iso-Butane	58.1	−11.7	462	2657	2872.8	1.80–8.44
n-Butene	56.1	−6.0	443	2542	2718.1	1.98–9.65
iso-Butene	56.1	−6.9	465	2485	2706.2	1.8–9.0

Source: Adapted from Speight, James G. 2011. Chapter 3 - Fuels for Fuel Cells, In: Dushyant Shekhawat, J.J. Spivey And David A. Berry, Editor(s), Fuel Cells: Technologies for Fuel Processing, (Amsterdam, Elsevier), 2011, Pages 29–48.

Table 17.9. Hydrocarbon synthesis product distributions (reduced iron catalyst)

Variable	Fixed bed	Fluid bed
Temperature	220–240 °C	320–340 °C
H_2/CO feed ratio	1.7	3.5
Product selectivity (%)		
Gases (C_1–C_4)	22	52.7
Liquids	75.7	39
Nonacid chemicals	2.3	7.3
Acids	—	1
Gases (% total product)		
C_1	7.8	13.1
C_2	3.2	10.2
C_3	6.1	16.2
C_4	4.9	13.2
Liquids (as % of liquid products)		
C_3–C_4 (liquified petroleum gas)	5.6	7.7
C_5–C_{11} (gasoline)	33.4	72.3
C_{12}–C_{20} (diesel)	16.6	3.4
Waxy oils	10.3	3
Medium wax (mp 60–65 °C)	11.8	—
Hard wax (mp 95–100 °C)	18	—
Alcohols and ketones	4.3	12.6
Organic acids	—	1
Paraffin/olefin ratio	~1:1	~0.25:1

Source: Larsen, John W. and Martin L. Gorbaty, Coal Structure and Reactivity. 2003. In: Robert A. Meyers, Editor-in-Chief, Encyclopedia of Physical Science and Technology (Third Edition), (New York, Academic Press), Pages 107–122.

Table 17.10. Properties of gasoline, natural gas, and hydrogen

Property	Gasoline	Natural Gas	Hydrogen
Density (g/cm^3)	0.73	0.78×10^{-3}	0.8×10^{-4}(Gas)
			0.71×10^{-1} (liquid)
Boiling point (°C)	38/204	−156	−253 (20K)
Lower heating valve			
Gravimetric (KJ/Kg)	4.45×10^4	4.80×10^4	12.50×10^4
Volumetric (KJ/m^3)	32.0×10^6	37.3×10^3	10.40×10^3 (Gas)
			8.52×10^6 (liquid)
Stoichiometric Composition in air (Volume %)	1.76	9.43	29.3
Flammable limits (% in air)	1.46–7.6	5–16	4.75
Flame speed (m/sec)	0.40	0.41	3.45
Flame temperature in air (°C)	2197	1875	2045
Ignition temperature (°C)	257	540	585
Flame Luminosity	High	Medium	Low

Source: Adapted from Garg, H. P., S. C. Mullick, and A. K. Bhargava. 1985. Solar thermal energy storage, (Dordrecht, D. Reidel Publishing).

Table 17.11. High-temperature, closed-loop chemical C-H-O reactions

Closed-loop system	Enthalpy[a]$\Delta_{H0 \text{ (kJ mol}^{-1})}$	Temperature range (K)
$CH_4 + H_2O \leftrightarrow CO + 3H_2$	206 (250)[b]	700–1200
$CH_4 + CO_2 \leftrightarrow 2CO + 2H_2$	247	700–1200
$CH_4 + 2H_2O \leftrightarrow CO_2 + 4H_2$	165	500–700
$C_6H_{12} \leftrightarrow C_6H_6 + 3H_2$	207	500–750
$C_7H_{14} \leftrightarrow C_7H_8 + 3H_2$	213	450–700
$C_{10}H_{18} \leftrightarrow C_{10}H_8 + 5H_2$	314	450–700

[a]Standard enthalpy for complete reaction.
[b]Including heat of evaporation of water.
Source: Adapted from Sørensen, Bent. 2012. Hydrogen and Fuel Cells-Emerging Technologies and Applications (Second Edition), (Boston, Academic Press).

Table 17.12. Properties of common heat transfer liquids[a]

Material	Density (kg/m²)	Specific heat (kJ/°C kg)	Boiling point (°C)
Water	~1000	~4.2	100
Therminol 66™ [b]	820 (250 °C)	2.4 (250 °C)	~330
Dowtherm A™ [c]	859 (250 °C)	2.2 (250 °C)	257
Hi Tec™ [d]	1890 (250 °C)	1.5 (250 °C)	--d
Caloria HT-43™ [e]	694 (250 °C)	3.0 (250 °C)	--
Diethylene glycol	1020 (150 °C)	2.8 (150 °C)	240
Tetraethylene glycol	1110 (66 °C)	2.5 (100 °C)	280

[a]Data collected directly from manufacturers.
[b]Trademark of Monsanto Chemical Co.
[c]Trademark of Dow Chemical Co; Dowtherm is a mixture of diphenyl and diphenyl oxide.
[d]Trademark of E.I. duPont; solidifies below approximately 190°C; does not boil or decompose below 500°; consists of 40% $NaNO_2$, 7% $NaNO_3$, 53% KNO_3.
[e]Trademark of ExxonMobil.
[f]At 1 atm
Source: Adapted from Goswami, Yogi D., Frank Kreith, and Jan F. Kreider. 2000. Principles of Solar Engineering, 2nd edition, (Philadelphia, Taylor & Francis).

Table 17.13. Specific heat capacity of selected substances

	Specific heat (J kg⁻¹ K⁻¹)	Density (kg m⁻³)	Heat capacity[a] (J m⁻³ K⁻¹)
Water[b]	4182	1000	4.18×10^6
Sandy soil, saturated	1480	2000	2.96×10^6
Sandy soil, dry	800	1600	1.28×10^6
Soil, inorganic	733	2600	1.91×10^6
Soil, organic	1921	1300	2.50×10^6
Peat soil, saturated	3650	1100	4.02×10^6
Peat soil, dry	1920	300	0.58×10^6
Snow, fresh	2090	100	0.21×10^6
Snow, old	2090	480	1.00×10^6
Ice	2100	920	1.93×10^6
Air[b]	1004	1.2	0.001×10^6

[a]Substances change their temperature by differing amounts for a given amount of heat, depending on their specific heat capacity and their density. The last column, the product of the former two quantities, describes the amount of heat that is necessary to raise the temperature of 1 m³ of a given substance by 1 K.
[b]Density depends on temperature. Values given are for 293 K.
Source: Adapted from Kleidon, A. 2008. Energy Balance, In: Sven Erik Jorgensen and Brian Fath, Editor(s)-in-Chief, Encyclopedia of Ecology, (Oxford, Academic Press), Pages 1276–1289.

Table 17.14. Pyrolysis methods

Method	Residence time	Temperature (°C)	Heating rate	Products
Carbonation	Days	402	Very low	Charcoal
Conventional	5–30 min	602	Low	Oil, gas, char
Fast	0.5–5 s	925	Very high	Bio-oil
Flash-liquid[a]	<1 s	<652	High	Bio-oil
Flash-gas[b]	<1 s	<652	High	Chemicals, gas
Hydro-pyrolysis[c]	<10 s	<502	High	Bio-oil
Methano-pyrolysis[d]	<10 s	>702	High	Chemicals
Ultra pyrolysis[e]	<0.5 s	1002	Very high	Chemicals, gas
Vacuum pyrolysis	2–30 s	402	Medium	Bio-oil

[a]Flash-liquid: liquid obtained from flash pyrolysis accomplished in a time of <1 s.
[b]Flash-gas: gaseous material obtained from flash pyrolysis within a time of <1 s.
[c]Hydropyrolysis: pyrolysis with water.
[d]Methanopyrolysis: pyrolysis with methanol.
[e]Ultra pyrolysis: pyrolysis with very high degradation rate.
Source: Adapted from Panwar, N.L., Richa Kothari, V.V. Tyagi. 2012. Thermo chemical conversion of biomass – Eco friendly energy routes, Renewable and Sustainable Energy Reviews, Volume 16, Issue 4, Pages 1801–1816.

Table 17.15. Characteristics of some pyrolysis processes

Pyrolysis process	Residence time	Heating rate	Final temperature (°C)	Products
Carbonization	Days	Very low	400	Charcoal
Conventional	5–30 min	Low	600	Char, bio-oil, gas
Fast	<2 s	Very high	~500	Bio-oil
Flash	<1 s	High	<650	Bio-oil, chemicals, gas
Ultra-rapid	<0.5 s	Very high	~1000	Chemicals, gas
Vacuum	2–30 s	Medium	400	Bio-oil
Hydropyrolysis	<10 s	High	<500	Bio-oil
Methano-pyrolysis	<10 s	High	>700	Chemicals

Source: Basu, Prabir. 2010. Chapter 2 - Biomass Characteristics, Biomass Gasification and Pyrolysis, (Boston, Academic Press), Pages 27–63.

Table 17.16. Exothermic reactions on pyrolysis of cellulose

Process	Reaction	Enthalpy, kJ/g-mol carbon converted at[a]	
		300 K	1000 K
Methanation	$CO + 3H_2 \longrightarrow CH_4 + H_2O$	−205	−226
	$CO_2 + 4H_2 \longrightarrow CH_4 + 2H_2O$	−167	−192
Methanol formation	$CO + 2H_2 \longrightarrow CH_3OH$	−92	−105
	$CO_2 + 3H_2 \longrightarrow CH_3OH + H_2O$	−50	−71
Char formation	$0.17C_6H_{10}O_5 \longrightarrow C + 0.85H_2O$	−81	−80
Water gas shift	$CO + H_2O \longrightarrow CO_2 + H_2$	−42	−33

[a]The standard enthalpy of formation of cellulose was calculated from its heat of combustion.
Source: Klass, Donald L. 1998. Biomass for Renewable Energy, Fuels, and Chemicals, (San Diego, Academic Press).

Earth Science

Figures

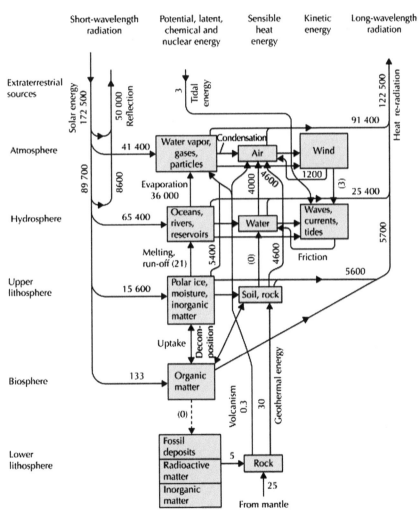

Figure 18.1. Schematic diagram of the Earth's energy flows without anthropogenic interference. The flows are in TW (10^{12} W). Numbers in parentheses are uncertain or rounded off. Sørensen, Bent. 2011. Renewable Energy (Fourth Edition), (Boston, Academic Press).

Handbook of Energy. http://dx.doi.org/10.1016/B978-0-08-046405-3.00018-8

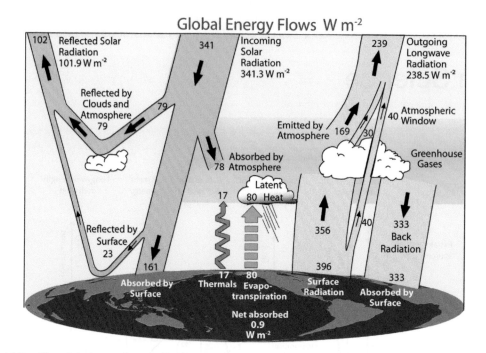

Figure 18.2. The global annual mean Earth's energy budget for the March 2000 to May 2004 period (W m⁻²). The broad arrows indicate the schematic flow of energy in proportion to their importance.
Source: Trenberth, Kevin E., John T. Fasullo, Jeffrey Kiehl. 2009. Earth's Global Energy Budget, Bulletin of the American Meteorological Society, Volume 90, Issue 3, 311-323.

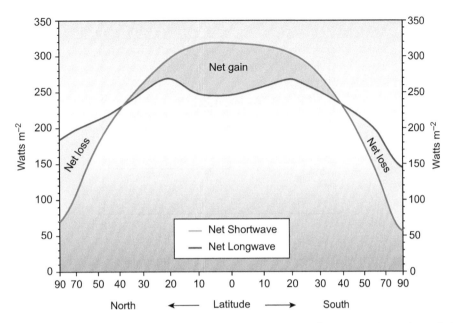

Figure 18.3. Earth's radiation balance. From 0 - 30° latitude North and South incoming solar radiation exceeds outgoing terrestrial radiation and a surplus of energy exists. The reverse holds true from 30 - 90° latitude North and South and these regions have a deficit of energy. Surplus energy at low latitudes and a deficit at high latitudes results in energy transfer from the equator to the poles. It is this meridional transport of energy that causes atmospheric and oceanic circulation. If there were no energy transfer the poles would be 25° Celsius cooler, and the equator 14° Celsius warmer.
Source: Adapted from Pidwirny, Michael, Energy balance of Earth, Encyclopedia of Earth, <http://www.eoearth.org/article/Energy_balance_of_Earth>.

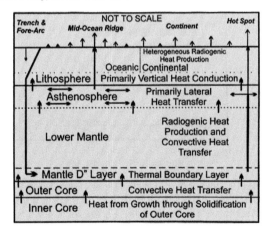

Figure 18.4. Schematic representation of heat sources and transport in the Earth and heat loss at the surface. The two primary heat sources are radiogenic heat production (~80%) and secular cooling of the earth (~20%). Lithospheric extension, magmatism, and upward water flow increase surface heat flow by advection. Subduction, compression, and downward water flow decrease surface heat flow unless accompanied by magmatism. Changes in surface temperature associated with climate change, erosion, and sedimentation cause transient changes in surface heat flow.
Source: Morgan, Paul. 2003. Heat Flow, In: Robert A. Meyers, Editor-in-Chief, Encyclopedia of Physical Science and Technology (Third Edition), (New York, Academic Press), Pages 265-278.

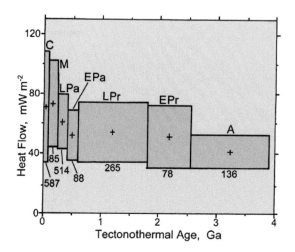

Figure 18.5. Average continental heat flow versus tectonothermal age in billions of years (Ga), the age of the last major tectonic or magmatic event at the heat flow site. Data are grouped in geological age ranges: C–Cenozoic, M–Mesozoic, LPa–Late Paleozoic, EPa–Early Paleozoic, LPr–Late Proterozoic, EPr–Early Proterozoic, A–Archean. Crosses are plotted at the mean heat flow and the midpoint of the age range. Box widths indicate age range and box heights indicate ± one standard deviation of the heat flow data about the mean. Numbers below the boxes indicate the number of data in each group.
Source: Morgan, Paul. 2003. Heat Flow, In: Robert A. Meyers, Editor-in-Chief, Encyclopedia of Physical Science and Technology (Third Edition), (New York, Academic Press), Pages 265-278.

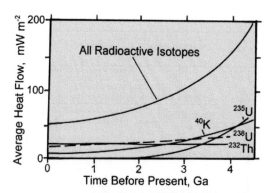

Figure 18.6. Average surface heat flow in mW m^{-2} from the Earth generated from each of the major radiogenic heat producing isotopes, and all of the isotopes, as a function of time before present in billions of years (Ga). An additional 20% of heat flow from the modern Earth is estimated to come from gradual cooling of the earth.
Source: Morgan, Paul. 2003. Heat Flow, In: Robert A. Meyers, Editor-in-Chief, Encyclopedia of Physical Science and Technology (Third Edition), (New York, Academic Press), Pages 265-278.

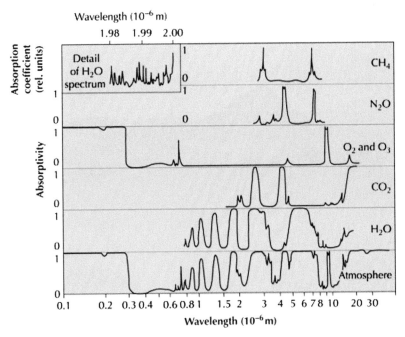

Figure 18.7. Spectral absorption efficiency of selected gaseous constituents of the atmosphere, and for the atmosphere as a whole (bottom).
Source: Sørensen, Bent. 2011. Renewable Energy (Fourth Edition), (Boston, Academic Press).

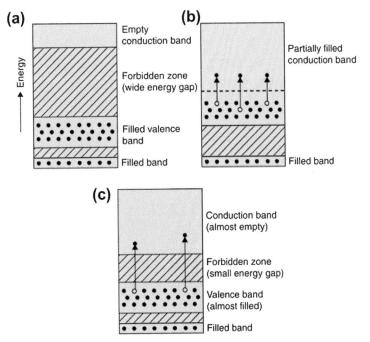

Figure 18.8. Schematic summary of the water cycle.
Source: Sørensen, Bent. 2011. Renewable Energy (Fourth Edition), (Boston, Academic Press).

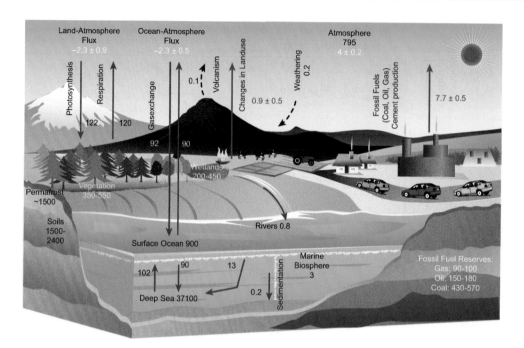

Figure 18.9. The global carbon cycle.
Source: Adapted from Intergovernmental Panel on Climate Change.

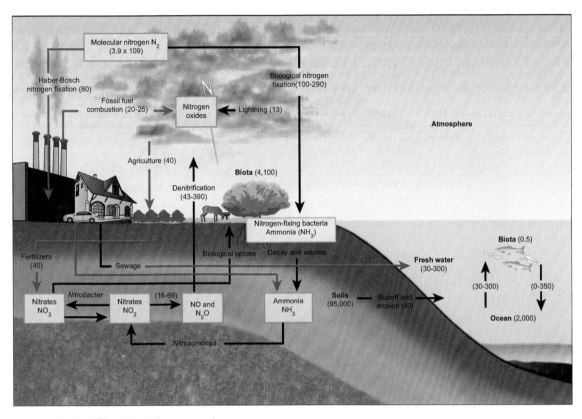

Figure 18.10. The global nitrogen cycle.
Source: Adapted from Data from Kaufmann, Robert K. and Cleveland, Cutler J. 2007. Environmental Science (McGraw-Hill, Dubuque, IA).

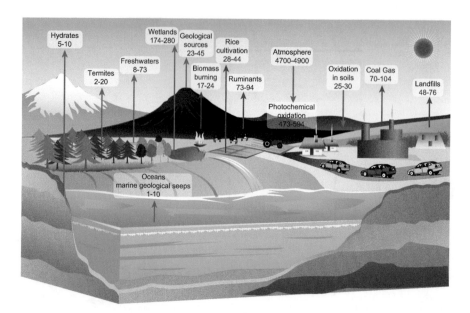

Figure 18.11.   The global methane cycle.
Source: Adapted from Intergovernmental Panel on Climate Change.

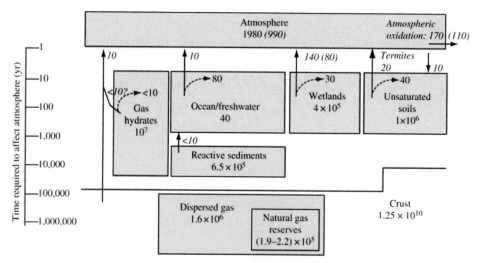

Figure 18.12.   Reservoirs (Tg CH$_4$ or Tg C) and fluxes (Tg CH$_4$ yr^{-1}) of the natural, pre-industrial methane carbon subcycle. Values for glacial periods, where available, are shown in parentheses. Values for reactive sediments, wetlands, unsaturated soils, and crustal reservoirs represent organic carbon that might be converted to methane, and are all given as Tg C. The vertical bar on the left shows the approximate time (in years) necessary for the different reservoirs to affect the atmosphere. Estimates of methane consumption within reservoirs are shown as dashed arrows. Gross production of methane within a reservoir can be calculated by adding the flux to the atmosphere and the consumption value. The flux and consumption values are rounded to the nearest 10 Tg CH$_4$ or Tg C.
Source: Sundquist, E.T. and K. Visser. 2003. The Geologic History of the Carbon Cycle, In: Heinrich D. Holland and Karl K. Turekian, Editor(s)-in-Chief, Treatise on Geochemistry, (Oxford, Pergamon), Pages 425-472.

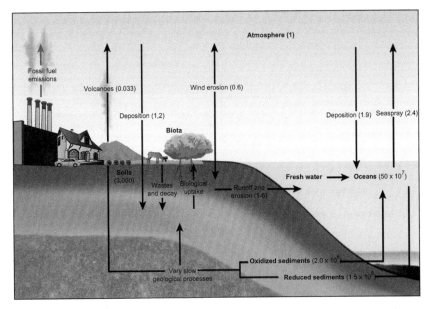

Figure 18.13.   Global sulfur cycle.
Source: Adapted from Data from Kaufmann, Robert K. and Cleveland, Cutler J. 2007. Environmental Science (McGraw-Hill, Dubuque, IA).

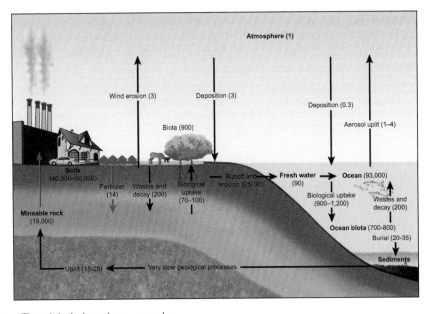

Figure 18.14.   The global phosphorous cycle.
Source: Adapted from Data from Kaufmann, Robert K. and Cleveland, Cutler J. 2007. Environmental Science (McGraw-Hill, Dubuque, IA).

Charts

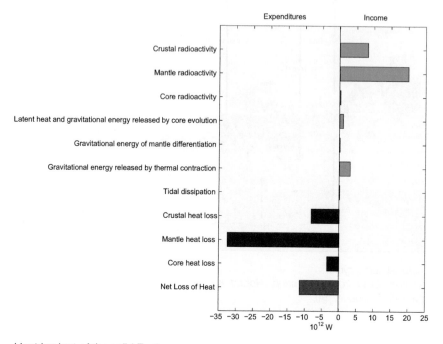

Chart 18.1.   Heat budget of the solid Earth.
Source: Data from Stacey, F.D. and P.M. Davis. 2008. Physics of the Earth, 4th ed., (Cambridge University Press, Cambridge, UK).

Tables

Table 18.1. Energy in the global atmosphere

Name	Symbol	Amount × 10^6 J m^{-2}	% of total
Internal energy	IE	1800	70
Potential energy	PE	700	27
Latent energy	LH	70	2.7
Kinetic energy	KE	1.3	0.05
Total energy	IE + PE + LH + KE	2571	100

Source: Adapted from Hartmann, Dennis L. 1994. Global Physical Climatology, (San Diego, Academic Press).

Table 18.2. Comparison of global energies in the solid Earth

	Source of Energy	Energy (10^{30} J)
1	Accretion of a homogeneous mass	219.0
2	Core separation less strain energy	13.9
3	Inner core formation	0.09
4	Mantle differentiation	0.03
5	Elastic strain	15.8
6	Radiogenic heat in 4.5×10^9 years	7.6
7	Residual stored heat	13.3
8	Heat loss in 4.5×10^9 years	13.4
9	Present rotational energy	0.2
10	Tidal dissipation in 4.5×10^9 years	~1.1

The first four entries give gravitational energy release with the resulting elastic strain energies subtracted. Strain enegy is listed seperately, so that gravitational energy is the sum of items 1-5. Tidal dissipation occurs mainly in the sea and does not influence the thermal state of the solid Earth. Source: Adapted from Stacey, F.D. and P.M. Davis. 2008. Physics of the Earth, 4th ed., (Cambridge University Press, Cambridge, UK).

Table 18.3. Surface components of the annual mean energy budget for the Earth

Global	Solar absorbed	Net solar	Solar reflected	LH evaporation	SH	Radiation up	Back radiation	Net LW	Net down
ISCCP-FD	70.8	165.7	22.8			393.9	345.4	48.5	
NRA	64.4	160.4	45.2	83.1	15.6	396.9	336.5	60.4	1.3
JRA	74.7	169.8	25.6	90.2	19.4	396.9	324.1	72.8	−12.6
Trenbeth et al. (2009)	78.2	161.2	23.1	80	17	396	333	63	0.9

Land	Solar absorbed	Net solar	Solar reflected	LH evaporation	SH	Radiation up	Back radiation	Net LW	Net down
ISCCP-FD	70.6	148.7	40.1			381.2	327.6	53.6	
NRA	59.1	155.1	70.3	50.2	26.3	371	296.8	74.1	4.5
JRA	71.9	155.8	51.6	39.4	27.4	374.4	287.4	87	2
Trenbeth et al. (2009)	78	145.1	39.6	38.5	27	383.2	303.6	79.6	0

Ocean	Solar absorbed	Net solar	Solar reflected	LH evaporation	SH	Radiation up	Back radiation	Net LW	Net down
ISCCP-FD	70.8	172	16.3			398.7	352	46.7	9.7
NRA	66.3	162.3	36.2	95	11.7	406.2	350.8	55.4	0.2
JRA	75.6	174.9	16.2	108.5	16.6	405	337.3	67.7	−17.9
HOAPS				103.6	14.6			56.1	
WHOI				93.8	10.8				
Trenbeth et al. (2009)	78.2	167.8	16.6	97.1	12	400.7	343.3	57.4	1.3

Surface components of the annual mean energy budget for the globe, global land, and global ocean, except for atmospheric solar radiation absorbed (Solar absorb, left column), for the clouds and the earth's radiant energy system period of Mar 2000 to May 2004 (W m⁻²). Included are the solar absorbed at the surface (Solar down), reflected solar at the surface (Solar reflected), surface latent heat from evaporation (LH evaporation), sensible heat (SH), LW radiation up at the surface (Radiation up), LW downward radiation to the surface (Back radiation), net LW (Net LW), (net long wave) and net energy absorbed at the surface (NET down). ISCCP = International Satellite Cloud Climatology Project; NRA = reanalysis of NCAR (National Centers for Atmospheric Research) and NCEP (National Center for Environmental Prediction); JRA = Japanese reanalysis; HOAPS = Hamburg Ocean Atmosphere Parameters and Fluxes from Satellite Data; WHOI = Woods Hole Oceanographic Institution. HOAPS version 3 covers 80°S–80°N and is for 1988 to 2005. For the ocean, the ISCCP-FD is combined with HOAPS to provide a NET value.

Source: Adapted from Trenbeth, Kevin E., John T. Fasullo, Jeffrey Kiehl, 2009: Earth's global energy budget. Bull. Amer. Meteor. Soc., 90, 311–323.

Table 18.4. Annual energy balance of the oceans and continents (W m^{-2})

Area	R$_s$	LH	SH	Δ F$_{eo}$	SH/LH
Europe	52	32	20	0	0.62
Asia	62	29	33	0	1.14
North America	53	30	22	0	0.74
South America	93	60	33	0	0.56
Africa	90	34	56	0	1.61
Australia	93	29	64	0	2.18
Antarctica	−16	0	−15	0	-
All land	65	33	32	0	0.96
Atlantic Ocean	109	95	11	3	0.11
Indian Ocean	113	102	9	1	0.09
Pacific Ocean	114	103	11	0	0.10
Arctic Ocean	−5	7	−7	−5	−1.00
All Oceans	109	98	11	0	0.11

R$_s$ = net radiative flux at surface; LH = latent heat flux; SH = sensible heat flux; ΔF$_{eo}$ = horizontal flux below surface.
Source: Adapted from Hartmann, Dennis L. 1994. Global Physical Climatology, (San Diego, Academic Press).

Table 18.5. Estimates of global primary exergy sources

Primary exergy source	Magnitude MJ/yr
Solar radiation[a]	3.33×10^{18} to 3.83×10^{18}
Moon gravity	$7.6E \times 10^{13}$ to 1.17×10^{14}
Geothermal heat	9.78×10^{14} to 1.98×10^{15}

[a]The exergy input of net solar radiation into the Earth system.
Source: Adapted from Liao, Wenjie, Reinout Heijungs, Gjalt Huppes. 2012. Natural resource demand of global biofuels in the Anthropocene: A review, Renewable and Sustainable Energy Reviews, Volume 16, Issue 1, Pages 996–1003.

Table 18.6. Physical properties of air at various temperatures

Temperature (K)	Density (kg/m³)	Dynamic Viscosity, $\mu.10^7$ (N.s/m²)	Kinematic Viscosity, $\gamma.10^6$ (m²/s)	Thermal Conductivity, $K_g.10^3$ (W/m.K)	Thermal Diffusivity, $\alpha.10^6$ (m²/s)	Prandtl Number[a]
100	3.5562	71.1	2	9.34	2.54	0.786
150	2.3364	103.4	4.426	13.8	5.84	0.758
200	1.7458	132.5	7.59	18.1	10.3	0.737
250	1.3947	159.6	11.44	22.3	15.9	0.72
300	1.1614	184.6	15.89	26.3	22.5	0.707
350	0.995	208.2	20.92	30	29.9	0.7
400	0.8711	230.1	26.41	33.8	38.3	0.69
450	0.774	250.7	32.39	37.3	47.2	0.686
500	0.6964	270.1	38.79	40.7	56.7	0.684
550	0.6329	288.4	45.57	43.9	66.7	0.683
600	0.5804	305.8	52.69	46.9	76.9	0.685
650	0.5356	322.5	60.21	49.7	87.3	0.69
700	0.4975	338.8	68.1	52.4	98	0.695
750	0.4643	354.6	796.37	54.9	109	0.702
800	0.4354	369.8	84.93	57.3	120	0.709
850	0.4097	384.3	93.8	59.6	131	0.716
900	0.3868	398.1	102.9	62	143	0.72
950	0.3666	411.3	112.2	64.3	155	0.723
1000	0.3482	424.4	121.9	66.7	168	0.726
1100	0.3166	449	141.8	71.5	195	0.728
1200	0.2902	473	162.9	76.3	224	0.728
1300	0.2679	496	185.1	82	238	0.719
1400	0.2488	530	213	91	303	0.703
1500	0.2322	557	240	100	350	0.685
1600	0.2177	584	268	106	390	0.688

[a]The ratio of momentum diffusivity (kinematic viscosity) to thermal diffusivity.

Table 18.7. Key constituents of air in the troposphere and stratosphere

Constituent	Fraction by volume in (or relative to) dry air	Significant absorption bands
N_2	78.10%	---
O_2	20.90%	UV-C, MW near 60 and 118 GHz, weak bands in VIS and IR
H_2O	(0–2%)	numerous strong bands throughout IR; also in MW, especially near 183 GHz
Ar and other inert gases	0.94%	---
CO_2	370 ppm	near 2.8, 4.3, and 15 μm
CH_4	1.7 ppm	near 3.3 and 7.8 μm
N_2O	0.35 ppm	4.5, 7.8, and 17 μm
CO	0.07 ppm	4.7 μm (weak)
O_3	$\sim 10^{-8}$	UV-B, 9.6 μm
$CFCl_3$, CF_2Cl_2, etc	$\sim 10^{-10}$	IR

Source: Adapted from Petty, Grant. 2006. Atmospheric Radiation (Second edition), (Madison, Wisconsin, Sundog Publishing).

Table 18.8. Standard atmosphere table

Height (m)	Temperature (C)	Pressure (hPa)	Density (kg/m3)
0	15	1013	1.2
1,000	8.5	900	1.1
2,000	2	800	1
3,000	−4.5	700	0.91
4,000	−11	620	0.82
5,000	−17.5	540	0.74
6,000	−24	470	0.66
7,000	−30.5	410	0.59
8,000	−37	360	0.53
9,000	−43.5	310	0.47
10,000	−50	260	0.41
11,000	−56.5	230	0.36
12,000	−56.5	190	0.31
13,000	−56.5	170	0.27
14,000	−56.5	140	0.23
15,000	−56.5	120	0.19
16,000	−56.5	100	0.17
17,000	−56.5	90	0.14
18,000	−56.5	75	0.12
19,000	−56.5	65	0.1
20,000	−56.5	55	0.088
21,000	−55.5	47	0.075
22,000	−54.5	40	0.064
23,000	−53.5	34	0.054
24,000	−52.5	29	0.046
25,000	−51.5	25	0.039

Continued

Table 18.8. Continued

Height (m)	Temperature (C)	Pressure (hPa)	Density (kg/m3)
26,000	−50.5	22	0.034
27,000	−49.5	18	0.029
28,000	−48.5	16	0.025
29,000	−47.5	14	0.021
30,000	−46.5	12	0.018
31,000	−45.5	10	0.015
32,000	−44.5	8.7	0.013
33,000	−41.7	7.5	0.011
34,000	−38.9	6.5	0.0096
35,000	−36.1	5.6	0.0082

Source: International Organization for Standardization.

Table 18.9. Alternative measures of greenhouse gases

1 pound methane, measured in carbon units (CH_4) = 1.333 pounds methane, measured at full molecular weight (CH_4)
1 pound carbon dioxide, measured in carbon units (CO_2-C) = 3.6667 pounds carbon dioxide, measured at full molecular weight (CO_2)
1 pound carbon monoxide, measured in carbon units (CO-C) = 2.333 pounds carbon monoxide, measured at full molecular weight (CO)
1 pound nitrous oxide, measured in nitrogen units (N_2O-N) = 1.571 pounds nitrous oxide, measured at full molecular weight (N_2O)

Source: Oak Ridge National Laboratory, Transportation Energy Data Book: Edition 30, <http://cta.ornl.gov/data/index.shtml>, accessed 11 June 2012.

Table 18.10. Basic parameters in terrestrial heat flow

Parameter	Working unit	Typical range
Vertical temperature gradient	°C/km ($mK\ m^{-1}$)	5–50
Thermal conductivity	$W\ m^{-1}\ K^{-1}$	1–4
Heat production	$\mu W\ m^{-3}$	0–8
Heat flow	$mW\ m^{-2}$	5–125

Source: Morgan, Paul. 2003. Heat Flow, In: Robert A. Meyers, Editor-in-Chief, Encyclopedia of Physical Science and Technology (Third Edition), (New York, Academic Press), Pages 265–278.

Table 18.11. **Major modes of heat loss from the Earth**

Component	Value
Heat loss though the continents	1.2×10^{13} W
Heat loss through the oceans	3.1×10^{13} W
Total-continent and oceans	4.2×10^{13} W
Heat loss by hydrothermal circulation	1.0×10^{13} W
Heat lost in plate creation	2.6×10^{13} W
Mean heat flow	
Continents	50 mW m^{-2}
Oceans	100 mW m^{-2}
Global	84 mW m^{-2}
Convective heat transport by surface plates[a]	~65% heat loss
Radioactive decay in crust	~17% heat loss

[a]Includes lithospheric creation in oceans and magmatic activity in continents.
Source: Morgen, Paul. 2003. Heat Flow, In: Robert A. Meyers, Editor-in-Chief, Encyclopedia of Physical Science and Technology (Third Edition), (New York, Academic Press), Pages 265–278.

Table 18.12. **Thermal resistance of stagnant air and surface resistance (m^2-K/W)**

	Heat flow direction		
Thickness of air (mm)	Sideways	Up	Down
5	0.11	0.11	0.11
7	0.13	0.13	0.13
10	0.15	0.15	0.15
15	0.17	0.16	0.17
25	0.18	0.16	0.19
50	0.18	0.16	0.21
100	0.18	0.16	0.22
300	0.18	0.16	0.23
Surface resistance			
Internal surface	0.12	0.11	0.16
External surface			0.044

Source: Kalogirou, Soteris A. 2009. Solar Energy Engineering, (Boston, Academic Press).

Table 18.13.   Average radiogeneic heat in geological materials[a]

	Material	Concentration (parts per million by mass)				Heat Production 10^{-12} W kg^{-1}
		U	Th	K	K/U	
Igneous rocks	Granites	4.6	18	33000	7000	1050
	Tholeitic basalts	0.75	2.5	12000	16000	180
	Alkali basalts	0.11	0.4	1500	13600	27
	Eclogites	0.035	0.15	500	1400	9.2
	Peridoites, dunites	0.006	0.02	100	17000	1.5
	Ordinary chrondites	0.020	0.070	400	20000	5.2
Meteorites	Carbonaceous chondrites	0.015	0.046	900	60000	5.8
	Iron meteorites	nil	nil	nil	-	$<3 \times 10^{-4}$
Moon	Apollo samples	0.23	0.85	590	2500	47
Global averages	Crust (2.8×10^{22} kg)	1.2	4.5	15500	13000	293
	Mantle (4.0×10^{24} kg)	0.025	0.087	70	2800	5.1
	Core	nil	nil	29	-	0.1
	Whole Earth	0.022	0.081	118	5400	4.7

[a]The total heat flux per unit mass of the Earth is 7.4×10^{-12} Wkg^{-1}.
Source: Adapted from Stacey, F.D. and P.M. Davis. 2008. Physics of the Earth, 4th ed., (Cambridge University Press, Cambridge, UK).

Table 18.14.   Radiative properties of natural surfaces

Surface type	Other specifications	Albedo[a]	Emissivity[b]
Water	Small zenith angle	0.03–0.10	0.92–0.97
	Large zenith angle	0.10–1.00	0.92–0.97
Snow	Old	0.40–0.70	0.82–0.89
	Fresh	0.45–0.95	0.90–0.99
Ice	Sea	0.30–0.45	0.92–0.97
	Glacier	0.20–0.40	
Bare sand	Dry	0.35–0.45	0.84–0.90
	Wet	0.20–0.30	0.91–0.95
Bare soil	Dry clay	0.20–0.40	0.95
	Moist clay	0.10–0.20	0.97
	Wet fallow field	0.05–0.07	
Paved	Concrete	0.17–0.27	0.71–0.88
	Black gravel road	0.05–0.10	0.88–0.95
Grass	Long (1 m)	0.16	0.90
	Short (0.02 m)	0.26	0.95
Agricultural	Wheat, rice, etc.	0.18–0.25	0.90–0.99
	Orchards	0.15–0.20	0.90–0.95
Forests	Deciduous	0.10–0.20	0.97–0.98
	Coniferous	0.05–0.15	0.97–0.99

[a]the ratio of reflected radiation from a surface to incident radiation upon it.
[b]the relative ability of its surface to emit energy by radiation, measured by the ratio of energy radiated by a particular material to energy radiated by a black body at the same temperature.
Source: Adapted from Arya, S. Pal. 2001. Introduction to Micrometeorology (Second Edition). (San Diego, Academic Press).

Table 18.15. Infared emissivities (%) of some surfaces[a]

Water and soil surfaces		Vegetation	
Water	92–96	Alfalfa, dark green	95
Snow, fresh fallen	82–99.5	Oak leaves	91–95
Snow, ice granules	89	Leaves and plants	
Ice	96	0.8 μm	5–53
Soil, frozen	93–94	1.0 μm	5–60
Sand, dry playa	84	2.4 μm	70–97
Sand, dry light	89–90	10.0 μm	97–98
Sand, wet	95		
Gravel, coarse	91–92	**Miscellaneous**	
Limestone, light gray	91–92	Paper, white	89–95
Concrete, dry	71–88	Glass pane	87–94
Ground, moist, bare	95–98	Bricks, red	92
Ground, dry plowed	90	Plaster, white	91
		Wood, planed oak	90
Natural surfaces		Paint, white	91–95
Desert	90–91	Paint, black	88–95
Grass, high dry	90	Paint, aluminum	43–55
Field and shrubs	90	Aluminum foil	1–5
Oak woodland	90	Iron, galvinized	13–28
Pine forest	90	Silver, highly polished	2
		Skin, human	~5

[a]Emissivity is the relative ability of its surface to emit energy by radiation, measured by the ratio of energy radiated by a particular material to energy radiated by a black body at the same temperature.
Source: Adapted from Hartmann, Dennis L. 1994. Global Physical Climatology, (San Diego, Academic Press).

Table 18.16. Albedos for various surfaces (%)[a]

Surface type	Range	Typical value
Water		
Deep water: low wind, low altitude	5–10	7
Deep water: high wind, high altitude	10–20	12
Bare surfaces		
Moist dark soil, high humus	5–15	10
Moist gray soil	10–20	15
Dry soil, desert	20–35	30
Wet sand	20–30	25
Dry light sand	30–40	35
Asphalt pavement	5–10	7
Concrete pavement	15–35	20
Vegetation		
Short green vegetation	10–20	17
Dry vegetation	20–30	25
Coniferous forest	10–15	12
Deciduous forest	15–25	17
Snow and ice		
Forest with surface snowcover	20–35	25
Sea ice, no snowcover	25–40	30
Old, melting snow	35–65	50
Dry, cold snow	60–75	70
Fresh, dry snow	70–90	80

[a]Albedo is the ratio of reflected radiation from a surface to incident radiation upon it.
Source: Adapted from Hartmann, Dennis L. 1994. Global Physical Climatology, (San Diego, Academic Press).

Table 18.17.   **Top of atmosphere annual mean radiation budget**

Global	Solar in	Solar reflected	Albedo (%)	ASR	OLR	NET down
ISCCP-FD	341.7	105.2	30.8	236.5	235.6	0.9
NRA	341.8	117	34.2	224.5	237.8	−13
JRA	339.1	94.6	27.9	244.5	253.6	−9.1
Trenbeth et al. (2009)	341.3	101.9	29.8	239.4	238.5	0.9
Land	Solar in	Solar reflected	Albedo (%)	ASR	OLR	NET down
ISCCP-FD	330.9	111.6	33.7	219.3	231.3	−12
NRA	330.6	116.4	35.2	214.2	234.7	−20.5
JRA	328.3	100.6	30.6	227.7	250.8	−23.1
Trenbeth et al. (2009)	330.2	113.4	34.4	216.8	232.4	−15.6
Ocean	Solar in	Solar reflected	Albedo (%)	ASR	OLR	NET down
ISCCP-FD	345.7	102.9	29.8	242.8	237.2	5.6
NRA	345.9	117.3	33.9	228.7	238.9	−10.2
JRA	343	92.5	27	250.5	254.7	−4.2
Trenbeth et al. (2009)	345.4	97.8	28.3	247.7	240.8	6.9

Top of atmosphere annual mean radiation budget quantities for the clouds and the earth's radiant energy system period of Mar 2000 to May 2004 for global, global land, and global ocean. The downward solar (Solar in), reflected solar (Solar reflected), and net (NET down) radiation are given with the absorbed solar radiation (ASR) and outgoing longwave radiation (OLR)(W m^{-2}), and albedo is given in percent. ISCCP = International Satellite Cloud Climatology Project; NRA = reanalysis of NCAR (National Centers for Atmospheric Research) and NCEP (National Center for Environmental Prediction); JRA = Japanese reanalysis.
Source: Adapted from Trenberth, Kevin E., John T. Fasullo, Jeffrey Kiehl, 2009: Earth's global energy budget. Bull. Amer. Meteor. Soc., 90, 311–323.

Table 18.18.   Direct normal solar irradiance at air mass 1.5[a]

λ (μm)	E (W/m²-μm)	λ (μm)	E (W/m²-μm)	λ (μm)	E (W/m²-μm)
0.305	3.4	0.71	1002.4	1.35	30.1
0.31	15.8	0.718	816.9	1.395	1.4
0.315	41.1	0.724	842.8	1.4425	51.6
0.32	71.2	0.74	971	1.4625	97
0.325	100.2	0.7525	956.3	1.477	97.3
0.33	152.4	0.7575	942.2	1.497	167.1
0.335	155.6	0.7625	524.8	1.52	239.3
0.34	179.4	0.7675	830.7	1.539	248.8
0.345	186.7	0.78	908.9	1.558	249.3
0.35	212	0.8	873.4	1.578	222.3
0.36	240.5	0.816	712	1.592	227.3
0.37	324	0.823	660.2	1.61	210.5
0.38	362.4	0.8315	765.5	1.63	224.7
0.39	381.7	0.84	799.8	1.646	215.9
0.4	556	0.86	815.2	1.678	202.8
0.41	656.3	0.88	778.3	1.74	158.2
0.42	690.8	0.905	630.4	1.8	28.6
0.43	641.9	0.915	565.2	1.86	1.8
0.44	798.5	0.925	586.4	1.92	1.1
0.45	956.6	0.93	348.1	1.96	19.7
0.46	990	0.937	224.2	1.985	84.9
0.47	998	0.948	271.4	2.005	25
0.48	1046.1	0.965	451.2	2.035	92.5
0.49	1005.1	0.95	549.7	2.065	56.3
0.5	1026.7	0.9935	630.1	2.1	82.7
0.51	1066.7	1.04	582.9	2.148	76.5
0.52	1011.5	1.07	539.7	2.198	66.4
0.53	1084.9	1.1	366.2	2.27	65
0.54	1082.4	1.12	98.1	2.36	57.6
0.55	1102.2	1.13	169.5	2.45	19.8
0.57	1087.4	1.137	118.7	2.494	17
0.59	1024.3	1.161	301.9	2.537	3
0.61	1088.8	1.18	406.8	2.941	4
0.63	1062.1	1.2	375.2	2.973	7
0.65	1061.7	1.235	423.6	3.005	6
0.67	1046.2	1.29	365.7	3.056	3
0.69	859.2	1.32	223.4	3.132	5

[a]Direct normal irradiance is the amount of solar radiation received per unit area by a surface that is always held perpendicular (or normal) to the rays that come in a straight line from the direction of the sun at its current position in the sky. The air mass (AM) coefficient defines the direct optical path length through the Earth's atmosphere, expressed as a ratio relative to the path length vertically upwards, i.e. at the zenith. "AM1.5" is a common reference in solar energy research and technology.

λ = wavelength; E = emissive power.

Source: National Renewable Energy Laboratory.

Table 18.19. Temperature and solar irradiation for select cities in the world

Country	Reference climate	Horizontal irradiation [kWh/m²/year]	Avg. Outside air temperature [°C]
Albania	Tirana	1604.3	13.5
Australia	Sydney	1674	18.1
Austria	Graz	1126	9.2
Barbados	Grantley Adams	2016.3	27.4
Belgium	Brussels	971.1	10
Brazil	Brasília	1792.5	22
Bulgaria	Sofia	1187.5	10.1
Canada	Montreal	1351.4	6.9
Chile	Santiago de Chile	1752.7	14.5
China	Shanghai	1281.9	17.1
Cyprus	Nicosia	1885.5	19.9
Czech Republic	Praha	998.4	7.9
Denmark	Copenhagen	989.4	8.1
Estonia	Tallin	960.2	5.3
Finland	Helsinki	948	4.6
France	Paris	1112.4	11
FYRM	Skopje	1380.8	12.5
Germany	Würzburg	1091.3	9.5
Greece	Athens	1584.6	18.5
Hungary	Budapest	1198.7	11
India	New-Delhi	1960.5	24.7
Ireland	Dublin	948.7	9.5
Israel	Jerusalem	2198	17.3
Italy	Bologna	1419	14.3
Japan	Tokyo	1175.2	16.7
Jordan	Amman	2145.4	17.9
Korea, South	Seoul	1161.1	12.7
Latvia	Riga	991.2	6.3
Lithuania	Vilnius	1001.2	6.2
Luxembourg	Luxembourg	1037.4	8.4
Malta	Luqa	1901.9	18.7
Mexico	Mexico City	1706.3	16.6
Namibia	Windhoek	2363	21
Netherlands	Amsterdam	999	10
New Zealand	Wellington	1401.2	13.6
Norway	Oslo	971.1	5.8
Poland	Warsaw	1024.2	8.1
Portugal	Lisbon	1686.4	17.4
Romania	Bucharest	1324.3	10.6
Slovakia	Bratislava	1213.8	10.3
Slovenia	Ljubjana	1114.6	9.8
South Africa	Johannesburg	2075.1	15.6
Spain	Madrid	1643.5	15.5
Sweden	Gothenburg	933.9	7.2
Switzerland	Zürich	1093.8	9.6

Continued

Table 18.19.   Continued

Country	Reference climate	Horizontal irradiation [kWh/m²/year]	Avg. Outside air temperature [°C]
Taiwan	Taipei	1372.2	20.8
Thailand	Bangkok	1764.8	29.1
Tunisia	Tunis	1808.2	19.3
Turkey	Ankara	1700.9	12
United Kingdom	London	942.6	12
United States	Indianapolis	1,492.3	11.3
	Los Angeles	1,799.80	17.2
Uruguay	Montevideo	1534.2	15.9
Zimbabwe	Harare	2017.1	18.9

Source: Weiss, Werner and Franz Mauthner. 2012. Solar Heat Worldwide, Markets and Contribution to the Energy Supply 2010, Edition 2012, (Gleisdorf, AEE - Institute for Sustainable Technologies; Paris, Solar Heating and Cooling Programme, International Energy Agency).

Table 18.20.   Roughness classification[a]

Surface type	Roughness length (m)
Sea, loose sand, and snow	≈0.0002
Concrete, flat desert, tidal flat	0.0002–0.0005
Flat snow field	0.0001–0.0007
Rough ice field	0.001–0.012
Fallow ground	0.001–0.004
Short grass and moss	0.008–0.03
Long grass and heather	0.02–0.06
Low mature agricultural crops	0.04–0.09
Low mature crops ("grain")	0.12–0.18
Continuous bushland	0.35–0.45
Mature pine forest	0.8–1.6
Dense low buildings ("suburb")	0.4–0.7
Large town	0.7–1.5
Tropical forest	1.7–2.3

[a]The roughness length is the height above the displacement plane at which the mean wind becomes zero when extrapolating the logarithmic wind-speed profile downward through the surface layer. It is a theoretical height that must be determined from the wind-speed profile, although there has been some success at relating this height to the arrangement, spacing, and physical height of individual roughness elements such as trees or houses.
Source: Adapted from Cataldo, J. and Zeballos, M. 2009. Roughness terrain consideration in a wind interpolation numerical model, Paper presented at 11th Americas Conference on Wind Energy, San Juan, Puerto Rico.

Table 18.21. **Classes of terrain roughness**

Roughness class	Character	Terrain	Obstacles	Farms	Buildings	Forest
0	Sea, lakes	Open water	–	–	–	–
1	Open landscape, with sparse vegetation and buildings	Plain to smooth hills	Only low vegetation	$0–3 \text{ km}^{-2}$	–	–
2	Countryside with a mix of open areas, vegetation, and buildings	Plain to hilly	Small woods, alleys are common	Up to 10 km^{-2}	Some villages and small towns	–
3	Small towns or countryside with many farms woods and obstacles	Plain to hilly	Many woods, vegetation, and alleys	Many farms $> 10 \text{ km}^{-2}$	Many villages, small towns, or suburbs	Low forest
4	Large cities or high forest	Plain to hilly	–	–	Large cities	High forest

Source: Adapted from Wizelius, T. 2012. 2.13 - Design and Implementation of a Wind Power Project, In: Ali Sayigh, Editor-in-Chief, Comprehensive Renewable Energy, (Oxford, Elsevier), Pages 391–430.

Table 18.22. **Classes of wind power density at 10 meters, 30 meters, and 50 meters elevation[a]**

Class	10 m (33 ft)		30 m (98 ft)		50 m (164 ft)	
	Wind power density (W/m²)	Speed[b] m/s (mph)	Wind power density (W/m²)	Speed[b] m/s (mph)	Wind power density (W/m²)	Speed[b] m/s (mph)
1	0–100	0–4.4 (0–9.8)	0–160	0–5.1	0–200	0–5.6 (0–12.5)
2	100–150	4.4–5.1 (9.8–11.5)	160–240	5.1–5.9 (11.4–13.2)	200–300	5.6–6.4 (12.5–14.3)
3	150–200	5.1–5.6 (11.5–12.5)	240–320	5.9–6.5 (13.2–14.6)	300–400	6.4–7.0 (14.2–15.7)
4	200–250	5.6–6.0 (12.5–13.4)	320–400	6.5–7.0 (14.6–15.7)	400–500	7.0–7.5 (15.6–16.8)
5	250–300	6.0–6.4 (13.4–14.3)	400–480	7.0–7.4 (15.7–16.6)	500–600	7.5–8.0 (16.8–17.9)
6	300–400	6.4–7.0 (14.3–15.7)	480–640	7.4–8.2 (16.6–18.3)	600–800	8.0–8.8 (17.9–19.7)
7	400–1000	7.0–9.4 (15.7–21.1)	640–1600	8.2–11.0 (18.3–24.7)	800–2000	8.8–11.9 (19.7–26.6)

[a]Vertical extrapolation of wind speed based on the 1/7 power law.
[b]Mean wind speed is based on Rayleigh speed distribution of equivalent mean wind power density. Wind speed is for standard sea-level conditions. To maintain the same power density, speed increases 3%/1000 m (5%/5000 ft) elevation.
Source: National Renewable Energy Laboratory.

Table 18.23. Geographic areas of high natural radiation background

Country	Area	Characteristics of area	Approximate Population	Absorbed dose rate in air[a] (nGy h[1])[b]
Brazil	Guarapari	Monazite sands; coastal areas	73,000	90–170 (streets) 90–90,000 (beaches)
	Mincas Gerais and Goias Pocos de Caldas Araxá	Volcanic Intrusives	350	2,800 average
China	Yanjiang Quandong	Monazite particles	80,000	370 average
Egypt	Nile delta	Monazite sands		20–400
France	Central region Southwest	Granitic, schistous, sandstone area Uranium minerals	7,000,000	200–4,000 10–10,000
India	Kerala and Madras Ganges delta	Monazite sands, coastal area 200 km long, 0.5 km wide	100,000	200–4,000 1,800 average 260–440
Iran (Islamic Rep of.)	Ramsar Mahallat	Spring waters	2,000	70–17,000 800–4 000
Italy	Lazio Campania Orvieto town South Toscana	Volcanic soil	5,100,000 5,600,000 21,000 100,000	180 average 200 average 560 average 150–200
Niue Island	Pacific	Volcanic Soil	4,500	1,100 maximum
Switzerland	Tessin, Alps, Jura	Gneiss, verucano, ^{226}Ra in karst soils	300,000	100–200

[a]Includes cosmic and terrestrial radiation.

[b]The Gray (G) is the SI derived unit of absorbed radiation dose of ionizing radiation, and is defined as the absorption of one joule of ionizing radiation by one kilogram of matter (usually human tissue).

Source: United Nations Scientific Committee on the Effects of Atomic Radiation (UNSCEAR), UNSCEAR 2000 Report Vol. 1: Sources and Effects of Ionizing Radiation. United Nations Scientific Committee on the Effects of Atomic Radiation, <http://www.unscear.org/unscear/en/publications/2000_1.html>, Accessed 28 April 2012.

Table 18.24. Radioactive elements in common rocks in the Earth's crust

Rock type	^{40}K Total K (%)	^{40}K Bq/kg	^{87}Rb ppm	^{87}Rb Bq/kg	^{232}Th ppm	^{232}Th Bq/kg	^{238}U ppm	^{238}U Bq/kg
Igneous rocks								
Basalt, crustal average	0.8	300	40	30	3–4	10–15	0.5–1	7–10
Mafic	0.3–1.1	70–400	10–50	1–40	1.6,2.7	7,10	0.5,0.9	7,10
Salic	4.5	1100–1500	170–200	150–180	16,20	60,83	3.9,4.7	50,60
Granite (crustal average)	>4	>1000	170–200	150–180	17	70	3	40
Sedimentary rocks								
Shale,								
Sandstones	2.7	800	120	110	12	50	3.7	40
Clean quatz	<1	<300	<40	<40	<2	<8	<1	<10
Dirty quartz	2?	400?	90?	80?	3–6?	10–25?	2–3?	40?
Arkose	2–3	600–900	80–120	80	2?	<8	1–2?	10–25?
Beach sands (unconsolidated)	<1	<300?	<40?	<40?	6	25	3	40
Carbonate rocks	0.3	70	10	8	2	8	2	25
Continental upper crust (average)	2.8	850	112	100	10.7	44	2.8	36
Soils	1.5	400	65	50	9	37	1–8	66

Question marks indicate estimates in the absence of measured values.
The becquerel (symbol Bq) is the SI-derived unit of radioactivity. One Bq is defined as the activity of a quantity of radioactive material in which one nucleus decays per second. The Bq unit is thus equivalent to an inverse second, s^{-1}.
Source: International Atomic Energy Agency. 2003. Extent of environmental contamination by naturally occurring radioactive material (NORM) and technological options for mitigation. Technical Reports Series No. 419.

Table 18.25. Natural radioactivity in the ocean

Nuclide	Activity used in calculation	Activity in Ocean Pacific	Activity in Ocean Atlantic	Activity in Ocean All Oceans
Uranium	0.9 pCi/L (33 mBq/L)	6×10^8 Ci (22 EBq)	3×10^8 Ci (11 EBq)	1.1×10^9 Ci (41 EBq)
Potassium-40	300 pCi/L (11 Bq/L)	2×10^{11} Ci (7400 EBq)	9×10^{10} Ci (3300 EBq)	3.8×10^{11} Ci (14000 EBq)
Tritium	0.016 pCi/L (0.6 mBq/L)	1×10^7 Ci (370 PBq)	5×10^6 Ci (190 PBq)	2×10^7 Ci (740 PBq)
Carbon-14	0.135 pCi/L (5 mBq/L)	8×10^7 Ci (3 EBq)	4×10^7 Ci (1.5 EBq)	1.8×10^8 Ci (6.7 EBq)
Rubidium-87	28 pCi/L (1.1 Bq/L)	1.9×10^{10} Ci (700 EBq)	9×10^9 Ci (330 EBq)	3.6×10^{10} Ci (1300 EBq)

The curie (symbol Ci) is a non-SI unit of radioactivity, and is defined as: 1 Ci = 3.7×10^{10} decays per second.
The becquerel (symbol Bq) is the SI-derived unit of radioactivity. One Bq is defined as the activity of a quantity of radioactive material in which one nucleus decays per second. The Bq unit is thus equivalent to an inverse second, s^{-1}.
Source: Adapted from The Radiation Information Network, Idaho State University, Radioactivity in Nature, <http://www.physics.isu.edu/radinf/natural.htm>, Accessed 21 November 2009.

Table 18.26.   Summary of concentrations of major radionuclides in major rock types and soil

Rock type	^{40}K		^{87}Rb		^{232}Th		^{238}U	
	Total K (%)	Bq/kg	ppm	Bq/kg	ppm	Bq/kg	ppm	Bq/kg
Igneous rocks								
Basalt (crustal average)	0.8	300	40	30	3–4	10–15	0.5–1	7–10
mafic	0.3–1.1	70–400	10–50	1–40	1.6,2.7	7,10	0.5,0.9	7,10
salic	4.5	1100–1500	170–200	150–180	16,20	60,83	3.9,4.7	50,60
Granite (crustal average)	>4	>1000	170–200	150–180	17	70	3	40
Sedimentary rocks								
Shale, sandstones	2.7	800	120	110	12	50	3.7	40
Clean quartz	<1	<300	<40	<40	<2	<8	<1	<10
Dirty quartz	2?	400?	90?	80?	3–6?	10–25?	2–3?	40?
Arkose	2–3	600–900	80–120	80	2?	<8	1–2?	10–25?
Beach sands (unconsolidated)	<1	<300?	<40?	<40?	6	25	3	40
Carbonate rocks	0.3	70	10	8	2	8	2	25
Continental upper crust (average)	2.8	850	112	100	10.7	44	2.8	36
Soils	1.5	400	65	50	9	37	1–8	66

Note: Question marks indicate estimates in the absence of measured values.

Source: International Atomic Energy Agency. 2003. Extent of environmental contamination by naturally occurring radioactive material (NORM) and technological options for mitigation. Technical Report Series no. 419.

Table 18.27. Naturally occurring radionuclides in mineral resources

Element/mineral	Source	Radioactivity
Aluminium	Ore	250 Bq/(kg U)
	Bauxitic limestone, soil	100–400 Bq/(kg Ra)
	Bauxitic limestone, soil	30–130 Bq/(kg Th)
	Tailings	70–100 Bq/(kg Ra)
Copper	Ore	30–100 000 Bq/(kg U)
	Ore	10–110 Bq/(kg Th)
Fluorspar	Mineral	Uranium series
	Tailings	4000 Bq/(kg Ra)
Iron		Uranium series
		Thorium series
Molybdenum	Tailings	Uranium series
Monazite	Sands	6000–20 000 Bq/(kg U)
		Thorium Series (4% by weight)
Natural gas	Gas, average for groups of US and Canadian wells	2–17 000 Bq/(m^3 Rn)
	Gas, average for groups of US and Canadian wells	0.4–54 000 Bq/(m^3 Rn)
	Scale, residue in pumps, vessels and residual gas pipelines	100–50 000 Bq/(kg^{210}Pb/^{210}Po)
Oil	Brines or produced water	Ranging from mBq to 100 Bq/(L Ra)
	Sludges (scales)	Ranging up to 70 000 Bq/(kg Ra) Typically 10^3–10^4 Bq/kg, ranging up to 4×10^6 Bq/(kg Ra)
Phosphate	Ore	100–4000 Bq/(kg U$_{natural}$)
	Ore	15–150 Bq/(kg Th$_{natural}$)
	Ore	600–3000 Bq/(kg Ra)
Potash		Thorium series ^{40}K
Rare earths		Uranium series
		Thorium series
Tantalum/niobium		Uranium series
		Thorium series
Tin	Ore and slag	1000–2000 Bq/(kg Ra)
Titanium (rutile, ilmenite)	Ore	30–750 Bq/(kg U)
	Ore	35–750 Bq/(kg Th)
Uranium	Ore	15 000 Bq/(kg Ra)
	Slimes	10^5 Bq/(kg Ra)
	Tailings	10 000–20 000 Bq/(kg Ra)
Vanadium		Uranium Series
Zinc		Uranium Series, Thorium Series
Zirconium	Sands	4000 Bq/(kg U)
	Sands	600 Bq/(kg Th)
		4000–7000 Bq/(g Ra)

The becquerel (symbol Bq) is the SI-derived unit of radioactivity. One Bq is defined as the activity of a quantity of radioactive material in which one nucleus decays per second. The Bq unit is thus equivalent to an inverse second, s^{-1}.
Source: International Atomic Energy Agency. 2003. Extent of environmental contamination by naturally occurring radioactive material (NORM) and technological options for mitigation. Technical Reports Series No. 419.

Table 18.28.   Primordial nuclides

Nuclide	Symbol	Half-life	Natural Activity
Uranium-235	^{235}U	7.04×10^8 yr	0.72% of all natural uranium
Uranium-238	^{238}U	4.47×10^9 yr	99.2745% of all natural uranium; 0.5 to 4.7 ppm total uranium in the common rock types
Thorium-232	^{232}Th	1.41×10^{10} yr	1.6 to 20 ppm in the common rock types with a crustal average of 10.7 ppm
Radium-226	^{226}Ra	1.60×10^3 yr	0.42 pCi/g (16 Bq/kg) in limestone and 1.3 pCi/g (48 Bq/kg) in igneous rock
Radon-222	^{222}Rn	3.82 days	Noble Gas; annual average air concentrations range in the US from 0.016 pCi/L (0.6 Bq/m^3) to 0.75 pCi/L (28 Bq/m^3)
Potassium-40	^{40}K	1.28×10^9 yr	soil - 1–30 pCi/g (0.037–1.1 Bq/g)

Primordial radionuclides are those left over from when the world and the universe were created, and which have not completely decayed due to their long half-life. They are typically long lived, with half-lives often on the order of hundreds of millions of years. Primordial radionuclides also include those generated from the primordial radionuclides U-238, U-235 and Th-232 of the associated decay chain. The becquerel (symbol Bq) is the SI-derived unit of radioactivity. One Bq is defined as the activity of a quantity of radioactive material in which one nucleus decays per second. The Bq unit is thus equivalent to an inverse second, s^{-1}. The curie (symbol Ci) is a non-SI unit of radioactivity defined as 1 Ci = 3.7 × 1010 decays per second.
Source: Adapted from the Radiation Information Network, Idaho State University, Radioactivity in Nature, <http://www.physics.isu.edu/radinf/natural.htm>.

Table 18.29.   Common cosmogenic nuclides

Nuclide	Symbol	Half-life	Source	Natural Activity
Carbon 14	^{14}C	5730 yr	Cosmic-ray interactions, ^{14}N(n,p)^{14}C	6 pCi/g (0.22 Bq/g) in organic material
Hydrogen (Tritium)	^{3}H	12.3 yr	Cosmic-ray interactions with N and O, spallation from cosmic-rays, ^{6}Li(n, alpha)3H	0.032 pCi/kg (1.2 x 10^{-3} Bq/kg)
Beryllium 7	^{7}Be	53.28 days	Cosmic-ray interactions with N and O	0.27 pCi/kg (0.01 Bq/kg)

Cosmic radiation originates directly or indirectly from extraterrestrial sources. The radiation is in many forms, from high speed heavy particles to high energy photons and muons. The upper atmosphere interacts with many of the cosmic radiations, and produces radioactive nuclides. Some have long half-lives, but the majority have shorter half-lives than the primordial nuclides. Some other cosmogenic radionuclides are ^{10}Be, ^{26}Al, ^{36}Cl, ^{80}Kr, ^{14}C, ^{32}Si, ^{39}Ar, ^{22}Na, ^{35}S, ^{37}Ar, ^{33}P, ^{32}P, ^{38}Mg, ^{24}Na, ^{38}S, ^{31}Si, ^{18}F, ^{39}Cl, ^{38}Cl, ^{34m}Cl.
Source: Adapted from The Radiation Information Network, Idaho State University, Radioactivity in Nature, <http://www.physics.isu.edu/radinf/natural.htm>.

Measurement

Tables

Table 19.1. Energy units and constants named after people

Quantitiy					
Unit	Symbol	Measured	Relation to SI or SI-Derived Units	Definition	Person
ampere	Å	electric current	1A	The constant current needed to produce a force of 2×10^{-7} newton per meter between two straight parallel conductors of infinite length and negligible circular cross-section placed one meter apart in a vacuum.	French physicist and mathematician André-Marie Ampère, (1775–1836), one of the main discoverers of electromagnetism.
becquerel	Bq	disintegration rate	s^{-1}	The activity of a quantity of radioactive material in which one nucleus decays per second.	French physicist Antoine Henri Becquerel (1852–1908), the discoverer of radioactivity, for which he won the 1903 Nobel Prize in Physics (along with Marie Curie and Pierre Curie)
biot	Bi	electric current	10A	A unit of electric current equal to 10 amperes.	French mathematician and physicist Jean-Baptiste Biot (1774–1862), one of the founders of the theory of electromagnetism.
clausius	Cl	entropy	$J \bullet K^{-1}$	A unit of entropy, a measure of the extent to which heat or energy in a physical system is not available for performing work. 1 Cl = 4.1868 $J \bullet K^{-1}$	German physicist Rudolf Clausius (1822–1888), who introduced and named the concept of entropy, and who first stated the basic ideas of the second law of thermodynamics.

Continued

Handbook of Energy. http://dx.doi.org/10.1016/B978-0-08-046405-3.00019-X

Table 19.1. Continued

Quantitiy					
Unit	Symbol	Measured	Relation to SI or SI-Derived Units	Definition	Person
coulomb	C	electric charge, quantity of electricity	s•A	One coulomb is the amount of charge accumulated in one second by a current of one ampere.	French physicist Charles-Augustin de Coulomb (1736–1806), who was the first to measure accurately the forces exerted between electric charges.
curie	Ci	radioactivity	3.7×10^{10} Bq	One curie was originally defined as the radioactivity of one gram of pure radium. In 1953 the curie was defined as 3.7×10^{10} atomic disintegrations per second, or 37 gigabecquerels (GBq), this being the best estimate of the activity of a gram of radium.	French physicists Marie Curie (1867–1934) and Pierre Curie (1859–1906), pioneers in the field of radioactivity. In 1903 the Curies and Henri Becquerel receive the Nobel Prize in Physics. Marie Curie also receives the receive the 1911 Nobel Prize in Chemistry.
debey	D	electric dipole moment	C•m	$1\ D = 1 \times 10^{-18}$ statcoulomb-centimeter.	Dutch physicist P.J.W. Debye (1884–1966), who was famous for his research on polar molecules. He is awarded the 1936 Nobel Prize in Chemistry "for his contributions to the study of molecular structure," primarily referring to his work on dipole moments and X-ray diffraction.
degree Celsius	°C	temperature	°C = K − 273.15	The freezing point of water (at one atmosphere of pressure) was originally defined to be 0 °C, while the boiling point is 100 °C. In the SI system, the Celsius scale is defined so that the temperature of the triple point of water (the temperature at which water can exist simultaneously in the gaseous, liquid, and solid states) is exactly 0.01 °C, and the size of the degree is 1/273.16 of the difference between this temperature and absolute zero.	Swedish astronomer and physicist Anders Celsius (1701–1744), who developed a similar scale. Celsius originally called his scale *centigrade* derived from the Latin centum, "hundred" and gradus, "step."

Table 19.1.   **Continued**

Quantitiy					
Unit	**Symbol**	**Measured**	**Relation to SI or SI-Derived Units**	**Definition**	**Person**
degree Farenheit	°F	temperature	$°F = K \times \frac{9}{5} - 459.67$	A traditional unit of temperature still used customarily in the United States in which the freezing point of water is 32 degrees Fahrenheit (°F) and the boiling point 212 °F (at standard atmospheric pressure), placing the boiling and freezing points of water exactly 180 degrees apart. A degree on the Fahrenheit scale is 1/180 of the interval between the freezing point and the boiling point.	The physicist and engineer Daniel Gabriel Fahrenheit (1686–1736) who determined the temperature scale now named after him, and who was a skilled instrument maker known for inventing the alcohol thermometer and mercury thermometer.
degree Rankine	R	temperature	$R = K \times \frac{9}{5}$	A traditional unit of absolute temperature. Zero on both the Kelvin and Rankine scales is absolute zero, but the Rankine degree is defined as equal to one degree Fahrenheit, rather than the one degree Celsius used by the Kelvin scale. This means the Rankine temperature is 459.67° plus the Fahrenheit temperature. 1 degree Rankine is equal to exactly 5/9 Kelvin	The British physicist and engineer William Rankine (1820-1872) who was a key figure in the development of the science of thermodynamics. Rankine developed a complete theory of the steam engine and ultimately of all heat engines.
einstein	E	quantity of light	1 mole of photons	A unit used in irradiance and in photochemistry. One einstein is defined as one mole of photons, regardless of their frequency, carried by a beam of monochromatic light. Therefore, the number of photons in an einstein is Avogadro's number.	German physicist Albert Einstein (1879–1955), who explained how light carries energy in a famous 1905 paper. Einstein is best known for his theories of special relativity and general relativity. He received the 1921 Nobel Prize in Physics, and often is regarded as the father of modern physics.

Continued

Table 19.1.   Continued

Quantitiy					
Unit	Symbol	Measured	Relation to SI or SI-Derived Units	Definition	Person
farad	F	capacitance	$C \bullet V^{-1}$	The charge in coulombs a capacitor will accept for the potential across it to change 1 volt.	British physicist Michael Faraday (1791–1867), who was known for his seminal work in electricity and electrochemistry. His inventions of electromagnetic rotary devices formed the foundation of electric motor technology. Some scholars refer to him as the best experimentalist in the history of science.
gauss	G	magnetic induction	10^{-4} T	A field of one gauss exerts, on a current-carrying conductor placed in the field, a force of 0.1 dyne per ampere of current per centimeter of conductor. One gauss represents a magnetic flux of one maxwell per square centimeter of cross-section perpendicular to the field.	German mathematician and astronomer Karl Friedrich Gauss (1777–1855) who contributed significantly to many fields, including number theory, statistics, analysis, differential geometry, geodesy, geophysics, electrostatics, astronomy, and optics. He is considered one of the formemost mathematicians in history.
gray	Gy	absorbed dose	$J \bullet kg^{-1}$	The SI unit of absorbed radiation dose due to ionizing radiation. One gray is the absorption of one joule of energy, in the form of ionizing radiation, by one kilogram of matter.	British physician L. Harold Gray (1905–1965), an authority on the use of radiation in the treatment of cancer, and a founder of the field of radiobiology.
henry	H	electrical inductance	$Wb \bullet A^{-1}$	If the rate of change of current in a circuit is one ampere per second and the resulting electromotive force is one volt, then the inductance of the circuit is one henry.	American physicist Joseph Henry (1797–1878), one of several scientists who discovered independently how magnetic fields can be used to generate alternating currents.

Table 19.1. Continued

Quantitiy					
Unit	Symbol	Measured	Relation to SI or SI-Derived Units	Definition	Person
hertz	Hz	frequency	s^{-1}	The number of complete cycles per second. The unit hertz is defined in the SI system such that the hyperfine splitting in the ground state of the cesium 133 atom is exactly 9,192,631,770 hertz.	German physicist Heinrich Rudolf Hertz (1857–1894), who proved that energy is transmitted through a vacuum by electromagnetic waves. He did so by building an apparatus to produce and detect VHF or UHF radio waves.
joule	J	energy, work, quantity of heat	$N\bullet m$	The work done by a force of one newton acting to move an object through a distance of one meter in the direction in which the force is applied.	British physicist James Prescott Joule (1818–1889), who demonstrated the equivalence of mechanical and thermal energy. He also worked with Lord Kelvin to develop the absolute scale of temperature, made observations on magnetostriction, and found the relationship between the current through a resistance and the heat dissipated.
kelvin	K	thermodynamic temperature		A unit increment of temperature, and one of the seven SI base units. The Kelvin scale is a thermodynamic (absolute) temperature scale where absolute zero, the theoretical absence of all thermal energy, is zero kelvin (0 K).	British physicist and engineer William Thomson, 1st Baron Kelvin (or Lord Kelvin) (1824–1907) who made important contributions to the mathematical analysis of electricity and to the formation of the first and second laws of thermodynamics, and who helped unify the discipline of physics in its modern form.

Continued

Table 19.1. Continued

Quantitiy					
Unit	Symbol	Measured	Relation to SI or SI-Derived Units	Definition	Person
lambert	L, LA, lam	luminance	lumen•cm^{-2}	The luminous intensity of a surface, measuring the intensity of the light emitted (or reflected) in all directions per unit of area of the surface. One lambert is the luminance of a surface that emits or reflects one lumen per square centimeter.	German mathematician, physicist and astronomer Johann Lambert (1728–1777), who showed that the illuminance of a surface is inversely proportional to the square of the distance from the light source. Lambert was also a gifted mathematician, and was the first to give a rigorous proof that π is irrational.
maxwell	M, Mx	magnetic flux	10^{-8} Wb	In a magnetic field with a strength of one gauss, one maxwell is the total flux across a surface of one square centimeter perpendicular to the field.	British physicist James Clerk Maxwell (1831–1879), who presented the unified theory of electromagnetism. His set of equations— Maxwell's equations— demonstrated that electricity, magnetism and even light are all manifestations of the electromagnetic field.
newton	N	force	m•kg•s^{-2}	The amount of force required to accelerate a mass of one kilogram at a rate of one meter per second per second.	Isaac Newton (1642–1727), the British mathematician, physicist, and natural philosopher. He was the first person to understand clearly the relationship between force (F), mass (m), and acceleration (a) expressed by the formula $F = ma$. He is widely considered one of the most influential men in history.
ohm	W	electric resistance	V/•A^{-1}	The resistance between two points of a conductor when a constant potential difference of 1 volt, applied to these points, produces in the conductor a current of 1 ampere, the conductor not being the seat of any electromotive force.	German physicist Georg Simon Ohm (1787–1854). Ohm's experiments defined the fundamental relationship among voltage, current, and resistance, which represents the beginning of electrical circuit analysis.

Table 19.1. Continued

Quantitiy					
Unit	Symbol	Measured	Relation to SI or SI-Derived Units	Definition	Person
pascal	Pa	pressure, stress	$N \bullet m^{-2}$	One kilogram per meter per second per second.	French philosopher and mathematician Blaise Pascal (1623–1662) who was the first person to use a barometer to measure differences in altitude. He also made important contributions to mathematics in the areas of conic sections, projective geometry and for the theory of probability.
planck	-	action	$J \bullet s$	The planck is equal to 1 joule second (an energy of one joule multiplied by a time of one second).	German physicist Max Planck (1858–1947), the originator of quantum theory and widely viewed as one of the most important physicists of the twentieth century. Planck was awarded the Nobel Prize in Physics in 1918.
rayleigh	R	luminous inensity	photons $m^{-2} \cdot sr^{-1}$	Used in astronomy and physics to measure the brightness of the night sky, auroras, etc.	English mathematician and physicist John William Strutt, the third Lord Rayleigh (1842–1919), co-discoverer of the element argon, an achievement for which he earned the Nobel Prize for Physics in 1904. He also discovered the phenomenon now called Rayleigh scattering, explaining why the sky is blue, and predicted the existence of the surface waves now known as Rayleigh waves.
röntgen	R	radiation exposure	$C \bullet kg^{-1}$	1 R is the amount of radiation required to liberate positive and negative charges of one electrostatic unit of charge (esu) in 1 cm^3 of dry air at standard temperature and pressure. $1R = 2.58 \times 10^{-4}$ C/kg.	German physicist Wilhelm Konrad Röntgen (1845–1923) who produced and detected electromagnetic radiation in a wavelength range today known as x-rays, which earned him the first Nobel Prize in Physics in 1901.

Continued

Table 19.1. Continued

Quantitiy					
Unit	Symbol	Measured	Relation to SI or SI-Derived Units	Definition	Person
rutherford	rd	disintegration rate	10^6 bq	1 rutherford represents 1 million radioactive disintegrations per second, or one megabecquerel (MBq).	New Zealand chemist and physicist Ernest Rutherford, known as the "father of nuclear physics" for his discovery that atoms have their positive charge concentrated in a very small nucleus. He was awarded the Nobel Prize in Chemistry in 1908.
siemens	S	electric conductance	Ω^{-1}	The siemens is the reciprocal of the ohm.	German electrical engineer, inventor and industrialist Werner von Siemens (1816–1892). The company he founded in 1847 is now Siemens AG, one of the largest electrotechnological firms of the world.
sievert	Sv	dose equivalent of radiation	$J \bullet kg^{-1}$	Attempts to reflect the biological effects of radiation as opposed to the physical aspects, which are characterised by the absorbed dose, measured in gray. The equivalent dose to a tissue is found by multiplying the absorbed dose, in gray, by a dimensionless "quality factor" Q, dependent upon radiation type, and by another dimensionless factor N, dependent on all other pertinent factors. N depends upon the part of the body irradiated, the time and volume over which the dose was spread, even the species of the subject.	Swedish medical physicist Rolf Sievert (1898–1966), who worked over many years to measure and standardize the radiation doses used in cancer treatment. He chaired the United Nations Scientific Committee on the Effects of Atomic Radiation, and invented a number of instruments for measuring radiation doses, the most widely known being the Sievert chamber.

Table 19.1. Continued

Quantitiy					
Unit	Symbol	Measured	Relation to SI or SI-Derived Units	Definition	Person
sverdrup	Sv	volume transport		One sverdrup equals one million cubic meters per second, which is also one cubic hectometer per second. The sverdrup is used almost exclusively in oceanography to measure the transport of ocean currents.	Norwegian oceanographer and Arctic explorer H. U. Sverdrup (1888–1957), who made important contributions to the understanding of ocean circulation.
telsa	T	magnetic flux density	$Wb \bullet m^{-2}$	One tesla is the field intensity generating one newton of force per ampere of current per meter of conductor. Equivalently, one tesla represents a magnetic flux density of one weber per square meter of area.	Serbian-American electrical engineer Nikola Tesla (1856-1943), best known for his many revolutionary developments in the field of electromagnetism in the late 19th and early 20th centuries. Tesla's patents and theoretical work formed the basis of modern alternating current (AC) electric power systems.
volt	V	electric charge, quantity of electricity	$W \bullet A^{-1}$	The value of the voltage across a conductor when a current of one ampere dissipates one watt of power in the conductor.	Italian scientist Count Alessandro Volta (1745–1827), known especially for the development of the first electric cell (battery).
watt	W	power, radiant flux	$J \bullet s^{-1}$	One watt is equal to a power rate of one joule of work per second of time. In terms of mechanical energy, one watt is the rate at which work is done when an object is moved at a speed of one meter per second against a force of one newton.	Scottish inventor and mechanical engineer James Watt (1736–1819) whose improvements to the steam engine were fundamental to the changes brought by the Industrial Revolution in both the Kingdom of Great Britain and the world.
weber	Wb	magnetic flux	$v \bullet s$	The weber is the magnetic flux which, linking a circuit of one turn, would produce in it an electromotive force of 1 volt if it were reduced to zero at a uniform rate in 1 second.	German physicist Wilhelm Eduard Weber (1804–1891), one of the early researchers of magnetism, and co-inventor of the electromagnetic telegraph.

Table 19.2.   Examples of parameters measurable by sensors

Measurement Type	Measurable Quantities
Thermal	Temperature, heat, heat flow, entropy, heat capacity, etc.
Radiation	Gamma rays, X-rays, ultraviolet, visible and IR light, microwaves, radio waves, etc.
Mechanical	Displacement, velocity, acceleration, force, pressure, mass flow, acoustic wavelength and amplitude, etc.
Magnetic	Magnetic field, flux, magnetic moment, magnetization, magnetic permeability, etc.
Chemical	Humidity, pH, concentrations, toxic and flammable materials, pollutants, etc.
Biological	Sugars, proteins, hormones, antigens, etc.

Pistoia, Gianfranco. 2009. Battery Operated Devices and Systems, (Amsterdam, Elsevier).

Table 19.3.   International energy conversions

To: From:	Terajoules	Giga-calories	Million tonnes of oil equivalent	Million Btu	Gigawatt-hours
	multiply by:				
Terajoules	1	238.8	2.388×10^{-5}	947.8	0.2778
Gigacalories	4.1868×10^{-3}	1	10^{-7}	3.968	1.163×10^{-3}
Million tonnes of oil equivalent	4.1868×10^{4}	107	1	3.968×10^{7}	11,630
Million Btu	1.0551×10^{-3}	0.252	2.52×10^{-8}	1	2.931×10^{-4}
Gigawatthours	3.6	860	8.6×10^{-5}	3412	1

Source: Davis, Stacy C., Susan W. Diegel, and Robert G. Boundy. 2009. Transportation Energy Data Book: Edition 28, Oak Ridge National Laboratory, Table B.7.

Table 19.4.   Distance and Velocity Conversions

1 in.	$= 83.33 \times 10^{-3}$ ft $= 27.78 \times 10^{-3}$ yd $= 15.78 \times 10^{-6}$ mile $= 25.40 \times 10^{-3}$ m $= 0.2540 \times 10^{-6}$ km	1 ft	$= 12.0$ in. $= 0.33$ yd $= 189.4 \times 10^{-3}$ mile $= 0.3048$ m $= 0.3048 \times 10^{-3}$ km
1 mile	$= 63360$ in. $= 5280$ ft $= 1760$ yd $= 1609$ m $= 1.609$ km	1 km	$= 39370$ in. $= 3281$ ft $= 1093.6$ yd $= 0.6214$ mile $= 1000$ m

1 ft/sec = 0.3048 m/s = 0.6818 mph = 1.0972 km/h
1 m/sec = 3.281 ft/s = 2.237 mph = 3.600 km/h
1 km/h = 0.9114 ft/s = 0.2778 m/s = 0.6214 mph
1 mph = 1.467 ft/s = 0.4469 m/s = 1.609 km/h

Table 19.5. Power unit conversions

Per second basis	TO					
From	hp	hp-metric	kW	kJ s^{-1}	Btu$_{IT}$ s^{-1}	kcal$_{IT}$ s^{-1}
Horsepower	1	1.014	0.746	0.746	0.707	0.1780
Metric horsepower	0.986	1	0.736	0.736	0.697	0.1757
Kilowatt	1.341	1.360	1	1	0.948	0.2388
kilojoule per sec	1.341	1.359	1	1	0.948	0.2388
Btu$_{IT}$ per sec	1.415	1.434	1.055	1.055	1	0.2520
Kilocalories $_{IT}$ per sec	5.615	5.692	4.187	4.187	3.968	1
Per hour basis	**TO**					
From	hp	hp- metric	kW	J hr^{-1}	Btu$_{IT}$ hr^{-1}	kcal$_{IT}$ hr^{-1}
Horsepower	1	1.014	0.746	268.5 x 10^4	2544	641.19
Metric horsepower	0.986	1	0.736	265.8 x 10^4	2510	632.42
kilowatt	1.341	1.360	1	360 x 10^4	3412	859.85
Joule per hr	3.73 x 10^{-7}	3.78 x 10^{-7}	2.78 x 10^{-7}	1	9.48 x 10^{-4}	2.39 x 10^{-4}
Btu$_{IT}$ per hr	3.93 x 10^{-4}	3.98 x 10^{-4}	2.93 x 10^{-4}	1055	1	0.2520
Kilocalories $_{IT}$ per hr	1.56 x 10^{-3}	1.58 x 10^{-3}	1.163 x 10^{-3}	4187	3.968	1

The subscript "IT" stands for International Table values, which are only slightly different from thermal values normally subscripted "th". The "IT" values are most commonly used in current tables and generally are not subscripted, but conversion calculators ususally include both.

Table 19.6. Pressure conversions

	Weight		Gas		Liquid	
	Pounds (lb)	Kilograms (kg)	Standard cubic feet (SCF)	Normal cubic meter (Nm^3)	Gallons (gal)	Liters (L)
1 lb	1.0	0.4536	192	5.047	1.6928	6.408
1 kg	2.205	1.0	423.3	11.126	3.733	14.128
1 SCF gas	0.005209	0.002363	1.0	0.02628	0.00882	0.0339
1 Nm^3 gas	0.19815	0.08988	38.04	1.0	0.3355	1.2699
1 gal liquid	0.5906	0.2679	113.41	2.981	1.0	3.785
1 L liquid	0.15604	0.07078	29.99	0.77881	0.2642	1.0

Table 19.7.   Units for magnetic properties

Quantity	Symbol	CGS and EMU Units	Conversion to SI	SI Units
Magnetic flux	Φ	Mx (*maxwell*), G cm^2	10^{-8}	Wb (*weber*), V s
Magnetic flux density, magnetic induction	$\vec{B}$	G (*gauss*)	10^{-4}	T (tesla),[a] Wb m^{-2}
Magnetic potential differ, magnetomotive force	U, F	Gb (*gilbert*)	$10/4\pi$	A (*ampere*)
Magnetic field strength, magnetizing force	$\vec{H}$	Oe (*oersted*), Gb cm^{-1}	$10^3/4\pi$	A m^{-1}
Vector potential	$\vec{A}$	Oe cm	10^{-6}	T m
Volume magnetization	$\vec{M}$	emu cm^{-3}	103	A m^{-1}
Mass magnetization	$\vec{\sigma}, \vec{M}_\rho$	emu g^{-1}	1	A m^2 kg^{-1}
Molar magnetization	$\vec{M}_{mol}$	emu mol^{-1}	10^{-3}	A m^2 mol^{-1}
Magnetic polarization, intensity of magnetization	$\vec{J}, \vec{I}$	emu cm^{-3}	$4\pi \times 10^{-4}$	T, Wb m^{-2}
Magnetic moment of the specimen	$\vec{m}, \vec{\mu}$	emu, erg G^{-1}	10^{-3}	A m^2, J T^{-1}
Magnetic dipole moment	$\vec{j}$	emu, erg G^{-1}	$4\pi \times 10^{-10}$	Wb m
Volume susceptibility	χ, κ	Dimensionless, emu cm^{-3}	4π	Dimensionless
Mass susceptibility	$\chi\rho, \kappa\rho$	cm^3 g^{-1}, emu g^{-1}	$4\pi \times 10^{-3}$	m^3 kg^{-1}
Molar susceptibility	χ_{mol}, κ_{mol}	cm^3 mol^{-1}, emu mol^{-1}	$4\pi \times 10^{-6}$	m^3 mol^{-1}
Permeability	μ	Dimensionless	$4\pi \times 10^{-7}$	H m^{-1}, Wb A^{-1} m^{-1}
Vacuum permeability	μ_0	1	$4\pi \times 10^{-7}$	H m^{-1}
Relative permeability	μ_r	Undefined	–	Dimensionless
Volume energy density	W	erg cm^{-3}	10^{-1}	J m^{-3}
Demagnetization factor	D, N	Dimensionless	$1/4\pi$	Dimensionless

[a]T = J m^{-2} A^{-1} = kg s^{-2} A^{-1}.

Source: Boca, Roman. 2012. A Handbook of Magnetochemical Formulae, (Oxford, Elsevier).

Table 19.8.   Some common SI units of electromagnetism

Symbol	Name of Quantity	Derived Units	Unit	Base Units
I	Electric current	ampere (SI base unit)	A	A (= W/V = C/s)
Q	Electric charge	coulomb	C	A·s
$U, \Delta V, \Delta\varphi; E$	Potential difference; Electromotive force	volt	V	$kg·m^2·s^{-3}·A^{-1}$ (= J/C)
$R; Z; X$	Electric resistance; Impedance; Reactance	ohm	Ω	$kg·m^2·s^{-3}·A^{-2}$ (= V/A)
ρ	Resistivity	ohm metre	$\Omega·m$	$kg·m^3·s^{-3}·A^{-2}$
P	Electric power	watt	W	$kg·m^2·s^{-3}$ (= V·A)
C	Capacitance	farad	F	$kg^{-1}·m^{-2}·s^4·A^2$ (= C/V)
E	Electric field strength	volt per metre	V/m	$kg·m·s^{-3}·A^{-1}$ (= N/C)
D	Electric displacement field	Coulomb per square metre	C/m^2	$A·s·m^{-2}$
ε	Permittivity	farad per metre	F/m	$kg^{-1}·m^{-3}·s^4·A^2$
χ_e	Electric susceptibility	(dimensionless)	-	-
$G; Y; B$	Conductance; Admittance; Susceptance	siemens	S	$kg^{-1}·m^{-2}·s^3·A^2$ (= Ω^{-1})
κ, γ, σ	Conductivity	siemens per metre	S/m	$kg^{-1}·m^{-3}·s^3·A^2$
B	Magnetic flux density, Magnetic induction	tesla	T	$kg·s^{-2}·A^{-1}$ (= Wb/m^2 = $N·A^{-1}·m^{-1}$)
ϕ	Magnetic flux	weber	Wb	$kg·m^2·s^{-2}·A^{-1}$ (= V·s)
H	Magnetic field strength	ampere per metre	A/m	$A·m^{-1}$
L, M	Inductance	henry	H	$kg·m^2·s^{-2}·A^{-2}$ (= Wb/A = V·s/A)
μ	Permeability	henry per metre	H/m	$kg·m·s^{-2}·A^{-2}$
x	Magnetic susceptibility	(dimensionless)	-	-

Source: International Union of Pure and Applied Chemistry. 1993. Quantities, Units and Symbols in Physical Chemistry, 2nd edition, (Oxfor, Blackwell Science).

Table 19.9.   Radiometric terminology and units

Term	Description	Units
Absorptance	The fraction of the incident radiation flux by the Earth's surface	Dimensionless
Albedo	The ratio of reflected to the incident radiation component by the surface of the Earth	Dimensionless
Diffuse solar radiation	The downward scattered and reflected short-wave radiation coming from the whole sky vault with the exception of the solid angle subtended by the Sun's disk	$W\ m^{-2}$ (instantaneous value) $Wh\ m^{-2}$ (integrated value over 1 h)
Direct solar radiation	The short-wave radiation emitted from the solid angle of the Sun's disk, comprising mainly unscattered and unreflected solar radiation	$W\ m^{-2}$ (instantaneous value) $Wh\ m^{-2}$ (integrated value over 1 h)
Global (or total) solar radiation	The sum of diffuse and direct short-wave radiation components	$W\ m^{-2}$ (instantaneous value) $Wh\ m^{-2}$ (integrated value over 1 h)

Continued

Table 19.9.   Continued

Term	Description	Units
IR (or terrestrial or long-wave or thermal) radiation	The radiation coming from the sky at wavelengths longer than about 4 μm	$W\,m^{-2}$ (instantaneous value)
		$Wh\,m^{-2}$ (integrated value over 1 h)
Irradiance	The radiant flux incident on a surface from all directions per unit area of this surface	$W\,m^{-2}$ (instantaneous value)
		$Wh\,m^{-2}$ (integrated value over 1 h)
Radiance	The radiant flux emitted by a unit solid angle of a source or scatterer incident on a unit area of a surface	$W\,m^{-2}$ (instantaneous value)
		$Wh\,m^{-2}$ (integrated value over 1 h)
Radiant flux	The amount of radiation coming from a source per unit time	W
Radiant intensity	The radiant flux leaving a source point per unit solid angle of space surrounding the point	$W\,sr^{-1}$
Reflectance/ transmittance	The fraction of radiant flux reflected by a surface or transmitted by a semitransparent medium	Dimensionless
Spectroradiometry	The radiant flux per unit wavelength	$W\,m^{-2}\,nm^{-1}$ or $W\,m^{-2}\,\mu m^{-1}$ (instantaneous value)
		$Wh\,m^{-2}\,nm^{-1}$ or $Wh\,m^{-2}\,\mu m^{-1}$ (integrated value over 1 h)

Source: Kambezidis, H.D. 2012. 3.02 - The Solar Resource, In: Ali Sayigh, Editor-in-Chief, Comprehensive Renewable Energy, (Oxford, Elsevier), 2012, Pages 27-84.

Table 19.10.   Unit-of-measure equivalents for electricity

Unit	Equivalent
Kilowatt (kW)	1,000 (One Thousand) Watts
Megawatt (MW)	1,000,000 (One Million) Watts
Gigawatt (GW)	1,000,000,000 (One Billion) Watts
Terawatt (TW)	1,000,000,000,000 (One Trillion) Watts
Gigawatt	1,000,000 (One Million) Kilowatts
Thousand Gigawatts	1,000,000,000 (One Billion) Kilowatts
Kilowatthours (kWh)	1,000 (One Thousand) Watthours
Megawatthours (MWh)	1,000,000 (One Million) Watthours
Gigawatthours (GWh)	1,000,000,000 (One Billion) Watthours
Terawatthours (TWh)	1,000,000,000,000 (One Trillion) Watthours
Gigawatthours	1,000,000 (One Million) Kilowatthours
Thousand Gigawatthours	1,000,000,000 (One Billion) Kilowatthours

Table 19.11.   Heat contents and carbon content coefficients of various fossil fuels

Fuel Type	Heat Content	Carbon(C) Content Coefficients	Carbon Dioxide (CO_2) per Physical Unit
Solid Fuels	**Million Btu/Metric Ton**	**kg C/Million Btu**	**kg CO_2/Metric Ton**
Anthracite coal	24.88	28.28	2,579.9
Bituminous coal	26.33	25.44	2,456.6
Sub-bituminous coal	18.89	26.50	1,835.9
Lignite	14.18	26.65	1,385.6
Coke	27.56	31.00	3,131.9
Unspecified coal	27.56	25.34	2,560.0
Gas Fuels	**Btu/Cubic Foot**	**kg C/Million Btu**	**kg CO_2/Cubic Foot**
Natural gas	1,026	14.46	0.0544
Liquid Fuels	**Million Btu/Petroleum Barrel**	**kg C/Million Btu**	**kg CO_2/Petroleum Barrel**
Motor gasoline	5.22	19.46	372.2
Distillate fuel oil	5.83	20.17	430.8
Residual fuel oil	6.29	20.48	472.1
Jet fuel	5.67	19.70	409.5
Aviation gasoline	5.05	18.86	349.0
LPG	3.55	16.83	219.3
Kerosene	5.67	19.96	415.1
Still gas	6.00	18.20	400.3
Petroleum coke	6.02	27.85	615.1
Pentanes plus	4.62	19.10	323.6
Unfinished oils	5.83	20.31	433.8

Note: For fuels with variable heat contents and carbon content coefficients, 2009 U.S. average values are presented. All factors are presented in gross calorific values (GCV) (i.e., higher heating values). Miscellaneous products includes all finished products not otherwise classified, (e.g., aromatic extracts and tars, absorption oils, ram-jet fuel, synthetic natural gas, naptha-type jet fuel, and specialty oils).
Source: United States Environmental Protection Agency.

Table 19.12.   Common quantities in CO_2 research

Quantity	Value
Solar constant	1.375 kW/m^2
Earth mass	5.976×10^{24}
Equatorial radius	6.378×10^6 m
Polar radius	6.357×10^6 m
Mean radius	6.371×10^6 m
Surface area	5.101×10^{14} m^2
Land area	1.481×10^{14} m^2
Ocean area	5.620×10^{14} m^2
Mean land elevation	840 m
Mean ocean depth	3730 m
Mean ocean volume	1.550×10^{18} m^3
Ocean mass	1.384×10^{21} kg
Mass of atmosphere	5.137×10^{18} kg
Equatorial surface gravity	8.780 m/s^2

Source: Adapted from Kreith, Frank, Ed. 2000. CRC Handbook of Thermal Engineering (Boca Raton, CRC Press).

Table 19.13.   Solar insolation conversion factors (multiply top row by factor to obtain side column)

	W/m²	kW•h/(m²*day)	sun hours/day	kWh/(m²•y)	kWh/(kWp•y)
W/m²	1	41.666666	41.666666	0.1140796	0.1521061
kW•h/(m²•day)	0.024	1	1	0.0027379	
sun hours/day	0.024	1	1	0.0027379	0.0036505
kWh/(m²•y)	8.765813	365.2422	365.2422	1	1.333333
kWh/(kWp•y)	6.57436	273.9316	273.9316	0.75	1

Table 19.14.   The solar constant in alternative units

Solar constant	Units
1366.1	W•m^{-2} [SI unit]
0.13661	W•cm^{-2}
136.61	mW•cm^{-2}
1.3661	x 10^6 erg*cm^{-2}s^{-1}
126.9	W•ft^{-2}
1.959	cal•cm^{-2}•min^{-1} ($\pm$ 0.03 cal•cm^{-2}•min^{-1})
0.0326	cal•cm^{-2}s^{-1}
433.4	Btu•ft^{-2}•h^{-1}
0.1202	Btu•ft^{-2}•s^{-1}
1.956	Langleys•min^{-1}

Notes:
The calorie is the thermochemical calorie-gram and is defined as 4.1840 absolute joules.
The Btu is the thermochemical British thermal unit and is defined by the relationship:
1 Btu (thermochemical)/(°F*lb) = 1 cal•g (thermochemical)/(°C•g).
The Langley, however, is defined in terms of the older thermal unit the calorie•g (mean); that is, 1 Langley = 1 cal•g (mean)•cm^{-2}; 1 cal•g (mean) = 4.19002 J.
Source: Solar Spectra: Standard Air Mass Zero, <http://rredc.nrel.gov/solar/spectra/am0/ASTM2000.html>.

Table 19.15.   List of uncertainty sources and total uncertainty associated with estimating solar radiation

Uncertainty source	Total hemispherical (%)	Direct normal (%)	Diffuse (%)
Measurement (site/model development/validation)	6	5	6
Station meteorological data bias	2	4	2
Station meteorological data random (2 s)	8	15	8
Total expanded uncertainty: meteorological model	10	16	10
Satellite data bias	0	4	1
Satellite data random	5	14	5
Total expanded uncertainty: satellite model	8	15	8

Source: Adapted from National Renewable Energy Laboratory, Solar Radiation Data Manual for Flat-Plate and Concentrating Collectors, <http://rredc.nrel.gov/solar/pubs/redbook/>, accessed 13 May 2012.

Table 19.16.   Motor vehicle fuel efficiency conversions

MPG	Miles/liter	Kilometers/L	L/100 kilometers
10	2.64	4.25	23.52
15	3.96	6.38	15.68
20	5.28	8.50	11.76
25	6.60	10.63	9.41
30	7.92	12.75	7.84
35	9.25	14.88	6.72
40	10.57	17.00	5.88
45	11.89	19.13	5.23
50	13.21	21.25	4.70
55	14.53	23.38	4.28
60	15.85	25.51	3.92
65	17.17	27.63	3.62
70	18.49	29.76	3.36
75	19.81	31.88	3.14
80	21.13	34.01	2.94
85	22.45	36.13	2.77
90	23.77	38.26	2.61
95	25.09	40.38	2.48
100	26.42	42.51	2.35
105	27.74	44.64	2.24
110	29.06	46.76	2.14
115	30.38	48.89	2.05
120	31.70	51.01	1.96
125	33.02	53.14	1.88
130	34.34	55.26	1.81
135	35.66	57.39	1.74
140	36.98	59.51	1.68
145	38.30	61.64	1.62
150	39.62	63.76	1.57
Formula	MPG/3.785	MPG/[3.785/1.609]	235.24/MPG

Source: Davis, S.C., et.al., *Transportation Energy Data Book: Edition 27*, Appendix B.13, ORNL-6981, Oak Ridge National Laboratory, Oak Ridge, TN. 2008.

PART III

APPLICATIONS

Consumption

Figures

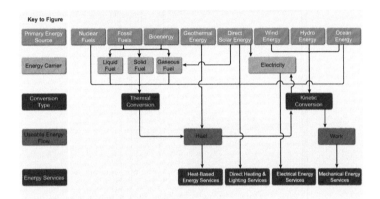

Figure 20.1. Pathways of energy from service to service.
Source: Adapted from Moomaw, W., F. Yamba, M. Kamimoto, L. Maurice, J. Nyboer, K. Urama, T. Weir, 2011: Introduction. In IPCC Special Report on Renewable Energy Sources and Climate Change Mitigation [O. Edenhofer, R. Pichs-Madruga, Y. Sokona, K. Seyboth, P. Matschoss, S. Kadner, T. Zwickel, P. Eickemeier, G. Hansen, S. Schlömer, C.von Stechow (eds)], Cambridge University Press, Cambridge, United Kingdom and New York, NY, USA.

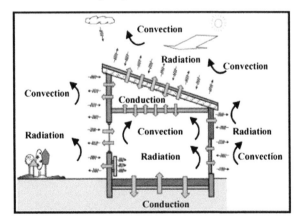

Figure 20.2. Schematic representation of heat transfers in a building.
Source: Kuznik, Frédéric, Damien David, Kevyn Johannes, Jean-Jacques Roux 2011. A review on phase change materials integrated in building walls, Renewable and Sustainable Energy Reviews, Volume 15, Issue 1, Pages 379-391.

Handbook of Energy. http://dx.doi.org/10.1016/B978-0-08-046405-3.00020-6

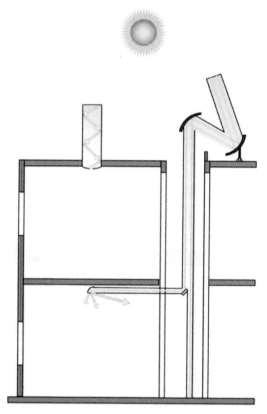

Figure 20.3.   Schematic diagram of light pipe systems, consisting of a skylight dome, a reflective tube, and a diffuser assembly.
Source: Shi, Long, Michael Yit Lin Chew. 2012. A review on sustainable design of renewable energy systems, Renewable and Sustainable Energy Reviews, Volume 16, Issue 1, Pages 192-207.

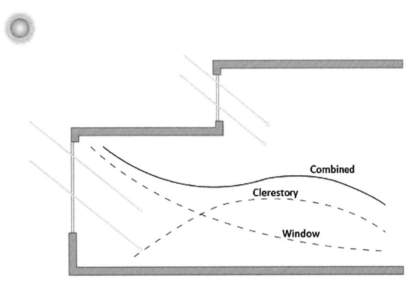

Figure 20.4. Daylight penetration resulting from the combination of a vertical clerestory and a side window. *Source: Shi, Long, Michael Yit Lin Chew. 2012. A review on sustainable design of renewable energy systems, Renewable and Sustainable Energy Reviews, Volume 16, Issue 1, Pages 192-207.*

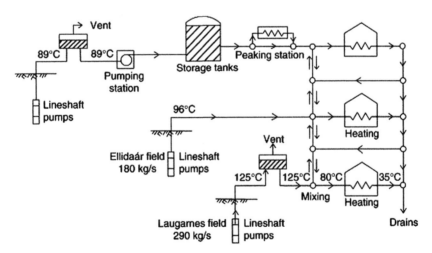

Figure 20.5. The Reykjavik, Iceland, district heating system supplies heat for more than 90 percent of the city's population The installed capacity of about 780 MWt is designed to meet the heating load to approximately −10 °C; during colder periods, the increased load is met by large storage tanks and an oil-fired booster station. *Source: Lund, John W. 2004. Geothermal Direct Use, In: Cutler J. Cleveland, Editor-in-Chief, Encyclopedia of Energy, (New York, Elsevier), Pages 859-873.*

Charts

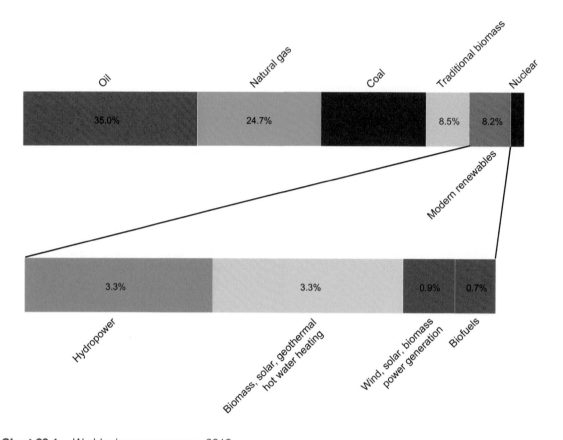

Chart 20.1. World primary energy use, 2010.
Source: Data from BP, Statistical Review of World Energy 2012, <http://www.bp.com/sectionbodycopy. do?categoryId=7500&contentId=7068481>; Renewable Energy Policy Network for the 21st Century, 2012 Global Status Report, <http://www.ren21.net/>.

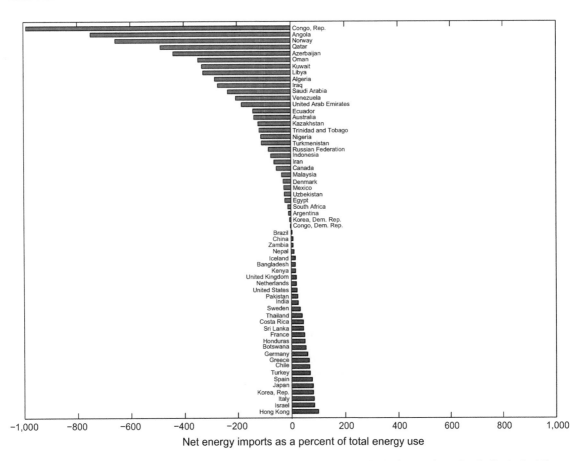

Chart 20.2. Net energy imports as a percenage of total energy use, 2010. A negative value indicate that the nation is a net exporter of energy.

Source: Data from United States Department of Energy, Energy Information Administration, International Energy Statistics, <http://www.eia.gov/coal/data.cfm#reserves>.

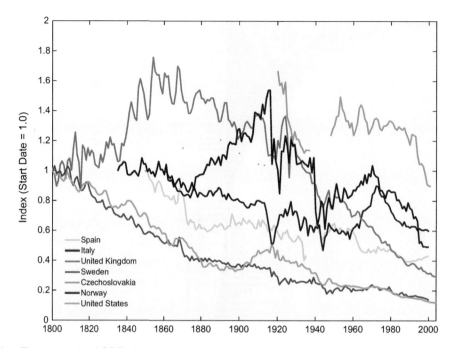

Chart 20.3.   The energy/real GDP ratio for selected nations.
Source: Sweden: Kander, Astrid. 2002. Economic growth, energy consumption and CO_2 emissions in Sweden 1800-2000, Lund Studies in Economic History 19, (Lund, Sweden, Lund University). Post-2000 data supplied by the author; UK: Fouquet, Roger. 2008. Heat, Power and Light: Revolutions in Energy Services.(Cheltenham, UK, and Northampton, MA, USA, Edward Elgar Publications); Netherlands and Spain: Gales, Ben, Kander, Astrid, Malanima, Paolo, and Rubio, Mar. 2007. North versus South: Energy transition and energy intensity in Europe over 200 years. European Review of Economic History, 11(02), 219-253; US: O'Connor, Peter and Cutler J. Cleveland. Energy use and economic output in the United States, 1780-2010. Department of Earth and Environment, Boston University. Italy: Malanima, Paolo. 2006. Energy Consumption in Italy in the 19th and 20th Centuries: A Statistical Outline. Naples: CNR-ISSM; Czechoslovakia: Kuskova, Petra, Simone Gingrich, and Fridolin Krausmann. 2008. Long term changes in social metabolism and land use in Czechoslovakia, 1830-2000: An energy transition under changing political regimes. Ecological Economics, 68, 394-407; Norway: Lindmark, Magnus. 2007. Estimates of Norwegian Energy consumption 1835-2000, Department of Economic History, Umeå University & Department of Economics, Norwegian School of Economics.

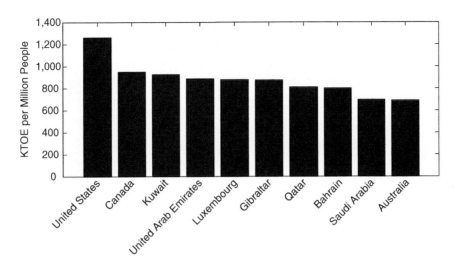

Chart 20.4. Motor gasoline consumption per capita, top 10 nations, 2010.
Source: Data from International Energy Agency (IEA), Energy statistics database, <http://www.iea.org/stats/index.asp>.

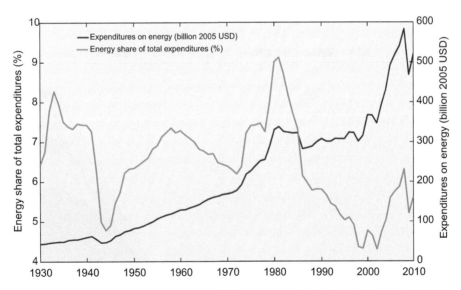

Chart 20.5. Personal consumption expenditures (PCEs) on energy in the United States, 1930-2010. PCEs measure the goods and services purchased by households and by nonprofit institutions serving households who are resident in the United States. Energy consists of gasoline and other energy goods and of electricity and natural gas.
Source: Data from United Sates Department of Commerce, Bureau of Economic Analysis, National Economic Accounts, <http://www.bea.gov/national/index.htm>.

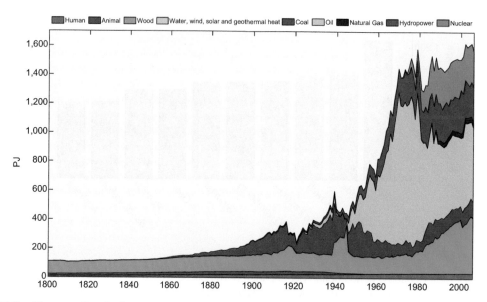

Chart 20.6. The quantity of primary energy use in Sweden by fuel type, 1800-2006.
Source: Data from Kander, Astrid. 2002. Economic growth, energy consumption and CO_2 emissions in Sweden 1800-2000, Lund Studies in Economic History 19, (Lund, Sweden, Lund University). Post-2000 data supplied by the author.

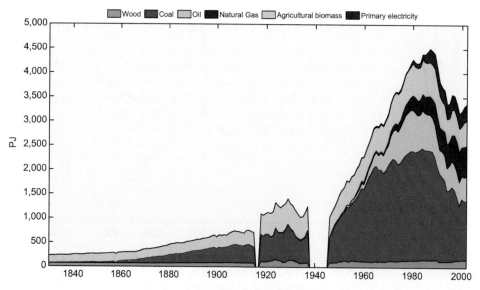

Chart 20.7. The quantity of primary energy use in Czechoslovakia, 1830-2000.
Source: Data from 1830-1917 refer to Bohemia+Moravia; data from 1919-1991 refer to Czechoslovakia; data from 1992-2003 refer to Czechia+Slovakia.
Source: Data from Kuskova, Petra, Simone Gingrich, and Fridolin Krausmann. 2008. Long term changes in social metabolism and land use in Czechoslovakia, 1830-2000: An energy transition under changing political regimes. Ecological Economics, 68, 394-407

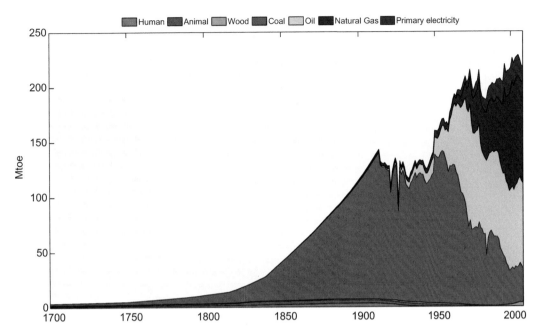

Chart 20.8. The quantity of primary energy use in the United Kingdom by fuel type, 1700-2008.
Source: Data from Fouquet, Roger. 2008. Heat, Power and Light: Revolutions in Energy Services. (Cheltenham, UK, and Northampton, MA, USA, Edward Elgar Publications).

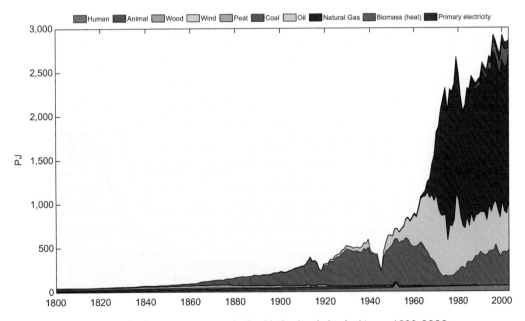

Chart 20.9. The quantity of primary energy use in the Netherlands by fuel type, 1800-2003.
Source: Datab from Gales, Ben, Kander, Astrid, Malanima, Paolo, and Rubio, Mar. 2007. North versus South: Energy transition and energy intensity in Europe over 200 years. European Review of Economic History, 11(02), 219-253.

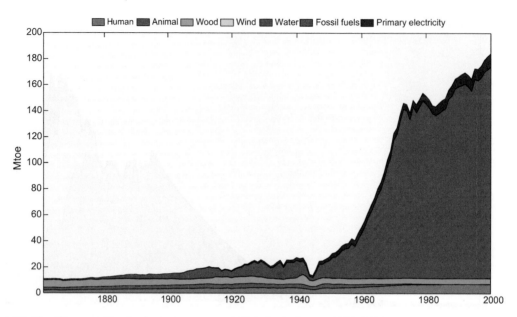

Chart 20.10. The quantity of primary energy use in Italy by fuel type, 1861-2000.
Source: Data from Malanima, Paolo. 2006. Energy Consumption in Italy in the 19th and 20th Centuries: A Statistical Outline. Naples: CNR-ISSM.

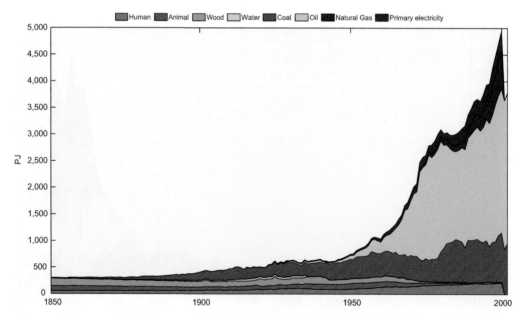

Chart 20.11. The quantity of primary energy use in Spain by fuel type, 1850-2000.
Source: Data from Gales, Ben, Kander, Astrid, Malanima, Paolo, and Rubio, Mar. 2007. North versus South: Energy transition and energy intensity in Europe over 200 years. European Review of Economic History, 11(02), 219-253.

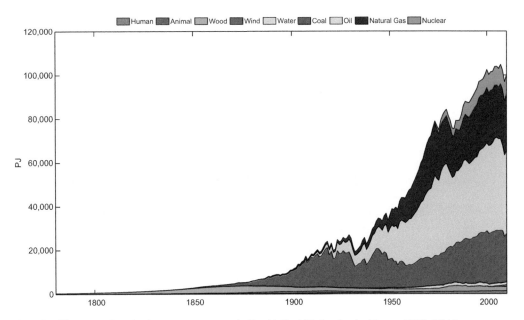

Chart 20.12. The quantity of primary energy use in the United States by fuel type, 1780-2010.
Source: Data from O'Connor, Peter and Cutler J. Cleveland. Energy use and economic output in the United States, 1780-2010. Department of Earth and Environment, Boston University.

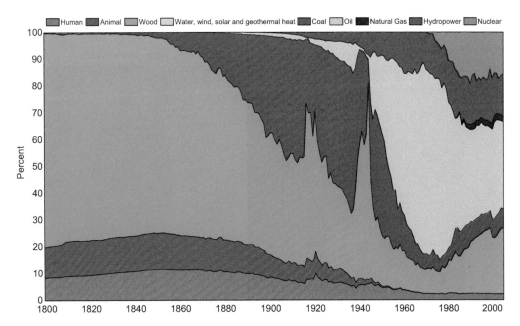

Chart 20.13. Fuel shares of primary energy use in Sweden, 1800-2006.
Source: Data from Kander, Astrid. 2002. Economic growth, energy consumption and CO$_2$ emissions in Sweden 1800-2000, Lund Studies in Economic History 19, (Lund, Sweden, Lund University). Post-2000 data supplied by the author.

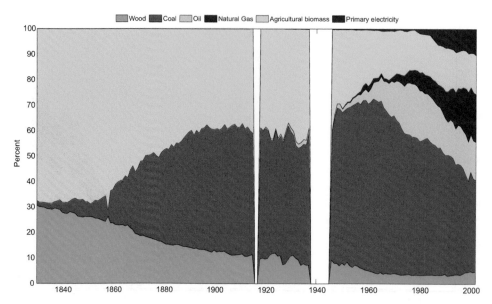

Chart 20.14. Fuel shares of primary energy use in Czechoslovakia, 1830-2000.
Source: Data from 1830-1917 refer to Bohemia+Moravia; data from 1919-1991 refer to Czechoslovakia; data from 1992-2003 refer to Czechia+Slovakia.
Source: Data from Kuskova, Petra, Simone Gingrich, and Fridolin Krausmann. 2008. Long term changes in social metabolism and land use in Czechoslovakia, 1830-2000: An energy transition under changing political regimes. Ecological Economics, 68, 394-408

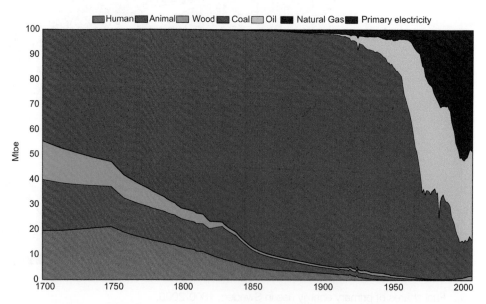

Chart 20.15. Fuel shares of primary energy use in the United Kingdom, 1700-2008.
Source: Data from Fouquet, Roger. 2008. Heat, Power and Light: Revolutions in Energy Services. (Cheltenham, UK, and Northampton, MA, USA, Edward Elgar Publications).

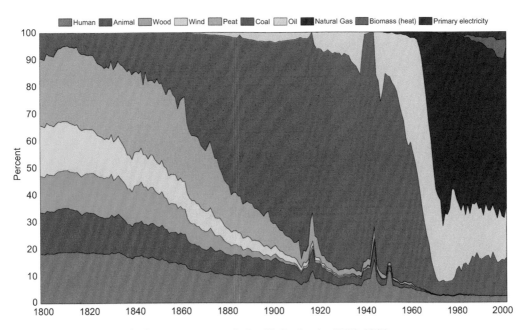

Chart 20.16. Fuel shares of primary energy use in the Netherlands, 1800-2003.
Source: Data from Gales, Ben, Kander, Astrid, Malanima, Paolo, and Rubio, Mar. 2007. North versus South: Energy transition and energy intensity in Europe over 200 years. European Review of Economic History, 11(02), 219-253.

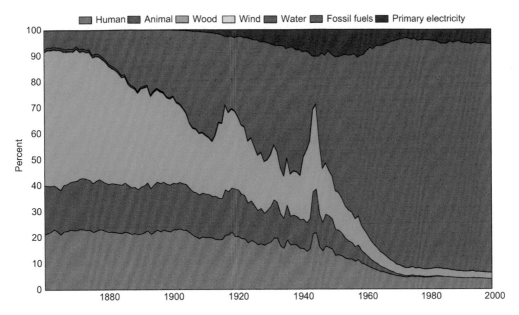

Chart 20.17. Fuel shares of primary energy use in Italy, 1861-2000.
Source: Data from Malanima, Paolo. 2006. Energy Consumption in Italy in the 19th and 20th Centuries: A Statistical Outline. Naples: CNR-ISSM.

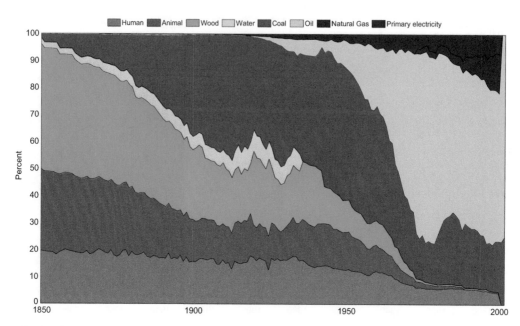

Chart 20.18.   Fuel shares of primary energy use in Spain, 1850-2000.
Source: Data from Gales, Ben, Kander, Astrid, Malanima, Paolo, and Rubio, Mar. 2007. North versus South: Energy transition and energy intensity in Europe over 200 years. European Review of Economic History, 11(02), 219-253.

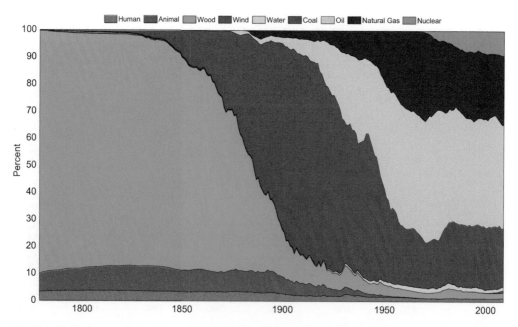

Chart 20.19.   Fuel shares of primary energy use in the United States, 1780-2010.
Source: Data from O'Connor, Peter and Cutler J. Cleveland. Energy use and economic output in the United States, 1780-2010. Department of Earth and Environment, Boston University.

Tables

Table 20.1. **World energy use and economic activity, 1800–2010**

	1800	1850	1900	1950	2000	2010	Change 1800-2010	Change 2000-2010
Primary energy (EJ)	20.5	27.4	50.8	113.9	432.8	502.0	2348%	16%
Population (Billions)	1.0	1.3	1.6	2.5	6.1	6.9	602%	14%
GDP-PPP (1990 GK$)	5.34E+11	9.02E+11	1.98E+12	5.37E+12	3.65E+13	5.11E+13	9463%	40%
Per capita energy (TJ/person)	20.9	21.8	31.0	45.3	71.5	72.8	249%	2%
Per capita GDP ($/person)	543.7	716.7	1,206.4	2,137.0	6,030.2	7,406.2	1262%	23%
Energy/GDP (MJ/$)	38.4	30.4	25.7	21.2	11.9	9.8	−74%	−17%

Source: 1800-2000: Grubler, Arnulf, Energy transitions, Encyclopedia of Earth, <http://www.eoearth.org/article/Energy_transitions>, accessed 15 June 2012; 2010: World Bank, World Development Indicators and Global Development Finance, statistical database, <http://databank.worldbank.org/data/Home.aspx>.

Table 20.2.   Global primary energy projections, 2020–2100, in Exajoules (EJ)

Organization and year	2020	2030	2050	2100
BP (2011)	565–635	600–760	NA	NA
EC (2006)	570–610	650–705	820–935	NA
EIA (2010)	600–645	675–780	NA	NA
IAEA (2009)	585–650	670–815	NA	NA
IEA (2011)	NA	730–816	NA	NA
IIASA (2007)	555–630	NA	800–1175	985–1740
Shell International (2008)	630–650	690–735	770–880	NA
WEC (2008)	615–675	700–845	845–1150	NA
Tellus Institute (2010)	504–644	489–793	425–1003	243–1200
ExxonMobil (2012)	667[a]		730[b]	

[a]For 2025.
[b]For 2040.
References:

- BP, BP Energy Outlook 2030, BP, London (2011).
- European Commission (EC), World Energy Technology Outlook-WETO H$_2$. Directorate General for Research: EUR 22038, EC, Brussels, Belgium (2006).
- Energy Information Administration (EIA), International energy outlook 2011, US Department of Energy, Washington, DC (2011).
- International Atomic Energy Agency (IAEA), Energy, electricity and nuclear power estimates for the period up to 2030. IAEA-RDS-1/29, IAEA, Vienna (2009).
- International Energy Agency (IEA), Key world energy statistics 2010, IEA/OECD, Paris (2010) [also earlier editions].
- IIASA: K. Riahi, A. Grubler, N. Nakicenovic, Scenarios of long-term socio-economic and environmental development under climate stabilization, Technol Forecast Soc Change, 74 (2007), pp. 887–935.
- Shell International BV, Shell energy scenarios to 2050, Shell International BV, The Hague, NL (2008).
- World Energy Council (WEC), 2010 survey of energy resources, WEC, London (2010).
- Tellus: P.D. Raskin, C. Electris, R.A. Rosen, The century ahead: searching for sustainability, Sustainability, 2 (2010), pp. 2626–2651.
- ExxonMobil, 2012 The Outlook for Energy: A View to 2040, ExxonMobil, Irving, TX.

Source: Adapted from Moriarty, Patrick, Damon Honnery. 2012. What is the global potential for renewable energy?, Renewable and Sustainable Energy Reviews, Volume 16, Issue 1, Pages 244–252.

Table 20.3.   Summary of global energy and exergy flows to the anthroposphere in 2008, in MJ/yr

Exergy resource	Energy flow	β value[a]	Exergy consumption
Coal	1.262 E+14	1.04	1.313 E+14
Oil	1.578 E+14	1.03	1.625 E+14
Gas	1.01 E+14	0.94	9.49 E+13
Nuclear	2.59 E+13	1	2.59 E+13
Geothermal	8.8 E+11	0.71	6.25 E+11
Tidal	1.6 E+10	1	1.6 E+10
Hydro	1.26 E+13	1	1.26 E+13
Biomass	5.36 E+13	1.05	5.63 E+13
Wind	1.9 E+12	1	1.9 E+12
Solar[b]	5.4 E+11	0.93	5.0 E+11

[a]β is the exergy-to-energy ratio of an energy resource.
[b]Solar energy collected directly by photovoltaic panels and solar thermal panels.
Source: Adapted from Liao, Wenjie, Reinout Heijungs, Gjalt Huppes. 2012. Natural resource demand of global biofuels in the Anthropocene: A review, Renewable and Sustainable Energy Reviews, Volume 16, Issue 1, Pages 996–1003.

Table 20.4. Primary energy consumption by region and fuel, 2010

Region	Oil	Natural gas	Coal	Nuclear energy	Hydroelectricity	Renewables	Total
North America	1,039.7	767.4	556.3	213.8	149.9	44.2	2,771.5
Total S. & Cent. America	282.0	132.9	23.8	4.9	157.2	11.1	611.9
Europe & Eurasia	922.9	1,023.5	486.8	272.8	195.9	69.6	2,971.5
Middle East	360.2	329.0	8.8	-	3.0	0.1	701.1
Africa	155.5	94.5	95.3	3.1	23.2	1.1	372.6
Asia Pacific	1,267.8	510.8	2,384.7	131.6	246.4	32.6	4,573.8
World	4,028.1	2,858.1	3,555.8	626.2	775.6	158.6	12,002.4

Soure: BP, Statistical Review of World Energy 2011.

Table 20.5. Primary energy consumption, top 10 nations 2010

Country	Oil	Natural gas	Coal	Nuclear energy	Hydroelectricity	Renewables	Total
China	428.6	98.1	1713.5	16.7	163.1	12.1	**2432.2**
United States	850.0	621.0	524.6	192.2	58.8	39.1	**2285.7**
Russian Federation	147.6	372.7	93.8	38.5	38.1	0.1	**690.9**
India	155.5	55.7	277.6	5.2	25.2	5.0	**524.2**
Japan	201.6	85.1	123.7	66.2	19.3	5.1	**500.9**
Germany	115.1	73.2	76.5	31.8	4.3	18.6	**319.5**
Canada	102.3	84.5	23.4	20.3	82.9	3.3	**316.7**
South Korea	105.6	38.6	76.0	33.4	0.8	0.5	**255.0**
Brazil	116.9	23.8	12.4	3.3	89.6	7.9	**253.9**
France	83.4	42.2	12.1	96.9	14.3	3.4	**252.4**
Top 10 total	2306.6	1494.9	2933.7	504.6	496.4	95.1	7831.3
World	4,028.1	2,858.1	3,555.8	626.2	775.6	158.6	12,002.4
Top 10 share of total	57.3%	52.3%	82.5%	80.6%	64.0%	60.0%	65.2%

Soure: BP, Statistical Review of World Energy 2011.

Table 20.6. Per capita energy use in high- and low-energy consumption nations, 1980–2009 (Million btu per person)

Country	1980	1990	2000	2009	Change 2000–09
10 Highest					
Qatar	904.6	767.8	1,028.4	1,229.6	19.6%
Bahrain	395.7	512.0	574.7	764.7	33.1%
Trinidad and Tobago	159.7	180.3	336.0	697.8	107.7%
United Arab Emirates	267.2	673.1	579.7	679.4	17.2%
Iceland	247.6	306.9	424.2	669.3	57.8%
Singapore	183.0	263.4	377.1	485.2	28.7%
Kuwait	351.1	208.4	460.8	462.3	0.3%
Norway	328.2	403.3	436.1	407.9	−6.5%
Canada	394.2	395.1	420.0	389.5	−7.3%
Luxembourg	389.6	377.9	350.7	352.6	0.5%
10 Lowest					
Uganda	1.3	1.1	1.4	1.4	2.2%
Somalia	2.9	2.1	1.4	1.3	−3.4%
Burkina Faso	1.0	0.8	1.4	1.3	−10.9%
Niger	1.5	1.9	1.4	1.2	−10.0%
Central African Republic	1.2	1.4	1.4	1.2	−14.3%
Rwanda	0.7	1.7	1.5	1.1	−23.2%
Mali	1.0	1.0	1.0	1.0	4.3%
Afghanistan	1.8	8.1	1.1	0.8	−25.8%
Burundi	0.4	1.0	1.0	0.7	−23.8%
Chad	0.9	0.8	0.4	0.4	16.9%
World average	63.6	65.6	64.9	71.3	9.9%

Source: United States Department of Energy, Energy Information Administration, International Energy Statistics, <http://www.eia.gov/countries/data.cfm>, accessed 3 June 2012.

Table 20.7. Personal consumption expenditures (PCE)[a] on energy in the U.S., 1930–2010

	1930	1950	1970	1990	2010
PCEs on energy[b]	44.0	83.2	170.5	308.6	520.7
Total PCEs	685.5	1,311.4	2,663.7	5,307.8	9,230.8
Energy share of total	6.4%	6.3%	6.4%	5.8%	5.6%

[a]Measures the goods and services purchased by households and by nonprofit institutions serving households that are located in the United States.
[b]Consists of gasoline and other energy goods and of electricity and natural gas.
Source: Data from United States Bureau of Economic Analysis.

Table 20.8. Energy service company (ESCO) and country indicators[a]

Country	Date of first ESCO	Age of ESCO market based on 2009	Number of ESCOs	Total value of ESCO projects in 2001 in millions of USD
Argentina	1990	19	5	1
Australia	1990	19	8	25
Austria	1995	14	25	7
Belgium	1990	19	4	n.a.
Brazil	1992	17	60	100
Bulgaria	1995	14	12	n.a.
Canada	1982	27	5	100
Chile	1996	13	3	0.2
China	1995	14	23	49.7
Columbia	1997	12	3	0.2
Côte d'Ivoire	2000	9	4	0.25
Czech Republic	1993	16	3	2
Egypt	1996	13	14	n.a.
Estonia	1986	23	20	3
Finland	2000	9	4	1
Germany	1995	14	500	150
Ghana	1996	13	3	0.1
Hungary	1990	19	20	n.a.
India	1994	15	8	1
Italy	1980	29	20	n.a.
Japan	1997	12	21	61.7
Jordan	1994	15	1	2
Kenya	1997	12	2	0.01
Korea	1992	17	158	20
Lithuania	1998	11	3	n.a.
Mexico	1998	11	7	n.a.
Morocco	1990	19	1	0.5
Nepal	2002	7	2	0.25
Philippines	1990	19	5	0.2
Poland	1995	14	8	30
Slovak Republic	1995	14	10	1.7
South Africa	1998	11	5	10
Sweden	1978	31	12	30
Switzerland	1995	14	50	13.5
Thailand	2000	9	6	6
Tunisia	2000	9	1	0.5
Ukraine	1996	13	5	2.5
United Kingdom	1980	29	20	n.a.

[a]Energy Service Companies (ESCOs) are private-sector instruments that offer energy-/emission-improvement (energy saving, energy efficiency, energy conservation and emission reduction) projects, or renewable-energy projects.
Source: Adapted from Okay, Nesrin, Ugur Akman. 2010. Analysis of ESCO activities using country indicators, Renewable and Sustainable Energy Reviews, Volume 14, Issue 9, Pages 2760–2771.

Table 20.9.   Syngas specifications for various applications[a]

Specification	Hydrogen or refinery use	Ammonia production	Methanol synthesis	Fischer-tropsch synthesis
Hydrogen content	>98%	75%	71%	60%
Carbon monoxide content	<10–50 ppm(v)	$[CO + CO_2]$ <20 ppm(v)	19%	30%
Carbon dioxide content	<10–50 ppm(v)		4–8%	
Nitrogen content	<2%	25%		
Other gases	N_2, Ar, CH_4	Ar, CH_4	N_2, Ar, CH_4	N_2, Ar, CH_4, CO_2
Balance		As low as possible	As low as possible	Low
H_2/N_2 ratio		~3		
H_2/CO ratio				0.6–2.0
$H_2/[2CO + 3CO_2]$ ratio			1.3–1.4	
Process temperature		350–550 °C	300–400 °C	200–350 °C
Process pressure	>50 bar	100–250 bar	50–300 bar	15–60 bar

[a]Syngas is a gas mixture that is composed of carbon monoxide, carbon dioxide, and hydrogen. The syngas is produced due to the gasification of a carbon containing fuel to a gaseous product that has some heating value.
Source: Adapted from Basu, Prabir. 2010. Biomass Gasification and Pyrolysis, (Boston, Academic Press), Pages 167–228.

Table 20.10.   Production and consumption of peat for fuel in 2008 (thousand metric tons)

Country	Producton	Consumption
Burundi	10	10
Total Africa	10	10
Falkland Islands	13	13
Total South America	13	13
Austria	1	1
Belarus	2,364	2,208
Estonia	214	294
Finland	4,971	7,959
Ireland	3,089	4,139
Kazakhstan		1
Latvia	11	9
Lithuania	67	38
Macedonia (Republic)	4	
Romania	10	39
Russian Federation	762	884
Sweden	837	1,201
Ukraine	358	340
United Kingdom	20	20
Total Europe	12,704	17,137
World	12,727	17,160

Data on production relate to peat produced for energy purposes; data on consumption (including imported peat) similarly relate only to fuel use Tonnages are generally expressed in terms of air-dried peat (35%-55% moisture content).
Source: Adapted form World Energy Council. 2010. Survey of World Energy Resources, (London, United Kingdom).

Table 20.11. Contribution of mining sectors to total final energy use by global mining industry

Sector	Electricity usage	Fossil fuel usage	Combustible waste and renewables	Heat	Percentage of total
Iron and steel	7%	46%	0.50%	1.10%	55%
Non-ferrous metals	6%	4%	0%	0.20%	10%
Non-metallic minerals	4%	24%	0.70%	0.30%	29%
Mining and quarrying	2%	3%	0%	0.20%	6%
Percentage of Total	19%	78%	1%	2%	100%

Source: Adapted from McLellan, B.C., G.D. Corder, D.P. Giurco, K.N. Ishihara. 2012. Renewable energy in the minerals industry: a review of global potential, Journal of Cleaner Production, Volume 32, Pages 32–44.

Table 20.12. Typical chemical exergy content of some fuels

Fuel	Exergy coefficient	Net heating value (kJ/kg)	Chemical exergy (kJ/kg)
Coal	1.088	21680	23587.84
Coke	1.06	28300	29998
Fuel oil	1.073	39500	42383.5
Natural gas	1.04	44000	45760
Diesel fuel	1.07	39500	42265
Fuelwood	1.15	15320	17641

Source: Ayres, R. U., & Warr, B. 2005. Accounting for growth: the role of physical work. Structural Change and Economic Dynamics, 16(2), 181–209.

Table 20.13. Carbon/hydrogen rations (C:H) of selected fuels

Fuel	C:H
Wood/biomass	0.8–1.1
Coal	1.4–6.8
Oil	0.9–1.1
Natural gas	0.25
Alcohol fuels	0.3–0.5
Hydrogen	0

Source: U.S.Department of Energy.

Table 20.14. Heating values of fuel gases

	HHV[a]	LHV[b]	HHV[a]	LHV[b]
	Btu ft^{-3} [c]		MJ Nm^{-3} [d]	
Hydrogen	325	275	12.75	10.79
Carbon monoxide	322	322	12.63	12.63
Methane	1013	913	39.74	35.81
Ethane	1792	1641	69.63	63.74
Propane	2590	2385	99.02	91.16
Butane	3370	3113	128.39	118.56

Note: Conversion factors for 1 MJ Nm^{-3} at 273.15K and 101.325 kPa. | 25.45 Btu ft^{-3} at 60 °F and 14.73 psia. Inverse 1 Btu ft^{-3} at 60 °F and 30 in. Hg. | 0.0393 MJ Nm^{-3}.
[a]Higher heating value.
[b]Lower heating value.
[c]Standard temperatures and pressure of dry gas are 60 °F and 14.73 psia (NIST, 2004).
[d]S.I. units.
Source: Adapted from Kreith Frank and D. Yogi Goswami. 2004. The CRC Handbook of Mechanical Engineering, Second Edition, (Boca Raton, CRC Press).

Table 20.15. Lower and higher heating values of gas, liquid and solid fuels

Fuels	Lower Heating Value (LHV) [1]			Higher Heating Value (HHV) [1]			Density
Gaseous Fuels @ 32 F and 1 atm	Btu/ft3 [2]	Btu/lb [3]	MJ/kg [4]	Btu/ft3 [2]	Btu/lb [3]	MJ/kg [4]	grams/ft3
Natural gas	983	20,267	47.141	1089	22,453	52.225	22.0
Hydrogen	290	51,682	120.21	343	61,127	142.18	2.55
Still gas (in refineries)	1458	20,163	46.898	1,584	21,905	50.951	32.8
Liquid Fuels	Btu/gal [2]	Btu/lb [3]	MJ/kg [4]	Btu/gal [2]	Btu/lb [3]	MJ/kg [4]	grams/gal
Crude oil	129,670	18,352	42.686	138,350	19,580	45.543	3,205
Conventional gasoline	116,090	18,679	43.448	124,340	20,007	46.536	2,819
Reformulated or low-sulfur gasoline	113,602	18,211	42.358	121,848	19,533	45.433	2,830
CA reformulated gasoline	113,927	18,272	42.500	122,174	19,595	45.577	2,828
U.S. conventional diesel	128,450	18,397	42.791	137,380	19,676	45.766	3,167
Low-sulfur diesel	129,488	18,320	42.612	138,490	19,594	45.575	3,206
Petroleum naphtha	116,920	19,320	44.938	125,080	20,669	48.075	2,745
NG-based FT naphtha	111,520	19,081	44.383	119,740	20,488	47.654	2,651

Table 20.15.   Continued

Fuels	Lower Heating Value (LHV) [1]			Higher Heating Value (HHV) [1]			Density
Residual oil	140,353	16,968	39.466	150,110	18,147	42.210	3,752
Methanol	57,250	8,639	20.094	65,200	9,838	22.884	3,006
Ethanol	76,330	11,587	26.952	84,530	12,832	29.847	2,988
Butanol	99,837	14,775	34.366	108,458	16,051	37.334	3,065
Acetone	83,127	12,721	29.589	89,511	13,698	31.862	2,964
E-Diesel Additives	116,090	18,679	43.448	124,340	20,007	46.536	2,819
Liquefied petroleum gas (LPG)	84,950	20,038	46.607	91,410	21,561	50.152	1,923
Liquefied natural gas (LNG)	74,720	20,908	48.632	84,820	23,734	55.206	1,621
Dimethyl ether (DME)	68,930	12,417	28.882	75,610	13,620	31.681	2,518
Dimethoxy methane (DMM)	72,200	10,061	23.402	79,197	11,036	25.670	3,255
Methyl ester (biodiesel, BD)	119,550	16,134	37.528	127,960	17,269	40.168	3,361
Fischer-Tropsch diesel (FTD)	123,670	18,593	43.247	130,030	19,549	45.471	3,017
Renewable Diesel I (Super Cetane)	117,059	18,729	43.563	125,294	20,047	46.628	2,835
Renewable Diesel II (UOP-HDO)	122,887	18,908	43.979	130,817	20,128	46.817	2,948
Renewable Gasoline	115,983	18,590	43.239	124,230	19,911	46.314	2,830
Liquid Hydrogen	30,500	51,621	120.07	36,020	60,964	141.80	268
Methyl tertiary butyl ether (MTBE)	93,540	15,094	35.108	101,130	16,319	37.957	2,811
Ethyl tertiary butyl ether (ETBE)	96,720	15,613	36.315	104,530	16,873	39.247	2,810
Tertiary amyl methyl ether (TAME)	100,480	15,646	36.392	108,570	16,906	39.322	2,913
Butane	94,970	19,466	45.277	103,220	21,157	49.210	2,213
Isobutane	90,060	19,287	44.862	98,560	21,108	49.096	2,118
Isobutylene	95,720	19,271	44.824	103,010	20,739	48.238	2,253

Continued

Table 20.15. Continued

Fuels	Lower Heating Value (LHV) [1]			Higher Heating Value (HHV) [1]			Density
Propane	84,250	19,904	46.296	91,420	21,597	50.235	1,920
Solid Fuels	**Btu/ton [2]**	**Btu/lb [5]**	**MJ/kg [4]**	**Btu/ton [2]**	**Btu/lb [5]**	**MJ/kg [4]**	
Coal (wet basis) [6]	19,546,300	9,773	22.732	20,608,570	10,304	23.968	
Bituminous coal (wet basis) [7]	22,460,600	11,230	26.122	23,445,900	11,723	27.267	
Coking coal (wet basis)	24,600,497	12,300	28.610	25,679,670	12,840	29.865	
Farmed trees (dry basis)	16,811,000	8,406	19.551	17,703,170	8,852	20.589	
Herbaceous biomass (dry basis)	14,797,555	7,399	17.209	15,582,870	7,791	18.123	
Corn stover (dry basis)	14,075,990	7,038	16.370	14,974,460	7,487	17.415	
Forest residue (dry basis)	13,243,490	6,622	15.402	14,164,160	7,082	16.473	
Sugar cane bagasse	12,947,318	6,474	15.058	14,062,678	7,031	16.355	
Petroleum coke	25,370,000	12,685	29.505	26,920,000	13,460	31.308	

Adapted from Table A.1, Wright, Lynn, Bob Boundy, Bob Perlack, Stacy Davis, Bo Saulsbury. 2006. Biomass Energy Data Book, Edition 1, Oak Ridge National Laborator,(Oak Ridge, TN)

Notes:

[1] The **lower heating value** (also known as net calorific value) of a fuel is defined as the amount of heat released by combusting a specified quantity (initially at 25 °C) and returning the temperature of the combustion products to 150 °C, which assumes the latent heat of vaporization of water in the reaction products is not recovered. The LHV are the useful calorific values in boiler combustion plants and are frequently used in Europe.

The **higher heating value** (also known as gross calorific value or gross energy) of a fuel is defined as the amount of heat released by a specified quantity (initially at 25 °C) once it is combusted and the products have returned to a temperature of 25 °C, which takes into account the latent heat of vaporization of water in the combustion products. The HHV are derived only under laboratory conditions, and are frequently used in the US for solid fuels.

[2] Btu = British thermal unit.

[3] The heating values for gaseous fuels in units of Btu/lb are calculated based on the heating values in units of Btu/ft3 and the corresponding fuel density values. The heating values for liquid fuels in units of Btu/lb are calculated based on heating values in units of Btu/gal and the corresponding fuel density values.

[4] The heating values in units of MJ/kg, are converted from the heating values in units of Btu/lb.

[5] For solid fuels, the heating values in units of Btu/lb are converted from the heating values in units of Btu/ton.

[6] Coal characteristics assumed by GREET for electric power production.

[7] Coal characteristics assumed by GREET for hydrogen and Fischer-Tropsch diesel production.

Source: GREET Transportation Fuel Cycle Analysis Model, GREET 1.8b, developed by Argonne National Laboratory, Argonne, IL, released May 8, 2008. http://www.transportation.anl.gov/software/GREET/index.html

Table 20.16. **Temperature ranges for various industrial processes**

Industry	Process	Temperature (°C)
Dairy	Pressurization	60–80
	Sterilization	100–120
	Drying	120–180
	Concentrates	60–80
	Boiler feedwater	60–90
Canned food	Sterilization	110–120
	Pasteurization	60–80
	Cooking	60–90
	Bleaching	60–90
Textile	Bleaching, dyeing	60–90
	Drying, degreasing	100–130
	Dyeing	70–90
	Fixing	160–180
	Pressing	80–100
Paper	Cooking, drying	60–80
	Boiler feedwater	60–90
	Bleaching	130–150
Chemical	Soaps	200–260
	Synthetic rubber	150–200
	Processing heat	120–180
	Pre-heating water	60–90
Meat	Washing, sterilization	60–90
	Cooking	90–100
Beverages	Washing, sterilization	60–80
	Pasteurization	60–70
Flours and by-products	Sterilization	60–80
Timber by-products	Thermodifusion beams	80–100
	Drying	60–100
	Pre-heating water	60–90
	Preparation pulp	120–170
Bricks and blocks	Curing	60–140
Plastics	Preparation	120–140
	Distillation	140–150
	Separation	200–220
	Extension	140–160
	Drying	180–200
	Blending	120–140

Kalogirou, Soteris A. 2009. Solar Energy Engineering, (Boston, Academic Press).

Table 20.17. Typical nameplate wattages for various household appliances

Appliance	Watts
Aquarium	50–1210
Clock radio	10
Coffee maker	900–1200
Clothes washer	350–500
Clothes dryer	1800–5000
Dishwasher	1200–2400
Dehumidifier	785
Electric blanket- *Single/Double*	60/100
Fans	
Ceiling	65–175
Window	55–250
Furnace	750
Whole house	240–750
Hair dryer	1200–1875
Heater *(portable)*	750–1500
Clothes iron	1000–1800
Microwave oven	750–1100
Personal computer	
CPU - awake / asleep	120/30 or less
Monitor - awake / asleep	150/30 or less
Laptop	50
Radio *(stereo)*	70–400
Refrigerator *(frost-free, 16 cubic feet)*	725
Televisions (color)	
19″	65–110
27″	113
36″	133
Flat screen	120
Toaster	800–1400
Toaster oven	1225
VCR/DVD	17–21/20–25
Vacuum cleaner	1000–1400
Water heater *(40 gallon)*	4500–5500
Water pump *(deep well)*	250–1100
Water bed *(with heater, no cover)*	120–380

http://www.energysavers.gov/your_home/appliances/index.cfm/mytopic=10040
United States Department of Energy, Office of Energy Efficiency and Renewable Energy (EERE), Estimating Appliance and Home Electronic Energy Use, <http://www.energysavers.gov/your_home/appliances/index.cfm/mytopic=10040>, accessed 22 June 2012.

Table 20.18.   Standby power consumption by appliances and equipment[a]

Product/Mode	Average (W)	Min (W)	Max (W)
Air Conditioner, room/wall			
Off	0.9	0.9	0.9
Charger, mobile phone			
On, charged	2.24	0.75	4.11
On, charging	3.68	0.27	7.5
Power supply only	0.26	0.02	1
Clock, radio			
On	2.01	0.97	7.6
Computer Display, CRT			
Off	0.8	0	2.99
On	65.1	34.54	124.78
Sleep	12.14	1.6	74.5
Computer Display, LCD			
Off	1.13	0.31	3.5
On	27.61	1.9	55.48
Sleep	1.38	0.37	7.8
Computer, desktop			
On, idle	73.97	27.5	180.83
Off	2.84	0	9.21
Sleep	21.13	1.1	83.3
Computer, notebook			
Fully on, charged	29.48	14.95	73.1
Fully on, charging	44.28	27.38	66.9
Off	8.9	0.47	50
Power supply only	4.42	0.15	26.4
Sleep	15.77	0.82	54.8
Fax, inkjet			
Off	5.31	0	8.72
On	6.22	2.89	14
Fax, laser			
Off	0	0	0
On	6.1	6.1	6.1
Ready	6.42	6.42	6.42
Heating, furnace central			
Off	4.21	0	9.8
On	339.71	70.5	796
Hub, USB			
Off	1.44	0.95	1.81
On	2.06	1.06	3.55
Modem, DSL			
Off	1.37	0.33	2.02
On	5.37	3.38	8.22
Modem, cable			
Off	3.84	1.57	6.62
On	6.25	3.64	8.62

Continued

Table 20.18.   Continued

Product/Mode	Average (W)	Min (W)	Max (W)
Standby	3.85	3.59	4.11
Multi-function Device, inkjet			
Off	5.26	0	10.03
On	9.16	3.9	17.7
Multi-function Device, laser			
Off	3.12	0	4.7
On	49.68	5	175
Night Light, interior			
Off	0.05	0	0.34
On	4.47	0	27.97
Ready	0.22	0	1.2
Phone, cordless			
Ready, handset	2.81	1.05	4.89
Ready, no handset	1.58	0.59	3.09
Active (talking)	1.9	0.59	3.38
Off	0.98	0.54	1.8
Phone, cordless with answering machine			
Ready, handset	4	2.15	7.4
Ready, no handset	2.82	1.72	4.7
Active (talking)	3.53	2.2	6.5
Off	2.92	0.9	7.4
Power Tool, cordless			
Ready, charged	8.34	1.82	14
Active	29.53	1.39	66
Ready	1.74	0	4.7
Printer, inkjet			
Off	1.26	0	4
On	4.93	1.81	22
Printer, laser			
Off	1.58	0	4.5
On	131.07	1.7	481.9
Range, gas			
Ready	1.13	0.7	1.7
Scanner, flatbed			
Off	2.48	0.27	8.2
On	9.6	1.71	15.6
Security Systems, home			
Ready	2.7	2.7	2.7
Set-top Box, DVR			
On, no recording	37.64	25.95	49.2
On, recording	29.29	27.27	31.3
Off	36.68	23.3	48.6
Set-top Box, digital cable with DVR			
Not recording, TV off	44.63	44.38	44.87

Table 20.18. Continued

Product/Mode	Average (W)	Min (W)	Max (W)
Not recording, TV on	44.4	44.2	44.6
Off by remote	43.46	43.3	43.61
Set-top Box, digital cable			
On, TV off	24.65	14.2	74.74
On, TV on	29.64	14.1	102.23
Off by remote	17.83	13.24	30.6
Off by switch	17.5	13.7	26.3
Set-top Box, satellite with DVR			
Not recording, TV off	28.35	25.8	30.9
Not recording, TV on	31.37	24.2	36.3
Off by remote	27.8	22	33.6
Set-top Box, satellite			
On, TV off	15.95	7.69	33.2
On, TV on	16.15	7.69	33.2
Off by remote	15.66	6.58	33.05
Off by switch	15.47	6.58	32.7
Speakers, computer			
On, no sound	4.12	0.69	9.84
Off	1.79	0	5.6
Stereo, portable			
CD, not playing	4.11	1.29	6.83
Cassette, not playing	2.42	1.16	5.92
CD playing	6.8	3.96	9.2
Off	1.66	0.7	5.44
Radio playing	3.3	1.36	8.25
Television, CRT			
Off by remote	3.06	0.3	10.34
Off by switch	2.88	0	16.1
Television, rear projection			
On	186.09	186.09	186.09
Off by remote	6.97	0.2	48.5
Off by switch	6.6	0.2	48.5
Timer, irrigation			
Off	2.75	1.5	5.9
Ready	2.84	1.5	5.9
Tuner, AM/FM			
On, not playing	9.48	5.08	16.4
On, playing	9.92	5.07	17.7
Off	1.12	0	3.37
Amplifier			
On, not playing	33.99	21.4	70.93
On, playing	39.16	21.11	69.3
Off	0.27	0	1.8
Audio Minisystem			
CD, not playing	13.99	1.67	36.95

Continued

Table 20.18. Continued

Product/Mode	Average (W)	Min (W)	Max (W)
Cassette, not playing	13.85	1.67	33.14
CD playing	19.09	5.2	41.2
Off	8.32	0.3	24.58
Radio playing	14.41	2.98	38
CD Player			
On, not playing	8.62	4	25.7
On, playing	9.91	5.8	25.6
Off	5.04	2	18.4
Caller ID Unit			
Ready	1.27	1.27	1.27
Clock			
On	1.74	0.99	3.61
Radio playing	2.95	1.7	4.2
Coffee Maker			
Off	1.14	0	2.7
Copier			
Off	1.49	0	2.97
On	9.63	3.6	14
DVD Recorder			
Off	0.75	0	1.5
DVD Player			
On, not playing	7.54	0.24	12.7
On, playing	9.91	5.28	17.17
Off	1.55	0	10.58
DVD/VCR			
On, not playing	13.51	8.48	20.5
On, playing	15.33	9.43	22.37
Off	5.04	0.09	12.7
Game Console			
Active	26.98	5.4	67.68
Off	1.01	0	2.13
Ready	23.34	2.12	63.74
Garage Door Opener			
Ready	4.48	1.8	7.3
Microwave Ovens			
Ready, door closed	3.08	1.4	4.9
Ready, door open	25.79	1.6	39
Cooking	1433	966.2	1723
Musical Instruments			
Off	2.82	1.2	4.2
Receiver (audio)			
On, not playing	37.61	17.1	65.2
Off	2.92	0	19.7
Subwoofer			
On, not playing	10.7	5.8	20.6

Table 20.18. **Continued**

Product/Mode	Average (W)	Min (W)	Max (W)
On, playing	12.42	5.9	20.6
Surge Protector			
Off	1.05	0	6.3
On	0.8	0	6.92
Telephone Answering Device			
Off	2.01	1.31	2.55
Ready	2.25	1.42	2.83
Television/VCR			
Off by remote	5.15	2.15	13.3
Off by switch	5.99	2.15	13.11
VCR			
On, not playing	7.77	3.8	11.62
Off	4.68071	1.2	9.9

[a]Standby power is electricity used by appliances and equipment while they are switched off or not performing their primary function. That power is consumed by power supplies (the black cubes—sometimes called "vampires"—converting AC into DC), the circuits and sensors needed to receive a remote signal, soft keypads and displays including miscellaneous LED status lights. Standby power use is also caused by circuits that continue to be energized even when the device is "off".

Source: Adapted from Standy Power, Energy Analysis Department, Environmental Energy Technologies Division Lawrence Berkeley National Laboratory, <http://standby.lbl.gov/standby.html>.

Conversion

Figures

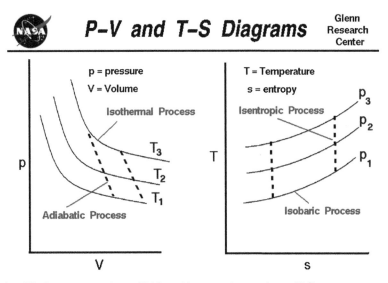

Figure 21.1. A simplified pressure-volume (P-V) and temperature-entropy (T-S).
Adapted from United States National Aeronautics and Space Administration (NASA), P-V and S-T diagrams, <http://www.grc.nasa.gov/WWW/k-12/airplane/pvtsplot.html>.

Handbook of Energy. http://dx.doi.org/10.1016/B978-0-08-046405-3.00021-8

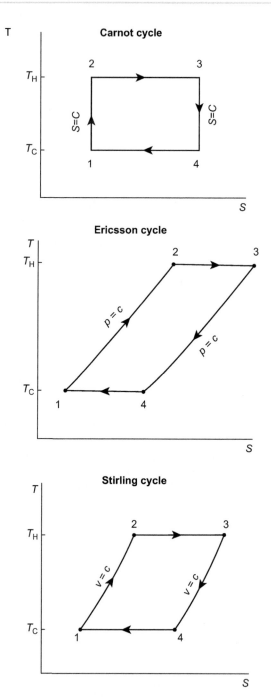

Figure 21.2. The Carnot, Ericsson, Stirling cycles.
Source: Adapted from Moran, Michael J. 2005. Engineering Thermodynamics, In: Frank Kreith and D. Yogi Goswami, Editors. CRC Handbook of Mechanical Engineering, Second Edition (Boca Raton, CRC Press).

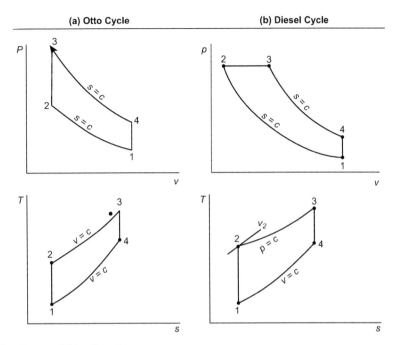

Figure 21.3. The Otto and Diesel cycles.
Source: Adapted from Moran, Michael J. 2005. Engineering Thermodynamics, In: Frank Kreith and D. Yogi Goswami, Editors. CRC Handbook of Mechanical Engineering, Second Edition (Boca Raton, CRC Press).

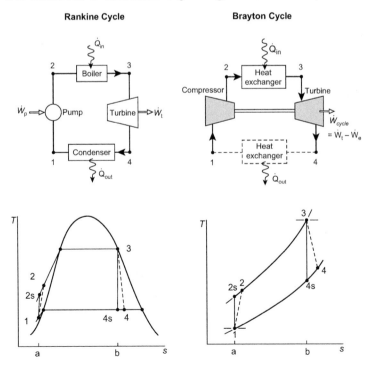

Figure 21.4. The Rankine and Brayton cycles.
Source: Adapted from Moran, Michael J. 2005. Engineering Thermodynamics, In: Frank Kreith and D. Yogi Goswami, Editors. CRC Handbook of Mechanical Engineering, Second Edition (Boca Raton, CRC Press).

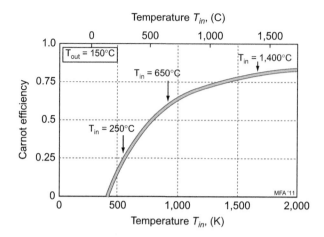

Figure 21.5. Carnot efficiency of heat engines as a function of temperature T_{in}, assuming $T_{out} = 150\,°C$. *Source: Ashby, Michael F. 2013. Materials and the Environment (Second Edition), (Boston, Butterworth-Heinemann).*

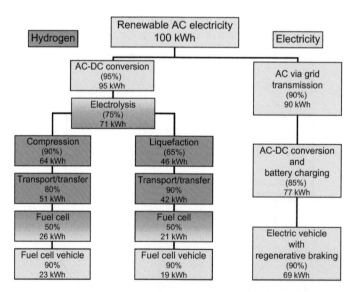

Figure 21.6. Comparison of the conversion efficiency for electric vehicles (EV) and fuel cell vehicles (FCV). *Source: Adapted from Bossel, Ulf. 2006. Does a Hydrogen Economy Make Sense? Proceedings of the IEEE, Vol. 94, No. 10, 1826-1837.*

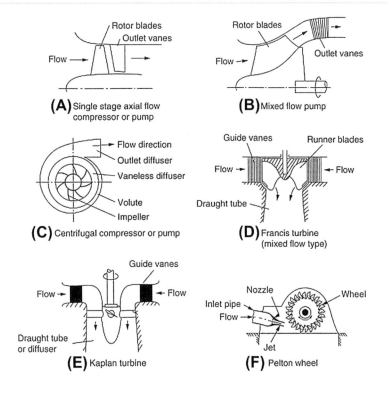

Figure 21.7. Examples of turbomachines.
Source: Dixon, S.L. and C.A. Hall. 2010. Fluid Mechanics and Thermodynamics of Turbomachinery (Sixth Edition), (Boston, Butterworth-Heinemann).

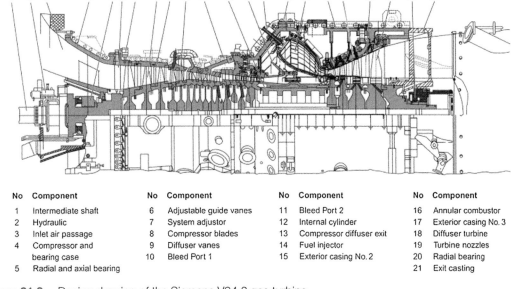

No	Component	No	Component	No	Component	No	Component
1	Intermediate shaft	6	Adjustable guide vanes	11	Bleed Port 2	16	Annular combustor
2	Hydraulic	7	System adjustor	12	Internal cylinder	17	Exterior casing No. 3
3	Inlet air passage	8	Compressor blades	13	Compressor diffuser exit	18	Diffuser turbine
4	Compressor and bearing case	9	Diffuser vanes	14	Fuel injector	19	Turbine nozzles
5	Radial and axial bearing	10	Bleed Port 1	15	Exterior casing No. 2	20	Radial bearing
						21	Exit casting

Figure 21.8. Design drawing of the Siemens V94.2 gas turbine.
Source: Boyce, Meherwan P. 2012. An Overview of Gas Turbines, Gas Turbine Engineering Handbook (Fourth Edition), (Oxford, Butterworth-Heinemann).

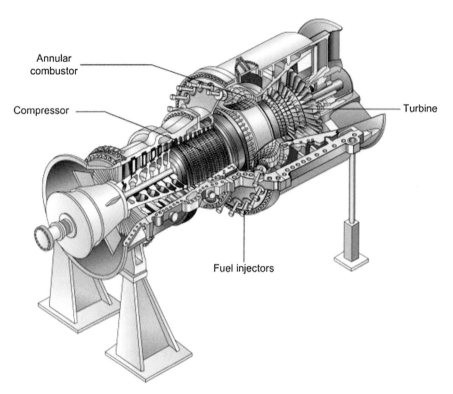

Figure 21.9. Cross-sectional representation of the Siemens V94.2 annular combustor-type gas turbine. This turbine consists of a 16-stage axial-flow compressor followed by an annular combustor and a four-stage reaction type axial-flow turbine, which drives both the axial-flow compressor and the generator.
Source: Boyce, Meherwan P. 2012. An Overview of Gas Turbines, Gas Turbine Engineering Handbook (Fourth Edition), (Oxford, Butterworth-Heinemann).

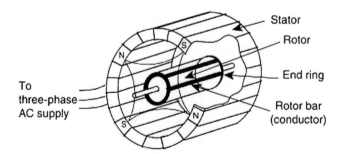

Figure 21.10. Schematic diagram of a squirrel-cage induction motor.
Source: de Almeida, Anibal and Steve Greenberg. 2004. Electric Motors, In: Cutler J. Cleveland, Editor-in-Chief, Encyclopedia of Energy, (New York, Elsevier), Pages 191-201.

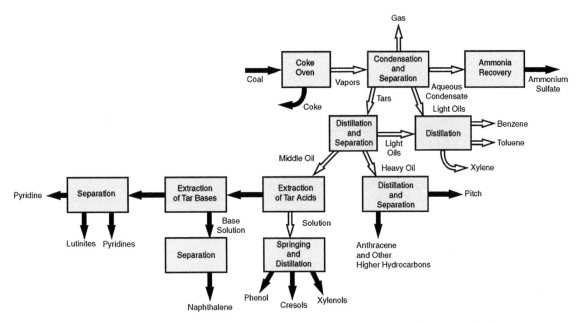

Figure 21.11. Byproducts recovered and processing performed to produce useful chemicals from coal.
Source: Miller, Bruce G. 2011. Clean Coal Engineering Technology, (Boston, Butterworth-Heinemann).

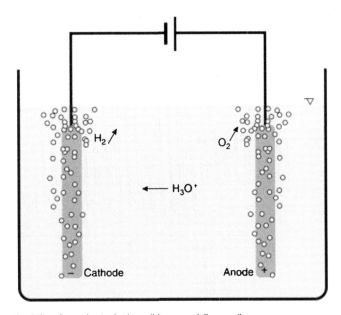

Figure 21.12. Basic principle of an electrolysis cell in an acidic medium.
Source: Smolinka, T. 2009. Water Electrolysis, In: Jürgen Garche, Editor-in-Chief, Encyclopedia of Electrochemical Power Sources, (Amsterdam, Elsevier), Pages 394-413.

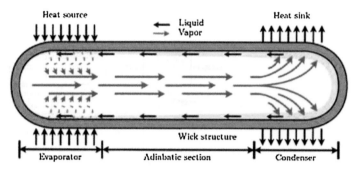

Figure 21.13. Schematic diagram of a conventional heat pipe.
Source: Zhang, Xingxing, Xudong Zhao, Stefan Smith, Jihuan Xu, Xiaotong Yu. 2012. Review of progress and practical application of the solar photovoltaic/thermal (PV/T) technologies, Renewable and Sustainable Energy Reviews, Volume 16, Issue 1, Pages 599-617.

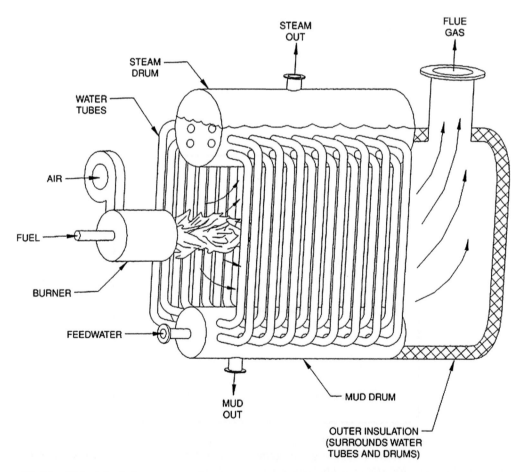

Figure 21.14. Watertube boiler cross section.
Source: Santoleri, Joseph J. 2003. Hazardous Waste Incineration, In: Robert A. Meyers, Editor-in-Chief, Encyclopedia of Physical Science and Technology (Third Edition), (New York, Academic Press), Pages 223-244.

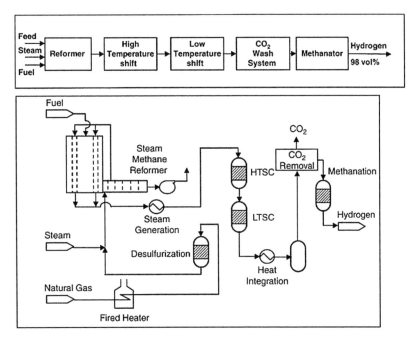

Figure 21.15. Steam reforming with carbon dioxide absorber and methanator. The final hydrogen purity ranges between 95 and 98%.
Source: Fahim, Mohamed A., Taher A. Alsahhaf, Amal Elkilani. 2010. Fundamentals of Petroleum Refining, (Amsterdam, Elsevier).

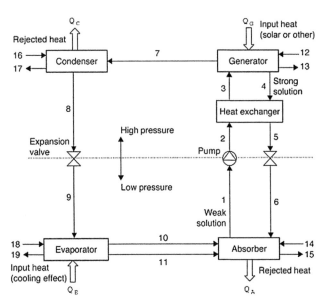

Figure 21.16. Schematic diagram of an absorption refrigeration system.
Source: Kalogirou, S.A. 2012. 3.01 - Solar Thermal Systems: Components and Applications – Introduction, In: Ali Sayigh, Editor-in-Chief, Comprehensive Renewable Energy, (Oxford, Elsevier), Pages 1-25.

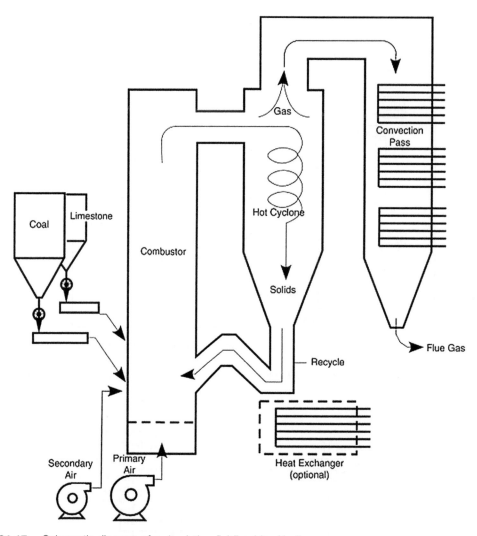

Figure 21.17. Schematic diagram of a circulating fluidized-bed boiler.
Source: Miller, Bruce G. 2011. Clean Coal Engineering Technology, (Boston, Butterworth-Heinemann).

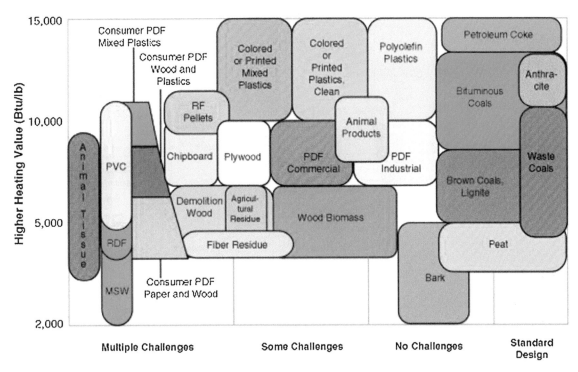

Figure 21.18. Applicable fuels for fluidized bed combustion.
Source: Miller, Bruce G. 2011. Clean Coal Engineering Technology, (Boston, Butterworth-Heinemann).

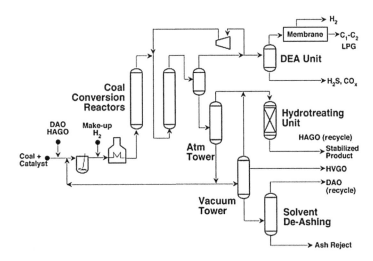

Figure 21.19. Hydrocarbon Technologies, Inc. coal liquefaction process.
Source: Miller, Bruce G. 2011. Clean Coal Engineering Technology, (Boston, Butterworth-Heinemann).

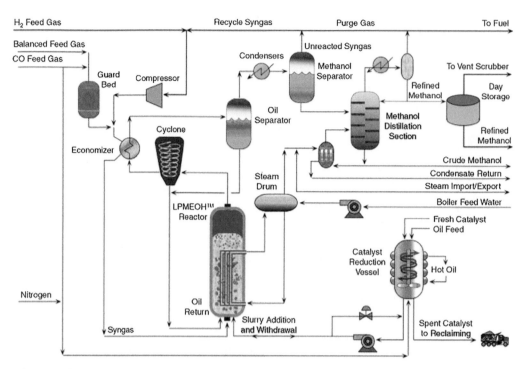

Figure 21.20. The production of methanol from coal-derived synthesis gas using the Liquid-Phase Methanol (LPMEOH) process, with the production of dimethyl ether (DME) as a mixed coproduct.
Source: United States Department of Energy, Office of Fossil Energy. 1999. Clean Coal Technology, Commercial-Scale Demonstration of the Liquid Phase Methanol (LPMEOH™) Process, Technical Report Number 11.

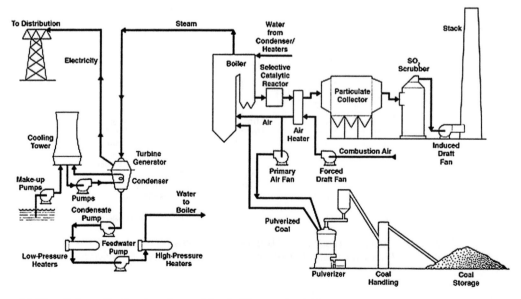

Figure 21.21. Schematic diagram of a coal-fired utility power plant.
Source: Miller, Bruce G. 2011. Clean Coal Engineering Technology, (Boston, Butterworth-Heinemann).

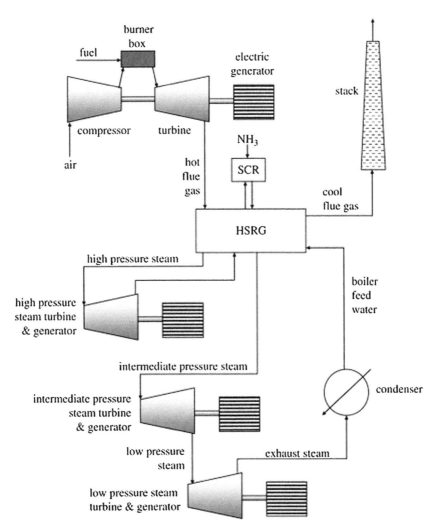

Figure 21.22. Schematic diagram of a combined cycle power plant. Steam from syngas processing is not shown. HRSG = heat recovery steam generator (HSRG).
Source: Bell, David and Brian Towler. 2011. Coal Gasification and Its Applications, (Boston, William Andrew Publishing).

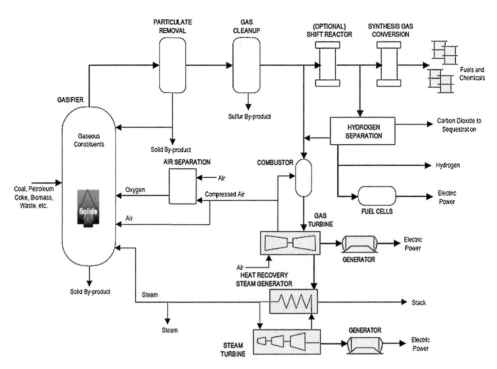

Figure 21.23. Gasification-based energy conversion.
Source: Speight, James G. 2011. Handbook of Industrial Hydrocarbon Processes, (Boston, Gulf Professional Publishing).

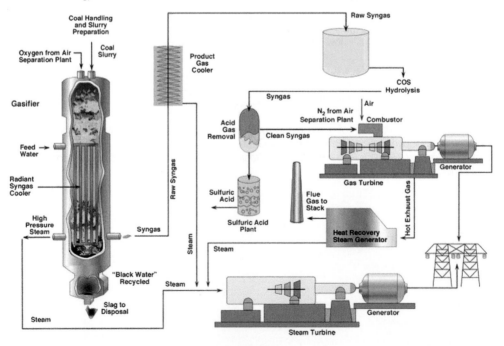

Figure 21.24. Integrated Gasification Combined Cycle (IGCC).
Source: Miller, Bruce G. 2011. Clean Coal Engineering Technology, (Boston, Butterworth-Heinemann).

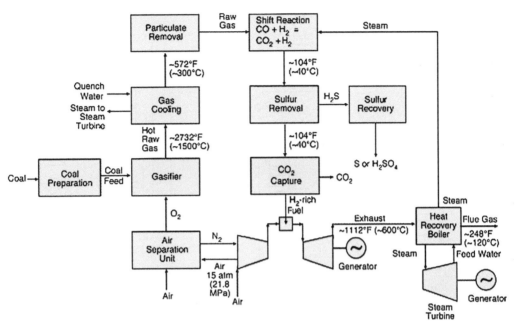

Figure 21.25. Simplified Integrated Gasification Combined Cycle process flowsheet with pre-combustion CO_2 capture and coal gas cleanup.
Source: Miller, Bruce G. 2011. Clean Coal Engineering Technology, (Boston, Butterworth-Heinemann).

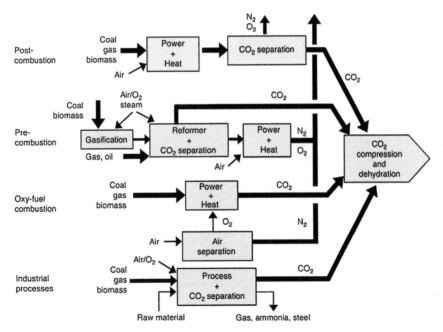

Figure 21.26. Principal routes for managing carbon dioxide emissions from power stations.
Source: Rand, D.A.J. and R.M. Del. 2009. Fuels - Hydrogen Production\Coal Gasification, In: Jürgen Garche, Editor-in-Chief, Encyclopedia of Electrochemical Power Sources, (Amsterdam, Elsevier), Pages 276-292.

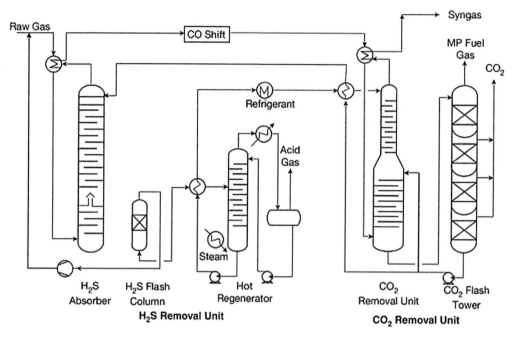

Figure 21.27. Rectisol process for H₂S and CO₂ removal.
Source: Miller, Bruce G. 2011. Clean Coal Engineering Technology, (Boston, Butterworth-Heinemann).

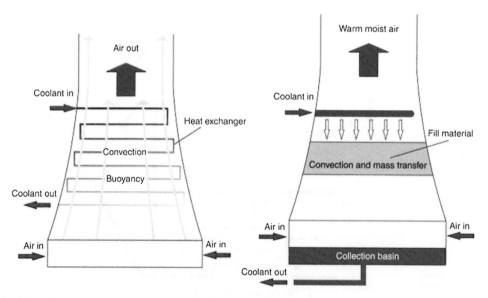

Figure 21.28. Natural draft cooling towers.
Source: Hoffschmidt, B.S. Alexopoulos, C. Rau, J. Sattler, A. Anthrakidis, C. Boura, B. O'Connor, P. Hilger. 2012. Concentrating Solar Power, In: Ali Sayigh, Editor-in-Chief, Comprehensive Renewable Energy, (Oxford, Elsevier), Pages 595-636.

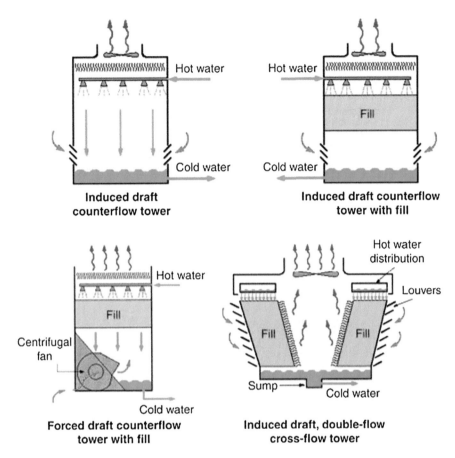

Figure 21.29. Wet cooling tower arrangements.
Source: Hoffschmidt, B.S. Alexopoulos, C. Rau, J. Sattler, A. Anthrakidis, C. Boura, B. O'Connor, P. Hilger. 2012. Concentrating Solar Power, In: Ali Sayigh, Editor-in-Chief, Comprehensive Renewable Energy, (Oxford, Elsevier), Pages 595-636.

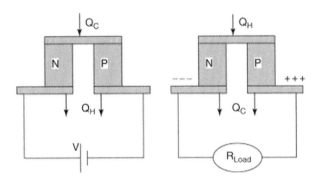

Figure 21.30. Schematic illustrations of thermoelectric couples in use for cooling (left) and power generation (right). In each case, QH and QC differ by the work done, either by the voltage source or on the load. For cooling, the heat sink is at the hotter temperature (TH), while the sink is at the lower temperature (TC) for power generation. *Source: Sharp, J.W. 2005. Thermoelectric and Energy Conversion Devices, In: Franco Bassani, Gerald L. Liedl, and Peter Wyder, Editor-in-Chief, Encyclopedia of Condensed Matter Physics, (Oxford,Elsevier), Pages 173-180.*

Charts

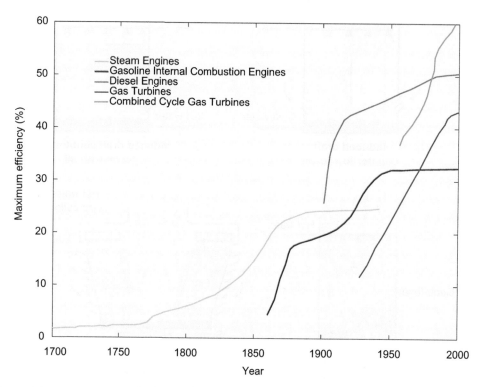

Chart 21.1. Maximum efficiency of inanimate prime movers, 1700-2000.
Source: Data from Smil, Vaclav. 2010. Energy Transitions. History, Requirements, Prospects, (Santa Barbara, Praeger).

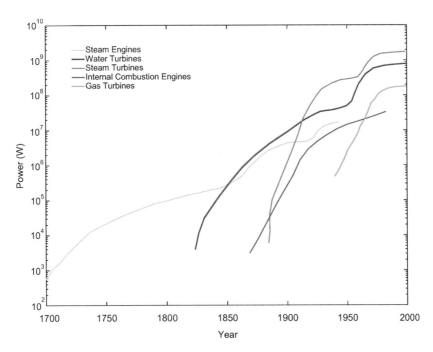

Chart 21.2. Maximum capacities of inanimate prime movers, 1700-2000.
Source: Data from Smil, Vaclav. 2010. Energy Transitions. History, Requirements, Prospects, (Santa Barbara, Praeger).

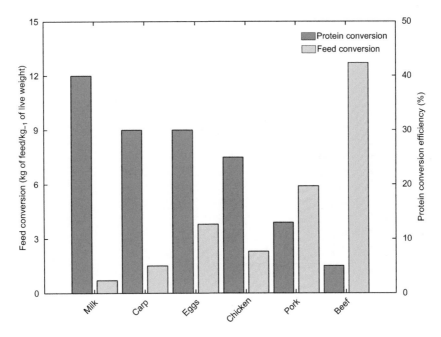

Chart 21.3. The efficiency of selected feed and protein conversions.
Source: Data from Smil, Vaclav. 2002. Nitrogen and food production: Proteins for human diets. Ambio 31:126-131.

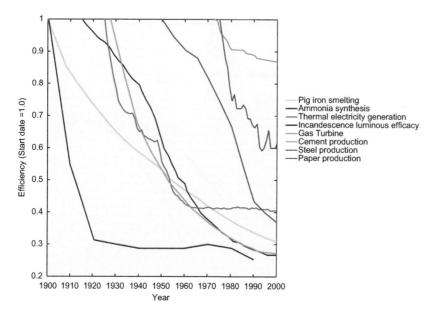

Chart 21.4. The index of energy effficiency for some important industrial processes. Efficiency is measured as energy input per unit output. The efficiency in the first year of each series set equal to 1.0. Thus, the efficiency of thermal electricity generation improved by about 60% from 1900 to 2000.
Source: Data compiled by authors from various sources.

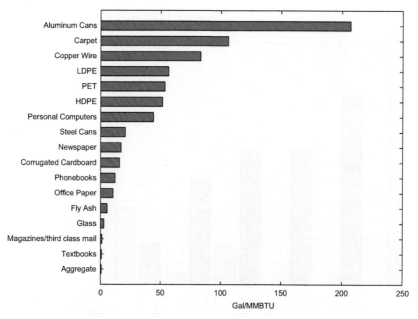

Chart 21.5. Energy savings per ton of recycled material, calculated as the reduction in energy use compared to using virgin materials. HDPE = high-density polyethylene; LDPE = low-density polyethylene; PET = polyethylene terephthalate.
Source: Data from Choate, Anne, Lauren Pederson, Jeremy Scharfenberg, Henry Ferland. 2005. Waste Management and Energy Savings: Benefits by the Numbers." (Washington, D.C., U.S. Environmental Protection Agency).

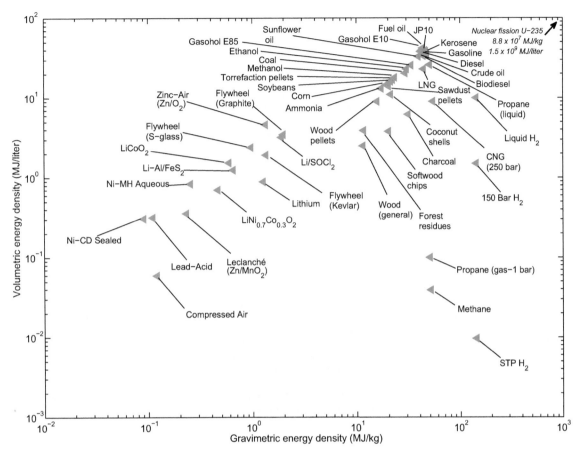

Chart 21.6. The volumetric and gravimetric density of selected energy sources and energy storage technologies.
Source: Data compiled by authors from various sources.

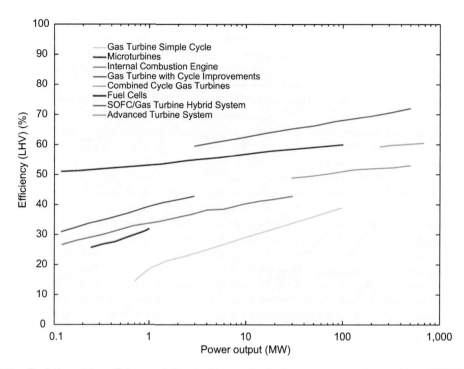

Chart 21.7. Evolution of the efficiency of chemical to mechanical energy conversion systems. SOFC = solid oxide fuel cell.

Source: Data from Ghoniem, Ahmed F. 2011. Needs, resources and climate change: Clean and efficient conversion technologies, Progress in Energy and Combustion Science, Volume 37, Issue 1, Pages 15-51.

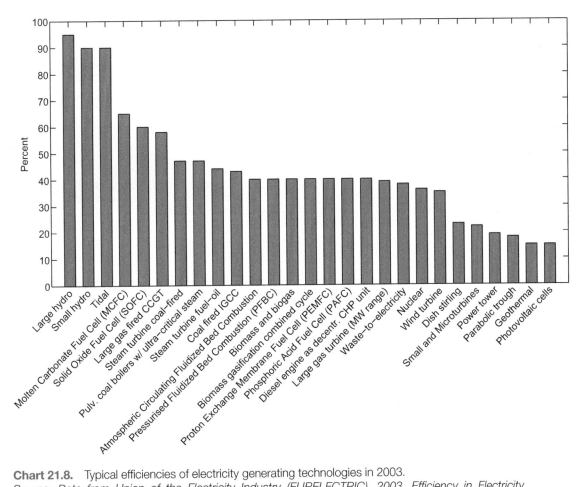

Chart 21.8. Typical efficiencies of electricity generating technologies in 2003.
Source: Data from Union of the Electricity Industry (EURELECTRIC). 2003. Efficiency in Electricity Generation (Essen, VGB PowerTech e.V.).

Chart 21.9. Estimated world energy use by electric motor-driven systems by end use sector.
Source: Data from Union of the Electricity Industry (EURELECTRIC). 2003. Efficiency in Electricity Generation (Essen, VGB PowerTech e.V.).

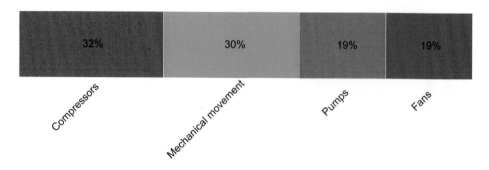

Chart 21.10. Estimated world energy use by electric motor-driven systems by end use application.
Source: Data from International Energy Agency (IEA). 2011. Walking the Torque: Proposed work plan for energy-efficiency policy opportunities for electric motor-driven systems, (Paris, IEA).

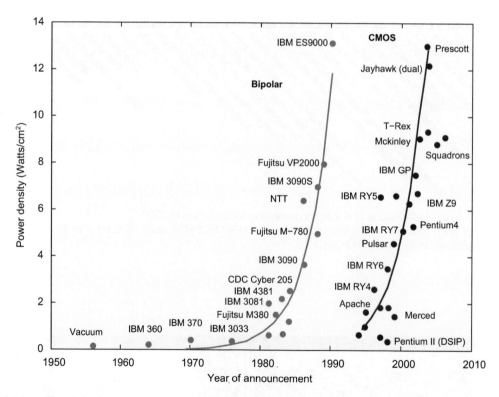

Chart 21.11. Trends in the power density of transistors.
Source: Data from Hutchinson, D. Dan. 2011. Power: Where It Matters, When It Matters, and When It Does Not, 2nd Berkeley Symposium on Energy Efficient Electronic Systems, University of California, Berkeley, November 3-4, 2011.

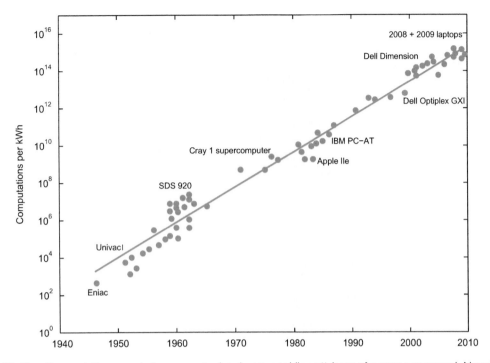

Chart 21.12. Computations made by computer hardware per kilowatt-hour of power consumed. Now known as "Koomey's law," the data indicate that the energy efficiency of computers has doubled nearly every eighteen months since the 1950s.
Source: Data from Koomey, Jonathan G., Stephen Berard, Marla Sanchez, and Henry Wong. 2011. Implications of Historical Trends in the Electrical Efficiency of Computing, IEEE Annals of the History of Computing. vol. 33, no. 3. July-September. pp. 46-54.

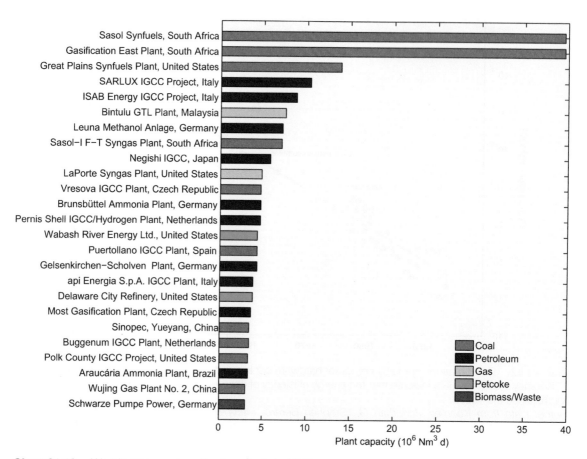

Chart 21.13. World's 25 largest gasification plants in 2010.
Source: Data from United States Department of Energy, National Energy Technology Laboratory, Worldwide Gasification Database, <http://www.netl.doe.gov/technologies/coalpower/gasification/database/database.html>.

Chart 21.14. Global syngas production by feedstock, 2010.
Source: Data from Gasification Technologies Council, <http://www.gasification.org>.

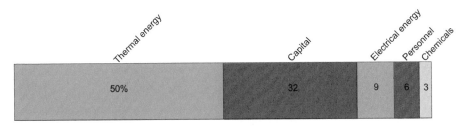

Chart 21.15. Typical cost structure for a thermal desalination plant.
Source: Data from Cooley, Heather, Peter H. Gleick, Gary Wolff. 2006. Desalination, with a grain of salt, (Oakland CA, Pacific Institute).

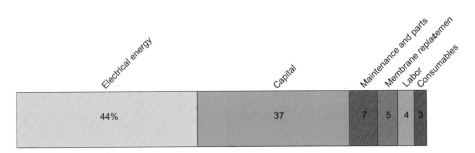

Chart 21.16. Typical cost structure for a seawater reverse osmosis (SWRO) desalination plant.
Source: Data from Cooley, Heather, Peter H. Gleick, Gary Wolff. 2006. Desalination, with a grain of salt, (Oakland CA, Pacific Institute).

Tables

Table 21.1. Common chemical reactions of combustion

Fuel	Reaction	Moles	Pounds	Heat of combustion (high) BTU/LB of fuel
Carbon (to CO)	$2C + O_2 = 2CO$	$2 + 1 = 2$	24 lbs + 23 lbs = 56 lbs	3,960
Carbon (to CO_2)	$C + O_2 = CO_2$	$1 + 1 = 1$	$12 + 32 = 44$	14,100
Carbon Monoxide	$2CO + O_2 = 2CO_2$	$2 + 1 = 2$	$56 + 32 = 88$	4,345
Hydrogen	$2H_2 + O_2 = 2H_2O$	$2 + 1 = 2$	$4 + 32 = 36$	61,100
Sulfur (to SO_2)	$S + O_2 = SO_2$	$1 + 1 = 1$	$32 + 32 = 64$	3,980
Methane	$CH_4 + 2O_2 = CO_2 + H_2O$	$1 + 2 = 1 + 2$	$16 + 64 = 80$	23,875
Acetylene	$2C_2H_2 + 5O_2 = 4CO_2 + 2H_2O$	$2 + 5 = 4 + 2$	$52 + 160 = 212$	21,500
Ethylene	$C_2H_4 + 3O_2 = 2CO_2 + 2H_2O$	$1 + 3 = 2 + 2$	$28 + 96 = 124$	21,635
Ethane	$2C_2H_6 + 7O_2 = 4CO_2 + 6H_2O$	$2 + 7 = 4 + 6$	$60 + 224 = 284$	22,325
Hydrogen to Sulfide	$2H_2S + 3O_2 = 2SO_2 + 2H_2O$	$2 + 3 = 2 + 2$	$68 + 96 = 164$	7,100

Source: Adapted from Babcock and Wilcox. 1978. Steam: Its Generation and Use, (Charlotte, Babcock and Wilcox).

Table 21.2. Basic gasification reactions

Reaction	Heating value (kJ/mol)
$2C + O_2 \Leftrightarrow 2CO$	246.4
$C + O2 \Leftrightarrow CO_2$	408.8
$CH_4 + H_2O \Leftrightarrow CO + 3H_2$	−206
$CH_4 + 2H_2O \Leftrightarrow CO_2 + 4H2$	−165
$C + CO_2 \Leftrightarrow 2CO$	−172
$C + H_2O \Leftrightarrow CO + H_2$	−131

Source: Adapted from Panwar, N.L., Richa Kothari, V.V. Tyagi. 2012. Thermo chemical conversion of biomass–Eco friendly energy routes, Renewable and Sustainable Energy Reviews, Volume 16, Issue 4, Pages 1801–1816.

Table 21.3. Common gasification processes

Process	Temperature (°C)	Pressure (Psia)	Gasification medium	Type of coal feed	Product gas composition[b] (dry basis)						Product gas calorific value[b] (Btu/scf)	Comments
					CO	CO$_2$	H$_2$	CH$_4$	C$_2$H$_6$	Others		
Lurgi[a]	980	400	O$_2$ + steam	Noncaking	68.5	29.5	40.4	9.4	–	1.2	302	
Winkler[a]	815	15	O$_2$ + steam	Any coal	33.4	20.5	41.9	3.1	–	1.1	275	
Synthane	980	1000	O$_2$ + steam	Any coal	16.7	28.9	27.8	24.5	0.8	1.3	405	
Hygas steam–oxygen	980	1000	O$_2$ + steam	Any coal	23.8	24.5	30.2	18.6	0.6	2.3	374	
Koppers–Totzek[a]	1510	20	O$_2$ + steam	Any coal	55.8	6.2	36.6	–		1.4	298	
Bi-gas	1480	1200	O$_2$ + steam	Any coal	21.5	29.3	32.1	15.6		1.5	367	
Texaco	1480	600	O$_2$	Any coal in the form of coal–water slurry	46.6	11.5	38.7	0.7		2.7	300	
Cogas	980	75	Steam, and air as heat source	Any coal	7.4	9.3	26.2	34	8.3	14.8	726	
Producer	980	15	Air + steam	Any coal	29	4	12	3	–	52	130	
Water gas	980	15	Air + steam	Any coal	41	5	49	0.5	–	4.5	300	
Westinghouse	1930	20	Air + steam	Any coal	19.2	9.4	14.4	2.8		54.2	140	Oxygen can also be used
U-gas	1040	350	Air + steam	Any coal	9.8	12	10.3	–		67.9	75	
Combustion	1930	20	Air	Any coal	24.4	4.1	10.7		–	–	125	
CO$_2$ acceptor	815	150	Air + steam	Low rank	17	6.6	53.8		20.9	0.4	440	Oxygen can also be used
Hydrane	815	>1000	H$_2$	Any coal	3.9	–	22	73.2	–	–	826	
Kellog salt	930	1200	O$_2$ + steam	Any coal	33.5	13.3	45	7.5	–	0.7	348	Oxygen can also be used

[a]Commercially available processes.
[b]Composition varies with coal used.
Source: Adapted from Pisupadti, Sarma V. 2003. Fuel Chemistry, In: Robert A. Meyers, Editor-in-Chief, Encyclopedia of Physical Science and Technology (Third Edition), (New York, Academic Press), Pages 253–274.

Table 21.4. Some significant gasification units in operation worldwide

Plant Owner	Country	Gasifier Technology	Total Gasifiers	Syngas Capacity (Nm³/d)	MWth Output	Feedstock	Products
Sasol Chemical Industries Ltd./ Sasol Ltd.	South Africa	Sasol Lurgi	17	7,100,100	970.6	Coal	FT liquids
Beijing No. 4 Chemical	China	GE Energy	1	320,000	43.7	Petroleum	Oxcochemicals
China National Petrochemical Corp./ Sinopec	China	GE Energy	1	210,000	28.7	Petroleum	Oxcochemicals
CNPC Ningxia Dayuan Refining & Chemical Ind. Co. Ltd.	China	GE Energy	3	2,500,000	341.8	Petroleum	Gases
China National Petrochemcial Corp./ Sinopec	China	GE Energy	3	2,100,000	286.6	Petroleum	Ammonia
Dalian Chemical Industrial Corp.	China	GE Energy	2	2,096,500	286.6	Petroleum	Ammonia
Lu Nan Chemical Industry Co./ CNTIC	China	GE Energy	2	525,000	71.8	Coal	Ammonia
Shanghai Coking & Chemical (Shanghai Pacific)	China	GE Energy	3	1,530,000	209.2	Coal	Methanol, town gas, acetic acid
Weihe Fertilizer Co.	China	GE Energy	3	2,040,000	278.9	Coal	Ammonia
Zhnhai Refining & Chemical Co.	China	GE Energy	3	2,100,000	287.1	Petroleum	Ammonia
Ube Ammonia Industry Co. Ltd.	Japan	GE Energy	4	2,150,000	293.9	Coal	Ammonia
Shell MDS (Malaysia) Sdn. Bhd.	Malayasia	Shell	6	7,552,000	1,032.40	Gas	Mid-distillates
Linde AG	Singapore	GE Energy	2	1,610,000	220.1	Petroleum	H_2, CO
Air Products & Chemicals, Inc.	United States	GE Energy	2	1,850,000	252.7	Gas	H_2, CO
Dakota Gasification Co.	United States	Sasol Lurgi	14	1,390,000,000	1,900.30	Coal	SNG, CO_2
Eastman Chemical Co.	United States	GE Energy	2	1,600,000	218.7	Coal	Acetic anhydride, methanol
Frontier Oil & Refining Co. (Texaco Inc.)	United States	GE Energy	1	80,559	11	Petcoke	Electricity, HP steam

Table 21.4. **Continued**

Plant Owner	Country	Gasifier Technology	Total Gasifiers	Syngas Capacity (Nm³/d)	MWth Output	Feedstock	Products
Valero Energy Corp.	United States	GE Energy	2	3,800,000	519.5	Petcoke	Electricity, steam
Motiva Enterprises LLC	United States	GE Energy	2	1,880,000	257	Petroleum	H₂
Global Energy, Inc.	United States	E-Gas	2	4,320,000	590.6	Petcoke	Electricity

Source: Adapted from Miller, Bruce G. 2011. Clean Coal Engineering Technology, (Boston,Butterworth-Heinemann).

Table 21.5. **Characteristics of generic gasifier technologies**

Gasifier Type	Fixed-Bed		Fluidized-Bed		Entrained-Flow
Ash Conditions	Dry Ash	Slagging	Dry Ash	Agglomerating	Slagging
Fuel Characteristics					
Fuel size limits	1/4–2 inches	1/4–2 inches	<1/4 inch	<1/4 inch	<0.005 inches
Acceptability of caking coal	Yes (with modifications)	Yes	Possibly	No, noncaking only	Yes
Preferred feedstock	Lignite, reactive bituminous coal, anthracite, wastes	Bituminous coal, anthracite, petcoke, wastes	Lignite, reactive bituminous coal, anthracite, wastes	Lignite, bituminous coal, anthracite, cokes, biomass, wastes	Lignite, reactive bituminous coal, anthracite, petcokes
Ash content limits	No limitation	<25% preferred	No limitation	No limitation	<25% preferred
Preferred ash melting temperature, °F	>2,200	<2,370	>2,000	>2,000	<2,372
Operating Characteristics					
Exit gas temperature, °F	Low[a] (800–1,200)	Low (800–1,200)	Moderate (1,700–1,900)	Moderate (1,700–1,900)	High (>2,300)
Gasification pressure, psig	435+	435+	15	15–435	<725
Oxidant requirement	Low	Low	Moderate	Moderate	High
Steam requirement	High	Low	Moderate	Moderate	Low
Unit capacities, MWth equivalent	10–350	10–350	100–700	20–150	Up to 700
Key Distinguishing Characteristics	Hydrocarbon liquids in raw gas		Large char recycle		Large amount of sensible heat energy in the hot raw gas
Key Technical Issue	Utilization of fines and hydrocarbon liquids		Carbon conversion		Raw gas cooling

[a]Fixed-bed gasifiers operating on low-rank coals have exit temperatures lower than 800 °F.
Source: Adapted from Ratafia-Brown, J., L. Manfredo, J. Hoffmann, and M. Ramezan. 2002. Major Environmental Aspects of Gasification-Based Power Generation Technologies, prepared for U.S. Department of Energy, Office of Fossil Energy, National Energy Technology Laboratory.

Table 21.6. Gasification-based energy conversion options

Resources	Gasifiers	Environmental control	Energy conversion	Products
		Particulate Removal and Recycle		
		Filtration		
		Water Scrubbing		
		Chloride and Alkali Removal		
Air/Oxygen	Oxygen-Blown	Water Scrubbing	Gas Turbine	Steam
Coal	Entrained Flow	**Acid Gas Removal**	Heat Recovery Steam	Electric Power
Biomass	Fluidized Bed	Amine Processes		Liquid Fuels
Petroleum Coke	Moving Bed	Rectisol, Selexol	Steam Turbine	Chemicals
Heavy Oil	Transport Reactor	**COS Hydrolysis**	Boiler	Methanol
RefineryWastes	Air-Blown	**Sulfur Recovery**	Syngas Conversion to Fuels & Chemicals	Hydrogen
MSW	Fluidized Bed	Claus Process		Ammonia/Fertilizers
Orimulsion	Spouting Bed	Scott Process	Catalytic Conversion	Slag
Other Wastes	Entrained Flow	Sulfuric Acid Plant	Shift Conversion	Sulfur/Sulfuric
	Transport Reactor	**Water Treatment**	Fischer-Tropsch	Acid
		Process Water, BFW	Fuel Cell	
		Tail Gas Treating	H_2 Turbine	
		Turbine NOx Control		
		Nitrogen/Steam Dilution		
		Syngas Mercury Capture		
		Syngas CO2 Capture		

MSW = municipal solid waste
Source: Adapted from Ratafia-Brown, J., L. Manfredo, J. Hoffmann, and M. Ramezan. 2002. Major Environmental Aspects of Gasification-Based Power Generation Technologies, prepared for U.S. Department of Energy, Office of Fossil Energy, National Energy Technology Laboratory.

Table 21.7. The coalification process

Materials	Partial processes	Main chemical reactions
Decaying Vegetation		
↓	Peatification	Bacterial and fungal life cycles
Peat		
↓	Lignification	Air oxidation, followed by decarboxylation and dehydration
Lignite		
↓	Bituminization	Decarboxylation and hydrogen disproportioning
Bituminous coal		
↓	Preanthracitization	Condensation to small aromatic ring systems
Semianthracite		
↓	Anthracitization	Condensation to small aromatic ring systems to larger ones; dehydrogenation
Anthracite	Graphitization	Complete carbonification

Source: Adapted from D.W. Van Krevelen. 1993. Coal: Typology–Physics–Chemistry–Constitution (third edition), (Elsevier Science).

Table 21.8. Comparison of characteristics of coal combustion methods

Variables	Fixed bed (stoker)	Fluidized bed	Suspension
	Combustion method		
Particle size			
Approximate top size	<2 inches	<0.2 inches	180 µm
Average size	0.25 inches	0.04 inches	45 µm
System/bed temperature	<1,500 °F	1,500–1,800 °F	>2,200 °F
Particle heating rate	≈1°/s	$10^3–10^4$ °/s	$10^3–10^6$ °/s
Reaction time			
Volatiles	≈100 seconds	10–50 seconds	<0.1 seconds
Char	≈1,000 seconds	100–500 seconds	<1 second
Reactive element description	Diffusion-controlled combustion	Diffusion-controlled combustion	Chemically controlled combustion

Source: Adapted from Miller, Bruce G. 2011. Clean Coal Engineering Technology, (Boston,Butterworth-Heinemann).

Table 21.9. Characteristics of select coal liquefaction processes

Process	I.G. Farben/ Bergius	SRC-I	SRC-II	H-Coal	EDS	Costeam	HTI DCL
Temperature (°F)							
Stage 1	900	840	860	840	700–900	750–840	800
Stage 2	750	NA[a]	NA	NA	500–840		?[b]
Pressure (atm)							
Stage 1	350–700	140	140	210	20–170	270	170
Stage 2	300	NA	NA	NA	80–210[c]	NA	?
Carbon content of coal (wt.%)	70–83	70–85	≈75	≈78	75–80	60–70	?
H_2 consumption (wt.% on coal)							
Stage 1	1	2.4	4.7	5.5	—	6.5+	?
Stage 2	4	NA	NA	NA	6[c]	NA	?
Scale of operation (short tons per day)	150–1,800	50	50	600	250	(5–10 lb/h)[d]	4,300[e]
Sulfur in coal (wt.%)	<3.0	3	3	1–3	1–3	1	?
Sulfur in product (wt.%)	0.1	0.7	0.3–0.7	0.3–0.7	0.3	low	low

[a]Not applicable.
[b]Unknown.
[c]Recycle oil hydrogenation.
[d]Small-scale, lb/h.
[e]Per train, 3 trains planned.
Source: Adapted from Miller, Bruce G. 2011. Clean Coal Engineering Technology, (Boston,Butterworth-Heinemann).

Table 21.10.   Typical coal liquefaction conditions

Reaction	Temperature (°C)	Pressure (psig)
Pyrolysis	500–650	50
Hydroliquefaction	400–480	1500–2500

Source: Larsen, John W. and Martin L. Gorbaty, Coal Structure and Reactivity. 2003. In: Robert A. Meyers, Editor-in-Chief, Encyclopedia of Physical Science and Technology (Third Edition), (New York, Academic Press), Pages 107–122.

Table 21.11.   Variation of hydroliquefaction of coal conversion with rank

	Batch hydroliquefaction	
Coal rank	Conversion to liquids and gas (% of coal, dry ash-free basis)	Barrels oil/ton coal
Lignite	86	3.54
High-volatile bituminous-1	87	4.68
High-volatile bituminous-2	92	4.9
Medium-volatile bituminous	44	2.1
Anthracite	9	0.44

Source: Adapted from Larsen, John W. and Martin L. Gorbaty, Coal Structure and Reactivity. 2003. In: Robert A. Meyers, Editor-in-Chief, Encyclopedia of Physical Science and Technology (Third Edition), (New York, Academic Press), Pages 107–122.

Table 21.12.   Comparison of Coal-to-Liquids (CTL) technologies

Plant type	DCL	ICL recycle	ICL once–through	Hybrid
Coal Consumption (STPD[b] dry basis)				
Coal Feed Rate to DCL	15,568	0	0	7,784
Coal Feed Rate to Gasifier	7,476	32,305	37,974	17,730
Total Coal Feed Rate	23,044	32,305	37,974	25,514
Liquid Product Capacity (BPD)				
Diesel	45,812	47,687	47,687	46,750
Naphtha	18,863	22,313	22,313	20,591
LPG	5,325	0	0	2,660
Total	70,000	70,000	70,000	70,000
Electric Power Capacity (MW)				
Gross	0	1,419	2,214	725
Parasitic	282	1,018	1,077	680
Net Export	0	399	1,139	45
Net Import	282	0	0	0
Energy Balance				
Total Energy Input (MM BTU/D)	653,057	831,012	976,855	656,323
Total Energy Output (MM BTU/D)	392,776	402,001	462,559	385,490

Table 21.12. Continued

Plant type	DCL	ICL recycle	ICL once-through	Hybrid
Overall Thermal Efficiency (%)	60.14	48.37	47.35	58.73
Coal Input (MM BTU/BBL product)	8.47	11.87	13.96	9.38
Product Yield (BBL of product /ST dry coal)	3.04	2.17	1.84	2.74
Carbon Balance				
Carbon in Product (% of input C)	53	34	29	45
Carbon in Slag/Ash (% of input C)	1	1	1	1
Carbon in CO_2 (% of input C)	46	65	70	54
Plant CO_2 Generation (lbs/bbl product)	783	1,557	1,972	1,010
Economics				
Relative Capital Cost	1.00	1.10	1.25	1.03

Table 21.13. Reactions in the Fischer–Tropsch process

Main reactions	
Paraffins	$(2n + 1)H_2 + nCO \rightarrow CnH2n + 2 + nH_2O$
Olefins	$2nH_2 + nCO \rightarrow CnH2n + nH_2O$
Water gas shift reaction	$CO + H_2O \rightleftharpoons CO_2 + H_2$
Side reactions	
Alcohols	$2nH_2 + nCO \rightarrow CnH2n + 2O + (n-1)H_2O$
Boudouard reaction	$2CO \rightarrow C + CO_2$
Catalyst modifications	
Catalyst oxidation/reduction	a. $MxOy + yH_2 \rightleftharpoons yH_2O + xM$
	b. $MxOy + yCO \rightleftharpoons y_2CO + xM$
Bulk carbide formation	$yC + xM \rightleftharpoons MxCy$

Source: Fahim, Mohamed A., Taher A. Alsahhaf, Amal Elkilani. 2010. Fundamentals of Petroleum Refining, (Amsterdam, Elsevier).

Table 21.14. Principal characteristics of different desalination processes[a]

Characteristics	Type of process		
	Phase change	Non-phase change	Hybrid
Nature	Thermal process: MED, MSF, MVC, TVC (evaporation and condensation)	Pressure/concentration gradient driven: RO (membrane separation), ED (electrochemical separation)	Thermal + membrane: membrane distillation, MSF/RO, MED/RO
Membrane pore size	-	0.1–3.5 nm	0.2–0.6 mm
Feed temperature	60–120 °C	<45 °C	40–80 °C
Cold water stream	May be required	-	20–25 °C
Driving force for separation	Temperature and concentration gradient	Concentration an pressure gradient	Temperature and concentration gradient
Energy	Thermal and mechanical	Mechanical and/or electrical	Thermal and mechanical
Form of energy	Steam, low-grade heat or waste heat and some mechanical energy for pumping	Requires prime quality mechanical/electrical energy derived from fossil fuels or renewable sources	Low-grade heat sources or renewable energy sources
Product quality	High quality distillate with TDS[b] <20 ppm	Potable water quality TDS <500 ppm	High quality distillate with TDS 20–500 ppm

[a]solar distillation (SD); multi-effect distillation (MED); multi-stage flash distillation (MSF); mechanical vapor compression (MVC); thermal vapor compression (TVC); electrodialysis (ED); reverse osmosis (RO); membrane distillation (MD).
[b]TDS = total dissolved solids.
Source: Adapted from Gude, Veera Gnaneswar, Nagamany Nirmalakhandan, Shuguang Deng. 2010. Renewable and sustainable approaches for desalination, Renewable and Sustainable Energy Reviews, Volume 14, Issue 9, Pages 2641–2654.

Table 21.15. Desalination costs for different desalination processes based on capacities[a]

Desalination process	Capacity (m³/day)	Energy source	Energy cost ($/kWh)	Desalinated water cost ($/m³)
Domestic applications				
Solar still	0.006	Solar		12.53
Solar still	0.009	Solar		10
Solar still	0.8	Solar		12.5
Solar still	1	Solar		12
MESS	1	Solar		50
PV-RO	1	Solar		12.05
PV-RO	1	Solar		3.73
MD	0.1	Solar		15
MD	0.5	Solar		18
MD	1	Geothermal		130
MSF	1	Solar		2.84
Solar still	5	Solar		0.52–2.99
ED	5	Electric		5
RO	10	Electric		4
MED	<100	Conventional		2.5–10
MVC	375	Conventional		2.9–3.8

Table 21.15. **Continued**

Desalination process	Capacity (m³/day)	Energy source	Energy cost ($/kWh)	Desalinated water cost ($/m³)
Small-scale applications				
Reverse osmosis	250	Diesel generators	0.07	3.21
Reverse osmosis	300	Diesel generators	0.06	1.82
Reverse osmosis	350	Diesel generators	0.06	1.36
Reverse osmosis	500	Diesel generators	0.07	2.94
Reverse osmosis	500	Diesel generators	0.06	1.42
Reverse osmosis	500	Diesel generators	0.06	1.25
Reverse osmosis	500	Diesel generators	0.06	2.57
Reverse osmosis	600	Diesel generators	0.06	2.95
MVC	1000	Conventional		1.51
MVC	1000–1200	Wind		2–2.6
MVC	1200	Conventional		3.22
Vapor compression	3000	Conventional		0.7
MFD	10,000	Conventional		0.88
MED	12,000–55,000	Conventional		0.95–1.95
MSF	20,000	Natural gas		2.02
Large-scale applications				
Dual-purpose MSF	20,000	Natural gas/steam	0.0001	0.08
Reverse osmosis	2000	Diesel generator	0.06	2.23
Reverse osmosis	5000	Diesel generators	0.06	1.54
Reverse osmosis	10,000	Diesel generators	0.05	1.18
Reverse osmosis	20,000	Diesel generators	0.05	1.04
Reverse osmosis	50,000	Diesel generators		0.86
Reverse osmosis	95,000	Conventional		0.83
Reverse osmosis	100,000	Conventional		0.43
Reverse osmosis	$100–320 \times 10^3$	Conventional		0.45–0.66
MED	$91–320 \times 103$	Conventional		0.52–1.01
MSF	$23–528 \times 103$	Conventional		0.52–1.75

[a]solar distillation (SD); multi-effect distillation (MED); multi-stage flash distillation (MSF); mechanical vapor compression (MVC); thermal vapor compression (TVC); electrodialysis (ED); reverse osmosis (RO); membrane distillation (MD).

[b]TDS = total dissolved solids.

Source: Adapted from Gude, Veera Gnaneswar, Nagamany Nirmalakhandan, Shuguang Deng. 2010. Renewable and sustainable approaches for desalination, Renewable and Sustainable Energy Reviews, Volume 14, Issue 9, Pages 2641–2654.

Table 21.16. Relative power requirements of desalination processes (assumed conversion efficiency of electricity generation of 30%)

Process	Heat input (kJ/kg of product)	Mechanical power input (kWh/m³)	Primary energy consumption (kJ/kg)
Multi-stage flash	294	2.5:4	338.4
Multiple-effect boiling	123	2.2	149.4
Vapor compression	N/A	8:16	192
Electro-dialysis	N/A	12	144
Solar still	2330	0.3	2333.6
Reverse osmosis	N/A	5:13	120
Energy recovery reverse osmosis	N/A	4:06	60

Source: Adapted from El-Ghonemy, A.M.K. 2012. Water desalination systems powered by renewable energy sources: Review, Renewable and Sustainable Energy Reviews, Volume 16, Issue 3, Pages 1537–1556.

Table 21.17. Renewable energy system desalination combinations

RES technology	Feedwater salinity	Desalination technology
Solar thermal	Seawater	Multiple-effect boiling (MEB)
	Seawater	Multi-stage flash (MSF)
Photovolatics	Seawater	Reverse osmosis (RO)
	Brackish water	Reverse osmosis (RO)
	Brackish water	Electrodialysis (ED)
Wind energy	Seawater	Reverse osmosis (RO)
	Brackish water	Reverse osmosis (RO)
	Seawater	Mechanical vapor compression (MVC)
Geothermal	Seawater	Multiple-effect boiling (MEB)

Source: Kalogirou, Soteris A. 2009. Solar Energy Engineering, (Boston, Academic Press).

Table 21.18. Energy requirements and greenhouse gas emissions for different desalination processes[a]

Process	Multi-effect solar still (MESS)	Multi-stage distillation (MSF)	Multi-effect distillation (MED)	MVC	MED-TVC	Reverse osmosis	ED
Energy requirements							
Thermal energy (kJ/kg)	1500	250–300	150–220	-	220–240	-	-
Electrical energy (kWh/m³)	0	3.5–5	1.5–2.5	11–12	1.5–2	5–9	2.6–5.5
GHG emissions (kg CO_2/m³ H_2O)							
Total electric equivalent (kWh/m³)	0	15–25	8–201	11–12	21.5–22	5–9	2.6–5.5
Maximum value	0	24	19.2	11.5	21	8.6	5.3

[a]solar distillation (SD); multi-effect distillation (MED); multi-stage flash distillation (MSF); mechanical vapor compression (MVC); thermal vapor compression (TVC); electrodialysis (ED); reverse osmosis (RO); membrane distillation (MD).
[b]TDS = total dissolved solids.
Source: Adapted from Gude, Veera Gnaneswar, Nagamany Nirmalakhandan, Shuguang Deng. 2010. Renewable and sustainable approaches for desalination, Renewable and Sustainable Energy Reviews, Volume 14, Issue 9, Pages 2641–2654.

Table 21.19. **Properties of potential working fluid candidates for organic Rankine cycles and supercritical rankine cycles**

Name	Molecular weight	Tc (K)	Pc (MPa)	Vapor C_p (J/kg K)	Latent heat L (kJ/kg)	ξ (J/kg K^2)[a]
Dichlorofluoromethane	102.92	451.48	5.18	339.85	216.17	−0.78
Chlorodifluoromethane	86.47	369.3	4.99	1069.13	158.46	−1.33
Trifluoromethane	70.01	299.29	4.83	3884.02	89.69	−6.49
Difluoromethane	52.02	351.26	5.78	2301.61	218.59	−4.33
Fluoromethane	34.03	317.28	5.9	3384.66	270.04	−7.20
Hexafluoroethane	138.01	293.03	3.05	4877.91	30.69	−5.54
2,2-Dichloro-1,1,1-trifluoroethane	152.93	456.83	3.66	738.51	161.82	0.26
2-Chloro-1,1,1,2-tetrafluoroethane	136.48	395.43	3.62	908.7	132.97	0.26
Pentafluoroethane	120.02	339.17	3.62	1643.89	81.49	−1.08
1,1,1,2-Tetrafluoroethane	102.03	374.21	4.06	1211.51	155.42	−0.39
1,1-Dichloro-1-fluoroethane	116.95	477.5	4.21	848.37	215.13	0
1-Chloro-1,1-difluoroethane	100.5	410.26	4.06	1036.52	185.69	0
1,1,1-Trifluoroethane	84.04	345.86	3.76	1913.97	124.81	−1.49
1,1-Difluoroethane	66.05	386.41	4.52	1456.02	249.67	−1.14
Ethane	30.07	305.33	4.87	5264.72	223.43	−8.28
Octafluoropropane	188.02	345.02	2.64	1244.87	58.29	0.45
1,1,1,2,3,3,3-Heptafluoropropane	170.03	375.95	3	1013	97.14	0.76
1,1,1,2,3,3-Hexafluoropropane	152.04	412.44	3.5	973.69	142.98	0.76
1,1,2,2,3-Pentafluoropropane	134.05	447.57	3.93	1011.26	188.64	0.6
1,1,1,3,3-Pentafluoropropane	134.05	427.2	3.64	980.9	177.08	0.19
Cyclopropane	42.08	398.3	5.58	1911.81	366.18	−1.54
Propane	44.1	369.83	4.25	2395.46	292.13	−0.79
Octafluorocyclobutane	200.03	388.38	2.78	896.82	93.95	1.05
Decafluorobutane	238.03	386.33	2.32	928.83	77.95	1.32
Dodecafluoropentane	288.03	420.56	2.05	884.25	86.11	1.56
Butane	58.12	425.13	3.8	1965.59	336.82	1.03
Isobutane	58.12	407.81	3.63	1981.42	303.44	1.03
Pentane	72.15	469.7	3.37	1824.12	349	1.51
Ammonia	17.03	405.4	11.33	3730.71	1064.38	−10.48
Water	18	647.1	22.06	1943.17	2391.79	−17.78
Carbon dioxide	44.01	304.13	7.38	3643.72	167.53	−8.27
Propene	42.08	365.57	4.66	2387.36	284.34	−1.77
Propyne	40.06	402.38	5.63	2100.54	431.61	−1.87
Benzene	78.11	562.05	4.89	1146.72	418.22	−0.70
Toluene	92.14	591.75	4.13	1223.9	399.52	−0.21

[a]$\xi = ds/dT$; the type of working fluid can be classified by the value of ξ, i.e. $\xi > 0$: a dry fluid (e.g. pentane), $\xi \approx 0$: an isentropic fluid (e.g. R11), and $\xi < 0$: a wet fluid (e.g. water).
Source: Adapted from Chen, Huijuan, D. Yogi Goswami, Elias K. Stefanakos. 2010. A review of thermodynamic cycles and working fluids for the conversion of low-grade heat, Renewable and Sustainable Energy Reviews, Volume 14, Issue 9, Pages 3059–3067.

Table 21.20.   Velocities in steam generating systems

Nature of service	Velocity ft/min
Air	
Air Heater	1000–5000
Coal-and-air lines, pulverized coal	3000–4500
Compressed-air lines	1500–2000
Forced draft ducts	1500–3600
Forced draft ducts, entrance to burners	1500–2000
Register grills	300–600
Ventilating ducts	1000–3000
Crude Oil Lines (6 to 30 in.)	60–360
Flue Gas	
Air heater	1000–5000
Boiler gas passes	3000–6000
Induced draft flues and breeching	2000–3500
Stacks and chimneys	2000–5000
Natural-Gas Lines (large interstate)	1000–1500
Steam	
Steam lines	
High pressure	8000–12000
Low pressure	12000–15000
Vacuum	20000–40000
Superheater tubes	2000–5000
Water	
Boiler circulation	70–700
Economizer tubes	150–300
Pressurized water reactors	
Fuel assembly channels	400–1300
Reactor coolant piping	2400–3600
Water lines, general	500–750

Source: Adapted from Babcock and Wilcox. 1978. Steam: Its Generation and Use, (Charlotte, Babcock and Wilcox).

Table 21.21.   Chronological advances in conversion technology

Prime mover	Date	Output in horsepower
Man pushing a lever	3000 B.C.	0.05
Ox pulling a load	3000 B.C.	0.5
Water turbine	1000 B.C.	0.4
Donkey mill	500 B.C.	0.5
Vertical water wheel	350 B.C.	3
Vitruvian water mill	50 B.C.	3
Eighteen-foot overshot water wheel	1200 A.D.	5
Post windmill	1400 A.D.	8
Turret windmill	1600 A.D.	14
Versailles waterworks	1600 A.D.	75
Savery's steam pump	1697 A.D.	1

Table 21.21. Continued

Prime mover	Date	Output in horsepower
Newcomen's steam engine	1712 A.D.	5.5
Watt's steam engine (land)	1800 A.D.	40
Steam engine (marine)	1837 A.D.	750
Steam engine (marine)	1843 A.D.	1,500
Water turbine	1854 A.D.	800
Steam engine (marine)	1900 A.D.	8,000
Steam engine (land)	1900 A.D.	12,000
Steam turbine	1906 A.D.	17,500
Steam turbine	1921 A.D.	40,000
Steam turbine	1943 A.D.	288,000
Coal-fired steam power plant	1973 A.D.	1,465,000
Nuclear power plant	1974 A.D.	1,520,000

Source: Adapted from Earl Cook, *Man, Energy, Society* (San Francisco, Freeman, 1976).

Table 21.22. Historical efficiencies of real fuel–air cycle engines

Year	Engine thermal efficiency[a]	After partial load losses[b] (25%)	After transmission losses (25%)
1902	4%	3.00%	2.30%
1912	5%	3.80%	2.80%
1923	7%	5.30%	3.90%
1935	10%	7.50%	5.60%
1960	20%	15%	11%
1989	28%	21%	16%
2000	32%	24%	18.00%

Source: Adapted from Williams, Eric, Benjamin Warr, and Robert U. Ayres. 2008. Efficiency Dilution: Long-Term Exergy Conversion Trends in Japan, Environmental Science & Technology, 42 (13), 4964–4970.

Table 21.23. Classification of chemical heat pumps

Operation	Batch		Continuous	
Reaction phase	**Gas-solid**		**Gas-liquid**	
Driving operation	Gas-liquid phase change	Chemical reaction	Distillation, membrane separation	Compressor
Driving force	Reaction pressure	Reaction pressure	Concentration	Reaction pressure
Reactor system	Reaction-Phase change	Reaction-reaction	Reaction-separation	Reaction-separation
Reaction example	$CaCl_2/NH_3$, CaO/H_2O, MgO/H_2O, Adsoprtion heat pump	$CaO/PbO/CO_2$, Metal hydrate/H_2	Acetone/H_2, Absorption heat pump	Isobutene/H_2O, benzene/H_2

Source: Adapted from Kato, Yukitaka. 2011. Possibility of Chemical Heat Pump Technologies, Paper presented at High Density Thermal Energy Storage Workshop, Arlington, VA, 31 January, 2011.

Table 21.24.   Number of ground source heat pumps in Europe

Country		Year		
		2000	2005	2008
Austria	AUT	19,000	35,810	48,641
Belgium	BEL	n.a.	6000	9500
Czech Republic	CZE	390	3727	9168
Denmark	DNK	250	6000	11,250
Estonia	EST	n.a.	3500	4874
Finland	FIN	10,000	29,106	46,412
France	FRA	4000	63,830	121,900
Germany	DEU	18,000	61,912	148,000
Hungary	HUN	20	230	4000
Ireland	IRL	n.a.	1500	9500
Italy	ITA	100	6000	12,000
Netherlands	NLD	900	1600	14,600
Norway	NOR	500	14,000	26,000
Poland	POL	4000	8100	11,000
Slovenia	SVN	66	300	3440
Spain	ESP	n.a.	n.a.	7000
Sweden	SWE	55,000	230,094	320,687
Switzerland	CHE	21,000	38,128	61,000
United Kingdom	GBR	40	550	10,350
Total		133,266	510,387	879,322

Source: Adapted from Bayer, Peter, Dominik Saner, Stephan Bolay, Ladislaus Rybach, Philipp Blum. 2012. Greenhouse gas emission savings of ground source heat pump systems in Europe: A review, Renewable and Sustainable Energy Reviews, Volume 16, Issue 2, Pages 1256–1267.

Table 21.25.   Typical motor systems, by type of drive machine

Drive machine	Service delivered by motor system
Compressors	Air: pneumatics, blow molding aeration, material transportation.
	Refrigeration: commercial and domestic refrigeration/freezing, heat pumps; gas liquefaction.
Mechanical movements	Stirrers, mixers, crushers, extruders, textiles, materials handling, lifts, escalators, machine tools.
Pumps	Water supply and treatment, hydraulic pumps, petrochemicals, food, boreholes. Circulators: in cooling, heating, cooling tower or chilling systems.
Fans	Ventilation, drying, boiler/industrial furnace combustion, equipment cooling.

Source: Adapted from International Energy Agency (IEA). 2011. Walking the Torque: Proposed work plan for energy-efficiency policy opportunities for electric motor-driven systems, (Paris, IEA).

Table 21.26. **Electricity end-use by country and estimated demand for all electric motor drive systems (EMDS) by sector**

| Country | Total electricity use | Electricity demand for all kinds of EMDS by sector (TWh/year) | | | | | | |
	TWh/yr	Industry	Commercial	Agricultural	Transport	Residential	Total motors	% total demand
United States	3722	632	498	0	4	297	1431	38.4%
European Union	2813	787	282	13	44	177	1303	46.3%
China	2317	1092	50	24	13	72	1251	54.0%
Japan	981	221	138	0	11	62	432	44.1%
Russia	681	244	43	4	52	25	367	53.9%
Canada	499	141	51	2	3	33	229	45.9%
India	506	157	15	24	6	24	226	44.7%
Korea, South	371	131	46	1	2	12	191	51.4%
Brazil	375	126	34	4	1	19	184	49.0%
South Africa	198	78	11	1	3	8	102	51.4%
Australia	210	65	19	0	2	14	99	47.3%
Mexico	199	77	8	2	1	11	98	49.4%
Taiwan	207	70	11	1	1	9	92	44.4%
Ukraine	130	47	8	1	6	6	68	52.3%
Turkey	141	46	14	1	0	8	68	48.4%
Thailand	128	41	16	0	0	6	62	48.8%
Iran	151	36	10	4	0	11	62	41.1%
Norway	108	34	8	0	1	7	51	47.3%
Indonesia	113	30	10	0	0	10	49	43.8%
Argentina	99	33	9	0	0	6	48	48.6%
Saudi Arabia	143	9	16	1	0	19	44	31.0%
Venezuela	81	28	7	0	0	5	40	49.7%
Pakistan	73	15	4	2	0	7	28	38.2%
Switzerland	58	13	6	0	2	4	26	44.3%
Vietnam	49	16	2	0	0	5	22	45.9%
Israel	46	8	6	1	0	3	18	38.6%
New Zealand	38	10	3	0	0	3	17	43.2%
Bangladesh	22	6	1	0	0	2	9	42.8%
Costa Rica	8	1	1	0	0	1	3	38.1%
Total (55 countries)	14,465	4193	1324	89	153	862	6621	45.8%
Share of motor electricity	100%	29%	90%	10%	10%	60%	46%	
World	15,660	4488	1412	101	159	948	7108	45.4%

Source: International Energy Agency (IEA). 2011. Walking the Torque: Proposed work plan for energy-efficiency policy opportunities for electric motor-driven systems, (Paris, IEA).

Table 21.27.   Matrix of waste-heat recovery devices and applications

Heat recovery device	Temp. range	Typical sources	Typical uses
Radiation Recuperator	H	Incinerator or boiler exhaust Soaking or annealing ovens	Combustion air preheat
Convective Recuperator	M-H	Soaking or annealing ovens, melting furnaces, afterburners, gas incinerators, radiant-tube burners, reheat furnaces	Combustion air preheat
Furnace Regenerator	H	Glass- and steel-melting furnaces	Combustion air preheat
Metallic Heat Wheel	L-M	Curing exhaust and drying ovens, boiler exhaust	Combustion air preheat, space heat
Hygroscopic Heat Wheel	L	Curing and drying ovens	Combustion air preheat, space heat
Ceramic Heat Wheel	M-H	Large boiler or incinerator exhaust	Combustion air preheat
Passive Regenerator	L-H	Drying, curing & baking ovens, exhaust from boilers, incinerators & turbines	Combustion air preheat, space heat
Finned-Tube Regenerator	L-M	Boiler exhaust	Boiler make-up water preheat
Shell & Tube Regenerator	L	Refrigeration condensates, waste steam, distillation condensates, coolants from engines, air compressors, bearings & lubricants	Liquid feed flows requiring heating
Heat Pipes	L-M	Drying, curing & baking ovens, waste steam, air dryers, kilns (secondary recovery), reverberatory furnaces (secondary recovery)	Combustion air preheat, boiler makeup water preheat, steam generation, domestic hot water, space heat
Waste Boiler Heat	M-H	Exhaust from gas turbines, reciprocating engines, incinerators, furnaces	Hot water or steam generation
Gas/Steam Turbines	M-H	High-pressure steam reduced for low-pressure application, waste steam	Generation of electrical or mechanical power

Source: Adapted from Pacific Gas and Electric. 1997. Industrial Heat-Recovery Strategies.

Table 21.28.   Energy content of wastes

Agriculture and foods	MJ/kg	Municipal solid waste	MJ/kg	Synthetics/ plastics	MJ/kg	Miscellaneous	MJ/kg
Animal fats	39.5	General municipal waste	13.1	Algea	15.6	Asphalt	39.8
Barley dust	24	Brown paper	1.9	Cellulose	17	Cellophane	14.9
Citrus rinds	4	Cardboard	18.2	Gelatin	18.4	Latex	23.3
Cocoa waste	18	Corrugated boxes	16.4	Gluten	24.2	Leather	18.9
Coffee grounds	22.8	Food fats	38.8	Melamine	20.1	Leather trimmings	17.8
Corn cobbs	17.5	Garbage	19.7	Naphthene	40.2	Leather shreddings	19.8
Corn shelled	19.9	Glass bottles	0.2	Nylon	27	Lignin	14
Cotton hulls	24.7	Magazine paper	12.7	Phenol-formaldehyde	35.1	Linoleum	17.9

Table 21.28.   Continued

Agriculture and foods	MJ/kg	Municipal solid waste	MJ/kg	Synthetics/ plastics	MJ/kg	Miscellaneous	MJ/kg
Cotton husks	18.4	Metal cans	1.7	Polethylene	46.5	Lubricants (spent)	27.9
Dry food waste	18.1	Paper food cartons	18	Polypropylene	46.5	Paint (waste)	19–29
Furfural	33.3	Rags	14	Polystyrene foam	41.9	Pigbristles	21.4
Grass (lawn)	4.7			Polyurethane	30.2	Pitch	35.1
Grape stalks	7.9	**Solvents**	**MJ/kg**	Polyvinyl acetate	22.8	Rubber	23.3
Starch	18.1	Acetone	30.8	Polyvinyl chloride	20	Tires	34.9
Straw (dry)	14	Benzene	41.6	Urea-formaldehyde	17.7	Black liquor	13.7
Sugar	16	Chloroform	3.1			Paper pellets	15.2
Sunflower husks	17.7	Dichlorobenzene	19			Peat	9.3
Bagasse	21.3	Diethyl ketone	40.5			Railroad ties	14.7
Feedlot manure	17.2	Diisopropyl ketone	38.4			Sludge wood	11.7
Rice hulls	15.4	Ethyl acetate	24.7			Solid byproducts	30
Rice straw	15.2	Ethanol	29.5			Spent sulfite liquor	14.8
		Ethylene dichloride	10.9			Utility poles	14.5
Wood	**MJ/kg**	Ethylene glycol	19.1				
Sawdust and chips, dry	17.4	Heptane	46.5				
30% wet	15.1	Isopropyl alcohol	33				
40% wet	12.8	methanol	22.6				
50% wet	10.5	Methyl butyl ketone	37.4				
		Methyl ethyl ketone	34				
Sewage	**MJ/kg**	Methyl isopropyl ketone	35.6				
Raw sewage	16.5	Toluene	42.3				
Sewage sludge	4.7	Xylene	42.8				

Source: U.S. Environmental Protection Agency.

Table 21.29.   Energy density for wind electricity

Scenario		h_f	Electricity production per turbine		GED	Input	NED	Energy output/ energy input
		h/y	MWh/y	GJ/y	GJ/ha/y	GJ/ha/y	GJ/ha/y	
LO		3634	5397	19,430	493	129	903	7.03
AV–		1886	3487	12,552	728	177	552	3.13
AV		3866	7176	25,835	1500	177	1323	7.49
AV+		5846	10,848	39,053	2271	177	2094	11.9
HI		4058	8401	30,244	3224	233	1910	8.21

h_f = number of full-load hours.

GED = Gross energy density = energy contained in the energy carrier per unit land area NED = net energy density = GED - energy required directly and indirectly to generate the energy carrier Low (LO), average (AV), and high (HI) scenarios reflect range of estimates regarding energy use and efficiency of conversion of biomass to biofuels, and variations in the energy requirement for wind turbine and solar cell manufacturing; geographical variations are excluded. The AV– and AV+ scenarios are used to determine the variations in energy density caused by geographical factors such as yield differences, as well as differences in wind speed and solar insolation.

Source: Adapted from McLellan, B.C., G.D. Corder, D.P. Giurco, K.N. Ishihara. 2012. Renewable energy in the minerals industry: a review of global potential, Journal of Cleaner Production, Volume 32, Pages 32–44.

Table 21.30.   Energy density for solar electricity

Scenario		kWh/kWp	kWh/m²/y	GED	NED	O/I
				GJ/ha/y	GJ/ha/y	
LO		1000	29.7	1069	233	1.28
AV–		600	21.8	784	39	1.05
AV		1000	36.6	1307	562	1.75
AV+		1500	54.5	1960	1215	2.63
HI		1000	47.9	1723	1069	2.64

GED = Gross energy density = energy contained in the energy carrier per unit land area.

NED = net energy density = GED - energy required directly and indirectly to generate the energy carrier.

Low (LO), average (AV), and high (HI) scenarios reflect range of estimates regarding energy use and efficiency of conversion of biomass to biofuels, and variations in the energy requirement for wind turbine and solar cell manufacturing; geographical variations are excluded. The AV– and AV+ scenarios are used to determine the variations in energy density caused by geographical factors such as yield differences, as well as differences in wind speed and solar insolation.

Source: Adapted from McLellan, B.C., G.D. Corder, D.P. Giurco, K.N. Ishihara. 2012. Renewable energy in the minerals industry: a review of global potential, Journal of Cleaner Production, Volume 32, Pages 32–44.

Table 21.31. Energy density of renewable fuels produced from biomass

	Scenario	GED GJ/ha/y	Energy inputs — Agriculture GJ/ha/y	Energy inputs — Conversion GJ/ha/y	NED GJ/ha/y	Energy Output/Energy Input
Biodiesel	LO	30.6	8.4	8	14.2	1.87
	AV−	12.2	4.1	3.2	5	1.68
	AV	30.6	6.8	8	15.9	2.08
	AV+	48.9	9.5	12.7	26.8	2.36
	HI	30.6	5	8	17.6	2.21
Bioethanol	LO	81.5	11	103	−32.9	0.68
	AV−	32.6	5.5	24.3	2.8	1.09
	AV	81.5	8.9	60.8	11.8	1.17
	AV+	130	13.4	97.3	20.7	1.19
	HI	81.5	6.7	36.5	38.3	1.92
Electricity	LO	13.8	2.2	–	8.2	2.46
	AV−	3.2	0.48	–	2.1	2.75
	AV	16.1	2.1	–	10.5	2.89
	AV+	32.2	4.1	–	21.1	2.91
	HI	18.4	2.1	–	12.9	3.33

GED = Gross energy density = energy contained in the energy carrier per unit land area.

NED = net energy density = GED - energy required directly and indirectly to generate the energy carrier.

Low (LO), average (AV), and high (HI) scenarios reflect range of estimates regarding energy use and efficiency of conversion of biomass to biofuels, and variations in the energy requirement for wind turbine and solar cell manufacturing; geographical variations are excluded. The AV− and AV+ scenarios are used to determine the variations in energy density caused by geographical factors such as yield differences, as well as differences in wind speed and solar insolation.

Source: Adapted from McLellan, B.C., G.D. Corder, D.P. Giurco, K.N. Ishihara. 2012. Renewable energy in the minerals industry: a review of global potential, Journal of Cleaner Production, Volume 32, Pages 32–44.

Table 21.32.   Power density of electricity-generating technologies[a]

Power Source	Power density (Wm^{-2})	
	Low	**High**
Natural gas	200	2000
Coal	100	1000
Solar-Photovoltaics	4	9
Solar-Concentrating solar power	4	10
Wind	0.5	1.5
Biomass	0.5	0.6

[a]Wm^{-2} of horizontal area of land or water surface occupied by an energy conversion system.
Source: Adapted from Smil, Vaclav. 2010. Power Density Primer, <http://www.vaclavsmil.com/publications/>, accessed 15 June 2012.

Table 21.33.   Efficiencies of common anthropogenic and natural energy conversions[a]

Conversion	Energy conversion[b]	Efficiencies
Large electricity generators	$m \rightarrow e$	98–99
Large power plant boilers	$c \rightarrow t$	90–98
Large electric motors	$e \rightarrow m$	90–97
Best home natural gas furnaces	$c \rightarrow t$	90–96
Dry-cell batteries	$c \rightarrow e$	85–95
Home gas furnace	$c \rightarrow t$	85
Human lactation	$c \rightarrow c$	75–85
Overshot waterwheels	$m \rightarrow m$	60–85
Small electric motors	$e \rightarrow m$	60–75
Best bacterial growth	$c \rightarrow c$	50–65
Glycolysis maxima	$c \rightarrow c$	50–60
Home oil furnace	$c \rightarrow t$	60
Fluidized bed coal combustion	$c \rightarrow e$	40–50
Integrated gasification combined cycle	$c \rightarrow e$	40–50
Large steam turbines	$t \rightarrow m$	40–45
Improved wood stoves	$c \rightarrow t$	25–45
Large gas turbines	$c \rightarrow m$	35–40
Fossil fuel electric generating plant	$c \rightarrow e$	30–40
Charcoal production—earth kiln	$c \rightarrow c$	20–40
Modern wind turbine	$m \rightarrow e$	40
Diesel engines	$c \rightarrow m$	30–35
Mammalian postnatal growth	$c \rightarrow c$	30–35
Best photovolatic cells	$r \rightarrow e$	20–30
Best large steam engines	$c \rightarrow m$	20–25
Internal combustion engines	$c \rightarrow m$	15–25
High-pressure sodium lamps	$e \rightarrow r$	15–20
Mammalian muscles	$c \rightarrow m$	15–20
Milk production	$c \rightarrow c$	15–20
Pregnancy	$c \rightarrow c$	10–20
Fluorescent lights	$e \rightarrow r$	10–20
Broiler production	$c \rightarrow c$	10–15

Table 21.33. Continued

Conversion	Energy conversion[b]	Efficiencies
Traditional stoves	$c \rightarrow t$	10–15
Beef production	$c \rightarrow c$	5–10
Steam locomotives	$c \rightarrow m$	3–6
Peak field photosynthesis	$r \rightarrow c$	4–5
Incandescent light bulbs	$e \rightarrow r$	2–5
Paraffin candles	$c \rightarrow r$	1–2
Most productive ecosystems	$r \rightarrow c$	1–2
Global photosynthetic mean	$r \rightarrow c$	0.3

[a]All ranges are the first-law efficiencies in percent.
[b]Energy labels: c = chemical, e = electrical, m = mechanical (kinetic), r = radiant (electromagnetic, solar), t = thermal.
Source: Adapted from Smil, Vaclav. 1991. General Energetics, (New York, Wiley).

Table 21.34. Power levels of continuous phenomena

	Power Ratings	
Energy Flows	Watts	Exponent
Luminosity of Milky Way galaxy	1	36
Global intercept of solar radiation	1.7	17
Wind-generated waves on the ocean	9	16
Solar radiation received by China	2	15
Estimated heat flux by Earth's atmosphere and oceans away from the equator towards the poles	1.4	15
Global gross primary productivity	1	14
Global Earth heat flow	4.2	13
Global human energy use	1.6	13
Jupiter's magnetic pull on its moon	2	12
Global earthquake activity	3	11
Planned total electric generating capacity of Three Gorges Dam	2.25	10
Florida Current between Miami and Mimini	2	10
World's largest coal-fired power plant	4.1	9
Rotating turbogenerator	1	9
Midsize nuclear reactor	5	8
Maximum power output of one GE90 jet engine on the Boeing 777	1.4	8
Gas pipeline compressors	2	7
Rate at which an average gasoline pump transfers chemical energy in fuel to a vehicle	1.6	7
Most powerful locomotive (GE AC6000 CW)	4.5	6
Electricity for a 30-story high-rise	1.5	6
Most powerful production car (SSC Ultimate Aero TT)	8.8	5
Most powerful truck (Terex TR100)	7.8	5
Energy needs of a typical supermarket	2	5
Large waterwheel	1	4
Japanese per capita energy use	3	3
Solar radiation received per square at outside of Earth's atmosphere	1.37	3
Transistor used in microwave communications	1.74	2

Continued

Table 21.34. Continued

	Power Ratings	
Energy Flows	Watts	Exponent
Output of 1 m² solar panel in full sunlight (approx. 12% efficiency)	1.2	2
Basal metabolism of a 70 kg man	8	1
Typical household incandescent light bulb	6	1
Typical household compact fluorescent light bulb	1.5	1
Flashlight	4	
Net productivity per square meter of tropical forest	1	0
Metabolic rate of neonate heart[b]	4	−1
Mean global rate of erosion per square meter	5	−2
Laser in a DVD player	5	−3
Sound produced during normal human speech	1.8	−5
Approximate consumption of a quartz wristwatch	3	−6
Cosmic microwave background radiation per square meter	1	−6
Power entering a human eye from a 100-watt lamp 1 km away	1.5	−10
Power lost by a proton in the Large Hadron Collider at 7000 GeV	1.84	−11
Average power use by human cell	1	−12
Power limit of power reception on digital spread-spectrum cell phones	1	−14
Minimum discernible signal of a FM radio receiver	2.5	−15
Galileo space probe's radio signal (when at Jupiter) as received on Earth	1	−23

Source: Adapted from Smil, Vaclav. 1991. General Energetics: Energy in the Biosphere and Civilization (New York, Wiley).

Table 21.35. **Power levels of ephemeral phenomena**

Energy Flows	Power Ratings (W)	
	Actual Multiple	Order of Magnitude
Luminosity of gamma-ray burst	1	45
Typical quasar	3.6	39
Output of the Sun	4	26
Detonation of Tsar Bomb, the most powerful human made device	5.4	24
Richter magnitude 8 earthquake	1.6	15
World's most powerful laser pulses	1.25	15
Firing of Z machine, largest X-ray generator	2.9	14
Large volcanic eruption	1	14
Giant lightning	2	13
Rainstorm's latent heat	1	12
First stage of the Saturn V rocket	1.9	11
Thunderstorm's kinetic energy	1	11
Large World War II bombing raid	2	10
Space shuttle at launch	1.2	10
Average U.S. tornado	1.7	9
Mount St. Helens seismic waves	5	8
Small avalanche with 500 m drop	1.1	7
Large coal unit train shuttle	5	6
Intercity truck trip	3	5
Gasoline for a 20 km drive	4	4
Elite power weight lifter performing a clean lift of 265 kg	2.6	3
Elite male sprinter (first 30 m)	2	3
Elite male track cyclist (sprint)	1.8	3
Elite female track cyclist (sprint)	1	3
Elite female hockey player	8	2
Elite male rower	6	2
Elite female rower	4	2
Machine-washing laundry	5	2
Record holder for one hour cycling record (56.4 km)	4.6	2
Output of a top rider in the Tour de France averaged over the course of the entire race	2.3	2
CD player spinning Mozart's last symphony	2.5	1
Small candle burning to the end	3	0
Hummingbird flight	7	−1

Source: Adapted from Smil, Vaclav. 1991. General Energetics: Energy in the Biosphere and Civilization (New York, Wiley).

Storage

Figures

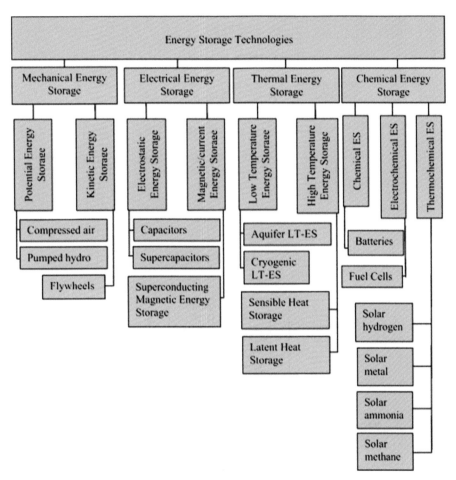

Figure 22.1. Classification of the major energy storage technologies.
Source: Evans, Annette, Vladimir Strezov, Tim J. Evans. 2012. Assessment of utility energy storage options for increased renewable energy penetration, Renewable and Sustainable Energy Reviews, Volume 16, Issue 6, Pages 4141-4147.

Handbook of Energy. http://dx.doi.org/10.1016/B978-0-08-046405-3.00022-X

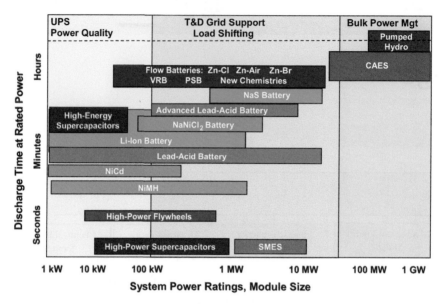

Figure 22.2. Comparison of energy storage technologies by power rating and discharge times.
Source: Adapted from Electric Power Research Institute, Electric Energy Storage Technology Options: A White Paper Primer on Applications, Costs, and Benefits, Technical Update, December 2010, (Palo Alto, EPRI).

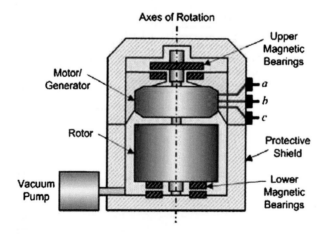

Figure 22.3. Structure of a conventional flywheel.
Source: Marcelo Gustavo Molina (2010). Dynamic Modelling and Control Design of Advanced Energy Storage for Power System Applications, Dynamic Modelling, Alisson V. Brito (Ed.), ISBN: 978-953-7619-68-8, InTech, Available from: http://www.intechopen.com/books/dynamic-modelling/dynamic-modelling-and-control-design-of-advanced-energy-storage-for-power-system-applications.

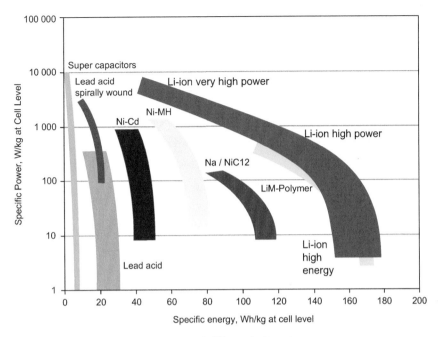

Figure 22.4. Specific energy and specific power of different battery types.
Source: Adapted from International Energy Agency (IEA). 2011. Technology Roadmap: Electric and plug-in hybrid electric vehicles, (Paris, IEA).

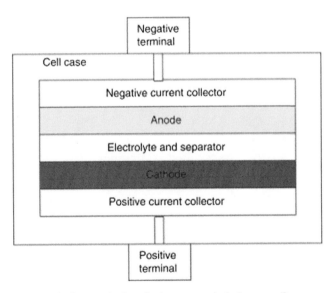

Figure 22.5. General scheme of a battery in the discharge mode (primary cell).
Source: Owens, B.B., P. Reale, B. Scrosati. 2009. Primary Batteries | Overview, In: Jürgen Garche, Editor-in-Chief, Encyclopedia of Electrochemical Power Sources, (Amsterdam, Elsevier), Pages 22-27.

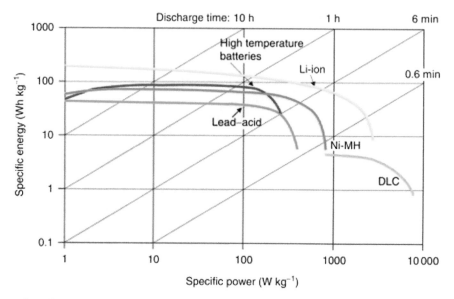

Figure 22.6.   Specific discharge energy vs specific discharge power for different traction-relevant electrochemical storage systems.
Source: Gutmann, G. 2009. Electric Vehicle: Batteries, In: Editor-in-Chief: Jürgen Garche, Encyclopedia of Electrochemical Power Sources, (Amsterdam, Elsevier), Pages 219-235.

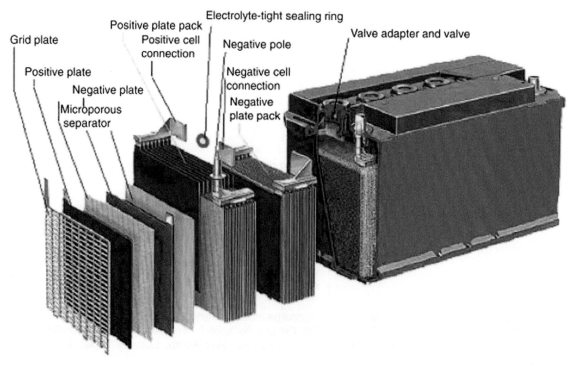

Figure 22.7.   Cross-section of lead-acid battery.
Source: Sloop, S.E., K. Kotaich, T.W. Ellis, R. Clarke. 2009. Recycling | Lead–Acid Batteries: Electrochemical, In: Jürgen Garche, Editor-in-Chief, Encyclopedia of Electrochemical Power Sources, (Amsterdam, Elsevier), Pages 179-187.

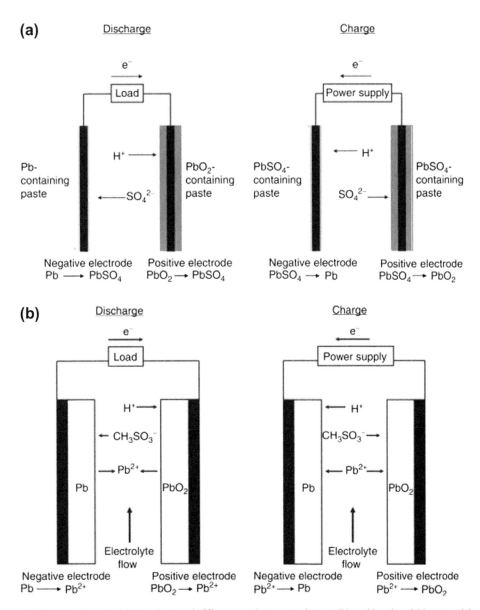

Figure 22.8. The principles of operation and differences between the traditional lead–acid battery (a) and the soluble lead flow battery (b), a type of redox flow batteries (RFB).
Source: Pletcher, D., F.C. Walsh, R.G.A. Wills. 2009. Secondary Batteries – Lead – Acid Systems | Flow Batteries, In: Jürgen Garche, Editor-in-Chief, Encyclopedia of Electrochemical Power Sources, (Amsterdam, Elsevier), Pages 745-749.

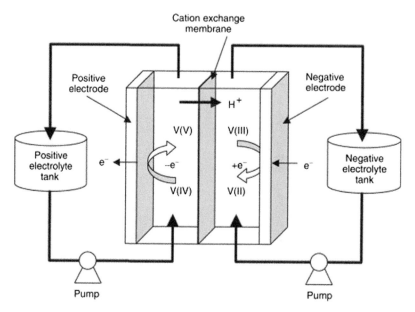

Figure 22.9. Principle of a redox flow battery (RFB), showing recirculation of an electrolyte through a cell compartment–tank loop with cell divided by a cation-exchange membrane. The cell is shown under charge. *Source: Watt-Smith, M.J., R.G.A. Wills, F.C. Walsh. 2009. Secondary Batteries – Flow Systems | Overview, In: Jürgen Garche, Editor-in-Chief, Encyclopedia of Electrochemical Power Sources, (Amsterdam, Elsevier), Pages 438-443.*

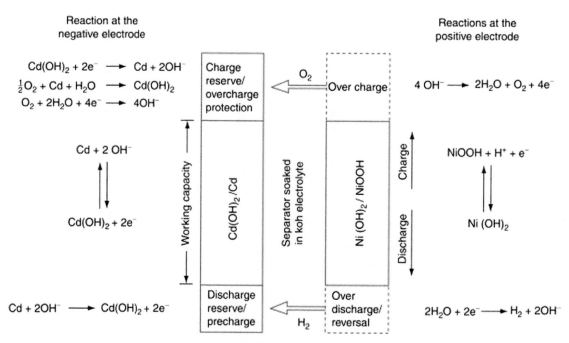

Figure 22.10. Operating principle of a sealed Ni–Cd cell. *Source: Shukla, A.K., S. Venugopalan, B. Hariprakash. 2009. Secondary Batteries – Nickel Systems | Nickel–Cadmium: Overview, In: Jürgen Garche, Editor-in-Chief, Encyclopedia of Electrochemical Power Sources, (Amsterdam, Elsevie), Pages 452-458.*

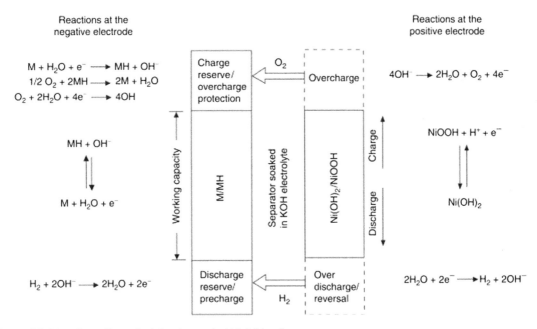

Reactions at the negative electrode

$$M + H_2O + e^- \longrightarrow MH + OH^-$$
$$1/2\, O_2 + 2MH \longrightarrow 2M + H_2O$$
$$O_2 + 2H_2O + 4e^- \longrightarrow 4OH^-$$

$$MH + OH^- \updownarrow M + H_2O + e^-$$

$$H_2 + 2OH^- \longrightarrow 2H_2O + 2e^-$$

Reactions at the positive electrode

$$4OH^- \longrightarrow 2H_2O + O_2 + 4e^-$$

$$NiOOH + H^+ + e^- \updownarrow Ni(OH)_2$$

$$2H_2O + 2e^- \longrightarrow H_2 + 2OH^-$$

Charge reserve/ overcharge protection — Overcharge — O_2

Working capacity

M/MH — Separator soaked in KOH electrolyte — $Ni(OH)_2/NiOOH$

Charge / Discharge

Discharge reserve/ precharge — Over discharge/ reversal — H_2

Figure 22.11. Operating principle of a sealed Ni–MH cell.
Source: Hariprakash, B., A.K. Shukla, S. Venugoplan. 2009. Secondary Batteries – Nickel Systems | Nickel–Metal Hydride: Overview, In: Jürgen Garche, Editor-in-Chief, Encyclopedia of Electrochemical Power Sources, (Amsterdam, Elsevier), Pages 494-501.

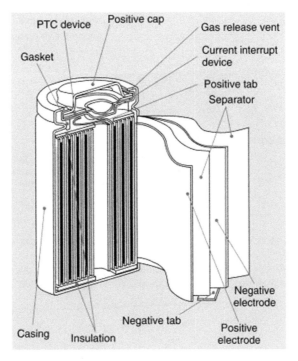

PTC device Positive cap Gas release vent Current interrupt device Positive tab Separator Gasket Negative electrode Negative tab Positive electrode Casing Insulation

Figure 22.12. Cross-section of a cylindrical Li-ion cell.
Source: Pistoia, Gianfranco. 2009. Battery Operated Devices and Systems, (Amsterdam, Elsevier).

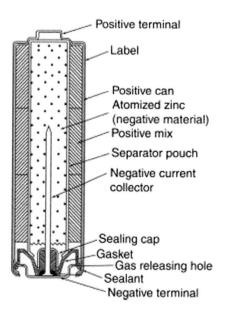

Figure 22.13. Cross section model of cylindrical zinc–alkaline manganese oxide battery.
Source: Takamura, T. 2009. Primary Batteries – Aqueous Systems | Alkaline Manganese–Zinc, In: Jürgen Garche, Editor-in-Chief, Encyclopedia of Electrochemical Power Sources, (Amsterdam, Elsevier), Pages 28-42.

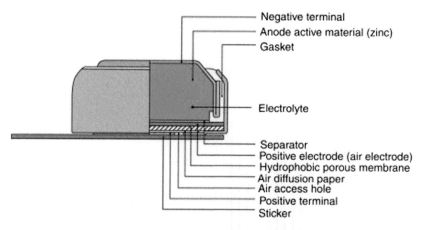

Figure 22.14. Typical cross-sectional view of a button-type zinc–air battery. Courtesy of Panasonic Corporation Energy Company.
Source: Arai, H., M. Hayashi, Primary Batteries – Aqueous Systems | Zinc–Air, In: Jürgen Garche, Editor-in-Chief, Encyclopedia of Electrochemical Power Sources, (Amsterdam, Elsevier), Pages 55-61.

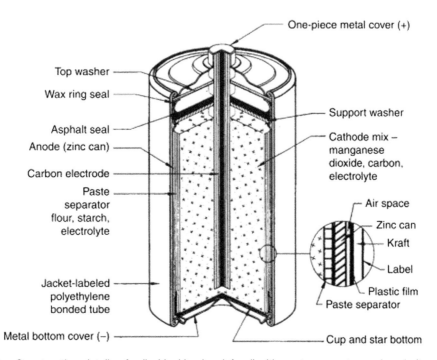

Figure 22.15. Construction details of cylindrical Leclanché cell with paste separator and asphalt seal.
Source: Kordesch, K., W. Taucher-Mautner. 2009. Primary Batteries – Aqueous Systems | Leclanché and Zinc–Carbon, In: Jürgen Garche, Editor-in-Chief, Encyclopedia of Electrochemical Power Sources, (Amsterdam, Elsevier), Pages 43-54.

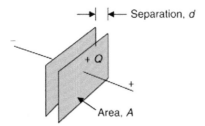

Figure 22.16. Simple electrostatic capacitor formed by separating two parallel conductors by a distance *d*. A dielectric material is commonly placed between the conductors to increase the capacitance.
Source: Miller, J.R. 2009. Capacitors | Overview, In: Jürgen Garche, Editor-in-Chief, Encyclopedia of Electrochemical Power Sources, (Amsterdam, Elsevier), Pages 587-599.

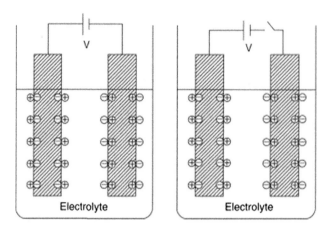

Figure 22.17. Electrochemical capacitor (EC) formed by placing two conductors in an electrolyte and applying a voltage (left). Charge separation occurs at the solid–liquid interface of both electrodes and persists after the voltage source is removed, creating two double-layer capacitors in series.
Source: Miller, J.R. 2009. Capacitors | Overview, In: Jürgen Garche, Editor-in-Chief, Encyclopedia of Electrochemical Power Sources, (Amsterdam, Elsevier), Pages 587-599.

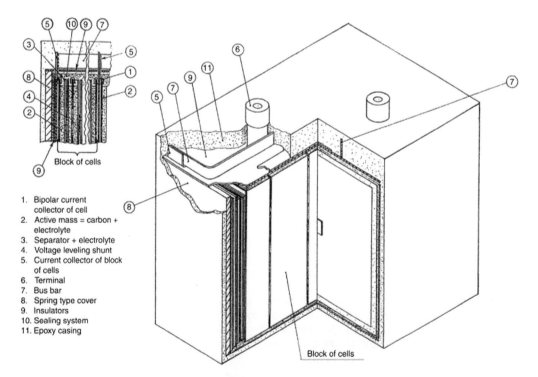

1. Bipolar current collector of cell
2. Active mass = carbon + electrolyte
3. Separator + electrolyte
4. Voltage leveling shunt
5. Current collector of block of cells
6. Terminal
7. Bus bar
8. Spring type cover
9. Insulators
10. Sealing system
11. Epoxy casing

Figure 22.18. Cutaway drawing of a multicell electrochemical capacitor (EC) constructed using bipolar design. Here capacitor cells are stacked face-to-face to form a block that meets the application voltages. Blocks are then parallel-connected to meet the energy requirements. Drawing courtesy of ELIT Company.
Source: Miller, J.R. 2009. Capacitors | Overview, In: Jürgen Garche, Editor-in-Chief, Encyclopedia of Electrochemical Power Sources, (Amsterdam, Elsevier), Pages 587-599.

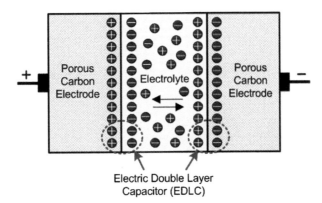

Figure 22.19. Schematic view of a super capacitor.
Source: Marcelo Gustavo Molina (2010). Dynamic Modelling and Control Design of Advanced Energy Storage for Power System Applications, Dynamic Modelling, Alisson V. Brito (Ed.), ISBN: 978-953-7619-68-8, InTech, Available from: http://www.intechopen.com/books/dynamic-modelling/dynamic-modelling-and-control-design-of-advanced-energy-storage-for-power-system-applications.

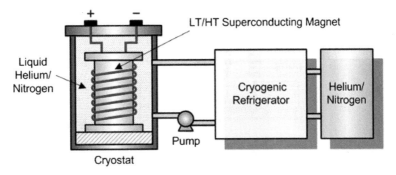

Figure 22.20. Basic structure of a superconducting magnetic energy storage (SMES) device.
Source: Marcelo Gustavo Molina (2010). Dynamic Modelling and Control Design of Advanced Energy Storage for Power System Applications, Dynamic Modelling, Alisson V. Brito (Ed.), ISBN: 978-953-7619-68-8, InTech, Available from: http://www.intechopen.com/books/dynamic-modelling/dynamic-modelling-and-control-design-of-advanced-energy-storage-for-power-system-applications.

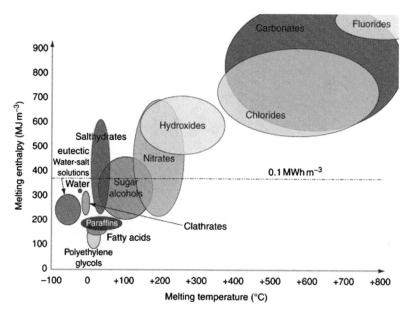

Figure 22.21.   Source: Classes of materials that can be used as phase change materials (PCMs) and their typical range of melting temperature and melting enthalpy.
Source: Cabeza, L.F. 2912. 3.07 - Thermal Energy Storage, In: Editor-in-Chief: Ali Sayigh, Comprehensive Renewable Energy, (Oxford, Elsevier), Pages 211-253.

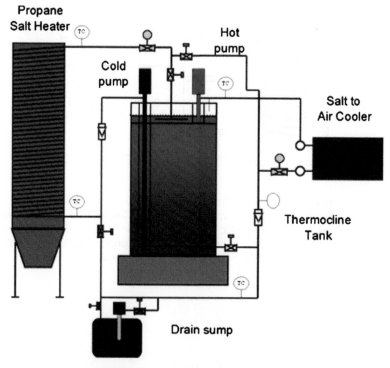

Figure 22.22.   A 2.5-MWhr thermal energy storage system with binary molten-salt fluid.
Source: United States Department of Energy, National Renewable Energy Laboratory, Parabolic Trough Thermal Energy Storage Technology, <http://www.nrel.gov/csp/troughnet/thermal_energy_storage.html>.

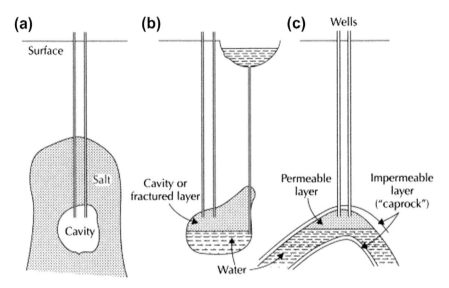

Figure 22.23.   Types of underground compressed air storage: (a) storage in salt cavity, (b) rock storage with compensating surface reservoir, and (c) aquifer storage.
Source: Sørensen, Bent. 2011. Renewable Energy (Fourth Edition), (Boston, Academic Press).

Charts

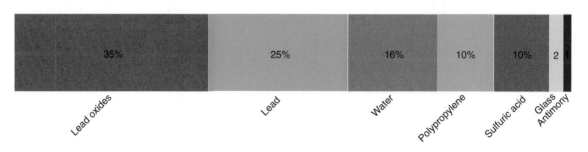

Chart 22.1. Composition for a representative lead-acid battery (%).
Source: Data from Sullivan, J.L. and L. Gaines. 2010. A Review of Battery Life-Cycle Analysis: State of Knowledge and Critical Needs, Center for Transportation Research Energy Systems Division, Argonne National Laboratory.

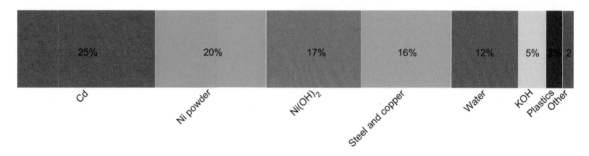

Chart 22.2. Typical composition of an automotive nickel-cadmium battery (%).
Source: Data from Sullivan, J.L. and L. Gaines. 2010. A Review of Battery Life-Cycle Analysis: State of Knowledge and Critical Needs, Center for Transportation Research Energy Systems Division, Argonne National Laboratory.

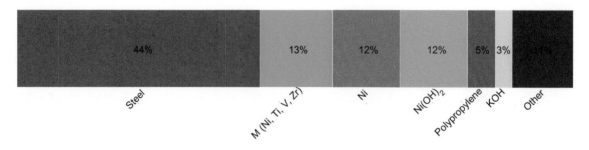

Chart 22.3. Composition of representative nickel-metal hydride battery (NiMH-AB$_2$).
Source: Data from Sullivan, J.L. and L. Gaines. 2010. A Review of Battery Life-Cycle Analysis: State of Knowledge and Critical Needs, Center for Transportation Research Energy Systems Division, Argonne National Laboratory.

Tables

Table 22.1.　Major storages of energy on the Earth

Reservoir	Volume (km³)[a]	Mass (Gt)	Temp. (K)	Heat Capacity (J/g x K)	Energy Content (EJ)
Atmosphere[b]	5.0×10^9	5.2×10^6	250	1.0	1.3×10^6
Land					
Surface[c]	3.0×10^4	3.3×10^4	290	3.7	3.5×10^4
Subsurface[e]	1.5×10^5	1.6×10^5	280	3.7	1.7×10^5
Oceans					
Surface[d]	3.3×10^7	3.3×10^7	280	4.2	4.0×10^7
Thermocline[d]	1.7×10^8	1.7×10^8	275	4.2	1.9×10^8
Deep[d]	1.2×10^9	1.7×10^9	270	4.2	1.4×10^9
Cryosphere[e]	5.1×10^7	4.7×10^7	265	2.1	2.6×10^7

[a]Volume is specified in order to estimate the reservoir's heat contents.
[b]The atmosphere is assumed to be 10 km thick (approximately the Troposphere) at density of 1.05 kg/m³.
[c]The land surface depth is taken to be 0.2 m thick for diurnal response, with a density of 1100 kg/m³, based on saturated sandy and clay soil with 80% saturated pore space. The deep soil layer is taken as 1.0 m thick, for seasonal variations.
[d]The oceans have a density of 1025 kg/m³ and average depths of about 100 m for the surface, 0.5 km for the thermocline, and 3.7 km for the deep oceans.
[e]The ice and snow reservoir has a density of 920 kg/m³ and an average depth of 2 km.
Source: Adapted from Shaw, Henry, Environmental Heat Transfer, in Kreith, Frank, Ed. 2000. CRC Handbook of Thermal Engineering (Boca Raton, CRC Press).

Table 22.2.　Product suitability for underground storage of liquid and gaseous fuels

Techniques	Products				
	Gas: Natural Gas/Compressed Air	Liquified Under Pressure Gas: Butane/Propane/ Propylene Ethane/ Ethylene	Liquified at Low Temperature Gas: Propane (−40°C)/ Ethylene (−105°C)/ Natural Gas (−162°C)	Liquids: Crude Oils/ Petroleum Products/Hot Water	Wastes: Industrial/ Nuclear
Aquifers	1	2	2	2	2
Leached cavities	1	1	2	1	1
Mined caverns	1	1	2	1	1
Disused mines	1	1	2	1	1
Cryogenic caverns	2	2	1	2	2

1 - Usable technique 2 - Non usable technique.
Source: Goel, R.K., Bhawani Singh, Jian Zhao. 2012. Underground Infrastructures, (Boston, Butterworth-Heinemann).

Table 22.3.   Physical properties of some sensible heat storage materials

Storage Medium	Temperature Range, °C	Density, kg/m³	Specific Heat, J/kg K	Energy Density, kWh/m³ K	Thermal Conductivity (W/m K)
Water	0–100	1000	4190	1.16	0.63 at 38 °C
50% ethylene glycol-50% water	0–100	1075	3,480	0.98	--
Dowtherm A (Dow Chemical, Co.)	12–260	867	2,200	0.53	0.122 at 260 °C
Therminol 66 (Monsanto Co.)	12–260	750	2,100	0.44	0.106 at 343 °C
Draw salt (50NaNO$_3$-50KNO$_3$)[a]	−9–343	1733	1,550	0.75	0.57
Molten salt (53KNO$_3$/40NANO$_3$/7NaNO$_3$)[a]	220–540	1680	1,560	0.72	0.61
Liquid Sodium	142–540	750	1,260	0.26	67.50
Cast Iron	100-760 m.p. (1150–1300)	7200	540	1.08	42.00
Taconite	--	3200	800	0.71	--
Aluminum	m.p. 660	2700	920	0.69	200
Fireclay	--	2100–2600	1000	0.65	1.0–1.5
Rock	--	1600	880	0.39	--

m.p. = melting point.
[a]Composition in percent by weight.
Source: Adapted from Goswami, D., Frank Kreith, and Jan F. Kreider. 2000. Principles of Solar Engineering, 2nd edition, (Philadelphia, Taylor & Francis).

Table 22.4.   Potential materials for thermochemical energy storage

Thermochemical Material	Solid Reactant	Working Fluid	Energy Storage Density (GJ/m³)	Charging Reaction Temperature (°C)
MgSO$_4$·7H$_2$O	MgSO$_4$	7H$_2$O	2.8	122
FeCO$_3$	FeO	CO$_2$	2.6	180
Ca(OH)$_2$	CaO	H$_2$O	1.9	479
Fe(OH)$_2$	FeO	H$_2$O	2.2	150
CaCO$_3$	CaO	CO$_2$	3.3	837
CaSO$_4$·2H$_2$O	CaSO$_4$	2H$_2$O	1.4	89

Source: Adapted from Abedin, Ali H. and Marc A. Rosen. 2011. A Critical Review of Thermochemical Energy Storage Systems, The Open Renewable Energy Journal,4: 42–46.

Table 22.5.   Comparison of thermal energy storage systems (TES)

Performance Parameter	Type of Thermal Energy Storage		
	Sensible TES	**Latent TES**	**Chemical TES (Sorption and Thermochemical)**
Temperature Range	Up to:		
	110 °C (water tanks)	20–40 °C (paraffins)	20–200 °C
	50 °C (aquifers and ground storage)	30–80 °C (salt hydrates)	
	400 °C (concrete)		
Storage density	Low (with high temperature interval):	Moderate (with low temperature interval):	Normally high: 0.5–3 GJ/m^3
	0.2 GJ/m^3 (for typical water tanks)	0.3–0.5 GJ/m^3	
Lifetime	Long	Often limited due to storage material cycling	Depends on reactant degradation and side reactions
Technology status	Available commerically	Avilable commercially for some temperatures and materials	Generally not available, but undergoing research and pilot project tests
Advantages	Low cost	Medium storage density	High storage density
	Reliable	Small volumes	Low heat losses (storage at ambient temperatures)
	Simple application with available materials	Short distance transport possibility	Long storage period
			Long distance transport possibility
			Highly compact energy storage
			High capital costs
			Technically complex
			High capital costs
Disadvantages	Significant heat losses	Low heat conductivity	Technically complex
	over time (depending on level of insulation)	Corrosivity of materials Significant heat losses	
	Large volume needed	(depending on level of insulation)	

Source: Adapted from Abedin, Ali H. and Marc A. Rosen. 2011. A Critical Review of Thermochemical Energy Storage Systems, The Open Renewable Energy Journal,4: 42–46.

Table 22.6.   Energy storage densities of different thermal energy storage (TES) methods

Type of storage technology	Material	Energy stored (MJ m⁻³)	Energy stored (kJ kg⁻¹)	Comments
Sensible heat	Granite	50	17	$\Delta T = 20\ °C$
	Water	84	84	$\Delta T = 20\ °C$
Latent heat	Water	306	330	$T_{melting} = 0\ °C$
	Paraffins	180	200	$T_{melting} = 5–130\ °C$
	Salt hydrates	300	200	$T_{melting} = 5–130\ °C$
	Salt	600–1,500	300–700	$T_{melting} = 300–800\ °C$
Chemical reactions	H_2 gas (oxidation)	11	120,000	300 K, 1 bar
	H_2 gas (oxidation)	2,160	120,000	300 K, 200 bar
	H_2 liquid (oxidation)	8,400	120,000	20 K, 1 bar
	Fossil gas	32		300 K, 1 bar
	Gasoline	33,000	43,000	
Electrical storage	Zn/Mn oxide battery		180	
	Pb battery		70–180	

Source: Adapted from Cabeza, L.F. 2012. Thermal Energy Storage, In: Ali Sayigh, Editor-in-Chief, Comprehensive Renewable Energy, (Oxford, Elsevier), Pages 211–253.

Table 22.7.   Thermal capacity at 20 °C of some common materials used in sensible thermal energy storage

Material	Density (kg m⁻³)	Specific heat (J kg⁻¹ K⁻¹)	Volumetric thermal capacity (×10⁶, J m⁻³ K⁻¹)
Clay	1458	879	1.28
Brick	1800	837	1.51
Sandstone	2200	712	1.57
Wood	700	2390	1.67
Concrete	2000	880	1.76
Glass	2710	837	2.27
Aluminum	2710	896	2.43
Iron	7900	452	3.57
Steel	7840	465	3.68
Gravelly earth	2050	1840	3.77
Magnetite	5177	752	3.89
Water	988	4182	4.17

Adapted from Cabeza, L.F. 2912. 3.07 - Thermal Energy Storage, In: Ali Sayigh, Editor-in-Chief, Comprehensive Renewable Energy, (Oxford, Elsevier), 2012, Pages 211–253.

Table 22.8. Potential storage media for solar thermal power plants

Storage Medium	Temperature		Average density	Average heat conductivity	Average heat capacity	Volume specific heat capacity
	Cold (°C)	Hot (°C)	(kg/m³)	(W/mK)	(kJ/kgK)	(kWh$_t$/m³)
Solid media						
Sand-rock-mineral oil	200	300	1,700	1	1.3	60
Reinforced concrete	200	400	2,200	1.5	0.85	100
NaCl (solid)	200	500	2,160	7	0.85	150
Cast iron	200	400	7,200	37	0.56	160
Cast steel	200	700	7,800	40	0.6	450
Silica fire bricks	200	700	1,820	1.5	1	150
Magnesia fire bricks	200	1,200	3,000	5	1.15	600
Liquid media						
Mineral oil	200	300	770	0.12	2.6	55
Synthetic oil	250	350	900	0.11	2.3	57
Silicone oil	300	400	900	0.1	2.1	52
Nitrite salts	250	450	1,825	0.57	1.5	152
Nitrate salts	265	565	1,870	0.52	1.6	250
Carbonate salts	450	850	2,100	2	1.8	430
Liquid sodium	270	530	850	71	1.3	80
Phase change media						
NaNo$_3$	308	2,257	0.5	200	125	0.2
KNO$_3$	333	2,110	0.5	267	156	0.3
KOH	380	2,044	0.5	150	85	1
Salt-ceramics	500-850	2,600	5	420	300	2
(NaCo$_3$-BaCO$_3$/MgO)						
NaCl	802	2,160	5	520	280	0.15
Na$_2$CO$_3$	854	2,533	2	276	194	0.2
K$_2$CO$_3$	897	2,290	2	236	150	0.6

Source: Adapted from Pilkington Solar International. 2000. Survey of Thermal Storage for Parabolic Trough Power Plants, NREL/SR-550-27925, (Golden, CO, National Renewable Energy Laboratory).

Table 22.9. Characteristics of thermal cells for stationary and vehicular applications

System	Voltage Range (V)	Operating Temperature (°C)	Cycle Life (cycles)	Specific Energy (Wh/kg)	Energy Density (Wh/L)	Specific Power (W/kg)
Li-Al/FeS	1.7–1.0	375 ÷ 500	1000	130	220	240
Li-Al/FeS$_2$	2.0–1.5	375 ÷ 450	1000	180	350	400
Na/S	2.0–1.8	300 ÷ 350	6000	1551	3001	2502
				1752	3502	
Na/NiCl$_2$	2.1–1.7	250 ÷ 300	2500	115	190	260
Li-Metal-Polymer	3.0–2.0	40 ÷ 60[a]	800	1401	1741	260
		60 ÷ 80[b]		1202	1602	

[a]Stationary applications.
[b]Traction.
Source: Adapted from Pistoia, Gianfranco. 2009. Battery Operated Devices and Systems, (Amsterdam, Elsevier).

Table 22.10. Comparison of different storage techniques for solar space heating and hot water production applications

	Sensible heat storage		Latent heat thermal storage material (PCM)
	Water	**Rock**	
Comparison between different heat storage media			
Operating temperature range	Limited (0–100 °C)	Large	Large, depending on the choice of the material
Specific heat	High	Low	Medium
Thermal conductivity	Low, convection effects improve the heat transfer rate	Low	Very low, insulating properties
Thermal storage capacity per unit mass and volume for small temperature differences	Low	Low	High
Stability to thermal cycling	Good	Good	Insufficient data
Availability	Overall	Almost overall	Dependent on the choice of material
Cost	Inexpensive	Inexpensive	Expensive
Comparison of heat transfer properties and lifetimes of different modes of thermal storage			
Required heat exchanger geometry	Simple	Simple	Complex
Temperature gradients during charging and discharging	Large	Large	Small
Thermal stratification effect	Existent works positively	Existent works positively	Generally nonexistent with proper choice of material
Simultaneous charging and discharging	Possible	Not possible	Possible with appropriate selection of heat exchanger
Integration with solar heating/ cooling systems	Direct integration with water systems	Direct integration with air systems	Indirect integration
Costs for pumps, fans, etc.	Low	High	Low
Corrosion with conventional materials of construction	Controlled w/corrosion inhibitors	Noncorrosive	Dependent on the choice of material
Lifetime	Long	Long	Short

PCM = phase change material.
Source: Adapted from Cabeza, L.F. 2912. 3.07 - Thermal Energy Storage, In: Ali Sayigh, Editor-in-Chief, Comprehensive Renewable Energy, (Oxford, Elsevier), 2012, Pages 211–253.

Table 22.11.   Energy storage characteristics by application (megawatt-scale) in electricity grid systems

Technology	Maturity	Capacity (MWh)	Power (MW)	Duration (hrs)	% Efficiency (total cycles)	Total Cost ($/kW)	(Cost $/kW-h)
Bulk Energy Storage to Support System and Renewables Integration							
Pumped Hydro	Mature	1680–5300	280–530	6–10	80–82 (>13000)	2500–4300	420–430
		5400–14,000	900–1400	6–10		1500–2700	250–270
Combustion turbine/ Compressed air (underground)	Demo	1440–3600	180	8	(>13000)	960	120
				20	(>13000)	1150	60
Compressed air (underground)	Commercial	1080	135	8	(>13000)	1000	125
		2700		20	(>13000)	1250	60
Sodium-Sulfur	Commercial	300	50	6	75 (4500)	3100–3300	520–550
Advanced Lead-Acid	Commercial	200	50	4	85–90 (2200)	1700–1900	425–475
	Commercial	250	20–50	5	85–90 (4500)	4600–4900	920–980
	Demo	400	100	4	85–90 (4500)	2700	675
Vanadium Redox	Demo	250	50	5	65–75 (>10000)	3100–3700	620–740
Zn/Br Redox	Demo	250	50	5	60 (>10000)	1450–1750	290–350
Fe/Cr Redox	R&D	250	50	5	75 (>10000)	1800–1900	360–380
Zn/air Redox	R&D	250	50	5	75 (>10000)	1440–1700	290–340
Energy Storage for ISO Fast Frequency Regulation and Renewables Integration							
Flywheel	Demo	5	20	0.25	85–97 (>100000)	1950–2200	7800–8800
Li-ion	Demo	0.25–25	1–100	0.25–1	87–92 (>100000)	1085–1550	4340–6200
Advanced Lead-Acid	Demo	0.25–50	1–100	0.25–1	75–90 (4500)	950–1590	2770–3800
Energy Storage for Utlity T&D Grid Support Applications							
CAES (aboveground)	Demo	250	50	5	See note 1 (>10000)	1950–2150	390–430
Advanced Lead-Acid	Demo	3.2–48	1–12	3.2–4	75–90 (4500)	2000–4600	625–1150
Sodium-Sulfur	Commercial	7.2	1	7.2	75 (4500)	3200–4000	445–555
Zn/Br Flow	Demo	5–50	1–10	5	60–65 (>10,000)	1670–2015	340–1350
Vanadium Redox	Demo	4–40	1–10	4	75 (>10000)	3000–3310	750–830
Fe/Cr Flow	R&D	4	1	4	75 (4500)	1200–1600	300–400
Zn/air	R&D	5.4	1	5.4	90–94 (4500)	1750–1900	325–350
Li-ion	Demo	4–24	1–10	2-4		1800–4100	900–1700

Continued

Table 22.11. Continued

Technology	Maturity	Capacity (MWh)	Power (MW)	Duration (hrs)	% Efficiency (total cycles)	Total Cost ($/kW)	(Cost $/kW-h)
Energy Storage for Commercial and Industrial Applications							
Advanced Lead-Acid	Demo-commercial	0.1–10	0.2–1	4–10	75–90 (4500)	2800–4600	700–460
Sodium-Sulfur	Commercial	7.2	1	7.2	75 (4500)	3200–4000	445–555
Zn/Br Flow	Demo	0.625	0.125	0.125	60–63 (>10000)	2420	485–440
		2.5	0.5	0.5		2200	
Vanadium Flow	Demo	0.6–4	0.6–4	0.2–1.2	65–70 (>10000)	4830–3020	1250–910
Li-ion	Demo	0.1–0.8	0.05–0.2	2–4	80–93 (4500)	3000–4400	950–1900

ISO = independent system operator; T&D = transmission and distribution.
Adapted from: Electric Power Research Institute. 2010. Electric Energy Storage Technology Options: A White Paper Primer on Applications, Costs, and Benefits, (Palo Alto, CA, Electric Power Research Institute).

Table 22.12. Types of energy storage applications in electricity grids

Value chain	Application	Description
Generation & System-level Applications	Wholesale Energy Services	Utility-scale storage systems for bidding into energy, capacity and ancillary services markets
	Renewables Integration	Utility-scale storage providing renewables time shifting, load and ancillary services for grid integratioo
	Stationary Storage for T&D Support	Systems for T&D system support, improving T&D system utilization factor, and T&D capital deferral
	Transportable Storage for T&D Support	Transportable storage systems for T&D system support and T&D deferral at multiple sites as needed
T&D System Applications	Distributed Energy Storage Systems	Centrally managed modular systems providing increased customer reliability, grid T&D support and potentially ancillary services
	ESCO Aggregated Systems	Residential-customer-sited storage aggregated and centrally managed to provide distribution system benefits
	C&I Power Quality and Reliability	Systems to provide power quality and reliability to commercial and industrial customers
	C&I Energy Management	Systems to reduce TOU energy charges and demand charges for C&I customers
	Home Energy Management	Systems to shift retail load to reduce TOU energy and demand charges
End- User Applications	Home Backup	Systems for backup power for home offices with high reliability value

T&D = Transmission and Distribution; C&I = Commercial and Industrial; ESCO = Energy Services Company; TOU = Time of Use.
Source: Adapted from Electric Power Research Institute. 2010. Electric Energy Storage Technology Options: A White Paper Primer on Applications, Costs, and Benefits, (Palo Alto, CA, Electric Power Research Institute).

Table 22.13. **Examples of grid-connected energy storage systems**

Project name	Location	Technology type	Rated power (kW)	Description	Benefits
Bath County Pumped Storage Station	Virginia	Open Loop Pumped Hydro	3,003,000	This project consists of a 3GW Pumped Hydro storage plant in Virginia that pumps water to an elevated reservoir at night and lets it run back down to generate electricity during the day.	Electric Energy Time Shift
Laurel Mountain	West Virginia	Lithium Ion Battery	32,000	AES installed a wind generation plant comprised of 98 MW of wind generation and 32 MW of integrated battery-based energy storage.	Frequency Regulation, Ramping
Angamos	Antofagasta, Chile	Lithium Ion Battery	20,000	This project will utilize 20MW of A123 lithium-ion batteries to supply a flexible and scalable emissions-free reserve capacity installation. The advanced energy storage installation provides critical contingency services to maintain the stability of the electric grid in Northern Chile, an important mining area.	Frequency Regulation, Electric Supply Reserve Capacity - Spinning,
Kahuku Wind Farm	Hawaii	Advanced Lead Acid Battery	15,000	Xtreme Power installed a 15 MW fully integrated energy storage and power management system designed to provide load firming for a 30 MW wind farm in Hawaii, as well as provide critical grid integration services.	Renewables Capacity Firming, Ramping, Voltage Support
Los Andes	Atacama, Chile	Lithium Ion Battery	12,000	The Los Andes project provides critical contingency services to maintain the stability of the electric grid in Northern Chile, an important mining area. The project continuously monitors the condition of the power system and if a significant frequency deviation occurs, such as the loss of a generator or transmission line, the Los Andes system provides up to 12MW of power nearly instantaneously. This output can be maintained for 20 minutes at full power, allowing the system operator to resolve the event or bring other standby units online.	Frequency Regulation, Electric Supply Reserve Capacity - Spinning

Continued

Table 22.13. Continued

Project name	Location	Technology type	Rated power (kW)	Description	Benefits
Johnson City	New York	Lithium Ion Battery	8,000	AES installed a bank of 800,000 A123 Lithium-ion batteries to perform frequency regulation for the New York ISO. The system was the largest Lithium-ion battery in commercial service on the US power grid when completed.	Frequency Regulation, Electric Supply Reserve Capacity - Spinning
Xcel and SolarTAC	Colorado	Advanced Lead Acid Battery	1,500	The project is designed to collect operational data on the integration of energy storage and solar energy systems at the Solar Technology Acceleration Center (SolarTAC).	Ramping , Renewables Capacity Firming, Frequency Regulation, Voltage Support
Glendale Water and Power - Peak Capacity Project	California	Ice Thermal Storage	1,500	Glendale Water and Power's (GWP) Ice Bear project installed Ice Thermal Energy storage units at 28 Glendale city buildings and 58 local small, medium sized, and large commercial businesses. A total of 180 Ice Bear units have been installed in Glendale since the program's inception.	Electric Energy Time Shift, Transmission Congestion Relief, Electric Supply Capacity
Kaheawa I Wind Project	Hawaii	Advanced Lead Acid Battery	1,500	Xtreme Power installed a 1.5 MW Dynamic Power Resource (DPR) as a demonstration project to perform Ramp Control for 3MW of the 30MW Kaheawa Wind Farm in Hawaii.	Renewables Capacity Firming
Lanai Sustainability Research	Hawaii	Advanced Lead Acid Battery	1,125	Xtreme Power deployed a 1.125 MVA Dynamic Power Resource (DPR) at the Lanai Sustainability Research's 1.5MW DC/1.2MW AC solar farm in order to double the output of the solar and control the ramp rate to +/- 360 kW/min.	Ramping , Renewables Capacity Firming, Frequency Regulation,

Table 22.13. Continued

Project name	Location	Technology type	Rated power (kW)	Description	Benefits
Painesville Municipal Power Vanadium Redox Battery Demonstration	Ohio	Vanadium Redox Flow Battery	1,080	This system is designed to demonstrate a 1.08 MW vanadium redox battery (VRB) storage system at the 32 MW municipal coal fired power plant in Painesville.	Load Following (Tertiary Balancing), Electric Energy Time Shift, Transmission Congestion Relief Electric Supply Reserve Capacity - Spinning, Voltage Support
Wind-to-Battery MinnWind Project	Minnesota	Sodium Sulfur Battery	1,000	The sodium-sulfur battery is commercially available and versions of this technology are in use elsewhere in the U.S. and other parts of the world, but this is the first U.S. application of the battery as a direct wind energy storage device. The project is being conducted in Luverne, Minn., about 30 miles east of Sioux Falls, S.D. The battery installation is connected to a nearby 11-megawatt wind farm owned by Minwind Energy, LLC.	Renewables Energy Time Shift, Ramping , Voltage Support, Frequency Regulation
Redding Electric Utilities - Peak Capacity, Demand Response, HVAC Replacement Program	California	Ice Thermal Storage	1,000	Ice Energy installed 1MW of Ice Bear Thermal Energy Storage Assets to assist Redding Electric Utility avoid procurement of high cost summer peak energy by shifting air conditioning load permanently to the night time hours when energy is more abundant and lower cost.	Electric Energy Time Shift, Transmission Congestion Relief, Electric Supply Capacity
Detroit Edison Community Energy Storage Project	Michigan	Lithium Ion Battery	1,000	This project is designed to demonstrate a proof of concept for aggregated Community Energy Storage Devices in a utility territory.	Voltage Support, Renewables Energy Time Shift, Frequency Regulation, Load Following (Tertiary Balancing), Grid-Connected Commercial (Reliability & Quality)

Source: United States Department of Energy, Energy Storage Database, <http://www.energystorageexchange.org>.

Table 22.14. General energy storage application requirements for electricity grids

Application	Description	Size	Duration	Cycles	Desired Lifetime
Wholesale Energy Services	Arbitrage	10–300 MW	2–10 hr	300–400/yr	15–20 yr
	Ancillary services[a]	a	a	a	a
	Frequency Regulation	1–100 MW	15 min	>8000/yr	15 yr
	Spinning reserve	10–100 MW	105 hr		20 yr
Renewables integration	Wind integration: ramp & voltage support	1–10 MW distributed 100–400 MW centralized	15 min	5000/yr 10,000 full energy cycles	20 yr
	Wind integration; off-peak storage.	100–400 MW	5–10 hr	300–500/yr	20 yr
	Photovoltaic Integration: time shift, voltage sag. Rapid demand support	1–2 MW	15 min–4 hr	>4000	15 yr
Stationary T&D Support	Urban and rural T&D deferral. Also ISO congestion mgt.	10–100 MW	2–6 hr	300–500/yr	15–20 yr
Transportable T&D Support	Urban and rural T&D deferral. Also ISO congestion mgt.	1–10 MW	2–6 hr	300–500/yr	15–20 yr
Distributed Energy Storage Systems (DESS)	Utility-sponsored; on utility side of meter, feeder line, substation, 75-85% ac-ac efficient	25–200 kW 1-phase 25–75 kW 3-phase Small footprint	2–4 hr	100–150/yr	10–15 yr
C&I Power Quality	Provide solutions ot avioid voltage sags and mometary outages.	50–500 kW 1000 kW	<15 min >15 min	<50/yr	10 yr
C&I Power Reliability	Provide UPS bridge to backup power, outage ride-through.	50–500 kW	>15 min	<50/yr	10 yr
C&I Energy Management	Reduce energy costs, increase reliability. Size varies by market segment	50–1000 kW Small Footprint 1 MW	3–4 hr 4–6 hr	400–1500/yr	15 yr
Home Energy Management	Efficiency, cost-savings	2–5 kW Small footprint	3–4 hr	150–400/yr	10–15 yr
Home Backup	Reliability	2–5 kW Small footprint	4–6 hr	150–400/yr	10–15 yr

T&D = Transmission and Distribution; C&I = Commercial and Industrial; ISO = Independent System Operator; UPS = uninterruptible power supply.
[a]Ancillary services encompass many market functions, such as black start capability and ramping services, that have a wide range of characteristics and requirements.
Source: Adapted from Electric Power Research Institute. 2010. Electric Energy Storage Technology Options: A White Paper Primer on Applications, Costs, and Benefits, (Palo Alto, CA, Electric Power Research Institute).

Table 22.15. Benefits of energy storage to the electrcity grid

Benefit	Discharge Duration[a]		Capacity (Power: kW, MW)	
	Low	High	Low	High
Electric Energy Time-shift	2	8	1 MW	500 MW
Electric Supply capacity	4	6	1 MW	500 MW
Load Following	2	4	1 MW	500 MW
Area Regulation	15 min.	30 min.	1 MW	40 MW
Electric Supply Reserve Capacity	1	2	1 MW	500 MW
Voltage Support	15 min.	1	1 MW	10 MW
Transmission Support	2 sec	5 sec.	10 MW	100 MW
Transmission Congestion Relief	3	6	1 MW	100 MW
T&D Upgrade Deferral 50th percentile	3	6	250 kW	5 MW
T&D Upgrade Deferral 90th percentile	3	6	250 kW	2 MW
Substation on-site Power	8	16	1.5 kW	5 W
Time-of-use Energy Cost Management	4	6	1 kW	1 MW
Demand Charge Management	5	11	50 kW	10 MW
Electric service Reliability	5 min.	1	0.2 kW	10 MW
Electric service Power Quality	10 sec.	1 min.	0.2 kW	10 MW
Renewables Energy Time-shift	3	5	1 kW	500 MW
Renewables capacity firming	2	4	1 kW	500 MW
Wind Generation Grid Integratlon, Short Duration	10 sec.	15 min.	0.2 kW	500 MW
Wind Generation Grid Integration, Lone Duration	1	6	0.2 kW	500 MW

Discharge duration indicates the amount of time that the storage must discharge at its rated output before charging.
Capacity indicates the range of storage system power ratings that apply for a given benefit. The benefit indicates the present worth of the respective benefit type for 10 years (2.5% inflation, 10% discount rate). Potential indicates the maximum market potential for the respective benefit over 10 years. Economy reflects the total value of the benefit given the maximum market potential.
[a]Hours unless indicated otherwise. Min = minutes Sec = seconds.
Source: Adapted from Eyer, Jim and Garth Corey, Energy Storage for the Electricity Grid: Benefits and Market Potential Assessment Guide, Sandia National Laboratories, Report SAND2010-0815, February 2010.

Table 22.16. Cradle-to-grave life-cycle energy (MJ/kg) results for five battery systems

Battery	Note	E_{mp}	E_{rcycl}	E_{mnf}	E_{ctg}
NiMH		108	19.6	8.1	119
					230
	AB_2				246
					195
	AB_5				263
		57			
		54–102	21–40	74–139	128–241
				14.6	
		86.5		105	191.5
PbA		25.1	8.4	11.3	36.4
				77	
		24.7			
		15–25	9.0–14.0	8.4–13	23.4–38
				16.6	

Continued

Table 22.16. Continued

Battery	Note	E_{mp}	E_{rcycl}	E_{mnf}	E_{ctg}
		16.8		6.7	23.5
		17.3	Included	8.81	26.1
NiCd		102.8			
		44.0		53.9	97.9
		44–60	22–30	46–63	90–123
Na/S		59.9			
		179		56	235
		82–93	30–34	62–70	144–163
Li-ion	NCA-G	93.3	4.8	32	125.3
	LMO-G	113	3.6	30	143
	NCA-G	53–80	25–37	96–144	149-224
		112.9		91.5	204.4
	NCA-G				222
	NCA-G				62.9

DOD = depth of discharge.
E_{ctg} = cradle-to-grave primary production energy for making battery.
E_{mp} = energy required to extract, and process materials used in battery production.
E_{mnf} = energy required to manufacture battery.
E_{rcycl} = energy required to recycle battery.
AB_2: includes titanium (Ti), zirconium (Zr), Ni, and vanadium (V).
AB_5: metals from the lanthanide series, or rare earths, including metals from lanthanum (atomic number = 57) to luterium (71),
NCA: a mixture of Ni, Co, and Al oxides.
LMO: manganese oxide.
Source: Adapted from Sullivan, J.L. and L. Gaines. 2010. A Review of Battery Life-Cycle Analysis: State of Knowledge and Critical Needs, Center for Transportation Research Energy Systems Division, Argonne National Laboratory.

Table 22.17. Best discharge, charge, and storage conditions for the main rechargeable batteries

	Lead-Acid	Ni–Cd	Ni–MH	Li-ion
Discharge	The limit is ~80% DOD	Can be discharged to 100% DOD	The limit is ~80% DOD; few deeper discharges are allowed	The safety circuit prevents full discharge. ~80% DOD is a safe limit
Typical charge methods	Constant voltage to 2.40 V, followed by float charging at 2.25 V. Float charge can be prolonged. Fast charge is not possible. Slow charge: 14 h. Rapid charge: 10 h	Constant current, followed by trickle charge. Fast-charge preferred to limit self-discharge. Slow charge:16 h. Rapid charge: 3 h. Fast charge: ~1 h	Constant current, followed by trickle charge. Slow charge not recommended. Heating when full charge is approached. Rapid charge: 3 h. Fast charge: ~1 h	Constant current to 4.1–4.2 V, followed by constant voltage. Trickle charge is not necessary. Rapid charge: 3 h. Fast charge recently reported: <1 h
Storage	To be stored at full charge. Storing below 2.10 V produces sulfation	To be stored at ~40% state of charge. Five years of storage (or more) possible at room temperature or below	To be stored at ~40% state of charge. Storage at low temperature is recommended, as this cell easily self-discharges above room temperature	To be stored at an intermediate DOD (3.7–3.8 V). Storing at full charge and above room temperature is to be avoided, as irreversible self-discharge occurs

Table 22.18. Metal hydride characteristics

Characteristic	Low Temperature				High Temperature		
	$Ti_2Ni-H_{2,5}$	$FeTi-H_2$	$VH-VH_2$	$LaNi_6-H_{6,7}$	Mg_2Cu-H_3	Mg_2Ni-H_4	$Mg-H$
Alloy mass that can absorb hydrogen							
Hydride mass equivalent to the energy in 0.264 gal (1L) of gasoline	1.61%	1.87%	1.92%	1.55%	2.67%	3.71%	8.25%
Alloy mass necessary to accumulate 5.5 lb (2.5 kg) of hydrogen	342 lb 155kg	295 lb 143 kg	286 lb 130 kg	355 lb 161 kg	No Data	149 lb 67.5 kg	79 lb 35 kg
Desorption temperature at 145 psig (10 barg)	93 °F 34 °C 307 K	125 °F 52 °C 325 K	127 °F 53 °C 326 K	163 °F 73 °C 346 K	604 °F 318 °C 591 K	662 °F 350 °C 623 K	683 °F 362 °C 635 K
Desorption temperature at 22 psig (1.5 barg)	26 °F −3 °C 270 K	44 °F 7 °C 280 K	59 °F 15 °C 288 K	70 °F 21 °C 295 K	480 °F 245 °C 522 K	512 °F 267 °C 540 K	565 °F 296 °C 569 K
Charging	Easy	No Data	No Data	Very Difficult	No Data	Difficult	Very Difficult
Safety	Safe	No Data	No Data	No Data	Highly Flammable	Safe	Highly Flammable

Source: U.S. Department of Energy, Office of Energy Efficiency and Renewable Energy, Hydrogen Fuel Cell Engines and Related Technologies Course Manual (2001), <http://www1.eere.energy.gov/hydrogenandfuelcells/tech_validation/h2_manual.html>.

Table 22.19. Composition of lithium-ion batteries

Component	Materials	Percentage
Cathodes		15–27
	Li_2CO_3	
	$LiCoO_2$	
	$LiMn_2O_4$	
	$LiNiO_2$	
	$LiFePO_4$	
	$LiCo_{1/3}Ni_{1/3}Mn_{1/3}O_2$	
	$LiNi_{0.8}Co_{0.15}Al_{0.05}O_2$	
Anodes		10–18
	Graphite (LiC_6)	
	$Li_4Ti_5O_{12}$	
Electrolyte		10–16
	Ethylene carbonate	
	Diethyl Carbonate	
	$LiPF_6$	
	$LiBF_4$	
	$LiClO_4$	
Separator	Polypropylene	3–5
Case	Steel	40

Source: Sullivan, J.L. and L. Gaines. 2010. A Review of Battery Life-Cycle Analysis: State of Knowledge and Critical Needs, Center for Transportation Research Energy Systems Division, Argonne National Laboratory.

Table 22.20. Basic characteristics of Lithium (Li)-ion batteries with different chemistries

System	Discharge Voltage (V)	Temperature Range (°C)	Specific Energy (Wh/kg)	Energy Density (Wh/L)	Cycles	Power
$LiCoO_2$	3.6	−20/60	140–190	360–500	800–1200	L-M
NCA[a]	3.5	−20/60	220–240	500–630	800–1200	L-M
NCM[b]	3.7	−20/60	100–150	230–400	500–700	M-H
Mn spinel[c]	3.7	−20/60	130–150	300–320	500–700	H
Fe phosphate[d]	3.3	−30/70	100–140	250–380	>1000	VH
Nexelion[e]	3.5	−20/60	160	480	~1000	M-H

(L=low, M=moderate, H=high, MH=moderately high, VH=very high).
[a]NCA, Ni-Co-Al.
[b]NCM, Ni-Co-Mn.
[c]$LiMn_2O_4$ doped with Mg
[d]Nano-sized, doped $LiFePO_4$.
[e]Sony's hybrid battery with Sn-Co-C as a negative and $LiCo_xNi_yMn_zO_2$ + $LiCoO_2$ as a positive.
Source: Adapted from Pistoia, Gianfranco. 2009. Battery Operated Devices and Systems, (Amsterdam, Elsevier).

Table 22.21. Characteristics of primary Lithium (Li) batteries

Attribute/Cell	Li/SO_2	$Li/SOCl_2$	Li/MnO_2	Li/CF_x
Average voltage (V)	2.7–2.9	3.4–3.6	2.8–3.0	2.6–2.8
Specific energy (Wh/kg)	260–280	450–600 (bobbin)	250–300 (bobbin)	200–250 (small)
		200–450 (spiral)	150–230 (spiral)	530–600 (large)
Energy density (Wh/L)	400–450	700–1100 (bobbin)	580–650 (bobbin)	580–635 (small)
		400–850 (spiral)	400–520 (spiral)	900–1050 (large)
Power density	High	Low/medium (bobbin)	Low/medium (bobbin)	Low
		Medium/high (spiral)	Medium/high (spiral)	
Temperature range (°C)	−55 to −70	−55 to −85 (standard)	−40 to −85 (laser)	−40 to −85
		−50 to −150 (h.t.)	−20 to −60 (crimp)	−40 to −125 (h.t.)
Shelf-life (years at room temperature)	10	10–15	10	15
Relative market	5	15	100	1
Relative cost (per kWh)	0.9	1.5	1	2

Source: Pistoia, Gianfranco. 2009. Battery Operated Devices and Systems, (Amsterdam, Elsevier).

Table 22.22. Characteristics of Lithium (LI)-ion batteries for stationary and traction applications

Application and Positive Electrode	Voltage Range (V)	Operating Temperature (°C)	Cycle Life (cycles)	Specific Energy (Wh/kg)	Energy Density (Wh/L)	Specific Power (W/kg)[a]
Stationary ($LiNi_{0.7}Co_{0.3}O_2$)	4.0–2.8	−20 ÷ 60	900[b]	128	197	
Traction ($LiNi_{1−x−y}Co_xMn_yO_2$)	4.0–2.8	−20 ÷ 50	570	150	252	490
Stationary ($Li_{1+x}Mn_2O_4$)	4.0–3.0	−20 ÷ 60	1200[b]	122	255	
Traction ($LiCr_xMn_{2−x}O_4$)	4.0–3.0	−20 ÷ 50	580	155	244	440

[a]Pulse discharge.
[b]Single cells.
Source: Pistoia, Gianfranco. 2009. Battery Operated Devices and Systems, (Amsterdam, Elsevier).

Table 22.23.   Characteristics of lead-acid and Nickel-Cadmium (Ni-Cd) batteries.

System	Voltage Range (V)	Operating Temperature (°C)	Cycle Life (cycles)[b]	Specific Energy (Wh/kg)	Energy Density (Wh/L)	Specific Power[c]	Self-Discharge (%/month)
Lead-Acid[a]							
Sealed	2.0–1.8	–40 to 60	250–500	30	90	H	4–6
SLI[d]	2.0–1.8	–40 to 55	200–500	35	70	H	3
Traction	2.0–1.8	–20 to 40	1500	25–30	80	MH	4–6
Stationary	2.0–1.8	–10 to 40	400	20–30	50–70	MH	2
Ni-Cd[a]							
Vented Pocket Plate	1.35–1.1	–20 to 45	500–2000	15–35	35–45	H	5
Vented Sintered Plate	1.35–1.1	–40 to 40	500–2000	30–40	60–100	H	10
FNC[d]	1.35–1.0	–50 to 60	500–3000+	10–40	15–80	L to VH	10–15
Sealed	1.35–1.0	–40 to 45	300–700	15–35	50–120	M to H	15–20

[a]Cell level.
[b]Dependent on DOD (depth of discharge).
[c]Pulse discharge (L=low, M=moderate, H=high, MH=moderately high, VH=very high).
[d]Starting, Lighting, Ignition.
[d]Fibre nickel cadmium.

Table 22.24.   Comparison of the main characteristics of aqueous primary batteries

	Leclanché (Zn/MnO₂)	Zinc Chloride (Zn/MnO₂)	Alkaline/Manganese Dioxide (Zn/MnO₂)	Silver Oxide (Zn/Ag₂O)	Zinc-Air (Zn/O₂)
System	Zinc/manganese dioxide	Zinc/manganese dioxide	Zinc/alkaline manganese dioxide	Zinc/silver oxide	Zinc/oxygen
Voltage per cell	1.5	1.5	1.5	1.5	1.4
Positive electrode	Manganese dioxide	Manganese dioxide	Manganese dioxide	Monovalent silver oxide	Oxygen
Electrolyte	Aqueous solution of NH_4Cl and $ZnCl_2$	Aqueous solution of $ZnCl_2$ (may contain some NH_4Cl)	Aqueous solution of KOH	Aqueous solution of KOH or NaOH	Aqueous solution of KOH
Overall reaction equations	$2MnO_2 + 2NH_4Cl + Zn \rightarrow ZnCl_2 \cdot 2NH_3 + Mn_2O_3 \cdot H_2O$	$8MnO_2 + 4Zn + ZnCl_2 \cdot 9H_2O \rightarrow 8MnOOH + ZnCl_2 \cdot 4ZnO \cdot 5H_2O$	$Zn + 2MnO_2 + 2H_2O \rightarrow Zn(OH)_2 + 2MnOOH$	$Zn + Ag_2O \rightarrow ZnO + 2Ag$	$2Zn + O_2 \rightarrow 2ZnO$
Typical commercial service capacities	Several hundred mAh	Several hundred mAh to 38 Ah	30 mAh to 45 Ah	5 to 190 mAh	30 to 1100 mAh
Specific energies (Wh/kg)	65 (cylindrical)	85 (cylindrical)	80 (button); 145 (cylindrical)	135 (button)	370 (button); 300 (prismatic)

Continued

Table 22.24.   Continued

	Leclanché (Zn/MnO$_2$)	Zinc Chloride (Zn/MnO$_2$)	Alkaline/ Manganese Dioxide (Zn/MnO$_2$)	Silver Oxide (Zn/Ag$_2$O)	Zinc-Air (Zn/O$_2$)
Energy densities(Wh/L)	100 (cylindrical)	165 (cylindrical)	360 (button); 400 (cylindrical)	530 (button)	1300 (button); 800 (prismatic)
Discharge curve	Sloping	Sloping	Sloping	Flat	Flat
Temperature range (storage)	−40 to 50 °C	−40 to 50 °C	−40 to 50 °C	−40 to 60 °C	−40 to 50 °C
Temperature range (operating)	−5 to 55 °C	−18 to 55 °C	−18 to 55 °C	−10 to 55 °C	−10 to 55 °C
Effect of temperature on service capacity	Poor low temperature	Good low temperature relative to Leclanché	Good low temperature	Low temperature depends upon construction	Good low temperature
Internal resistance	Moderate	Low	Very low	Low	Low
Gassing	Medium	Higher than Leclanché	Low	Very low	Very low
Cost (initial)	Low	Low to medium	Medium plus	High	High
Cost (operating)	Low	Low to medium	Low to high	High	High
Capacity loss per year @ 0 °C	3%	2%	1%	1%	NA
Capacity loss per year @ 20 °C	6%	5%	3%	3%	5% (sealed)
Capacity loss per year @ 40 °C	20%	16%	8%	7%	NA

Source: Pistoia, Gianfranco. 2009. Battery Operated Devices and Systems, (Amsterdam, Elsevier).

Table 22.25.   Advantages, disadvantages, and main applications of aqueous primary batteries

Battery Type	Dry Manganese	Alkaline Manganese	Silver Oxide	Zinc Air
Designation	R	LR	SR	PR
Nominal voltage	1.5 V	1.5 V	1.55 V	1.4 V
Advantages	Wide applications	For heavy and continuous use	Flat discharge curve	For heavy and continuous use
	Excellent anti-leakage	High reliability	Superior long-term reliability	Stable discharge curve
	No Hg, no Cd	No Hg, no Cd	High-energy density	Excellent anti-leakage
	Lowest cost	Higher capacity than dry manganese	High power	High-energy density
Disadvantages	Low-energy density	More expensive than dry manganese but better cost/performance ratio at high rates	Expensive but cost-effective in button cells	Limited power
	Poor performance at low temperature and high rate		Good low-temperature performance	Performance affected by environment: flooding, drying out
				Short activated life

Table 22.25. Continued

Battery Type	Dry Manganese	Alkaline Manganese	Silver Oxide	Zinc Air
Applications	Radios	Cordless phones	Wristwatches	Telecommunications
	Headphone stereos	Headphones	Cameras	Hearing aids
	Remote controllers	CD players	Calculators	Glucose meters
	Transceivers	LCD TVs	Hearing aids	
	Flashlights	Lanterns	Glucose meters	
	Clocks	Electric shavers		
	Calculators	Remote controllers		

Source: Adapted from Pistoia, Gianfranco. 2009. Battery Operated Devices and Systems, (Amsterdam, Elsevier).

Table 22.26. Characteristics of aqueous secondary batteries (except Pb-acid and Ni–Cd)

System	V Range (V)	Operating Temperature (°C)	Cycle Life (cycles)[a]	Specific Energy (Wh/kg)	Energy Density (Wh/L)	Specific Power[b]	Self-Discharge (%/month)
Ni-MH[c]	1.4–1.2	−30 to 65	800–1200	60–80	200–270	M to H	15–25
Ni-H2[c]	1.5–1.2	−10 to 30	>2000	45–55	65–80	M	60
Ni-Fe[c]	1.5–1.2	−10 to 60	2000–3000	30	60	L to M	25–30
Ni-Zn[d]	1.9–1.5	−20 to 50	300–600	60	100–120	H	15–20
Zn-Air[d]	1.2–1.0	0 to 45	20–30	150–200	160–220	L to M	–
Zn/AgO[c]	1.8–1.5	−20 to 60	50–80	90–100	180	VH	5
Zn/Br2[d]	1.9–1.6	10 to 50	>500	65–70	60–70	L to M	12–15
VRB[d]	1.5–1.1	10 to 50	3000	10–30	10–30	M to H	5–10

[a]Battery.
[b]Pulse discharge (L=low, M=moderate, H=high, MH=moderately high, VH=very high).
[c]Cell
[d]Dependent on depth-of-discharge
Source: Pistoia, Gianfranco. 2009. Battery Operated Devices and Systems, (Amsterdam, Elsevier).

Table 22.27. Advantages, disadvantages, and main applications of secondary batteries

Battery Type	Sealed Lead-Acid	Nickel–Cadmium	Nickel-Metal Hydride	Lithium Ion
Nominal voltage	2.0 V	1.2 V	1.2 V	3.7 V
Advantages	For heavy duty use	For heavy duty use	For heavy duty use	For heavy duty use
	Superior long-term reliability	High mechanical strength	No heavy metals	High 3.7 V voltage
	Economical	High efficiency charge	Relatively high capacity	No memory effect
	Easy to recycle	Charge cycle: 500 times	Charge cycle: 500 times	Low self-discharge
		Easy to recycle		
Disadvantages	Relatively low cycle life	Low energy	More expensive than Ni–Cd	Expensive
	Low energy	Memory effect	Very high self-discharge	Potential safety problems
	High self-discharge in flooded batteries	Toxicity		Requires control of charge/disch. limits

Continued

Table 22.27.　Continued

Battery Type	Sealed Lead-Acid	Nickel–Cadmium	Nickel-Metal Hydride	Lithium Ion
		High self-discharge espec. in sealed cells		Degrades at high temperature
Applications	Automot. applications	Portable OA equipment	Portable OA equipment	Portable OA equipment
	Portable AV equipment	Portable AV equipment	Portable AV equipment	Military and space appl.
	Lighting equipment	Power tools	Power tools	Many consumer devices
	Stationary applications	Medical instruments	Medical instruments	Candidate for next-generation HEV
		Stationary applications	Hybrid cars	Power tools
		Space applications		

Source: Adapted from Pistoia, Gianfranco. 2009. Battery Operated Devices and Systems, (Amsterdam, Elsevier).

Table 22.28.　Comparison of technical characteristics of energy storage systems: energy and power density, lifetime, and cycle life

Systems	Energy and power density				Life time and cycle life	
	Wh/kg	W/kg	Wh/L	W/L	Life time (years)	Cycle life (cycles)
Pumped-hydro	0.5–1.5		0.5–1.5		40–60	
Compressed air (CAES)	30–60		3–6	0.5–2.0	20–40	
Lead-acid	30–50	75–300	50–80	10–400	5–15	500–1000
NiCd	50–75	150–300	60–150		10–20	2000–2500
NaS	150–240	150–230	150–250		10–15	2500
Sodium nickel chloride battery (ZEBRA)	100–120	150–200	150–180	220–300	10–14	2500+
Li-ion	75–200	150–315	200–500		5–15	1000–10,000+
Fuel cells	800–10,000	500+	500–3000	500+	5–15	1000+
Metal-Air	150–3000		500–10,000			100–300
Polysulphide bromide battery	10–30		16–33		5–10	12,000+
ZnBr	30–50		30–60		5–10	2000+
Polysulphide bromide battery	–	–	–	–	10–15	
Solar fuel	800–100,000		500–10,000		–	–
SMES[a]	0.5–5	500–2000	0.2–2.5	1000–4000	20+	100,000+
Flywheel	10–30	400–1500	20–80	1000–2000	~15	20,000+
Capacitor	0.05–5	~100,000	2–10	100,000+	~5	50,000+
Super-capacitor	2.5–15	500–5000	10–30	100,000+	20+	100,000+
AL-TES[b]	80–120		80–120		10–20	
CES[c]	150–250	10–30	120–200		20–40	
HT-TES[d]	80–200		120–500		5–15	

[a]Superconducting magnetic energy storage
[b]Aquiferous low-temperature thermal energy storage
[c]Cryogenic energy storage
[d]High temperature thermal energy storage
Source: Adapted from Chen, Haisheng, Thang Ngoc Cong, Wei Yang, Chunqing Tan, Yongliang Li, Yulong Ding. 2009. Progress in electrical energy storage system: A critical review, Progress in Natural Science, Volume 19, Issue 3, Pages 291-312.

Table 22.29. Comparison of technical characteristics of energy storage systems: power rating, discharge times, duration, and cost

Systems	Power rating and discharge time		Storage duration		Capital cost		
	Power rating	Discharge time	Self discharge per day	Suitable storage duration	$/kW	$/kWh	¢/kWh-Per cycle
Pumped-hydro	100–5000 MW	1–24 h+	Very small	Hours–months	600–2000	5–100	0.1–1.4
Compressed air (CAES)	5–300 MW	1–24 h+	Small	Hours–months	400–800	2–50	2–4
Lead-acid	0–20 MW	Seconds–hours	0.1–0.3%	Minutes–days	300–600	200–400	20–100
NiCd	0–40 MW	Seconds–hours	0.2–0.6%	Minutes–days	500–1500	800–1500	20–100
NaS	50 kW–8 MW	Seconds–hours	~20%	Seconds–hours	1000–3000	300–500	8–20
Sodium nickel chloride battery (ZEBRA)	0–300 kW	Seconds–hours	~15%	Seconds–hours	150–300	100–200	5–10
Li-ion	0–100 kW	Minutes–hours	0.1–0.3%	Minutes–days	1200–4000	600–2500	15–100
Fuel cells	0–50 MW	Seconds–24 h+	Almost zero	Hours–months	10,000+		6000–20,000
Metal-Air	0–10 kW	Seconds–24 h+	Very small	Hours–months	100–250	10–60	
Polysulfide bromide battery	30 kW–3 MW	Seconds–10 h	Small	Hours–months	600–1500	150–1000	5–80
ZnBr	50 kW–2 MW	Seconds–10 h	Small	Hours–months	700–2500	150–1000	5–80
Polysulfide bromide battery	1–15 MW	Seconds–10 h	Small	Hours–months	700–2500	150–1000	5–80
Solar fuel	0–10 MW	1–24 h+	Almost zero	Hours–months	–	–	–
SMES[a]	100 kW–10 MW	Milliseconds–8 s	10–15%	Minutes–hours	200–300	1000–10,000	
Flywheel	0–250 kW	Milliseconds–15 min	100%	Seconds–minutes	250–350	1000–5000	3–25
Capacitor	0–50 kW	Milliseconds–60 min	40%	Seconds–hours	200–400	500–1000	
Super-capacitor	0–300 kW	Milliseconds–60 min	20–40%	Seconds–hours	100–300	300–2000	2–20
AL-TES[b]	0–5 MW	1–8 h	0.50%	Minutes–days		20–50	
CES[c]	100 kW–300 MW	1–8 h	0.5–1.0%	Minutes–days	200–300	3–30	2–4

Continued

Table 22.29. **Continued**

Systems	Power rating and discharge time		Storage duration		Capital cost		
	Power rating	Discharge time	Self discharge per day	Suitable storage duration	$/kW	$/kWh	¢/kWh-Per cycle
HT-TES[d]	0–60 MW	1–24 h+	0.05–1.0%	Minutes–months		30–60	

[a]Superconducting magnetic energy storage.
[b]Aquiferous low-temperature thermal energy storage.
[c]Cryogenic energy storage.
[d]High temperature thermal energy storage.
Source: Adapted from Chen, Haisheng, Thang Ngoc Cong, Wei Yang, Chunqing Tan, Yongliang Li, Yulong Ding. 2009. Progress in electrical energy storage system: A critical review, Progress in Natural Science, Volume 19, Issue 3, Pages 291–312.

Materials

Figures

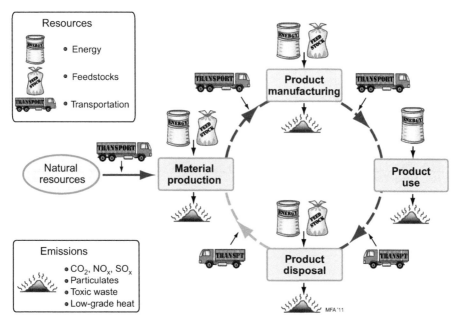

Figure 23.1. The material life cycle. Ore and feedstock are mined and processed to yield a material. *Source: Ashby, Michael F. 2013. Materials and the Environment (Second Edition), (Boston, Butterworth-Heinemann).*

Handbook of Energy. http://dx.doi.org/10.1016/B978-0-08-046405-3.00023-1

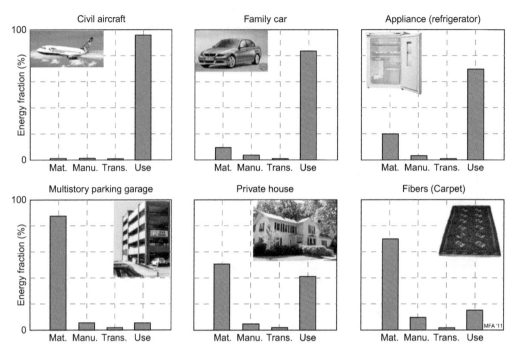

Figure 23.2. Approximate values for the energy consumed at each phase for selected products. The disposal phase is not shown because there are many alternatives for each product.
Source: Ashby, Michael F. 2013. Materials and the Environment (Second Edition), (Boston, Butterworth-Heinemann).

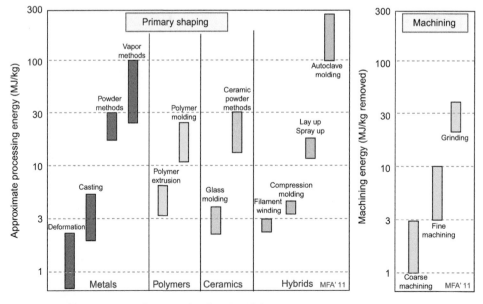

Figure 23.3. Approximate processing energies for materials.
Source: Ashby, Michael F. 2013. Materials and the Environment (Second Edition), (Boston, Butterworth-Heinemann).

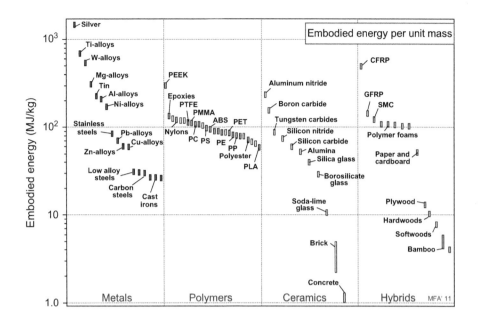

Figure 23.4. Embodied energies of materials per unit mass.
Source: Ashby, Michael F. 2013. Materials and the Environment (Second Edition), (Boston, Butterworth-Heinemann).

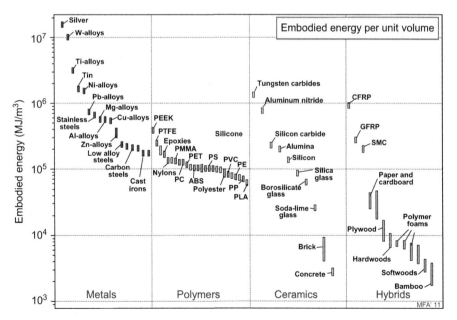

Figure 23.5. Embodied energies of materials per unit volume.
Source: Ashby, Michael F. 2013. Materials and the Environment (Second Edition), (Boston, Butterworth-Heinemann).

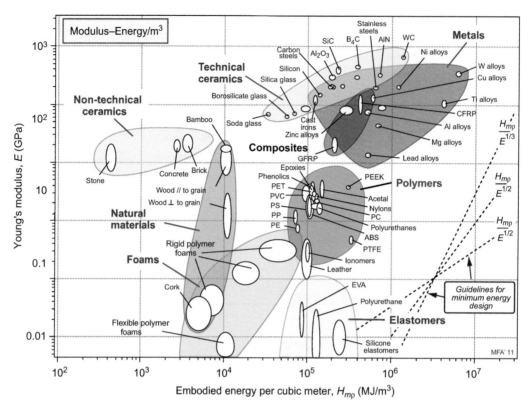

Figure 23.6. Young's modulus, a measure of the stiffness of an elastic material, and embodied energy per unit volume for selected materials.
Source: Ashby, Michael F. 2013. Materials and the Environment (Second Edition), (Boston, Butterworth-Heinemann).

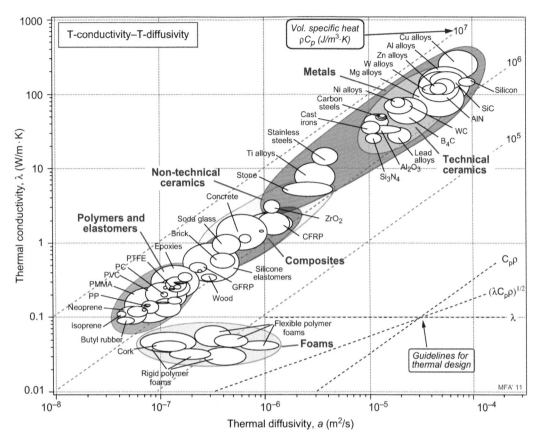

Figure 23.7. The Thermal conductivity-Thermal diffusivity chart with contours of volumetric specific heat: the one for minimum thermal loss.
Source: Ashby, Michael F. 2013. Materials and the Environment (Second Edition), (Boston, Butterworth-Heinemann).

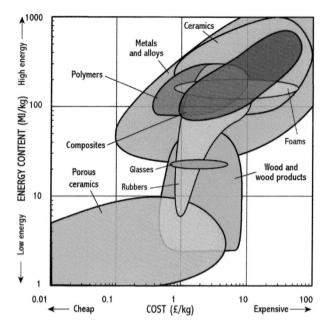

Figure 23.8. The relation between the embodied energy (MJ/kg) in broad classes of materials and their monetary cost (£/kg).
Source: Ashby, Mike, and Shercliff, Hugh, Material selection and processing, Department of Engineering, Cambridge University, <http://www-materials.eng.cam.ac.uk/mpsite/default.html>.

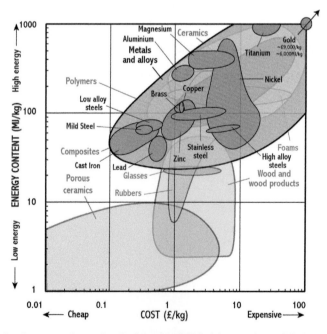

Figure 23.9. The relation between the embodied energy (MJ/kg) in metals and their monetary cost (£/kg).
Source: Ashby, Mike, and Shercliff, Hugh, Material selection and processing, Department of Engineering, Cambridge University, <http://www-materials.eng.cam.ac.uk/mpsite/default.html>.

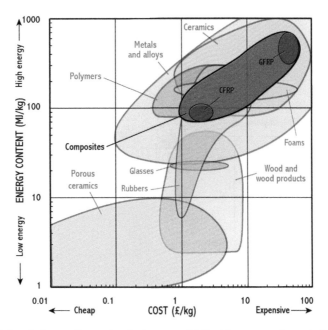

Figure 23.10. The relation between the embodied energy (MJ/kg) in composite materials and their monetary cost (£/kg).
Source: Ashby, Mike, and Shercliff, Hugh, Material selection and processing, Department of Engineering, Cambridge University, <http://www-materials.eng.cam.ac.uk/mpsite/default.html>.

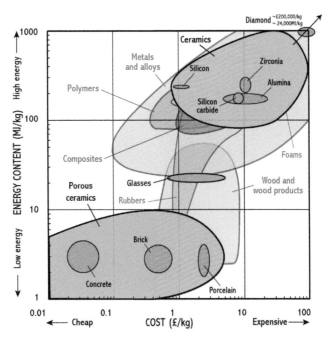

Figure 23.11. The relation between the embodied energy (MJ/kg) in ceramic materials and their monetary cost (£/kg).
Source: Ashby, Mike, and Shercliff, Hugh, Material selection and processing, Department of Engineering, Cambridge University, <http://www-materials.eng.cam.ac.uk/mpsite/default.html>.

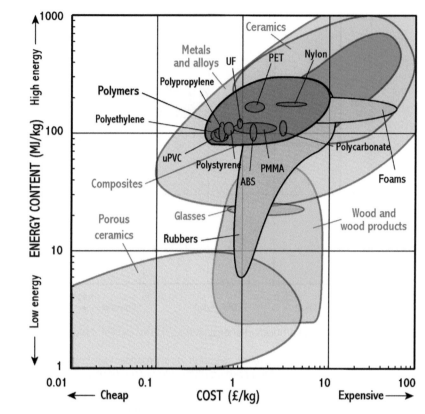

Figure 23.12.   The relation between the embodied energy (MJ/kg) in polymers and their monetary cost (£/kg).
Source: Ashby, Mike, and Shercliff, Hugh, Material selection and processing, Department of Engineering, Cambridge University, <http://www-materials.eng.cam.ac.uk/mpsite/default.html>.

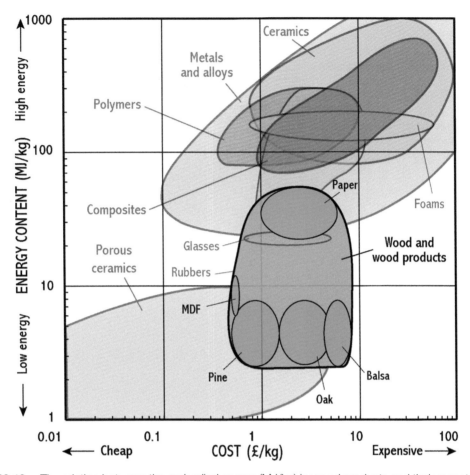

Figure 23.13. The relation between the embodied energy (MJ/kg) in wood products and their monetary cost (£/kg).
Source: Ashby, Mike, and Shercliff, Hugh, Material selection and processing, Department of Engineering, Cambridge University, <http://www-materials.eng.cam.ac.uk/mpsite/default.html>.

Tables

Table 23.1. Properties of Common Solids

Material	Specific gravity	Specific heat		Thermal conductivity	
		Btu/lbm•deg R	kJ/kg•K	Btu/hr•ft•deg F	W/m•K
Asbestos cement board	1.4	0.2	0.837	0.35	0.607
Asbestos millboard	1	0.2	0.837	0.08	0.14
Asphalt	1.1	0.4	1.67		
Brick, common	1.75	0.22	0.92	0.42	0.71
Brick, hard	2	0.24	1	0.75	1.3
Chalk	2	0.215	0.9	0.48	0.84
Charcoal, wood	0.4	0.24	1	0.05	0.088
Coal, bituminous	1.2	0.33	1.38		
Concrete, Light	1.4	0.23	0.962	0.25	0.42
Concrete, stone	2.2	0.18	0.753	1	1.7
Corkboard	0.2	0.45	1.88	0.025	0.04
Earth, dry	1.4	0.3	1.26	0.85	1.5
Fiberboard, light	0.24	0.6	2.51	0.035	0.05
Fiber hardboard	1.1	0.5	2.09	0.12	0.2
Glass, window	2.5	0.2	0.837	0.55	0.96
Gypsum board	0.8	0.26	1.09	0.1	0.17
Hairfelt	0.1	0.5	2.09	0.03	0.05
Ice	0.9	0.5	2.09	1.25	2.2
Leather, dry	0.9	0.36	1.51	0.09	0.2
Limestone	2.5	0.217	0.908	1.1	1.9
Magnesia (85%)	0.25	0.2	0.837	0.04	0.071
Marble	2.6	0.21	0.879	1.5	2.6
Mica	2.7	0.12	0.502	0.4	0.71
Paper	0.9	0.33	1.38	0.07	0.1
Paraffin Wax	0.9	0.69	2.89	0.15	0.2
Plaster, light	0.7	0.24	1	0.15	0.2
Plaster, sand	1.8	0.22	0.92	0.42	0.71
Plastics, solid	1.2	0.4	1.67	0.11	0.19
Porcelain	2.5	0.22	0.92	0.9	1.5
Sandstone	2.3	0.22	0.92	1	1.7
Sawdust	0.15	0.21	0.879	0.05	0.08
Vermiculite	0.13	0.2	0.837	0.035	0.058
Wood, balsa	0.16	0.7	2.93	0.03	0.05
Wood, oak	0.7	0.5	2.09	0.1	0.17
Wood, white pine	0.5	0.6	2.51	0.07	0.12
Wool, felt	0.3	0.33	1.38	0.04	0.071
Wool, loose	0.1	0.3	1.26	0.02	0.3

Source: Adapted from Norton, Paul, Appendices, in Kreith, Frank, Ed. 2000. CRC Handbook of Thermal Engineering (Boca Raton, CRC Press).

Table 23.2. Typical properties of commerical metals and alloys

Common name and classification	Thermal conductivity Btu/hr ft °F	Specific gravity	Coefficient of linear expansion μ in./ in. °F	Electrical resistivity microhm-cm	Approximate melting point °F
Ingot iron (included for comparison)	42	7.86	6.8	9	2800
Plain carbon steel AISI-SAE 1020	30	7.86	6.7	10	2760
Stainless steel type 304	10	8.02	9.6	72	2600
Cast gray iron ASTM A48-48, Class 25	26	7.2	6.7	67	2150
Malleable iron ASTM A47	--	7.32	6.6	30	2250
Ductile cast iron ASTM A339, A395	19	7.2	7.5	60	2100
Ni-resist cast iron, type 2	23	7.3	9.6	170	2250
Cast 28-7 alloy (HD) ASTM A297-63T	1.5	7.6	9.2	41	2700
Inconel X, annealed	9	8.25	6.7	122	2550
Haynes Stellite alloy 25 (L605)	5.5	9.15	7.61	88	2500
Aluminum alloy 3003, rolled ASTM B221	90	2.73	12.9	4	1200
Aluminum alloy 2017, annealed ASTM B221	95	2.8	12.7	4	1185
Copper ASTM B152, B134, B135	225	8.91	9.3	1.7	1980
Yellow brass (high brass) ASTM B36, B135, B135	69	8.47	10.5	7	1710
Aluminum bronze ASTM B169, alloy A; ASTM B124, B150	41	7.8	9.2	12	1900
Beryllium copper 25, ASTM B194	7	8.25	9.3	--	1700
Nickel silver 18%, alloy A (wrought) AST< B194	19	8.8	9	29	2030
Red brass (cast) ASTM B30, No. 4A	42	8.7	10	11	1825
Chemical lead	20	11.35	16.4	21	621
Antimonial lead (hard lead)	17	10.9	15.1	23	554
Solder 50–50	26	8.89	13.1	15	420
Magnesium alloy AZ31B	45	1.77	14.5	9	1160
Nickel ASTM B160, B161, B162	35	8.89	6.6	10	2625
Cupronicle 55–45 (Constantan)	13	8.9	8.1	49	2300
Commercial titanium	10	5	4.9	80	3300
Zinc ASTM B69	62	7.14	18	6	785
Zirconium, commerical	10	6.5	2.9	41	3350

Source: Adapted from Norton, Paul, Appendices, in Kreith, Frank, Ed. 2000. CRC Handbook of Thermal Engineering (Boca Raton, CRC Press).

Table 23.3. Thermal properties of metals and alloys

Material	k, Btu/(hr)(ft)(°F)				c, Btu/(lb_m)(°F)	p, lb_m/ft³	a, ft²/hr
	32°F	212°F	572°F	932°F	32°F	32°F	32°F
Metals							
Aluminum	117	119	133	155	0.208	169	3.33
Copper, pure	224	218	212	207	0.091	558	4.42
Gold	169	170			0.030	1203	4.68
Iron, pure	35.8	36.6			0.104	491	0.70
Lead	20.1	19	18		0.030	705	0.95
Magnesium	91	92			0.232	109	3.60
Mercury	4.8				0.033	849	0.17
Nickel	34.5	34	32		0.103	555	0.60
Silver	242	238			0.056	655	6.6
Tin	36	34			0.054	456	1.46
Zinc	65	64	59		0.091	446	1.60
Alloys							
Brass, 70% Cu, 30% Zn	56	60	66		0.092	532	1.14
Bronze, 75% Cu, 25% Sn	15				0.082	540	0.34
Cast iron							
Plain	33	31.8	27.7	24.8	0.11	474	0.63
Alloy	30	28.3	27		0.10	455	0.66
Constantan, 60% Cu, 40% Ni	12.4	12.8			0.10	557	0.22
18-8 Stainless steel, Type 304	8	9.4	10.9	12.4	0.11	488	0.15
Steel, mild, 1% C	26.5	26	25	22	0.11	490	0.49

Source: Adapted from Goswami, D., Frank Kreith, and Jan F. Kreider. 2000. Principles of Solar Engineering, 2nd edition, (Philadelphia, Taylor & Francis)

Table 23.4. Electron properties in common metals

Metal	Electron concentration N_e (10^{28} m⁻³)	Fermi energy E_F (10^{-19} J)	Fermi velocity v (10^6 m s⁻¹)	Sommerfeld parameter γ (J m⁻³ K⁻²)	Electrical conductivity, 300K σ (10^5 Ω⁻¹ m⁻¹)	Thermal conductivity, 300 K k (W m⁻¹ K⁻¹)	Electron Mean Free Path, 300 K Λ (nm)
Rb	1.15	2.96	0.81	46	0.8	58	16
Ag	5.85	8.78	1.39	62.8	6.21	429	49
Au	5.9	8.83	1.39	71.4	4.55	317	32
Cu	8.45	11.2	1.57	97.5	5.88	401	26
In	11.49	13.8	1.74	322	1.14	82	1.5
Pb	13.2	15	1.82	653	0.48	35	0.3
Al	18.06	18.6	2.02	405	3.65	237	2.9

Source: Adapted from Goodson, Kenneth E., Thermal Conduction in Electronic Microstructures, in Kreith, Frank, Ed. 2000. CRC Handbook of Thermal Engineering (Boca Raton, CRC Press).

Table 23.5. Thermal properties of selected nonmetals

Material	Average temperature, °F	k, Btu/(hr)(ft)(°F)	c, Btu/(lb$_m$)(°F)	p, lb$_m$/ft3	a, ft2/hr
Insulating materials					
Asbestos	32	0.087	0.25	36	~0.01
	392	0.12		36	~0.01
Cork	86	0.025	0.04	10	~0.006
Cotton, fabric	200	0.046			
Diatomaceous earth, powdered	100	0.030	0.21	14	~0.01
	300	0.036			
	600	0.046			
Molded pipe covering	400	0.051		26	
	1600	0.088			
Glass woll					
Fine	20	0.022			
	100	0.031		1.5	
	200	0.043			
Packed	20	0.016			
	100	0.022		6.0	
	200	0.029			
Hair felt	100	0.027		8.2	
Kaolin insulating firebrick	932	0.15		27	
	2102	0.26			
Kaolin insulating firebrick	392	0.05		19	
	1400	0.11			
Rock wool	20	0.017		8	
	200	0.030			
Rubber	32	0.087	0.48	75	0.0024
Building Materials					
Brick					
Fire-clay	392	0.58	0.20	144	0.02
	1832	0.95			
Masonry	70	0.38	0.20	106	0.018
Zirconia	392	0.84		304	
	1832	1.13			
Chrome brick	392	0.82		246	
	1832	0.96			
Concrete					
Stone	~70	0.54	0.20	144	0.019
10% Moisture	~70	0.70		140	~0.025
Glass, window	~70	~0.45	0.2	170	0.013
Limestone, dry	70	0.40	0.22	105	0.017
Sand (dry)	68	0.2		95	

Source: Adapted from Goswami, D., Frank Kreith, and Jan F. Kreider. 2000. Principles of Solar Engineering, 2nd edition, (Philadelphia, Taylor & Francis)

Table 23.6.   Physical and thermal properties of common liquids

Common name	Density, kg/m³	Specific Heat, kJ/kg•K	Viscosity, N•s/m²	Thermal Conductivity, W/m•K	Freezing Point K	Latent heat of fusion, kJ/kg	Boiling Point, K	Latent heat of Evaporation kJ/kg	Coefficient of cubical expansion per K
Acetic Acid	1049	2.18	0.001155	0.171	290	181	391	402	0.0011
Acetone	784.6	2.15	0.000316	0.161	179	98.3	329	518	0.0015
Alcohol, ethyl	785.1	2.44	0.001095	0.171	158.6	108	351.46	846	0.0011
Alcohol, methyl	786.5	2.54	0.00056	0.202	175.5	98.8	337.8	1100	0.0014
Alcohol, propyl	800	2.37	0.00192	0.161	146	86.5	371	779	
Ammonia (aqua)	823.5	4.38		0.353					
Benzene	873.8	1.73	0.000601	0.144	278.68	126	353.3	390	0.0013
Bromine		0.473	0.00095	0.147	245.84	66.7	331.6	193	0.0012
Carbon tetrachloride	1584	0.866	0.00091	0.104	250.35	174	349.6	194	0.0013
Chloroform	1465	1.05	0.00053	0.118	209.6	77	334.4	247	0.0013
Decane	726.3	2.21	0.000859	0.147	243.5	201	447.2	263	
Dodecane	754.6	2.21	0.001374	0.14	247.18	216	489.4	256	
Ether	713.5	2.21	0.000223	0.13	157	96.2	307.7	372	0.0016
Ethylene glycol	1097	2.36	0.0162	0.258	260.2	181	470	800	

Material									
Flourine refrig R-11	1476	0.87	0.00042	0.093	162		297	180	
Flourine refrig R-12	1311	0.971		0.071	115	34.4	243.4	165	
Flourine refrig R-22	1194	1.26		0.086	113	183	232.4	232	
Glycerine	1259	2.62	0.95	0.287	264.8	200	563.4	974	0.00054
Heptane	679.5	2.24	0.000376	0.128	182.54	140	371.5	318	
Hexane	654.8	2.26	0.000297	0.124	178	152	341.84	365	
Iodine		2.15		0.145	386.6	62.2	457.5	164	
Kerosene	820.1	2.09	0.00164					251	
Linseed oil	929.1	1.84	0.0331		253		560		
Mercury		0.139	0.00153		234.3	11.6	630	295	0.00018
Octane	698.6	2.15	0.00051	0.131	216.4	181	398	298	0.00072
Phenol	1072	1.43	0.008	0.19	316.2	121	455		0.0009
Propane	493.5	2.41	0.00011		85.5	79.9	231.08	428	
Propylene	514.4	2.85	0.00009		87.9	71.4	225.45	342	
Propylene glycol	965.3	2.5	0.042		213		460	914	
Sea water	1025	3.76–4.1			270.6				
Toluene	862.3	1.72	0.00055	0.133	178	71.8	383.6	363	
Turpentine	868.2	1.78	0.001375	0.121	214		433	293	0.00099
Water	997.1	4.18	0.00089	0.609	273	333	373	2260	0.0002

Source: Adapted from Norton, Paul, Appendices, in Kreith, Frank, Ed. 2000. CRC Handbook of Thermal Engineering (Boca Raton, CRC Press).

Table 23.7.   Molecular thermal properties of natural materials

Material	Condition	Mass density (kg m^{-3} × 10^3)	Specific heat (J kg^{-1} K^{-1} × 10^3)	Heat capacity (J m^{-3} K^{-1} × 10^6)	Thermal conductivity (W m^{-1} K^{-1})	Thermal diffusivity (m^2 s^{-1} × 10^{-6})
Air	20 °C, Still	0.0012	1.01	0.0012	0.025	20.5
Water	20 °C, Still	1.00	4.18	4.18	0.57	0.14
Ice	0 °C, Pure	0.92	2.10	1.93	2.24	1.16
Snow	Fresh	0.10	2.09	0.21	0.08	0.38
Snow	Old	0.48	2.09	0.84	0.42	0.05
Sandy soil	Fresh	1.60	0.80	1.28	0.30	0.24
(40% pore space)	Saturated	2.00	1.48	2.96	2.20	0.74
Clay soil	Dry	1.60	0.89	1.42	0.25	0.18
(40% pore space)	Saturated	2.00	1.55	3.10	1.58	0.51
Peat soil	Dry	0.30	1.92	0.58	0.06	0.10
(80% pore space)	Saturated	1.10	3.65	4.02	0.50	0.12
Rock	Solid	2.70	0.75	2.02	2.90	1.43

Source: Adapted from Arya, S. Pal. 2001. Introduction to Micrometeorology (Second Edition). (San Diego, Academic Press).

Table 23.8.   Thermal properties of plastics

Material	Formula	Coefficient of thermal expansion ×10⁻⁶ K⁻¹	Lower working temperature °C	Specific heat J K⁻¹ kg⁻¹	Thermal conductivity W m⁻¹ K⁻¹	Upper working temperature °C
Cellulose Acetate	CA	80–180	−20	1200–1900	0.16–0.36 @23°C	55–95
Cellulose Acetate Butyrate	CAB	140	<−40	–	0.16–0.32 @23°C	60–100
Ethylene-Chlorotrifluoroethylene copolymer	E-CTFE	80	−75	–	0.16 @23°C	130–170
Ethylene-Tetrafluoroethylene Copolymer	ETFE	90–170	<−100	1900–2000	0.24 @23°C	150–160
Fluorinated Ethylene Propylene Copolymer	FEP	83–104	−250	1100	0.19–0.24 @23°C	150–20
Polyacrylonitrile-butadienestyrene	ABS	80	–	–	0.17 @23°C	70–100
Polyamide - Nylon 6	PA 6	95	−40	1700	0.24–0.28 @23°C	80–160
Polyamide - Nylon 6, 6	PA 6,6	90	−30	1670	0.25 @23°C	80–80
Polyamide - Nylon 6, 6-30% Carbon Fiber Reinforced	PA 6,6 - 30% CFR	14	–	–	0.51	120–200
Polyimide	PI	30–60	−270	1090	0.10–0.35 @23°C	250–320
Polymethylmethacrylate	PMMA, Acrylic	70–77	−40	1400–1500	0.17–0.19 @23°C	50 to 90
Polymethylpentene	TPX®	117	−20 to−40	2000	0.17 @23°C	75–15
Polyoxymethylene - Copolymer	Acetal - Copolymer POMC	80–120	160	1500	0.23–0.3 @23°C	80–20
Polyoxymethylene - Homopolymer	Acetal- Copolymer POMC	122	170	1500	0.22–0.24 @23°C	80–120
Polyphenyleneoxide	PPO (modified), PPE (modified)	60	137	–	0.22 @23°C	80–120
Polyphenyleneoxide (modified), 30% Glass Fiber Reinforced	PPO 30% GFR	25–30	165	–	0.28 @23°C	90–160
Polyphenylenesulfide - 40% Glass Fiber Reinforced	PPS - 40% GFR	22–35	>260	–	0.29–0.45 @23°C	200–260
Polyphenylsulfone	PPSu	55	–	–	0.35	180–210
Polypropylene	PP	100–180	100–105	1700–900	0.1–0.22 @23°C	90–120
Polystyrene	PS	30–210	90	1200	0.1–0.13 @23°C	50–95
Polystyrene - Cross-linked	PS - X - Linked	70–90	–	–	0.17 @23°C	93
Polysulphone	PSu	56	–	–	0.26	150–180
Polytetrafluoroethylene	PTFE	100–160	120	1000	0.25 @23°C	180–260
Polytetrafluoroethylene filled with Glass	PTFE 25% GF	75–100	–	–	0.33–0.42 @23°C	260

Source: Adapted from Professional Plastics, <http://www.professionalplastics.com/>, accessed 6 May 2012.

Table 23.9. Thermal properties of building materials

Material	Density (kg/m³)	Thermal Conductivity W/mK)	Specific Heat (J/kgK)
WALLS			
Asbestos cement sheet	700	0.36	1050
Asbestos cement decking	1500	0.36	1050
Brickwork (outer leaf)	1700	0.84	800
Brickwork (inner leaf)	1700	0.62	800
Cast concrete (dense)	2100	1.4	840
Cast concrete (lightweight)	1200	0.38	1000
Concrete block (medium weight)	1400	0.51	1000
Fiberboard	300	0.06	1000
Plasterboard	950	0.16	840
Stone (Artificial)	1750	1.3	1000
Stone (Limestone)	2180	1.5	910
Tile hanging	1900	0.84	800
SURFACE FINISHES			
External rendering	1300	0.5	1000
Plaster (dense)	1300	0.05	1000
Plaster (lightweight)	600	0.16	1000
ROOFS			
Aerated concrete slab	500	0.16	840
Asphalt	1700	0.5	1000
Felt/Bitumen layers	1700	0.5	1000
Screed	1200	0.41	840
Stone chippings	1800	0.96	1000
Tile	1900	0.84	800
Wood wool slab	500	0.1	1000
FLOORS			
Cast concrete	2000	1.13	1000
Metal tray	7800	50	480
Timber flooring	650	0.14	1200
Wood blocks	650	0.14	1200
INSULATION			
Expanded polystyrene slab	25	0.035	1400
Glass fiber quilt	12	0.04	840
Glass fiber slab	25	0.035	1000
Mineral fiber slab	30	0.035	1000
Phenolic foam	30	0.04	1400
Polyurethane board	30	0.025	1400
Urea formaldehyde foam	10	0.04	1400

Source: Adapted from Martin A. Wilkinson, Department of Architecture and Civil Engineering, University of Bath, United Kingdom.

Table 23.10. Embodied energy of assemblies in the U.S.[a, b]

	Embodied energy (MMBtu/SF)	CO_2 equivalent emissions (lbs/SF)
Embodied Energy of Windows[c]		
Window Type		
Aluminium	0.59	71.24
PVC-clad Wood	0.37	62.15
Wood	0.33	51.83
Vinyl (PVC)	0.49	82.31
Curtainwall Viewable Glazing	0.27	61.6
Curtainwall		
Opaque glazing (with insulated backpan)	0.18	32.16
Spandrel panel (with insulated backpan)	0.1	9.53
Embodied Energy of Studded Exterior Walls		
Exterior Wall Type		
2×6 Steel Stud Wall[d]		
16" OC with brick cladding	0.15	20.68
24" OC with brick cladding	0.15	19.48
16" OC with wood cladding (pine)	0.06	7.82
24" OC with wood cladding (pine)	0.06	6.61
16" OC with steel cladding (26 ga)	0.17	37.02
2×6 Wood Stud Wall[e]		
16" OC with brick cladding	0.16	18.88
16" OC with PVC cladding	0.10	9.63
24" OC with steel cladding	0.18	35.04
24" OC with stucco cladding	0.09	8.85
24" OC with wood cladding (pine)	0.07	5.83
Structural Insulated Panel (SIP)[f]		
with Brick cladding	0.20	20.73
with Steel cladding	0.22	37.07
with Stucco cladding	0.13	10.88
with PVC cladding	0.14	11.48
with Wood cladding	0.11	7.86
Embodied Energy of Concrete Exterior Walls		
Exterior Wall Type		
8" Concrete Block		
with Brick cladding + rigid insulation + vapor barrier	0.22	32.04
+ Gypsum board + latex paint	0.24	33.08
with Stucco cladding + rigid insulation + vapor barrier + gypsum board + latex paint	0.16	23.24
6" Cast-In-Place Concrete[g]		
with Brick cladding	0.22	33.78
with Steel cladding	0.24	50.12
with Stucco cladding	0.14	23.93
with 1" rigid insulation + 2×6 steel stud wall (24" OC) + batt insulation	0.11	20.93

Continued

Table 23.10.   Continued

	Embodied energy (MMBtu/SF)	CO$_2$ equivalent emissions (lbs/SF)
8" Concrete Tilt-Up		
with Steel cladding[g]	0.24	50.25
with Stucco cladding[g]	0.15	24.06
with 2x6 steel stud wall (24" OC) + batt insulation	0.11	21.05
Insulated Concrete Forms		
with Steel cladding + gypsum board + latex paint	0.28	57.78
with PVC cladding + gypsum board + latex paint	0.20	32.19
with Wood cladding + gypsum board + latex paint	0.17	28.57
Embodied Energy of Wood-Based Roof Assemblies		
Glulam Joist with Plank Decking[h]		
with EPMD membrane	0.2	18.13
with PVC membrane	0.17	14.93
with Modified bitumen membrane	0.14	12.88
with 4-Ply built-up roofing	0.81	63.75
with Steel roofing	0.16	15.02
Wood I-Joist with WSP Decking[i]		
with PVC membrane	0.11	8.70
with 4-Ply built-up roofing	0.75	57.52
Solid Wood Joist with WSP Decking[i]		
with Modified bitumen membrane	0.10	6.77
Wood Cord / Steel Web Truss with WSP Decking[i]		
with Modified bitumen membrane	0.10	9.71
Wood Truss (Flat) with WSP Decking[i]		
with Modified bitumen membrane	0.09	7.10
Wood Truss (4:12 Pitch) with WSP Decking[i]		
with 30-yr Fiberglass Shingles	0.08	6.97
with Clay Tile	0.22	22.07
Embodied Energy of Other Roof Assemblies		
Concrete Flat Plate Slab[h]		
with EPDM membrane	0.30	47.55
with PVC membrane	0.27	44.34
with Modified bitumen membrane	0.25	42.29
with 4-Ply built-up roofing	0.91	93.17
with Steel Roofing	0.26	44.44
Precast Double-T[h]		
with EPDM membrane	0.18	23.78
with PVC membrane	0.15	20.57
with Modified bitumen membrane	0.13	18.52
with 4-Ply built-up roofing	0.80	69.39
with Steel Roofing	0.15	20.66

Table 23.10. Continued

	Embodied energy (MMBtu/SF)	CO$_2$ equivalent emissions (lbs/SF)
Open-Web Steel Joist[i]		
with Steel decking and EPDM membrane	0.19	20.29
with Steel decking and modified Bitumen membrane	0.14	15.03
with Steel decking and 4-ply built-up roofing	0.81	65.90
with Wood decking and modified bitumen membrane	0.14	11.89
with Wood decking and 4-ply built-up roofing	0.80	62.77
Embodied Energy of Interior Wall Assemblies[k]		
Wood stud (16" OC) + gypsum board	0.03	2.49
Wood stud (24" OC) + gypsum board	0.03	2.42
Wood stud (24" OC) + 2 gypsum boards[l]	0.05	4.08
Steel stud (24" OC) + 2 gypsum boards[l]	0.05	4.84
6" Concrete block + gypsum board	0.11	15.89
6" Concrete block	0.09	14.22
Clay brick (4") unpainted	0.11	13.37
Embodied Energy of Floor Structures		
Floor Structure with Interior Ceiling Finish of Gypsum Board, Latex Paint		
Concrete flat plate and slab column system 25% flyash	0.15	31.98
Precast double-T concrete system	0.08	17.73
Glulam joist and plank decking	0.07	6.41
Wood chord and steel web truss system	0.06	6.49
Wood I-joist and OSB decking system	0.05	3.72
Open web steel joist with steel decking system and concrete topping	0.09	12.67
Wood truss and OSB decking system	0.06	4.35
Open web steel joist with 3/4" OSB flooring system	0.06	5.01
Floor Structure without Interior Ceiling Finish		
Concrete flat plate and slab column system 25% flyash	0.14	30.94
Concrete hollow core slab	0.06	14.14
Open web steel joist with 3/4" OSB flooring system	0.05	3.96
Embodied energy of column and beam assemblies[m]		

Column type	Beam type		
Concrete	Concrete	0.13	20.17
Concrete	Steel I-beam	0.09	11.42
Hollow structural steel	Glulam	0.02	1.68
Hollow structural steel	Structural composite lumber	0.02	2.38
Glulam	Glulam	0.03	2.64
Glulam	Structural composite lumber	0.03	1.92
Steel I-beam	Steel I-beam	0.09	8.19

Continued

Table 23.10.   Continued

Embodied energy of column and beam assemblies[m]			
Column type	Beam type		
Steel I-beam	Structural composite lumber	0.02	1.64
Built-up softwood	Glulam	0.03	2.41
Built-up softwood	Structural composite lumber	0.02	1.7

Notes:

[a]Embodied Energy: Energy use includes extraction, processing, transportation, construction, and disposal of each material.

[b]Assumes a Low rise building. Values are general estimations for the U.S. 60 year building lifetime.

[c]Low-e glass.

[d]Includes cladding, 1" rigid insulation sheathing, batt insulation, vapor barrier, gypsum board, and latex paint.

[e]Includes cladding, wood structural panel (WSP) sheathing, batt insulation, vapor barrier, gypsum board, and latex paint.

[f]Includes cladding, vapor barrier, gypsum board, and latex paint.

[g]Includes cladding, 4" rigid insulation, vapor barrier, gypsum board, and latex paint unless otherwise described.

[h]Includes membrane, 8" rigid insulation, vapor barrier, and latex paint.

[i]Includes membrane, 9.5" batt insulation, vapor barrier, gypsum board, and latex paint. WSP = wood structural panel.

[j]Includes membrane, 8" rigid insulation, vapor barrier, gypsum board, and latex paint.

[k]All interior walls include latex paint on each side unless noted otherwise

[l]Rounding obscures difference in embodied energy figure: wood stud wall is 7% lower than steel stud wall.

[m]Bay size: 30 by 30 feet. Column Height: 10 feet.

Source: Adapted from U.S. Department of Energy, Office of Energy Efficiency and Renewable Energy, 2009 Buildings Energy Data Book, <http://buildingsdatabook.eren.doe.gov/>

Table 23.11.   Embodied energies for selected containers made from virgin material

Container type	Material	Mass (grams)	Embodied energy (MJ/kg)	Energy/liter (MJ/liter)
PET 400 ml bottle	PET	25	84	5.3
PE 1 liter milk bottle	High density PE	38	81	3.8
Glass 750 ml bottle	Soda glass	325	15.5	6.7
Al 440 ml can	5,000 series Al alloy	20	208	9.5
Steel 440 ml can	Plain carbon steel	45	32	3.3

Source: Ashby, Michael F. 2013. Materials and the Environment (Second Edition), (Boston, Butterworth-Heinemann).

Table 23.12.   Conductance and resistance value for exterior siding materials

Material	Description	Conductivity k, Btu/(hr)(ft2) (°F/in.)	Thickness, in.	Conductance C, Btu/(hr) (ft2)(°F)	Resistance R, 1/[Btu/(hr) (ft²)(°F)]
Brick	Common	5	4	1.25	0.80
Brick	Face	9	4	2.27	0.44
Stucco		5	1	5.0	0.20
Asbestos cement shingles				4.76	0.21
Wood shingles	16-7 1/2 in. exposure			1.15	0.87
Wood shingles	Double 16-22 in. exposure			0.84	1.19
Wood shingles	plus 5/16 in. Insulated backerboard			0.71	1.40
Asphalt roll siding				6.50	0.15
Asphalt insulating siding			1/2	0.69	1.46
Wood	Drop siding, 1 × 8 in.			1.27	0.79
Wood	Bevel, 1/2 × 8 in. lapped			1.23	0.81
Wood	Bevel, 3/4 ×10 in. lapped			0.95	1.05
Wood	Plywood, 3/8 in. lapped			1.59	0.59
Hardboard	Medium density	0.73	1/4	2.94	0.34
	Tempered	1	1/4	4.00	0.25
Plywood lap siding			3/8	1.79	0.56
Plywood flat siding			3/8	2.33	0.43

Source: Adapted from Goswami, D., Frank Kreith, and Jan F. Kreider. 2000. Principles of Solar Engineering, 2nd edition, (Philadelphia, Taylor & Francis)

Table 23.13.   Natural radioactivity in building materials

Material	Uranium		Thorium		Potassium	
	ppm	mBq/g (pCi/g)	ppm	mBq/g (pCi/g)	ppm	mBq/g (pCi/g)
Granite	4.7	63 (1.7)	2	8 (0.22)	4	1184 (32)
Sandstone	0.45	6 (0.2)	1.7	7 (0.19)	1.4	414 (11.2)
Cement	3.4	46 (1.2)	5.1	21 (0.57)	0.8	237 (6.4)
Limestone concrete	2.3	31 (0.8)	2.1	8.5 (0.23)	0.3	89 (2.4)
Sandstone concrete	0.8	11 (0.3)	2.1	8.5 (0.23)	1.3	385 (10.4)
Dry wallboard	1	14 (0.4)	3	12 (0.32)	0.3	89 (2.4)
By-product gypsum	13.7	186 (5.0)	16.1	66 (1.78)	0.02	5.9 (0.2)
Natural gypsum	1.1	15 (0.4)	1.8	7.4 (0.2)	0.5	148 (4)
Wood	-	-	-	-	11.3	3330 (90)
Clay Brick	8.2	111 (3)	10.8	44 (1.2)	2.3	666 (18)

The becquerel (symbol Bq) is the SI-derived unit of radioactivity. One Bq is defined as the activity of a quantity of radioactive material in which one nucleus decays per second. The Bq unit is thus equivalent to an inverse second, s^{-1}. The curie (symbol Ci) is a non-SI unit of radioactivity defined as 1 Ci = 3.7×10^{10} decays per second.
Source: Adapted from The Radiation Information Network, Idaho State University, Radioactivity in Nature, <http://www.physics.isu.edu/radinf/natural.htm>.

Table 23.14.   Radioactivity in various materials

Material	Radioactivity
1 adult human (100 Bq/kg)	7000 Bq
1 kg of coffee	1000 Bq
1 kg superphosphate fertilizer	5000 Bq
The air in a 100 sq meter Australian home (radon)	3000 Bq
The air in many 100 sq meter European homes (radon)	up to 3000 Bq
1 household smoke detector (with americium)	30,000 Bq
Radioisotope for medical diagnosis	70 million Bq
Radioisotope source for medical therapy	100,000,000 million Bq (100 TBq)
1 kg 50-year old vitrified high-level nuclear waste	10,000,000 million Bq (10 TBq)
1 luminous Exit sign (1970s)	1,000,000 million Bq (1 TBq)
1 kg uranium	25 million Bq
1 kg uranium ore (Canadian, 15%)	25 million Bq
1 kg uranium ore (Australian, 0.3%)	500,000 Bq
1 kg low level radioactive waste	1 million Bq
1 kg of coal ash	2000 Bq
1 kg of granite	1000 Bq

The becquerel (symbol Bq) is the SI-derived unit of radioactivity. One Bq is defined as the activity of a quantity of radioactive material in which one nucleus decays per second. The Bq unit is thus equivalent to an inverse second, s–1.
Source: World Nuclear Association,

Table 23.15. General characteristics of engineering materials

Characteristic	Metals	Ceramics	Polymers
Chemical resistance	Low to medium	Excellent	Good
Creep resistance	Poor to medium	Excellent	Poor
Density	High	Medium	Low
Electrical conductivity	High	Very low	Very low
Hardness	Medium	High	Low
Machinability	Good	Poor	Good
Malleability	High	Nil	High
Melting point	Low to high	High	Low
Stiffness	High	High	Low
Strength	High	Very high	Low
Thermal conductivity	Medium to high	Medium but often decreasing rapidly with temperature	Very low
Thermal expansion	Medium to high	Low to medium	Very high
Thermal shock resistance	Good	Generally poor	Good within limited temperature ranges

Source: Adapted from ASM International, <http://www.asminternational.org>.

Table 23.16. Production, energy use, and carbon emissions from the global minerals industry

Sector	Production (Mt)	Energy use[a] (PJ)	CO_2 emissions[b] (Mt CO_2)	Energy ratio (GJ/t production)	Carbon ratio[b] (t CO_2/t production)
Iron and steel	1330	25,043	2551	18.8	1.92
Non-ferrous metals	73	4638	508	60.1	6.58
Non-metallic minerals	3921	13,155	1120	3.4	0.29
Mining and quarrying[c]	20,000–40,000	2651	241	0.07–0.13	0.006–0.012

[a]Final energy consumption. (In the case of iron and steel, this includes coal used in blast furnaces and coke ovens.)
[b]CO_2 emissions and carbon ratio are based only on the combustion of fuel. This is particularly significant in the case of non-metallic minerals, in which cement production dominates. (Non-energy CO_2 emissions from cement production are estimated at 1500 Mt CO_2).
[c]There is significant uncertainty about the production figures for mining and quarrying. Estimates of aggregate production vastly outweigh the production of other products however these figures are difficult to verify due to poor global reporting of production.
Source: Adapted from McLellan, B.C., G.D. Corder, D.P. Giurco, K.N. Ishihara. 2012. Renewable energy in the minerals industry: a review of global potential, Journal of Cleaner Production, Volume 32, Pages 32–44.

Table 23.17.   Thermal energy use for the top 20 mineral-producing countries

Country	Minerals industry thermal energy usage (PJ/yr)				Ranking of total thermal energy use in minerals
	Iron and steel	Non-ferrous metals	Non-metallic minerals	Total	
Australia	a	225	95	320	14
Belgium	112	a	a	112	21
Brazil	658	107	298	1063	6
Canada	183	37	7	227	17
Chile	a	b	a	b	
China	10,266	680	5491	16,437	1
France	234	a	132	366	12
Germany	549	36	235	820	8
Iceland	a	0	a	0	
India	1219	23	403	1646	4
Indonesia	a	8	219	227	18
Iran	30	a	a	30	24
Italy	224	a	284	508	10
Japan	1333	a	266	1599	5
Kazakhstan	a	16	a	16	25
Korea	595	a	a	595	9
Mexico	212	8	169	389	11
Morocco	a	a	30	30	23
Norway	a	4	a	4	26
Peru	a	b	a	b	
Poland	136	17	97	251	16
Russia	1812	0	599	2411	2
Saudi Arabia	a	a	b	b	
South Africa	190	a	a	190	20
Spain	99	12	224	336	13
Turkey	161	a	55	217	19
Ukraine	836	a	a	836	7
United Kingdom	187	a	72	259	15
USA	939	262	1011	2211	3
Venezuela	a	32	a	32	22
Zambia	a	2	a	2	27
Other	368	84	265	717	–
World	21,672	1797	11,558	35,027	–

"b" indicates that data has not been reported; electrical energy usage only shown (and only included in Total) for sectors in which the country is in the top 20 producers – others marked with "a").
Source: Adapted from McLellan, B.C., G.D. Corder, D.P. Giurco, K.N. Ishihara. 2012. Renewable energy in the minerals industry: a review of global potential, Journal of Cleaner Production, Volume 32, Pages 32–44.

Table 23.18. Energy and CO_2 emissions in minerals production in leading producing nations

	Iron and Steel			Non-ferrous Metals			Non-metallic minerals		
	% Production	% Energy	% Emissions	% Production	% Energy	% Emissions	% Production	% Energy	% Emissions
Australia				7%	9%	11%	1%	1%	1%
Canada	1%	1%	1%	6%	5%	3%	1%	0%	0%
Chile				7%	0%	0%			
China	38%	46%	52%	28%	34%	46%	45%	47%	55%
India	4%	5%	5%	3%	1%	0%	5%	3%	3%
Korea	4%	3%	3%						
Peru				4%	0%	0%			
Poland	1%	1%	1%	1%	1%	0%	1%	1%	1%
Russia	5%	8%	6%	7%	8%	7%	1%	5%	3%
South Africa	1%	1%	1%	2%	1%	1%			
Ukraine	3%	4%	3%						
United States of America	7%	5%	5%	7%	12%	10%	6%	9%	8%
Other	63%	73%	76%	53%	62%	68%	59%	65%	70%
World	100%	100%	100%	100%	100%	100%	100%	100%	100%

Source: Adapted from McLellan, B.C., G.D. Corder, D.P. Giurco, K.N. Ishihara. 2012. Renewable energy in the minerals industry: a review of global potential, Journal of Cleaner Production, Volume 32, Pages 32–44.

Table 23.19. Electricity use the top 20 minerals producing countries

Country	Minerals industry electricity usage (PJ/yr)				Ranking of total electricity use in minerals
	Iron and steel	Non-ferrous metals	Non-metallic minerals	Total	
Australia	a	177	15	193	8
Belgium	23	a	a	23	23
Brazil	99	141	30	269	5
Canada	38	209	8	255	6
Chile	a	b	b	b	
China	1330	904	706	2940	1
France	44	a	41	85	14
Germany	102	62	46	211	7
Iceland	a	42	a	42	20
India	b	b	b	b	
Indonesia	a	b	b	b	
Iran	b	a	b	b	
Italy	78	a	50	128	11
Japan	227	a	82	309	4
Kazakhstan	a	34	a	34	22
Korea	159	a	a	159	9
Mexico	30	3	21	54	17
Morocco	a	a	9	9	25
Norway	a	83	a	83	15
Peru	a	b	a	b	
Poland	24	11	15	50	18
Russia	220	359	71	651	3
Saudi Arabia	a	a	2	2	26
South Africa	16	58	a	74	16
Spain	63	41	46	150	10
Turkey	58	a	32	90	13
Ukraine	105	a	a	105	12
United Kingdom	18	a	28	45	19
USA	287	293	152	732	2
Venezuela	a	41	a	41	21
Zambia	a	14	a	14	24
Other	135	45	a	180	
World	3372	2841	1548	7761	

"b" indicates that data has not been reported; electrical energy usage only shown (and only included in Total) for sectors in which the country is in the top 20 producers – others marked with "a").

Source: Adapted from McLellan, B.C., G.D. Corder, D.P. Giurco, K.N. Ishihara. 2012. Renewable energy in the minerals industry: a review of global potential, Journal of Cleaner Production, Volume 32, Pages 32–44.

Table 23.20. Energy and carbon intensity of precious metals

Metal	Embodied energy (MJ/kg)	Carbon footprint (kg/kg)	Important applications
Silver	1.43×10^3–1.55×10^3	95–105	Photosensitive compounds, dentistry
Gold	240×10^3–265×10^3	14,000–15,900	Corrosion-free contacts, dentistry
Palladium	5.1×10^3–5.9×10^3	404–447	Catalysis, hydrogen purification
Rhodium	13.5×10^3–14.9×10^3	1,000–1,200	Catalysis
Platinum	257×10^3–284×10^3	14,000–15,500	Catalysis, electrodes, contacts
Iridium	2×10^3–2.2×10^3	157–173	Catalysis, electrodes, high-temperature igniters

Souce: Adaped from Ashby, Michael F. 2013. Materials and the Environment (Second Edition), (Boston, Butterworth-Heinemann).

Table 23.21. Summary of isotopes of natural uranium

Isotope	Percent in natural uranium	# protons	# neutrons	Radioactivity[a]	Alpha energies (MeV)	Half-life (years)
Uranium-238	99.2745	92	146	48.9	4.196 (77%)	4.46 billion
					4.147 (23%)	
Uranium-235	0.720	92	143	2.2	4.597 (5%)	704 million
					4.395 (55%)	
					4.370 (6%)	
					4.364 (11%)	
					4.216 (5.7%)	
Uranium-234	0.0055	92	142	48.9	4.776 (72.5%)	245,000

[a]Percent of total uranium in rocks and soil.
Source: Craft, Elena S., Abu-Qare, Aquel W., Flaherty, Meghan M., Garofolo, Melissa C., Rincavage, Heather L. and Abou-Donia, Mohamed B. 2004. Depleted and natural uranium: chemistry and toxicological effects, Journal of Toxicology and Environmental Health, Part B, 7: 4, 297–317.

Table 23.22. Typical uranium concentrations (parts per million)

Very high-grade ore (Canada) - 20% U	200,000
High-grade ore - 2% U,	20,000
Low-grade ore - 0.1% U,	1,000
Very low-grade ore[a] (Namibia) - 0.01% U	100
Granite	4–5
Sedimentary rock	2
Earth's continental crust (av)	2.8
Seawater	0.003

[a]Where uranium is at low levels in rock or sands (certainly less than 1000 ppm) it needs to be in a form which is easily separated for those concentrations to be called "ore" - that is, implying that the uranium can be recovered economically. This means that it needs to be in a mineral form that can easily be dissolved by sulfuric acid or sodium carbonate leaching.
Source: World Nuclear Association,<http://www.world-nuclear.org/>

Table 23.23. Uranium production by nation, 2000–2010

Country	Production (metric tons U)			% change
	2000	2005	2010	2009–10
Australia	7,609	9,516	5,900	−22
Brazil	50	110	148	196
Canada	10,590	11,628	9,783	−8
China	500	750	827	65
Czech Rep	507	408	254	−50
France	320	7	7	−98
Germany	28	94	0	na
Hungary	10	0	0	na
India	200	230	400	100
Kazakhstan	1,740	4,357	17,803	923
Malawi	0	0	670	na
Namibia	2,714	3,147	4,496	66
Niger	2,900	3,093	4,198	45
Pakistan	23	45	45	96
Portugal	10	0	0	na
Romania	50	90	77	54
Russia	2,500	3,431	3,562	42
South Africa	878	674	583	−34
Spain	251	0	0	na
Ukraine	500	800	850	70
USA	1,456	1,039	1,660	14
Uzbekistan	2,350	2,300	2,400	2
Total	35,186	41,179	53,663	53

Source: World Nuclear Association.

Table 23.24. World production, reserves, and requirements for uranium.

	Production[b] (metric tons U)	Known Recoverable Resources[a] (metric tons U)	Requirements[c] (metric tons U)
Australia	5,900	1,673,000	0
Kazakhstan	17,803	651,800	0
Russia	3,562	480,300	4,500
South Africa	583	295,600	290
Canada	9,783	485,300	1,600
USA	1,660	207,400	16,160
Brazil	148	278,700	0
Namibia	4,496	284,200	0
Niger	4,198	272,900	0
Ukraine	850	105,000	2,480
Jordan	0	111,800	0
Uzbekistan	2,400	114,600	0

Table 23.24. Continued

	Production[b] (metric tons U)	Known Recoverable Resources[a] (metric tons U)	Requirements[c] (metric tons U)
India	400	80,200	930
China	827	171,400	330
Mongolia	0	49,300	0
Germany	0	0	2,600
Sweden	0	10,000	1,685
Spain	0	11,300	680
United Kingdom	0	0	1,215
Finland	0	1,100	460
Czech Republic	254	500	590
Belgium	0	0	835
Korea, Rep. of	0	0	3,400
France	0	100	9,000
Other	409	119,500	14,975
World total	53,663	5,404,000	61,730

[a]Reasonably Assured Resources plus Inferred Resources, to US$ 130/kg U, as of 2009.
[b]For 2010.
[c]For 2009.
Source: OECD Nuclear Energy Agency and the International Atomic Energy Agency. 2010. Uranium 2009: Resources, Production and Demand, (Paris, OECD).

Table 23.25. Largest uranium mines in 2010

Mine	Country	Main Owner	Type	Production (metric tons U)	% of world
McArthur River	Canada	Cameco	underground	7,654	14
Ranger	Australia	ERA (Rio Tinto 68%)	open pit	3,216	6
Rossing	Namibia	Rio Tinto (69%)	open pit	3,077	6
Kraznokamensk	Russia	ARMZ	underground	2,920	5
Arlit	Niger	Somair/Areva	open pit	2,650	5
Tortkduk	Kazakhstan	Katco JV/Areva	in-situ leaching	2,439	5
Olympic Dam	Australia	BH BIlliton	by-product/ underground	2,330	4
Budenovskoye 2	Kazakhstan	Karatau JV/ Kazatomoprom	in-situ leaching	1,708	3
South Inkai	Kazakhstan	Bedpak Dala JV/ Uranium One	in-situ leaching	1,701	3
Inkai	Kazakhstan	Inkai JV/Cameco	in-situ leaching	1,642	3
Top 10 Total				29,337	55%

Source : Adapted from World Nuclear Association, <http://www.world-nuclear.org/info/inf23.html>.

Table 23.26.   Radon releases in airborne effluents and collective dose from uranium mining and milling

Source	Release per unit production (GBq t $^{-1}$)	Release rate per unit area (Bq s^{-1} m^{-2})	Normalized release[a] [TBq (Gwa)$^{-1}$]	Nomalized collective effective dose [man Sv (Gwa)$^{-1}$][b]
Mining	300		75	0.19
Milling	13		3	0.0075
Mill tailings				
Operational mill		10	3[c]	0.04[d]
Closed mill		1	0.3[c]	7.5[e]

The becquerel (symbol Bq) is the SI-derived unit of radioactivity. One Bq is defined as the activity of a quantity of radioactive material in which one nucleus decays per second. The Bq unit is thus equivalent to an inverse second, s^{-1}. The curie (symbol Ci) is a non-SI unit of radioactivity defined as 1 Ci = 3.7 × 10^{10} decays per second. The sievert (Sv) is the International System of Units (SI) derived unit of dose equivalent radiation that takes into account the relative biological effectiveness of different forms of ionizing radiation. One millisievert (mSv) corresponds to 10 ergs of energy of gamma radiation transferred to one gram of living tissue.
[a]Normalization basis; production. 250 t (Gw a)$^{-1}$; tailings, 1 ha (GWa)$^{-1}$.
[b]Dose coefficient: 0.0025 man Sv TBq $^{-1}$.
[c]Normalized release rate: TBq a^{-1} (GWa)$^{-1}$.
[d]Assuming release period of five years.
[e]Assuming release period of 10,000 years and unchanging population density.
Source: United Nations Scientific Committee on the Effects of Atomic Radiation (UNSCEAR), UNSCEAR 2000 Report vol. 1: Sources and Effects of Ionizing Radiation. United National Scientific Committee on the Effects of Atomic Radiation.

Table 23.27.   World lithium resource and reserve estimates

Li Resources[a] million tonnes	Reference	Li Reserves[b] million tonnes	Reference
13.8	USGS (2009)	4.1	USGS (2009)
19.2	Tahil (2008)	4.6	Tahil (2008)
29.9	Evans 2008)	29.4	Yaksic and Tilton (2009)
33	USGS (2011)	13	USGS (2011)
34.5	Evans (2010)	39.4	Clarke and Harben (2009)
64	Yaksic and Tilton (2009)		

[a]The concentration of naturally occurring solid, liquid, or gaseous material in or on the Earth's crust in such form and amount that economic extraction of a commodity from the concentration is currently or potentially feasible.
[b]That part of the reserve base that could be economically extracted or produced at the time of determination. (Reserve Base. — That part of an identified resource that meets specified minimum physical and chemical criteria related to current mining and production practices, including those for grade, quality, thickness, and depth).

References:
U.S. Geological Survey. 2009.Mineral commodity summaries 2009: Lithium. Reston, VA, USA: U.S. Geological Survey.
U.S. Geological Survey. 2011. Mineral commodity summaries 2011: Lithium. Reston, VA, USA: U.S. Geological Survey.
Tahil, W. 2008. The trouble with lithium 2: Under the microscope. Martainville, France: Meridian International Research.
Evans, R. K. 2008. An abundance of lithium: Part. two. <www.evworld.com/library/KEvans_LithiumAbunance_pt2.pdf>.
Evans, R. K. 2010. The lithium-brine reserve conundrum. Northern Miner. 96(35): 2–6.
Yaksic, A. and J. Tilton. 2009. Using the cumulative availability curve to assess the threat of mineral depletion: The case of lithium. Resources Policy 34(4): 185–194
Clarke, G. M. and P. W. Harben. 2009. Lithium availability wall map LAWM.
Source: Addapted from Gruber, Paul W. Pablo A. Medina, Gregory A. Keoleian, Stephen E. Kesler, Mark P. Everson, Timothy J. Wallington. 2011. Global Lithium Availability A Constraint for Electric Vehicles? Journal of Industrial Ecology Volume 15, Issue 5, pages 760–775.

Table 23.28.   Major deposits of in-situ lithium (Li) resources

Deposit	Country	Type	Average concentration (%Li)	In-situ resource (metric tons Li)
Uyuni	Bolivia	Brine	0.0532	10.2
Atacama[a]	Chile	Brine	0.14	6.3
Kings Mountain Belt	U.S.	Pegmatite	0.68	5.9
Qaidam	China	Brine	0.03	2.02
Kings Valley	U.S.	Sedimentary rock	0.27	2.0
Zabuye[a]	China	Brine	0.068	1.53
Manono/Kitolo	Congo	Pegmatite	0.58	1.145
Ricon	Argentina	Brine	0.033	1.118
Brawley	U.S.	Brine	-	1.0
Jadar Valley	Serbia	Sedimentary rock	0.0087	0.99
Hombre Muerto[a]	Argentina	Brine	0.052	0.8
Smackover	U.S.	Brine	0.0146	0.75
Gajika	China	Pegmatite	-	0.591
Greenbushes[a]	Australia	Pegmatite	1.59	0.56
Beaverhill	Canada	Brine	-	0.515
Yichun[a]	China	Pegmatite	-	0.325
Salton Sea	U.S.	Brine	0.02	0.316
Silver Peak[a]	U.S.	Brine	0.02	0.3
Kolmorzerskoe	Russia	Pegmatite	-	0.288
Maerking[a]	China	Pegmatite	-	0.225
Maricunga	Chile	Brine	0.092	0.22
Jaijika[a]	China	Pegmatite	0.59	0.204
Daoxian	China	Pegmatite	-	0.182
Dangxiongcuo[a]	CHina	Brine	0.04	0.181
Olaroz	Argentina	Brine	0.07	0.156
Other (producing)[a]	8 deposits in Brazil, Canada, China, Portugal	Pegmatite	-	0.147
Goltsovoe	Russia	Pegmatite	-	0.139
Polmostundrovskoe	Russia	Pegmatite	-	0.139
Ulug-Tanzek	Russia	Pegmatite	-	0.139
Urikskoe	Russia	Pegmatite	-	0.139
Korale	Austria	Pegmatite	-	0.1
Mibra	Brazil	Pegmatite	-	0.1
Bikita	Zimbabwe	Pegmatite	1.4	0.0567[b]
Dead Sea	Israel	Brine	0.001	-
Great Salt Lake	U.S.	Brine	0.004	-
Searles Lake	U.S.	Brine	0.005	-
Total				38.68

[a]Producing.

[b]Lowest estimate in the literature, although some estimates for Bikita are over 100,000 metric tons Li.

Kesler, Mark P. Everson, Timothy J. Wallington. 2011. Global Lithium Availability A Constraint for Electric Vehicles? Journal of Industrial Ecology, Volume 15, Issue 5, pages 760–775.

Source: Adapted from Gruber, Paul W. Pablo A. Medina, Gregory A. Keoleian, Stephen E.

Table 23.29. Properties of materials for electrochromic devices[a]

	Material	Method of preparation	Thickness (nm)	Type of coloration	Charge capacity (mC cm^{-2})	Luminous transmittance (%) (Bleached/colored)	Stability durability
Electrochromic layer	WO_3	e-gun, sputtering, chemical methods (sol–gel, electrodeposition)	350–500	Cathodic	20–40	80/10	Stable, more than 5,000 voltammetric cycles
	MoO_3	Thermal evaporation, sputtering, chemical methods	300–400	Cathodic	~20	85/20	Unstable above 5,000 cycles
	Prussian Blue	Chemical methods, mostly electrodeposition	300–600	Anodic, also used as IS layer	~20	70/10	Stable up to 20,000 cycles, degrades at rest
Transparent conductor	SnO_2:F (TFO)	Spray pyrolysis	>1000	NA	NA	~90	Hard coating, stable up to 350 °C
	In_2O_3:Sn (ITO)	Spray pyrolysis	>1000	NA	NA	~90	Hard coating, stable up to 350 °C
	ZnS/Ag/ZnS	e-gun, sputtering	40/10/40	NA	NA	~85	Soft coating, stable up to 259 °C, optical interference problems
Ion storage-protective layer	V_2O_5	e-gun, sputtering, chemical methods	300–500	Anodic	30 (maximum)	70/60	Stable up to 500 °C, degrade with moisture, unstable due to phase transitions
	CeO_2	e-gun, sputtering, chemical methods	150–500	Passive	10	90	Stable
	CeO_2-TiO_2	Sputtering, chemical methods	150–450	Passive	20–50	80	10% reduction of charge capacity after 300 cycles
	NiO	Sputtering (low yield), chemical methods	200–400	Anodic	1	70/50	Stable up to 1,200 cycles
	MgF_2	e-gun, sputtering, chemical methods	150–200	Passive	5	95	Stable, optically neutral

[a]Electrochromism involves electroactive materials that show a reversible colorchange when a small DC voltage is applied. Electrochromic devices change light transmission properties in response to voltage and thus allow control over the amount of light and heat passing through. In electrochromic windows, the electrochromic material changes its opacity. IS = insulator-semiconductor.

Source: Adapted from Leftheriotis, G. and P. Yianoulis. 2012. 3.10 - Glazings and Coatings, In: Editor-in-Chief: Ali Sayigh, Comprehensive Renewable Energy, (Oxford, Elsevier), Pages 313–355.

Table 23.30.   Characteristics of materials used in flywheel rotors

Material	Density (kg/m³)	Tensile strength (MPa)	Max energy density (for 1 kg)	Cost ($/kg)
Monolithic material	7700	1520	0.19 MJ/kg = 0.05 kWh/kg	1
4340 Steel				
Composites				
E-glass	2000	100	0.05 MJ/kg = 0.014 kWh/kg	11
S2-glass	1920	1470	0.76 MJ/kg = 0.21 kWh/kg	24.6
Carbon T1000	1520	1950	1.28 MJ/kg = 0.35 kWh/kg	101.8
Carbon AS4C	1510	1650	1.1 MJ/kg = 0.30 kWh/kg	31.3

Bolund, Björn, Hans Bernhoff, Mats Leijon. 2007. Flywheel energy and power storage systems, Renewable and Sustainable Energy Reviews, Volume 11, Issue 2, Pages 235–258.

Table 23.31.   Selected automotive materials and their properities

Material option	Density	Failure strength (MPa)	Energy content (MJ/kg)	Mass index (kg/m³ MPa²ᐟ³)	Energy index (MJ/m³ MPa²ᐟ³)
1015 Steel	7850	328	66	165	10,893
6061-T6 aluminum	2700	270	285	65	18,420
Titanium alloy	4480	845	1000	50	50,143
Epoxy-kelvar composite	1325	460	500	22	11,118

Source: Adapted from Mayyas, Ahmad, Ala Qattawi, Mohammed Omar, Dongri Shan. 2012. Design for sustainability in automotive industry: A comprehensive review, Renewable and Sustainable Energy Reviews, Volume 16, Issue 4, Pages 1845–1862

Table 23.32.   Average material consumption for a domestic light vehicle in the United States, model years 1995 and 2009

Material	1995		2009	
	Pounds	Percentage	Pounds	Percentage
Regular steel	1,630.0	44.1%	1,501.0	38.3%
High and medium strength steel	324.0	8.8%	524.0	13.4%
Stainless steel	51.0	1.4%	69.0	1.8%
Other steels	46.0	1.2%	31.0	0.8%
Iron castings	466.0	12.6%	206.0	5.3%
Aluminum	231.0	6.3%	324.0	8.3%
Magnesium castings	4.0	0.1%	12.0	0.3%
Copper and brass	50.0	1.4%	63.0	1.6%
Lead	33.0	0.9%	45.0	1.1%
Zinc castings	19.0	0.5%	9.0	0.2%
Powder metal parts	29.0	0.8%	41.0	1.0%
Other metals	4.0	0.1%	5.0	0.1%
Plastics and plastic composites	240.0	6.5%	384.0	9.8%
Rubber	149.0	4.0%	212.0	5.4%
Coatings	23.0	0.6%	34.0	0.9%
Textiles	42.0	1.1%	53.0	1.4%
Fluids and lubricants	192.0	5.2%	219.0	5.6%
Glass	97.0	2.6%	93.0	2.4%
Other materials	64.0	1.7%	90.0	2.3%
Total	3,694.0	100.0%	3,915.0	100.0%

Source: Oak Ridge National Laboratory, Transportation Energy Data Book: Edition 30, <http://cta.ornl.gov/data/index.shtml>.

Table 23.33.   Energy impacts of recyling

Material	Recycled fraction in current supply (%)	Embodied energy, virgin material (MJ/kg)	Embodied energy, recycled material (MJ/kg)	Ratio of recycled to virgin energies (%)
Aluminum	36	210	26	12
Steel	42	26.5	7.3	27
Copper	42	58	13.5	23
Lead	72	27	7.4	27
PET	21	85	39	46
PP	5	74	50	67
Glass	24	10.5	8.2	78
Paper	72	45	20	44

Ashby, Michael F. 2013. Materials and the Environment (Second Edition), (Boston, Butterworth-Heinemann).

Table 23.34. Energy and CO_2 reductions per ton recycled material

Materials	Grade	% Reduction of Energy[a]	Million BTUs	Equivalent in Barrels of Oil	Tons CO_2 Reduced
Aluminum		95	196	37.2	13.8
Paper[b]	Newsprint	45	20.9	3.97	−0.03
	Print/Writing	35	20.8	3.95	−0.03
	Linerboard	26	12.3	2.34	0.07
	Boxboard	26	12.8	2.43	0.04
Glass	Recycle	31	4.74	0.9	0.39
	Reuse	328	50.18	9.54	3.46
Steel		61	14.3	2.71	1.52
Plastic	PET	57	57.9	11	0.985
	PE	75	56.7	10.8	0.346
	PP	74	53.6	10.2	1.32

PE = polyethylene; PET = polyethylene terephthalate; PP = polypropylene.
[a]Relative to energy required for virgin production.
[b]Energy calculations for paper recycling count unused wood as fuel.
Source: Adapted from Hershkowitz, Allen. 1997. Too Good To Throw Away: Recycling's Proven Record, (Washington, D.C., Natural Resouces Defense Council).

Table 23.35. Important environmental aspects of materials: aluminum alloys

Stage	Quantity
Raw material	
Global production, main component	37×10^6 metric ton/yr
Reserves	2.03×10^9 metric ton
Embodied energy, primary production	200–220 MJ/kg
CO_2 footprint, primary production	11–13 kg/kg
Water usage	495–1490 l/kg
Processing	
Casting energy	11–12.2 MJ/kg
Casting CO_2 footprint	0.82–0.91 kg/kg
Deformation processing energy	3.3–6.8 MJ/kg
Deformation processing CO_2 footprint	0.19–0.23 kg/kg
End of life	
Embodied energy, recycling	22–39 MJ/kg
CO_2 footprint, recycling	1.9–2.3 kg/kg
Recycle fraction in current supply	41–45%

Source: Adapted from Ashby, Michael F. 2013. Materials and the Environment (Second Edition), (Boston, Butterworth-Heinemann).

Transportation

Figures

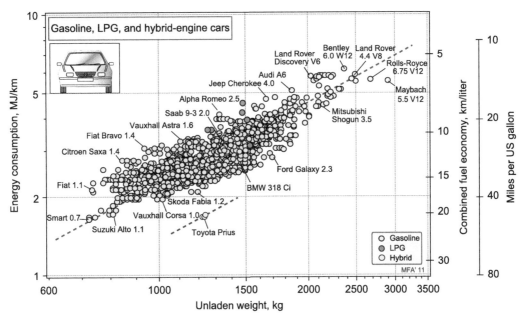

Figure 24.1. Energy consumption and vehicle weight for gasoline-powered cars.
Source: Ashby, Michael F. 2013. Materials and the Environment (Second Edition), (Boston, Butterworth-Heinemann).

Handbook of Energy. http://dx.doi.org/10.1016/B978-0-08-046405-3.00024-3

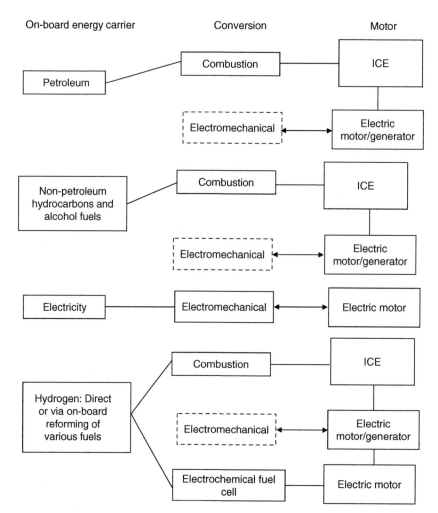

Figure 24.2. Possible combinations of on-board fuels and conversion technologies for personal transportation.
Source: Kundu, Arunabha, Yong Gun Shul, Dong Hyun Kim, Methanol Reforming Processes, In: T.S. Zhao, K.-D. Kreuer and Trung Van Nguyen, Editor(s), Advances in Fuel Cells, (Amsterdam, Elsevier Science), Volume 1, Pages 419-472.

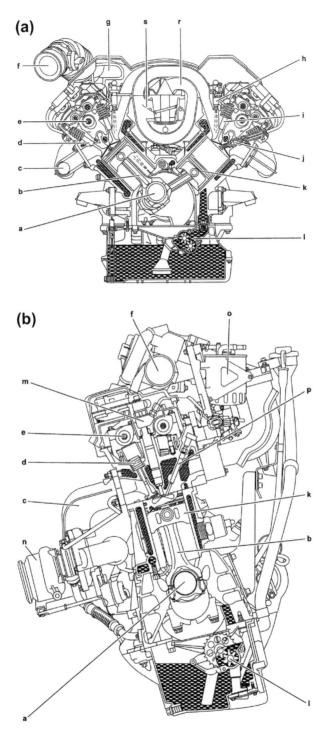

Figure 24.3. (a) Spark-ignition engine (SI, or otto-engine) (Mercedes-Benz). a, Crankshaft; b, connecting rod; c, exhaust pipe; d, outlet valve; e, camshaft; f, air intake; g, air filter; h, fuel-injection nozzle; i, inlet valve; j, spark plug; k, piston with rings; l, oil pump; r, switchover induction system; s, switchover induction system door. (b) Direct-injection diesel engine for passenger car (Mercedes-Benz). See legend to part (A); m, common-rail injection system; n, turbocharger; o, fuel filter; p, glow plug.
Source: Förster, Hans Joachim, Hermann Gaus. 2003. Automobile Technology, In: Robert A. Meyers, Editor-in-Chief, Encyclopedia of Physical Science and Technology (Third Edition), (New York, Academic Press), Pages 805-837.

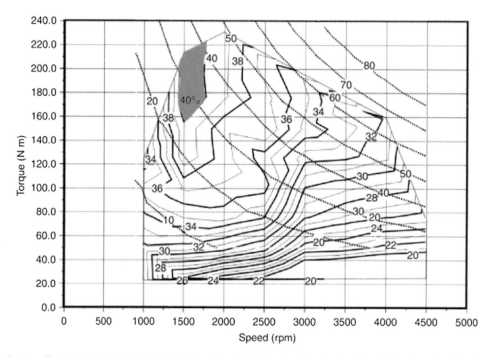

Figure 24.4. Typical efficiency map of an internal combustion engine (turbo direct injection diesel) with contour lines of constant efficiency and hyperbolas of constant power in kW.
Source: Kabza, H. 2009. Hybrid Electric Vehicles: Overview, In: Jürgen Garche, Editor-in-Chief, Encyclopedia of Electrochemical Power Sources, (Amsterdam, Elsevier), Pages 249-268.

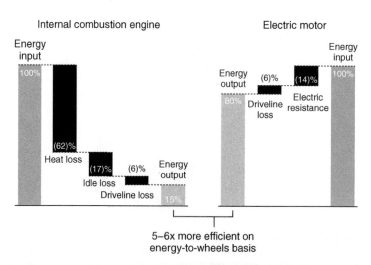

Figure 24.5. Comparison of an internal combustion engine (ICE) with an electric motor on tank-to-wheels basis. Note: Excluded are the inefficiences associated with the generation of electricity or the production of fuel for the ICE.
Source: Pistoia, Gianfranco. 2009. Battery Operated Devices and Systems, (Amsterdam, Elsevier).

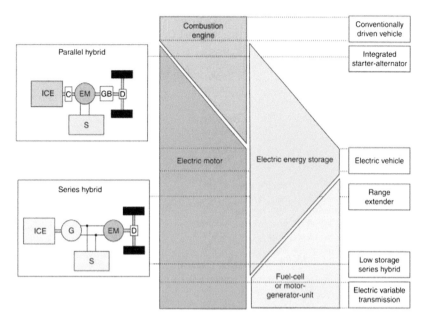

Figure 24.6. Structural transitions between different drivetrains. ICE, internal combustion engine; EM, electric motor.
Source: Kabza, H. 2009. Hybrid Electric Vehicles: Overview, In: Jürgen Garche, Editor-in-Chief, Encyclopedia of Electrochemical Power Sources, (Amsterdam, Elsevier), Pages 249-268.

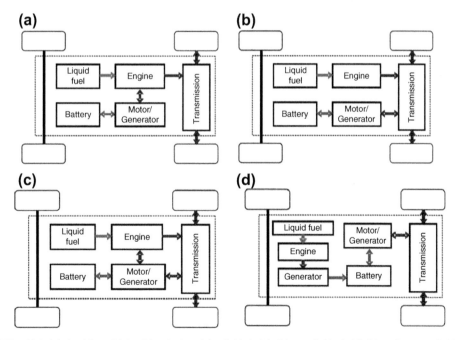

Figure 24.7. Hybrid electric vehicle drive trains: (a) mild hybrid; (b) parallel hybrid; (c) series–parallel hybrid; and (d) series hybrid.
Source: Dicks, A.L. 2012. PEM Fuel Cells: Applications, In: Ali Sayigh, Editor-in-Chief, Comprehensive Renewable Energy, (Oxford, Elsevier), Pages 183-225.

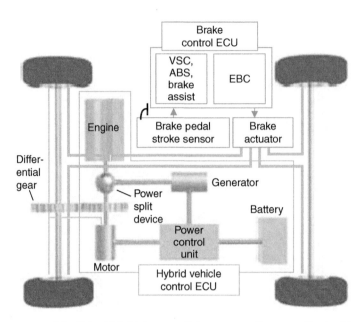

Figure 24.8. Components of the Prius Hybrid Synergy Drive system. Regenerative braking follows the route (in the lower half of the figure): front wheels→electric motor→power control unit→battery. The hydraulic braking components are in the upper half of the figure. ECU, electronic control unit; VSC, vehicle stability control; ABS, anti-lock braking system.
Source: Pistoia, Gianfranco. 2009. Battery Operated Devices and Systems, (Amsterdam, Elsevier).

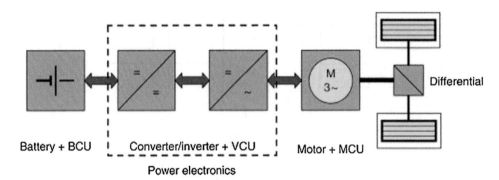

Figure 24.9. Battery electric vehicle (BEV) drivetrain structure. MCU = motor control unit; BCU = battery control unit.
Source: Gutmann, G. 2009. Electric Vehicle: Batteries, In: Editor-in-Chief: Jürgen Garche, Encyclopedia of Electrochemical Power Sources, (Amsterdam, Elsevier), Pages 219-235.

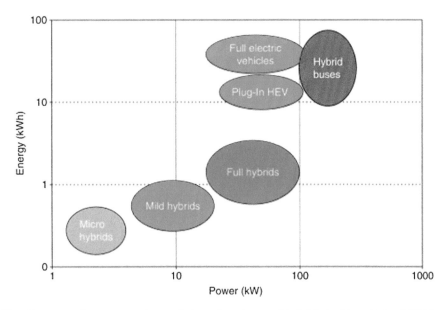

Figure 24.10. Energy and power requirements in various types of hybrid electric vehicles (HEVs). *Source: Köhler, U. 2009. Hybrid Electric Vehicles: Batteries, In: Jürgen Garche, Editor-in-Chief, Encyclopedia of Electrochemical Power, (Amsterdam, Elsevier), Pages 269-285.*

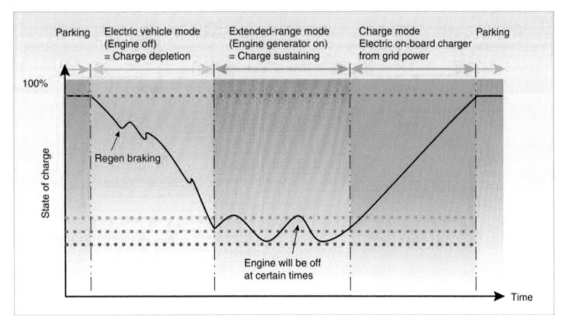

Figure 24.11. Battery state-of-charge (SoC) in different extended-range electric vehicle (E-REV) operation modes. *Source: Mettlach, H. 2009. Hybrid Electric Vehicle: Plug-In Hybrids, In: Jürgen Garche, Editor-in-Chief, Encyclopedia of Electrochemical Power Sources, (Amsterdam, Elsevier), Pages 286-291.*

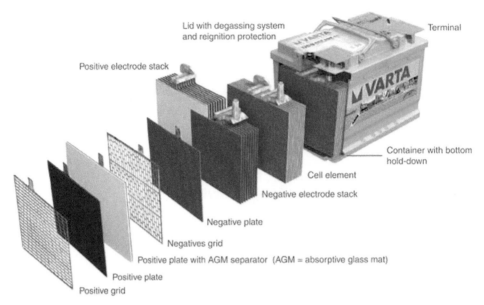

Figure 24.12. 12 volt lead–acid battery (Johnson Controls – VARTA) used in hybrid electric vehicles based on absorptive glass mat technology.
Source: Köhler, U. 2009. Hybrid Electric Vehicles: Batteries, In: Jürgen Garche, Editor-in-Chief, Encyclopedia of Electrochemical Power Sources, (Amsterdam, Elsevier), Pages 269-285.

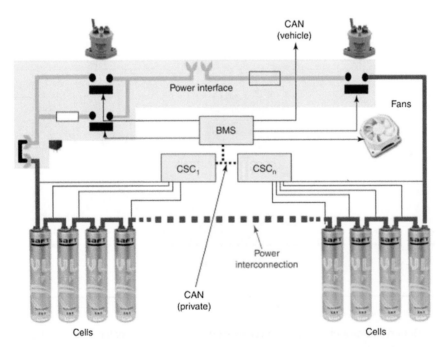

Figure 24.13. Battery management system (BMS) for Li-ion batteries (schematic). CSC, cell supervisory circuit.
Source: Köhler, U. 2009. Hybrid Electric Vehicles: Batteries, In: Jürgen Garche, Editor-in-Chief, Encyclopedia of Electrochemical Power Sources, (Amsterdam, Elsevier), Pages 269-285.

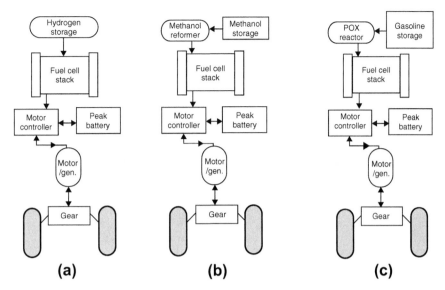

Figure 24.14. Three different configurations for a fuel cell vehicle: (a) off-board hydrogen production, (b) and (c) on-board hydrogen production.
Source: Kundu, Arunabha, Yong Gun Shul, Dong Hyun Kim, Methanol Reforming Processes, In: T.S. Zhao, K.-D. Kreuer and Trung Van Nguyen, Editor(s), Advances in Fuel Cells, (Amsterdam, Elsevier Science), Volume 1, Pages 419-472.

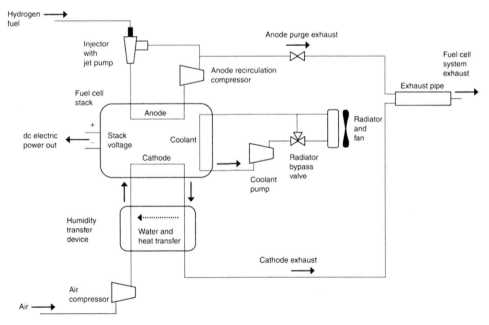

Figure 24.15. Schematic diagram of generic fuel cell system configuration in a vehicle, showing cathode air supply system, fuel supply system, and cooling system.
Source: Hochgraf, C. 2009. Electric Vehicles: Fuel Cells, In: Jürgen Garche, Editor-in-Chief, Encyclopedia of Electrochemical Power Sources, (Amsterdam, Elsevier), Pages 236-248.

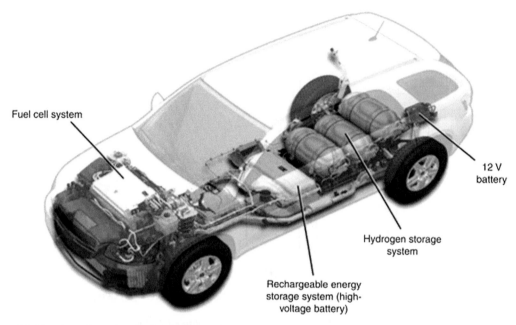

Figure 24.16. Location of major fuel cell propulsion system components in a vehicle.
Source: Hochgraf, C. 2009. Electric Vehicles: Fuel Cells, In: Jürgen Garche, Editor-in-Chief, Encyclopedia of Electrochemical Power Sources, (Amsterdam, Elsevier), Pages 236-248.

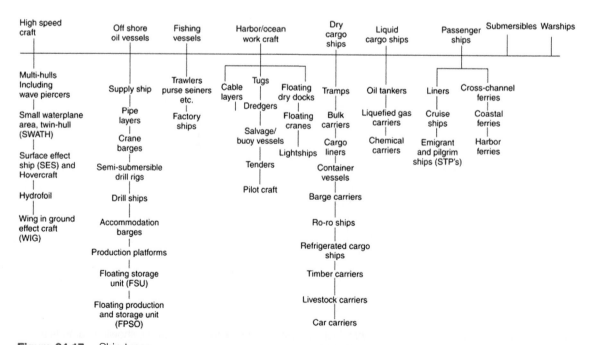

Figure 24.17. Ship types.
Source: Eyres, D.J. and G.J. Bruce. 2012. Ship Construction (Seventh Edition), (Oxford, Butterworth-Heinemann).

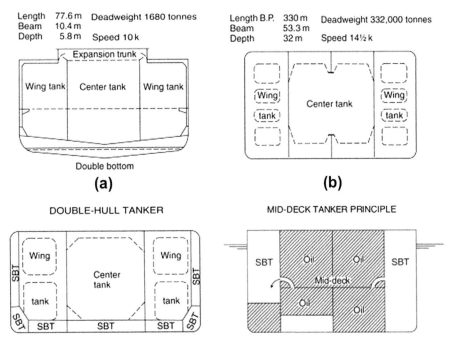

Figure 24.18. Oil tankers. In the double hull design (a), the bottom and sides of the ship have two complete layers of watertight hull surface: one outer layer forming the normal hull of the ship, and a second inner hull which is some distance inboard, typically by a few feet, which forms a redundant barrier to seawater in case the outer hull is damaged and leaks. The mid-deck design (b) includes an additional deck placed at about the middle of the draft of the ship that limits spills if the tanker is damaged.
Source: Eyres, D.J. and G.J. Bruce. 2012. Ship Construction (Seventh Edition), (Oxford, Butterworth-Heinemann).

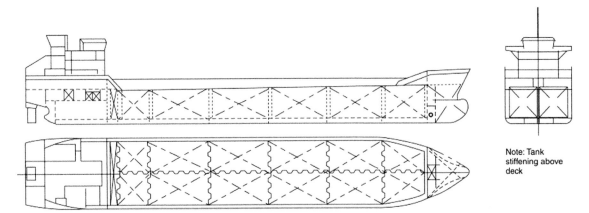

Figure 24.19. Oil products/chemical tanker of 12,700 metric tons deadweight. The 130 meter length overall by 20 meter breadth by 9.75 meter depth vessel has 12 cargo tanks, equipped to carry six different types or grades of cargo simultaneously. Each tank has a deepwell pump rated at 300 cubic meters per hour and the ship's maximum discharge rate is 1800 cubic meters per hour with six lines working at the same time.
Source: Eyres, D.J. and G.J. Bruce. 2012. Ship Construction (Seventh Edition), (Oxford, Butterworth-Heinemann).

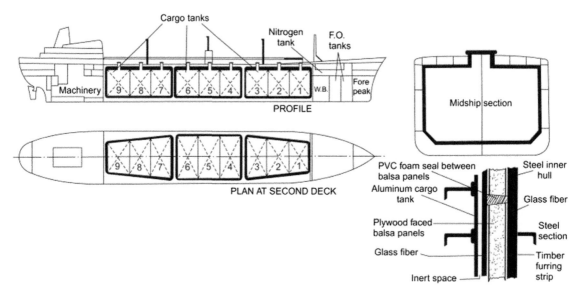

Figure 24.20. Liquefied natural gas (LNG) carrier.
Source: Eyres, D.J. and G.J. Bruce. 2012. Ship Construction (Seventh Edition), (Oxford, Butterworth-Heinemann).

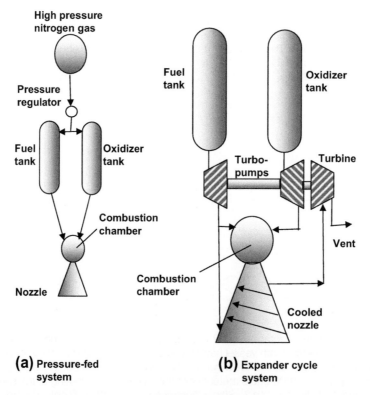

Figure 24.21. Basic propellant feed systems for liquid rocket motors.
Source: Sforza, Pasquale M. 2012. Theory of Aerospace Propulsion, (Boston, Butterworth-Heinemann).

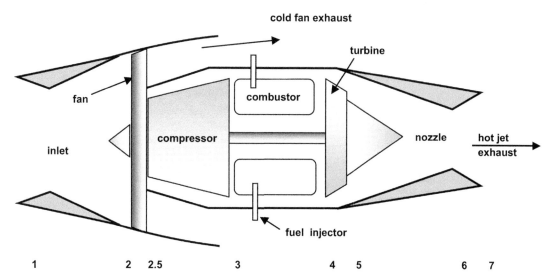

Figure 24.22. Schematic diagram of a typical turbofan engine showing required turbomachinery components and usual station numbering scheme. The combustor burns the injected fuel supplying heat to the central flow passing out the nozzle, and the fan accelerates the cold outer flow of the turbofan.
Source: Sforza, Pasquale M. 2012. Theory of Aerospace Propulsion, (Boston, Butterworth-Heinemann).

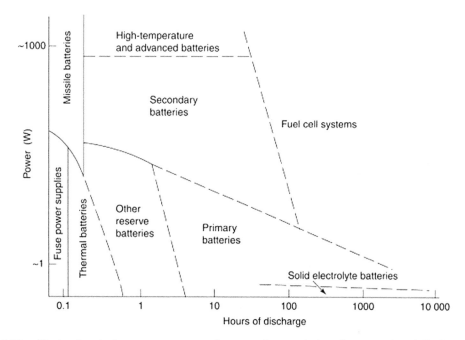

Figure 24.23. Electrochemical energy sources and energy storage devices for space travel. Performance envelope of electrochemical energy sources for different power output levels and space mission durations.
Source: Rao, G.M., R.C. Pandipati. 2009. Satellites: Batteries, In: Jürgen Garche, Editor-in-Chief, Encyclopedia of Electrochemical Power, (Amsterdam, Elsevier), Pages 323-337.

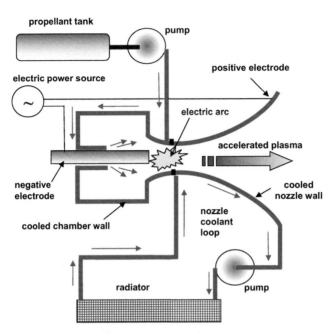

Figure 24.24. Schematic diagram of an electrothermal propulsion device used in space travel, the arcjet.
Source: Sforza, Pasquale M. 2012. Theory of Aerospace Propulsion, (Boston, Butterworth-Heinemann).

Charts

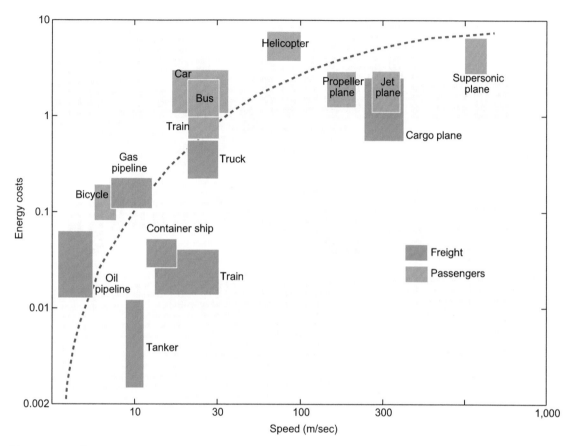

Chart 24.1.   Speed of transport mode versus energy use.
Source: Data from Chapman, J.D. 1989. Geography and Energy: Commercial Energy Systems and National Policies, (New York, Longman Scientific & Technical).

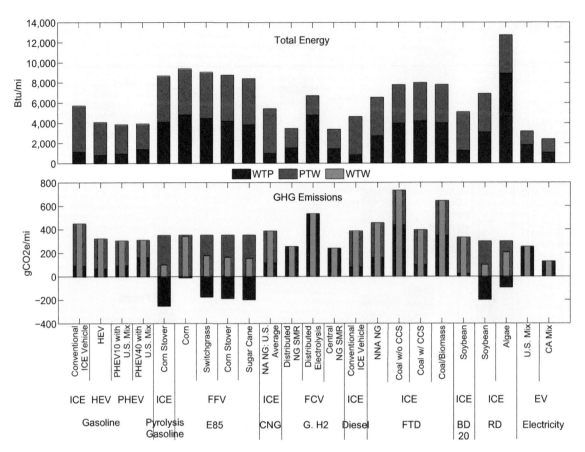

Chart 24.2. Wells-to-wheels energy use and greenhouse emissions of various transportation vehicles and fuels. WTP = well-to-pump; WTW = well-to-wheels; ICE = internal combustion engine; EV = electric vehicle; HEV = hybrid electric vehicle; PHEV = Plug-in Hybrid Electric Vehicle; NA = North America; NNA = non-North America; NG = Natural gas; SMR = Steam methane reforming; CCS = Carbon capture and sequestration; FTD = Fischer Tropsch diesel; G.H2 = Gaseous hydrogen; CNG = Compressed natural gas; BD = Biodiesel; RD = Renewable diesel; E85 = a mixture of 85% ethanol and 15% gasoline (by volume); FFV = flexible fuel vehicle. *Source: Data from Argonne National Laboratory, Transportation Technology R&D Center, The Greenhouse Gases, Regulated Emissions, and Energy Use in Transportation (GREET) Model, GREET 1 2012 version.*

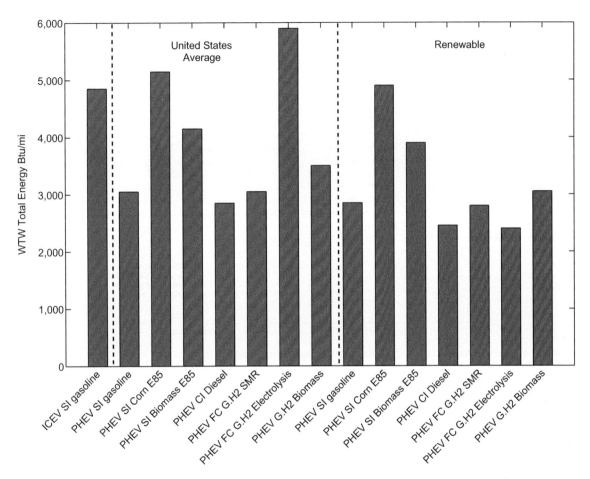

Chart 24.3. Well-to-wheels energy use for plug-in hybrid electric vehicles (PHEVs). The "United States Average" results assume that electricity is generated from the average fuel mix at U.S. power plants. The "Renewable" results assume all electricity is generated by renewable sources. ICEV = internal combustion engine vehicle; SI = spark ignition; H2 = hydrogen; FC = fuel cell; SMR = steam methane reformation; E85 = a blend of 85% ethanol and 15% reformulated gasoline.
Source: Data from Elgowainy, A., A. Burnham, M. Wang, J. Molburg, and A. Rousseau. 2009. Well-to-Wheels Energy Use and Greenhouse Gas Emissions Analysis of Plug-in Hybrid Electric Vehicles, Center for Transportation Research Energy Systems Division, Argonne National Laboratory.

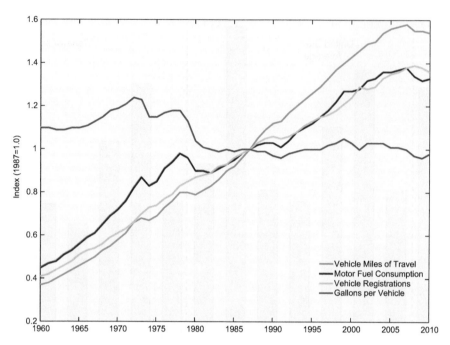

Chart 24.4. Vehicle registrations, fuel consumption, and vehicle miles of travel in the United States, 1960-2010. *Source: Data from United States Department of Transportation, Federal Highway Administration, Highway Statistics 2010.*

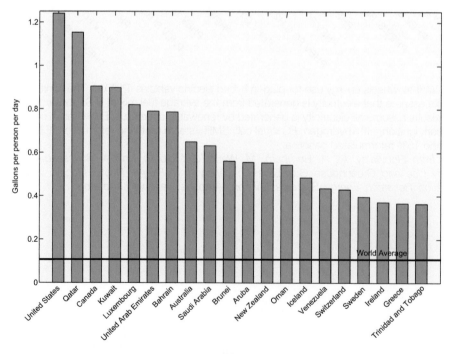

Chart 24.5. Per capita gasoline use, top 20 nations, 2008. *Source: Data from United States Department of Energy, Energy Information Administration, International Energy Statistics, <http://www.eia.gov/countries/data.cfm>.*

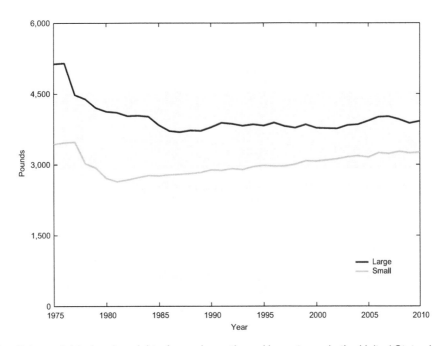

Chart 24.6. Sales-weighted curb weight of new domestic and import cars in the United States by size class and model years, 1975–2010.
Source: Data from Oak Ridge National Laboratory, Transportation Energy Data Book: Edition 30, <http://cta.ornl.gov/data/index.shtml>.

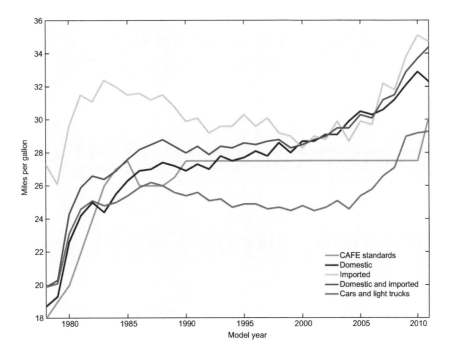

Chart 24.7. Corporate Average Fuel Economy (CAFE) in the United States, 1978-2011.
Source: Data from Oak Ridge National Laboratory, Transportation Energy Data Book: Edition 30, <http://cta.ornl.gov/data/index.shtml>.

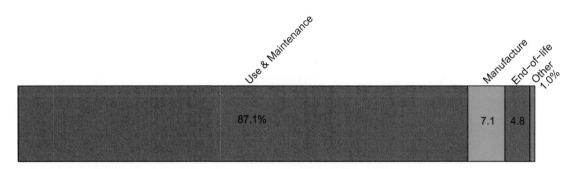

Chart 24.8. The direct and indirect energy used in an automobile life cycle.
Source: Data from Mayyas, Ahmad, Ala Qattawi, Mohammed Omar, Dongri Shan. 2012. Design for sustainability in automotive industry: A comprehensive review, Renewable and Sustainable Energy Reviews, Volume 16, Issue 4,Pages 1845-1862.

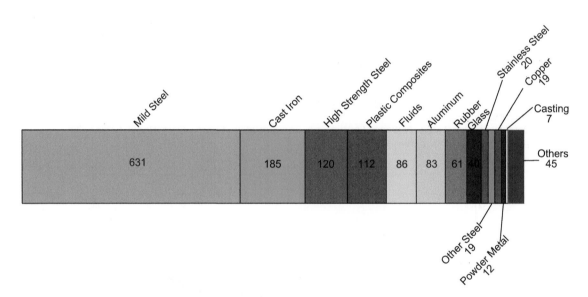

Chart 24.9. A typical material composition of an automobile (kg).
Source: Data from Omar, Mohammed A. 2011. The Automotive Body Manufacturing Systems and Processes, (New York, John Wiley & Sons).

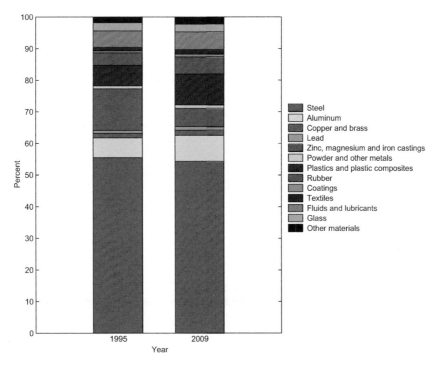

Chart 24.10. Average material consumption for a domestic light vehicle in the United States, model years 1995 and 2009.
Source: Data from Oak Ridge National Laboratory, Transportation Energy Data Book: Edition 30, <http://cta.ornl.gov/data/index.shtml>.

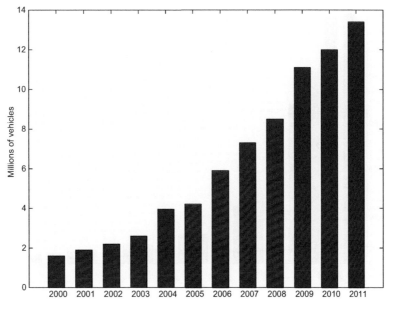

Chart 24.11. Number of natural gas vehicles (NGVs) in the world, 2000-2011.
Source: Data from NGVA Europe, Worldwide NGV Statistics, <http://www.ngvaeurope.eu/worldwide-ngv-statistics>.

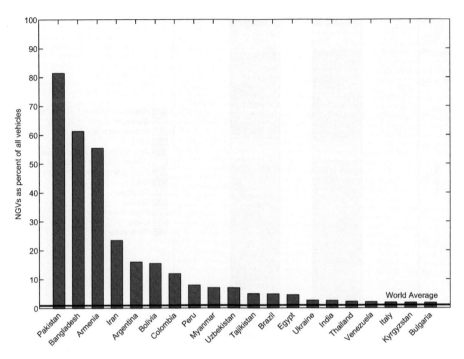

Chart 24.12. Natural gas vehicles share of total vehicles, 2011.
Source: Data from NGVA Europe, Worldwide NGV Statistics, <http://www.ngvaeurope.eu/worldwide-ngv-statistics>.

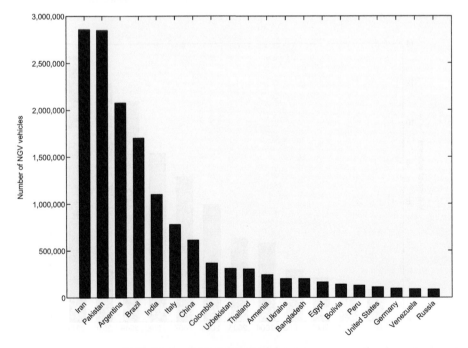

Chart 24.13. Global natural gas vehicle use, top 20 nations, 2011.
Source: Data from NGVA Europe, Worldwide NGV Statistics, <http://www.ngvaeurope.eu/worldwide-ngv-statistics>.

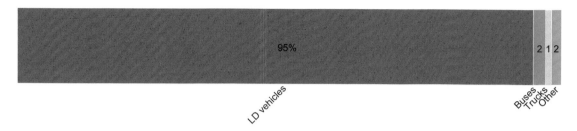

Chart 24.14. Global natural gas vehicle use by type of vehicle, 2011. LD = light duty.
Source: Data from NGVA Europe, Worldwide NGV Statistics, <http://www.ngvaeurope.eu/worldwide-ngv-statistics>.

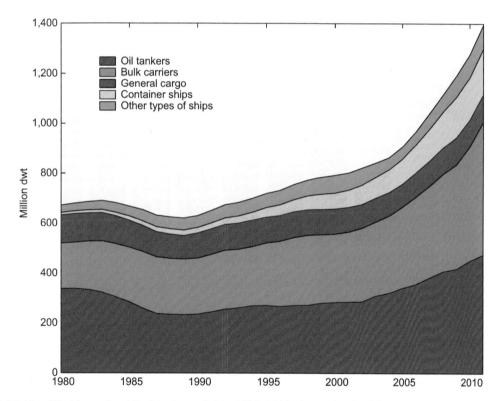

Chart 24.15. World merchant fleet by type of ship, 1980-2010. dwt = deadweight, a measure of how much weight a ship is carrying or can safely carry; sum of the weights of cargo, fuel, fresh water, ballast water, provisions, passengers, and crew.
Source: Data from United Nations Conference on Trade and Development (UNCTAD), UNCTAD Statistics, <http://unctad.org/en/pages/Statistics.aspx>.

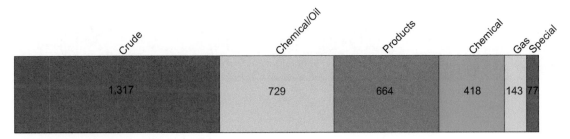

Chart 24.16. Number of tankers by type of product carried.
Source: Data from INTERTANKO,
<http://www.intertanko.com/Book-Shop/Annual-Report-and-Industry-Facts/INTERTANKO-Tanker-Facts-2010/>.

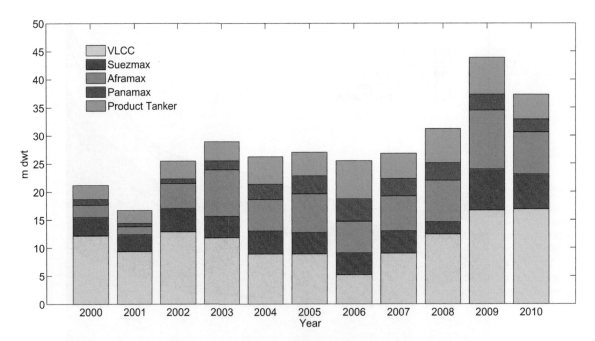

Chart 24.17. Tanker deliveries by year and size, 2000-2010. VLCC=very large crude carrier.
Source: Data from INTERTANKO,
<http://www.intertanko.com/Book-Shop/Annual-Report-and-Industry-Facts/INTERTANKO-Tanker-Facts-2010/>.

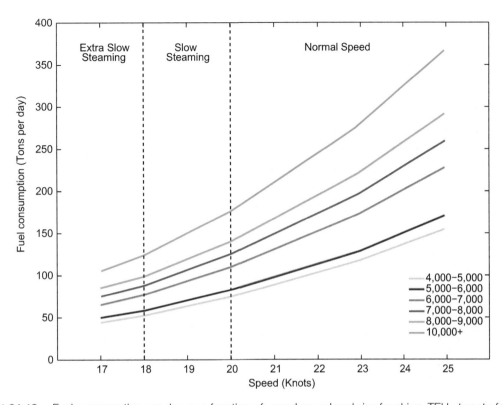

Chart 24.18. Fuel consumption per day as a function of vessel speed and size for ships. TEU = twenty-foot equivalent unit. One TEU represents (approximately) the cargo capacity of a standard intermodal container, 20 feet (6.1 m) long and 8 feet (2.44 m) wide.
Source: Data from Notteboom, T. and P. Carriou. 2009. Fuel surcharge practices of container shipping lines: Is it about cost recovery or revenue making?. Proceedings of the 2009 International Association of Maritime Economists (IAME) Conference, June, Copenhagen, Denmark; Rodrigue, Jean-Paul, The Geography of transport systems, Dept. of Global Studies & Geography , Hofstra University, New York, <http://people.hofstra.edu/geotrans/index.html>.

Tables

Table 24.1.   Vehicle energy flows (%)

	City	Highway	Combined
Fuel tank	100	100	100
Engine losses	74–75	65–69	70–72
Thermal (radiator, exhaust heat)	63–64	56–60	60–62
Combustion	3	3	3
Pumping	5	3	4
Friction	3	3	3
Parasitic losses (water pump, alternator, etc.)	6–7	3–4	5–6
Power to wheels	14–16	20–26	17–21
Dissipated as:			
Wind resistance	4	13–16	8–10
Rolling resistance	4–5	6–8	5–6
Braking	6–7	2	4–5
Drivetrain losses	4–5	5–7	5–6
Idle losses	6	0	3

Source: United States Environmental Protection Agency, <http://www.fueleconomy.gov>, accessed 12 June 2102.

Table 24.2.   Comparison of the characteristics of various transportation fuels

Fuel	Energy Density	Production cost with oil at USD 100/bbl	Distribution infrastructure	Current production and retail availability for vehicles	Compatibility with existing ICE vehicles	Typical greenhouse gas emissions
Gasoline	High	Moderate	Complete	Complete	Complete	High
Distillate	High	Moderate	Complete	Complete	Complete	High
Jet fuel	High	Moderate	Complete	Complete	Complete	High
HFO	High	Moderate	Complete	Complete	Complete	High
CTL diesel	High	Moderate-high	Compatible with existing	Very low	Complete	Very high (high with CCS)
GTL diesel	High	Moderate-high	Compatible with existing	Very low	Complete	High (even with CCS)
Grain ethanol	Medium	Moderate-high	Partial	Low-moderate	Partial	Moderate-high
Cane ethanol	Medium	Low-moderate	Partial	Low-moderate	Partial	Low
Advanced ligno-cellulosic ethanol	Medium	High	Partial	None	Partial	Low
Oil-seed biodiesel	High	Moderate-high	Partial	Low-moderate	Partial	Moderate
Advanced BTL diesel	High	High	Compatible with existing	None	Complete	Low

Table 24.2.   Continued

Fuel	Energy Density	Production cost with oil at USD 100/bbl	Distribution infrastructure	Current production and retail availability for vehicles	Compatibility with existing ICE vehicles	Typical greenhouse gas emissions
CNG	Low	Low-moderate	Partial	Very low	Requires conversion	Moderate-high
LPG	Low	Low-moderate	Partial	Very low	Requires conversion	Moderate-high
Methanol from NG	Low	Moderate	Very low	Very low	Requires conversion	Moderate-high
DME from NG	Medium	Moderate	Very low	Very low	Requires conversion	Moderate-high
H2 from fossil sources	Low	Moderate	Very low	Very low	Requiresconversion	Moderate-high
H2 from renewable sources	Low	High	Very low	None	Requiresconversion	Very low
Electricity/fossil	Low	Low	Widespread	Very low	Incompatible	Moderate-
Electricity/ renewable	Low	Moderate	Widespread	Very low	Incompatible	Very low

Classifications are indicative, based on current characteristics and estimates, and apply only to near term. There may be situations and regions in which these classifications do not apply.

HFO = heavy fuel oil; CTL = coal-to-liquids; GTL = gas-to-liquids; CNG = compressed natural gas; LPG = liquefied petroleum gas; BTL = biomass-to-liquid; NG = natural gas; CCS = carbon capture and storage; ICE = internal combustion engine; DME = dimethyl ether

Source: Adapted from International Energy Agency (IEA). 2009. Transport Energy and Sustainabilty, (Paris, IEA).

Table 24.3. Comparison of properties for various transportation fuels

	Gasoline	No. 2 Diesel	Biodiesel	Compressed Natural Gas(CNG)	Electricity	Ethanol	Hydrogen	Liquified Natural Gas (LNG)	Liquified Petroleum Gas (LPG)	Methanol
Chemical Structure	C_4 to C_{12}	C_8 to C_{25}	Methyl esters of C_{12} to C_{22} fatty acids	CH_4 (83–99%), $_2H_6$ (1–13%)	N/A	CH_3CH_2OH	H_2	CH_4	C_3H_8 (majority) and C_4H_{10} (minority)	CH_3OH
Cetane Number	N/A	40–55	48–65	N/A	N/A	0–54	N/A	N/A	N/A	N/A
Pump Octane Number[a]	84–93	N/A	N/A	120+	N/A	110	130+	120+	105	112
Main Fuel Source	Crude Oil	Crude Oil	Fats and oils from sources such as soy beans, waste cooking oil, animal fats, and rapeseed; many other feedstocks possible	Underground reserves	Coal, nuclear, natural gas, oil, hydroelectric biomass, solar, wind, geothermal	Starches, sugars and cellulose	Natural gas, methanol, and electrolysis of water.	Underground reserves	A by-product of petroleum refining or natural gas processing	Natural gas, coal, or, woody biomass
Energy Content (Lower Heating Value)	116,090 Btu/gal	128,450 Btu/gal	119,550 Btu/gal for B100	20,268 Btu/lb[c]	3,414 Btu/kWh	76,330 Btu/gal for E100	51,585 Btu/lb[c]	74,720 Btu/gal	84,950 Btu/gal	57,250 Btu/gal
Energy Content (Higher Heating Value)	124,340 Btu/gal	137,380 Btu/gal	127,960 Btu/gal for B100	22,453 Btu/lb[c]	3,414 Btu/kWh	84,530 Btu/gal for E100	61,013 Btu/lb[c]	84,820 Btu/gal	91,410 Btu/gal	65,200 Btu/gal
Energy Contained in Various Alternative Fuels as Compared to One Gallon of Gasoline[b]	100%	1 gallon of diesel has 113% of the energy of one gallon of gasoline.	B100 has 103% of the energy in one gallon of gasoline or 93% of the energy of one gallon of diesel. B20 has 109% of the energy of one gallon of gasoline or 99% of the energy of one gallon of diesel.	5.66 pounds or 126.67 cu. ft. of CNG has 100% of the energy of one gallon of gasoline.[c]	33.70 kWh has 100% of the energy of one gallon of gasoline.	1 gallon of E85 has 77% of the energy of one gallon of gasoline.[d]	1 kg or 2.198 lbs. of H2 has 100% of the energy of one gallon of gasoline.[c]	1 gallon of LNG has 64% of the energy of one gallon of gasoline.	1 gallon of propane has 73% of the energy of one gallon of gasoline.	1 gallon of methanol has 49% of the energy of one gallon of gasoline.

Maintenance Issues								
	Hoses and seals may be affected by higher-percent blends, lubricity is improved over that of conventional diesel fuel.	High-pressure tanks require periodic inspection and certification.	Service requirements are less than with gasoline or diesel. No tune-ups, oil changes, timing belts, water pumps, radiators, or fuel injectors are required. However, it is likely that the battery will need replacement before the vehicle is retired.	Special lubricants may be required. Practices are very similar, if not identical, to those for conventionally fueled operations.	When hydrogen is used in fuel cell applications, maintenance should be very minimal.	High-pressure tanks require periodic inspection and certification.	Some fleets report service lives that are 2–3 years longer, as well as extended intervals between required maintenance.	Special lubricants must be used as directed by the supplier and M-85-compatible replacement parts must be used.

[a] Pump octane number is the average of the research octane number and motor octane number.

[b] Energy comparisons are given in percent energy content on a gallon-to-gallon basis unless other units are given.

[c] Due to the infinite temperature and pressure combinations of gaseous fuels and their effect on fuel density, ft^3 units are not given. Most of these fuels are dispensed by Coriolis flow meters, which track fuel mass and report fuel dispensed on a "gallon of gasoline-equivalent" (GGE) basis.

[d] The ethanol content of E85 is usually lower than 85% for two reasons: 1) fuel ethanol contains 2–5% gasoline as a denaturant and 2) fuel ethanol content is lowered to 70% in the winter in cold climates to facilitate cold starts. When the actual composition of E85 is accounted for, the lower heating value of E85 varies from 82,970 Btu/gal to 89,650 Btu/gal, which is 72% to 77% the heat content of gasoline.

Source: U.S Department of Energy, Alternative Fuels and Advanced Vehicles Data Center, Fuel Properties database, <http://www.afdc.energy.gov/afdc/fuels/properties.html>.

Table 24.4.   Transportation fuels and their production processes

Fuel	Feedstock	Process[a]
Liquid petroleum fuels: gasoline, diesel, kerosene, jet fuel	Oil from both conventional sources and non-conventional sources such as heavy crudes, oil sands, and oil shales	Refining
Liquid synthetic fuels	Natural gas, coal	Gasification/FT (with or without CCS)
Biodiesel	Oil-seed crops	Esterification, hydrogenation
Ethanol	Grain crops	Saccharification and distillation
Ethanol	Sugar crops (cane)	Distillation
Advanced biodiesel (and other distillate fuels)	Biomass from crops or waste products	Gasification/FT (with or without CCS)
Compressed natural gas	Natural gas	Gasification/FT (with or without CCS)
Electricity	Coal, gas,oil,nuclear, renewables	Different mixes in different regions (with or without CCS)
Hydrogen (H_2)	Natural gas	
	Electricity	Electrolysis at point of use
	Direct production w/ renewables (solar, wind), or nuclear energy	High-temperature process

[a]FT = Fischer-Tropsch synthesis; CCS = carbon capture and storage.
Source: Adapted from International Energy Agency (IEA). 2009. Transport Energy and Sustainabilty, (Paris, IEA).

Table 24.5.   Energy equivalency of transportation fuels[a]

	Hydrogen (kg)	Natural Gas (million cubic feet)	Crude Oil (barrel)	Conventional Gasoline (gallon)	Reformulated Gasoline (RFG) (gallon)	California RFG (gallon)	U.S. Conventional Diesel (gallon)	Low-Sulfur Diesel (gallon)
Hydrogen, 1 kg =	1	0.000117	0.0211	0.992	1.014	1.011	0.896	0.889
Natural Gas, 1 million cubic feet =	8538	1	180.5	8468	8653	8628	7653	7591
Crude Oil, 1 barrel =	47.30	0.00554	1	46.91	47.94	47.80	42.40	42.06
Conventional Gasoline, 1 gallon =	1.008	0.000118	0.0213	1	1.022	1.019	0.904	0.897
Reformulated Gasoline (RFG), 1 gallon =	0.987	0.000116	0.0209	0.979	1	0.997	0.884	0.877
California RFG, 1 gallon =	0.989	0.000116	0.0209	0.981	1.003	1	0.887	0.880
U.S. Conventional Diesel, 1 gallon =	1.116	0.000131	0.0236	1.106	1.131	1.127	1	0.992
Low-Sulfur Diesel, 1 gallon =	1.125	0.000132	0.0238	1.115	1.140	1.137	1.008	1

[a]The shows the amount of each fuel necessary to provide the same energy as 1 kg of hydrogen, 1 million cubic feet natural gas, 1 barrel of crude oil, or 1 gallon of other fuels, based on lower heating values.

Source: U.S. Department of Energy, Hydrogen Resource Center, <http://hydrogen.pnl.gov/cocoon/morf/hydrogen/article/401>.

Table 24.6.   General properties of oxygenated fuels

	Chemical formula	Mol. wt.	Composition (wt%)			Specific gravity	Boiling point (°C)	Autoignition temperature (°C)	Flash point (°C)	Heating values (kJ/kg)		Flammability limits (vol%)
			C	H	O					LHV	HHV	
Methanol	CH_3OH	32.04	37.5	12.6	49.9	0.796	65	464	11	20,100	22,900	7.3–36.0
Ethanol	C_2H_5OH	46.07	52.2	13.1	34.7	0.794	78	423	13	27,000	29,800	4.3–19.0
Dimethyl ether	$(CH_3)_2O$	46.07	52.2	13.1	34.7	0.61	−24	350		28,900	31,700	
Biodiesel	C12–C22	292	77	12	11	0.88	315–350		100–170	37,500	40,200	

Source: Adapted from Speight, James G. 2011. Chapter 3 - Fuels for Fuel Cells, In: Dushyant Shekhawat, J.J. Spivey And David A. Berry, Editor(s), Fuel Cells: Technologies for Fuel Processing, (Amsterdam, Elsevier), 2011, Pages 29–48.

Table 24.7.   Typical Properties of oxygenates for motor gasoline

	Ethanol	MTBE	ETBE	TAME
Chemical formula	CH_3CH_2OH	$CH_3OC(CH_3)_2$	$CH_3CH_2OC(CH_3)_3$	$(CH)_3CCH_2OCH_3$
oxygen content, percent by weight	34.73	18.15	15.66	15.66
Octane, (R+M)/2	115	110	111	105
Blending vapor pressure, RVP	18	8	4	1.5

MTBE = Methyl tert-butyl ether; ETBE = Ethyl tert-butyl ether; TAME = tert-Amyl methyl ether.
Source: Adapted from National Petroleum Council, U.S. Petroleum Refining: Meeting Requirements for Cleaner Fuels and Refineries (Washington, DC, August 1993) Appendix L.

Table 24.8.   Comparison of ExxonMobil methanol-to-gasoline (MTG) properties to typical 2005 U.S. gasolines

	MTG gasoline	Summer gasoline	Winter gasoline
Oxygen, wt.%	0	0.95	1.08
API gravity	61.8	58.4	61.9
Aromatics, vol.%	26.5	27.7	24.7
Olefins (alkenes), vol.%	12.6	12	11.6
Reid vapor pressure, psi	9	8.3	12.12
50% boiling point, °F	201	211.1	199.9
90% boiling point, °F	320	330.7	324.1
Sulfur, ppm	0	106	97
Benzene, vol. %	0.3	1.21	1.15

Source: Adapted from Bell, David A., Brian F. Towler, Maohong Fan. 2011. Methanol and Derivatives, In: David Bell and Brian Towler, Coal Gasification and Its Applications, (Boston, William Andrew Publishing), Pages 353–371.

Table 24.9.   Typical properties of biodiesel compared to petroleum diesel and renewable diesel

Property	No. 2 petroleum diesel	Biodiesel (FAME)[a]	Renewable diesel[b]
Carbon, wt.%	86.8	76.2	84.9
Hydrogen, wt.%	13.2	12.6	15.1
Oxygen, wt.%	0	11.2	0.0
Specific Gravity	0.85	0.88	0.78
Cetane no.	40–35	45–55	70–90
T °C	300–330	330–360	290–300
Viscosity, mm²/s. @ 40 °C	2–3	4–5	3–4
Energy content (LHV)			
Mass basis, MJ/kg	43	39	44
Mass basis, BTU/lb.	18,500	16,600	18,900
Vol. basis, 1000 BTU/gal	130	121	122

[a]Transportation fuel consisting of fatty acid methyl esters.
[b]Renewable diesel fuel (also known as Green Diesel) is produced by catalytic hydroprocessing of the same triglyceride feedstocks used to produce biodiesel. In this process, an alcohol is not required, the products are hydrocarbons rather than fatty acid alkyl esters, and no glycerol byproduct is formed. The general term "biodistillate" is used to refer to both biodiesel and renewable diesel.
Source: Adapted from Hoekman, S. Kent, Amber Broch, Curtis Robbins, Eric Ceniceros, Mani Natarajan. 2012. Review of biodiesel composition, properties, and specifications, Renewable and Sustainable Energy Reviews, Volume 16, Issue 1, Pages 143–169.

Table 24.10. Fuel property comparison for ethanol, gasoline, and No. 2 Diesel

Property	Ethanol	Gasoline	No. 2 Diesel
Chemical Formula	C_2H_5OH	C_4 to C_{12}	C_3 to C_{25}
Molecular Weight	46.07	100–105	≈200
Carbon	52.2	85–88	84–87
Hydrogen	13.1	12–15	33–16
Oxygen	34.7	0	0
Specific gravity, 60 °F/60 °F	0.796	0.72–0.78	0.81–0.89
Density, lb/gal @ 60 °F	6.61	6.0–6.5	6.7–7.4
Boiling temperature, °F	172	80–437	370–650
Reid vapor pressure, psi	2.3	8–15	0.2
Research octane no.	108	90–100	--
Motor octane no.	92	81–90	--
(R + M)/2	100	86–94	N/A
Cetane no.(1)	--	5–20	40–55
Fuel in water, volume %	100	Negligible	Negligible
Water in fuel, volume %	100	Negligible	Negligible
Freezing point, °F	–173.2	–40	–40–30[b]
Centipoise @ 60 °F	1.19	0.37–0.44[a]	2.6–4.1
Flash point, closed cup, °F	55	-45	165
Autoignition temperature, °F	793	495	≈600
Higher (liquid fuel-liquid water) Btu/lb	12,800	18,800–20,400	19,200–20000
Lower (liquid fuel-water vapor) Btu/lb	11,500	18,000–19,000	18,000–19,000
Higher (liquid fuel-liquid water) Btu/gal	84,100	124,800	138,700
Lower (liquid fuel-water vapor) Btu/gal @ 60 °F	76,000[a]	115,000	128,400
Mixture in vapor state, Btu/cubic foot @ 68 °F	92.9	95.2	96.9[c]
Fuel in liquid state, Btu/lb or air	1,280	1,290	–
Specific heat, Btu/lb °F	0.57	0.48	0.43
Stoichiometric air/fuel, weight	9	14.7[a]	14.7
Volume % fuel in vaporized stoichiometric mixture	6.5	2	–

[a]Calculated.
[b]Pour Point, ASTM D 97.
[c]Based on Cetane.
Source: U.S. Department of Energy, Office of Energy Efficiency and Renewable Energy, Alternative Fuels Data Center,
<http://www.eere.energy.gov/afdc/altfuel/fuel_properties.html>.

Table 24.11. International standard (EN 14214) requirements for biodiesel

Property	Units	Lower limit	Upper limit
Ester Content	% (m/m)	96.5	
Density at 15°C	kg/m3	860	900
Viscosity at 40°C	mm2/s	3.5	5
Flash point	°C	>101	
Sulfur content	mg/kg		10
Tar remnant (at 10% distillation remnant)	% (m/m)		0.3
Cetane number		51	
Sulfated ash content	% (m/m)		0.02
Water content	mg/kg		500
Total contamination	mg/kg		24
Copper band corrosion (3h at 50°C)	rating	Class 1	Class 1
Oxidation stability at 110°C	hours	6	
Acid value	mg KOH/g		0.5
Iodine value			120
Linoleic acid methyl ester	% (m/m)		12
Polyunsaturated (>4 double bonds) methylester	% (m/m)		1
Methanol content	% (m/m)		0.2
Monoglyceride content	% (m/m)		0.8
Diglyceride content	% (m/m)		0.2
Triglyceride content	% (m/m)		0.2
Free glyceride	% (m/m)		0.02
Total glyceride	% (m/m)		0.25
Alkali metals (Na + K)	mg/kg		5
Phosphorus content	mg/kg		10

Source: European Committee for Standardization.

Table 24.12. ASTM D7467 specification for diesel blends B6 to B20

Property	Test Method	Grade		
		B6 to B20 S15	B6 to B20 S500[i]	B6 to B20 S5000[k]
Acid Number, mg KOH/g, max.	D664	0.3	0.3	0.3
Viscosity, mm2/s at 40 °C	D445	1.9–4.1[a]	1.9–4.1[a]	1.9–4.1[a]
Flash Point, °C, min	D93	52[b]	52[b]	52[b]
Cloud Point, °C, max	D2500	[c]	[c]	[c]
Sulfur Content, (µg/g)[d]				
mass %, max.	D5453	15	–	–
mass %, max.	D2622	–	0.05	–
mass %, max.	D129	–	–	0.50
Distillation Temperature, °C, 90% evaporated, max.	D86	343	343	343
Ramsbottom carbon residue on 10% bottoms, mass %, max	D524	0.35	0.35	0.35
Cetane Number, min	D613[f]	40[g]	40[g]	40[g]
One of the following must be met:				
(1) Cetane index, min.	D976–80[e]	40	40	40
(2) Aromaticity, vol %, max.	D1319–88+B15	35	35	–
Ash Content, mass %, max	D482	0.01	0.01	0.01
Water and Sediment, vol %, max	D2709	0.05	0.05	0.05
Copper Corrosion, 3 h @ 50 °C, max.	D130	No. 3	No. 3	No. 3
Biodiesel Content, % (V/V)	DXXXX[h]	6-20	6-20	6-20
Oxidation Stability, hours, min.	EN14112	6	6	6
Lubricity, HFRR @ 60 °C, micron, max.	D6079	520[j]	520[j]	520[j]

[a] If Grade No. 1-D or blends of Grade No. 1-D and Grade No. 2-3 diesel fuel are used, the minimum viscosity shall be 1.3 mm²/s.

[b] If Grade No. 1-D or blends of Grade No. 1-D and Grade No. 2-D diesel fuel are used, or a cloud point of less than –12°C is specified, the miniumum flash point shall be 38 °C.

[c] It is unrealistic to specify low-temperature properties that will ensure satisfactory operation at all ambient conditions. However, satisfactory operation below the cloud poing (or wax appearance point) may be achieved depending on equipment design, operating conditions, and the use of flow-improver additives as described in X3.1.1.2. Appropriate low-temperature operability properties should be agreed upon between the fuel supplier and purchaser for the intended use and expected ambient temperatures. Test Methods D4539 and D6371 may be useful to estimate vehicle low temperature operability limits when flow improvers are used, but their use with Bxx blends from a full range of biodiesel feedstock sources has not been validated. The tenth percentile miniumum air temperatures may be used to estimate expected regional target temperatures for use with Test Methods D2500, D4539, and D6371.

[d] Other sulfur limits can apply in selected areas in the United States and in other countries.

[e] These test methods are specified in 40 CFR Part 80

[f] Calculated cetane index approximation, Test Method D4737, is not applicable to biodiesel blends.

[g] Low ambient temperatures as wella s engine operation at high altitudes may require the use of fuels with higher cetane ratingss. If the diesel fuel is qualified under Table 1 of D 975 for cetane, it is not necessary to measure the cetane number of the blend. This is because the cetane number of the individual blend components will be at least 40, so the resulting blend will also be at least 40 cetane number.

[h] Where specified, the blend level shall be +/- 2% volume unless a different tolerance is agreed to by the purchaser and the supplier. If the diesel fuel is qualified under Table 1 of D975 for lubricity, it is not necessary to measure the lubricity of the blend. This is because the individual blend components will be at least 520 microns, so the resulting blend will also be at least 520 microns.

[i] Under U.S. regulations, if Grades B20 S500 are sold for tax-exempt purposes, then, at or beyond terminal storage tanks, it is required by 26 CFR Part 48 to contain the dye Solvent Red 164 at a concentration spectrally equivalent to 3.8 lb per thousand barrels of the solid dye standard Solvent Red 164, or the tax must be colected.

[k] Under U.S. regulations, Grades B20 S5000 are required by 40 CFR part 80 to contain a sufficient amount of the dye solvent Red 164 so its presence is visually apparent. At or beyond terminal storage tanks; they are required by 36 CFR Part 48 to contain the dye Solvent Red 164 at a concentration spectrally equivalent to 3.9 lb per thousand barrels of the solid dye standard Solvent Red 26.

Table 24.13. Comparison of different technologies to produce biodiesel

Variable	Alkali catalysis (NaOH or KOH)	Lipase catalysis	Supercritical alcohol	Acid catalysis (H$_2$SO$_4$ or H$_2$SO$_3$
Reaction temperature/°C	25–120	30–40	239–385	55–80
Free fatty acid in raw materials	Saponified products	Methyl esters	Esters	Esters
Water in raw materials	Interference with reaction	No influence		Interference with reaction
Yield of methyl esters	Normal	Higher	Good	Normal
Recovery of glycerol	Difficult	Easy		Difficult
Purification of methyl esters	Repeated washing	None		Repeated washing
Production cost of catalyst	Cheap	Relatively expensive	Medium	Cheap
Amount of catalyst used	0.005–1% w/w			0.5–3.5 mol %

Source: Adapted from Earle, M.J., Plechkova, N.V., Seddon, K.R. Green synthesis of biodiesel using ionic liquids (2009) Pure and Applied Chemistry, 81 (11), pp. 2045–2057.

Table 24.14. Heating value of diesel and some biodiesel (B100) fuels

	Heating Value	
Fuel	Btu/lb	MJ/kg
Typical No. 2 Diesel	18,300	42.57
Soy Methyl Ester	15,940	37.08
Canola Methyl Ester	15,861	36.89
Lard Methyl Ester	15,841	36.85
Edible Tallow Methyl Ester	15,881	36.94
Inedible Tallow Methyl Ester	15,841	36.85
Low Free Fatty Acid Yellow Grease Methyl Ester	15,887	36.95
High Free Fatty Acid Yellow Grease Methyl Ester	15,710	36.54

Source: Adapted from National Renewable Energy Laboratory. 2009. Biodiesel handling and use guide (Fourth Edition) <www.nrel.gov; http://www.nrel.gov/vehiclesandfuels/pdfs/43672.pdf>.

Table 24.15.   Car registrations for selected countries, 1950–2009 (thousands)

Country	1950	1960	1970	1980	1990	2000	2009	Change 1990-2009
Argentina	318	474	1,482	3,112	4,284	5,060	6,465	50.9%
Brazil	a	a	a	a	12,127	15,393	23,612	94.7%
Canada[b]	1,913	4,104	6,602	10,256	12,622	16,832	19,877	57.5%
China	a	a	a	351	1,897	3,750	25,301	1233.7%
France	a	4,950	11,860	18,440	23,550	28,060	31,050	31.8%
Germany[c]	a	4,856	14,376	23,236	35,512	43,772	41,738	17.5%
India	a	a	a	a	2,300	5,150	10,400	352.2%
Indonesia	a	a	a	a	1,200	650	5,005	317.1%
Japan	43	457	8,779	23,660	34,924	52,437	58,020	66.1%
Malaysia	a	a	a	a	1,811	4,213	7,375	307.2%
Pakistan	a	a	a	a	738	375	460	−37.7%
Russia	a	a	a	a	a	20,353	33,187	a
United Kingdom	2,307	5,650	11,802	15,438	22,528	27,185	31,036	37.8%
United States	40,339	61,671	89,244	121,601	143,550	127,721	132,424	−7.8%
U.S. Percentage of World	76.0%	62.7%	46.1%	38.0%	32.3%	23.3%	19.4%	
World	53,051	98,305	193,479	320,390	444,900	548,558	681,154	53.1%

[a]Data are not available.
[b]Data from 2000 and later are not comparable to prior data. Canada reclassified autos and trucks prior to 2000.
[c]Data for 1990 and prior include West Germany only. Kraftwagen are included with automobiles.
Source: Oak Ridge National Laboratory, Transportation Energy Data Book: Edition 30, <http://cta.ornl.gov/data/index.shtml>.

Table 24.16.   Consumption of motor gasoline, top 20 nations, 2008 (1000 barrels per day)

Country	1990	2000	2008	Rank 2008	Rank 1990	Change 2000-08
United States	7,235	8,472	8,989	1	1	6.1%
China	455	891	1,437	2	7	61.2%
Japan	762	999	982	3	2	-1.6%
Mexico	469	534	776	4	6	45.1%
Russia	--	542	738	5	--	--
Canada	583	657	715	6	4	8.9%
Germany	724	665	480	7	3	-27.8%
Iran	142	263	410	8	14	56.1%
United Kingdom	563	499	388	9	5	-22.1%
Saudi Arabia	160	228	375	10	13	64.3%
Brazil	164	579	327	11	12	-43.6%
Australia	298	308	325	12	10	5.4%
Indonesia	110	214	285	13	15	33.4%
Venezuela	165	210	274	14	11	30.4%
Italy	336	398	262	15	9	-34.0%
India	83	151	257	16	17	70.1%
France	422	318	206	17	8	-35.3%
South Africa	107	175	187	18	16	6.8%
Malaysia	65	142	186	19	19	31.2%
Korea, South	65	170	172	20	18	0.9%
World	17,227	19,887	21,323			7.2%

Source: United States Department of Energy, Energy Information Administration, International Energy Statistics, <http://www.eia.gov/countries/data.cfm>.

Table 24.17.   Consumption of jet fuel, top 20 nations, 1990-2008 (1000 barrels per day)

	1990	2000	2008	Rank 2008	Rank 1990	Change 2000-08
United States	1,522	1,725	1,539	1	1	−10.8%
United Kingdom	142	233	261	2	3	12.0%
China	51	119	251	3	9	111.3%
Russia	--	189	248	4	--	31.1%
Japan	154	207	223	5	2	8.0%
Germany	117	154	192	6	4	25.0%
France	81	140	153	7	6	9.1%
Spain	52	94	121	8	8	28.8%
Singapore	39	49	120	9	14	145.4%
Canada	87	110	119	10	5	8.8%
Australia	54	91	107	11	7	17.8%
Hong Kong	38	57	100	12	15	75.2%
India	37	53	95	13	16	78.2%
Korea, South	46	69	91	14	10	31.8%
Brazil	na	65	91	15	na	39.6%
Italy	43	77	88	16	12	13.2%
United Arab Emirates	18	21	84	17	24	299.7%
Netherlands	34	70	79	18	18	12.7%
Thailand	41	57	77	19	13	36.2%
Mexico	35	54	64	20	17	17.0%
World	3,806	4,551	5,270			15.8%

na = not available.
Source: United States Department of Energy, Energy Information Administration, International Energy Statistics, <http://www.eia.gov/countries/data.cfm>.

Table 24.18. Global natural gas vehicle use, top 20 nations, 2011

Country	Natural Gas Vehicles	N/m³ monthly sales average	Refuelling stations
Iran	2,859,386	547,500,000	1,800
Pakistan	2,850,500	245,750,000	3,330
Argentina	2,077,581	225,455,000	1,913
Brazil	1,702,790	165,210,000	1,792
India	1,100,376	163,210,000	724
Italy	779,09	72,500,000	860
China	611,900		2,300
Colombia	365,168	45,000,000	651
Uzbekistan	310,000		175
Thailand	305,290	285,170,000	470
Armenia	244,000	26,520,000	345
Ukraine	200,019	83,000,000	294
Bangladesh	200,000	91,550,000	600
Egypt	165,392	38,000,000	146
Bolivia	140,400	26,278,135	156
Peru	129,981	17,584,416	179
United States	112,000	105,000,000	1,100
Germany	96,215	26,300,000	903
Venezuela	90,000	8,152,054	166
Russia	86,012	30,400,000	247
World	14,805,457	2,364,506,605	20,354

Source: NGV Journal, The NGC Statistics, <http://www.ngvjournal.dreamhosters.com/en/statistics>.

Table 24.19. Energy Intensity of passenger transportation modes

	Btu per vehicle-mile	Btu per passenger-mile
Bicycling (moderate pace)		180
Walking (moderate pace)		360
Bus (motorcoach)		750
Trolley bus[c,e]		1,321
Van pool[c]		1,354
Car (hybrid; 1.57 persons)		1,659
Motorcycles [a,b]	2,224	1,853
Car pool-2 person[c]		2,492
Rail		
Intercity rail [a,b]	54,585	2,516
Transit [a,b]	62,833	2,577
Commuter rail [a,b]	90,328	2,638
Air carriers [a,b]	301,684	3,153
Car (average trip, 1.57 persons) [a,b]	5,517	3,514
Personal truck [a,b]	6,788	3,946
Bus (transit) [a,b]	39,408	4,315
Car (single person)[c]		4,983
Ferry boat[c]		10,987
Demand response [a,b,d]	16,771	16,429

[a]U.S. 2007 value.
[b]Source: Davis, Stacy C., Susan W. Diegel, and Robert G. Boundy. 2009. Transportation Energy Data Book: Edition 28, Oak Ridge National Laboratory.
[c]M.J. Bradley & Associates. 2007. Comparison of Energy Use & CO2 Emissions From Different Transportation Modes, Manchester, NH.
[d]A transit mode that includes passenger cars, vans, and small buses operating in response to calls from passengers to the transit operator who dispatches the vehicles. The vehicles do not operate over a fixed route on a fixed schedule. Also be known as paratransit or dial-a-ride.
[e]A transit mode that uses electric-powered rubber-tired vehicles for fixed route scheduled service within an urban area, and usually operated in mixed traffic on city streets. Electricity to power the vehicles is drawn from overhead wires installed along the route.

Table 24.20.   Energy and carbon intensity of freight transport[a]

	MJ/t-km	t CO_2e/t-km $\times 10^6$
inland water (U.S.)	0.3	21
rail (U.S.)	0.3	18
truck (U.S.)	2.7	180
air (U.S.)	10.0	680
oil pipeline (U.S.)	0.2	16
gas pipeline (U.S.)	1.7	180
international air	10.0	680
international water container	0.2	14
international water bulk	0.2	11
international water tanker	0.1	7

[a]Adapted from Weber, Christopher L. and H. Scott Matthews, Food-Miles and the Relative Climate Impacts of Food Choices in the United States. Environmental Science & Technology, 2008, 42 (10), 3508–3513.
Note: CO_2 equivalent (CO_2-e) is based on the global warming potentials (GWP) of greenhouse gases. The GWP of a gas is the warming caused over a 100-year period by the emission of one ton of the gas relative to the warming caused over the same period by the emission of one ton of CO_2.

Table 24.21.   Emission factors (g/mile) and fuel economy (mile per gasoline equivalent gallon) of selected vehicle technologies

Vehicle technology	VOC Exhaust	NO_x	PM_{10} Exhaust	$PM_{2.5}$ Exhaust	CO	Fuel economy
GV	0.1	0.07	0.01	0.01	3.49	23.5
DV	0.06	0.08	0.01	0.01	0.53	28.2
E85 FFV	0.1	0.07	0.01	0.01	3.49	23.5
Gasoline HEV	0.05	0.06	0.01	0.01	3.49	34.8
Diesel HEV	0.05	0.07	0.01	0.01	0.53	37.6
EV	0	0	0	0	0	82.3
H_2 FCV	0	0	0	0	0	54

FFV: flexible-fuel vehicle; GV: gasoline vehicle; EV: electric vehicle; HEV: hybrid electric vehicle; FCV: fuel cell vehicle.
Source: Adapted from Huo, Hong, Ye Wu, Michael Wang. 2009. Total versus urban: Well-to-wheels assessment of criteria pollutant emissions from various vehicle/fuel systems, Atmospheric Environment, Volume 43, Issue 10, Pages 1796–1804.

Table 24.22. Life cycle greenhouse gas emissions for plug-in hybrid electric vehicles sold in the United States, model years 2011 and 2012

Vehicle	EPA rated All-electric range	EPA rated combined fuel economy[a]	Lower emission electric grids			U.S. national average electric mix	Higher emission electric grids		
			Alaska (Juneau)	California (San Francisco)	Mid-Atlantic South (Washington, D.C.)		Southeast (Atlanta)	Midwest (Des Moines)	Rocky Mountains (Denver)
Mitsubishi i-MiEV	62 mi (100 km)	112 mpg-e (30 kW-hrs/100 miles)	80 g/mi (50 g/km)	100 g/mi (62 g/km)	160 g/mi (99 g/km)	200 g/mi (124 g/km)	230 g/mi (143 g/km)	270 g/mi (168 g/km)	290 g/mi (180 g/km)
Ford Focus Electric	76 mi (122 km)	105 mpg-e (32 kW-hrs/100 miles)	80 g/mi (50 g/km)	110 g/mi (68 g/km)	170 g/mi (106 g/km)	210 g/mi (131 g/km)	250 g/mi (155 g/km)	280 g/mi (174 g/km)	310 g/mi (193 g/km)
BMW ActiveE	94 mi (151 km)	102 mpg-e (33 kW-hrs/100 miles)	90 g/mi (56 g/km)	110 g/mi (68 g/km)	180 g/mi (112 g/km)	220 g/mi (137 g/km)	250 g/mi (155 g/km)	290 g/mi (180 g/km)	320 g/mi (199 g/km)
Nissan Leaf	73 mi (117 km)	99 mpg-e (34 kW-hrs/100 miles)	90 g/mi (56 g/km)	120 g/mi (75 g/km)	190 g/mi (118 g/km)	230 g/mi (143 g/km)	260 g/mi (162 g/km)	300 g/mi (186 g/km)	330 g/mi (205 g/km)
Chevrolet Volt	35 mi (56 km)	94 mpg-e (36 kW-hrs/100 miles)	170 g/mi (106 g/km)	190 g/mi (118 g/km)	230 g/mi (143 g/km)	260 g/mi (162 g/km)	290 g/mi (180 g/km)	310 g/mi (193 g/km)	330 g/mi (205 g/km)
Smart ED	63 mi (101 km)	87 mpg-e (39 kW-hrs/100 miles)	100 g/mi (62 g/km)	130 g/mi (81 g/km)	210 g/mi (131 g/km)	260 g/mi (162 g/km)	300 g/mi (186 g/km)	350 g/mi (218 g/km)	380 g/mi (236 g/km)
Coda	88 mi (142 km)	73 mpg-e (46 kW-hrs/100 miles)	120 g/mi (76 g/km)	160 g/mi (99 g/km)	250 g/mi (155 g/km)	300 g/mi (186 g/km)	350 g/mi (218 g/km)	410 g/mi (255 g/km)	440 g/mi (273 g/km)
Average U.S. new car (Chevrolet Impala)	Gasoline only	22 mpg	Total emissions: 500 g/mi (311 g/km) Upstream: 100 g/mi (62 g/km) and tailpipe: 400 g/mi (249 g/km)						

[a]mpg-e = Miles per gallon gasoline equivalent. Used by the U.S. Environmental Protection Agency (EPA) to compare energy consumption of alternative fuel vehicles, plug-in electric vehicles and other advanced technology vehicles with the fuel economy of conventional internal combustion vehicles. One gallon of gasoline is equivalent to 33.7 kilowatt-hours of electrical energy stored in a vehicle's battery pack.
Data from United States Environmental Protection Agency, <http://www.fueleconomy.gov/>. Table format adapted from Wikipedia, "Electric car," http://en.wikipedia.org/wiki/Electric_car>.

Table 24.23.   Energy and CO$_2$ rating of cars as a function of their mass

	Energy per km•mass	Energy per km•mass	dH_{km}/dm MJ/km•kg
	$(H_{km}$ in MJ/km, m kg)	$(CO_2/km$ in g/km, m in kg)	$(m = 1000$ kg)
Gasoline power	$H_{km} \approx 3.7 \times 10^{-3}\ m^{0.93}$	$CO_2/km \approx 0.25\ m^{0.93}$	2.1×10^{-3}
Diesel power	$H_{km} \approx 2.8 \times 10^{-3}\ m^{0.93}$	$CO_2/km \approx 0.21\ m^{0.93}$	1.6×10^{-3}
LPG power	$H_{km} \approx 3.7 \times 10^{-3}\ m^{0.93}$	$CO_2/km \approx 0.17\ m^{0.93}$	2.2×10^{-3}
Hybrid power	$H_{km} \approx 2.3 \times 10^{-3}\ m^{0.93}$	$CO_2/km \approx 0.16\ m^{0.93}$	1.3×10^{-3}

Source: Ashby, Michael F. 2013. Materials and the Environment (Second Edition), (Boston, Butterworth-Heinemann).

Table 24.24.   Selected technologies that improve motor vehicle fuel efficiency

Technology	Average Efficiency Increase	Description
Variable Valve Timing & Lift	5%	Improves engine efficiency by optimizing the flow of fuel & air into the engine for various engine speeds.
Cylinder Deactivation	7.5%	Saves fuel by deactivating cylinders when they are not needed.
Turbochargers & Superchargers	7.5%	Increase engine power, allowing manufacturers to downsize engines without sacrificing performance or to increase performance without lowering fuel economy.
Integrated Starter/Generator (ISG) Systems	8%	Automatically turn the engine on/off when the vehicle is stopped to reduce fuel consumed during idling.
Direct Fuel Injection (w/ turbocharging or supercharging)	11–13%	Delivers higher performance with lower fuel consumption.
Continuously Variable Transmissions (CVTs)	6%	Infinite number of "gears", providing seamless acceleration and improved fuel economy.
Automated Manual Transmissions (AMTs)	7%	Combine the efficiency of manual transmissions with the convenience of automatics (gears shift automatically).

Source: United States Environmental Protection Agency, <http://www.fueleconomy.gov>.

Table 24.25.   Most fuel effcient cars sold in the United States, model year 2012

Rank	Model	MPG Combined	MPG City/Highway
	Including electric (EV) and plug-in hybrid electric (PHEV) vehicles		
1	Mitsubishi i-MiEV	112	126/99
	Electric Vehicle, Auto (A1)		
2	Ford Focus EV	105	110/99
	Electric Vehicle, Auto (variable gear ratios)		
3	Nissan Leaf	99	106/92
	Electric Vehicle, Auto (A1)		
4	Coda Automotive Coda	73	77/68
	Electric Vehicle, Auto (A1)		
5 (tie)	Azure Dynamics Transit Connect Van	62	62/62
	Electric Vehicle, Auto		
	Azure Dynamics Transit Connect Wagon	62	62/62
	Electric Vehicle, Auto		
6	Chevrolet Volt (Ranked by combined gas/electricity rating of 60 MPGe)[a]	60	58/62
	PHEV, 4 cyl, 1.4 L, Auto (variable gear ratios), Premium		
7	Toyota Prius c	50	53/46
	Hybrid, 4 cyl, 1.5 L, Auto (variable gear ratios), Regular		
8	Toyota Prius	50	51/48
	Hybrid, 4 cyl, 1.8 L, Auto (variable gear ratios), Regular		
9	Honda Civic Hybrid	44	44/44
	Hybrid, 4 cyl, 1.5 L, Auto (variable gear ratios), Regular		
10	Toyota Prius V	42	44/40
	Hybrid, 4 cyl, 1.8 L, Auto (variable gear ratios), Regular		
	Excluding electric (EV) and plug-in hybrid electric (PHEV) vehicles		
1	Toyota Prius c	50	53/46
	Hybrid, 4 cyl, 1.5 L, Auto (variable gear ratios), Regular		
2	2012 Toyota Prius	50	51/48
	Hybrid, 4 cyl, 1.8 L, Auto (variable gear ratios), Regular		

Continued

Table 24.25.   Continued

Rank	Model	MPG Combined	MPG City/Highway
	Including electric (EV) and plug-in hybrid electric (PHEV) vehicles		
3	2012 Honda Civic Hybrid	44	44/44
	Hybrid, 4 cyl, 1.5 L, Auto (variable gear ratios), Regular		
4	2012 Toyota Prius V	42	44/40
	Hybrid, 4 cyl, 1.8 L, Auto (variable gear ratios), Regular		
5	2012 Lexus CT 200h	42	43/40
	Hybrid, 4 cyl, 1.8 L, Auto (variable gear ratios), Regular		
6 (tie)	2012 Honda Insight	42	41/41
	Hybrid, 4 cyl, 1.3 L, Auto (variable gear ratios), Regular		
	2012 Honda Insight	42	41/41
	Hybrid, 4 cyl, 1.3 L, Auto (AV-S7), Regular		
7	2012 Toyota Camry Hybrid LE	41	43/39
	Hybrid, 4 cyl, 2.5 L, Auto (variable gear ratios), Regular		
8	2012 Toyota Camry Hybrid XLE	40	40/38
	Hybrid, 4 cyl, 2.5 L, Auto (variable gear ratios), Regular		
9 (tie)	2012 Ford Fusion Hybrid	39	41/36
	Hybrid, 4 cyl, 2.5 L, Auto (variable gear ratios), Regular		
	2012 Lincoln MKZ Hybrid	39	41/36
	Hybrid, 4 cyl, 2.5 L, Auto (variable gear ratios), Regular		
10	2012 Scion iQ	37	36/37
	4 cyl, 1.3 L, Auto (variable gear ratios), Regular		

[a]mpg-e =Miles per gallon gasoline equivalent . Used by the U.S. Environmental Protection Agency (EPA) to compare energy consumption of alternative fuel vehicles, plug-in electric vehicles and other advanced technology vehicles with the fuel economy of conventional internal combustion vehicles expressed as miles per US gallon. One gallon of gasoline is equivalent to 33.7 kilowatt-hours of electrical energy stored in a vehicle's battery pack.

Source: United States Environmental Protection Agency, <http://www.fueleconomy.gov/>, accessed 12 June 2012.

Table 24.26. Fuel quality standards for hydrogen in the state of California[a]

Specification	Value
Hydrogen Fuel Index (minimum, %)[b]	99.97
Total Gases (maximum, ppmv)[c]	300
Water (maximum, ppmv)	5
Total Hydrocarbons (maximum, ppmv)[d]	2
Oxygen (maximum, ppmv)	5
Helium (maximum, ppmv)	300
Nitrogen and Argon (maximum, ppmv)	100
Carbon Dioxide (maximum, ppmv)	2
Carbon Monoxide (maximum, ppmv)	0.2
Total Sulfur Compounds (maximum, ppmv)	0.004
Formaldehyde (maximum, ppmv)	0.01
Formic Acid (maximum, ppmv)	0.2
Ammonia (maximum, ppmv)	0.1
Total Halogenated Compounds (maximum, ppmv)	0.05
Particulate Size (maximum, μm)	10
Particulate Concentration (maximum, μg/L @ NTP)[e]	1

[a]California Department of Food and Agriculture Division of Measurement Standards. "Hydrogen Fuel Standards"
<http://www.cdfa.ca.gov/dms/hydrogenfuel/HydrogenFuelFinalText.pdf>, accessed 23 May 2012.
[b]The Hydrogen Fuel Index is the value obtained with the value of Total Gases measured in % subtracted from 100%.
[c]Total Gases = Sum of all impurities listed on the table except particulates.
[d]Total Hydrocarbons may exceed 2 ppmv only due to the presence of methane, provided that the total gases do not exceed 300 ppmv.
[e]μg/L @ NTP = micrograms per liter of hydrogen fuel at 0°C and 1 atmosphere pressure.

Table 24.27. Driving cycle tests performed by the U.S Environmental Protection Agency

	Test Schedule				
	City	Highway	High Speed	Air Conditioning (AC	Cold Temp
Trip type	Low speeds in stop-and-go urban traffic	Free-flow traffic at highway speeds	Higher speeds; harder acceleration & braking	AC use under hot ambient conditions	*City* test w/ colder outside temperature
Top speed	56 mph	60 mph	80 mph	54.8 mph	56 mph
Average speed	20 mph	48 mph	48 mph	22 mph	20 mph
Max. acceleration	3.3 mph/sec	3.2 mph/sec	8.46 mph/sec	5.1 mph/sec	3.3 mph/sec
Simulated distance	11 mi.	10 mi.	8 mi.	3.6 mi.	11 mi.
Time	31 min.	12.5 min.	10 min.	9.9 min.	31 min.
Stops	23	None	4	5	23
Idling time	18% of time	None	7% of time	19% of time	18% of time
Engine startup	Cold	Warm	Warm	Warm	Cold
Lab temperature	68–86° F	68–86° F	68–86° F	95° F	20° F
Vehicle air conditioning	Off	Off	Off	On	Off

[a] EPA tests vehicles by running them through a series of driving routines, also called cycles or schedules, that specify vehicle speed for each point in time during the laboratory tests.
Source: Oak Ridge National Laboratory, Transportation Energy Data Book: Edition 30, <http://cta.ornl.gov/data/index.shtml>.

Table 24.28.   Global merchant fleet by flag of registration and by type of ship, 1980-2011[a]

	1980	1990	2000	2011	Change 2000-11
Oil tankers	337,896	235,785	283,066	474,846	67.8%
Bulk carriers	181,880	223,619	274,445	532,039	93.9%
General cargo[b]	112,841	100,457	101,520	108,971	7.3%
Container ships	10,290	22,346	63,580	183,859	189.2%
Other types of ships	29,236	47,770	71,160	96,027	34.9%

[a]thousand dwt.
[b]including passenger/cargo combined.
dwt = deadweight tons: the number of tons of 2,240 pounds that a vessel can transport of cargo, stores and bunker fuel. It is the difference between the number of tons of water a vessel displaces "light" and the number of tons it displaces when submerged to the "load line."
Source: UNCTADstat, <http://unctadstat.unctad.org/ReportFolders/reportFolders.aspx?sCS_referer=&sCS_ChosenLang=en>, accessed 10 June 2012.

Table 24.29.   World's largest tanker companies

10 Largest Independent Tanker Companies			10 Largest Oil Co/State Tanker Companies		
Owner	No. Tankers	m dwt		No. Tankers	m dwt
Frontline Ltd. (Bermuda)	82	19.1	Sovcomflot Group (Russia)	136	11.7
Misui OSK Lines Ltd. (Japan)	105	14.4	AET/MISC Group (Malaysia)	78	10.2
Teekay Corp. (Bermuda)	111	12.6	China Shipping Group Co	85	7.5
Overseas Shipholding Group (USA)	101	11.9	National Shipping SA (Argentina)	45	6.5
NITC (Iran)	45	10.7	Saudi Aramco (Saudi Arabia)	23	5.7
Nippon Yusen Kaisha (Japan)	52	10.3	India Govt.	43	5.3
Maran Tankers Management (Greece)	37	8.1	COSCO (China)	32	5.2
Euronav NV (Belgium)	33	7.8	BP Plc (United Kingdom)	33	3.2
Tanker Pacific Management (Singapore)	49	6.2	Kuwait Petroleum Corp.	18	3.1
Dr. Peters Group (Germany)	30	5.8	Petrobras (Venezuala)	42	2.4
Totals	624	103.3	Totals	535	60.8
Independent Fleet	4787	378.3	Total Oil/State Fleet	723	74.7

dwt = deadweight tons: the number of tons of 2,240 pounds that a vessel can transport of cargo, stores and bunker fuel. It is the difference between the number of tons of water a vessel displaces "light" and the number of tons it displaces when submerged to the "load line."
Source: http:INTERTANKO, //www.intertanko.com, accessed 5 May 2012.

Table 24.30. Tanker size categories

Classification	Size (dwt)[a]	Remarks
Handy	10,000–30,000	
Handymax	30,000–50,000	suited for small ports with length and draft restrictions and also lacking transshipment infrastructure; carry grains and minor bulks including steel products, forest products and fertilizers; small-sized oil tankers are in this size range
Panamax	60,000–80,000	largest dry bulk ships capable of transiting the Panama Canal; carry coal, grain and, to a lesser extent, minor bulks, including steel products, forest products and fertilizers; also used in iron ore trade in recent years
Capesize	80,000–200,000	incapable of using the Panama or Suez canals due to size (over 1000' in length); transit via Cape Horn (South America) or the Cape of Good Hope (South Africa); serve deepwater terminals handling raw materials, such as iron ore and coal; trading between Australia/Indonesia and Asia, Brazil to the EU/Asia, and South Africa to the EU/Asia
Aframax	80,000–120,000	name derives from Average Freight Rate Assessment (AFRA) system; largely used in the basins of the Black Sea, the North Sea, the Caribbean Sea, the China Sea and the Mediterranean; used for oil transport when volumes are modest or port size/traffic prevents larger tankers from being used
Suezmax	120,000–200,000	largest vessels capable of transiting the Suez Canal Canal; usually refers to oil tankers
Very Large Crude Carrier (VLCC)	200,000–315,000	very large crude oil carriers that transport crude oil from the Persian Gulf, West Africa, the North Sea and Prudhoe Bay to destinations in the United States, Mediterranean Europe and Asia.
Ultra Large Crude Carrier (ULCC)	320,000–550,000	largest man-made vessels that move; transport oil; sail the longest routes, typically from the Persian Gulf to Europe, the U.S. and Asia; require custom-built terminals for loading and unloading.

[a]dwt = deadweight tons: the number of tons of 2,240 pounds that a vessel can transport of cargo, stores and bunker fuel. It is the difference between the number of tons of water a vessel displaces "light" and the number of tons it displaces when submerged to the "load line."
Source: American Bureau of Shipping.

Table 24.31. Deliveries of tankers , 2000-2010. Units are m dwt

Segment	2000	2005	2010	Change 2000-10
VLCC	12.2	8.9	16.9	38.5%
Suezmax	3.3	3.8	6.2	87.9%
Aframax	2.2	6.9	7.5	240.9%
Panamax	1	3.2	2.3	130.0%
Product tanker	2.5	4.2	4.4	76.0%
Total	21.2	27	37.3	75.9%

dwt = deadweight tons: the number of tons of 2,240 pounds that a vessel can transport of cargo, stores and bunker fuel. It is the difference between the number of tons of water a vessel displaces "light" and the number of tons it displaces when submerged to the "load line."
VLCC = Very Large Crude Carrier (200,000–320,000 dwt).
Suezmax (120,000–200,000 dwt).
Aframax (80,000–120,000 dwt).
Panamax (60,000–80,000 dwt).
Product tanker (10,000–60,000 dwt).
Source: Adapted from INTERTANKO,
http://www.intertanko.com/Book-Shop/Annual-Report-and-Industry-Facts/INTERTANKO-Tanker-Facts-2010/>.

Table 24.32.   General properties of representative rocket propellants

Type	Propellant	Energy Source	Isp (s)[a] Vacuum	$F_{(lbs)}$ Range[b]	Specific Gravity	Advantages	Disadvantages
Solid	Organic polymers, ammonium perchlorate, and powdered Al	Chemical	280–300	10 to 10^6	1.8	Simple, reliable, low cost	Limited performance, safety issues
Cold gas	N_2, NH_3, He, freon	High pressure	50–75	0.01 to 50	1.46	Simple, reliable, low cost	Low performance, high weight
Liquid	H_2O_2	Exothermic decomposition	150	0.01–0.1	1	Simple, reliable, low cost	Low performance, higher weight
monopropellant	N_2H_4		200				
Liquid	RP-1/O_2	Chemical	270–360	1 to 10^6	1	High performance	Complicated
bipropellant	H_2/O_2	Chemical	360–450	1 to 10^6	1.26	Very high performance	Cryogenic, complicated
	UMDH/N_2O_4	Chemical	270–340	1 to 10^6	1.14	Storable, good performance	Complicated
	N_2H_4/F_2	Chemical	425	1 to 10^6	1.1	Very high performance	Toxic, dangerous, complicated

[a]specific impulse.
[b]thrust.
Source: Adapted from Sforza, Pasquale M. 2012. Theory of Aerospace Propulsion, (Boston, Butterworth-Heinemann).

Table 24.33.   Rocket propellant performance comparison

Oxidizer	Fuel	Mixture ratio	Specific impulse (s, sea level)	Density impulse (kg s l⁻¹, S.L.)
Liquid oxygen	Liquid hydrogen	5	381	124
	Liquid methane	2.77	299	235
	Ethanol + 25% water	1.29	269	264
	Kerosene	2.29	289	294
	Hydrazine	0.74	303	321
	MMH	1.15	300	298
	UDMH	1.38	297	286

Specific impulses are theoretical maximum assuming 100% efficiency; actual performance will be less. All mixture ratios are optimum for the operating pressures indicated, unless otherwise noted. LO2/LH2 and LF2/LH2 mixture ratios are higher than optimum to improve density impulse. FLOX-70 is a mixture of 70% liquid fluorine and 30% liquid oxygen. Where kerosene is indicated, the calculations are based on n-dodecane.
Source: Burke, K. 2012. Current Perspective on Hydrogen and Fuel Cells, In: Ali Sayigh, Editor-in-Chief, Comprehensive Renewable Energy, (Oxford, Elsevier), Pages 13–43.

Table 24.34.   Typical Performance of Electric Propulsion Systems used in space travel.

Type	Propellant	Energy Source	I_{sp}, v_{ac} (s)[a]	Thrust (N)	Density (kg/m³)
Resistojet	N_2, NH_3, N_2O_4, H_2	Resistive $\eta = 0.9$	150–700	0.005–0.5	280, 600, 1000, 19
Arcjet	NH_3, H_2, N_2H_4	Arc heat $\eta = 0.3$	450–1500	0.05–5	600, 19, 1000
Ion	Hg, A, Xe, Cs	Electrostatic $\eta = 0.75$	2000–6000	$5 \times 10^{-6} - 0.5$	13,500, 440, 273, 187
MHD	A	Magnetic	2000	25–200	440
Pulsed plasma	Teflon	Magnetic	1500	$5 \times 10^{-6} - 5 \times 10^{-3}$	220

[a]specific impulse.
Source: Adapted from Sforza, Pasquale M. 2012. Theory of Aerospace Propulsion, (Boston, Butterworth-Heinemann).

Table 24.35. Radioisotope Power Systems (RPS) for Space Exploration

Name and Model	Used on (Number of RTGs per User)	Maximum Output		Maximum Fuel Used (kg)	RPS Mass (kg)
		Electrical (W)	Heat (W)		
SNAP-3B	Transit-4A/B (1)	2.7	52.5	~0.2	2
SNAP-9A	Transit 5BN-1/2/3 (1)	25	525	~1	12
SNAP-19	Nimbus B1 (2)	40.3	525	~1	14
	Nimbus III (2)				
	Pioneer 10/11 (4)				
Modified SNAP-19	Viking 1/2 (2)	42.7	525	~1	15
SNAP-27	Apollo 12-17 ALSEP (1)	73	1480	3.8	20
MHW-RTG	LES-8/9 (2)	470	2400	~4.5	38
	Voyager 1/2 (3)				
GPHS-RTG	Galileo (2)	285	4500	7.6	56
	Ulysses (1)				
	Cassini (3)				
	New Horizons (1)				

ALSEP, Apollo Lunar Surface Experiments Package; GPHS, General Purpose Heat Source; LES, Lincoln Experimental Satellite; MHW, Multi-hundred Watt; SNAP, Systems for Nuclear Auxiliary Power.
Source: Radioisotope Power Systems Committee, National Research Council, Radioisotope Power Systems: An Imperative for Maintaining U.S. Leadership in Space Exploration, The National Academies Press, 2009.

Table 24.36. Possible *in situ* nuclear thermal propulsion propellant sources in space travel

Propellant	Source	I_{sp} Potential (s)[a]
CO_2	Martian atmosphere, Martian frost, Earth	160–380
CH_4	Asteroids, Phobos and Deimos, Earth, outer planets	460–670
H_2	Lunar polar ice, lunar silane, NEO asteroids, Earth, outer planets	800–1200
NH_3	Earth, outer planets	350–700
H_2O	Lunar ice, Martian ice, planetary moons	160–240

[a]specific impulse.
Source: Adapted from Sforza, Pasquale M. 2012. Theory of Aerospace Propulsion, (Boston, Butterworth-Heinemann).

Table 24.37.   **Specifications for solar-powered aircarft built by AeroVironment, Inc**

	Solar Challenger	Pathfinder	Pathfinder-Plus	Centurion
Wingspan (ft)	46.5	98.4	121	206
Length (ft)	30.3	12	12	12
Wing Chord (ft)	5.8	8	8	8
Gross weight (lbs)	336	560	700	~1,900
Payload (lbs)	150	100	150	100-600
Airspeed (mph)	25–34	17–20	17–20	17–21
Power (Watts)	2,700	7,500	12,500	31,000
Motors (kW)/ (n motors)	2.7/1	1.25/6	1.5/8	2.2/14
Primary materials	Composites, plastic, foam			

Source: Dryden Flight Research Center, NASA.

Lighting/HVAC/Refrigeration

Charts

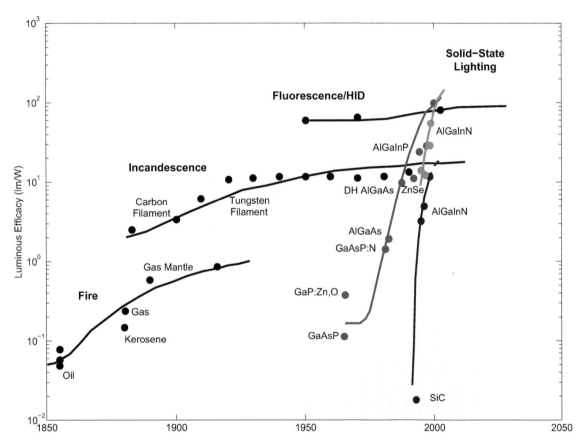

Chart 25.1. Trends in luminous efficacy for various lighting technologies. Luminous efficacy is measured in units of lumens (lm), a measure of light that factors in the human visual response to various wavelengths, per watt (W). The red, green, blue and white curves represent their respective solid-state lighting, light-emitting diodes (SSL-LED). HID=high intensity discharge.
Source: Data from Tsao, Jeff T. 2004. Solid State Lighting, IEEE Circuits and Devices Magazine, May/June, pages 28-37.

Handbook of Energy. http://dx.doi.org/10.1016/B978-0-08-046405-3.00025-5

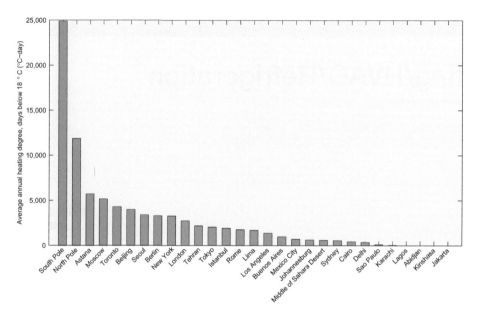

Chart 25.2. Annual average heating degree-days below 18 °C for selected cities/regions (22-year annual average, July 1983 - June 2005).
Source: Data from United States National Aeronautics and Space Administration (NASA), Surface meteorology and Solar Energy database, <http://eosweb.larc.nasa.gov/sse/>.

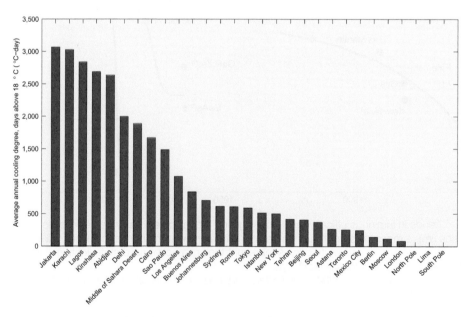

Chart 25.3. Annual average cooling degree-days above 18 °C for selected cities/regions (22-year annual average, July 1983 - June 2005).
Source: Data from United States National Aeronautics and Space Administration (NASA), Surface meteorology and Solar Energy database, <http://eosweb.larc.nasa.gov/sse/>.

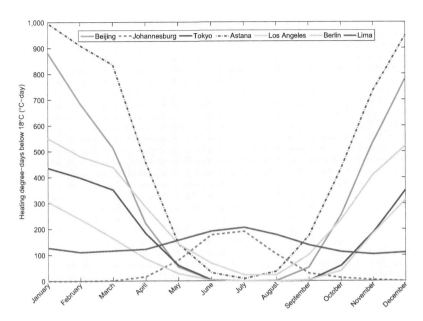

Chart 25.4. Monthly average heating degree-days below 18 °C for selected cities (22-year annual average, July 1983 - June 2005).
Source: Data from United States National Aeronautics and Space Administration (NASA), Surface meteorology and Solar Energy database, <http://eosweb.larc.nasa.gov/sse/>.

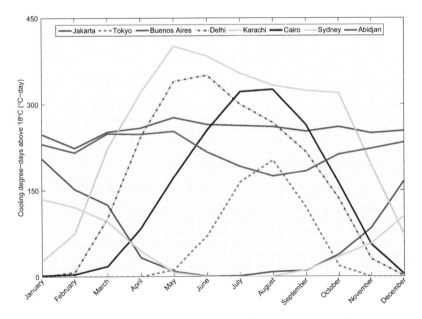

Chart 25.5. Monthly average cooling degree-days above 18 °C for selected cities (22-year annual average, July 1983 - June 2005).
Source: Data from United States National Aeronautics and Space Administration (NASA), Surface meteorology and Solar Energy database, <http://eosweb.larc.nasa.gov/sse/>.

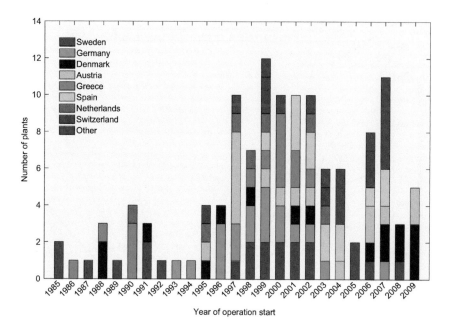

Chart 25.6. Large-scale solar heating and cooling systems in operation in Europe, 1985–2009.
Source: Data from Weiss, Werner and Franz Mauthner. 2012. Solar Heat Worldwide, Markets and Contribution to the Energy Supply 2010, Edition 2012, (Gleisdorf, AEE - Institute for Sustainable Technologies; Paris, Solar Heating and Cooling Programme, International Energy Agency).

Tables

Table 25.1. Comparison of different forms of lighting

Lighting Type	Efficacy (lumens/ watt)	Lifetime (hours)	Color Rendition Index (CRI)	Color Temperature (K)	Indoors/ Outdoors
Incandescent					
Standard "A" bulb	10–17	750–2500	98–100 (excellent)	2700–2800 (warm)	Indoors/ outdoors
Energy-Saving Incandescent (or Halogen)	12–22	1,000–4,000	98–100 (excellent)	2900–3200 (warm to neutral)	Indoors/ outdoors
Reflector	12–19	2000–3000	98–100 (excellent)	2800 (warm)	Indoors/ outdoors
Fluorescent					
Straight tube	30–110	7000–24,000	50–90 (fair to good)	2700–6500 (warm to cold)	Indoors/ outdoors
Compact fluorescent lamp (CFL)	50–70	10,000	65–88 (good)	2700–6500 (warm to cold)	Indoors/ outdoors
Circline	40–50	12,000			Indoors
High-Intensity Discharge					
Mercury vapor	25–60	16,000–24,000	50 (poor to fair)	3200–7000 (warm to cold)	Outdoors
Metal halide	70–115	5000–20,000	70 (fair)	3700 (cold)	Indoors/ outdoors
High-pressure sodium	50–140	16,000–24,000	25 (poor)	2100 (warm)	Outdoors
Light-Emitting Diodes					
Cool White LEDs	60–92	25,000–50,000	70–90 (fair to good)	5000 (cold)	Indoors/ outdoors
Warm White LEDs	27–54	25,000–50,000	70–90 (fair to good)	3300 (neutral)	Indoors/ outdoors
Low-Pressure Sodium	60–150	12,000–18,000	-44 (very poor)		Outdoors

Source: United States Department of Energy, Office of Energy Efficiency and Renewable Energy (EERE), Types of Lighting, <http://www.energysavers.gov/your_home/lighting_daylighting/index.cfm/mytopic=12030>.

Table 25.2. Comparison of LED, CFL and incandescent lamps

	Light emitting diodes (LEDs)	Compact fluorescent lamps (CFLs)	Traditional incandescent
Light bulb projected lifespan	50,000 hours	10,000 hours	1,200 hours
Watts per bulb (equiv. 60 watts)	10	14	60
Cost per bulb	35.95	3.95	1.25
KWh of electricity used over 50,000 hours	300 500	700	3000
Cost of electricity (@ 0.10per KWh)	50	70	300
Bulbs needed for 50k hours of use	1	5	42
Equivalent 50k hours bulb expense	35.95	19.75	52.5
Total cost for 50k hours	85.75	89.75	352.5

Source: Data from United States Department of Energy, Office of Energy Efficiency and Renewable Energy (EERE), How Energy-Efficient Light Bulbs Compare with Traditional Incandescents, <http://www.energysavers.gov/your_home/lighting_daylighting/index.cfm/mytopic=12060>.

Table 25.3. Trends in recommended light levels

Task	Recommended lumens per square foot		
	1972	1987	2000
Hospital operating table	2500	2500	300-1000
Basketball, college and pro	50	50	80-125
Open parking	1	0.2-0.9	0.2-0.5
Office accounting	150	50-100	30
Retail feature displays	500	150-500	100
Retail circulation	30	10-30	10
Active storage, small items	50	20-50	30
Kitchen general lighting	50	20-50	30
Reading	70	20-50	30

Source: Adapted from Horner, Pamela. 2008. What Have We Accomplished in Lighting Efficiency?, Regulatory & Industry Relations, OSRAM SYLVANIA.

Table 25.4. Trends in recommended light power densities

Area	Watts per square foot		
	1989	1999	2004
Warehouse	1.03	1.2	0.8
Sports Arena	2.07	1.5	1.1
School/University	1.29	1.5	1.1
Retail	2.25	1.9	1.5
Office	1.26	1.3	1
Manufacturing Facility	0.96	2.2	1.3
Library	1.29	1.5	1.3
Hotel	1.15	1.7	1
Hospital	1.44	1.6	1.2
Dining: Family	1.37	1.9	1.6

Source: Adapted from Horner, Pamela. 2008. What Have We Accomplished in Lighting Efficiency?, Regulatory & Industry Relations, OSRAM SYLVANIA.

Table 25.5. Common light levels outdoors

Condition	Illumination	
	(foot candle)	lux
Sunlight	10,000	107,527
Full Daylight	1,000	10,752
Overcast Day	100	1,075
Very Dark Day	10	107
Twilight	1	10.8
Deep Twilight	0.1	1.08
Full Moon	0.01	0.108
Quarter Moon	0.001	0.0108
Starlight	0.0001	0.0011
Overcast Night	0.00001	0.0001

Source: Adapted from Engineering Toolbox, Illuminance - Recommended Light Levels,
<http://www.engineeringtoolbox.com/light-level-rooms-d_708.html>.

Table 25.6. First and second law efficiencies of space heating technologies (T_h = 20 °C, T_c = 7 °C)

Technology	1st law efficiency	2nd law efficiency
Hand fired coal fire	45%	2.10%
Wood fire	80%	3.50%
Oil or gas fired furnace	60%–75%	2.6–3.3%
Kerosene/gas stove	100%	4.40%
Electric resistance heater (40% electricity generation efficiency)	100%	4.4%(1.8%)
Heat pump (COP = 3.2, 40% electricity efficiency)	300%	14.2% (5.7%)

Source: Warr, Benjamin, Robert Ayres, Nina Eisenmenger, Fridolin Krausmann, Heinz Schandl. 2010. Energy use and economic development: A comparative analysis of useful work supply in Austria, Japan, the United Kingdom and the US during 100 years of economic growth, Ecological Economics, Volume 69, Issue 10, Pages 1904-1917

Table 25.7. Heating degree days in the United States by region and month, 2011

Date	New England	Middle Atlantic	E N Central	W N Central	South Atlantic	E S Central	W S Central	Mountain	Pacific	US
12/03/11	136	136	192	231	111	144	111	194	90	149
11/05/11	168	152	136	140	112	102	63	145	65	117
10/01/11	7	15	62	55	10	22	5	14	7	26
04/02/11	196	201	218	195	133	115	53	110	55	147
03/05/11	245	203	217	255	105	78	50	178	116	164
02/05/11	309	274	317	350	168	180	199	282	122	244
01/01/11	247	244	242	283	192	175	118	256	150	211

Source: United States Department of Energy, National Renewable Enegy Laboratory

Table 25.8. Internal sensible heat gains that offset the heating requirements of buildings

Type	Magnitude (W or J/s)
Incandescent lights	total W
Fluorescent lights	total W
Electric motors	746 X (hp/efficiency)
Natural gas, stove	8.28 X m³/hr
Appliances	total W
A dog	50-90
People	
Sitting	70
Walking	75
Dancing	90
Working hard	170
Sunlight	Solar heat gain x fenestration transmittance x shading factor[a]

[a]Shading factor is the amount of a window not in a shadow expressed as a decimal between 1.0 and 0.0
Source: Adapted from Kreith, Frank and D. Yogi. Goswami, Eds. 2005. CRC Handbook of Mechanical Engineering, Second Edition (Boca Raton, CRC Press).

Table 25.9.   Total emittance of selected surface

	Temperature °C	Total Normal Emittance[a]
Alumina, Flame Sprayed	−25	0.8
Aluminum foil		
As received	20	0.04
Bright dipped	20	0.025
Aluminum, vacuum deposited	20	0.025
Hard-anodized	−25	0.84
Highly polished plate, 98.3% pure	225–575	0.039–0.057
Commercial sheet	100	0.09
Rough polish	100	0.18
Heavily oxidized	95–500	0.20–0.31
Antimony, polished	35–260	0.28–0.31
Asbestos	35–370	0.93–0.94
Beryllium	150	0.18
	370	0.21
	600	0.3
Beryllium, anodized	150	0.9
	370	0.88
	600	0.82
Bismuth, bright	75	0.34
Black paint		
Parson's optical black	−25	0.95
Black silicone	−25–750	0.93
Black epoxy paint	−25	0.89
Black enamel paint	95–425	0.81–0.80
Brass, polished	40–315	0.1
Rolled plate, natural surface	22	0.06
Dull plate	50–350	0.22
Oxidized by heating at 600 °C	200–600	0.61–0.59
Carbon, graphitized	100–320	0.76–0.75
	320–500	0.75–0.71
Candle soot	95–270	0.952
Graphite, pressed, filed surface	250–510	0.98
Chromium, polished	40–1100	0.08–0.36
Copper, electroplated	20	0.03
Carefully polished electrolytic copper	80	0.018
Polished	115	0.023
Plate heated at 600 °C	200–600	0.57
Molten copper	1075–1275	0.16–0.13
Glass, Pyrex, lead, and soda	260–540	0.95–0.85
Gypsum	20	0.903
Gold, pure, highly polished	225–625	0.018–0.035
Inconel X, oxidized	−25	0.71
Lead, pure (99.96%), unoxidized	125–225	0.057–0.075
Gray oxidized Oxidized at 150°C	25 200	0.28 0.63
Magnesium oxide	275–825	0.55–0.20

Table 25.9.   Continued

	Temperature °C	Total Normal Emittance[a]
	900–1705	0.2
Magnesium, polished	35–260	0.07–0.13
Mercury	0–100	0.09–0.12
Molybdenum, polished	35–260	0.05–0.08
	540–1370	0.10–0.18
	2750	0.29
Nickel, electroplated	20	0.03
Polished	100	0.072
Platinum, pure, polished	225–625	0.054–0.104
Silica, sintered, powdered, fused silica	35	0.84
Silicon carbide	150–650	0.83–0.96
Silver, polished, pure	40–625	0.020–0.032
Stainless steel		
Type 312, heated 300 h at 260°C	95–425	0.27–0.32
Type 301 with Armco black oxide	−25	0.75
Type 410, heated to 700 °C in air	35	0.13
Type 303, sandblasted	95	0.42
Titanium, 75A	95–425	0.10–0.19
75A, oxidized 300 h at 450 °C	35–425	0.21–0.25
Anodized	−25	0.73
Tungsten, filament, aged	27–3300	0.032–0.35
Zinc, pure, polished	225–325	0.045–0.053
Galvanized sheet	100	0.21

[a](energy emitted from a surface/energy emitted by a black surface at same temperature), with direction perpendicular to the surface
Source: Adapted from Modest, Michael F., Radiation, in Kreith, Frank and D. Yogi. Goswami, Eds. 2005. CRC Handbook of Mechanical Engineering, Second Edition (Boca Raton, CRC Press).

Table 25.10. Thermophysical properties of some heat-pipe fluids

(°C)	Heat (kJ/kg)	Density (kg/m³)	Density (kg/m³)	Conductivity (W/m°C)	Viscosity (cP)	Viscosity (cP × 10²)	Pressure (bars)	Specific Heat (kJ/kg°C)	Tension (N/m ×10²)
				Methanol					
-50	1194	843.5	0.01	0.21	1.7	0.72	0.01	1.2	3.26
50	1125	764.1	0.77	0.202	0.399	1.04	0.55	1.54	2.01
150	850	653.2	15.9	0.193	0.138	1.38	8.94	1.92	1.04
				Water					
20	2448	998	0.02	0.603	1	0.96	0.02	1.81	7.28
100	2258	958	0.6	0.68	0.28	1.27	1.01	2.01	5.89
200	1967	865	7.87	0.659	0.14	1.65	16.19	2.91	3.89
				Potassium					
350	2093	763.1	0.002	51.08	0.21	0.15	0.01	5.32	9.5
500	2040	725.4	0.031	45.08	0.17	0.17	0.05	5.32	8.44
800	1913	665.4	0.716	34.81	0.11	0.2	1.55	5.32	6.32
850	1883	653.1	1.054	33.31	0.1	0.21	2.34	5.32	5.92

Source: Adapted from Kreith, Frank, Ed. 2000. CRC Handbook of Thermal Engineering (Boca Raton, CRC Press).

Table 25.11. Energy density of different materials combined with water for cooling applications[a]

Absorption	Solid-Gas Reaction	Adsorption
NH_3 - H_2O: 110 Wh/kg	H_2O - Na_2S: 353 Wh/kg	H_2O - zeolite 4A: 23 Wh/kg
H2O - NaOH: 277 Wh/kg	H_2O - $MgCl_2$: 233 Wh/kg	H_2O - silica gel: 40 Wh/kg
		H_2O - $CaCl_2$: 271 Wh/kg
		H_2O - LiCl: 197 Wh/kg

[a]For the weight of the pairs when fully loaded with water.
Adapted from Bales, Chris, Ed. 2005. Thermal Properties of Materials for Thermo-chemical Storage of Solar Heat, A Report of IEA Solar Heating and Cooling programme - Task 32 "Advanced storage concepts for solar and low energy buildings," IEA Solar Heating and Cooling Programme, (Paris, IEA).

Table 25.12. Flammability and environmental characteristics of selected refrigerants

Refrigerant	Chemical Formula	ODP[a]	GWP[b] (CO_2 at 100 yr)	Atmos. Life (yr)	TLV[c] (ppm)	LFL[d] (%)
R-11 (CFC)	CCl_3F	1	4000	50	1000	None
R-12 (CFC)	CCl_2F_2	1	8500	102	1000	None
R-22 (HCFC)	$CHClF_2$	0.055	1700	13.3	1000	None
R-123 (HCFC)	$CHCl_2CF_3$	0.02	93	1.4	10	None
R-134a (HFC)	CH_2FCF_3	0	1300	14	1000	None
R-401A	R-22/152a/124	0.037	--	--	--	None
R-401a	R-125/143a/134a	0	3260	--	1000	None
R-407C	R-32/125/134a	0	--	--	1000	None
R-401A	R-32/125	0	--	--	1000	None
R-717 (Inorganic)	NH_3	0	0	<1	25	14.8
R-744 (Inorganic)	CO_2	0	1	50-200	5000	None
R-170 (Hydrocarbon)	C_2H_6	0	3	--	--	3.3
R-290 (Hydrocarbon)	C_3H_8	0	3	--	--	2.1

[a]ODP is the ratio of the impact on ozone of a chemical compared to the impact of a similar mass of CFC-11.
[b]GWP is the ratio of the warming caused by a substance to the warming caused by a similar mass of carbon dioxide.
[c]Vapor toxicity is expressed via the threshold limit value or "TLV." It is defined as the time weighted average exposure that a worker can have to a substance for a normal 40-h work week without any adverse health effects.
[d]Lower Flammability Limit: lower end of the concentration range of a flammable solvent at a given temperature and pressure for which air/vapor mixtures can ignite.
Source: Adpated from Barron, Randall F., Cryogenic Systems, in Kreith, Frank, Ed. 2000. CRC Handbook of Thermal Engineering (Boca Raton, CRC Press).

Table 25.13.   Performance characteristics of several high temperature refrigerants per ton of refrigeration

Refrigerant	Evap. Press. (psia)	Cond. Press. (psia)	Net Refrig. Effect (Btu/lb)	Mass Flow Rate (lb/minper ton)	Suction Vapor Specific Volume (ft3/lb)	Volume Rate Leaving Evap. (ft3/min)	Compr. Discharge Temp. (°F)	COP
R-11	2.9	18.3	67.21	2.98	12.24	36.43	110	5.02
R-123	2.3	15.9	61.19	3.27	14.08	46.02	94	4.84
R-12	26.5	108	50.25	3.98	1.46	5.83	100	4.75
R-134a	23.8	111.6	64.77	3.09	1.95	6.02	108	4.41
R-22	43	172.9	69.9	2.86	1.34	3.55	128	4.67
R-125	58.9	228.1	37.69	5.31	0.628	3.33	108	3.67
R-717	34.2	168.8	474.2	0.422	8.18	3.45	210	4.77
R-744	332.4	1045.4	57.75	3.46	0.264	0.914	156	2.81
R-170	236.4	674.7	69.27	2.89	0.534	1.54	123	2.72
R-290	42.4	156.8	169.6	1.66	2.46	4.09	98	4.41

Source: Adapted from Barron, Randall F., Cryogenic Systems, in Kreith, Frank, Ed. 2000. CRC Handbook of Thermal Engineering (Boca Raton, CRC Press).

PART IV

IMPACTS

Environment

Figures

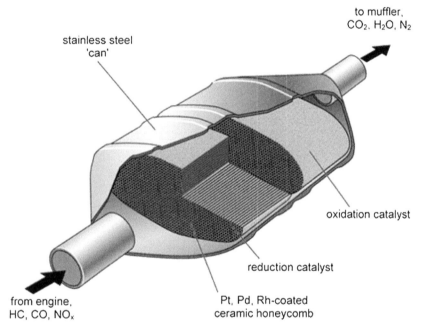

Figure 26.1. The automobile catalytic converter, which catalyzes the reduction of NO_x and hydrocarbons and the oxidation of carbon monoxide. It operates at ~200 °C during start-up and 800°C during cruising. Converters are typically oval, ~0.35 m long and ~0.25 m across the oval (sometimes round).
Source: Crundwell, Frank K., Michael S. Moats, Venkoba Ramachandran, Timothy G. Robinson, William G. Davenport. 2011. Platinum-Group Metals, Production, Use and Extraction Costs, Extractive Metallurgy of Nickel, Cobalt and Platinum Group Metals, (Oxford, Elsevier).

Handbook of Energy. http://dx.doi.org/10.1016/B978-0-08-046405-3.00026-7

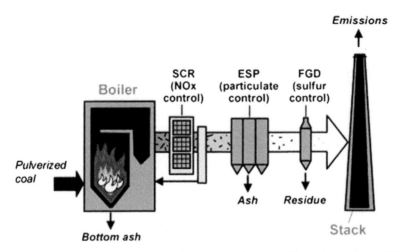

Figure 26.2. Typical pollution controls on a conventional pulverized coal-fired steam plant. ESP = electrostatic precipitator; FGD = flue gas desulfurization; SCR = selective catalytic reduction.
Source: United States Environmental Protection Agency

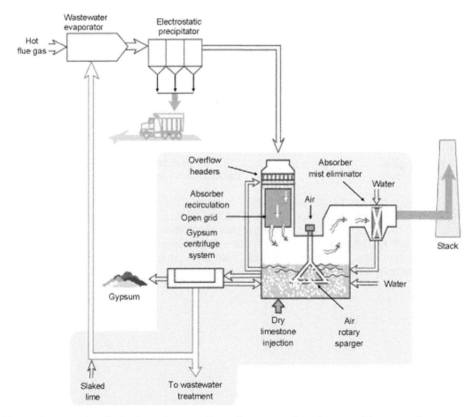

Figure 26.3. Flue gas desulfurization process using a limestone slurry in a scrubber vessel.
Source: Perry, Mildred B. 2004. Clean Coal Technology, In: Cutler J. Cleveland, Editor-in-Chief, Encyclopedia of Energy, (New York, Elsevier), Pages 343-357.

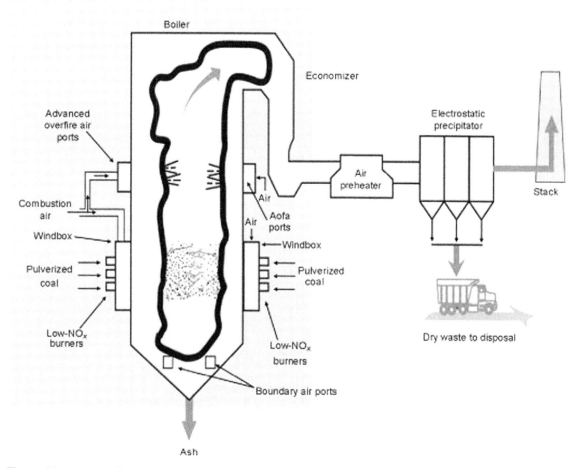

Figure 26.4. Low-NO$_x$ burners with advanced overfire air (AOFA) installation on a wall-fired boiler. *Source: Perry, Mildred B. 2004. Clean Coal Technology, In: Cutler J. Cleveland, Editor-in-Chief, Encyclopedia of Energy, (New York, Elsevier), Pages 343-357.*

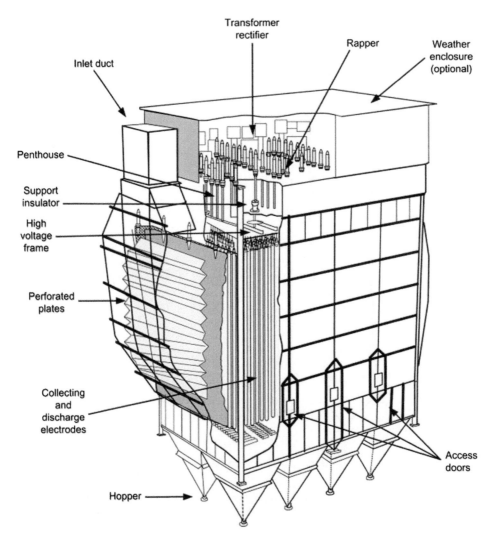

Figure 26.5. Typical electrostatic precipitator.
Source: Sadeghbeigi, Reza. 2012. Fluid Catalytic Cracking Handbook (Third Edition), (Oxford, Butterworth-Heinemann).

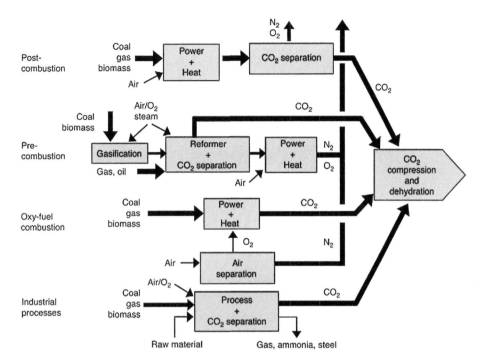

Figure 26.6. Principal routes for managing carbon dioxide emissions from power stations.
Source: Rand, D.A.J. and R.M. Del. 2009. Fuels - Hydrogen Production | Coal Gasification, In: Jürgen Garche, Editor-in-Chief, Encyclopedia of Electrochemical Power Sources, (Amsterdam, Elsevier), Pages 276-292.

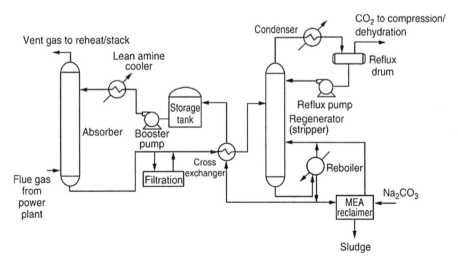

Figure 26.7. Process flow diagram for the amine separation process used in the post-combustion capture of CO_2. MEA = monoethanolamine
Source: Herzog, Howard and Dan Golomb. 2004. Carbon Capture and Storage from Fossil Fuel Use, In: Cutler J. Cleveland, Editor-in-Chief, Encyclopedia of Energy, (New York, Elsevier), Pages 277-287.

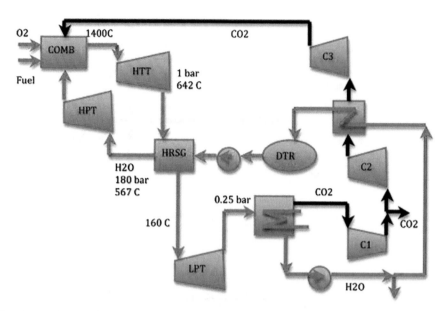

Figure 26.8. An oxyfuel combustion process for CO_2 capture that utilizes the Graz cycle. HTT: high-temperature turbine, LPT: low-pressure turbine, HPT: High-pressure turbine, C1–C3: CO_2 compressors.
Source: Ghoniem, Ahmed F. 2011. Needs, resources and climate change: Clean and efficient conversion technologies, Progress in Energy and Combustion Science, Volume 37, Issue 1, Pages 15-51.

Charts

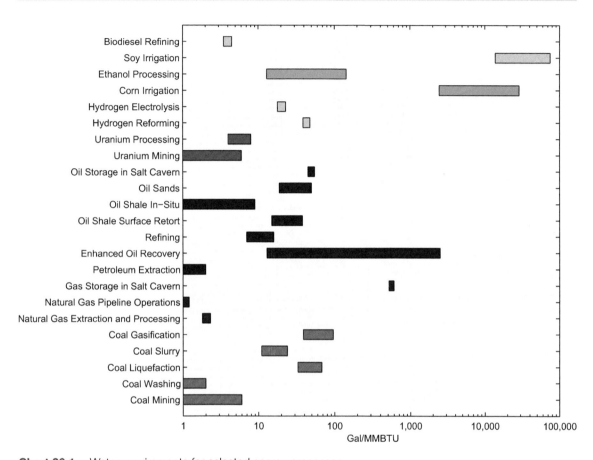

Chart 26.1.   Water requirements for selected energy processes.
Source: Data from United States Department of Energy, Office of Environmental Management. 2007. Energy Demands on Water Resources.

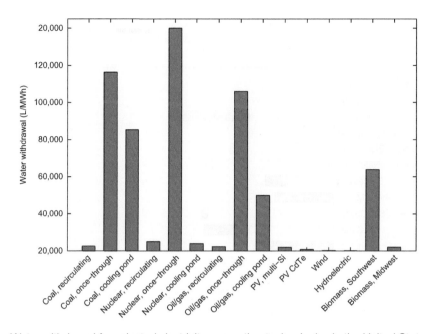

Chart 26.2. Water withdrawal for selected electricity generation technologies in the United States.
Source: Data from Fthenakis, Vasilis, Hyung Chul Kim. 2010. Life-cycle uses of water in U.S. electricity generation, Renewable and Sustainable Energy Reviews, Volume 14, Issue 7, Pages 2039-2048.

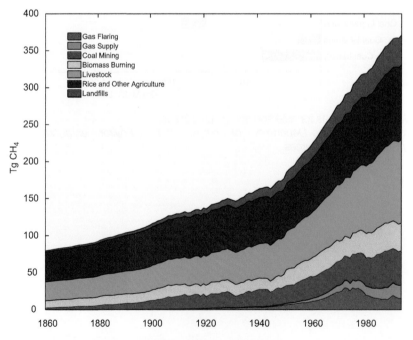

Chart 26.3. Anthropogenic methane emissions by source, 1860-1994.
Source: Data from Stern D. I. and R. K. Kaufmann. 1998. Estimates of Global Anthropogenic Methane Emissions: 1860-1994, Trends Online: A Compendium of Data on Global Change, Carbon Dioxide Information Analysis Center, Oak Ridge National Laboratory, U.S. Department of Energy, Oak Ridge, TN.

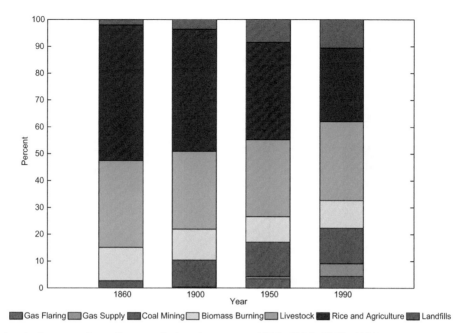

Chart 26.4. Anthropogenic methane emissions by source, 1860, 1900, 1950, 1990.
Source: Data from Stern D. I. and R. K. Kaufmann. 1998. Estimates of Global Anthropogenic Methane Emissions: 1860-1994, Trends Online: A Compendium of Data on Global Change, Carbon Dioxide Information Analysis Center, Oak Ridge National Laboratory, U.S. Department of Energy, Oak Ridge, TN.

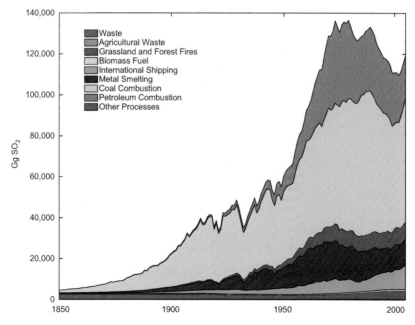

Chart 26.5. Anthropogenic sulfur dioxide emissions by source, 1850-2005.
Source: Data from Smith, S. J., J. van Aardenne, Z. Klimont, R. J. Andres, A. Volke, and S. Delgado Arias. 2011. Anthropogenic sulfur dioxide emissions: 1850–2005, Atmos. Chem. Phys., 11, 1101-1116.

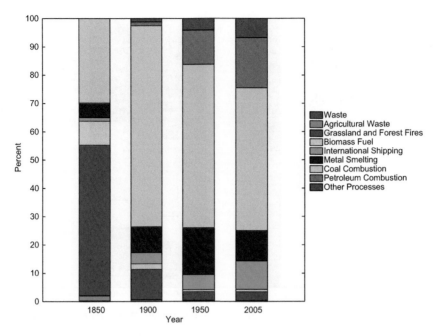

Chart 26.6. Anthropogenic sulfur dioxide emissions by source, 1850, 1900, 1950, 2005.
Source: Data from Smith, S. J., J. van Aardenne, Z. Klimont, R. J. Andres, A. Volke, and S. Delgado Arias. 2011. Anthropogenic sulfur dioxide emissions: 1850–2005, Atmos. Chem. Phys., 11, 1101-1116.

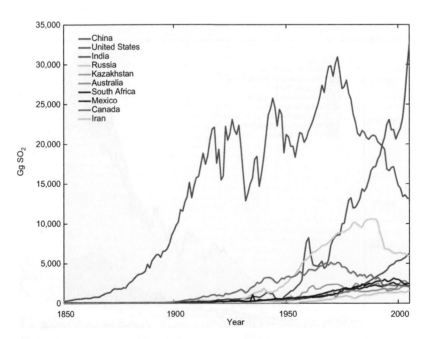

Chart 26.7. Anthropogenic sulfur dioxide emissions, top 10 emitting nations, 1850-2005.
Source: Data from Smith, S. J., J. van Aardenne, Z. Klimont, R. J. Andres, A. Volke, and S. Delgado Arias. 2011. Anthropogenic sulfur dioxide emissions: 1850–2005, Atmos. Chem. Phys., 11, 1101-1116.

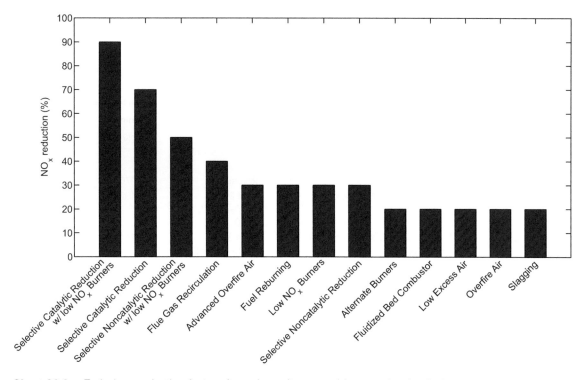

Chart 26.8.   Emissions reduction factors for various nitrogen oxide control technologies.
Source: Data from United States Environmental Protection Agency.

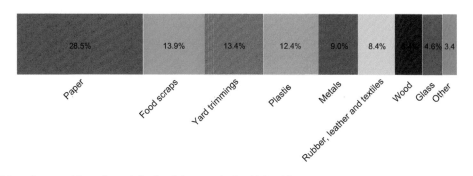

Chart 26.9.   Composition of municipal solid waste in the United States, 2010.
Source: Data from United States Environmental Protection Agency.

Tables

Table 26.1. Some significant oil spills

Spill / Vessel	Location	Dates	Oil released (metric tons)	
			Minimum	Maximum
Lakeview Gusher	United States, Kern County, California	March 14, 1910–September 10, 1911	1,230,000	1,230,000
Gulf War oil spill	Iraq, Persian Gulf	23-Jan-91	270,000	820,000
Deepwater Horizon	United States, Gulf of Mexico	April 20, 2010–July 15, 2010	492,000	627,000
Ixtoc I oil spill	Mexico, Bay of Campeche, Gulf of Mexico	June 3, 1979–March 23, 1980	454,000	480,000
Atlantic Empress	Trinidad and Tobago	19-Jul-79	287,000	287,000
Fergana Valley	Uzbekistan	2-Mar-92	285,000	285,000
Nowruz Field Platform	Iran, Persian Gulf	4-Feb-83	260,000	260,000
ABT Summer	Angola, 700 nmi (1,300 km; 810 mi) offshore	28-May-91	260,000	260,000
Castillo de Bellver	South Africa, Saldanha Bay	6-Aug-83	252,000	252,000
Amoco Cadiz	France, Brittany	16-Mar-78	223,000	227,000
MT Haven	Italy, Mediterranean Sea near Genoa	11-Apr-91	144,000	144,000
Odyssey	Canada, 700 nmi (1,300 km; 810 mi) off Nova Scotia	10-Nov-88	132,000	132,000
Torrey Canyon	United Kingdom, Isles of Scilly	18-Mar-67	80,000	119,000
Sea Star	Iran, Gulf of Oman	19-Dec-72	115,000	115,000
Exxon Valdez	United States, Prince William Sound, Alaska	24-Mar-89	37,000	104,000
Irenes Serenade	Greece, Pylos	23-Feb-80	100,000	100,000
Urquiola	Spain, A Coruña	12-May-76	100,000	100,000
Greenpoint, Brooklyn oil spill	United States, Newtown Creek, Greenpoint, Brooklyn, New York	1940s–present	55,200	97,400
Sea Empress	United Kingdom, Pembrokeshire	15-Feb-96	40,000	72,000
Prestige oil spill	Spain, Galicia	13-Nov-02	63,000	63,000
Santa Barbara oil spill	United States, Santa Barbara, California	28-Jan-69	10,000	14,000

Source: Deepwater Horizon oil spill, : Encyclopedia of Earth, <http://www.eoearth.org/article/Deepwater_Horizon_oil_spilltopic=50364>; Wikipedia, List of oil spills, <en.wikipedia.org/wiki/List_of_oil_spills>.

Table 26.2. Air pollutant emission factors for fuel combustion in the United States

	VOCs	NOₓ	PM₁₀	PM₂.₅	CO
Unit: grams(g)/10⁶Btu of fuel burned					
NG-fired large industrial boilers	1.6	52	3.2	3.2	16.4
NG-fired small industrial boilers	2.4	25	3	3	28.8
Diesel-fired industrial boilers	1.2	82	42.5	38	16.7
Diesel-fired stationary reciprocating engines	47.7	463	39.5	35.5	229.4
Diesel-powered farming tractors	47.7	463	39.5	35.5	229.4
Coal-fired industrial boilers	2.1	120	85	45	76.2
Biomass-fired industry boilers	5.3	110	12.7	6.3	76.8
Unit: g/kilowatt-hour(kWh) of electricity generated					
Oil-fired utility boilers	0.03	1.72	0.025	0.019	0.2
NG-fired utility boilers	0.01	0.41	0.014	0.014	0.18
NG-fired simple-cycle turbines	0.03	0.42	0.016	0.016	0.21
NG-fired combined-cycle turbines	0.03	0.09	0.016	0.016	0.21
Coal-fired utility boilers	0.01	0.87	0.073	0.037	0.19

Source: Huo, Hong, Ye Wu, Michael Wang. 2009. Total versus urban: Well-to-wheels assessment of criteria pollutant emissions from various vehicle/fuel systems, Atmospheric Environment, Volume 43, Issue 10, Pages 1796–1804.

Table 26.3. Carbon dioxide uncontrolled emission factors (pounds of CO_2 per million Btu)

Fuel	Emissions Factor (Pounds of CO_2 Per Million Btu)
Bituminous Coal	205.3
Distillate Fuel Oil	161.386
Geothermal	16.59983
Jet Fuel	156.258
Kerosene	159.535
Lignite Coal	215.4
Municipal Solid Waste	91.9
Natural Gas	117.08
Petroleum Coke	225.13
Propane Gas	139.178
Residual Fuel Oil	173.906
Synthetic Coal	205.3
Subbituminous Coal	212.7
Tire-Derived Fuel	189.538
Waste Coal	205.3
Waste Oil	210

Source: United States Department of Energy, Information Energy Administration.

Table 26.4. Sulfur dioxide uncontrolled emission factors

Fuel	Emissions units[a]	Combustion system type/Firing configuration							
		Cyclone boiler	Fluidized bed boiler	Opposed firing boiler	Spreader stoker boiler	Tangential boiler	All Other boiler types	Combustion turbine	Internal combustion engine
Agricultural Byproducts	Lbs per ton	0.08	0.01	0.08	0.08	0.08	0.08	NA	NA
Blast Furnace Gas	Lbs per MMCF	0.6	0.06	0.6	0.6	0.6	0.6	0.6	0.6
Bituminous Coal	Lbs per ton	38	3.8	38	38	38	38	NA	NA
Black Liquor	Lbs per ton	7	0.7	7	7	7	7	NA	NA
Distillate Fuel Oil	Lbs per MG	157	15.7	157	157	157	157	140	140
Jet Fuel	Lbs per MG	157	15.7	157	157	157	157	140	140
Kerosene	Lbs per MG	157	15.7	157	157	157	157	140	140
Landfill Gas	Lbs per MMCF	0.6	0.06	0.6	0.6	0.6	0.6	0.6	0.6
Lignite Coal	Lbs per ton	30	3	30	30	30	30	NA	NA
Municipal Solid Waste	Lbs per ton	1.7	0.17	1.7	1.7	1.7	1.7	NA	NA
Natural Gas	Lbs per MMCF	0.6	0.06	0.6	0.6	0.6	0.6	0.6	0.6
Petroleum Coke	Lbs per ton	39	3.9	39	39	39	39	NA	NA
Propane Gas	Lbs per MMCF	0.6	0.06	0.6	0.6	0.6	0.6	0.6	0.6
Residual Fuel Oil	Lbs per MG	157	15.7	157	157	157	157	NA	NA
Synthetic Coal	Lbs per ton	38	3.8	38	38	38	38	NA	NA
Sludge Waste	Lbs per ton	2.8	0.28	2.8	2.8	2.8	2.8	NA	NA
Subbituminous Coal	Lbs per ton	35	3.5	35	38	35	35	NA	NA
Tire-Derived Fuel	Lbs per ton	38	3.8	38	38	38	38	NA	NA
Waste Coal	Lbs per ton	30	3	30	30	30	30	NA	NA
Wood Waste Liquids	Lbs per MG	157	15.7	157	157	157	157	140	140
Wood Waste Solids	Lbs per ton	0.29	0.08	0.29	0.08	0.29	0.29	NA	NA
Waste Oil	Lbs per MG	147	14.7	147	147	147	147	NA	NA

[a]Lbs = pounds, MMCF = million cubic feet, MG = thousand gallons.
Source: United States Department of Energy, Information Energy Administration.

Table 26.5.   Anthopogenic sulfur dioxide emissions, top 20 nations, 1850–2005

	1850	1900	1950	1990	2005	Rank 2005	Rank 1950	Change 1990–2005
China	161.6	147.8	1,070.0	17,194.2	32,673.4	1	9	90.0%
USA	306.9	9,146.0	21,215.1	20,986.5	13,106.4	2	1	-37.5%
India	35.0	102.1	424.6	3,301.7	6,274.5	3	22	90.0%
Russia	39.7	277.7	2,565.1	10,631.9	5,974.8	4	5	-43.8%
Kazakhstan	0.6	5.0	429.9	2,426.9	2,580.9	5	20	6.3%
Australia	9.2	124.1	418.0	1,594.0	2,522.1	6	23	58.2%
South Africa	1.2	29.5	588.2	2,283.5	2,476.7	7	16	8.5%
Mexico	1.5	51.8	658.1	2,729.2	2,145.2	8	14	-21.4%
Canada	4.1	198.9	3,297.8	3,079.5	2,024.2	9	4	-34.3%
Iran	1.4	3.1	178.3	1,213.6	1,598.0	10	32	31.7%
Indonesia	3.1	13.0	42.2	348.8	1,535.0	11	54	340.1%
Turkey	4.8	41.1	148.6	1,573.2	1,483.9	12	35	-5.7%
Brazil	1.7	3.8	211.7	1,643.3	1,437.8	13	29	-12.5%
Saudi Arabia	0.3	0.4	58.3	810.0	1,378.8	14	48	70.2%
Ukraine	5.1	132.3	1,607.3	4,121.6	1,342.6	15	6	-67.4%
Spain	21.7	321.9	454.6	2,161.8	1,259.1	16	19	-41.8%
Poland	20.9	360.3	1,044.1	3,435.9	1,232.4	17	10	-64.1%
Chile	26.1	48.7	811.8	2,324.4	1,158.8	18	13	-50.1%
Peru	0.2	9.1	97.4	674.8	1,110.6	19	39	64.6%
Serbia and Montenegro	1.2	12.9	396.0	1,083.4	1,077.0	20	24	-0.6%
World	4,605	22,702	60,188	131,413	119,602			-9.0%

Source: Smith, S. J., J. van Aardenne, Z. Klimont, R. J. Andres, A. Volke, and S. Delgado Arias. 2011. Anthropogenic sulfur dioxide emissions: 1850–2005, Atmos. Chem. Phys., 11, 1101–1116.

Table 26.6.   Global anthropgenic sulfur dioxide emissions, by source, 1850–2005 (Gg SO_2)

Source	1850	1900	1950	1990	2005	Change 1990–2005
Waste	11	19	31	49	54	8.9%
Ag Waste Burning	83	110	141	211	205	–3.0%
Grassland and Forest Fires	2,447	2,447	1,859	3,357	3,836	14.3%
Biomass fuel	395	457	471	794	982	23.7%
International Shipping	61	892	3,247	7,041	12,078	71.6%
Metal Smelting	234	2,062	9,908	13,470	12,768	–5.2%
Other Process	0	272	2,473	7,948	8,070	1.5%
Coal Combustion	1,374	16,137	34,801	67,483	60,307	–10.6%
Petroleum Combustion	0	306	7,258	31,060	21,302	–31.4%
Total	4,605	22,702	60,188	131,413	119,602	–9.0%

Source: Smith, S. J., J. van Aardenne, Z. Klimont, R. J. Andres, A. Volke, and S. Delgado Arias. 2011. Anthropogenic sulfur dioxide emissions: 1850–2005, Atmos. Chem. Phys., 11, 1101–1116.

Table 26.7. Summary of commercially available and demonstrated processes for SO$_2$ removal

Process	Efficiency	Capital Cost	Reagent Cost	Complexity	Comments
Wet limestone	High	High	Low	High	Low reagent cost offsets high cost of operation and maintenance in very large systems.
Wet soda ash/ caustic	High	Moderate	High	Moderate	No slurry or solids handling. Very effective for smaller systems.
Lime spray dryer	Moderate	Moderate	Moderate	Moderate	Mature technology widely used for industrial applications. Subject to deposits accumulation during upset conditions.
Circulating lime reactor	Moderate to high	Moderate	Moderate	Moderate	High solids circulation rate prevents deposits accumulation, enabling slightly higher reactivity or lower reagent cost than lime spray dryer.
Sodium bicarbonate/ trona injection[a]	Moderate	Low	High	Low	Lower cost of trona an advantage in HCl applications, but mostly offset by lower reactivity in SO applications.

[a]trona (trisodium hydrogendicarbonate dihydrate), $Na_3(CO_3)(HCO_3) \bullet 2H2O$.
Source: Adapted from Schnelle, Karl B. and Charles A. Brown. 2001. Air Pollution Control Technology Handbook (Boca Raton, CRC Press).

Table 26.8. Global nitrous oxide (N$_2$O) emissions, 1500–1994 (Tg N$_2$O–N yr^{-1})

Year	Natural	Energy	Industry	Biomass Burning	Agriculture	Total Anthropogenic	Total Global	Net additions to Atmosphere
1500	10.4	0	0	0.1	0.5	0.6	11	0
1600	10.4	0	0	0.1	0.6	0.7	11	0
1700	10.3	0	0	0.1	0.6	0.7	11	0
1800	9.9	0	0	0.1	1	1.1	11	0
1850	9.6	0	0	0.1	1.3	1.4	11	0
1900	9.6	0	0	0.1	1.6	1.7	11.3	0.3
1930	9.6	0	0	0.2	2.2	2.4	12	1
1950	9.6	0.1	0	0.2	2.9	3.1	12.8	1.8
1960	9.6	0.2	0.1	0.2	3.4	3.9	13.5	2.5
1970	9.6	0.3	0.3	0.3	4.2	5.1	14.7	3.7
1975	9.6	0.4	0.3	0.3	4.8	5.8	15.4	4.4
1980	9.6	0.5	0.4	0.3	5.3	6.5	16.1	5.1
1985	9.6	0.6	0.4	0.4	5.8	7.2	16.9	5.9
1990	9.6	0.7	0.5	0.5	6.2	8	17.6	6.6
1994	9.6	0.9	0.3	0.6	6.2	8	17.7	6.7

Natural emissions include N$_2$O from soils under natural vegetation, oceans, aquatic systems, and formation in the atmosphere. Emissions from energy, industry, and biomass burning are assumed to be 0 before 1900 except 0.1 Tg N yr^{-1} from biomass burning. Net additions to atmosphere are total emissions minus 11 Tg N (1500–1800 emissions).
Source: Adapted from Kroeze, Carolien, Arvin Mosier, Lex Bouwman. 1999. Closing the global N$_2$O budget: A retrospective analysis 1500–1994, Global Biogeochemical Cycles, 13(1): 1–8.

Table 26.9. Nitrogen oxides uncontrolled emission factors

Fuel and EIA fuel code	Emissions units[a]	Combustion system type/firing configuration							
		Cyclone boiler	Fluidized bed boiler	Opposed firing boiler	Spreader stoker boiler	Tangential boiler	All Other boiler types	Combustion turbine	Internal combustion engine
Agricultural Byproducts	Lbs per ton	1.2	1.2	1.2	1.2	1.2	1.2	NA	NA
Blast Furnace Gas	Lbs per MMCF	15.4	15.4	15.4	15.4	15.4	15.4	30.4	256.55
Bituminous Coal	Lbs per ton	33	5	12 [31]	11	10.0 [14.0]	12.0 [31.0]	NA	NA
Black Liquor	Lbs per ton	1.5	1.5	1.5	1.5	1.5	1.5	NA	NA
Distillate Fuel Oil	Lbs per MG	24	24	24	24	24	24	122	443.8
Jet Fuel	Lbs per MG	24	24	24	24	24	24	118	432
Kerosene	Lbs per MG	24	24	24	24	24	24	118	432
Landfill Gas	Lbs per MMCF	72.44	72.44	72.44	72.44	72.44	72.44	144	1215.22
Lignite Coal	Lbs per ton	15	3.6	6.3	5.8	7.1	6.3	NA	NA
Municipal Solid Waste	Lbs per ton	5	5	5	5	5	5	NA	NA
Natural Gas	Lbs per MMCF	280	280	280	280	170	280	328	2768
Other Biomass Gas	Lbs per MMCF	112.83	112.83	112.83	112.83	112.83	112.83	313.6	2646.48
Other Biomass Liquids	Lbs per MG	19	19	19	19	19	19	NA	NA

Continued

Table 26.9. Continued

Fuel and EIA fuel code	Emissions units[a]	Combustion system type/firing configuration							
		Cyclone boiler	Fluidized bed boiler	Opposed firing boiler	Spreader stoker boiler	Tangential boiler	All Other boiler types	Combustion turbine	Internal combustion engine
Other Biomass Solids	Lbs per ton	2	2	2	2	2	2	NA	NA
Other Gases	Lbs per MMCF	152.82	152.82	152.82	152.82	152.82	152.82	263.82	2226.41
Other	Lbs per MMCF	280	280	280	280	170	280	328	2768
Petroleum Coke	Lbs per ton	21	5	21	21	21	21	NA	NA
Propane Gas	Lbs per MMCF	215	215	215	215	215	215	330.75	2791.22
Residual Fuel Oil	Lbs per MG	47	47	47	47	32	47	NA	NA
Synthetic Coal	Lbs per ton	33	5	12 [31]	11	10.0 [14.0]	12.0 [31.0]	NA	NA
Sludge Waste	Lbs per ton	5	5	5	5	5	5	NA	NA
Subbituminous Coal	Lbs per ton	17	5	7.4 [24]	8.8	7.2	7.4 [24.0]	NA	NA
Tire-Derived Fuel	Lbs per ton	33	5	12 [31]	11	10.0 [14.0]	12.0 [31.0]	NA	NA
Waste Coal	Lbs per ton	15	3.6	6.3	5.8	7.1	6.3	NA	NA
Wood Waste Liquids	Lbs per MG	5.43	5.43	5.43	5.43	5.43	5.43	NA	NA
Wood Waste Solids	Lbs per ton	2.51	2	2.51	1.5	2.51	2.51	NA	NA
Waste Oil	Lbs per MG	19	19	19	19	19	19	NA	NA

[a]Lbs = pounds, MMCF = million cubic feet, MG = thousand gallons.
Factors for Wet-Bottom Boilers are in Brackets; All Other Boiler Factors are for Dry-Bottom.
Source: United States Department of Energy, Information Energy Administration.

Table 26.10. Global mercury emissions by natural and anthropogenic sources

Source	Emissions Mg yr^{-1}
Natural (for 2008)	
Oceans	2682
Lakes	96
Forests	342
Tundra/Grassland/Savannah/Prairie/Chaparral	448
Desert/Metalliferous/Non-vegetated Zones	546
Agricultural areas	128
Evasion after mercury depletion events	200
Biomass burning	675
Volcanoes and geothermal areas	90
Total-natural	**5207**
Anthropogenic	
Coal and oil combustion	810
Non-ferrous metal prod.	310
Pig iron and steel prod.	43
Cement production	236
Caustic soda production	163
Mercury production	50
Artisanal gold mining prod.	400
Waste disposal	187
Coal bed fires	32
VCM production	24
Other	65
Total-Anthropogenic	**2320**

Source: Adapted from Pirrone, N., S. Cinnirella, X. Feng, R. B. Finkelman, H. R. Friedli, J. Leaner, R. Mason, A. B. Mukherjee, G. B. Stracher, D. G. Streets, and K. Telmer. 2010. Global mercury emissions to the atmosphere from anthropogenic and natural sources, Atmos. Chem. Phys. Discuss., 10, 4719–4752.

Table 26.11. Global emissions of total mercury from major anthropogenic sources by region (Mg yr^{-1})

	SC[a]	NFMP	PISP	CP	CSP	MP	GP	WD	O	T	Ref. year
S. Africa	32.6	0.3	1.3	3.8	0	0	0.3	0.6	1.3	40.2	2004
China	268	203.3	8.9	35	0	27.5	44.7	14.1	7.6	609.1	2003
India	124.6	15.5	4.6	4.7	6.2	0	0.5	77.4	7.5	240.9	2004
Australia	2.2	11.6	0.8	0.9	0	0	0.3	0.2	0.6	16.6	2005
Europe	76.6	18.7	0	18.8	6.3	0	0	10.1	14.7	145.2	2005
Russia	46	5.2	2.6	3.9	2.8	0	4.3	3.5	1.5	69.8	2005
N. America	65.2	34.7	12.8	15.1	10.3	0	0	13	1.7	152.8	2005
S. America	8	13.6	1.8	6.4	2.2	0	16.2	0	1.5	49.7	2005
Total	623.2	302.9	32.8	88.6	27.8	27.5	66.3	118.9	36.4	1324.3	
Rest of the world	186.8	7.1	10.4	147.1	135.1	22.5	333.7	68.5	28.2	939.4	2006
Total	810	310	43.2	235.7	162.9	50	400.4	187.4	64.6	2319.7[b]	

[a]SC, Stationary combustion; NFMP, Non-ferrous metal production; PISP, Pig iron and steel production; CP, Cement production; CSP, Caustic soda production; MP, Mercury production; GP, Gold production; WD, Waste disposal; CB, Coal-bed fires; VCM, Vinyl chloride monomer production; O, Other; T, Total.
[b]This sum considers also CB and VCM estimates, which account for 32.0 Mgyr^{-1} and 24 Mgyr^{-1}, respectively. Totals for countries do not include these values.
Source: Adapted from Pirrone, N., S. Cinnirella, X. Feng, R. B. Finkelman, H. R. Friedli, J. Leaner, R. Mason, A. B. Mukherjee, G. B. Stracher, D. G. Streets, and K. Telmer. 2010. Global mercury emissions to the atmosphere from anthropogenic and natural sources, Atmos. Chem. Phys. Discuss., 10, 4719–4752.

Table 26.12. Power plants releasing the most carbon dioxide in 2007

	Company	Facility name	CO$_2$ emissions 10^6 tons CO$_2$	Energy Generated TWh	Carbon Intensity lbs CO$_2$/MWh	Country	State
1	TAIWAN POWER CO	TAICHUNG	39.7	39.2	2,022.0	Taiwan (China)	Taichung County
2	BOT ELEKTROWNIA BELCHATOW SA	BELCHATOW	34.6	32.8	2,110.0	Poland	Lodz
3	HUADIAN POWER INTL CORP	ZOUXIAN	33.4	34.5	1,940.4	China	Shandong
4	RWE POWER AG	NIEDERAUSSEM	30.4	29.6	2,056.1	Germany	Nordrhein-Westfalen
5	MAILIAO POWER CO	MAILIAO FP	29.9	32.9	1,817.0	Taiwan (China)	Yunlin County
6	KOREA WESTERN POWER (KOWEPO)	TAEAN	29.5	32.4	1,819.1	South Korea	Ch'ungch'ong-namdo
7	CHUBU ELECTRIC POWER CO	HEKINAN	27.5	30	1,831.2	Japan	Aichi
8	VATTENFALL EUROPE AG	JANSCHWALDE	27.4	25.8	2,123.5	Germany	Brandenburg

Table 26.12. Continued

	Company	Facility name	CO₂ emissions 10⁶ tons CO₂	Energy Generated TWh	Carbon Intensity lbs CO₂/ MWh	Country	State
9	GEORGIA POWER CO	SCHERER	27.2	26.5	2,058.7	United States	Georgia
10	ESKOM	KENDAL	26.8	25.5	2,103.5	South Africa	Mpumalanga
11	HUANENG POWER INTERNATIONAL	HUANENG YUHUAN	26.4	28.8	1,837.9	China	Zhejiang
12	PT INDONESIA POWER - SURALAYA	SURALAYA	25.8	24.5	2,110.1	Indonesia	Banten
13	KOREA EAST-WEST POWER CO	TANGJIN	24.7	26.7	1,849.5	South Korea	Ch'ungch'ong-namdo
14	DATANG INTL POWER GEN CO	TUOKETUO-1	24.7	23.3	2,120.0	China	Inner Mongolia
15	ESKOM	MAJUBA	24.4	23	2,121.9	South Africa	Mpumalanga
16	RWE POWER AG	FRIMMERSDORF	24.1	21.2	2,272.2	Germany	Nordrhein-Westfalen
17	ESKOM	MATIMBA	24	22.6	2,125.6	South Africa	Mpumalanga
18	KOREA SOUTHEAST POWER CO	SAMCHONPO	24	23.7	2,030.7	South Korea	Kyongsang-namdo
19	ALABAMA POWER CO	MILLER	23.7	23	2,066.0	United States	Alabama
20	KOREA MIDLAND POWER (KOMIPO)	PORYONG	23.5	27.8	1,687.5	South Korea	Ch'ungch'ong-namdo

Source: Carbon Monitoring for Action (CARMA), power plant database, <http://carma.org/>.

Table 26.13. Geologic CO_2 storage potential by region and type of storage (Pg C)

	Coal basins	Depleted oil pools	Gas basins	Deep Saline formations (On-shore)	Deep Saline formations (Off-shore)	Total
USA	16	3	10	745	248	1,022
Canada	1	0	1	273	68	344
Western Europe	1	2	11	20	39	73
Japan	0	0	0	0	0	0.3
Australia and New Zealand	8	0	3	56	130	196
Former Soviet Union	5	6	70	101	378	560
China	4	1	2	90	9	106
Middle East	0	9	52	61	4	126
Africa	2	4	17	32	63	118
Latin America	1	4	13	51	15	85
Southeast Asia	7	1	8	33	49	97
Eastern Europe	1	0	2	29	3	35
Korea	0	0	0	0	0	0.3
India	2	0	2	51	51	105
Total capacity, all grades:	48	31	190	1,540	1,057	2,867

Source: Adapted from Dooley, J.J., S.H. Kim, J.A. Edmonds, S.J. Friedman, M.A. Wise. 2005. A first-order global geological CO_2-storage potential supply curve and its application in a global integrated assessment model, In: E.S. Rubin, D.W. Keith, C.F. Gilboy, M. Wilson, T. Morris, J. Gale and K. Thambimuthu, Editors, Greenhouse Gas Control Technologies 7, (Oxford, Elsevier Science Ltd), Pages 573–581.

Table 26.14. Cost of CO_2 avoidance from coal combustion with different capture technologies ($/metric ton)

	Membrane	Adsorbent
Flourine solvent, SC[b]	60.8	60.8
Advanced flourine solvent, SC	52.6	52.6
MHI KS-1 solvent, SC[c]	54	54.1
TDA, SC[e]		46.3
TDA, USC[e]		41.3
TDA, AUSC[e]		38.1
TDA, AUSC, adv. compression[e]		34.9
MTR, AUSC, adv. compression, low risk[ef]		30.6
MTR, SC[d]	42.3	
MTR, USC[d]	37.6	
MTR, AUSC[d]	33.4	
MTR, AUSC, adv. compression[d]	31.7	
MTR, AUSC, adv. compression, low risk[df]	27.5	

[a]Assumes a mid-Western, Greenfield site using Illinois No. 6 coal. The nominal plant capacity for all cases is 550 MW with all plants using an evaporative cooling tower. The plants are designed for 90 percent CO_2 capture; capacity factor = 85 percent. The cost for transport, storage, and monitoring is not included.
SC = supercritical steam cycle.
AUSC = advanced ultra-supercritical steam cycle.
[b]Fluor Econamine technology, <http://www.fluor.com/econamine/
[c]Mitsubishi Heavy Industries, Ltd CO2 recovery, <http://www.mhi.co.jp/en/products/category/co2_recovery_plants.html>.
[d]Advanced membrane technology by MTR, Inc., <http://www.mtrinc.com/>.
[e]Advanced adsorbent process by TDA Research, Inc., <http://www.tda.com/>
[f]low risk finance structure for investor-owned utilities.
Source: Data from United States Department of Energy, National Energy Technology Laboratory. 2012. Current and Future Technologies for Power Generation with Post-Combustion Carbon Capture, DOE/NETL-2012/1557.

Table 26.15. Particulate emissions from biomass combustion systems

Appliance	Fuel	Environmental control	Efficiency to end-use %	Particulate emission gMWh⁻¹
Fireplace	Hardwood		7	31000
Woodstove	Hardwood		54	6048
Boiler	Bagasse		55	4929
Boiler	Bagasse	Multiclone	55	3501
Woodstove	Hardwood	Catalyst	68	1500
3R stove	Eucalyptus		18	1267
Pellet stove	Hardwood		78	468
Boiler	Bagasse	Scrubber	55	476
Boiler	Mixed wood	ESP	23	191
Gas burner	Biogas		57	88
Haslev CHP	Straw	Filter	84	88
Combined Cycle	Natural gas		58	4.5

Source: Overend, Ralph P. 2004. Chapter 3 - Heat, Power and Combined Heat and Power, In: Ralph E.H. Sims, Editor(s), Bioenergy Options for a Cleaner Environment, (Oxford, Elsevier), Pages 63–102.

Table 26.16. Comparison of particulate removal systems in electric power generation

Type of Collector	Particle Size Range (µm)	Removal Efficiency	Space Required	Max. Temp. (°C)	Pressure Drop (cm H₂O)
Baghouse (cotton bags)	0.1–0.1	Fair	Large	80	10
	1.0–10.0	Good	Large	80	10
	10.0–50.0	Excellent	Large	80	10
Baghouse (Dacron, nylon, Orlon)	0.1–1.0	Fair	Large	120	12
	1.0–10.0	Good	Large	120	12
	10.0–50.0	Excellent	Large	120	12
Baghouse (glass fiber)	0.1–1.0	Fair	Large	290	10
	1.0–10.0	Good	Large	290	10
	10.0–50.0	Good	Large	290	10
Baghouse (Teflon)	0.1–1.0	Fair	Large	260	20
	1.0–10.0	Good	Large	260	20
	10.0–50.0	Excellent	Large	260	20
Electrostatic precipitator	0.1–1.0	Excellent	Large	400	1
	1.0–10.0	Excellent	Large	400	1
	10.0–50.0	Good	Large	400	1
Standard cyclone	0.1–1.0	Poor	Large	400	5
	1.0–10.0	Poor	Large	400	5
	10.0–50.0	Good	Large	400	5
High-efficiency cyclone	0.1–1.0	Poor	Moderate	400	12
	1.0–10.0	Fair	Moderate	400	12
	10.0–50.0	Good	Moderate	400	12

Continued

Table 26.16. Continued

Type of Collector	Particle Size Range (μm)	Removal Efficiency	Space Required	Max. Temp. (°C)	Pressure Drop (cm H$_2$O)
Spray tower	0.1–1.0	Fair	Large	540	5
	1.0–10.0	Good	Large	540	5
	10.0–50.0	Good	Large	540	5
Impingement scrubber	0.1–1.0	Fair	Moderate	540	10
	1.0–10.0	Good	Moderate	540	10
	10.0–50.0	Good	Moderate	540	10
Venturi scrubber	0.1–1.0	Good	Small	540	88
	1.0–10.0	Excellent	Small	540	88
	10.0–50.0	Excellent	Small	540	88
Dry scrubber	0.1–1.0	Fair	Large	500	10
	1.0–10.0	Good	Large	500	10
	10.0–50.0	Good	Large	500	10

Source: Adapted from Kreider, Jan F. 2005. Air Pollution Control, In Kreith, Frank and D. Yogi. Goswami, Eds. CRC Handbook of Mechanical Engineering, Second Edition (Boca Raton, CRC Press).

Table 26.17. Technologies to control organic vapors from point sources

Device	Inlet Concentration PPMV	Efficiency	Advantages	Disadvantages
Absorption	250	90%	Especially good for inorganic acid gases	Limited applicability
	1000	95%		
	5000	98%		
Adsorption	200	50%	Low capital investment	Selective applicability
	1000	90–95%	Good for solvent recovery	Moisture and temperature constraints
	5000	98%		
Condensation	500	50%	Good for product or solvent recovery	Limited applicability
	10,000	95%		
Thermal incineration	20	95%	High destruction efficiency	No organics can be recovered
	100	99%	Wide applicability Can recover heat energy	Capital intensive
Catalytic incineration	50	90%	High destruction efficiency	No organics can be recovered
	100	>95%	Can be less expensive than thermal incineration	Technical limitations that can poison
Flares		>98%	High destruction efficiency	No organics can be recovered
				Large emissions only

Source: Adapted from Schnelle, Karl B. and Charles A . Brown. 2001. Air Pollution Control Technology Handbook (Boca Raton, CRC Press); based on data from United States Environmental Protection Agency. 1991. Handbook: Control Technologies for Hazardous Air Pollutants, EPA/625/6–91/014.

Table 26.18. Physical properties of selected polynuclear aromatic compounds

	Naphthalene	Anthracene	Phenanthrene	Benzo[a]pyrene
Molecular weight (g/mole)	128.16	178.23	178.23	252.3
Melting point (°C)	80.28	216.4	100.5	179
Boiling point (°C)	217.95	340	338	310–312
Solubility (aqueous, mg/L)	30	0.005	1.28	3.8×10^{-3}
Vapor pressure (Torr)	0.08	5.63×10^{-6}	1.250×10^{-4}	5.25×10^{-9}
Henry's constant (atm-m^3/mol)	4.27×10^{-4}	1.8×10^{-6}	2.800×10^{-4}	5.53×10^{-7}
Molar volume (cm^3/mole)	148	197	199	263
Heat of vaporization (kJ/mol)	43.2	52.4	52.7	71.7
Molecular volume (Angstroms^3)	26.9	170.3	169.5	228.6
Molecular surface area (Angstroms^2)	55.8	202.2	198	225.6

Source: Adapted from Speight, James G. 2011. Handbook of Industrial Hydrocarbon Processes, (Boston, Gulf Professional Publishing).

Table 26.19. Typical potential air emissions from fossil fuel electricity generation

Technology/Fuel	Inputs	Air Emissions
Steam cycle/pulverized coal	coal, demineralized water, auxiliary fuel (fuel oil, natural gas, briquettes), lubricants, degreasers, water treatment chemicals	NO_X, SO_X, particulates (including PM_{10}), fugitive dust, trace metals, OCs, CL_2
Steam cycle/natural gas	natural gas, auxiliary fuel (fuel oil, distillate, LPG), demineralized water, cooling water, lubricants, degreasers, water treatment chemicals	NO_X, CO, SO_X, (very low), PM_{10}, OCs, CL_2 and trace metals
Steam cycle/oil	fuel oil, auxiliary fuel (natural gas, distillate, LPG), demineralised water, lubricants, degreasers	NO_X, CO, SO_X, particulates (including PM_{10}), OCs, CL_2 and trace metals
Gas turbine/natural gas	natural gas, auxiliary fuel (distillate, LPG), lubricants, degreasers	NO_X, CO, SO_X, (very low), OCs, and trace metals
Gas turbine/distillate	distillate, auxiliary fuel (LPG), lubricants, degreasers	NO_X, CO, SO_X, particulates (including PM_{10}), OCs, and trace metals
Internal combustion engine/natural gas, LPG	natural gas (or LPG), auxiliary fuel (distillate), lubricants, degreasers, coolants	NO_X, CO, SO_X, (very low), OCs, PM_{10}, and trace metals
Internal combustion engine/distillate	distillate, lubricants, degreasers, coolant	NO_X, CO, SO_X, OCs, and trace metals, PM_{10}

Cl_2: chlorine.
CO: carbon monoxide.
NO_X: oxides of nitrogen.
OCs: organic compounds; include volatile organic compounds, and polycyclic aromatic hydrocarbons.
SO_X: oxides of sulfur.
Source: Environment Australia, Emission Estimation Manual for Fossil Fuel Electric Power Generation.

Table 26.20.   Potential typical land emissions from fossil fuel electricity generation

Technology/Fuel	Inputs	Potential land emissions
Steam cycle/pulverized coal, natural gas, oil	Coal, demineralised water, auxiliary fuel (fuel oil, natural gas, briquettes), lubricants, degreasers, water treatment chemicals	ash, oil/chemical spills, metals, wastes
Gas turbine/natural gas, distillate	natural gas, auxiliary fuel (distillate, LPG), lubricants, degreasers	oil spills, wastes
Internal combustion engine/natural gas, LPG, distillate	natural gas (or LPG), auxiliary fuel (distillate), lubricants, degreasers, coolants	oil spills, wastes

Source: Environment Australia, Emission Estimation Manual for Fossil Fuel Electric Power Generation.

Table 26.21.   Potential typical water emissions from fossil fuel electricity generation

Technology/Fuel	Inputs	Potential water emissions
Steam cycle/pulverized coal, natural gas, oil	Coal, demineralised water, auxiliary fuel (fuel oil, natural gas, briquettes), lubricants, degreasers, water treatment chemicals/effluent, detergents	Chlorine, acids, alkalis, suspended solids, nitrogen, phosphorous, trace metals, oil spills, degreasers, detergents
Gas turbine/natural gas, distillate	natural gas, auxiliary fuel (distillate, LPG), lubricants, degreasers, detergents, cooling system inhibitors	oil spills, degreasers, detergents, cooling system inhibitors
Internal combustion engine/natural gas, LPG, distillate	natural gas (or LPG), auxiliary fuel (distillate), lubricants, degreasers, coolants	oil spills, detergents, waste coolant, degreasers,

Source: Environment Australia, Emission Estimation Manual for Fossil Fuel Electric Power Generation.

Table 26.22.   Connections between the energy sector and water availability and quality

Energy element	Connection to water quantity	Connection to water quality
Energy Extraction and Production		
Oil and Gas Exploration	Water for drilling, completion, and fracturing	Impact on shallow groundwater quality
Oil and Gas Production	Large volume of produced, impaired water[a]	Produced water can impact surface and groundwater
Coal and Uranium Mining	Mining operations can generate large quantities of water	Tailings and drainage can impact surface water and ground-water
Electric Power Generation		
Thermo-electric (fossil, biomass, nuclear)	Surface water and groundwater for cooling[b] and scrubbing	Thermal and air emissions impact surface waters and ecology
Hydro-electric	Reservoirs lose large quantities to evaporation	Can impact water temperatures, quality, ecology
Solar PV and Wind	None during operation; minimal water use for panel and blade washing	

Table 26.22. Continued

Energy Element	Connection to Water Quantity	Connection to Water Quality
Refining and Processing		
Traditional Oil and Gas Refining	Water needed to refine oil and gas	End use can impact water quality
Biofuels and Ethanol	Water for growing and refining	Refinery waste-water treatment
Synfuels and Hydrogen	Water for synthesis or steam reforming	Wastewater treatment
Energy Transportation and Storage		
Energy Pipelines	Water for hydrostatic testing	Wastewater requires treatment
Coal Slurry Pipelines	Water for slurry transport; water not returned	Final water is poor quality; requires treatment
Barge Transport of Energy	River flows and stages impact fuel delivery	Spills or accidents can impact water quality
Oil and Gas Storage Caverns	Slurry mining of caverns requires large quantities of water	Slurry disposal impacts water quality and ecology

[a]Impaired water may be saline or contain contaminants.
[b]Includes solar and geothermal steam-electric plants.
Source: Adapted from United States Department of Energy, Office of Environmental Management. 2007. Energy Demands on Water Resources.

Table 26.23. Water withdrawal during fuel acquisition and preparation for thermoelectric fuel cycles in the United States

Fuel cycle	Stage	Withdrawal – on-site (L/MWh)	Withdrawal – upstream (L/MWh)
Coal	Eastern underground mining[a]	190	507
	Eastern surface mining[b]	38[c]	148
	Western surface mining[d]	N/A	11
	US coal mining	106	N/A
	Beneficiation (Material fractionation)	>45	53
	Transportation – train	N/A	26–38
	Transportation – slurry pipeline	450	3100
	Construction – coal-power plant	N/A	11–45
Nuclear	Uranium mining	38	15
	Milling	19	68
	Conversion	15	8
	Enrichment – diffusion	79	1150
	Enrichment – centrifuge	8	102
	Fuel fabrication	3	0.4
	Power plant construction – PWR	N/A	19
	Power plant construction – BWR	N/A	38
	Spent fuel disposal	N/A	19
Natural gas	Extraction – on shore	130	300
	Extraction – off shore	0.8	0.4
	Purification	64	N/A
	Pipeline transportation	1.5	38
	Storage – underground	N/A	15
	Power plant environmental control	N/A	890

PWR = pressurized water reactor; BWR = boiling water reactor; N/A = not available.
[a]Including coal washing.
[b]Seam thickness = 0.9 m.
[c]Washing only.
[d]Seam thickness = 7 m.
Source: Adapted from Fthenakis, Vasilis, Hyung Chul Kim. 2010. Life-cycle uses of water in U.S. electricity generation, Renewable and Sustainable Energy Reviews, Volume 14, Issue 7, Pages 2039–2048.

Table 26.24. Water consumption during fuel acquisition and preparation of thermoelectric fuel cycles in the United States[a]

Fuel cycle	Stage	Consumption (L/MWh)
Coal	Surface mining	11–53
	Underground mining	30–200
	Washing	30–64
	Beneficiation	42–45
	Transportation – slurry pipeline	420–870
Nuclear	Surface uranium mining	200
	Underground uranium mining	4
	Milling	83–100
	Conversion	42

Table 26.24.   Continued

Fuel cycle	Stage	Consumption (L/MWh)
	Enrichment – diffusion	45–130
	Enrichment – centrifuge	4–19
	Fabrication	11
Natural gas	Extraction – on shore	negligible
	Extraction – off shore	negligible
	Purification	57
	Pipeline transportation	30

[a]Upstream water consumption not included.
Source: Adapted from Fthenakis, Vasilis, Hyung Chul Kim. 2010. Life-cycle uses of water in U.S. electricity generation, Renewable and Sustainable Energy Reviews, Volume 14, Issue 7, Pages 2039–2048.

Table 26.25.   Water withdrawal factors for PV technologies[a]

Type	On-site (L/MWh)	Upstream (L/MWh)	Note
Multi-Si	200	1470	Efficiency = 13.2%
Mono-Si	190	1530	Efficiency = 14%
Frame	N/A	64	Based on multi-Si PV
CdTe	0.8	575	Efficiency = 10.9%
BOS	1.5	210	Based on ground-mount multi-Si PV

[a]for manufacturing the devices, and building the power plants (insolation = 1800 kWh/m2/year; lifetime = 30 years; performance ratio = 0.8).
Source: Adapted from Fthenakis, Vasilis, Hyung Chul Kim. 2010. Life-cycle uses of water in U.S. electricity generation, Renewable and Sustainable Energy Reviews, Volume 14, Issue 7, Pages 2039–2048.

Table 26.26.   Water withdrawal factors of the wind-fuel cycle

Type	Upstream (L/MWh)	Capacity factor
Off shore, Denmark	230	0.29
On land, Denmark	170	0.25
Off shore, Denmark	170	0.46
On shore, Denmark	320	0.32
On land, Italy	250	0.19
On shore, Spain	210	0.23

Source: Adapted from Fthenakis, Vasilis, Hyung Chul Kim. 2010. Life-cycle uses of water in U.S. electricity generation, Renewable and Sustainable Energy Reviews, Volume 14, Issue 7, Pages 2039–2048.

Table 26.27.   Water demand of biomass/bioenergy production

Energy type	Biomass	On-site (L/GJ)[a]	Water use type	Upstream (L/GJ)[a]	Assumptions
Electricity	Hybrid Poplar, US	0	W/C	52	Rain-fed; yield = 13.5 t/ha/year; gasification combined cycle with 37.2% efficiency
Electricity	Herbaceous perennials, Southwestern US, irrigation	121,000	W	310	Yield = 27 t/ha/year; steam power plant with 25% efficiency
Electricity	Maize, global average	20,000	C	N/A	Total biomass yields are used
Electricity	Sugar beet, global average	27,000	C	N/A	Total biomass yields are used
Electricity	Soybean, global average	95,000	C	N/A	Total biomass yields are used
Electricity	Jatropha, global average	231,000	C	N/A	Total biomass yields are used
Ethanol	Corn, US	350–12,100	W	N/A	Irrigation + fuel conversion; crop yield = 142 bushel per acre, ethanol yield = 10.2 L per bushel
Ethanol	Corn, US	270–8600	C	N/A	Irrigation + fuel conversion; crop yield = 142 bushel per acre, ethanol yield =10.2 L per bushel
Ethanol	Switchgrass, US	50–260	W/C	N/A	Rain-fed; fuel conversion only; yield = 9.0–15.7 dry metric tons per hectare
Ethanol	Corn, Illinois	505	W	N/A	Corp yield =10.2 t/ha/year; ethanol yield = 9.7 L per bushel
Ethanol	Corn, Iowa	170	W	N/A	Yield = 9.2 t/ha/year; ethanol yield = 9.7 L per bushel
Ethanol	Corn, Nebraska	18,700	W	N/A	Yield = 8.8 t/ha/year; ethanol yield = 9.7 L per bushel
Ethanol	Corn, US	130–56,800	C	N/A	
Ethanol	Sugar beet, global average	35,000	C	N/A	Total biomass yields are used
Biodiesel	Soybean, global average	217,000	C	N/A	Total biomass yields are used
Biodiesel	Rapeseed, global average	245,000	C	N/A	Total biomass yields are used

W: withdrawal; C: consumption; W/C: consumption is equal to withdrawal.

[a]GJ instead of MWh were used to represent both electrical- and thermal-end-use energy.

Source: Adapted from Fthenakis, Vasilis, Hyung Chul Kim. 2010. Life-cycle uses of water in U.S. electricity generation, Renewable and Sustainable Energy Reviews, Volume 14, Issue 7, Pages 2039–2048.

Table 26.28. Estimated water needs for drilling and fracturing wells in selected shale gas plays in the United States

	Volume of Drilling Water per well (gal)	Volume of Fracturing Water per well (gal)	Total Volume of Water per well (gal)
Barnett Shale	400,000	2,300,000	2,700,000
Fayetteville Shale	60,000[a]	2,900,000	3,060,000
Haynesville Shale	1,000,000	2,700,000	3,700,000
Marcellus Shale	80,000*	3,800,000	3,880,000

[a]Drilling performed with an air "mist" and/or water-based or oil-based muds for deep horizontal well completions.
These volumes are approximate and may vary substantially between wells.
Source: Ground Water Protection Council. 2009. Modern Shale Gas Development in the United States: A Primer, Prepared for U.S. Department of Energy, Office of Fossil Energy and National Energy Technology Laboratory.

Table 26.29. Estimated time of restoration for select ecosystem types

Ecosystems	Time (years)
Arable land, pioneer vegetation	<5
Species-poor meadows, mature pioneer vegetation	5–25
Species-poor immature shrubs and hedgerows, oligotrophic vegetation, species-rich marshland, meadows, dry meadows and heathland	25–50
Species-rich forests, shrubs, hedgerows	50–200
Immature peat bogs, old dry meadows and heathland	200–1,000
Mature peat bogs, old growth forests	1,000–10,000

Source: Adapted from Fthenakis, Vasilis, Hyung Chul Kim. 2009. Land use and electricity generation: A life-cycle analysis, Renewable and Sustainable Energy Reviews, Volume 13, Issues 6–7, Pages 1465–1474.

Table 26.30. Composition of municpal solid waste by world region (percent)

Region	Food waste	Paper/cardboard	Wood	Textiles	Rubber/leather	Plastic	Metal	Glass	Other
Asia									
Eastern Asia	26.2	18.8	3.5	3.5	1	14.3	2.7	3.1	7.4
South-Central Asia	40.3	11.3	7.9	2.5	0.8	6.4	3.8	3.5	21.9
South-Eastern Asia	43.5	12.9	9.9	2.7	0.9	7.2	3.3	4	16.3
Western Asia & Middle East	41.1	18	9.8	2.9	0.6	6.3	1.3	2.2	5.4
Africa									
Eastern Africa	53.9	7.7	7	1.7	1.1	5.5	1.8	2.3	11.6
Middle Africa	43.4	16.8	6.5	2.5		4.5	3.5	2	1.5
Northern Africa	51.1	16.5	2	2.5		4.5	3.5	2	1.5
Southern Africa	23	25	15						
Western Africa	40.4	9.8	4.4	1		3	1		
Europe									
Eastern Europe	30.1	21.8	7.5	4.7	1.4	6.2	3.6	10	14.6
Northern Europe	23.8	30.6	10	2		13	7	8	
Southern Europe	36.9	17	10.6						
Western Europe	24.2	27.5	11						
Oceania									
Australia and New Zealand	36	30	24						
Rest of Oceania	67.5	6	2.5						
Americas									
North America	33.9	23.2	6.2	3.9	1.4	8.5	4.6	6.5	9.8
Central America	43.8	13.7	13.5	2.6	1.8	6.7	2.6	3.7	12.3
South America	44.9	17.1	4.7	2.6	0.7	10.8	2.9	3.3	13
Caribbean	46.9	17	2.4	5.1	1.9	9.9	5	5.7	3.5

Source: Intergovernmental Panel on Climate Change. 2006. Guidelines for National Greenhouse Gas Inventories.

Table 26.31. Direct land transformation by coal mining

Region	Land transformation (m²/1000 t), facilities and waste disposal	Land transformation (m²/1000 t), excavated	Total land transformation[a] (m²/GWh)	Average seam thickness (m)
Western U.S.	0.31	250	140	7
Eastern U.S.—case 1	1.8	650	310	1.8
Eastern U.S.—case 2	1.8	3040	1450	0.9
Northern Appalachia	na	760	350	1–1.6
Central Appalachia	na	780	360	0.8–1.5
Southern Appalachia	na	1240	570	0.8
Wyoming	na	90	43	9.1
Kansas	na	1820	840	0.5
U.S. average	na	na	400	na
Eastern U.S.	0.8	4.5	2.3	na
Eastern Europe	156	0	67	na
Northern Appalachia[b]	na	1010	470	1.4–1.9
Central Appalachia[b]	na	1110	510	1–1.6
Southern Appalachia[b]	na	1050	480	1.4
Utah[b]	na	520	240	2.9
U.S. average	na	na	200	na

[a]Based on electricity conversion efficiency of 35%; loss during preparation = 25%.
[b]Area undermined.
Source: Adapted from Fthenakis, Vasilis, Hyung Chul Kim. 2009. Land use and electricity generation: A life-cycle analysis, Renewable and Sustainable Energy Reviews, Volume 13, Issues 6–7, Pages 1465–1474.

Table 26.32. Direct land transformation for power plant operation in the United States

Plant type	Size (MW)	Land cover (acre)	Lifetime (years)	Land transformation (m²/GWh)
Eastern U.S.	500	849	30	32.6
Western U.S.	500	156	30	6
FBD[a]—bituminous	831	246	20	10.8
FBD[b]—western subbituminous	536	442	20	20.1
U.S. average[b]	1000	500	30	9.1[c]

[a]Fluidized bed combustion.
[b]Normalized for 1000 MW capacity.
[c]Facilities only, excluding waste storage yard.
Source: Adapted from Fthenakis, Vasilis, Hyung Chul Kim. 2009. Land use and electricity generation: A life-cycle analysis, Renewable and Sustainable Energy Reviews, Volume 13, Issues 6–7, Pages 1465–1474.

Table 26.33. Direct land transformation of hydroelectric power plants

Location/type	Capacity (MW)	Area (10^4 m²)	Lifetime electricity generation (GWh)	Lifetime (years)	Land transformation (m²/GWh)
Lake Powell Reservoir	1296	65,313	277,500	50	2,350
U.S., generic reservoir	114	75,000	30,000	30	25,000
Canada, reservoir	na	na	na	30	3,700
Canada, run-of-river	na	na	na	30	3

Source: Adapted from Fthenakis, Vasilis, Hyung Chul Kim. 2009. Land use and electricity generation: A life-cycle analysis, Renewable and Sustainable Energy Reviews, Volume 13, Issues 6–7, Pages 1465–1474.

Table 26.34. Direct land transformation by solar electric power plants

Type	System efficiency[a] (%)	Packing factor	Insolation (kWh/m²/year)	Plant lifetime (years)	Land transformation (m²/GWh)
Multi-Si PV[b]—case 1	10.6	2.5	1800	30	438
Multi-Si PV[b]—case 2	10.6	2.5	2400	30	329
Multi-Si PV[b]—case 3	10.6	2.5	2400	60	164
PV with 25° tilt	9.5	2.1	1770	30	402
PV with 1-axis tracker	9.5	2.8	2050	30	463
Concentrator PV—case 1	20.2	3.5	2500[c]	30	229
Concentrator PV—case 2	13.8	5	2200[c]	30	549
Solar thermal, tower	8.5	5	2700[c]	30	552
Solar thermal, parabolic trough	10.7	3.4	2900[c]	30	366

[a]Module efficiency times performance ratio.
[b]Optimum tilt.
[c]Direct Normal Insolation (DNI) with tracker.
Source: Adapted from Fthenakis, Vasilis, Hyung Chul Kim. 2009. Land use and electricity generation: A life-cycle analysis, Renewable and Sustainable Energy Reviews, Volume 13, Issues 6–7, Pages 1465–1474.

Table 26.35. Direct land transformation by wind farms based on 30-year plant lifetime

Location	Capacity factor	Area (10^4 m²/MW)	Land transformation (m²/GWh)
U.S.—case 1	0.26[a]	19[c]	2780
U.S.—case 2	0.36[b]	19[c]	2040
U.S., California	0.24	6.5	1030
Denmark, Tændpibe	0.2	17	3230
Denmark, Velling Mærsk-Tændpibe	0.2	12	2280
Germany, Fehmarn	0.2	11	2090

[a]Based on a wind speed of class 4, i.e., 5.8 m/s at 10 m.
[b]Based on a wind speed of class 6, i.e., 6.7 m/s at 10 m.
[c]Based on an array of 25 turbines by two rows. Each turbine is separated by 2.5 rotor diameters side-by-side and the rows are positioned 20 rotor diameters apart.
Source: Adapted from Fthenakis, Vasilis, Hyung Chul Kim. 2009. Land use and electricity generation: A life-cycle analysis, Renewable and Sustainable Energy Reviews, Volume 13, Issues 6–7, Pages 1465–1474.

Table 26.36. Direct land transformation during production of energy from biomass (normalized for 1 year timeframe)

Energy type	Biomass	Yield (t/10^4 m^2/year)	Land transformation (m^2/GJ)
Ethanol and biodiesel	Corn and soybean	5.2	199
Ethanol	Corn	7.8	121
Ethanol	Corn and corn stover[a]	11.6	87
Ethanol	Corn and corn stover[b]	13.8	74
Ethanol	Corn	7.5	154
Ethanol	Corn	7.5	150
Ethanol	Corn	8.8	125
Ethanol	Corn	7.7	142
Ethanol	Corn	8.7	133
Electricity	Willow, high-pressure gasification	15	104
Electricity	Willow, low-pressure gasification	15	101
Electricity	Willow, direct fire	15	136
Electricity	Hybrid poplar	13.4	114
Electricity	Gasification	11.2	126
Electricity	Direct fire	11.2	193
Electricity	Co-firing	9.4	175

[a]50% corn stover removal.
[b]70% corn stover removal and use of wheat as winter cover crop.
Source: Adapted from Fthenakis, Vasilis, Hyung Chul Kim. 2009. Land use and electricity generation: A life-cycle analysis, Renewable and Sustainable Energy Reviews, Volume 13, Issues 6–7, Pages 1465–1474.

Table 26.37. Typical landfill gas components

Component	Percent by volume	Characteristics
methane	45–60	Methane is a naturally occurring gas. It is colorless and odorless. Landfills are a significant source of anthropogenic methane emissions.
carbon dioxide	40–60	Carbon dioxide is naturally found at small concentrations in the atmosphere (0.03%). It is colorless, odorless, and slightly acidic.
nitrogen	2–5	Nitrogen comprises approximately 79% of the atmosphere. It is odorless, tasteless, and colorless.
oxygen	0.1–1	Oxygen comprises approximately 21% of the atmosphere. It is odorless, tasteless, and colorless.
ammonia	0.1–1	Ammonia is a colorless gas with a pungent odor.
non-methane organic compounds (NMOCs)	0.01–0.6	NMOCs are organic compounds (i.e., compounds that contain carbon). (Methane is an organic compound but is not considered an NMOC.) NMOCs may occur naturally or be formed by synthetic chemical processes. NMOCs most commonly found in landfills include acrylonitrile, benzene, 1,1-dichloroethane, 1,2-cis dichloroethylene, dichloromethane, carbonyl sulfide, ethyl-benzene, hexane, methyl ethyl ketone, tetrachloroethylene, toluene, trichloroethylene, vinyl chloride, and xylenes.
sulfides	0–1	Sulfides (e.g., hydrogen sulfide, dimethyl sulfide, mercaptans) are naturally occurring gases that give the landfill gas mixture its rotten-egg smell. Sulfides can cause unpleasant odors even at very low concentrations.
hydrogen	0–0.2	Hydrogen is an odorless, colorless gas.
carbon monoxide	0–0.2	Carbon monoxide is an odorless, colorless gas.

Source: Adapted from United States Enviironmental Protection Agency

Table 26.38. Common landfill gas components and their odor thresholds

Component	Odor Description	Odor Threshold (parts per billion)
Hydrogen Sulfide	Strong rotten egg smell	0.5 to 1
Ammonia	Pungent acidic or suffocating odor	1,000 to 5,000
Benzene	Paint-thinner-like odor	840
Dichloroethylene	Sweet, ether-like, slightly acrid odor	85
Dichloromethane	Sweet, chloroform-like odor	205,000 to 307,000
Ethylbenzene	Aromatic odor like benzene	90 to 600
Toluene	Aromatic odor like benzene	10,000 to 15,000
Trichloroethylene	Sweet, chloroform-like odor	21,400
Tetrachloroethylene	Sweet, ether-or chloroform-like odor	50,000
Vinyl Chloride	Faintly sweet odor	10,000 to 20,000

Source: United States Center for Disease Control.

Health and Safety

Charts

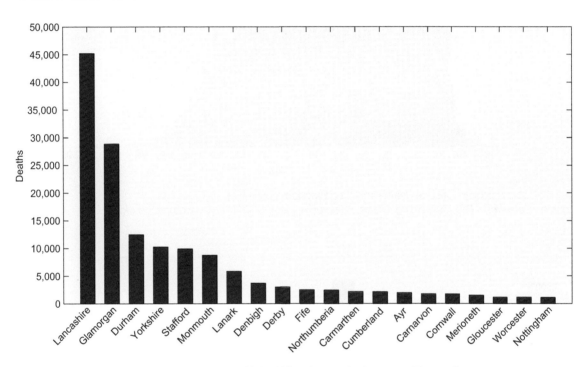

Chart 27.1. County distribution of deaths in United Kingdom coal mines, top 20 counties, 1700–2000.
Source: Data from The Coalmining History Resource Centre, <http://www.cmhrc.co.uk>.

Handbook of Energy. http://dx.doi.org/10.1016/B978-0-08-046405-3.00027-9

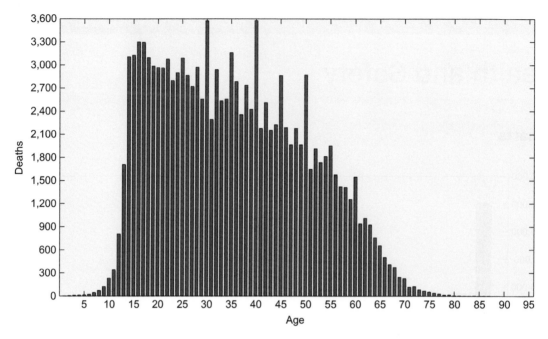

Chart 27.2. Age distribution of deaths in United Kingdom coal mines, 1700–2000.
Source: Data from The Coalmining History Resource Centre, <http://www.cmhrc.co.uk>.

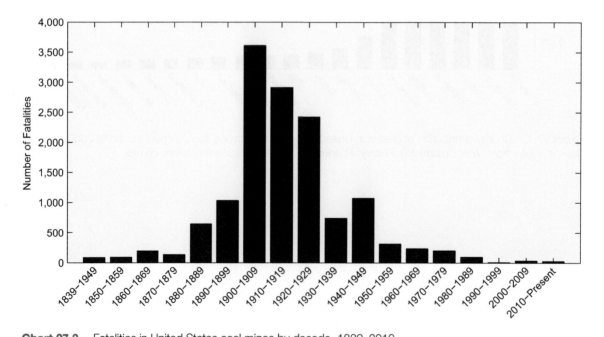

Chart 27.3. Fatalities in United States coal mines by decade, 1839–2010.
Source: Data from United States Department of Labor, Mine Safety and Health Administration.

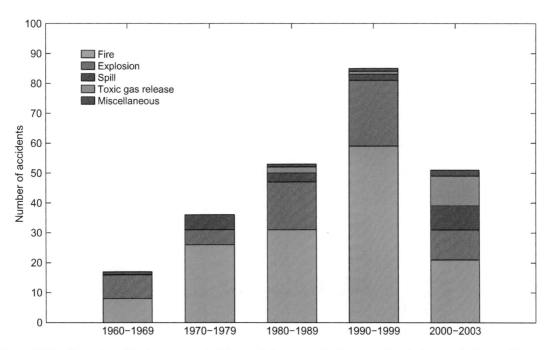

Chart 27.4. Type of accident associated with tank failures in refineries and chemical plants in the world, 1960–2003.
Source: Data from Chang, James I., Cheng-Chung Lin. 2006. A study of storage tank accidents, Journal of Loss Prevention in the Process Industries, Volume 19, Issue 1, Pages 51-59.

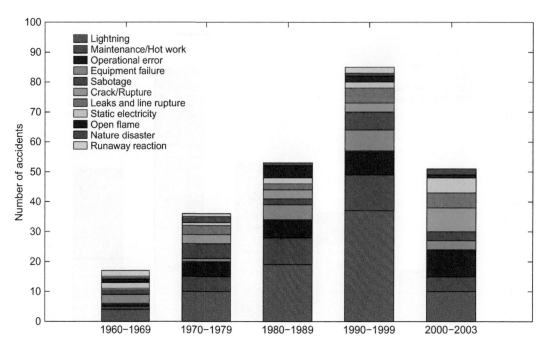

Chart 27.5. Cause of accidents with oil and chemical storage tanks in refineries and chemical plants in the world, 1960–2003.
Source: Data from Chang, James I., Cheng-Chung Lin. 2006. A study of storage tank accidents, Journal of Loss Prevention in the Process Industries, Volume 19, Issue 1, Pages 51-59.

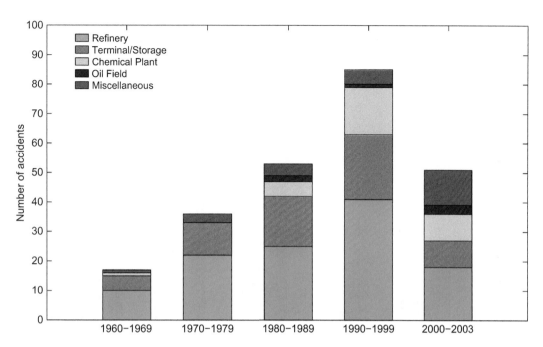

Chart 27.6. Type of contents associated with tank failures in refineries and chemical plants, 1960–2003. *Source: Data from Chang, James I., Cheng-Chung Lin. 2006. A study of storage tank accidents, Journal of Loss Prevention in the Process Industries, Volume 19, Issue 1, Pages 51-59.*

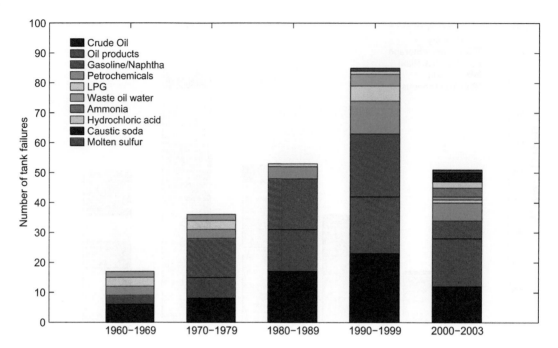

Chart 27.7. Type of facility where accidents occur with storage tanks in refineries and chemical plants in the world, 1960–2003.
Source: Data from Chang, James I., Cheng-Chung Lin. 2006. A study of storage tank accidents, Journal of Loss Prevention in the Process Industries, Volume 19, Issue 1, Pages 51-59.

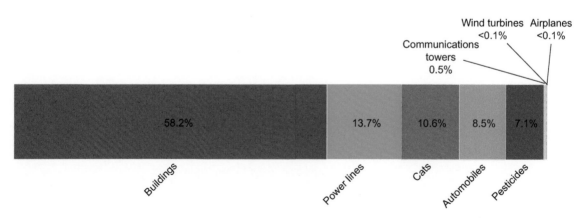

Chart 27.8. Estimates of annual anthropogenic causes of bird mortality in the United States.
Source: Data from Erickson, Wallace P., Gregory D. Johnson and David P. Young Jr. 2005. A Summary and Comparison of Bird Mortality from Anthropogenic Causes with an Emphasis on Collisions, USDA Forest Service Gen. Tech. Rep. PSW-GTR-191.

Tables

Table 27.1. Some notable energy accidents that caused fatalities

Facility	Year	Region/State	Country	Description	Approximate Fatalities
Hydroelectric	1975	Henan Province	China	Shimantan Dam fails and releases 15,738 billion tons of water causing widespread flooding that destroys 18 villages, 1500 homes, and induces disease epidemics and famine	171,000
Hydroelectric	1963	Longarone	Italy	Landslides associated with earthquake activity cause dam to overflow, resulting in flooding	2,500
Hydrolectric	1979	Gujarat	India	Abnormal rainfall and inadequate spillway capacity cause dam failure and massive flood	2,000
Hydrolectric	1943	Ruhr valley	Germany	British air force attacks Nazi hydroelectric plants, seriously damaging two dams	1,650
Coal mine	1942	Benxi, Liaoning	China	A gas and coal-dust explosion kills 1,549 people, 34% of the miners working that day, making it this worst disaster in the history of coal mining	1,549
Coal mine	2006	Pas-de-Calais	France	An explosion caused by the ignition of coal dust that swept through the underground mine	1,099
Oil pipeline	1998	Niger Delta	Nigeria	Petroleum pipeline ruptures and explodes, destroying two villages and hundreds of villagers scavenging gasoline	1,078
Hydroelectric	1980	Orissa	India	Officals intentionally open floodagtes during heavy rainfall event	1,000
Coal mine	1917	Fushun	China	Explosion in coal mine	917
Coal mine	1914	Mitsubishi Hōjō, Kyūshū	Japan	Ignition of methane gas causes explosion and fire	686
Coal mine	1960	Datong, Shanxi	China	An explosion at the Laobaidong mine in Datong, Shanxi province, kills 684 miners	684
Oil pipeline	1989	Ufa	Russia	Sparks from passing trains ignite gas leaking from petroleum pipeline, causing multiple explosions that derail both trains	643
Hydrolecetric	1928	Los Angeles County, California	USA	Failure of concrete gravity-arch dam designed to supply drinking water and generate power	600
Coal mine	1935		China	Flood in Zichuan coal mine	536
Oil storage tank	1984	Mexico City	Mexico	A leak at the PEMEX LPG Terminal in San Juan Ixhuatepec causes a flammable gas cloud to ignite, destroying 5 city blocks and forcing the evacuation of more than 200,000 residents	514
Oil pipeline	1984	Cubatão, São Paulo	Brazil	Pipeline rupture leads to ignition of gasoline	508

Continued

Table 27.1. Continued

Facility	Year	Region/State	Country	Description	Approximate Fatalities
Natural gas storage	1983	Nile River	Egypt	Explosion during transportation of LPG	500
Oil storage	1994	Durunkah	Egypt	Train carrying fuel oil derails as it enters an oil depot during a heavy downpour	500
Coal mine	1928		China	Flood in Fushun coal mine	482
Coal mine	1972	Wankie	Rhodesia	A methane explosion followed is followed by a coal dust explosion	472
Oil pipeline	2006	Lagos	Nigeria	Oil pipeline explodes, causing widespread fires that destroy more than 300 homes	466
Coal mine	1963	Kyushu	Japan	Explosion and carbon monoxide create deadly conditions	458
Coal mine	1920	Kailuan	China	Explosion in coal mine	451
Coal mine	1913	Cardiff	Wales	Mine shaft completely collapses	439
Coal mine	1960	Coalbrook	South Africa	A cave-in traps miners and exposes then to lethal levels of methane	437

Table 27.2. The International Nuclear and Radiological Event Scale (INES)

INES Level	People and Environment	Radiological Barriers and Control	Defense-in-Depth
Major Accident Level 7	• Major release of radio active material with widespread health and environmental effects requiring implementation of planned and extended countermeasures.		
Serious Accident Level 6	• Significant release of radioactive material likely to require implementation of planned countermeasures.		
Accident with Wider Consequences Level 5	• Limited release of radioactive material likely to require implementation of some planned countermeasures. • Several deaths from radiation.	• Severe damage to reactor core. • Release of large quantities of radioactive material within an installation with a high probability of significant public exposure. This could arise from a major criticality accident or fire.	

Table 27.2. Continued

INES Level	People and Environment	Radiological Barriers and Control	Defense-in-Depth
Accident with Local Consequences Level 4	• Minor release of radioactive material unlikely to result in implementation of planned countermeasures other than local food controls. • At least one death from radiation.	• Fuel melt or damage to fuel resulting in more than 0.1% release of core inventory. • Release of significant quantities of radioactive material within an installation with a high probability of significant public exposure.	
Serious Incident Level 3	• Exposure in excess of ten times the statutory annual limit for workers. • Non-lethal deterministic effect(e.g., burns) from radiation.	• Exposure rates of more than 1 Sv/h in an operating area. • Severe contamination in an area not expected by design, with a low probability of significant public exposure.	• Near accident at a nuclear power plant with no safety provisions remaining. • Lost or stolen highly radioactive sealed source. • Misdelivered highly radioactive sealed source without adequate procedures in place to handle it.
Incident Level 2	• Exposure of a member of the public in excess of 10 mSv. • Exposure of a worker in excess of the statutory annual limits.	• Radiation levels in an operating area of more than 50 mSv/h. • Significant contamination within the facility into an area not expected by design.	• Significant failures in safety provisions but with no actual consequences. • Found highly radioactive sealed orphan source, device or transport package with safety provisions intact. • Inadequate packaging of a highly radioactive sealed source.
Anomaly Level 1			• Overexposure of a member of the public in excess of statutory annual limits. • Minor problems with safety components with significant defense-in-depth remaining. • Low activity lost or stolen radioactive source, device or transport package.

Source: International Atomic Energy Agency, The International Nuclear and Radiological Event Scale, <http://www-ns.iaea.org/tech-areas/emergency/ines.htm>.

Table 27.3. Examples of events at nuclear facilities classified according to the International Nuclear and Radiological Event Scale (INES)

INES Level	People and environment	Radiological barriers and control	Defense-in-depth
7	Chernobyl, 1986 — Widespread health and environmental effects. External release of a significant fraction of reactor core inventory. Fukushima, 2011— Significant release of radioactive material resulting from damage to backup power and containment systems caused by the Tōhoku earthquake and tsunami.		
6	Kyshtym, Russia, 1957 — Significant release of radioactive material to the environment from explosion of a high activity waste tank.		
5	Windscale Pile, UK, 1957 — Release of radioactive material to the environment following a fire in a reactor core.	Three Mile Island, USA, 1979 — Severe damage to the reactor core.	
4	Tokaimura, Japan, 1999 — Fatal overexposures of workers following a criticality event at a nuclear facility.1980	Saint Laurent des Eaux, France, 1980—Melting of one channel of fuel in the reactor with no release outside the site.	
3	No example available.	Sellafield, UK, 2005 — Release of large quantity of radioactive material, contained within the installation.	Vandellos, Spain, 1989 — Near accident caused by fire resulting in loss of safety systems at the nuclear power station.
2	Argentina, 2005 — Overexposure of a worker at a power reactor exceeding the annual limit.	Cadarache, France, 1993 — Spread of contamination to an area not expected by design.	Forsmark, Sweden, 2006 — Degraded safety functions for common area cause failure in the emergency power supply system at nuclear power plant.
1			Breach of operating limits at a nuclear facility.

BWR = boiling water reactor; GCHWR = gas-cooled heavy-water reactor; GCR = gas-cooled reactor; PWR = pressurized water reactor; RMBK = Russian reaktor bolshoy moshchnosty kanalny, high-power channel reactor; VVER = Russian vodo-vodyanoi energetichesky reactor, water-water power reactor.

Source: Adapted from International Atomic Energy Agency, The International Nuclear and Radiological Event Scale, <http://www-ns.iaea.org/tech-areas/emergency/ines.htm>.

Table 27.4. Nuclear reactors closed following an accident or serious incident

Country	Reactor	Type	MWe net	Years operating	Shut down	Reason
Germany	Greifswald 5	VVER-440/V213	408	0.5	1989	Partial core melt
	Gundremmingen A	BWR	237	10	1977	Botched shutdown
Japan	Fukushima Daiichi 1	BWR	439	40	2011	Core melt from cooling loss
	Fukushima Daiichi 2	BWR	760	37	2011	Core melt from cooling loss

Table 27.4. Continued

Country	Reactor	Type	MWe net	Years operating	Shut down	Reason
	Fukushima Daiichi 3	BWR	760	35	2011	Core melt from cooling loss
	Fukushima Daiichi 4	BWR	760	32	2011	Damage from hydrogen explosion
Slovakia	Bohunice A1	Prot GCHWR	93	4	1977	Core damage from fueling error
Spain	Vandellos 1	GCR	480	18	1990	Turbine fire
Switzerland	St Lucens	Exp GCHWR	8	3	1966	Core melt
Ukraine	Chernobyl 4	RBMK LWGR	925	2	1986	Fire and meltdown
USA	Three Mile Island 2	PWR	880	1	1979	Partial core melt

BWR = boiling water reactor; GCHWR = gas-cooled heavy-water reactor; GCR = gas-cooled reactor; PWR = pressurized water reactor; RMBK = Russian *reaktor bolshoy moshchnosty kanalny,* high-power channel reactor; VVER = Russian *vodo-vodyanoi energetichesky reactor,* water-water power reactor
Source: Adapted from World Nuclear Association, <http://www.world-nuclear.org/info/inf19.html>.

Table 27.5. Radiation doses and their effects

Dose	Description
2 mSv/yr	Typical background radiation experienced by everyone (average 1.5 mSv in Australia, 3 mSv in North America).
1.5 to 2.0 mSv/yr	Average dose to Australian uranium miners, above background and medical.
2.4 mSv/yr	Average does to US nuclear industry employees.
Up to 5 mSv/yr	Typical incremental dose for aircrew in middle latitudes
9 mSv/yr	Exposure by airline crew flying the New York-Tokyo polar route.
10 mSv/yr	Maximum actual dose to Australian uranium miners.
20 mSv/yr	Current limit (averaged) for nuclear industry employees and uranium miners
50 mSv/yr	Former routine limit for nuclear industry employees. It is also the dose rate which arises from natural background levels in several places in Iran, India, and Europe.
	Allowable short-term dose for emergency workers (IAEA)
100 mSv/yr	Lowest level at which any increase in cancer is clearly evident. Above this, the probability of cancer occurrence (rather than the severity) is assumed to increase with dose. Allowable short term dose for emergency workers taking vital remedial actions (IAEA)
250 mSv	Allowable short-term dose for workers controlling the 2011 Fukushima accident
250 mSv/yr	Natural background level at Ramsar in Iran, with no identified health effects.
350 mSv/lifetime	Criterion for relocating people after Chernobyl accident.
500 mSv	Allowable short-term dose for emergency workers taking life-saving actions (IAEA)
1,000 mSv cumulative	Would probably cause a fatal cancer many years later in 5 of every 100 persons exposed to it (i.e. if the normal incidence of fatal cancer were 25%, this dose would increase it to 30%).
1,000 mSv single dose	Causes (temporary) radiation sickness (Acute Radiation Syndrome) such as nausea and decreased white blood cell count, but not death. Above this, the severity of illness increases with dose.
5,000 mSv single dose	Would kill about half those receiving it within a month.
10,000 mSv single dose	Fatal within a few weeks.

The sievert (Sv) is the International System of Units (SI) derived unit of dose equivalent radiation that takes into account the relative biological effectiveness of different forms of ionizing radiation. One millisievert (mSv) corresponds to 10 ergs of energy of gamma radiation transferred to one gram of living tissue.
Source: Adapted from World Nuclear Association, <http://www.world-nuclear.org/>.

Table 27.6. Annual average doses and ranges of individual doses of ionizing radiation by source (Millisieverts[a])

Source or Mode	Annual average dose (worldwide)	Typoical range of individual doses	Comments
Natural sources of exposure			
Inhalation (radon gas)	1.26	0.2–10	The dose is much higher in some dwellings.
External terrestrial	0.48	0.3–1	The dose is higher in some locations.
Ingestion	0.29	0.2–1	
Cosmic radiation	0.39	0.3–1	The dose increases with altitude.
Total natural	**2.42**	**1-13**	**Sizeable population groups receive 10-20 millisieverts (mSv).**
Artificial sources of exposure			
Medical diagnosis (not therapy)	0.6	0-several tens	The averages for different levels of health care range from 0.03 to 2.0 mSv; averages for some countries are higher than that due to natural sources; individual doses depend on specific examinations.
Atmospheric nuclear testing	0.005	Some higher doses around test sites	The average has fallen from a peak of 0.11 mSv in 1963.
Occupational exposure	0.005	0–20	The average dose to all workers is 0.7 mSv. Most of the average dose and most high exposures are due to natural radiation (specifically radon in mines).
Chernobyl accident	0.002[b]	In 1986, the average dose to more than 300, 000 recovery workers was nearly 150 mSv and more than 350,000 other individuals received doses greater than 10 mSv.	The average in the Northern Hemisphere has decreased from a maximum of 0.04 mSv in 1986. Thyroid doses were much higher.
Nuclear fuel cycle (public exposure)	0.0002[b]	Doses are up to 0.02 mSv for critical groups at 1 km from some nuclear reactor sites.	
Total artificial	**0.61**	**From essentially zero to several tens.**	**Individual doses depend primarily on medical treatment, occupational exposure and proximity to test or accident sites.**

Notes:
[a]Unit of measurement of effective dose.
[b]Globally dispersed radionuclides. The value for the nuclear fuel cycle represents the maximum per capita annual dose to the public in the future, assuming the practice continues for 100 years, and derives mainly from globally dispersed, long-lived radionuclides released during reprocessing of nuclear fuel and nuclear power plant operation.
Source: Report of the United Nations Scientific Committee on the Effects of Atomic Radiation, Fifty-sixth session, (10–18 July 2008).

Table 27.7. Doses of ionizing radiation to individuals

Dose	Effect
Worldwide average annual effective dose	
0.4 mSv/year (range is 0.3 to 1.0)[a]	External exposure-cosmic rays
0.5 mSv/year (range is 0.3 to 0.6)[b]	External exposure-Terrestrial gamma rays
1.2 mSv/year (range is 0.2 to 10)[c]	Internal exposure-Inhalation (mainly radon)
0.3 mSv/year (range is 0.2 to 0.8)[d]	Internal exposure-Ingestion
2.4 mSv/year (range is 1–10)	Total exposure-global average from all sources
3.6 mSv/year	Average American total exposure from all sources
1,300 mSv/year	Kerala, India, resident-Concentrated radioactive material in the soil
1.79 mSv/year	Colorado state resident-High altitude above sea level
1.0 mSv/year	Boston, Massachusetts, resident
0.92 mSv/year	Louisiana state resident-Low altitude above sea level
5 mSv/patient	Anyone near a patient released after a nuclear medicine test. Guidance for medical facilities. Quantity depends on the quantity of radioactive material
0.4 mSv	A person who gets a full set of dental x rays
1.5–2.0 mSv/year	Average dose to Australian uranium miners, above background and medical.
2.4 mSv/year	Average dose to US nuclear industry employees
0.05 mSv/flight	A flight attendant flying from New York to Los Angeles
0.0006 mSv/year	Wearing a luminous wristwatch (LCD)
0.00008 mSv/year	Smoke detector in home
1.0 mSv/year	Wearing a plutonium powered pacemaker
0.02-0.03 mSv/yr	Watching a color TV set
0.00009 mSv/yr	A person who lives within 50 miles of a nuclear power plant
0.0003 mSv/yr	A person who lives within 50 miles of a coal-fired power plant
~.01 mSv/yr for each 5 stories above the ground floor	A person who lives in a multi-storied apartment building
Less than 001 mSv /truck	A person who watches a truck carrying nuclear waste pass by
10 mSv/year	Maximum actual dose to Australian uranium miners
20 mSv/year	Current limit (averaged) for nuclear industry employees and uranium miners
50 mSv/year	Former routine limit for nuclear industry employees. It is also the dose rate which arises from natural background levels in several places in Iran, India and Europe
100 mSv/year	Lowest level at which any increase in cancer is clearly evident. Above this, the probability of cancer occurrence (rather than the severity) increases with dose
Chernobyl Accident	
~100 mSv	Average accumulated doses for 600,000 rescue workers (1986–1989)
33 mSv	Average accumulated doses for 116,000 evacuees from highly-contaminated zone (1986)
>50 mSv	Average accumulated doses for 270,000 residents in areas of "strict radiological control" near the accident (1986–2005)
10–20 mSv	Average accumulated doses for 5,000,000 residents of other 'contaminated' areas (1986–2005)

Continued

Table 27.7.   Continued

Dose	Effect
less than 1 mSv/year	Current annual effective doses from the Chernobyl fallout for most of the five million people esiding in contaminated areas of Belarus, Russia and Ukraine
more than 1 mSv per year	Current annual effective doses from the Chernobyl fallout for about 100,000 residents of the more contaminated areas
1,000 mSv/cumulative	Would probably cause a fatal cancer many years later in 5 of every 100 persons exposed to it (ie. if the normal incidence of fatal cancer were 25%, this dose would increase it to 30%)
1,000 mSv/single dose	Causes (temporary) radiation sickness such as nausea and decreased white blood cell count, but not death. Above this, severity of illness increases with dose
5,000 mSv/single dose	Would kill about half those receiving it within a month
10,000 mSv/single dose	Fatal within a few weeks

"Dose" is a broad term that is often used to mean either absorbed dose, or dose equivalent, depending on the context. The absorbed dose is measured in both a traditional unit called a rad and an International System (S.I.) unit called a gray (Gy). Both grays and rads are physical units (1 Gy = 100 rad) that measure the concentration of absorbed energy. The absorbed dose is the amount of energy absorbed per kilogram of absorber. Physical doses from different radiations are not biologically equivalent. For this reason, a unit called the dose equivalent, which considers both the physical dose and the radiation type, is used in radiation safety dosimetry. The unit of dose equivalent is called the rem in traditional units and the sievert (Sv) in S.I. units (1 rem = 0.01 Sv).

[a]Range from sea level to high ground elevation.
[b]Depending on radionuclide composition of soil and building materials.
[c]Depending on indoor accumulation of radon gas.
[d]Depending on radionuclide composition of foods and drinking water.

Sources: U.S. Department of Health and Human Services, Agency for Toxic Substances and Disease Registry (ATSDR). Health Statement for Ionizing Radiation, <http://www.atsdr.cdc.gov/toxprofiles/phs149.html>, Accessed 20 November 2009 World Nuclear Association, Radiation and Nuclear Energy, <http://www.world-nuclear.org/info/inf05.html>, Accessed 20 November 2009 U.S. Environmental Protection Agency, Calculate Your Radiation Dose, <http://www.epa.gov/rpdweb00/understand/calculate.html>, Accessed 20 November 2009. The Chernobyl Forum: 2003–2005 (Second revised version), Chernobyl's Legacy: Health, Environmental and Socio-Economic Impacts and Recommendations to the Governments of Belarus, the Russian Federation and Ukraine, <http://www.who.int/ionizing_radiation/chernobyl/chernobyl_digest_report_EN.pdf>, Accessed 21 November 2009. United Nations Scientific Committee on the Effects of Atomic Radiation. 2000. Report of the United Nations Scientific Committee on the Effects of Atomic Radiation to the General Assembly, Volume 1, Sources and Effects of Ionizing Radiation, <http://www.unscear.org/unscear/en/publications/2000_1.html>.

Table 27.8.   Normalized collective effective dose to members of the public from radionuclides released in effluents from the nuclear fuel cycle[a]

Source	Normalized collective effective dose [man Sv (GW a)$^{-1}$]				
	1970–1979	1980–1984	1985–1989	1990–1994	1995–1997
Local and regional component					
Mining	0.19	0.19	0.19	0.19	0.19
Milling	0.008	0.008	0.008	0.008	0.008
Mine and mill tailings (releases over five years)	0.04	0.04	0.04	0.04	0.04
Fuel fabrication	0.003	0.003	0.003	0.003	0.003
Reactor operation Atmospheric Aquatic	2.8 0.4	0.7 0.2	0.4 0.06	0.4 0.05	0.4 0.04

Table 27.8. Continued

Source	Normalized collective effective dose [man Sv (GW a)$^{-1}$]				
Reprocessing Atmospheric Aquatic	0.3 8.2	0.1 1.8	0.06 0.11	0.03 0.10	0.04 0.09
Transportation	<0.1	<0.1	<0.1	<0.1	<0.1
Total (rounded)	12	3.1	0.97	0.92	0.91
Solid waste disposal and global component					
Mine and mill tailings (releases of radon over 10,000 years)	7.5	7.5	7.5	7.5	7.5
Reactor operation Low-level waste disposal Intermediate-level waste disposal	0.00005 0.5	0.00005 0.5	0.00005 0.5	0.00005 0.5	0.00005 0.5
Reprocessing solid waste disposal	0.05	0.05	0.05	0.05	0.05
Globally dispersed radionuclides (truncated to 10,000 years)	95	70	50	40	40
Total (rounded)	100	80	60	50	50

[a]Analysis is based on reported releases per unit electrical energy generated and presently adopted dose coefficients. These results may, therefore, differ somewhat from earlier evaluations by the Committee.
Source: United Nations Scientific Committee on the Effects of Atomic Radiation (UNSCEAR), UNSCEAR 2000 Report Vol. 1: Sources and Effects of Ionizing Radiation.

Table 27.9. Occupational radiation exposures

Source	Number of monitored workers (thousands)	Average annual effective dose (mSv)
Man-made sources		
Nuclear fuel cycle (including uranium mining)	800	1.8
Industrial uses of radiation	700	0.5
Defense activities	420	0.2
Medical uses of radiation	2, 320	0.3
Education/veterinary	360	0.1
Total from man-made sources	4, 600	0.6
Enhanced natural sources		
Air travel (crew)	250	3.0
Mining (other than coal)	760	2.7
Coal mining	3910	0.7
Mineral processing	300	1.0
Above ground workplaces (radon)	1, 250	4.8
Total from natural sources	6, 500	1.8

The sievert (Sv) is the International System of Units (SI) derived unit of dose equivalent radiation that takes into account the relative biological effectiveness of different forms of ionizing radiation. One millisievert (mSv) corresponds to 10 ergs of energy of gamma radiation transferred to one gram of living tissue.
Source: United Nations Scientific Committee on the Effects of Atomic Radiation (UNSCEAR), UNSCEAR 2000 Report Vol. 1: Sources and Effects of Ionizing Radiation. United Nations Scientific Committee on the Effects of Atomic Radiation, <http://www.unscear.org/unscear/en/publications/2000_1.html>, accessed 29 April 2012.

Table 27.10. Annual estimated average effective dose equivalent received by a member of the population of the United States

Source	Average annual effective dose equivalent (mrem)	Average annual effective dose equivalent (µSv)
Inhaled (Radon and Decay Products)	2000	200
Other internally deposited radionuclides	390	39
Terrestrial radiation	280	28
Cosmic radiation	270	27
Cosmogenic radioactivity	10	1
Rounded total from natural sources	3000	300
Rounded total from artificial sources	600	60
Total	3600	360

The sievert (Sv) is the International System of Units (SI) derived unit of dose equivalent radiation that takes into account the relative biological effectiveness of different forms of ionizing radiation. One millisievert (mSv) corresponds to 10 ergs of energy of gamma radiation transferred to one gram of living tissue. The roentgen equivalent or rem is a unit of radiation dose equivalent. It is the product of the absorbed dose in rads and a weighting factor that accounts for the effectiveness of the radiation to cause biological damage.
Source: The Radiation Information Network, Idaho State University, Radioactivity in Nature, <http://www.physics.isu.edu/radinf/natural.htm>.

Table 27.11. Worldwide average annual exposures from the commercial nuclear fuel cycle [a]

Practice	Monitored workers [b] (thousands)	Average annual collective effective dose (man Sv)	Average annual collective effective dose per unit energy generated (man Sv per GW a) 1975–1979	Average annual effective dose to monitored workers	Distribution ratio[c]	
					NR$_{15}$[d]	SR$_{15}$
Mining[e, f]	240	1300	5.7	5.5	0.37	0.69
Milling[e,f]	12	120	0.52	10	0.41	0.76
Enrichment[e]	11	5.3	0.02	0.5	0	0
Fuel fabrication	20	36	0.59	1.8	0.012	0.38[i]
Reactor operation	150	600	11	4.1	0.078[h]	0.60[j]
Reprocessing[g]	7.2	53	0.7	7.3	0.16	0.29[g]
Research	120	170	1	1.4	0.035	0.42
Total	560	2300	20	4.1	0.2	0.63
1990–1994						
Mining	69 (62)	310	1.72	4.5 (5.0)	0.1	0.32
Milling[e, f]	6	20	0.11	3.3	0	0.01
Enrichment[e]	13	1	0.02	0.12	0	0
Fuel fabrication	21 (11)	22	0.1	1.03 (2.0)	0.01	0.11
Reactor operation	530 (300)	900	3.9	1.4 (2.7)	0.00[h]	0.08
Reprocessing[g, k]	45 (24)	67	3	1.5 (2.8)	0	0.13
Research	120 (36)	90	1	0.78 (2.5)	0.01	0.22
Total	800 (450)	1400	9.8	1.75 (3.1)	0.01	0.11

Notes: FBR = fast breeder reactor; GCR = gas-cooled reactor; HTGR = high temperature gas-cooled reactor; HWR = heavy water reactor; LWGR = light water graphite reactor; LWR = light water reactor
[a]The data are annual values averaged over the indicated periods.

Table 27.11.   Continued

[b]Data in parentheses relate to data for measurably exposed workers.

[c]The values of the distribution ratios should only be considered indicative of worldwide levels as they are based, in general, on data from far fewer countries than the data for number of workers and collective doses.

[d]This ratio applies to monitored workers.

[e]Also include uranium obtained or processed for purposes other than the commercial nuclear fuel cycle.

[f]For 1985-1989 the data for mining and milling (except for NR and SR) have been modified from those reported by using a conversion factor of 5.6 mSv WLM^{-1} for exposure to radon daughters (10 mSvWLM^{-1} used in the reported data). The ratios NR$_{15}$ and SR$_{15}$ are averages of reported data in which, in general, the previously used conversion factor has been applied. The tabulated ratios are thus strictly for a value of E somewhat less than 15 mSv. The relationship between the reported and revised data is not linear because exposure occurs from other than just inhalation of radon progeny. For 19901994 a conversion factor of 5.0 mSv WLM^{-1} for exposure to radon daughters is used.

[g]Also includes the reprocessing of some fuel from the defence nuclear fuel cycle.

[h]Does not include data for LWGRs, FBRs and HTGRs.

[i]Ratio applies to LWR and HWR fuels only, as data for other fuels are not available; the ratio would be smaller if all fuel types were included.

[j]Does not include data for GCRs, LWGRs, FBRs and HTGRs.

[k]In the absence of sufficient data on equivalent electrical energy generated from reporting countries for 1990-1994, the Committee has taken the normalized average annual collective effective per unit energy generated to be the same as that for the previous period.

Source:United Nations Scientific Committee on the Effects of Atomic Radiation (UNSCEAR), UNSCEAR 2000 Report Vol. 1: Sources and Effects of Ionizing Radiation.

Table 27.12.   Natural radioactivity in the human body[a]

Nuclide	Total mass of nuclide found in the body	Total activity of nuclide found in the body	Daily intake of nuclides
Uranium	90 µg	30 pCi (1.1 Bq)	1.9 µg
Thorium	30 µg	3 pCi (0.11 Bq)	3 µg
Potassium-40	17 mg	120 nCi (4.4 kBq)	0.39 mg
Radium	31 pg	30 pCi (1.1 Bq)	2.3 pg
Carbon-14	22 ng	0.1 µCi (3.7 kBq)	1.8 ng
Tritium	0.06 pg	0.6 nCi (23 Bq)	0.003 pg
Polonium	0.2 pg	1 nCi (37 Bq)	~0.6 fg

The becquerel (symbol Bq) is the SI-derived unit of radioactivity. One Bq is defined as the activity of a quantity of radioactive material in which one nucleus decays per second. The Bq unit is thus equivalent to an inverse second, s−1. The curie (symbol Ci) is a non-SI unit of radioactivity defined as 1 Ci = 3.7 × 1010 decays per second.

[a]Based on a 70,000 gram adult.

Source: Adapted from The Radiation Information Network, Idaho State University, Radioactivity in Nature, <http://www.physics.isu.edu/radinf/natural.htm>.

Table 27.13.   **Uranium toxicity on various body systems**

Body system	Human studies	Animal studies	In vitro
Renal	Elevated levels of protein excretion, urinary catalase and diuresis	Damage to Proximal convoluted tubules, necrotic cells cast from tubular epithelium, glomerular changes	No studies
Brain/CNS	Decreased performance on neurocognitive tests	Acute cholinergic toxicity; Dose-dependent accumulation in cortex, midbrain, and vermis; Electrophysiological changes in hippocampus	No studies
DNA	Increased reports of cancers	Increased urine mutagenicity and induction of tumors	Binucleated cells with micronuclei, Inhibition of cell cycle kinetics and proliferation; Sister chromatid induction, tumorigenic phenotype
Bone/muscle	No studies	Inhibition of periodontal bone formation; and alveolar wound healing	No studies
Reproductive	Uranium miners have more first born female children	Moderate to severe focal tubular atrophy; vacuolization of Leydig cells	No studies
Lungs/respiratory	No adverse health effects reported	Severe nasal congestion and hemorrage, lung lesions and fibrosis, edema and swelling, lung cancer	No studies
Gastrointestinal	Vomiting, diarrhea, albuminuria	n/a	n/a
Liver	No effects seen at exposure dose	Fatty livers, focal necrosis	No studies
Skin	No exposure assessment data available	Swollen vacuolated epidermal cells, damage to hair follicles and sebaceous glands	No studies
Tissues surrounding embedded DU fragments	Elevated uranium urine concentrations	Elevated uranium urine concentrations, perturbations in biochemical and neuropsychological testing	No studies
Immune system	Chronic fatigue, rash, ear and eye infections, hair and weight loss, cough. May be due to combined chemical exposure rather than DU alone	No studies	No studies
Eyes	No studies	Conjunctivitis, irritation inflammation, edema, ulceration of conjunctival sacs	No studies
Blood	No studies	Decrease in RBC count and hemoglobin concentration	No studies
Cardiovascular	Myocarditis resulting from the uranium ingestion, which ended 6 months after ingestion	No effects	No studies

Source: Craft, Elena S., Abu-Qare, Aquel W., Flaherty, Meghan M., Garofolo, Melissa C., Rincavage, Heather L. and Abou-Donia, Mohamed B.'DEPLETED AND NATURAL URANIUM: CHEMISTRY AND TOXICOLOGICAL EFFECTS', Journal of Toxicology and Environmental Health, Part B, 7: 4, 297–317.

Table 27.14. **Health effects of selected power plant pollutants**

	Human Toxicity		
Substance	Acute	Chronic	Comments
Sulfur dioxide	Lung irritant, triggers asthma, low birth weight in infants.	Reduces lung function, associated with premature death.	Also contributes to acid rain and poor visibility.
Nitrogen oxides	Changes lung function, increases respiratory illness in children.	Increases susceptibility to respiratory illnesses and causes permanent alteration of lung.	Forms ozone smog and acid rain. Ozone is associated with asthma, reduced lung function, adverse birth outcomes and allergen sensitization.
Particulate Matter	Asthma attacks, heart rate variability, heart attacks.	Cardiovascular disease, pneumonia, chronic obstructive pulmonary disease, premature death.	
Hydrogen chloride	Inhalation causes coughing, hoarseness, chest pain, and inflammation of respiratory tract.	Chronic occupational exposure is associated with gastritis, chronic bronchitis, dermatitis, photo sensitization in workers.	
Hydrogen Fluoride	Inhalation causes severe respiratory damage, severe irritation and pulmonary edema.		Very high exposures through drinking water or air can cause skeletal fluorosis.
Arsenic	Ingestion and inhalation: affects the gastrointestinal system and central nervous system.	Known human carcinogen with high potency. Inhalation causes lung cancer; ingestion causes lung, skin, bladder and liver cancer. The kidney is affected following chronic inhalation and oral exposure.	
Cadmium	Inhalation exposure causes bronchial and pulmonary irritation. A single acute exposure to high levels of cadmium can result in long-lasting impairment of lung function.	Probable human carcinogen of medium potency. The kidney is the major target organ in humans following chronic inhalation and oral exposure.	Other effects noted from chronic inhalation exposure are bronchiolitis and emphysema.
Chromium	High exposure to chromium VI may result in renal toxicity, gastrointestinal hemorrhage and internal hemorrhage.	Known human carcinogen of high potency.	Chronic effects from industrial exposures are inflammation of the respiratory tract, effects on the kidneys, liver, and gastrointestinal tract.
Mercury	Inhalation exposure to element- al mercury results in central nervous system effects and effects on gastrointestinal tract and respiratory system.	Methyl mercury ingestion causes developmental effects. Infants born to women who in- gested methylmercury may perform poorly on neurobehavorial tests	The major effect from chronic exposure to inorganic mercury is kidney damage.

Source: Adapted from Clean Air Task Force. 2001. Cradle to Grave: The Environmental Impacts from Coal, (Boston, MA), based on data from Agency for Toxic Substances and Disease Registry Online, ToxFAQs, Division of Toxicology, Atlanta, Georgia; U.S. EPA, 2000. Integrated risk information system (IRIS). Online. Office of Health and Environmental Assessment, Cincinnati, Ohio.

Table 27.15.   Potential explosion hazards from common landfill gas components[a]

Component	Potential to Pose an Explosion Hazard
Methane	Methane is highly explosive when mixed with air at a volume between its LEL of 5% and its UEL of 15%. At concentrations below 5% and above 15%, methane is not explosive. At some landfills, methane can be produced at sufficient quantities to collect in the landfill or nearby structures at explosive levels.
Carbon dioxide	Carbon dioxide is not flammable or explosive.
Nitrogen dioxide	Nitrogen dioxide is not flammable or explosive.
Oxygen	Oxygen is not flammable, but is necessary to support explosions.
Ammonia	Ammonia is flammable. Its LEL is 15% and its UEL is 28%. However, ammonia is unlikely to collect at a concentration high enough to pose an explosion hazard.
non-methane organic compounds (NMOCs)	Potential explosion hazards vary by chemical. For example, the LEL of benzene is 1.2% and its UEL is 7.8%. However, benzene and other NMOCs alone are unlikely to collect at concentrations high enough to pose explosion hazards.
Hydrogen sulfide	Hydrogen sulfide is flammable. Its LEL is 4% and its UEL is 44%. However, in most landfills, hydrogen sulfide is unlikely to collect at a concentration high enough to pose an explosion hazard.

[a]LEL = Lower Explosive Limit; HEL = Higher Explosive Limit.
Source: United States Center for Disease Control.

Table 27.16.   Incident records and estimated production for substances used in PV operation and regulated under the RMP rule in the U.S[a]

Substance	Source	Total average (1994–2000) US production (1000 metric tons yr^{-1})	Number of incidents (employees and public) in the RMP database (1994–2004)		
			Incidents	Injuries	Deaths
Toxic					
Arsine (AsH$_3$)	GaAs chemical vapor deposition (CVD)	23	2	1	0
Boron trichloride (BCl$_3$)	p-type dopant for epitaxial silicon	NA	0	0	0
Boron trifluoride (BF$_3$)	p-type dopant for epitaxial silicon	NA	1	1	0
Diborane (B$_2$H$_6$)	a-Si dopant	~50	0	0	0
Hydrochloric acid (HCl)	Cleaning agent for c-Si	3,500	28	12	1
Hydrogen fluoride (HF)	Etchant for c-Si	190	165 (57)[b]	209 (70)[b]	1
Hydrogen selenide (H$_2$Se)	CIGS selenization	N/A	4	17	0
Hydrogen sulfide (H$_2$S)	CIS sputtering	>110	40	47	1
Phosphine (PH$_3$)	a-Si dopant	N/A	0	0	0
Flammable					
Dichlorosilane (SiH$_2$Cl$_2$)	a-Si and c-Si deposition	N/A	2	0	0
Hydrogen (H$_2$)	a-Si deposition/GaAs	18,000 m^3	57	65	4
Silane (SiH$_4$)	a-Si deposition	8	5	2	0
Trichlorosilane (SiHCl$_3$)	Precursor of c-Si	110	14	14	2

a-Si, amorphous silicon; c-Si, crystalline silicon; CIGS, copper indium gallium selenide; CIS, copper indium diselenide; RMP, risk management program.
[a]Under the authority of section 112(r) of the Clean Air Act, the Chemical Accident Prevention Provisions require facilities that produce, handle, process, distribute, or store certain chemicals to develop a Risk Management Program, prepare a Risk Management Plan (RMP), and submit the RMP to the U.S. Environmental Protection Agency.
[b]Number excludes incidents in petroleum refineries.
Source: Adapted from Fthenakis, V.M. and H.C. Kim. 2012. 2012. 1.08 – Environmental Implacts of Photovoltaic Life Cycles, In: Editor-in-Chief: Ali Sayigh, Comperehensive Renewable Energy (Oxford Elsevier) pages 47-72

Table 27.17.   Safety-related properties of hydrogen and other fuels

Property	Hydrogen	Methanol	Methane	Propane	Gasoline	Unit
Minimum energy for ignition	0.02	–	0.29	0.25	0.24	10^{-3} J
Flame temperature	2045		1875		2200	°C
Auto-ignition temperature in air	585	385	540	510	230–500	°C
Maximum flame velocity	3.46	–	0.43	0.47		m s^{-1}
Range of flammability in air	4–75	7–36	5–15	2.5–9.3	1.0–7.6	vol.%
Range of explosivity in air	13–65	6.3–13.5	1.1–3.3	vol.%		
Diffusion coefficient in air	0.61	0.16	0.2	0.1	0.05	10^{-4} m^2 s^{-1}

Source: Adapted from Sørensen, Bent. 2012. Hydrogen and Fuel Cells-Emerging Technologies and Applications (Second Edition), (Boston, Academic Press).

Table 27.18.   Type of facility where accidents occur with oil and chemical storage tanks, 1960–2003

Year	Refinery	Terminal/Storage	Chemical Plant[a]	Oil Field	Miscellaneous[b]	Total
1960–1969	10	5	1	0	1	17
1970–1979	22	11	0	0	3	36
1980–1989	25	17	5	2	4	53
1990–1999	41	22	16	1	5	85
2000–2003	18	9	9	3	12	51
Total	116	64	31	6	25	242

[a]Petrochemical plants included.
[b]Other industrial facilities such as power, gas, pipeline, fertilizer, and plating plants.
Source: Chang, James I. 2006. Cheng-Chung Lin, A study of storage tank accidents, Journal of Loss Prevention in the Process Industries, Volume 19, Issue 1, Pages 51-59

Table 27.19.   Cause of accidents with oil and chemical storage tanks, 1960–2003

Year	1960–1969	1970–1979	1980–1989	1990–1999	2000–2003	Total
Lightning	4	10	19	37	10	80
Maintenance/hot work	1	5	9	12	5	32
Operational error	1	5	6	8	9	29
Equipment failure	3	1	5	7	3	19
Sabotage	2	5	2	6	3	18
Crack/rupture	0	3	3	3	8	17
Leaks and line rupture	0	3	2	5	5	15
Static electricity	2	1	2	2	5	12
Open flame	1	0	4	2	1	8
Natural disaster	1	2	1	1	2	7
Runaway reaction	2	1	0	2	0	5
Total	17	36	53	85	51	242

Source: Chang, James I. 2006. Cheng-Chung Lin, A study of storage tank accidents, Journal of Loss Prevention in the Process Industries, Volume 19, Issue 1, Pages 51–59.

Table 27.20.   Type of accident associated with tank failures in refineries and chemical plants, 1960–3003

Year	Fire	Explosion	Spill	Toxic gas Release	Misc.	Total
1960–1969	8	8	0	0	1[a]	17
1970–1979	26	5	5	0		36
1980–1989	31	16	3	2	1[a]	53
1990–1999	59	22	2	1	1[b]	85
2000–2003	21	10	8	10	2[c]	51
Total	145	61	18	13	5	242

[a]Tank body distortion.
[b]Personal fall.
[c]1 Person fell and 1 person was electrified to death.
Source: Chang, James I. 2006. Cheng-Chung Lin, A study of storage tank accidents, Journal of Loss Prevention in the Process Industries, Volume 19, Issue 1, Pages 51–59

Table 27.21. Type of contents associated with tank failures in refineries and chemical plants, 1960–2003

Year	Crude oil	Oil products[a]	Gasoline / Naphtha	Petro-chemicals	LPG[b]	Waste oil water	Ammonia	Hydrochloric acid	Caustic soda	Molten sulfur	Total
1960–1969	6	3	0	3	3	2	0				17
1970–1979	8	7	13	3	3	2	0				36
1980–1989	17	14	17	4	1	0	0				53
1990–1999	23	19	21	11	5	4	0	1		1	85
2000–2003	12	16	6	6	1	1	3	2	3	1	51
Total	66	59	55	27	15	9	3	3	3	2	242

[a]Fuel oil, diesel, kerosene, lubricants.
[b]Propane and butane included.
Chang, James I. 2006. Cheng-Chung Lin, A study of storage tank accidents, Journal of Loss Prevention in the Process Industries, Volume 19, Issue 1, Pages 51-59.

Table 27.22.   Estimates of anthropogenic causes of bird mortality in the United States

Mortality Source	Annual mortality rate	Percent composition
Buildings	550 million	58.20%
Power lines	130 million	13.70%
Cats	100 million	10.60%
Automobiles	80 million	8.50%
Pesticides	67 million	7.10%
Communications towers	4.5 million	0.50%
Wind turbines	28.5 thousand	<0.1%
Airplanes	25 thousand	<0.1%
Other sources (oil spills, oil seeps, fishing by-catch, etc.)	not calculated	not calculated

Source: Erickson, Wallace P., Gregory D. Johnson and David P. Young Jr. 2005. A Summary and Comparison of Bird Mortality from Anthropogenic Causes with an Emphasis on Collisions, USDA Forest Service Gen. Tech. Rep. PSW-GTR-191.

Table 27.23.   Studies of Bird, Bat, and Raptor Fatality Rates Associated with wind turbines, by Region, in the United States

Region	Location and year	Number of turbines	Fatalities per turbine, per year — Birds	Bats	Raptors
Pacific NW	Stateline, OR - 2003	181	1.93	1.12	0.06
	Nine Canyon, OR - 2003	37	3.59	3.21	0.07
	Klondike, OR - Phase I - 2003	16	1.16[a]	1.16	0
	Vansycle, OR - 2000	38	0.63	0.74	0
West	Foote Creek Rim, WY - 2003	69	1.5	1.34	0.03
	National Wind Tech Center, CO - 2003	Varies	0	0	0
California	Altamont Pass, CA - 2003	5,400	0.19	***	***
	Altamont Pass, CA - 2004	5,400	0.87	0.004	0.24
	Altamont Pass and Solano County, CA - 1992	7,340	***	***	0.058 (1989)
					0.025 (1990)
	Altamont Pass, CA - 1991	3,000	***	***	0.047[b]
	Montezuma Hills, CA - 1992	600	0.074[b]	***	0.047[b]
Midwest	Buffalo Ridge, MN - P1 - 2000	73	0.98	0.26	***
	Buffalo Ridge, MN - P2 - 2000	143	2.27	1.78	***
	Buffalo Ridge, MN - P3 - 2000	138	4.45	2.04	***
	Buffalo Ridge, MN - 2000	73	0.33–0.66	***	***
	Buffalo Ridge, MN - (Bats) - 2004	281	***	3.02 (2001)	***
				1.3 (2002)	
	Northeastern, WI - 2002	31	1.29	4.26	0
	Top of Iowa - 2004	89	0.12[c]	1.88[c]	***
Northeast	Searsburg, VT - 2002	11	0	***	0
Appalachian Mt.	Mountaineer, WV - 2004	44	4.04[d]	47.53	***

Table 27.23.   Continued

Region	Location and year	Number of turbines	Fatalities per turbine, per year		
Region	Location and year	Number of turbines	Birds	Bats	Raptors
Region	Tennessee - 2005	3	7.28	20.8	***
	Mountaineer, WV - 2005	44	***	38[e]	***
	Meyersdale, PA - 2005	20	***	23[e]	***

Some of the studies that presented a bird/turbine/year mortality rate also included raptors in that calculation. With the exception of the studies conducted in the Appalachian region, most of the studies listed were designed and timed to focus on bird mortality. Bats were found only incidentally to the study objectives; therefore, rates of bat mortality reported from those studies may not represent a reliable measure.
[a]Fatality rate applies to small birds only.
[b]Fatality rate not adjusted for both searcher efficiency and scavenging rate.
[c]Fatality rate represents number of birds and bats killed per turbine per 8-month study period.
[d]Fatality rate represents number of bats killed per turbine per 7-month study period.
[e]Fatality rate represents number of birds and bats killed per turbine per 6-week study period; however, bat mortality has been shown to be concentrated in the season during which these study periods took place.
***indicates that the study authors did not calculate a mortality rate for that category.

Table 27.24.   Hypotheses regarding possible mechanisms of bat attraction to or failure to detect wind turbines

Linear corridor hypothesis	Many species of bats (especially red and hoary bats) are known to use linear corridors during migration and while foraging. Wind farms in forested regions can be developed along natural corridors such as ridge tops or corridors are created when access roads are constructed. If bats use such corridors where wind turbines are located, they may increase the chance of collision during migration or while foraging.
Acoustic failure hypothesis	Either migrating or foraging bats may fail to acoustically detect wind turbines, particularly moving blades. If the smooth cylindrical turbine masts are not detected by echolocating bats, then bats may collide directly with and be killed by these structures during flight. The functional range of echolocation by North American bats typically varies from 3–5 m. Migrating bats flying at a velocity of 5 m/s would have less than a second to respond to a wind turbine.
Visual failure hypothesis	Rotating rotor blades are subject to motion smear, thus making them difficult for organisms to see and respond appropriately. This hypothesis relates more to birds, but bats do use vision and bats may fail to visually detect wind turbine rotor blades.
Roost attraction hypothesis	Bats may be attracted to wind turbines because the tall, white turbine masts are perceived as potential roosts. During migration in late summer and fall, bats seek shelter during the day, following night-time travel. Bats may mistake the large, white turbine masts for potential tree roosts and thus increase their susceptibility to collision at turbines.
Light attraction hypothesis	Bats may be attracted to the lights placed on wind turbines. Currently, these lights range from red lights or stroboscopic lights placed on alternative turbines, as recommended by the Federal Aviation Administration.
Acoustic attraction hypothesis	Bats may be attracted to sounds (audible and/or ultrasonic) produced by wind turbines. The uniform constant sounds made by the turbine generator and/or the variable "swishing" sounds made by rotating blades may attract bats and increase their risk of collision.
Motion attraction hypothesis	Curious bats may be attracted to the movement of rotating turbine blades. By investigating the moving blades, bats increase their risk of collision.

Continued

Table 27.24. Continued

Insect concentration hypothesis	Flying insects rise in altitude with warm daily air masses and may become concentrated, particularly along ridge tops on certain nights. If the activity of migrating and locally foraging bats increases in response to high insect concentrations they increase their exposure to turbines and possible collision.
Insect attraction hypothesis.	Flying insects may be attracted to the white turbine masts at night and then get trapped in the downstream wake of the rotors. Bats respond to these concentrations of insects in the wake and collide with the turbine in the process of feeding

Source: Adapted from Arnett, Edward B.,Wallace P. Erickson,Jessica Kerns, Jason Horn. 2005. Relationships between Bats and Wind Turbines in Pennsylvania and West Virginia: An Assessment of Fatality Search Protocols, Patterns of Fatality, and Behavioral Interactions with Wind Turbines, Bats and Wind Energy Cooperative, <www.batsandwind.org/pdf/ar2004.pdf>.

Climate Change

Figures

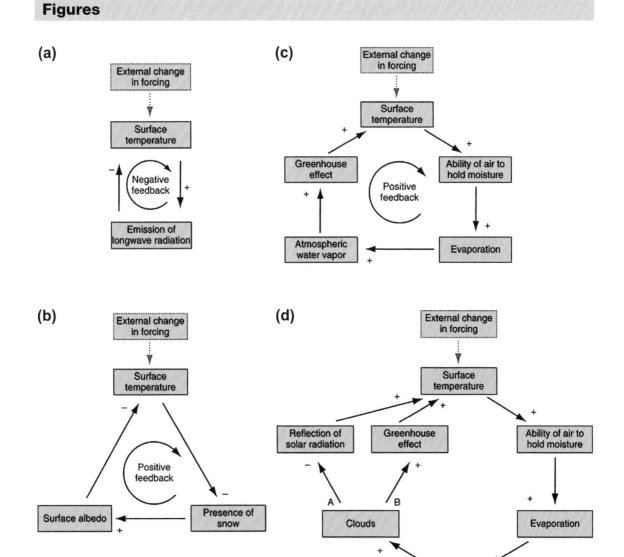

(Continued)

Handbook of Energy. http://dx.doi.org/10.1016/B978-0-08-046405-3.00028-0

Figure 28.1. Examples of important climate system feedbacks that shape the response of the global energy balance and surface temperature to external change. The diagrams show the variables involved in four important feedback processes. The (+/–) signs at the arrows indicate positive/negative influences. (a) The thermal radiation feedback. An external change in forcing that would increase surface temperature would also increase the emission of long-wave radiation (a '+' influence). An increased emission would result in a lower surface temperature (a '–' influence). The enhanced emission of long-wave radiation therefore counteracts the initial change, resulting in a negative feedback loop. The same line of reasoning also applies for an external change that would reduce surface temperature. (b) The snow/ice albedo feedback. An external change that warms the surface reduces the presence of snow, lowers the surface albedo, thereby amplifying the warming (a positive feedback). (c) The water vapor feedback. An external change that warms the surface heats the lower atmosphere. Since warmer air can hold more moisture, this enhances surface evaporation and the amount of water vapor in the atmosphere. More water vapor results in a stronger atmospheric greenhouse effect, thereby amplifying the initial change (a positive feedback). (d) Two types of cloud feedbacks. Continuing from the water vapor feedback, more water vapor in the atmosphere can result in more clouds. Depending on the balance of increased cloud cover on shortwave reflection (path A) or increased greenhouse forcing (path B), cloud feedbacks can form both positive and negative feedback loops on surface temperature.
Source: Kleidon, A. 2008. Energy Balance, In: Sven Erik Jorgensen and Brian Fath, Editors-in-Chief, Encyclopedia of Ecology, (Oxford, Academic Press), Pages 1276-1289.

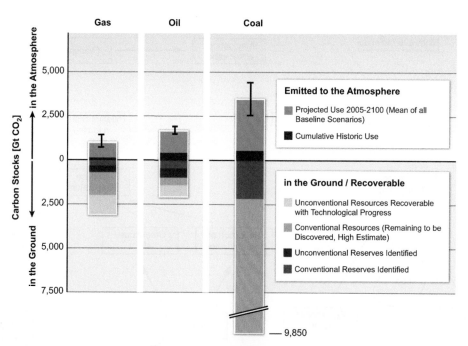

Figure 28.2. Carbon stocks in the atmosphere and in the ground.
Source: Adapted from Moomaw, W., F. Yamba, M. Kamimoto, L. Maurice, J. Nyboer, K. Urama, T. Weir, 2011: Introduction. In IPCC Special Report on Renewable Energy Sources and Climate Change Mitigation [O. Edenhofer, R. Pichs-Madruga, Y. Sokona, K. Seyboth, P. Matschoss, S. Kadner, T. Zwickel, P. Eickemeier, G. Hansen, S. Schlömer, C.von Stechow (eds)], Cambridge University Press, Cambridge, United Kingdom and New York, NY, USA.

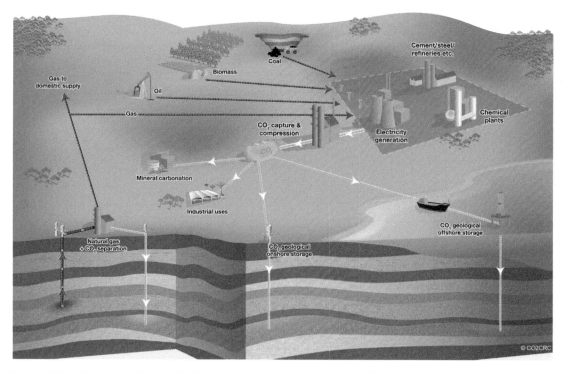

Figure 28.3. Possible pathways for the capture and geologic storage of carbon dioxide.
Source: Cooperative Research Centre for Greenhouse Gas Technologies (CO2CRC), Australian Government's Cooperative Research Centres program, <http://www.co2crc.com.au>.

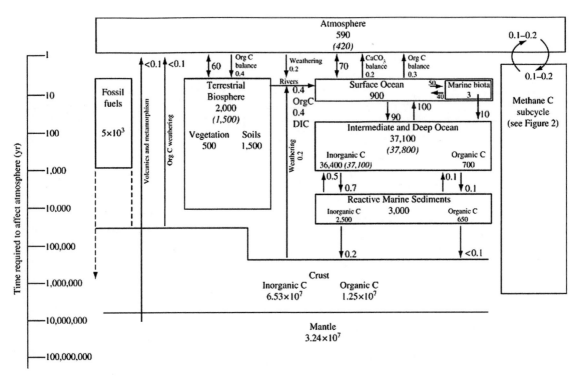

Figure 28.4. Reservoirs (PgC; boxes) and fluxes (PgC yr⁻¹; arrows) of the pre-industrial (1750 AD) global carbon cycle. Values for glacial periods, where available, are shown in parentheses. The vertical bar on the left shows the approximate time (in years) necessary for the different reservoirs to affect the atmosphere. Atmospheric "balance" fluxes are shown to indicate the small net atmospheric exchange required to maintain a steady state with respect to sedimentation of organic carbon and calcium carbonate. The terrestrial biosphere and oceanic reservoir values are rounded to the nearest 100 PgC, total reactive marine sediments to the nearest 1,000 PgC, and all other reservoirs to the number of significant figures shown. All fluxes are rounded to one significant figure.

Source: Sundquist, E.T. and K. Visser. 2003. The Geologic History of the Carbon Cycle, In: Heinrich D. Holland and Karl K. Turekian, Editors-in-Chief, Treatise on Geochemistry, (Oxford, Pergamon), Pages 425-472.

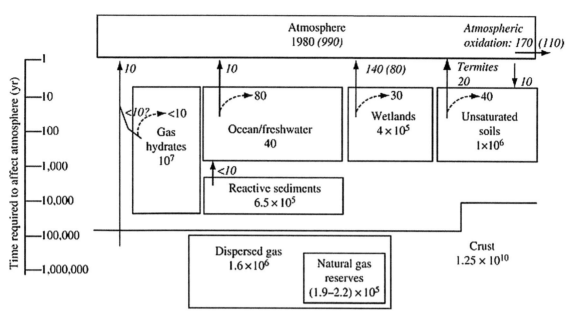

Figure 28.5.   Reservoirs (Tg CH$_4$ or Tg C) and fluxes (Tg CH$_4$ yr^{-1}) of the pre-industrial (1750 AD) methane carbon subcycle. Values for glacial periods, where available, are shown in parentheses. Values for reactive sediments, wetlands, unsaturated soils, and crustal reservoirs represent organic carbon that might be converted to methane, and are all given as Tg C. The vertical bar on the left shows the approximate time (in years) necessary for the different reservoirs to affect the atmosphere. Estimates of methane consumption within reservoirs, are shown as dashed arrows. Gross production of methane within a reservoir can be calculated by adding the flux to the atmosphere and the consumption value. The flux and consumption values are rounded to the nearest 10 Tg CH$_4$ or Tg C.
Source: Sundquist, E.T. and K. Visser. 2003. The Geologic History of the Carbon Cycle, In: Heinrich D. Holland and Karl K. Turekian, Editors-in-Chief, Treatise on Geochemistry, (Oxford, Pergamon), Pages 425-472.

Charts

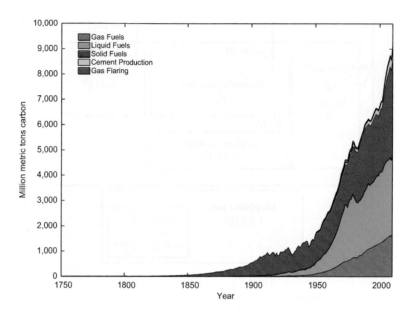

Chart 28.1. Global emissions of carbon dioxide from fossil fuels, by source, 1751–2008.
*Source: Data from United States Department of Energy, Carbon Dioxide Information Analysis Center
(CDIAC), <http://cdiac.ornl.gov/>.*

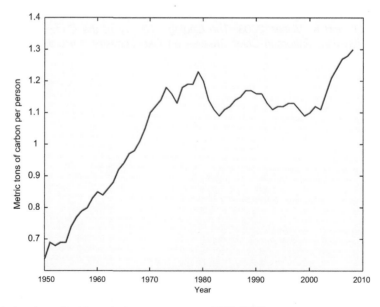

Chart 28.2. Global carbon dioxide emissions per person, 1950–2009.
*Source: Data from United States Department of Energy, Carbon Dioxide Information Analysis Center
(CDIAC), <http://cdiac.ornl.gov/>.*

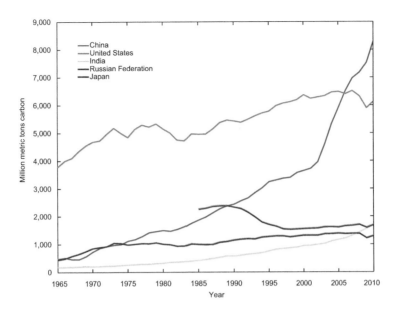

Chart 28.3. Total carbon dioxide emissions, top 5 emitting nations, 1965–2010.
Source: Data from BP, Statistical Review of World Energy 2012, <http://www.bp.com/ sectionbodycopy.do?categoryId=7500&contentId=7068481>.

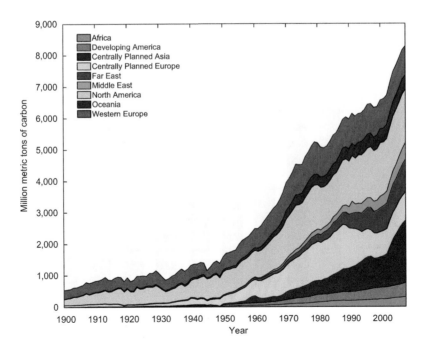

Chart 28.4. Fossil fuel carbon dioxide emissions by region,1900–2008.
Source: Data from United States Department of Energy, Carbon Dioxide Information Analysis Center (CDIAC), <http://cdiac.ornl.gov/>.

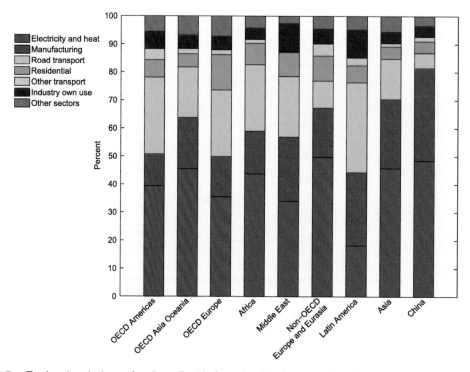

Chart 28.5. Regional emissions of carbon dioxide from fossil fuel combustion by sector, 2009.
OECD = Organisation for Economic Co-operation and Development.
Source: Data from International Energy Agency (IEA), Energy statistics database, <http://www.iea.org/stats/index.asp>.

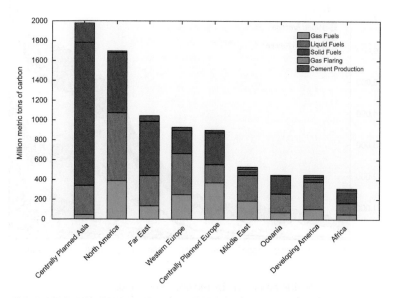

Chart 28.6. Fossil-fuel carbon dioxide emissions by region and by source, 2008.
Source: Data from United States Department of Energy, Carbon Dioxide Information Analysis Center (CDIAC), <http://cdiac.ornl.gov/>.

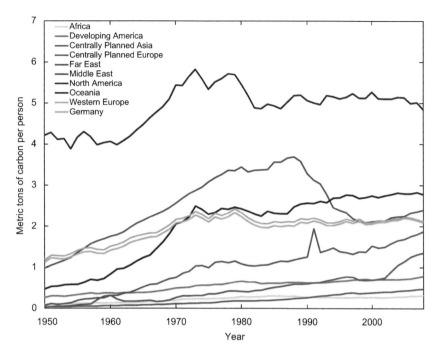

Chart 28.7.   Per capita fossil fuel carbon dioxide emissions by region, 1950–2008.
Source: Data from United States Department of Energy, Carbon Dioxide Information Analysis Center (CDIAC), <http://cdiac.ornl.gov/>.

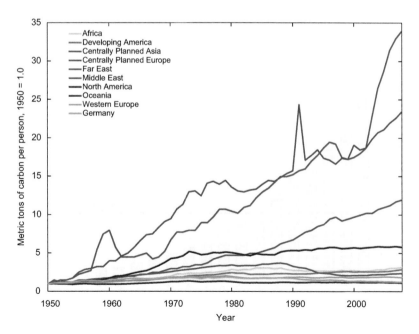

Chart 28.8.   Index of per capita fossil fuel carbon dioxide emissions by region, 1950–2008.
Source: Data from United States Department of Energy, Carbon Dioxide Information Analysis Center (CDIAC), <http://cdiac.ornl.gov/>.

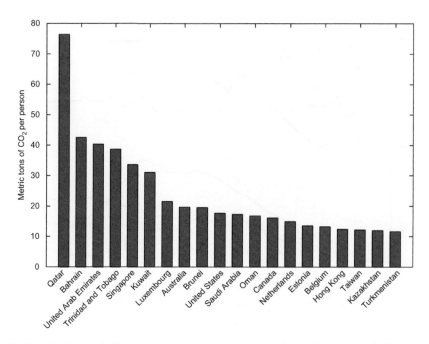

Chart 28.9.   Carbon dioxide emissions per person, 20 most carbon-intensive nations, 2009.
Source: Data from United States Department of Energy, Energy Information Administration, International Energy Statistics, <http://www.eia.gov/countries/data.cfm>.

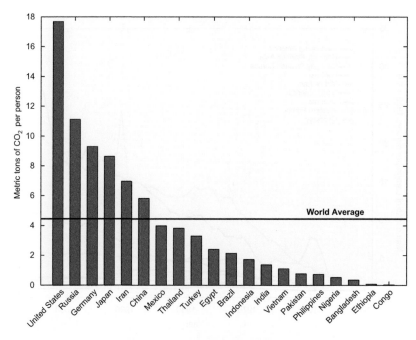

Chart 28.10.   Carbon dioxide emissions per person, 20 most populous nations, 2009.
Source: Data from United States Department of Energy, Energy Information Administration, International Energy Statistics, <http://www.eia.gov/countries/data.cfm>.

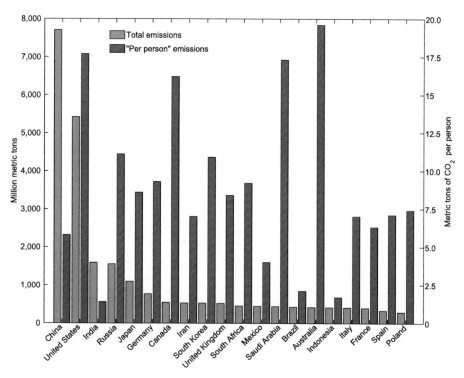

Chart 28.11. Total carbon emissions and carbon emissions per person, top 20 nations, 2009.
Source: Data from International Energy Agency (IEA), Energy statistics database, <http://www.iea.org/stats/index.asp>.

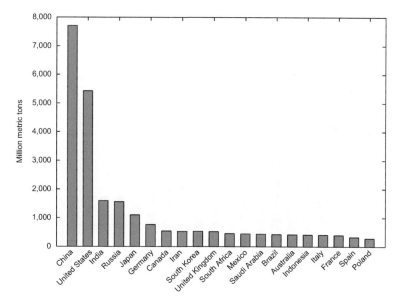

Chart 28.12. Carbon dioxide emissions from the consumption of energy, top 20 nations, 2009.
Source: Data from United States Department of Energy, Energy Information Administration, International Energy Statistics, <http://www.eia.gov/countries/data.cfm>.

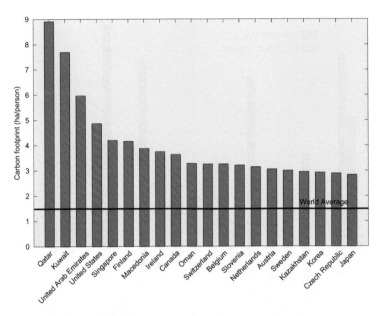

Chart 28.13.   Carbon footprint top 20 nations, 2011. Carbon footprint is defined as the demand on biocapacity required to sequester (through photosynthesis) the carbon dioxide emissions from fossil fuel combustion. Although fossil fuels are extracted from the Earth's crust and are not regenerated in human time scales, their use demands ecological services if the resultant carbon dioxide is not to accumulate in the atmosphere. The footprint therefore includes the biocapacity - typically that of unharvested forests - needed to absorb that fraction of fossil carbon dioxide which is not absorbed by the ocean.
Source: Data from Global Footprint Network, National Footprint Accounts 2011 Edition, <www.footprintnetwork.org/atla>.

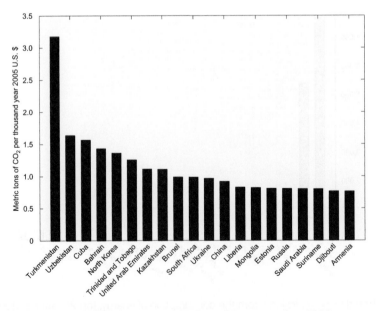

Chart 28.14.   Carbon intensity of GDP, 20 most carbon-intensive economies, 2009.
Source: Data from United States Department of Energy, Energy Information Administration, International Energy Statistics, <http://www.eia.gov/countries/data.cfm>.

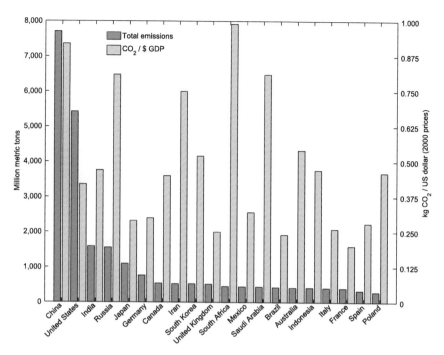

Chart 28.15. Total carbon dioxide emissions from fossil fuel use and the carbon intensity of GDP for the 20 largest emitters, 2009.
Source: Data from United States Department of Energy, Energy Information Administration, International Energy Statistics, <http://www.eia.gov/countries/data.cfm>.

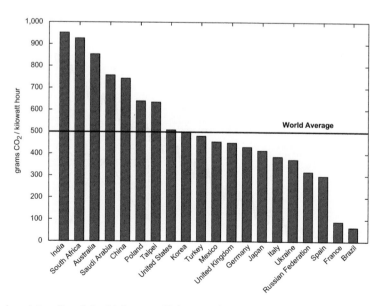

Chart 28.16. Carbon intensity of electricity use, 20 largest electricity users, 2009.
Source: Data from International Energy Agency (IEA), Energy statistics database, <http://www.iea.org/stats/index.asp>.

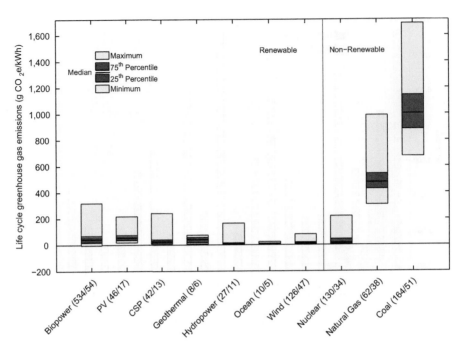

Chart 28.17. Life cycle greenhouse gas emissions from electricity generation. Each category displays (published data count/number of references used).
Source: Data from United States Department of Energy, National Renewable Energy Laboratory, LCA Harmonization database, <http://en.openei.org/apps/LCA/>.

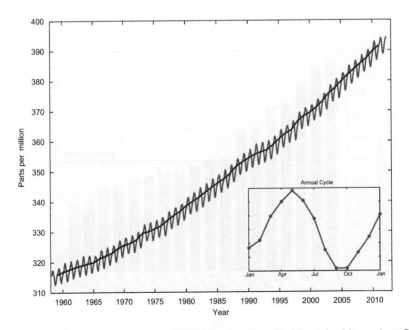

Chart 28.18. Monthly mean atmospheric concentrations of carbon dioxide at the Mauna Loa Observatory.
Source: Data from United States National Oceanic and Atmospheric Administration, Trends in Atmospheric Carbon Dioxide, <http://www.esrl.noaa.gov/gmd/ccgg/trends/>.

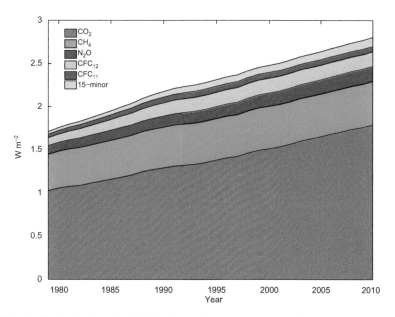

Chart 28.19. Global radiative forcing by individual greenhouse gases, 1979–2010.
Source: Data from United States Department of Energy, National Oceanic and Atmospheric Administration, Earth System Research Laboratory, Annual Greenhouse Gas Index, <http://www.esrl.noaa.gov/gmd/aggi/>.

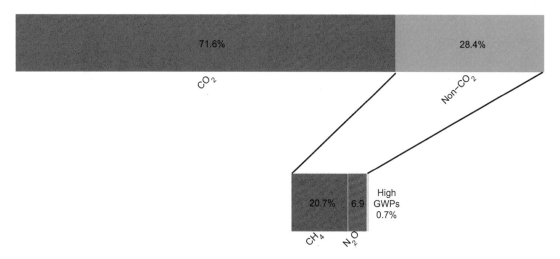

Chart 28.20. Contribution of anthropogenic greenhouse gas emissions to global radiative forcing (W/m²).
Source: Data from IPCC, 2007: Climate Change 2007: The Physical Science Basis. Contribution of Working Group I to the Fourth Assessment Report of the Intergovernmental Panel on Climate Change [Solomon, S., D. Qin, M. Manning, Z. Chen, M. Marquis, K.B. Averyt, M. Tignor and H.L. Miller (eds.)]. Cambridge University Press, Cambridge, United Kingdom and New York, NY, USA, 996 pp.

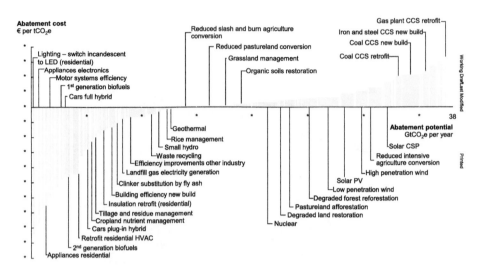

Chart 28.21. Global greenhouse gas abatement curve. The curve shows the estimated cost and potential of various measures to reduce greenhouse gas emissions. The vertical dimension represents the cost per ton of CO2e, while the horizontal dimension represents the potential quantity of CO2-e that could be abated by each measure. *Source: McKinsey & Company. 2010. Version 2.1 of the global greenhouse gas abatement cost curve, <http://www.mckinsey.com/client_service/sustainability/latest_thinking/costcurves>. Reproduced with permission of publisher.*

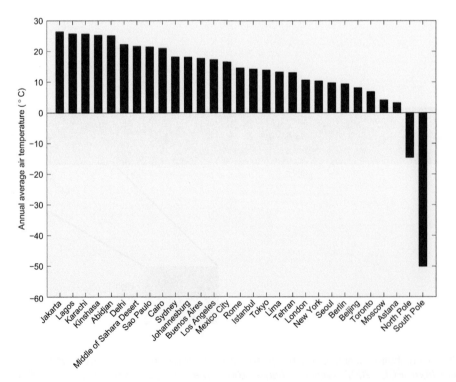

Chart 28.22. Average annual air temperature for selected cities/regions (22-year annual average, July 1983 - June 2005). *Source: Data from United States National Aeronautics and Space Administration (NASA), Surface meteorology and Solar Energy database, <http://eosweb.larc.nasa.gov/sse/>.*

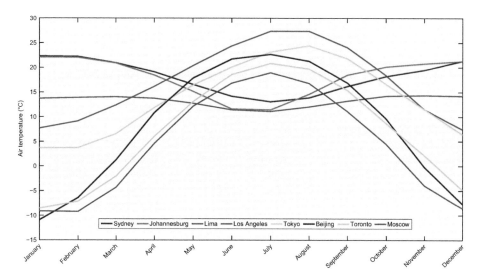

Chart 28.23. Monthly average air temperature for selected cities (22-year annual average, July 1983 - June 2005). *Source: Data from United States National Aeronautics and Space Administration (NASA), Surface meteorology and Solar Energy database, <http://eosweb.larc.nasa.gov/sse/>.*

Tables

Table 28.1. Contribution of human energy use and land use change to global greenhouse gas (GHG) emissions, 1990–2005

	1990	1995	2000	2005
Energy Contribution to Non-CO$_2$ GHG emissions (MtCO$_2$e)[a]	2,316.2	2,331.7	2,401.2	2,699.0
Total Non-CO$_2$GHG emissions (MtCO$_2$e)	9,909.0	9,942.0	10,059.8	10,927.6
Energy share non-CO$_2$ GHG emissions	23.4%	23.5%	23.9%	24.7%
Net carbon emissions from land use change (Tg C)[b]	1643.7	1561.6	1409.9	1467.3
Carbon emissions from fossil fuels (million metric tonnes C)[c]	5994	6225	6524	7766
Total anthropogenic carbon emissions	7637.7	7786.6	7933.9	9233.3
Energy share carbon emissions	78.5%	79.9%	82.2%	84.1%

[a]Source: United States Environmental Protection Agency. 2011. DRAFT: Global Anthropogenic Non-CO$_2$ Greenhouse Gas Emissions: 1990–2030, (Washington, D.C., EPA). Gases included are the direct non-CO$_2$ GHGs covered by the United Nations Framework Convention on Climate Change (UNFCCC): methane (CH$_4$), nitrous oxide (N$_2$O), and the high global warming potential (high-GWP) gases: hydrofluorocarbons (HFCs), perfluorocarbons (PFCs), and sulfur hexafluoride (SF$_6$). In addition, nitrogen fluoride (NF$_3$) is considered. Compounds covered by the Montreal Protocol are not included, although many of them are also high-GWP gases. Emissions of non-CO$_2$ gases are converted to a CO$_2$-equivalent (CO$_2$e) basis using the 100-year GWPs published in the IPCC's Second Assessment Report (SAR) (IPCC, 1996).
[b]Source: Houghton, R.A. 2008. Carbon Flux to the Atmosphere from Land-Use Changes: 1850-2005. In TRENDS: A Compendium of Data on Global Change. Carbon Dioxide Information Analysis Center, Oak Ridge National Laboratory, U.S. Department of Energy, Oak Ridge, Tenn., U.S.A.
[c]Source: Boden, T.A., G. Marland, and R.J. Andres. 2011. Global, Regional, and National Fossil-Fuel CO2 Emissions. Carbon Dioxide Information Analysis Center, Oak Ridge National Laboratory, U.S. Department of Energy, Oak Ridge, Tenn., U.S.A. doi 10.3334/CDIAC/00001_V2011.

Table 28.2. Global radiative forcing 1980–2010 (W m^{-2})

Gas	1980	1990	2000	2010	Change 1980–2010	Change 2000–2010
CO$_2$	1.056	1.29	1.512	1.791	69.6%	18.5%
CH$_4$	0.426	0.472	0.495	0.504	18.3%	1.8%
N$_2$O	0.104	0.129	0.151	0.175	68.3%	15.9%
CFC12	0.096	0.154	0.173	0.17	77.1%	−1.7%
CFC11	0.042	0.065	0.066	0.06	42.9%	−9.1%
15-minor[a]	0.034	0.065	0.083	0.105	208.8%	26.5%
Total	1.76	2.18	2.48	2.81	59.7%	13.3%
AGGI[b]	0.81	1	1.14	1.29	59.3%	13.2%

[a]15 minor long-lived halogenated gases (CFC-113, CCl4, CH3CCl3, HCFCs 22, 141b and 142b, HFCs 134a, 152a, 23, 143a, and 125, SF6, and halons 1211, 1301 and 2402). Except for the HFCs and SF6, which do not contain chlorine or bromine, these gases are also ozone-depleting gases and are regulated by the Montreal Protocol.
[b]The Annual Greenhouse Gas Index (AGGI) is the ratio of the total radiative forcing due to long-lived greenhouse gases for any year for which adequate global measurements exist to that which was present in 1990. 1990 was chosen because it is the baseline year for the Kyoto Protocol.
Source: National Oceanic and Atmospheric Administration, Earth System Research Laboratory, Global Monitoring Division , The NOAA Annual Greenhouse Gas Index (AGGI), <http://www.esrl.noaa.gov/gmd/aggi/>.

Table 28.3. Carbon dioxide emissions from fossil fuel use by region, 1970–2010 (Million metric tons carbon dioxide)

	1970	1980	1990	2000	2010	2010 share of total	Change 1970–2010	Change 2000–2010
North America	5,133	5,843	6,229	7,348	7,197	21.7%	40.2%	−2.1%
South & Central America	382	611	707	978	1,272	3.8%	232.5%	30.0%
Europe & Eurasia	6,725	8,328	8,644	7,044	7,164	21.6%	6.5%	1.7%
Middle East	214	388	745	1,157	1,913	5.8%	795.8%	65.3%
Africa	235	432	666	811	1,077	3.2%	358.8%	32.8%
Asia Pacific	2,305	3,720	5,621	8,239	14,536	43.8%	530.7%	76.4%
World	14,992.5	19,322.4	22,613.2	25,576.9	33,158.4	100.0%	121.2%	29.6%

Carbon emissions reflect only those through consumption of oil, gas and coal, and are based on standard global average conversion factors: oil - 73,300 kg CO_2 per TJ (3.07 metric tons per metric ton of oil equivalent); natural gas - 56,100 kg CO_2 per TJ (2.35 metric tons per metric ton of oil equivalent); coal - 94,600 kg CO_2 per TJ (3.96 metric tons per metric ton of oil equivalent). As such, these numbers may diverge from offical national estimates that are based on more specific emission conversion factors.
Source: BP Statistical Review of World Energy (2011).

Table 28.4. Carbon dioxide emissions per capita from fosssil fuels for the 10 most populous nations, 1980–2009

	Population in 2009 millions	Population Rank 2009	Per capita emissions[a]				Change in Per Capita Emissions	
			1980	1990	2000	2009	1980–2009	2000–2009
China	1,323.6	1	1.5	2.0	2.3	5.8	295.9%	158.2%
India	1,156.9	2	0.4	0.7	1.0	1.4	223.4%	38.0%
United States	307.0	3	21.0	20.2	20.8	17.7	−15.9%	−14.9%
Indonesia	240.3	4	0.6	0.9	1.2	1.7	204.0%	39.2%
Brazil	198.7	5	1.5	1.6	2.0	2.1	41.8%	9.5%
Pakistan	181.5	6	0.4	0.6	0.7	0.8	95.7%	8.2%
Bangladesh	153.7	7	0.1	0.1	0.2	0.4	313.0%	61.4%
Nigeria	149.2	8	0.9	0.9	0.7	0.5	−43.3%	−20.1%
Russia	140.0	9	--	--	10.6	11.1		4.8%
Japan	127.1	10	8.1	8.5	9.5	8.6	6.6%	−8.9%
World	6,776.9		4.1	4.1	3.9	4.5	8.0%	14.4%

[a]Metric tons of carbon dioxide per person.
Source: United States Energy Information Administration, International Energy Statistics, <http://www.eia.gov/countries/data.cfm>.

Table 28.5. Emissions and intensities of carbon dioxide from the consumption of energy, 2009

Country	Total Emissions[a]	World Rank	Emissions Per $ GDP[b]	World Rank	Emissions Per person[c]	World rank
China	7,706.8	1	0.92	12	5.8	52
United States	5,424.5	2	0.42	54	17.7	10
India	1,591.1	3	0.47	45	1.4	108
Russia	1,556.7	4	0.81	16	11.1	22
Japan	1,098.0	5	0.29	93	8.6	35
Germany	765.6	6	0.3	89	9.3	28
Canada	541.0	7	0.45	48	16.2	13
Iran	528.6	8	0.75	22	7	44
South Korea	528.1	9	0.52	37	10.9	23
United Kingdom	519.9	10	0.25	111	8.4	39
South Africa	451.2	11	0.99	10	9.2	30
Mexico	443.6	12	0.32	85	4	68
Saudi Arabia	438.2	13	0.81	17	17.3	11
Brazil	425.2	14	0.24	115	0.24	94
Australia	417.7	15	0.54	34	19.6	8
Indonesia	414.9	16	0.47	43	1.7	101
Italy	407.9	17	0.26	101	7	43
France	396.7	18	0.2	122	6.3	46
Spain	329.9	19	0.28	96	7.1	42
Poland	285.8	20	0.46	47	7.4	40

[a]Million metric tons.
[b]Carbon intensity using purchasing power parities (metric tons of carbon dioxide per thousand year 2005 U.S. dollars)
[c]Metric tons carbon dioxide per person.
Source: Energy Information Administration, U.S. Department of Energy.

Table 28.6. The carbon footprint[a] of nations, regions and income classes in 2011.

Country/region	Population (millions)	Carbon footprint (ha/person)
World	6,739.6	1.5
Income Class[b]		
High-income countries	1,037.0	3.4
Middle-income countries	4,394.1	0.8
Low-income countries	1,297.5	0.2
Region		
North America		
EU	495.1	2.4
Other Europe	238.1	2.2
Middle East/Central Asia	382.6	1.4
Asia-Pacific		
South America	390.1	0.6
Central America/Caribbean	66.8	0.6
Africa	938.4	0.3
Country (10 largest footrpints)		
Qatar	1.4	8.9
Kuwait	2.5	7.7
United Arab Emirates	8.1	6.0
USA	305.0	4.9
Singapore	4.8	4.2
Finland	5.3	4.2
Macedonia	2.1	3.9
Ireland	4.4	3.7
Canada	33.3	3.6
Oman	2.6	3.3

[a]The quantity of land required to sequester (through photosynthesis) that fraction of carbon dioxide (CO_2) emissions from fossil fuel combustion which is not absorbed by the ocean.
[b]Word Bank classification.
Source: Global Footprint Network, National Footprint Accounts 2011 Edition, <www.footprintnetwork.org/atla>.

Table 28.7. Carbon intensity of primary energy use for 10 largest energy users, 1980–2009

Country	Primary energy use EJ	Energy use rank 2010	Carbon intensity of primary energy use[a]				Change in intensity	
			1980	1990	2000	2010	1980–2009	2000–2009
China	101.8	1	86.1	86.3	84.2	81.8	−4.9%	−2.8%
US	95.7	2	68.0	66.1	65.8	64.2	−5.5%	−2.5%
Russian Federation	28.9	3	–	65.0	60.2	58.8		−2.3%
India	21.9	4	75.5	76.8	76.9	77.8	3.0%	1.1%
Japan	21.0	5	67.8	64.0	61.7	62.4	−8.0%	1.2%
Germany	13.4	6	75.7	70.4	64.9	61.9	−18.2%	−4.5%

Continued

Table 28.7.   Continued

Country	Primary energy use EJ	Energy use rank 2010	Carbon intensity of primary energy use[a]				Change in intensity	
			1980	1990	2000	2010	1980–2009	2000–2009
Canada	13.3	7	52.0	46.9	46.8	45.6	−12.3%	−2.5%
South Korea	10.7	8	78.2	67.6	66.6	67.1	−14.3%	0.6%
Brazil	10.6	9	50.5	45.8	45.2	43.6	−13.6%	−3.4%
France	10.6	10	62.8	45.0	40.4	38.2	−39.2%	−5.5%
Total World	502.5		69.7	66.6	65.1	66.0	-5.3%	1.3%

[a]Metric tons CO_2/TJ

Carbon emissions reflect only those through consumption of oil, gas and coal, and are based on standard global average conversion factors: oil - 73,300 kg CO_2 per TJ (3.07 metric tons per metric ton of oil equivalent); natural gas - 56,100 kg CO_2 per TJ (2.35 metric tons per metric ton of oil equivalent); coal - 94,600 kg CO_2 per TJ (3.96 metric tons per metric ton of oil equivalent). As such, these numbers may diverge from offical national estimates that are based on more specific emission conversion factors.

Source: BP Statistical Review of World Energy (2011); United States Energy Information Administration, International Energy Statistics, <http://www.eia.gov/countries/data.cfm>.

Table 28.8.   Carbon intensity of GDP for 10 largest economies, 1980–2009

Country	GDP in 2009 (billion constant 2005 intl. $)[b]	GDP Rank 2010	Carbon intensity of GDP [a] [b]				Change in intensity	
			1980	1990	2000	2009	1980–2009	2000–2009
United States	12,703.5	1	0.82	0.63	0.52	0.42	−48.5%	−19.3%
China	8,262.9	2	2.78	1.79	0.83	0.92	−66.9%	10.3%
Japan	3,746.7	3	0.46	0.32	0.33	0.29	−36.6%	−11.4%
India	3,458.3	4	0.47	0.55	0.56	0.47	−0.8%	−15.9%
Germany	2,635.3	5	NA	NA	0.35	0.30		−14.7%
United Kingdom	1,979.5	6	0.56	0.42	0.30	0.25	−55.6%	−18.2%
Russian Federation	1,932.4	7	NA	NA	1.23	0.81		−34.4%
France	1,895.4	8	0.43	0.25	0.23	0.20	−52.6%	−10.9%
Brazil	1,823.8	9	0.21	0.23	0.26	0.24	14.8%	−7.0%
Italy	1,616.5	10	NA	0.32	0.29	0.26		−10.1%
World	64,465.5		0.81	0.60	0.49	0.46	−42.8%	−5.4%

[a]Metric tons of carbon dioxide per thousand year 2005 U.S. dollars.
[b]Based on Purchasing Power Parity (PPP).
Carbon emissions reflect only those through consumption of oil, gas and coal, and are based on standard global average conversion factors: oil - 73,300 kg CO_2 per TJ (3.07 metric tons per metric ton of oil equivalent); natural gas - 56,100 kg CO_2 per TJ (2.35 metric tons per metric ton of oil equivalent); coal - 94,600 kg CO_2 per TJ (3.96 metric tons per metric ton of oil equivalent). As such, these numbers may diverge from offical national estimates that are based on more specific emission conversion factors.
Source: GDP: World Bank, World Development Indicators, <http://data.worldbank.org/data-catalog/world-development-indicators>, accessed 9 May 2012; Carbon Intensity of GDP: United States Energy Information Administration, International Energy Statistics, <http://www.eia.gov/countries/data.cfm>.

Table 28.9. Top 10 nations in carbon dioxide emissions from fossil fuel use, 1970–2010[a]

Country	1970	1980	1990	2000	2010	% of World Total 2010	World Rank 2010	World Rank 1970[b]	Change 1970–2010	Change 2000–2010
China	737.8	1,499.7	2,459.3	3,659.3	8,332.5	25%	1	4	1029.4%	127.7%
US	4,682.8	5,158.9	5,444.6	6,377.0	6,144.9	19%	2	1	31.2%	-3.6%
India	210.0	324.2	581.4	952.8	1,707.5	5%	3	10	713.0%	79.2%
Russian Federation[b]	NA	NA	2,343.4	1,563.0	1,700.2	5%	4	NA	NA	8.8%
Japan	856.8	1,008.4	1,158.2	1,327.1	1,308.4	4%	5	3	52.7%	-1.4%
Germany	1,058.3	1,126.2	1,030.5	902.5	828.2	2%	6	2	-21.7%	-8.2%
South Korea	48.0	126.3	255.0	527.0	715.8	2%	7	29	1390.8%	35.8%
Canada	363.7	474.8	494.5	592.1	605.1	2%	8	7	66.4%	2.2%
Saudi Arabia	66.1	112.4	239.1	329.5	562.5	2%	9	23	750.8%	70.7%
Iran	71.5	112.1	196.7	329.6	557.7	2%	10	22	680.3%	69.2%
Top 10	8,094.9	9,942.9	14,202.6	16,560.0	22,462.7	67.7%			121.2%	29.6%
Total World	14,992.5	19,322.4	22,613.2	25,576.9	33,158.4					

[a]Million metric tons carbon dioxide unless otherwise indicated.

[b]The former Soviet Union emitted 682.8 million metric tons of carbon dioxide from fossil fuel use in 1970, which would have ranked #5 in the world. Carbon emissions reflect only those through consumption of oil, gas and coal, and are based on standard global average conversion factors: oil - 73,300 kg CO_2 per TJ (3.07 metric tons per metric ton of oil equivalent); natural gas - 56,100 kg CO_2 per TJ (2.35 metric tons per metric ton of oil equivalent); coal - 94,600 kg CO_2 per TJ (3.96 metric tons per metric ton of oil equivalent). As such, these numbers may diverge from offical national estimates that are based on more specific emission conversion factors.

Source: BP Statistical Review of World Energy (2011).

Table 28.10. Top 10 emitters of Nitrous Oxide (N₂O) 1970–2008ᵃ

						Rank	Rank	Change
	1970	1980	1990	2000	2008	2008	1970	1970–2008
China	1,003	1,215	1,070	1,095	1,036	1	2	3.3%
United States	449	765	1,098	1,383	1,764	2	1	293.1%
India	404	511	518	328	232	3	6	−42.7%
Brazil	347	460	506	545	619	4	4	78.5%
Central African Republic	332	246	281	189	206	5	15	−37.9%
Indonesia	272	370	547	696	764	6	10	180.8%
Sudan	262	267	229	174	167	7	22	−36.3%
Russian Federation	209	239	238	172	154	8	3	−26.4%
Congo	169	222	203	244	186	9	5	10.0%
Australia	167	252	324	306	329	10	9	96.2%
World Total	7,180	8,810	9,520	9,410	10,600	NA	NA	47.6%

ᵃUnits are Gg N₂O unless otherwise indicated
Source: European Commission, Joint Research Centre (JRC)/PBL Netherlands Environmental Assessment Agency. Emission Database for Global Atmospheric Research (EDGAR), release version 4.2.< http://edgar.jrc.ec.europe.eu, 2010>.

Table 28.11. Top 10 emitters of Methane (CH₄) 1990–2005ᵃ

							Change
Country	1990	1995	2000	2005	Rank 2005	Rank 1990	1990–2005
China	727.1	798.7	767.8	879.0	1	1	20.9%
India	576.2	599.3	614.4	626.9	2	4	8.8%
United States	636.8	635.8	614.6	599.9	3	2	−5.8%
Russia	582.7	446.2	440.5	468.2	4	3	−19.6%
Brazil	283.3	285.8	280.7	413.3	5	5	45.9%
Indonesia	189.4	199.6	199.0	220.0	6	6	16.2%
Nigeria	116.3	121.0	139.1	157.1	7	9	35.0%
Mexico	110.7	128.8	146.6	154.4	8	10	39.5%
Australia	122.1	116.1	123.9	112.9	9	8	−7.5%
Kuwait	48.6	83.3	86.2	104.8	10	27	115.6%
World	6304.2	6322.4	6342.4	6837.0			8.5%

ᵃUnits are Mt CO₂ e unless otherwise indicated.
Source: United States Environmental Protection Agency. 2011. DRAFT: Global Anthropogenic Non-CO₂ Greenhouse Gas Emissions: 1990–2030, (Washington, D.C., EPA).

Table 28.12.   Methane emissions by region and source, 2000 (Tg)

Source	Canada	USA	Oecd Europe	Oceania	Japan	Eastern Europe	Former USSR	Latin America	Africa	Middle East	South Asia	East Asia	Southeast Asia	Grand Total
Biofuel combustion	0.02	0.36	0.13	0.04	0.00	0.10	0.32	0.52	4.18	0.15	4.24	3.04	1.79	14.90
Industry & transport	0.04	0.45	0.35	0.03	0.12	0.20	0.53	0.16	0.10	0.13	0.09	1.09	0.12	3.41
Fossil fuel prod. & trans.	2.56	22.84	4.24	1.70	0.71	3.83	25.45	6.31	5.72	10.61	2.40	13.23	4.70	93.81
Industrial processess	0.01	0.06	0.14	0.01	0.12	0.04	0.12	0.05	0.01	0.01	0.03	0.26	0.00	0.86
Agriculture	0.95	7.19	7.08	3.88	0.35	1.77	5.10	20.32	11.97	2.00	30.60	23.77	12.84	127.83
Biomass burning	0.22	0.35	0.07	1.04	0.01	0.07	0.69	7.85	8.88	0.11	0.27	0.12	2.35	22.03
Waste handling	1.28	9.78	4.35	0.61	1.52	1.22	3.82	8.36	4.65	1.67	8.98	8.04	3.87	58.15
Grand Total	5.09	41.04	16.36	7.30	2.84	7.24	36.03	43.57	35.50	14.68	46.60	49.54	25.68	320.98

Source: Emissions Database for Global Atmospheric Research (EDGAR).

Table 28.13.   Estimates for natural (pre-industrial) fluxes and reservoirs of methane

Reservoir	Size (Tg CH$_4$ or TgC)	Size (Pg C)
Atmosphere	1980 Tg CH$_4$	1.5
Oceans	22–65 Tg CH$_4$	0.016–0.049
Wetlands	$(2.2–4.9) \times 10^5$ Tg C	224–489
Reactive marine sediments (org. C)	6.5×10^5 Tg C	650
Non-wetland soils	9.7×10^5 Tg C	970
Geological sources		
Crust	1.25×10^{10} Tg C	1.25×10^7
Hydrates	5×10^5–2.4×10^7 Tg CH$_4$	4×10^2–1.8×10^4
Dispersed gas in sedimentary basins	1.6×10^6 Tg CH$_4$	1.2×10^3
Natural gas reserves (part of dispersed gas in sed. basins)	$(2.5–2.9) \times 10^5$ Tg CH$_4$	190–216
Flux	**Size (Tg CH$_4$ yr^{-1})**	**Size[a] (Pg C yr^{-1})**
Oceans	0.4–20	0–0.01
Marine sediments	0.4–12.2	0–0.01
Wetlands	92–260	0.07–0.19
Termites	2–22	0–0.02
Wild fires	2	0
Geological sources (Hydrates, volcanoes, natural gas seeps, geothermal)	5–65	0–0.05
Total source	159–290	0.12–0.22

[a]Fluxes less than 0.01 Pg C yr^{-1} are rounded to 0.

Source: Adapted from Sundquist, E.T. and K. Visser. 2003. The Geologic History of the Carbon Cycle, In: Heinrich D. Holland and Karl K. Turekian, Editor(s)-in-Chief, Treatise on Geochemistry, (Oxford, Pergamon), Pages 425–472.

Table 28.14.   Estimates for natural, pre-industrial reservoirs and geologic fluxes of carbon

Reservoir	Size (Pg C)
Atmosphere	590
Oceans	$(3.71–3.90) \times 10^4$
Surface layer—inorganic	700–900
Deep layer—inorganic	$(3.56–3.80) \times 10^4$
Total organic	685–700
Aquatic biosphere	1–3
Terrestrial biosphere and soils	$2.0–2.3 \times 10^3$
Vegetation	500–600
Soil	$(1.5–1.7) \times 10^3$
Reactive marine sediments	3,000
Inorganic	2,500
Organic	650
Crust	$(7.78–9.0) \times 10^7$
Sedimentary carbonates	6.53×10^7
Organic carbon	1.25×10^7
Mantle	3.24×10^8

Table 28.14. Continued

Reservoir	Size (Pg C)
Fossil fuel reserves and resources	$(4.22–5.68) \times 10^3$
Oil (conventional and unconventional)	636–842
Natural gas (conventional and unconventional)	483–564
Coal	$(3.10–4.27) \times 10^3$
Flux	**Size (Pg C yr^{-1})**
Carbonate burial	0.13–0.38
Organic carbon burial	0.05–0.13
Rivers (dissolved inorganic carbon)	0.39–0.44
Rivers (total organic carbon)	0.30–0.41
Rivers-dissolved organic carbon	0.21–0.22
Rivers-particular organic carbon	0.17–0.30
Volcanism	0.04–0.10
Mantle exchange	0.022–0.07

Source: Adapted from Sundquist, E.T. and K. Visser. 2003. The Geologic History of the Carbon Cycle, In: Heinrich D. Holland and Karl K. Turekian, Editor(s)-in-Chief, Treatise on Geochemistry, (Oxford, Pergamon), Pages 425–472.

Table 28.15. Carbon masses in the major environmental reservoirs

Reservoir	Mass of carbon	
	grams C	mols C
Atmosphere		
CO_2 (at pre-industrial 280ppmv)	6×10^{17}	5×10^{16}
Ocean		
Dissolved inorganic (DIC)	3.74×10^{19}	3.11×10^{18}
Dissolved organic (DOC)	1×10^{18}	8.33×10^{16}
Particulate organic (POC)	3×10^{16}	2.50×10^{15}
Ocean biota	3×10^{15}	2.50×10^{14}
Land biota		
Phytomass	7×10^{17}	$6 \pm 1 \times 10^{16}$
Bacteria and fungi	3×10^{15}	2.50×10^{14}
Animals	1 to 2×10^{15}	1.25×10^{14}
Land		
Soil humus	1.92×10^{18}	$1.6 +/-3 \times 10^{17}$
Reactive fraction of humus		2×10^{16}
Dead organic matter, litter, peat	2.5×10^{17}	2.08×10^{16}
Inorganic soil ($CaCO_3$)	7.2×10^{17}	6×10^{16}
Sediments		
Carbonates	6.53×10^{22}	5.44×10^{21}
Organic matter	1.25×10^{22}	1.05×10^{21}
Continental crust	2.576×10^{21}	2.14×10^{20}
Oceanic crust	9.200×10^{20}	7.66×10^{19}
Upper mantle	$(8.9 \text{ to } 16.6) \times 10^{22}$	$\sim1.1 \times 10^{22}$

Source: Adapted from MacKenzie, Fred T. 2010. Our Changing Planet: An Introduction to Earth System Science and Global Environmental Change (Prentice-Hall).

Table 28.16. CO$_2$ capture capacities of different solid sorbents

Solid Sorbent	Capacity (g CO$_2$/g sorbent)	Gas composition
Aminated SBA-15[a]	0.07~0.18	10% CO$_2$, 88% He with ~2% H$_2$O
SBA-HA[b]	0.09	15% CO$_2$, 85% N$_2$ with 20 mL/min rate of H$_2$O
Tertiary amine	0.13	10% CO$_2$, 88% He with ~2% H$_2$O
K-Li$_2$ZrO$_3$/ Li$_2$ZrO$_3$	0.22~0.29	10% CO$_2$, 90% N$_2$
MCM-41[c]	0.09	15% CO$_2$, 85% N$_2$, 4% O$_2$
K- HTlc[d]	0.44	15% CO$_2$, 75% N$_2$ with 10% H$_2$O
Hydrotalcites (HTlc)	0.04	11% CO$_2$, 89% N$_2$
Mg-Al-CO$_3$ HTlc	0.02	20% CO$_2$, 80% N$_2$
Zeolite 13X	0.22	15% CO$_2$, 85% N$_2$
Zeolite/activated carbon	0.22	15% CO$_2$, 85% N$_2$
Activated carbon	0.18	17% CO$_2$, 79% N$_2$, 4% O$_2$
Basic alumina	0.03	100% CO$_2$
Metal–organic frameworks (MOFs)	1.47~1.5	-

[a]SBA-15 denotes a type of mesoporous silica (Santa Barbara Amorphous type material, or SBA-15).
[b]SBA-HA Hyperbranched mesoporous aminosilica.
[c]MCM-41, a type of mesoporous silica (Mobil Crystalline of Materials, or MCM-41).
[d]K-HTlc potassium modified HTlc.
Source: Adapted from Kwon, Soonchul, Maohong Fan, Herbert F.M. DaCosta, Armistead G. Russell, Kathryn A. Berchtold, Manvendra K. Dubey. 2011. CO$_2$ Sorption, In: David Bell and Brian Towler, Coal Gasification and Its Applications, (Boston, William Andrew Publishing), Pages 293–339.

Table 28.17. Physical properties of different solid sorbents

Solid sorbent	Bulk density (kg/m^3)	BET surface area (m^2/g)	Pore volume (ml/g)
Tertiary amine (DBU)	-	369	1.1
Aminated SBA-15	-	200~230	0.72
Zeolite 13X	689	726	0.25
Zeolite/activated carbon	500	620	0.19
Li$_2$ZrO$_3$	-	5	0.02
Activated carbon	860	1300	0.6–0.8
Hydrotalcites (HTlc)	440	271	0.55
Mg-Al-CO$_3$ HTlc	-	184	0.31
Serpentine	2500	330	0.23
Metal–organic frameworks (MOFs)	-	345~2833	–

The Brunauer-Emmett-Teller (BET) equation is used to determine the surface area from the physical adsorption of a gas (e.g., N$_2$ or CO$_2$) on a solid surface.
Source: Adapted from Kwon, Soonchul, Maohong Fan, Herbert F.M. DaCosta, Armistead G. Russell, Kathryn A. Berchtold, Manvendra K. Dubey. 2011. CO$_2$ Sorption, In: David Bell and Brian Towler, Coal Gasification and Its Applications, (Boston, William Andrew Publishing), Pages 293–339.

Table 28.18. Comparison of selected physical properties of CO_2 and N_2 at 1 atm and 1000 K

Physical process	Physical property	Unit	CO_2	N_2	CO_2/N_2
Thermodynamic	Density	kg/m³	0.5362	0.3413	1.57
	Specific heat capacity	kJ/kgK	1.2343	1.1674	1.06
	Volumetric heat capacity	kJ/m³K	0.662	0.398	1.66
	Gas-water interfacial tension[a]	N/m	71.03	71.98	0.987
Momentum transfer	Kinematic viscosity	m²/s	7.69E-05	1.20E-04	0.631
Heat transfer	Thermal conductivity	W/mK	7.06E-02	6.60E-02	1.07
	Thermal diffusivity	m²/s	1.10E-04	1.70E-04	0.644
	Absorptivity/ emissivity		>0	≈0	-
Mass transfer	Mass diffusivity[b]	m²/s	9.80E-05	1.30E-04	0.778

[a]Water gas interfacial tensions are evaluated for CO_2/water and air/water interfaces at 298.15 K.
[b]Mass diffusivity refers to the binary diffusion of O_2 in CO_2 and nitrogen.
Source: Adapted from Chen, Lei, Sze Zheng Yong, Ahmed F. Ghoniem. 2012. Oxy-fuel combustion of pulverized coal: Characterization, fundamentals, stabilization and CFD modeling, Progress in Energy and Combustion Science, Volume 38, Issue 2, Pages 156-214.

Table 28.19. Lifetimes, radiative efficiencies, and direct (except for CH_4) global warming potentials relative to CO_2

Industrial designation or common name	Chemical formula	Lifetime (years)	Radiative efficiency (W m⁻² ppb⁻¹)	Global warming potential for given time horizon			
				SAR[b] (100-yr)	20-yr	100-yr	500-yr
Carbon dioxide	CO_2	See below[a]	1.4×10^{-5}	1	1	1	1
Methane[c]	CH_4	12	3.7×10^{-4}	21	72	25	7.6
Nitrous oxide	N_2O	114	3.03×10^{-3}	310	289	298	153
CFC-11	CCl_3F	45	0.25	3,800	6,730	4,750	1,620
CFC-12	CCl_2F_2	100	0.32	8,100	11,000	10,900	5,200
CFC-113	CCl_2FCClF_2	85	0.3	4,800	6,540	6,130	2,700
Halon-1301	$CBrF_3$	65	0.32	5,400	8,480	7,140	2,760
Halon-1211	$CBrClF_2$	16	0.3		4,750	1,890	575
Halon-2402	$CBrF_2CBrF_2$	20	0.33		3,680	1,640	503
Carbon tetrachloride	CCl_4	26	0.13	1,400	2,700	1,400	435
Methyl chloroform	CH_3CCl_3	5	0.06		506	146	45
HCFC-22	$CHClF_2$	12	0.2	1,500	5,160	1,810	549
HCFC-141b	CH_3CCl_2F	9.3	0.14		2,250	725	220

Continued

Table 28.19.   Continued

Industrial designation or common name	Chemical formula	Lifetime (years)	Radiative efficiency (W m^{-2} ppb^{-1})	Global warming potential for given time horizon			
				SAR[b] (100-yr)	20-yr	100-yr	500-yr
HCFC-142b	CH_3CClF_2	17.9	0.2	1,800	5,490	2,310	705
HFC-23	CHF3	270	0.19	11,700	12,000	14,800	12,200
HFC-125	CHF_2CF_3	29	0.23	2,800	6,350	3,500	1,100
HFC-134a	CH_2FCF_3	14	0.16	1,300	3,830	1,430	435
HFC-143a	CH_3CF_3	52	0.13	3,800	5,890	4,470	1,590
HFC-152a	CH_3CHF_2	1.4	0.09	140	437	124	38
Sulphur hexafluoride	SF_6	3,200	0.52	23,900	16,300	22,800	32,600

[a]The CO_2 response function is based on the revised version of the Bern Carbon cycle model (Bern2.5CC.) See Joos, F. et al. 2001, 'Global warming feedbacks on terrestrial carbon uptake under the Intergovernmental Panel on Climate Change (IPCC) emission scenarios', Global Biogeochemical Cycles, Vol. 15(4), pp. 891 – 907.
[b]The IPCC Third Assessment Report (1995)
[c]The perturbation lifetime for methane is 12 years as in the IPCC Third Assessment Report. The GWP for methane includes indirect effects from enhancements of ozone and stratospheric water vapor.
Source: Adapted from Solomon, S., D. Qin, M. Manning, Z. Chen, M. Marquis, K.B. Averyt, M. Tignor and H.L. Miller (eds.). 2007. Contribution of Working Group I to the Fourth Assessment Report of the Intergovernmental Panel on Climate Change, 2007, (Cambridge University Press, Cambridge, United Kingdom and New York, NY, USA.

Table 28.20.   Greenhouse gas intensity for electricity generation

Electricity technology	Energy intensity (kWh$_{th}$/kWh$_{el}$)	Greenhouse gas intensity (g CO_2-e/kWh$_{el}$)[a]
Light water reactors[b]	0.18 (0.16–0.40)	60 (10–130)
Heavy water reactors[b]	0.20 (0.18–0.35)	65 (10–120)
Black coal (new subcritical) [b]	2.85 (2.70–3.17)	941 (843–1171)
Black coal (supercritical) [b]	2.62 (2.48–2.84)	863 (774–1046)
Brown coal (new subcritical) [b]	3.46 (3.31–4.06)	1175 (1011–1506)
Natural gas (open cycle) [b]	3.05 (2.81–3.46)	751 (627–891)
Natural gas (combined cycle) [b]	2.35 (2.20–2.57)	577 (491–655)
Wind turbines[b]	0.066 (0.041–0.12)	21 (13–40)
Photovolatics[b]	0.33 (0.16–0.67)	106 (53–217)
Hydroelectricity (run-of-river) [b]	0.046 (0.020–0.137)	15 (6.5–44
Geothermal (low temperature) [c, f]		~1
Geothermal (high temperature) [c, f]		91–122
Solar thermal[d]		60–90
Solar pond[d]		6

Table 28.20.   Continued

Electricity technology	Energy intensity (kWh$_{th}$/kWh$_{el}$)	Greenhouse gas intensity (g CO$_2$-e/kWh$_{el}$)[a]
Wood cogeneration[e]		92–156
Oil[e]		519–1190

[a]A greenhouse gas's CO$_2$ equivalent (CO$_2$-e) is based on its global warming potential (GWP). The GWP of a gas is the warming caused over a 100-year period by the emission of one ton of the gas relative to the warming caused over the same period by the emission of one ton of CO$_2$.

[b]Lenzen, M. (2008) Life cycle energy and greenhouse gas emissions of nuclear energy: A review. Energy Conversion and Management 49, 2178–2199.

[c]Fridleifsson, I.B., R. Bertani, E. Huenges, J. W. Lund, A. Ragnarsson, and L. Rybach 2008. The possible role and contribution of geothermal energy to the mitigation of climate change. In: O. Hohmeyer and T. Trittin (Eds.) IPCC Scoping Meeting on Renewable Energy Sources, Proceedings, Luebeck, Germany, 20–25 January 2008, 59–80.

[d]Dey C, Lenzen M. Greenhouse gas analysis of electricity generation systems. In: Proceedings of the ANZSES Solar 2000 Conference, 29 November–1 December, Griffith University, Queensland, Australia; 2000. p. 658–68.

[e]Dones, Roberto Thomas Heck, Stefan Hirschberg, Greenhouse Gas Emissions from Energy Systems, Comparison and Overview, In: Cutler J. Cleveland, Editor-in-Chief, Encyclopedia of Energy, Elsevier, New York, 2004, Pages 77–95.

[f] CO$_2$ emissions only.

PART V

CORRELATIONS

Business and Economics

Charts

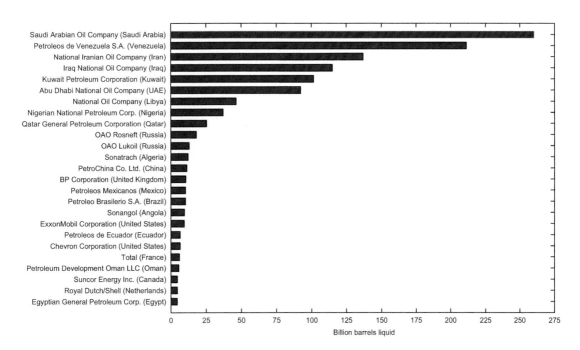

Chart 29.1. Companies holding the largest oil reserves, 2010.
Source: Data from PetroStrategies, Inc., <http://www.petrostrategies.org>.

Handbook of Energy. http://dx.doi.org/10.1016/B978-0-08-046405-3.00029-2

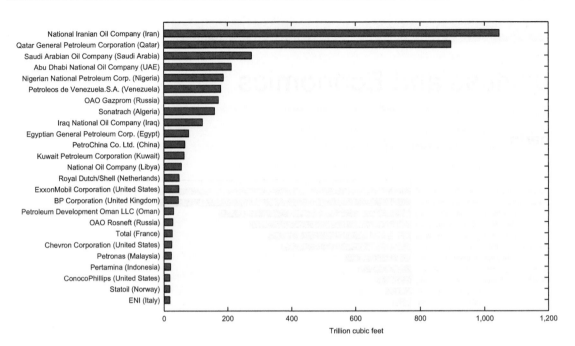

Chart 29.2. Companies holding the largest natural gas reserves, 2010.
Source: Data from PetroStrategies, Inc., <http://www.petrostrategies.org>.

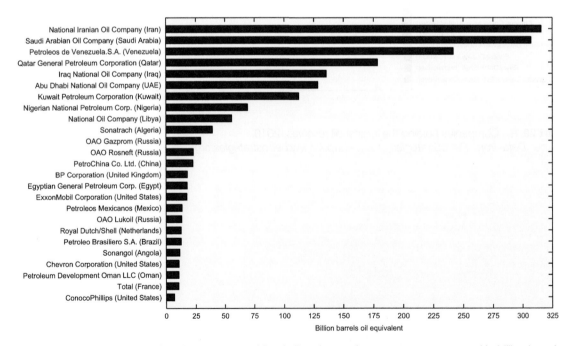

Chart 29.3. Companies holding the largest combined oil and natural gas reserves, measured in billion barrels oil equivalent, 2010.
Source: Data from PetroStrategies, Inc., <http://www.petrostrategies.org>.

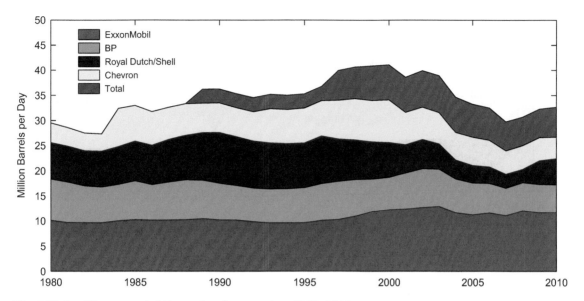

Chart 29.4. Oil reserves held by major oil companies, 1980–2010.
Source: Data from Organization of Petroleum Exporting Countries, Statistical database,
<http://www.opec.org/library/Annual%20Statistical%20Bulletin/interactive/current/FileZ/Main.htm>.

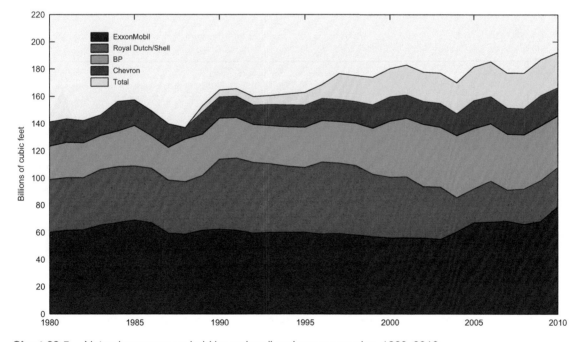

Chart 29.5. Natural gas reserves held by major oil and gas companies, 1980–2010.
Source: Data from Organization of Petroleum Exporting Countries, Statistical database,
<http://www.opec.org/library/Annual%20Statistical%20Bulletin/interactive/current/FileZ/Main.htm>.

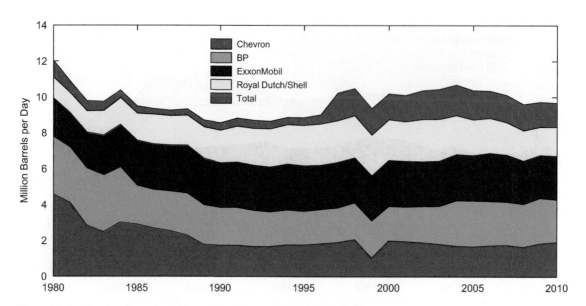

Chart 29.6. Crude oil produced by major oil companies, 1980–2010.
Source: Organization of Petroleum Exporting Countries, Statistical database, <http://www.opec.org/library/Annual%20Statistical%20Bulletin/interactive/current/FileZ/Main.htm>.

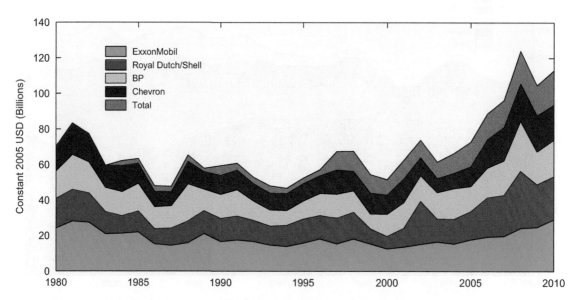

Chart 29.7. Capital and exploratory expenditures of the major oil companies, 1980–2010.
Source: Data from Organization of Petroleum Exporting Countries, Statistical database, <http://www.opec.org/library/Annual%20Statistical%20Bulletin/interactive/current/FileZ/Main.htm>.

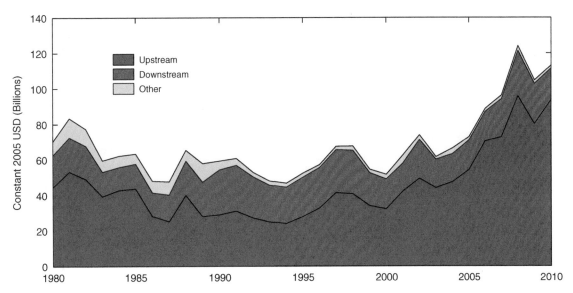

Chart 29.8. Upstream and downstream expenditures of the major oil companies, 1980–2010 (constant 2005 US$).
Source: Data from Organization of Petroleum Exporting Countries, Statistical database,
<http://www.opec.org/library/Annual%20Statistical%20Bulletin/interactive/current/FileZ/Main.htm>.

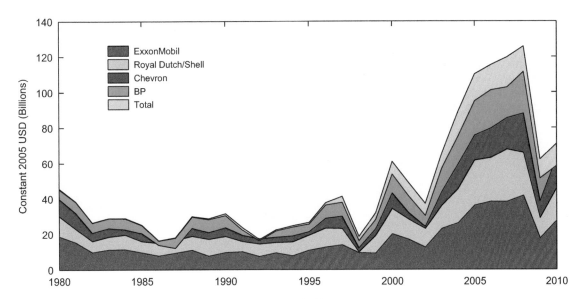

Chart 29.9. Net income of major oil companies, 1980–2010.
Source: Data from Organization of Petroleum Exporting Countries, Statistical database,
<http://www.opec.org/library/Annual%20Statistical%20Bulletin/interactive/current/FileZ/Main.htm>.

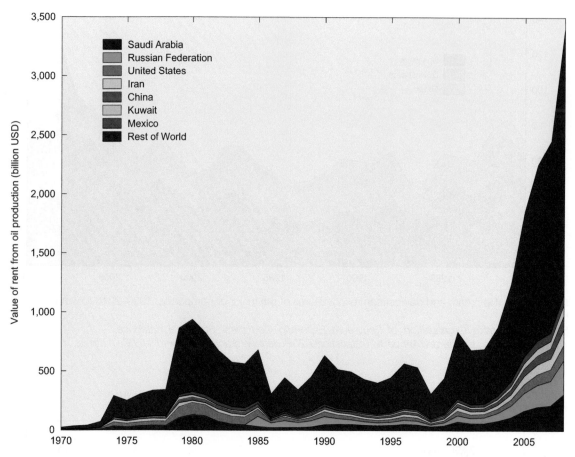

Chart 29.10. Rent earned from crude oil production, top nations, 1970–2008. Rents = Unit Rents *
Production. Unit Rents = Unit Price – Unit Cost.
*Source: Data from World Bank, Changing Wealth of Nations, 2010 update, <http://data.worldbank.org/
data-catalog/wealth-of-nations>.*

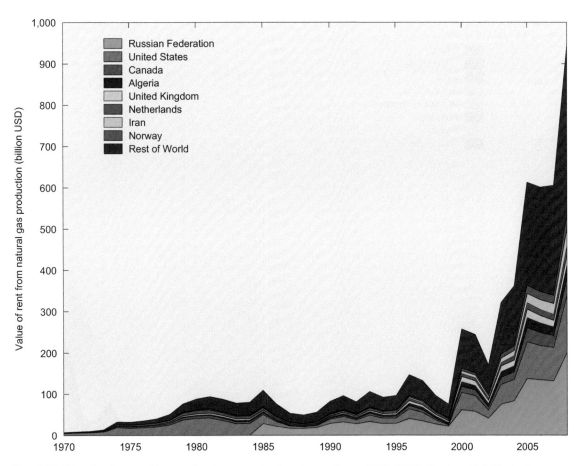

Chart 29.11. Rent earned from natural gas production, top nations, 1970–2008. Rents = Unit Rents * Production. Unit Rents = Unit Price – Unit Cost.
Source: Data from World Bank, Changing Wealth of Nations, 2010 update, <http://data.worldbank.org/ data-catalog/wealth-of-nations>.

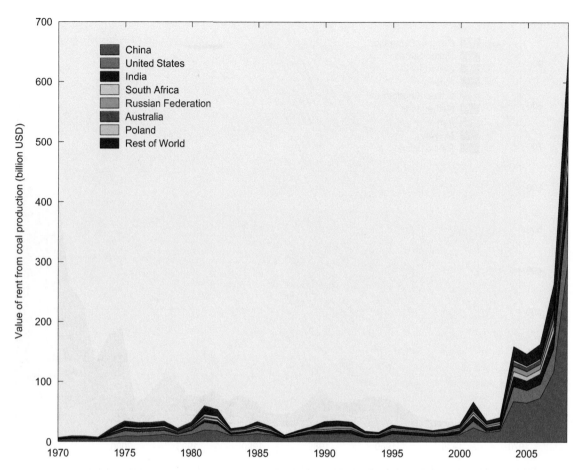

Chart 29.12. Rent earned from coal production, top nations, 1970–2008. Rents = Unit Rents * Production. Unit Rents = Unit Price – Unit Cost.
Source: Data from World Bank, Changing Wealth of Nations, 2010 update, <http://data.worldbank.org/ data-catalog/wealth-of-nations>.

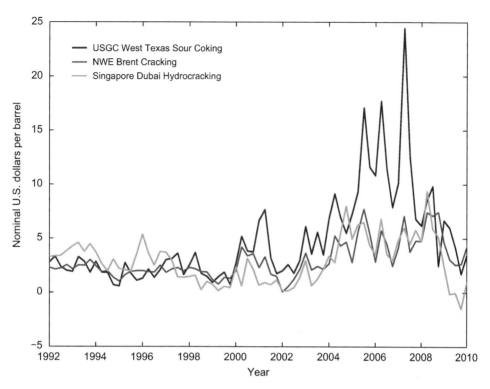

Chart 29.13. Oil refinery margins, 1992–2010. Refining margin is the difference in value between the products produced by a refinery and the value of the crude oil used to produce them.
Source: Data from BP, Statistical Review of World Energy 2012, <http://www.bp.com/sectionbodycopy.do?categoryId=7500&contentId=7068481>.

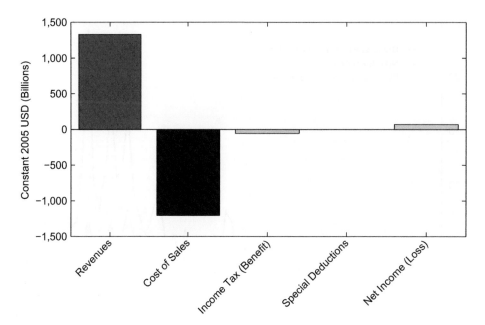

Chart 29.14. Revenue, operating costs, deductions, taxation and net income of the major oil companies, 2010.
Source: Data from Organization of Petroleum Exporting Countries, Statistical database, <http://www.opec.org/library/Annual%20Statistical%20Bulletin/interactive/current/FileZ/Main.htm>.

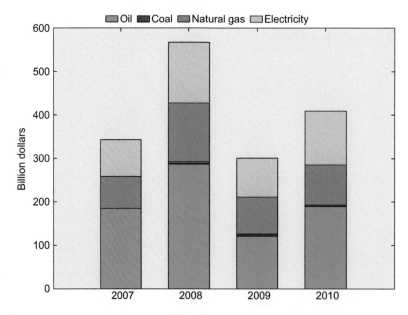

Chart 29.15. Global energy subsides by fuel, 2007–2010.
Source: International Energy Agency (IEA). 2011. IEA analysis of fossil-fuel subsidies, 4 October 2011 (Paris, IEA).

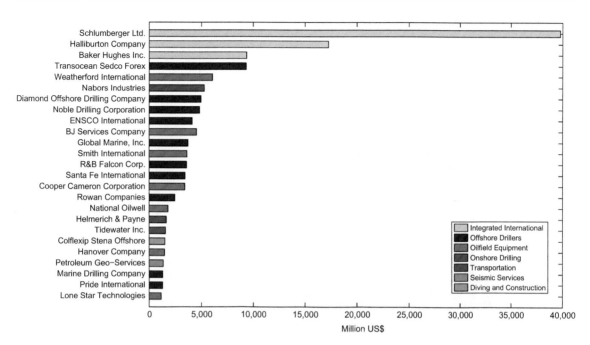

Chart 29.16. The world's 25 largest oil and gas service companies based on market capitalization, 2011. *Source: Data from PetroStrategies, Inc., <http://www.petrostrategies.org>.*

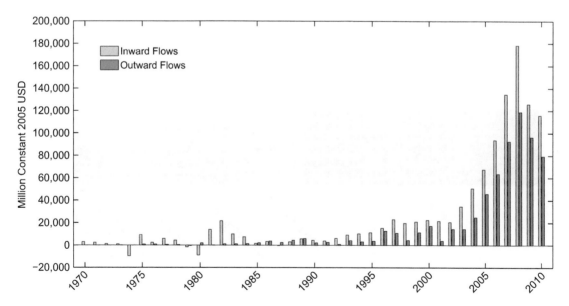

Chart 29.17. Foreign direct investment by the major petroleum exporting nations, 1970–2010. The group of major petroleum and gas exporters consists of countries whose share of petroleum and gas was not less than 50 per cent of their total exports, and whose exports of these products amounted to at least 1 per cent of petroleum and gas world share for the period 2004–2006. *Source: Data from United Nations Conference on Trade and Development (UNCTAD), UNCTAD Statistics, <http://unctad.org/en/pages/Statistics.aspx>.*

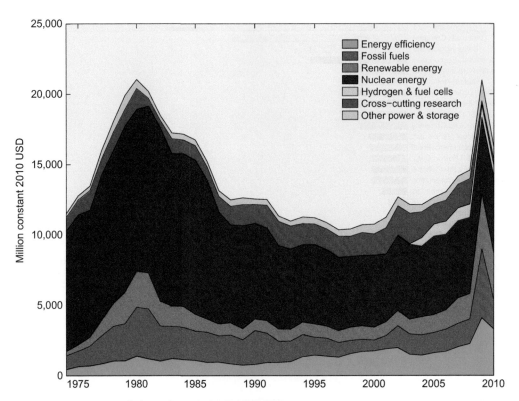

Chart 29.18. Energy research and development expenditures in IEA countries 1974–2010.
Source: Data from International Energy Agency (IEA), Energy statistics database, <http://www.iea.org/stat s/index.asp>.

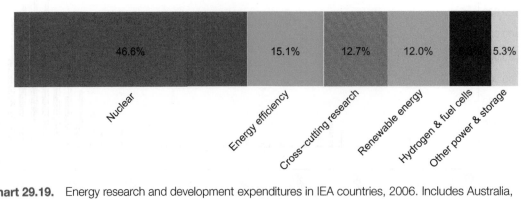

Chart 29.19. Energy research and development expenditures in IEA countries, 2006. Includes Australia, Austria, Belgium, Canada, Denmark, Finland, France, Germany, Greece, Hungary, Ireland, Italy, Japan, Korea, Luxembourg, the Netherlands, New Zealand, Norway, Portugal, Spain, Sweden, Switzerland, Turkey, the United Kingdom and the United States. Due to missing data, the Czech Republic, Poland and the Slovak Republic are not included.
Source: Data from International Energy Agency (IEA), Energy statistics database, <http://www.iea.org/stats/ index.asp>.

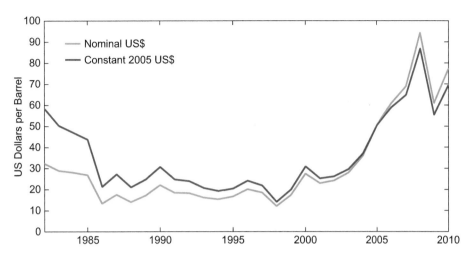

Chart 29.20. Spot OPEC reference basket prices, 1982–2010, a weighted average of prices for petroleum blends produced by the Organization of Petroleum Exporting Countries (OPEC) countries. It is used as an important benchmark for crude oil prices. As of June 2005, the basket is defined by: Saharan Blend (Algeria), Girassol (Angola), Oriente (Ecuador), Iran Heavy (Islamic Republic of Iran), Basra Light (Iraq), Kuwait Export (Kuwait), Es Sider (Libya), Bonny Light (Nigeria), Qatar Marine (Qatar), Arab Light (Saudi Arabia), Murban (UAE) and Merey (Venezuela).
Source: Data from Organization of Petroleum Exporting Countries, Statistical database, <http://www.opec.org/ library/Annual%20Statistical%20Bulletin/interactive/current/FileZ/Main.htm>.

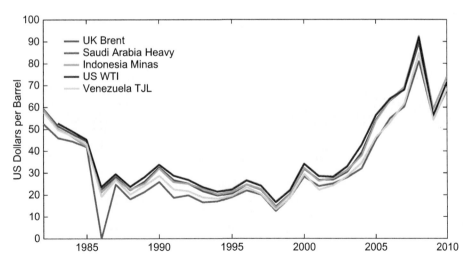

Chart 29.21. Spot crude oil prices, for US WTI, UK Brent, Saudi Arabia Heavy, Indonesia Minas, and Venezuela TJL, 1982–2010.
Source: Data from Organization of Petroleum Exporting Countries, Statistical database, <http://www.opec.org/library/Annual%20Statistical%20Bulletin/interactive/current/FileZ/Main.htm>.

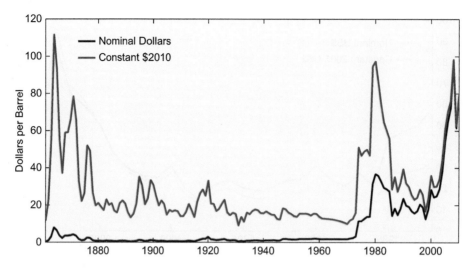

Chart 29.22.   Crude oil prices in the United States, 1859–2010.
Source: Data from United States Department of Energy, Energy Information Administration.

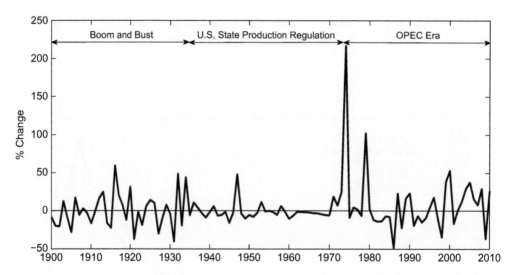

Chart 29.23.   Annual percent change in the real price of crude oil in the United States 1900–2010. Beginning in the 1930s, oil production was regulated by agencies in the major oil-producing states to dampen price fluctuations. By the 1970s, US production was declining and spare productive capacity was gone. Huge reserves among members of the Organization of Petroleum Exporting Countries (OPEC) gave OPEC more market control. One result was the return of more price volatility.
Source: Data from United States Department of Energy, Energy Information Administration.

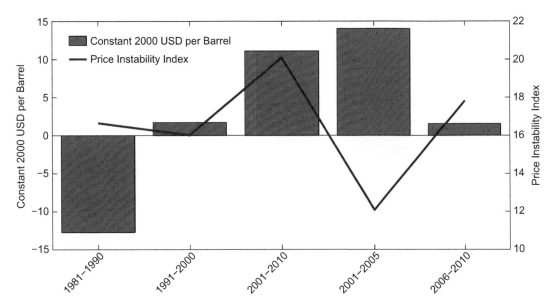

Chart 29.24. The price of crude oil and an index of oil price instability, 1981–2010. Instability is measured by the deviation of price from its long run trend.
Source: Data from United Nations Conference on Trade and Development (UNCTAD), UNCTAD Statistics, <http://unctad.org/en/pages/Statistics.aspx>.

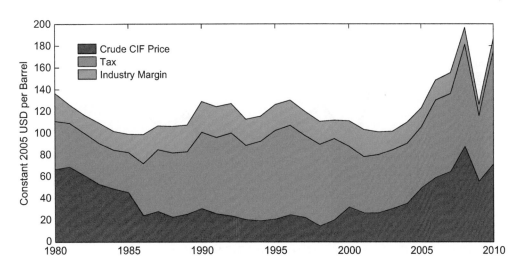

Chart 29.25. Composite barrel of oil and its components in G7 nations, 1980–2010. The CIF price (i.e. cost, insurance and freight price) is the price of a barrel delivered at the frontier of the importing country, including any insurance and freight charges incurred to that point.
Source: Data from Organization of Petroleum Exporting Countries, Statistical database, <http://www.opec.org/library/Annual%20Statistical%20Bulletin/interactive/current/FileZ/Main.htm>.

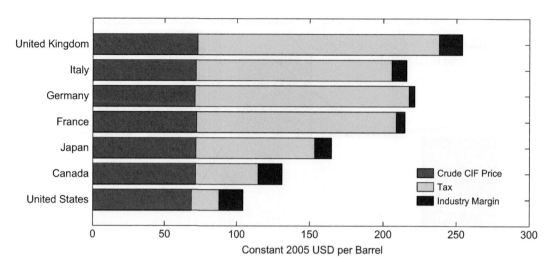

Chart 29.26. Composite barrel of oil and its components in major consuming nations, 2010. The CIF price (i.e. cost, insurance and freight price) is the price of a barrel delivered at the frontier of the importing country, including any insurance and freight charges incurred to that point.
Source: Data from Organization of Petroleum Exporting Countries, Statistical database,
<http://www.opec.org/library/Annual%20Statistical%20Bulletin/interactive/current/FileZ/Main.htm>.

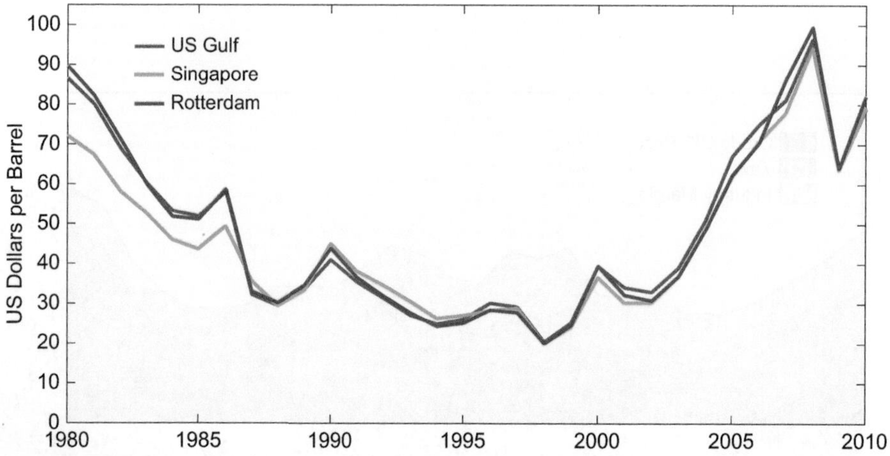

Chart 29.27. Spot prices of premium gasoline in major markets, 1980–2010.
Source: Data from Organization of Petroleum Exporting Countries, Statistical database, <http://www.opec.org/library/Annual%20Statistical%20Bulletin/interactive/current/FileZ/Main.htm>.

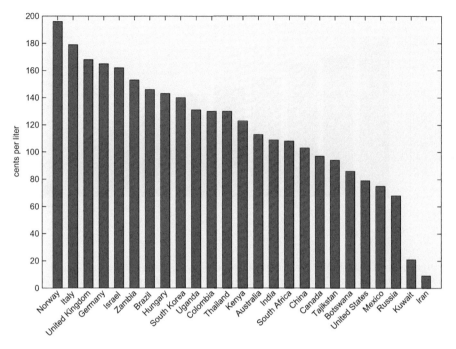

Chart 29.28. International gasoline prices, June 2012.
Source: Data from http://www.mytravelcost.com/petrol-prices/.

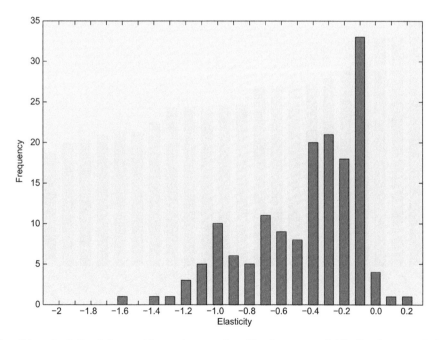

Chart 29.29. Price elasticity of demand for motor gasoline. The frequency distribution is generated from a meta-analysis of 312 elasticity observations from 43 primary studies.
Source: Data from Brons, Martijn, Peter Nijkamp, Eric Pels, Piet Rietveld. 2008. A meta-analysis of the price elasticity of gasoline demand. A SUR approach, Energy Economics, Volume 30, Issue 5, Pages 2105-2122.

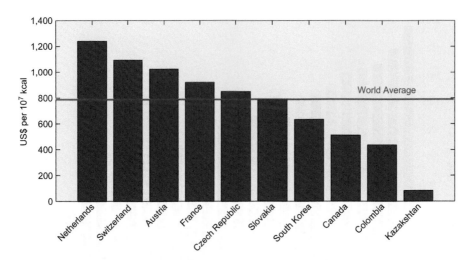

Chart 29.30. The price of natural gas to residential consumers in selected countries, 2009.
Source: Data from United States Department of Energy, Energy Information Administration, International Energy Statistics, <http://www.eia.gov/countries/data.cfm>.

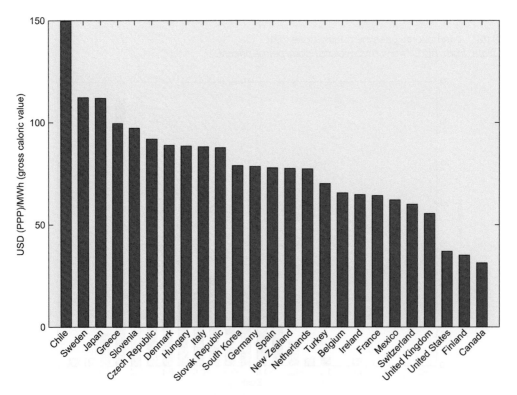

Chart 29.31. International prices paid by households for natural gas, 2010.
Source: Data from United States Department of Energy, Energy Information Administration, International Energy Statistics, <http://www.eia.gov/coal/data.cfm#reserves>.

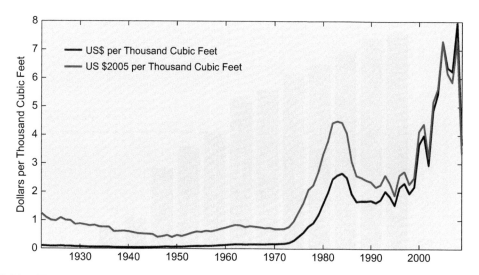

Chart 29.32. The wellhead price of natural gas in the United States, nominal and constant $2005 per thousand cubic feet.
Source: Data from United States Department of Energy, Energy Information Administration, <http://www.eia.gov/naturalgas/>.

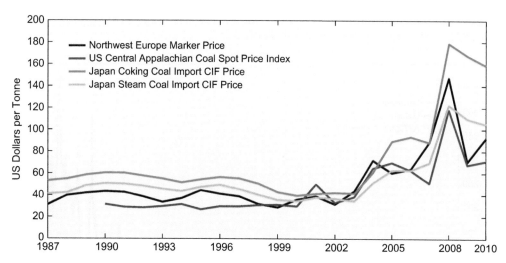

Chart 29.33. International prices for coal, 1987–2010. Coking coal is used in the smelting of iron ore; steam coal is that used by power plants and industrial steam boilers to produce electricity or process steam. Spot price is the price that is quoted for immediate (spot) settlement (payment and delivery). The CIF price (i.e. cost, insurance and freight price) is the price of a good delivered at the frontier of the importing country, including any insurance and freight charges incurred to that point.
Source: Data from United States Department of Energy, Energy Information Administration, International Energy Statistics, <http://www.eia.gov/countries/data.cfm>.

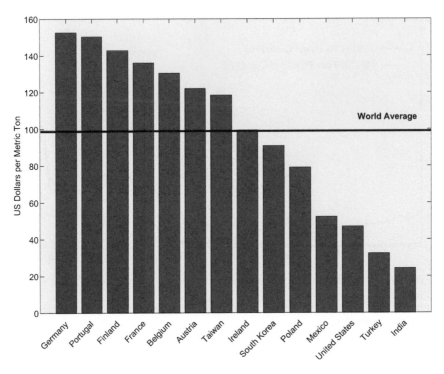

Chart 29.34. Coal prices in selected countries, 2010.
Source: Data from United States Department of Energy, Energy Information Administration, International Energy Statistics, <http://www.eia.gov/countries/data.cfm>.

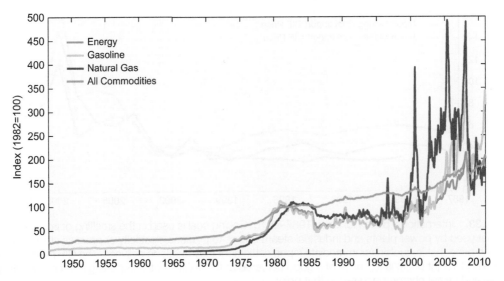

Chart 29.35. The Producer Price Index for natural gas, motor gasoline, all energy commodities, and all commodities in the United States, 1947–2011.
Source: Data from United States Department of Labor Bureau of Labor Statistics, Producer Price Indexes, <http://www.bls.gov/ppi/>.

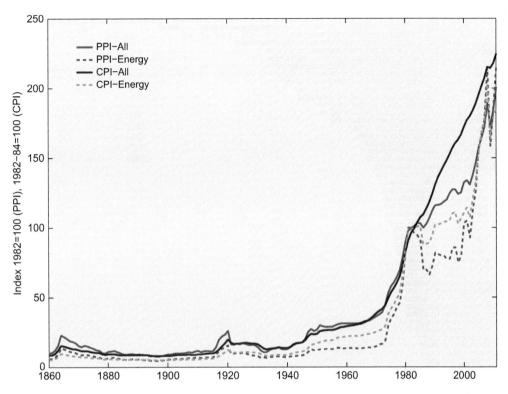

Chart 29.36. Consumer and Producer Indexes for energy and all commodities in the United States, 1860–2010.
Source: Data from Carter, Susan B., Gartner, Scott Sigmund, Haines, Michael R., Imstead, Alan L., Sutch, Richard, Wright, Gavin. 2006. Historical Statistics of the United States: Millennial Edition, (New York, Cambridge University Press); United States Department of Labor Bureau of Labor Statistics, <http://www.bls.gov/ppi/>.

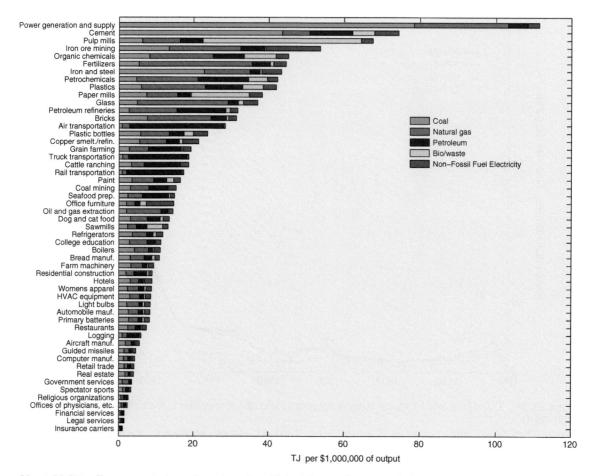

Chart 29.37. Energy cost of goods and services United States. Energy includes both direct and indirect energy use.

Source: Data from Carnegie Mellon University Green Design Institute. 2012 Economic Input-Output Life Cycle Assessment (EIO-LCA) US 2002 (428) model [Internet], <http://www.eiolca.net/>.

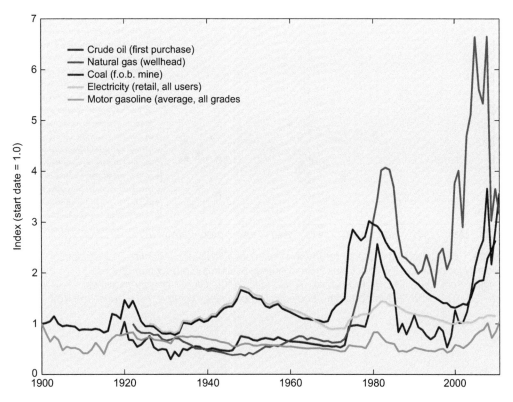

Chart 29.38. Index of real prices of energy in the United States, 1900–2011.
Source: Data from Carter, Susan B., Gartner, Scott Sigmund, Haines, Michael R., Imstead, Alan L., Sutch, Richard, Wright, Gavin. 2006. Historical Statistics of the United States: Millennial Edition, (New York, Cambridge University Press); United States Department of Energy, Energy Information Admnistration.

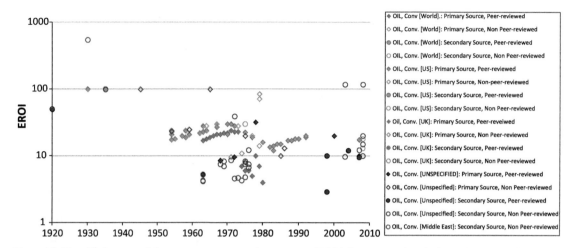

Chart 29.39. Estimates of the energy return on investment (EROI) for conventional oil production. EROI is the ratio of energy produced to the direct plus indirect energy used in the production process.
Source: Data from Dale, M., S. Krumdieck, P. Bodger. 2012. Global energy modelling — A biophysical approach (GEMBA) Part 2: Methodology, Ecological Economics, Volume 73, Pages 158-167.

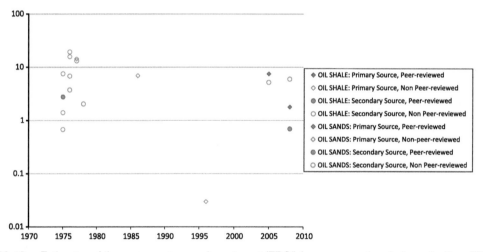

Chart 29.40. Estimates of the energy return on investment (EROI) for unconventional oil production. EROI is the ratio of energy produced to the direct plus indirect energy used in the production process.
Source: Data from Dale, M., S. Krumdieck, P. Bodger. 2012. Global energy modelling — A biophysical approach (GEMBA) Part 2: Methodology, Ecological Economics, Volume 73, Pages 158-167.

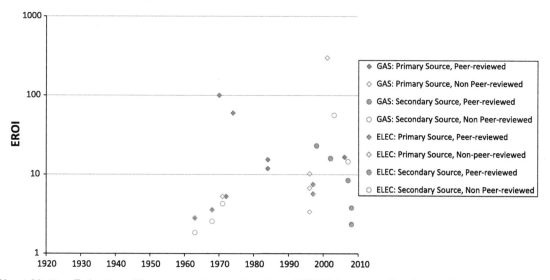

Chart 29.41. Estimates of the energy return on investment (EROI) for conventional natural gas production. EROI is the ratio of energy produced to the direct plus indirect energy used in the production process. CtX = clean coal technology.
Source: Data from Dale, M., S. Krumdieck, P. Bodger. 2012. Global energy modelling — A biophysical approach (GEMBA) Part 2: Methodology, Ecological Economics, Volume 73, Pages 158-167.

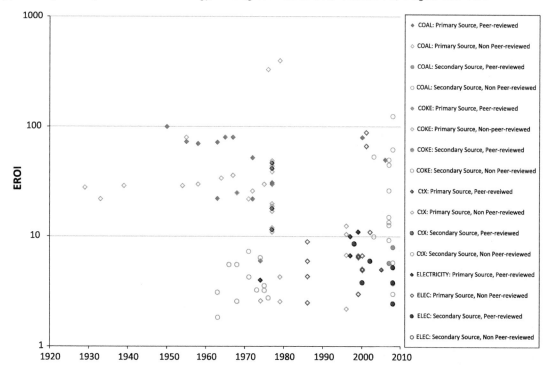

Chart 29.42. Estimates of the energy return on investment (EROI) for coal production. EROI is the ratio of energy produced to the direct plus indirect energy used in the production process.
Source: Data from Dale, M., S. Krumdieck, P. Bodger. 2012. Global energy modelling — A biophysical approach (GEMBA) Part 2: Methodology, Ecological Economics, Volume 73, Pages 158-167.

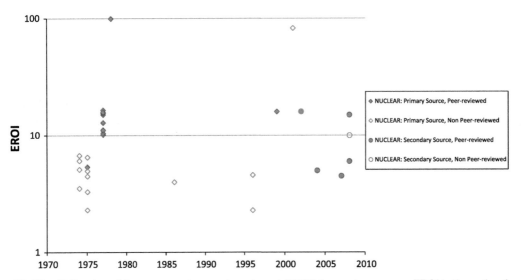

Chart 29.43. Estimates of the energy return on investment (EROI) for nuclear power. EROI is the ratio of energy produced to the direct plus indirect energy used in the production process.
Source: Data from Dale, M., S. Krumdieck, P. Bodger. 2012. Global energy modelling — A biophysical approach (GEMBA) Part 2: Methodology, Ecological Economics, Volume 73, Pages 158-167.

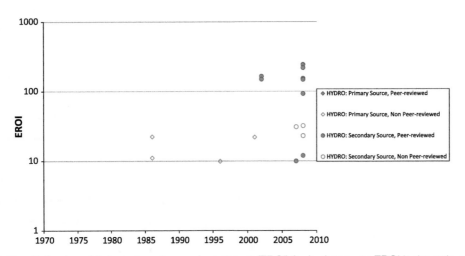

Chart 29.44. Estimates of the energy return on investment (EROI) for hydropower. EROI is the ratio of energy produced to the direct plus indirect energy used in the production process.
Source: Data from Dale, M., S. Krumdieck, P. Bodger. 2012. Global energy modelling — A biophysical approach (GEMBA) Part 2: Methodology, Ecological Economics, Volume 73, Pages 158-167.

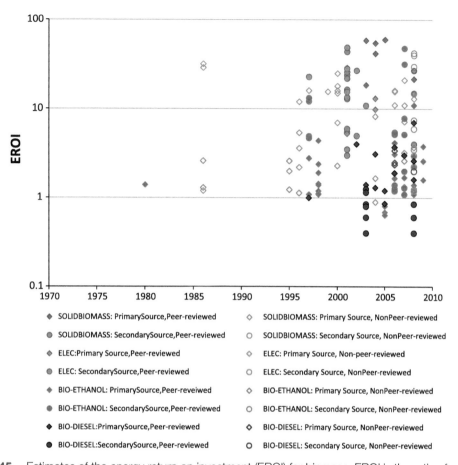

Chart 29.45. Estimates of the energy return on investment (EROI) for biomass. EROI is the ratio of energy produced to the direct plus indirect energy used in the production process.
Source: Data from Dale, M., S. Krumdieck, P. Bodger. 2012. Global energy modelling — A biophysical approach (GEMBA) Part 2: Methodology, Ecological Economics, Volume 73, Pages 158-167.

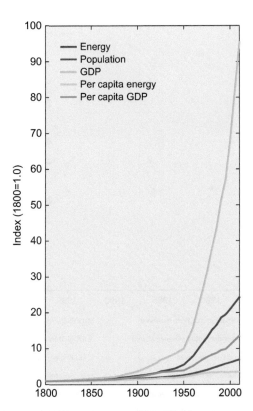

Chart 29.46. World energy use, population and GDP, 1800–2010.
Source: Grubler, Arnulf, Energy transitions, Encyclopedia of Earth, <http://www.eoearth.org/article/
Energy_transitions>, accessed 15 June 2012; World Bank, World Development Indicators and
Global Development Finance, statistical database, <http://databank.worldbank.org/data/Home.aspx>,
accessed 15 June 2012;BP, Statistical Review of World Energy 2012, <http://www.bp.com/
sectionbodycopy.do?categoryId=7500&contentId=7068481>.

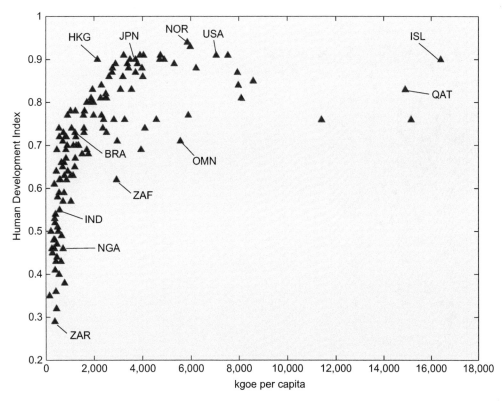

Chart 29.47.   Energy use per capita and the Human Development Index, 2010. QAT = Qatar; CHN = China; IND = India; BRA = Brazil; ISL = Iceland; JPN = Japan; NGA = Nigeria; HKG = Hong Kong; OMN = Oman; ZAF = South Africa; ZAR = Congo, Dem. Rep.
Source: Data from United Nations Development Programme, Human Development Report 2011, <http://hdr.undp.org/en/>; orld Bank, World Development Indicators and Global Development Finance, statistical database, <http://databank.worldbank.org/data/Home.aspx>.

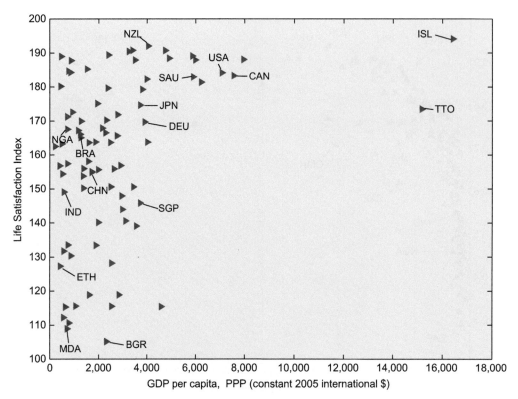

Chart 29.48. Energy use per capita and the Happy Index by nation. The Happy Index is dervied from the World Values Survey and the European Values Survey which asks individuals in various countries how happy they are. A higher value indicates a greater state of happiness.
Source: Data from Helliwell, John, Richard Layard and Jeffrey Sachs, Editors. 2012. World Happiness Report, (New York, Earth Institute,Columbia University); World Bank, World Development Indicators and Global Development Finance, statistical database, <http://databank.worldbank.org/data/Home.aspx>.

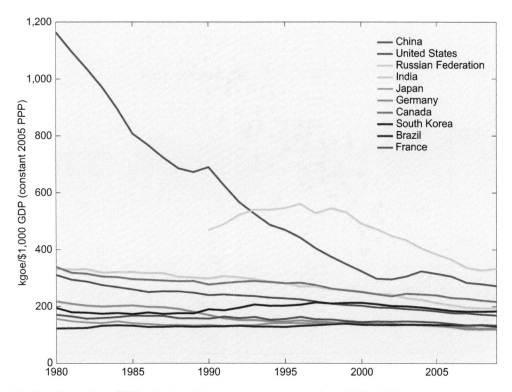

Chart 29.49.   Energy/real GDP ratio for 10 largest energy using nations 1980–2009.
Source: Data from World Bank, World Development Indicators and Global Development Finance, statistical database, <http://databank.worldbank.org/data/Home.aspx>.

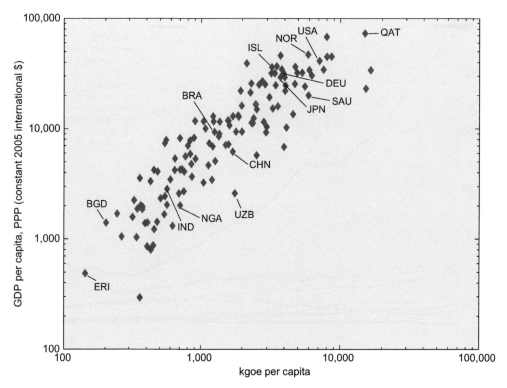

Chart 29.50. Energy/real GDP ratio and energy use per capita for selected nations, 2009. ERI = Eritrea; BGD = Bangladesh; UZB = Uzbekistan; QAT = Qatar; DEU = Germany; CHN = China; IND = India; BRA = Brazil; ISL = Iceland; JPN = Japan; NOR = Norway; SAU = Saudi Arabia; NGA = Nigeria.
Source: Data from World Bank, World Development Indicators and Global Development Finance, statistical database, <http://databank.worldbank.org/data/Home.aspx>.

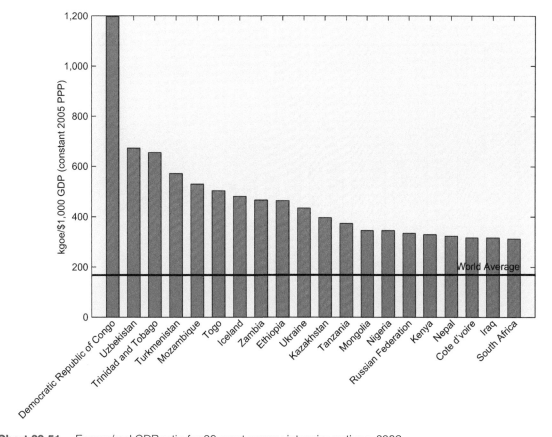

Chart 29.51. Energy/real GDP ratio for 20 most energy-intensive nations, 2009.
Source: Data from World Bank, World Development Indicators and Global Development Finance,
statistical database, <http://databank.worldbank.org/data/Home.aspx>.

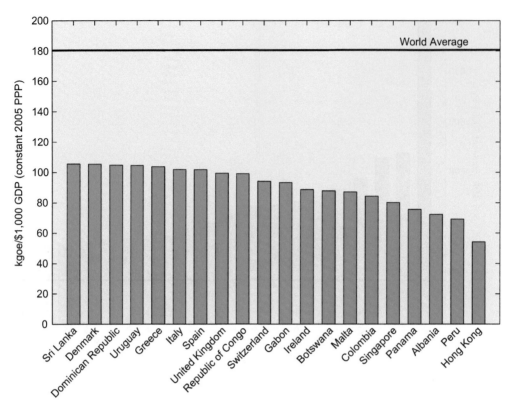

Chart 29.52. Energy/real GDP ratio for 20 least energy-intensive nations, 2009.
Source: Data from World Bank, World Development Indicators and Global Development Finance, statistical database, <http://databank.worldbank.org/data/Home.aspx>.

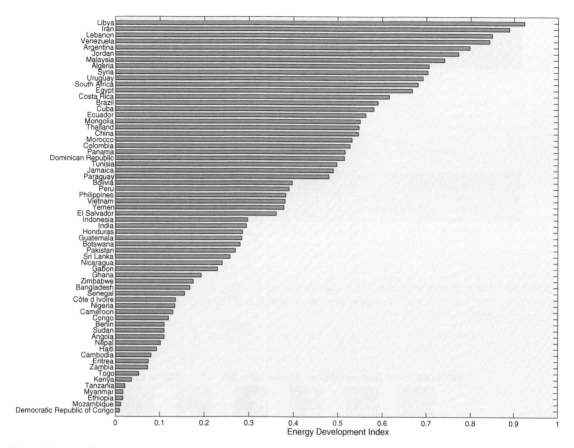

Chart 29.53. The Energy Development Index (EDI), 2011. The EDI is designed to track progress in a country's or region's transition to the use of modern fuels. The Index has four components: (1) per capita commercial energy consumption; (2) Per capita electricity consumption in the residential sector; (3) Share of modern fuels in total residential sector energy use, and (4) Share of population with access to electricity. A higher value indicates a greater degree of energy development.

Source: Data from International Energy Agency (IEA). 2011. World Energy Outlook, (Paris, IEA).

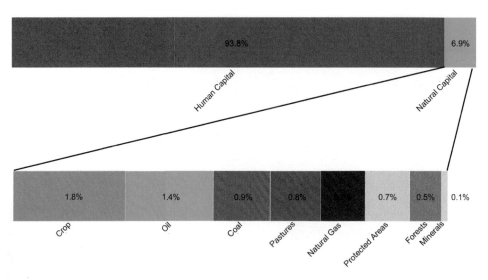

Chart 29.54. Value of human and natural capital for the world, 2005.
Source: Data from World Bank, Changing Wealth of Nations, 2010 update, <http://data.worldbank.org/data-catalog/wealth-of-nations>.

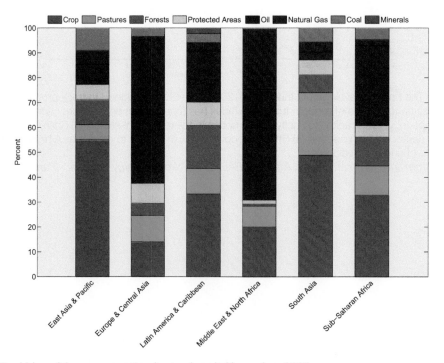

Chart 29.55. Value of the components of natural capital by region, 2005.
Source: Data from World Bank, Changing Wealth of Nations, 2010 update, <http://data.worldbank.org/data-catalog/wealth-of-nations>.

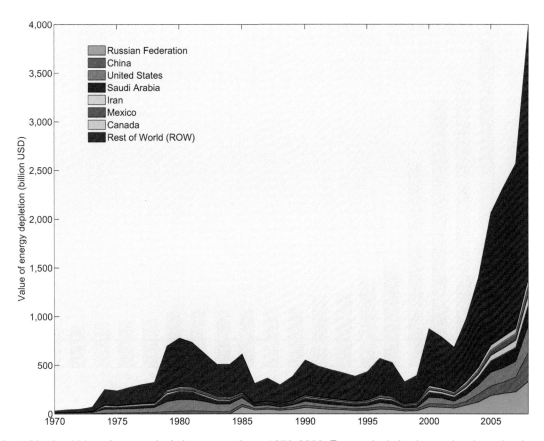

Chart 29.56.   Value of energy depletion, top nations, 1970-2008. Energy depletion is equal to the ratio of present value of rents, discounted at 4%, to exhaustion time of the resource. It covers crude oil, natural gas, and coal. Exhaustion time = Min (25 years, Reserves/Production). Rents = Unit Rents * Production. Unit Rents = Unit Price – Unit Cost.

Source: Data from World Bank, Changing Wealth of Nations, 2010 update, <http://data.worldbank.org/ data-catalog/wealth-of-nations>.

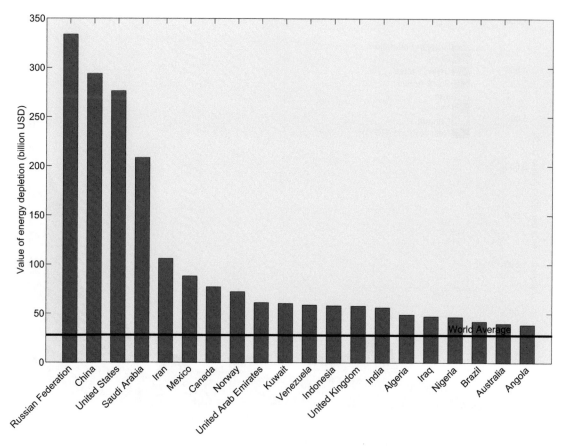

Chart 29.57. Value of the depletion of nonrenewable energy resources, top 20 nations, 2008. Energy depletion is equal to the ratio of present value of rents, discounted at 4%, to exhaustion time of the resource. It covers crude oil, natural gas, and coal. Exhaustion time = Min (25 years, Reserves/Production). Rents = Unit Rents * Production. Unit Rents = Unit Price – Unit Cost.
Source: Data from World Bank, Changing Wealth of Nations, 2010 update, <http://data.worldbank.org/data-catalog/wealth-of-nations>.

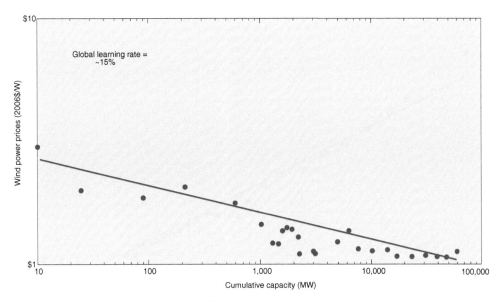

Chart 29.58. Learning curve for wind energy. The learning rate describes the cost reduction for every doubling of output.
Source: Data from Nemet, Gregory F. 2009. Interim monitoring of cost dynamics for publicly supported energy technologies, Energy Policy, Volume 37, Issue 3, Pages 825-835.

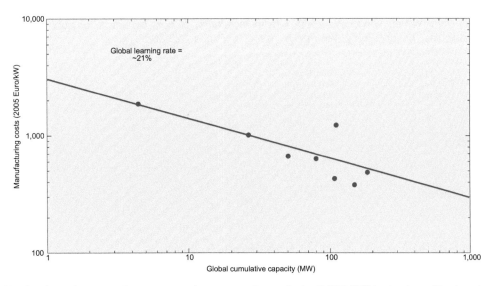

Chart 29.59. Learning curve for proton exchange membrane fuel cell (PEMFC) technology. The learning rate describes the cost reduction for every doubling of output.
Source: Data from Schoots, K., Kramer, G.J., and van der Zwaan, B.C.C., 2010. Technology learning for fuel cells: An assessment of past and potential cost reductions, Energy Policy, vol. 38(6), pages 2887-2897.

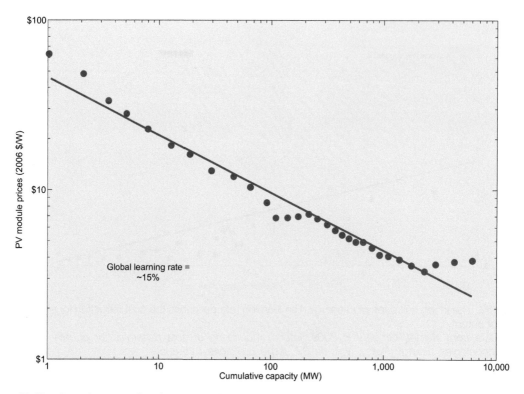

Chart 29.60. Learning curve for photovoltaic (PV) energy. The learning rate describes the cost reduction for every doubling of output.
Source: Data from Nemet, Gregory F. 2009. Interim monitoring of cost dynamics for publicly supported energy technologies, Energy Policy, Volume 37, Issue 3, Pages 825-835.

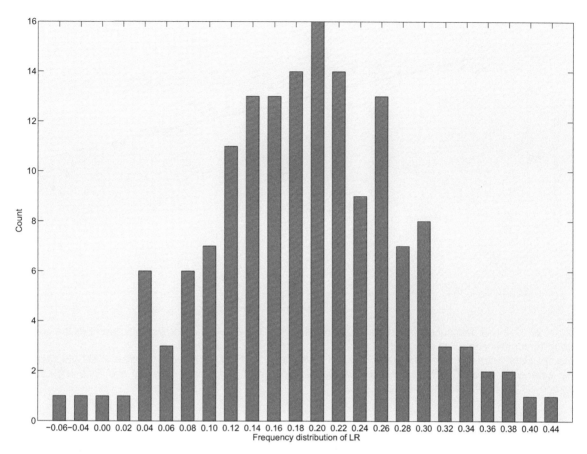

Chart 29.61. Frequency distribution of learning rates calculated in 156 learning curve studies for photovoltaics. The learning rate describes the cost reduction for every doubling of output.
Source: Data from Nemet, G.F., D. Husmann. 2012. Historical and Future Cost Dynamics of Photovoltaic Technology, In: Ali Sayigh, Editor-in-Chief, Comprehensive Renewable Energy, (Oxford, Elsevier) Pages 47-72.

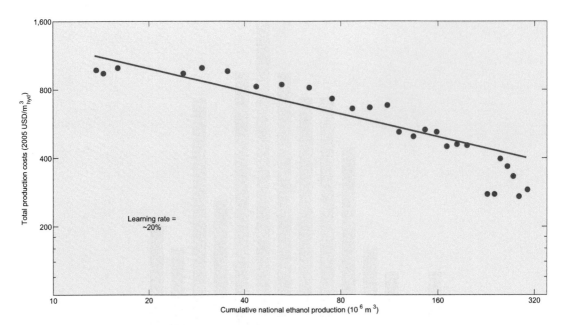

Chart 29.62. Learning curve for hydrated ethanol in Brazil, 1975-2004. The learning rate describes the cost reduction for every doubling of output.
Source: Data from van den Wall Bake, J.D., M. Junginger, A. Faaij, T. Poot, A. Walter. 2009. Explaining the experience curve: Cost reductions of Brazilian ethanol from sugarcane, Biomass and Bioenergy, Volume 33, Issue 4, Pages 644-658.

Tables

Table 29.1. Fifty largest energy corporations in the world, 2011

Rank	Company Name	Market Capitalization ($US bn)	Primary Business	HQ Country
1	ExxonMobil	406.3	Integrated IOC	US
2	PetroChina	276.6	Integrated NOC	China
3	Royal Dutch Shell	234.6	Integrated IOC	Netherlands
4	Chevron	211.9	Integrated IOC	US
5	Petrobras	156.3	Integrated NOC	Brazil
6	BP	135.5	Integrated IOC	UK
7	Gazprom	122.6	Integrated NOC	Russia
8	Total SA	121	Integrated IOC	France
9	Sinopec	97.4	Integrated NOC	China
10	ConocoPhillips	96.8	Integrated IOC	US
11	Schlumberger	91.7	Oilfield Services	US
12	Ecopetrol	88	Integrated NOC	Colombia
13	Eni	83.1	Integrated IOC	Italy
14	Statoil	81.9	Integrated NOC	Norway
15	CNOOC	78.1	Integrated NOC	China
16	Occidental	76.1	E&P	US
17	Rosneft	74.7	Integrated NOC	Russia
18	BG	72.5	Integrated IOC	UK
19	GDF SUEZ	61.7	Gas/Utilities	France
20	Suncor	45.5	Integrated IOC	Canada
21	LUKOIL	43.5	Integrated IOC	Russia
22	E.ON	43.2	Gas/Utilities	Germany
23	Reliance	42.7	R&M	India
24	ONGC	41.3	Integrated NOC	India
25	Enterprise	41.1	Midstream/Infrastructure	UA
26	Canadian Natural	41.1	E&P	Canada
27	TNK-BP	39.6	Integrated IOC	Russia
28	Anadarko	38	E&P	US
29	Imperial Oil	37.8	Integrated IOC	Canada
30	Repsol YPF	37.6	Integrated IOC	Spain
31	NOVATEK	37.2	E&P	Russia
32	BHP Billiton	a	Mining/E&P	Australia
33	Apache	34.8	E&P	US
34	Halliburton	31.8	Oilfield Services	US
35	Sasol	30.7	Integrated IOC	South Africa
36	Formosa Petrochemical	29.5	R&M	Taiwan
37	TransCanada	29.5	Midstream/Infrastructure	Canada
38	Nat'l Oilwell Varco	28.8	Equipment	US
39	Surgutneftegaz	28.2	Integrated IOC	Russia
40	Enbridge	28.1	Midstream/Infrastructure	Canada
41	Kinder Morgan	27.1	Midstream/Infrastructure	US

Continued

Table 29.1.   Continued

Rank	Company Name	Market Capitalization ($US bn)	Primary Business	HQ Country
42	EOG Resources	26.5	E&P	US
43	Woodside	25.3	E&P	Australia
44	Cenovus	25.1	Integrated IOC	Canada
45	Devon	25	E&P	US
46	OGX	23.6	E&P	Brazil
47	Centrica	23.2	Gas/Utilities	UK
48	Husky	23.1	Integrated IOC	Canada
49	INPEX	23	E&P	Japan
50	Tenaris	21.9	Equipment	Luxembourg

IOC = Integrated Oil Company NOC = National Oil Company; E&P = Exploration and Production: R&M = Refining and Marketing
[a]BHP Billiton is ranked using an estimated value for its E&P business based on reserves and production (#24). Prior listings ranked the company based on its total market capitalization.

Table 29.2.   Principal operations of the major oil companies, 1980–2010 (Thousand barrels/day)

	1980	1990	2000	2010
BP				
Crude Oil & NGL Reserves[a]	8,234	7,313	6,508	5,559
Natural Gas Reserves[b]	24,566	30,046	41,100	37,809
Crude Oil Produced	3,222	2,104	1,928	2,374
Crude Oil Processed	2,865	2,783	2,928	2,426
Refined Products Sold	3,558	3,835	5,859	5,927
Natural Gas Sold[c]	3,366	5,152	7,609	8,401
ExxonMobil				
Crude Oil & NGL Reserves	10,116	10,181	12,171	11,673
Natural Gas Reserves	59,997	62,249	55,866	78,815
Crude Oil Produced	2,134	2,491	2,553	2,422
Crude Oil Processed	6,108	4,952	5,642	5,253
Refined Products Sold	7,207	7,283	7,993	6,414
Natural Gas Sold	10,248	9,743	10343	12148
Total SA				
Crude Oil & NGL Reserves	n.a.	2,731	6,960	5,987
Natural Gas Reserves	n.a.	5,156	20,705	25,788
Crude Oil Produced	927	411	1,433	1,340
Crude Oil Processed	941	832	2,411	2,009
Refined Products Sold	1,309	1,201	3695	3776
Natural Gas Sold	564	1,487	3,758	5,648
Royal Dutch/Shell				
Crude Oil & NGL Reserves	7,223	10,107	6,907	5,179
Natural Gas Reserves	38,867	51,575	44,685	28,983
Crude Oil Produced	1,170	1,820	2,274	1,619
Crude Oil Processed	3,382	3,218	2,923	3197
Refined Products Sold	4,342	4,962	5,574	6,460
Natural Gas Sold	5,959	6,068	8,212	9,305

Table 29.2. Continued

	1980	1990	2000	2010
Chevron				
Crude Oil & NGL Reserves	3,961	5,909	8,519	4,270
Natural Gas Reserves	17,711	15,581	17,844	20,755
Crude Oil Produced	4,579	1,745	1997	1923
Crude Oil Processed	4,719	3,285	2,540	1,894
Refined Products Sold	5,225	4,680	5,188	3,113
Natural Gas Sold	4,584	6,521	10,866	10,425
Total Majors				
Crude Oil & NGL Reserves	29,534	36,241	41,065	32,668
Natural Gas Reserves	141,141	164,607	180,200	192,150
Crude Oil Produced	12,032	8,571	10,185	9,678
Crude Oil Processed	18,015	15,070	16,444	14,779
Refined Products Sold	21,641	21,961	28,309	25,690
Natural Gas Sold	24,721	28,971	40,788	45,927

[a]Million barrels, as at year end.
[b]Billions of cubic feet, as at year end.
[c]Million cubic feet daily
Source: Organization of Petroleum Exporting Countries

Table 29.3. Gasoline prices[a] in selected countries, 1990–2010

	Current dollars per gallon							Average annual percentage change
	1990	1995	2000	2005	2007	2009	2010[b]	1990–2010
China	c	1.03	c	1.7	2.29	3.27	c	c
Japan	3.16	4.43	3.65	4.28	4.49	4.86	5.93	3.20%
France[d]	3.63	4.26	3.8	5.46	6.6	6.35	6.72	3.10%
United Kingdom[d]	2.82	3.21	4.58	5.97	7.15	5.86	6.81	4.50%
Germany	2.65	3.96	3.45	5.66	6.88	6.81	6.86	4.90%
Canada	1.87	1.53	1.86	2.89	3.59	3.15	3.72	3.50%
United States[e]	1.16	1.15	1.51	2.27	2.8	2.34	2.72	4.30%

	Constant 2010 dollars[f] per gallon							Average annual percentage change
	1990	1995	2000	2005	2007	2009	2010[b]	1990–2010
China	c	1.47	c	1.9	2.41	3.33	c	c
Japan	5.27	6.34	4.62	4.78	4.73	4.94	5.93	0.60%
France[d]	6.06	6.1	4.81	6.09	6.94	6.45	6.72	0.60%
United Kingdom[d]	4.7	4.59	5.8	6.66	7.52	5.96	6.81	1.90%
Germany	4.42	5.67	4.37	6.31	7.24	6.92	6.86	2.30%

Continued

Table 29.3. Continued

| | Constant 2010 dollars[f] per gallon | | | | | | | Average annual percentage change |
	1990	1995	2000	2005	2007	2009	2010[b]	1990–2010
Canada	3.12	2.19	2.36	3.22	3.78	3.2	3.72	0.90%
United States[e]	1.94	1.65	1.91	2.54	2.94	2.38	2.72	1.80%

Comparisons between prices and price trends in different countries require care. They are of limited validity because of fluctuations in exchange rates; differences in product quality, marketing practices, and market structures; and the extent to which the standard categories of sales are representative of total national sales for a given period.

[a]Prices represent the retail prices (including taxes) for regular unleaded gasoline, except for France and the United Kingdom which are premium unleaded gasoline.

[b]3rd quarter 2010.

[c]Data are not available.

[d]Premium gasoline.

[e]These estimates are international comparisons only.

[f]Adjusted by the U.S. Consumer Price Inflation Index.

Table 29.4. Diesel fuel prices[a] for selected countries, 1998–2010

| | Current dollars per gallon | | | | | | | | Average annual percentage change |
	1998	2000	2003	2004	2005	2007	2009	2010[b]	1998–2010
China	c	c	1.32	1.47	1.69	2.42	3.23	c	c
Japan	2.25	2.85	2.76	3.08	3.45	3.82	4.19	5.04	7.0%
France	2.71	2.95	3.39	4.16	4.81	5.66	5.27	5.59	6.2%
United Kingdom	4.10	4.66	4.82	5.68	6.25	7.34	6.14	6.95	4.5%
Germany	2.45	2.79	3.79	4.41	5.01	6.06	5.73	5.94	7.7%
United States[d]	1.04	1.50	1.51	1.81	2.40	2.88	2.46	2.94	9.0%
	Constant 2010 dollars[e] per gallon								Average annual percentage change
	1998	2000	2003	2004	2005	2007	2009	2010[b]	1998–2010
China	c	c	1.57	1.70	1.89	2.55	3.28	c	c
Japan	3.01	3.61	3.27	3.56	3.85	4.02	4.26	5.04	4.4%
France	3.62	3.73	4.02	4.80	5.37	5.95	5.35	5.59	3.7%
United Kingdom	5.49	5.90	5.71	6.56	6.98	7.72	6.24	6.95	2.0%
Germany	3.28	3.54	4.49	5.09	5.59	6.37	5.83	5.94	5.1%
United States[d]	1.40	1.89	1.79	2.08	2.68	3.03	2.50	2.94	6.4%

Comparisons between prices and price trends in different countries require care. They are of limited validity because of fluctuations in exchange rates; differences in product quality, marketing practices, and market structures; and the extent to which the standard categories of sales are representative of total national sales for a given period.

[a]Prices represent the retail prices (including taxes) for automotive diesel fuel for non-commercial (household) use.

[b]3rd quarter 2010.

[c]Data are not available.

[d]These estimates are for international comparisons only and do not necessarily correspond to gasoline price estimates in other sections of the book.

[e]Adjusted by the U.S. Consumer Price Inflation Index. Source: Oak Ridge National Laboratory, Transportation Energy Data Book: Edition 30, <http://cta.ornl.gov/data/index.shtml>.

Table 29.5. The price of compressed natural gas (CNG) compared to other fuels in selected nations

Country	Fuel prices expressed in Euros					Date		CNG price as % of other fuel price	
	CNG (Euro/Nm³)	CNG price per liter gasoline equivalent	Gasoline (Euro/liter)	CNG price per liter diesel eqivalent	Diesel (Euro/liter)	Month	Year	Gasoline	Diesel
Argentina	0.26	0.23	1.00	0.27	1.00	December	2011	27	27
Bangladesh	0.18	0.16	0.49	0.18	0.34	November	2008	38	54
Pakistan	0.55	0.50	0.71	0.56	0.55	September	2011	71	103
Iran	0.23	0.21	0.29	0.24	0.12	July	2011	73	203
Brazil	0.71	0.64	1.20	0.73	0.77	October	2011	61	95
Italy[a]	0.64	0.59	1.69	0.66	1.66	December	2011	35	39
U.S.A.	0.46	0.42	0.71	0.47	0.78	October	2011	59	61
Japan	0.79	0.72	1.34	0.81	1.16	July	2011	54	70
Germany[a]	0.74	0.66	1.60	0.76	1.49	December	2011	47	51

Default values used: CH_4 content in natural gas is 97%. CNG energy content is 9.6 kWh/Nm³. Gasoline energy content is 8.8 kWh/litre. Diesel energy content is 9.85 kWh/litre. LPG energy content is 6.46 kWh/litre
[a]In these countries sales are measured in kg. The conversion factor depends on the normal density of gaseous natural gas in each country. The default value used is 0.73 kg/Nm³.
Source: NGVA Europe, Worldwide NGV Statistics, <http://www.ngvaeurope.eu/worldwide-ngv-statistics>.

Table 29.6. Energy costs of selected goods and services in the United States per $1,000,000 of output[a]

	Total energy	Coal	Natural gas	Petroleum	Bio/waste	Non-Fossil Fuel Electricity	Energy Intensity
Good or Service	TJ	TJ	TJ	TJ	TJ	TJ	MJ/$
Power generation and supply	111	78.4	24.7	5.14	0.097	3.06	111.0
Cement	74.4	43.6	7.16	11.4	5.79	6.41	74.4
Pulp mills	67.6	6.21	10.1	6.12	42	3.13	67.6
Iron ore mining	53.7	13.3	19.2	6.41	0.285	14.4	53.7
Organic chemicals	45.3	8.11	16.9	8.43	8.39	3.44	45.3
Fertilizers	44.6	5.28	30.2	4.86	0.908	3.39	44.6
Iron and steel	43.3	22.8	12.1	2.72	0.458	5.29	43.3
Petrochemicals	42.3	4.62	16.5	13.5	4.93	2.78	42.3
Plastics	42	5.94	17.1	10	5.41	3.51	42.0
Paper mills	38.2	7.33	8.31	3.72	15.4	3.46	38.2
Glass	37.1	4.78	24.3	2.8	1.32	3.85	37.1
Petroleum refineries	31.7	2.59	12.9	13	1.15	2.09	31.7
Bricks	31.4	7.52	17	4.14	0.589	2.15	31.4
Air transportation	28.4	0.78	2.01	24.9	0.2	0.503	28.4
Plastic bottles	23.7	5.75	7.63	4.08	2.46	3.75	23.7
Copper smelt./refin.	21.4	5.48	7.07	3.64	0.846	4.31	21.4
Grain farming	19.3	2.7	5.06	9.01	0.334	2.22	19.3
Truck transportation	18.8	0.92	1.56	15.5	0.181	0.585	18.8

Continued

Table 29.6. Continued

	Total energy	Coal	Natural gas	Petroleum	Bio/waste	Non-Fossil Fuel Electricity	Energy Intensity
Cattle ranching	18.7	3.31	3.36	9.57	0.228	2.23	18.7
Rail transportation	17.4	0.804	1.2	14.5	0.244	0.642	17.4
Paint	16.5	3.43	5.77	3.53	1.92	1.84	16.5
Coal mining	15.4	3.03	4.9	5.04	0.161	2.22	15.4
Seafood prep.	14.9	2.38	3.83	7.1	0.466	1.16	14.9
Office furniture	14.7	1.99	2.22	1.4	1.65	7.45	14.7
Oil and gas extraction	14.5	1.99	9.19	1.73	0.198	1.42	14.5
Dog and cat food	13.5	3.07	4.45	3.49	0.788	1.7	13.5
Sawmills	13.2	2.34	2.3	2.84	4.21	1.52	13.2
Refrigerators	11.8	3.6	3.78	1.74	0.827	1.8	11.8
College education	11.2	2.81	4.7	2.01	0.186	1.51	11.2
Boilers	11.1	4.16	3.69	1.15	0.332	1.73	11.1
Bread manuf.	10.8	3.04	3.71	2.06	0.803	1.24	10.8
Farm machinery	9.29	3.25	3.01	1.23	0.354	1.45	9.3
Residential construction	8.91	1.91	2.1	3.47	0.539	0.892	8.9
Hotels	8.85	3.06	2.23	1.32	0.267	1.98	8.9
Women's apparel	8.71	2.32	2.93	1.51	0.534	1.43	8.7
HVAC equipment	8.53	2.93	2.62	1.15	0.399	1.43	8.5
Light bulbs	8.42	2.16	3.21	0.947	0.56	1.54	8.4
Automobile manuf.	8.33	2.56	2.63	1.29	0.435	1.41	8.3
Primary batteries	8.2	2.64	2.51	1.1	0.576	1.37	8.2
Restaurants	7.41	2.43	1.96	1.49	0.302	1.22	7.4
Logging	5.95	0.933	1.26	3.12	0.145	0.489	6.0
Aircraft manuf.	5.56	1.79	1.6	0.948	0.249	0.96	5.6
Guided missiles	4.55	1.44	1.31	0.773	0.233	0.803	4.6
Computer manuf.	4.28	1.26	1.18	0.815	0.272	0.75	4.3
Retail trade	4.12	1.47	0.892	0.803	0.107	0.852	4.1
Real estate	3.97	1.7	0.873	0.435	0.047	0.913	4.0
Government services	3.5	0.977	1.76	0.478	0.087	0.192	3.5
Spectator sports	3.29	1.07	0.796	0.683	0.113	0.635	3.3
Religious organizations	2.58	0.636	0.66	0.821	0.108	0.352	2.6
Offices of physicians, etc.	2.35	0.689	0.613	0.529	0.132	0.385	2.4
Financial services	1.53	0.449	0.337	0.425	0.063	0.258	1.5
Legal services	1.52	0.446	0.351	0.401	0.071	0.249	1.5
Insurance carriers	1.07	0.289	0.258	0.26	0.076	0.183	1.1

[a]The direct plus indirect energy required to produce a good or service in 2002.
Source: Carnegie Mellon University Green Design Institute. 2012 Economic Input-Output Life Cycle Assessment (EIO-LCA) US 2002 (428) model [Internet], <http://www.eiolca.net/>.

Table 29.7. Estimates of the long-run direct rebound effect[a] for household energy services in the OECD[b]

End-use	Range of values in (%)	'Best guess' (%)	No. of studies	Degree of confidence
Personal automotive transport	3–87	10–30	17	High
Space heating	0.6–60	10–30	9	Medium
Space cooling	1–26	1–26	2	Low
Other consumer energy services	0–41	<20	3	Low

[a]An increase in energy use caused directly or indirectly by improved energy efficiency.
[b]Organisation for Economic Co-operation and Development
Source: Adapted from Sorrell, Steve, John Dimitropoulos, Matt Sommerville. 2009. Empirical estimates of the direct rebound effect: A review, Energy Policy, Volume 37, Issue 4, Pages 1356-1371.

Table 29.8. Estimates of the per unit external costs of energy activity in the U.S[a]

Energy-Related Activity			Climate Damages (per ton CO_2-eq)[c]		
Fuel Type	Nonclimate Damage	CO_2-eq Intensity	At $10	At $20	At $30
Electricity generation (coal)	3.2 cents/kWh	2 lb/kWh	1 cent/kWh	3 cents/kWh	10 cents/kWh
Electricity generation (natural gas)	0.16 cents/kWh	1 lb/kWh	0.5 cent/kWh	1.5 cents/kWh	5 cents/kWh
Transportation[b]	1.2 to >1.7 cents/VMT	0.3 to >1.3 lb/VMT	0.15 to >.65 cent/kWh	0.45 to >.2 cents/kWh	1.5 to >6 centskWh
Heat production (natural gas)	11 cents/MCF	140 lb/MCF	70 cents/MCF	210 cents/MCF	700 cents/MCF

CO_2-eq, carbon dioxide equivalent; VMT, vehicle miles traveled; MCF, thousand cubic feet; E85, ethanol 85% blend; HEV, hybrid electric vehicle; CNG, compressed natural gas; RFG, reformulated gasoline.
Committee on Health, Environmental, and Other External Costs and Benefits of Energy Production and Consumption; National Research Council, Hidden Costs of Energy: Unpriced Consequences of Energy Production and Use, (Washington, National Academy Press).
[a]Based on emission estimates for 2005. Damages are expressed in 2007 U.S. dollars. Damages that have not been quantified and monetized are not included.
[b]Transportation fuels include E85 herbaceous, E85 corn stover, hydrogen gaseous, E85 corn, diesel with biodiesel, grid-independent HEV, griddependent HEV, electric vehicle, CNG, conventional gasoline and RFG, E10, low-sulfur diesel, tar sands.
[c]Often called the "social cost of carbon."
Source: Adapted from Committee on Health, Environmental, and Other External Costs and Benefits of Energy Production and Consumption; National Research Council, Hidden Costs of Energy: Unpriced Consequences of Energy Production and Use, (Washington, National Academy Press).

Table 29.9. **Value of rent earned from coal production top 20 nations, 1970–2008**

	1970	1980	1990	2000	2008	Change 2000-08	Cumulative 1970-2008
China	1.64	11.65	12.98	12.47	294.80	2263.4%	987.52
United States	1.70	6.60	1.08	2.34	102.42	4275.3%	331.24
India	0.41	2.65	4.25	4.89	57.74	1081.7%	247.62
South Africa	0.25	2.08	2.10	2.24	26.03	1063.8%	124.43
Russian Federation	0.00	0.00	5.12	2.57	30.97	1107.0%	113.92
Australia	0.14	0.52	0.04	0.07	33.89	46853.3%	78.46
Poland	0.55	2.62	0.95	0.59	10.37	1669.2%	69.28
Indonesia	0.00	0.01	0.11	0.69	30.22	4276.7%	67.17
Germany	0.71	2.10	0.54	0.30	6.78	2133.8%	53.17
Kazakhstan	0.00	0.00	2.37	1.08	12.21	1027.8%	47.31
Ukraine	0.00	0.00	2.64	0.88	6.87	678.7%	35.01
Korea, Dem. Rep.	0.10	0.83	0.60	0.31	4.17	1224.5%	25.91
United Kingdom	0.43	1.05	0.00	0.06	1.61	2812.1%	20.51
Colombia	0.01	0.02	0.00	0.00	7.67		14.60
Vietnam	0.01	0.10	0.06	0.12	4.44	3513.4%	12.86
Canada	0.03	0.14	0.02	0.03	4.16	13935.6%	11.01
Czech Republic	0.00	0.30	0.02	0.03	2.13	6157.0%	10.44
Korea, Rep.	0.06	0.35	0.22	0.04	0.31	601.8%	7.38
Turkey	0.02	0.08	0.07	0.09	2.29	2446.0%	6.92
Greece	0.01	0.08	0.08	0.09	1.82	1926.4%	6.35
Rest of World	1.80	6.43	7.60	4.37	95.02	2076.8%	373.92
World	6.49	32.54	34.12	29.53	651.23	2105.6%	2,326.37

Source: World Bank, Changing Wealth of Nations, 2010 update, <http://data.worldbank.org/data-catalog/wealth-of-nations>.

Table 29.10. **Value of rent eaned from crude oil production top 20 nations, 1970–2008**

Country	1970	1980	1990	2000	2008	Change 2000-08	Cumlative Rent 1970-2008
Saudi Arabia	1.6	131.8	48.9	76.9	308.3	301.0%	2,607.1
Russian Federation			70.1	52.0	287.4	452.9%	1,790.8
United States	3.3	102.0	40.5	38.9	133.1	242.0%	1,665.9
Iran	1.5	19.0	22.1	32.3	127.8	295.8%	1,059.8
China	0.2	27.0	19.2	27.7	118.7	328.3%	880.8
Kuwait	1.2	21.9	6.6	18.6	91.8	392.9%	633.6
Mexico	0.2	27.3	19.2	26.5	91.2	244.3%	841.5
Venezuela	1.5	29.6	16.2	28.8	83.9	191.3%	804.8
United Arab Emirates	0.3	21.1	12.6	19.0	80.0	321.0%	598.3
Iraq	0.6	34.2	14.7	23.2	75.3	224.5%	574.3
Nigeria	0.4	26.1	12.3	19.5	62.9	222.6%	599.4
Canada	0.4	15.5	7.9	12.3	61.4	399.9%	433.0

Table 29.10.   Continued

Country	1970	1980	1990	2000	2008	Change 2000–08	Cumlative Rent 1970–2008
Norway	0.0	5.8	10.9	26.3	61.3	133.1%	514.1
Brazil	0.1	2.3	4.2	10.3	56.4	449.0%	286.7
Angola	0.0	1.9	3.0	5.8	56.4	874.7%	221.0
Libya	1.2	22.8	9.4	11.7	52.5	347.8%	448.1
Algeria	0.4	12.2	5.1	7.5	41.6	452.9%	299.8
United Kingdom	0.0	20.0	12.0	19.7	40.4	105.1%	506.9
Kazakhstan			3.1	5.4	36.3	569.6%	176.6
Azerbaijan			1.7	2.3	26.6	1048.6%	95.4
World	14.6	608.3	407.5	565.0	2,244.8	297.3%	17,976.3

Source: World Bank, Changing Wealth of Nations, 2010 update, <http://data.worldbank.org/data-catalog/wealth-of-nations>.

Table 29.11.   Value of rent eaned from natural gas production top 20 nations, 1970–2008.

	1970	1980	1990	2000	2008	Change 2000-08	Cumulative 1970-2008
Russian Federation	0.00	0.00	28.87	61.98	200.38	2.23	1,335.16
United States	5.04	41.42	10.74	41.57	137.91	2.32	1,112.09
Canada	0.49	5.94	2.66	14.62	43.16	1.95	311.90
Algeria	0.02	1.44	2.86	11.25	30.35	1.70	222.28
United Kingdom	0.11	3.36	2.29	12.84	23.66	0.84	220.54
Netherlands	0.28	7.38	3.05	6.83	22.92	2.35	215.51
Iran	0.12	0.41	1.23	7.05	41.16	4.84	191.06
Norway	0.00	2.44	1.35	6.03	33.20	4.50	174.48
Indonesia	0.01	1.51	1.85	7.30	22.82	2.13	148.45
Saudi Arabia	0.02	1.08	1.64	5.38	25.14	3.67	146.67
Turkmenistan	0.00	0.00	3.84	5.03	21.67	3.31	131.23
Malaysia	0.00	0.25	1.02	6.16	21.18	2.44	125.83
Uzbekistan	0.00	0.00	1.84	6.03	20.71	2.43	122.79
Mexico	0.10	2.34	1.32	4.22	16.48	2.91	112.14
China	0.03	1.30	0.75	3.08	24.58	6.98	106.85
United Arab Emirates	0.01	0.71	1.10	4.80	17.35	2.61	105.60
Australia	0.01	0.84	1.13	4.13	16.07	2.89	97.78
Qatar	0.01	0.32	0.36	3.27	28.44	7.70	94.56
Argentina	0.05	0.93	0.99	4.53	14.42	2.18	93.91
Venezuela	0.07	1.58	1.38	3.94	9.16	1.32	83.81
World	7.60	87.97	82.10	257.49	942.89	2.66	6,199.30

Source: World Bank, Changing Wealth of Nations, 2010 update, <http://data.worldbank.org/data-catalog/wealth-of-nations>.

Table 29.12.   Estimates of the external costs of electricity generated from fossil fuels

Specific Type	Author	Fuel & Country Scope	Period	Billion $	US c/ kWh
External Costs	ExternE	Weighted Average of Coal, Oil, Gas in EU	2001	33-59	2.2-3.9
External Costs	Owen	Weighted Average of Coal, Oil, Gas in EU	2004	48	3.2
External Costs	EEA	Solid Fuels in the EU-15	2001	23-42	3.3-5.9
External Costs	EEA	Oil and Gas in the EU-15	2001	11-20	1.3-2.4
External Costs	Badcock & Lenzen	Global Coal	2007	227-1890	2.9-23.8
External Costs	ATSE	Brown coal generation in Australia	NA	NA	4.3
External Costs	ATSE	Gas Generation in Australia	NA	NA	1.6
External Costs	NRC	Coal generation in the U.S.	2005	89-280	4.2-13.2
External Costs	NRC	Gas Generation in the U.S.	2005	6-47	0.7-5.2

ATSE. (2009). The Hidden Costs of Electricity: Externalities of power generation in Australia. Parkville: Australian Academy of Technological Sciences and Engineering; EEA: European Environment Agency; ; externE Project. 2003. External Costs: Research results on socio-environmental damages due to electricity and transport (Luxembourg, Office for Official Publications of the European Communities). ; Badcock, J., & Lenzen, M. (2010, September). Subsidies for Electricity-generating Technologies: A review. Energy Policy 38(9), 5038-5047; Owen, A. D. (2006). Renewable energy: Externality costs as market barriers. Energy Policy 34 (2006) 632-642; NRC: National Research Council. (2009). Hidden Costs of Energy: Unpriced consequences of energy production and use. Washington: The National Academies Press.

Source: Adapted from Global Subsidies Initiative. 2011. Subsidies and External Costs in Electric Power Generation: A comparative review of estimates, <http://www.iisd.org/gsi/sites/default/files/power_gen_subsidies.pdf>.

Table 29.13.   Estimates of the external cost of electricity generated from nuclear energy

Specific Type	Author	Fuel & Country Scope	Period	Billion $	US c/ kWh
Health, envt, & climate change	ExternE	EU-15	2001	1.5-5.1	0.2-0.6
	EEA	EU-15	2001	2.5	0.3
	Owen	EU-15	2001	2.2	0.3
	Badcock & Lenzen	Global Generation	2007	Nov-31	0.4-1.2
	Badcock & Menzen	Global Generation	1960-2007	250-741	NA

EEA: European Environment Agency; ; externE Project. 2003. External Costs: Research results on socio-environmental damages due to electricity and transport (Luxembourg, Office for Official Publications of the European Communities). ; Badcock, J., & Lenzen, M. (2010, September). Subsidies for Electricity-generating Technologies: A review. Energy Policy 38(9), 5038-5047; Owen, A. D. (2006). Renewable energy: Externality costs as market barriers. Energy Policy 34 (2006) 632-642.

Source: Adapted from Global Subsidies Initiative. 2011. Subsidies and External Costs in Electric Power Generation: A comparative review of estimates, <http://www.iisd.org/gsi/sites/default/files/power_gen_subsidies.pdf>.

Table 29.14. Estimates of the external costs of electricity generated from renewable energy.

Specific Type	Author	Fuel & Country Scope	Period	Billion $	US c/ kWh
Health, envt, & climate change	ExternE	EU-15	2001	3.7-4.0	0.4-0.44
	EEA	EU-15	2001	1.8-2.4	0.3-0.4
	Owen	EU-15	2004	3	0.5
	Babcock & Lenzen	Global Generation	2007, 2008	1.3-10.4	0.2-3.2

EEA: European Environment Agency; ; externE Project. 2003. External Costs: Research results on socio-environmental damages due to electricity and transport (Luxembourg, Office for Official Publications of the European Communities). ; Badcock, J., & Lenzen, M. (2010, September). Subsidies for Electricity-generating Technologies: A review. Energy Policy 38(9), 5038-5047; Owen, A. D. (2006). Renewable energy: Externality costs as market barriers. Energy Policy 34 (2006) 632-642.
Source: Adapted from Global Subsidies Initiative. 2011. Subsidies and External Costs in Electric Power Generation: A comparative review of estimates, <http://www.iisd.org/gsi/sites/default/files/power_gen_subsidies.pdf>.

Table 29.15. Estimates of the external cost of electricity generation in European Union countries (€ cent per kWh)

Country	Coal & lignite	Peat	Oil	Gas	Nuclear	Biomass	Hydro	PV	Wind
Austria				1-3		2-3	0.1		
Belgium	4-15			1-2	0.5				
Germany	3-6		5-8	1-2	0.2	3		0.6	0.05
Denmark	4-7			2-3		1			0.1
Spain	5-8			1-2		3.5			0.2
Finland	2-4	2-5				1			
France	7-10		8-11	2-4	0.3	1	1		
Greece	5-8		3-5	1		0-0.8	1		0.25
Ireland	6-8	3-4							
Italy			3-6	2-3			0.3		
Netherlands	3-4			1-2	0.7	0.5			
Norway				1-2		0.2	0.2		0-0.25
Portugal	4-7			1-2		1.2	0.03		
Sweden	2-4					0.3	0-0.7		
United Kingdom	4-7		3-5	1-2	0.25	1			0.15

Source: Adapted from externE Project. 2003. External Costs: Research results on socio-environmental damages due to electricity and transport (Luxembourg, Office for Official Publications of the European Communities).

Table 29.16. Global levelized costs of electricity ($ MWh^{-1}) by generating type

Generating type	Midpoint	Low	High
Wind onshore	68.08	36.39	168.71
Wind offshore	78.54	59.09	144.38
Solar thermal	193.64	193.64	315.2
Solar photovoltaic	192.21	141.1	2195.39
Small-scale run-of-river hydro	108.28	46.45	283.02

Continued

Table 29.16. **Continued**

Generating type	Midpoint	Low	High
Large-scale hydro	53.12	53.12	99.33
Nuclear	30.71	24.34	80.26
Coal (lignite)	39.35	34.4	75.35
Coal (high quality)	31.9	30.3	80.85
Coal (integrated coal gas)	44.73	31.94	69.15
Gas (CCGT)	54.62	44.69	73.24
Gas (open)	54.64	54.64	57.33
CHP (using CCGT)	55.12	33.11	94.65
CHP (using coal)	39.09	29.25	54.87
CHP (using other fuels)	40.01	34.4	116.42
Waste incineration	11.39	−4.68	61.19
Biomass	48.74	43.64	117.59

CCGT = combined cycle gas turbine; CHP = combined heat and power. The costs include capital, operation and maintenance, and fuel costs over the lifetime of a power plant, discounted to the present and 'levelized' over the expected output of the generating source over its lifetime. Values are in 2008 US dollars. The midpoint value is based on a 5% discount rate, as is the low value (except in the case of high-quality coal); the high value is derived using a 10% discount rate.
Source: Adapted from van Kooten GC and Timilsina GR. (2009. Wind Power Development: Economics and Policies, 32pp. Policy Research Working Paper 4868. Washington, DC: The World Bank, Development Research Group, Environment and Energy Team.

Table 29.17. **Estimated Levelized Cost[a] of new electricity generation in 2016**

Plant Type	U.S. Average Levelized Cost for Plants Entering Service in 2016					
	(2009 USD/MWh)					
	Capacity Factor (%)	Levelized Capital Cost	Fixed O&M	Variable O&M (including fuel)	Transmission Investment	Total System Levelized Cost
Conventional Coal	85	65.3	3.9	24.3	1.2	94.8
Advanced Coal	85	74.6	7.9	25.7	1.2	109.4
Advanced Coal with CCS	85	92.7	9.2	33.1	1.2	136.2
Natural Gas Fired						
Conventional Combined Cycle	87	17.5	1.9	45.6	1.2	66.1
Advanced Combined Cycle	87	17.9	1.9	42.1	1.2	63.1
Advanced CC with CCS	87	34.6	3.9	49.6	1.2	89.3
Conventional Combustion Turbine	30	45.8	3.7	71.5	3.5	124.5
Advanced Combustion Turbine	30	31.6	5.5	62.9	3.5	103.5
Advanced Nuclear	90	90.1	11.1	11.7	1	113.9
Wind	34	83.9	9.6	0	3.5	97
Wind — Offshore	34	209.3	28.1	0	5.9	243.2
Solar PV	25	194.6	12.1	0	4	210.7

Table 29.17.   Continued

| Plant Type | Capacity Factor (%) | U.S. Average Levelized Cost for Plants Entering Service in 2016 (2009 USD/MWh) | | | | |
		Levelized Capital Cost	Fixed O&M	Variable O&M (including fuel)	Transmission Investment	Total System Levelized Cost
Solar Thermal	18	259.4	46.6	0	5.8	311.8
Geothermal	92	79.3	11.9	9.5	1	101.7
Biomass	83	55.3	13.7	42.3	1.3	112.5
Hydro	52	74.5	3.8	6.3	1.9	86.4

[a]Levelized cost represents the present value of the total cost of building and operating a generating plant over an assumed financial life and duty cycle, converted to equal annual payments and expressed in terms of real dollars to remove the impact of inflation. CC = combined cycle; CCS = carbon capture and storage; M&O = maintenance and operation.
Source: U.S. Energy Information Administration, Levelized Cost of New Generation Resources in the Annual Energy Outlook 2011, <http://205.254.135.7/oiaf/aeo/electricity_generation.html>.

Table 29.18.   General objectives of subsidies to electricity generation and use

Generation type	Objective of subsidies
Renewable	• Environmental improvement (reductions in CO_2 and local pollution)
	• Boost the national economy and create jobs in high technology and growth industries
	• Improve energy security through diversification and reduced dependence on imports
	• Widen access to energy and realize related social benefits, particularly in rural areas
	• Stimulate cost reductions in renewable energy technologies
Fossil Fuel	• Welfare improvements through lower costs to consumers
	• Widen access to energy and realize related social benefits, particularly in rural areas
	• Subsidies for switching across fossil fuels (e.g. from coal to gas)
	• Development of local fossil-fuel sources (particularly otherwise stranded assets)
	• Stimulate national economy (or segments thereof) through lower costs to business
Nuclear Energy	• Reduce CO_2 emissions from combustion of fossil-fuel emissions
	• Support the creation of a strong nuclear industry and viable nuclear fuel cycle
	• Improve energy security through diversification and reduced dependence on imports
Any or All Types	• Stimulate and support economic growth
	• Meet growing consumer and industry demand
	• Generate employment and social benefits

Source: Global Subsidies Initiative. 2011. Subsidies and External Costs in Electric Power Generation: A comparative review of estimates, <http://www.iisd.org/gsi/sites/default/files/power_gen_subsidies.pdf>.

Table 29.19.   Examples of subsidies to electricity generation

	Direct and Indirect Transfer of Funds and Liabilities	Government Revenue Foregone	Provision of Goods or Services	Income or Price Support
R&D	• Renewable: Grants for Marine R&D (New Zealand)	• Renewable: Deduction of expenses for income tax (China)	• Nuclear: funding for national laboratories (U.S.)	
	• Renewable: Loan guarantees for new technology research (U.S.)	• Renewable: Tax deductions for R&D (Australia)	• Renewable: Promotion of domestic industry (Germany)	
	• Fossils: Funding for Clean Coal R&D (Australia)		• Renewable: Funding of research institues (Australia)	
	• Nuclear: Funding for R&D (France)			
Investment	• Renewable: Reduced rates of interest on loans (Germany, India)	• Nuclear: Accelerated depreciation (U.S.)	• Renewable: Exemption from planning requirements (France)	• Renewable: Priority access to grid (Czech Republic)
	• Fossil: Low interest loan for power plant build (Indonesia)	• Nuclear: Property tax abatements (U.S.)		• Renewable: Obligatory connection to grid (Bulgaria)
	• Fossil: Guarantees for raising construction capital (South Africa)	• Renewable: Invetsment tax credits (U.S.)		• Renewable: Refund of grid connection costs (Poland)
	• Fossil: Govt funding for new investment (South Africa)	• Renewable: Reduced rate import duty for solar components (India)		
	• Nuclear: Subsidized liabilities (U.K.)	• Renewable: Tax credit for Solar PV (Sweden)		
Generation	• All: Losses in generation, transmission and distribution (Cuba)	• Nuclear: Production tax credit (U.S.)	• All: Provision of grid infrastructure and service below cost	• Fossil: Gas provided to generators below cost (Iran)
			• Renewable: Grid strengthening	• Fossil: Income support to gas generation (Australia)
			• Nuclear: underpriced cooling water (USA)	• Renewable: Feed in tariffs for production (Germany)
			• Nuclear: subsidized heavy water (India)	
Consumption	• All: Absence of penalties for non-payment of bills	• All: VAT reductions for electricity sales		• All: Below cost provision to end users (Iran, Bangladesh)

Table 29.19.   Continued

	Direct and Indirect Transfer of Funds and Liabilities	Government Revenue Foregone	Provision of Goods or Services	Income or Price Support
	• All: Theft from network (Dominican Republic)	• Renewable: Reduced VAT on wind power (China)		• Renewable : Portfolio Standards (U.K., Chile, Italy)
				• Renewable: Obligatory long term PPAs (France)
Decommissioning	• Nuclear: subsidized waste management facilities (South Korea)	• Nuclear: Decomissioning funds tax exempt (Japan)		
Throughout	• Nuclear: public funds for education (U.K.)	• Renewable: Payroll subsidy for investors (Greece)	• Fossil: Oversight of new power stations (China)	
	• Fossil: subsidies for health programs			
	• Nuclear: cap onaccident liabilities			

Source: Global Subsidies Initiative. 2011. Subsidies and External Costs in Electric Power Generation: A comparative review of estimates, <http://www.iisd.org/gsi/sites/default/files/power_gen_subsidies.pdf>.

Table 29.20.   Estimates of subsidies to electricity generated from fossil fuels

Category	Specific Type	Author	Fuel & Country Scope	Period	Billion $	US c/ kWh
Financial	All subsidies captured through price gap	IEA	Global Fossil fuel Generation	2009	95	0.74
	Subsidies to electricity end users (EU)	EEA	Electrical power in the EU	2001	5.5	0.3
	Direct subsidies and tax incentives	Badcock & Menzen	Global Coal Generation	1974-2007	536	0.3
	Input price, tax incentives	Reidy	Electrical power in Australia	2006	0.5-1	0.3-0.6
	Direct subsidies, tax, and federal support	EIA	U.S. Fossil Fuel Generation	2007	2.7	0.1
R&D	National Government Expenditure	Badcock & Lenzen	Global Coal Generation	1974-2007	32.6	0.02
	IEA Member Goverment Expenditure	IEA Database	Fossil Fuel Combustion & network	2007	0.4	0.01
	IEA Member Goverment Expenditure	IEA Database	Fossil Fuel Combustion & network	1990-2007	16.5	0.02
	Federal Government Expenditure	EIA	U.S. Fossil Fuel Generation	2007	0.6	0.02

IEA: International Energy Agency; Badcock, J., & Lenzen, M. (2010, September). Subsidies for Electricity-generating Technologies: A review. Energy Policy 38(9), 5038-5047; Reidy, C. (2007). Energy and transport subsidies in Australia: 2007 update. Sydney: Institute for Sustainable Futures.
Source: Adapted from Global Subsidies Initiative. 2011. Subsidies and External Costs in Electric Power Generation: A comparative review of estimates, <http://www.iisd.org/gsi/sites/default/files/power_gen_subsidies.pdf>.

Table 29.21.   Estimates of subsidies to electricity generated from nuclear energy

Category	Specific Type	Author	Fuel & Country Scope	Period	Billion $	US c/ kWh
Financial	Direct Price Support	Koplow	Existing Reactors, US	NA	NA	0.7-5.7
	Direct Price Support	Koplow	New Reactors, U.S.	NA	NA	4.2-11.4
	On and off budget support	EEA	Nuclear, EU-15	2001	2	0.2
	Direct Price Support & tax incentive	Badcock & Menzen	Nuclear, Global	1960-2007	324	NA
R&D	National Government Expenditure	IEA	Nuclear Fission	2008	5.5	0.2
	National Government Expenditure	WNA	Nuclear Fission	2005	c. 3	0.1
	National Government Expenditure	Badcock & Lenzen	Global	1960-2007	178	NA
	US Government Expenditure	EIA	U.S.	1988-2007	13.5	0.1
	US Government Expenditure	EIA	U.S.	2007	0.9	0.1
	France Government Expenditure	IEA	France	2009	0.7	0.2

IEA: International Energy Agency; Badcock, J., & Lenzen, M. (2010, September). Subsidies for Electricity-generating Technologies: A review. Energy Policy 38(9), 5038-5047; WNA: World Nuclear Association; Koplow: Doug Koplow, Earthtarck, <http://www.earthtrack.net/. Source: Adapted from Global Subsidies Initiative. 2011. Subsidies and External Costs in Electric Power Generation: A comparative review of estimates, <http://www.iisd.org/gsi/sites/default/files/power_gen_subsidies.pdf>.

Table 29.22.   Estimates of subsidies to electricity generated from renewable sources

Category	Specific Type	Author	Fuel & Country Scope	Period	Billion $	US c/ kWh
Financial	Direct Price Support	IEA	Global, non-hydro renewables	2008	26.6	4.7
	Direct Price Support	CEER	All Renewables, EU (16 states)	2009	26.5	5.2-10.9
	Direct Price Support	EURElectric	All Renewables, EU 15	2001	3	3.3
	On and off budget support	EEA	All Renewables, EU	2001	4.8	5.6
	Direct Price Support & tax incentive	Babcock & Lenzen	Wind, Global	1975-2007	65	8.8
	"Material" Financial Subsidies	Vivid Economics	Japan	2009	0.6-0.7	2.3-2.8
			South Korea	2009	0.26-0.32	12.3-14.8
			Australia	2009	0.3-0.5	5.9-8.9
			China	2009	1.7-2.1	6-7.6
			U.K.	2009	1.4-1.5	6.9-7.3
			Germany	2009	7.9-9.9	9.8-12.3
			U.S.	2009	2.7-3.1	1.7-2.1

Table 29.22.   Continued

Category	Specific Type	Author	Fuel & Country Scope	Period	Billion $	US c/ kWh
R&D	National Government Expenditure	IEA / BNEF	Global, excluding hydro > 50 MW	2009	c. 2.1	0.1-0.6
	National Government Expenditure	UNEP / SEFI	Global, excluding large hydro	2010	c. 3.3	0.1-0.4
	IEA Member Goverment Expenditure	IEA Database	IEA Members	1974-2007	22.8	0-0.5
	National Government Expenditure	Babcock & Menzen	Global	1974-2007	25.6	0-0.5

IEA: International Energy Agency; CEER: Council of European Energy Regulators; EEA: European Environment Agency; Badcock, J., & Lenzen, M. (2010, September). Subsidies for Electricity-generating Technologies: A review. Energy Policy 38(9), 5038-5047; UNEP: United Nations Environmental Programme; BNEF: Bloomberg New Energy Finance; SEFI: UNEP Sustainable Energy Finance Initiative
Source: Adapted from Global Subsidies Initiative. 2011. Subsidies and External Costs in Electric Power Generation: A comparative review of estimates, <http://www.iisd.org/gsi/sites/default/files/power_gen_subsidies.pdf>.

Table 29.23.   Common forms of government interventions in energy markets

Intervention type	Description
Access[a]	Policies governing the terms of access to domestic onshore and offshore resources (e.g., leasing)
Cross-subsidy[a,b]	Policies that reduce costs to particular types of customers or regions by increasing charges to other customers or regions
Direct spending[b]	Direct budgetary outlays for an energy-related purpose
Government ownership[b]	Government ownership of all or a significant part of an energy enterprise or a supporting service organization
Import/Export restriction[a]	Restrictions on the free market flow of energy products and services between countries
Information[b]	Provision of market-related information that would otherwise have to be purchased by private market participants
Lending[b]	Below-market provision of loans or loan guarantees for energy-related activities
Price controls[a]	Direct regulation of wholesale or retail energy prices
Purchase requirements[a]	Required purchase of particular energy commodities, such as domestic coal, regardless of whether other choices are more economically attractive
Research and Development[b]	Partial or full government funding for energy-related research and development
Regulation[a]	Government regulatory efforts that substantially alter the rights and resposibilities of various parties in energy markets or that exempt certain parties from those changes
Risk[b]	Government-provided insurance or indemnification at below-market prices
Tax[a,b]	Special tax levies or exemptions for energy-related activities

[a]Can act either as a subsidy or a tax depending on program specifics and one's position in the marketplace.
[b]Interventions included within the realm of fiscal subsidies.
Source: Adapted from Koplow, Doug. 2004. Subsidies to Energy Industries, In: Cutler J. Cleveland, Editor-in-Chief, Encyclopedia of Energy, (New York, Elsevier), Pages 749-764.

Table 29.24. Matrix of transfers associated with subsidies to the energy industries, with examples of specific support policies

Transfer mechanism (how a transfer is created)	Statutory or formal incidence (to whom and what a transfer is first given)					
	Direct consumption		Output returns	Enterprise income	Cost of intermediate inputs	Cost of production factors
	Unit of consumption	Household or enterprise income				
Direct transfer of funds	Unit subsidy	Government-subsidized life-line electricity rate	Per-tonne subsidy for metallurgical coal	Operating grant to coal-mining company	Input subsidy for electricity used in mining	Capital grant linked to acquisition of mining-related capital
Transfer of risk to government	Price-triggered subsidy	Means-tested cold-weather grant	Government expenditure on coal buffer stock	Government limit on producer liability for mining accidents	Security guarantee for coal trains	Credit guarantee linked to acquisition of mining-related capital
Tax revenue foregone	Excise-tax concession on fuel	Tax deduction related to energy purchases that exceed given share of income	Production tax credit for making liquid fuels from coal	Reduced rate of income tax on coal-mining companies	Reduction in excise tax on fuel used by mining machines	Tax credit for investment in mining equipment
Other government revenue foregone	Under-pricing of access to a natural resource harvested by final consumer		Reduced royalty payments on access to coal deposits		Under-pricing of a good, government service or access to a natural resource	Under-pricing of access to government land used for storage of coal
Induced transfers	Regulated price; cross subsidy	Mandated life-line electricity rate	Import tariff or export subsidy on coal	Monopoly concession to coal company	Export restriction on domestically produced coal	Wage controls on mining labour

Source: Adapted from IEA, OPEC, OECD, WORLD BANK, Analysis of the Scope of Energy Subsidies and Suggestions for the G-20 Initiative, Prepared for submission to the G-20 Summit Meeting, Toronto (Canada), 26-27 June 2010.

Table 29.25.   Selected energy funds, June 2012

Fund	Net assets	Top holdings
Dow Jones U.S. Energy Sector Index Fund	770M	Exxon Mobil, Chevron, Schlumberger NV
S&P Global Energy Sector Index Fund	1075M	Exxon Mobil, Chevron,BP
Putnam Global Natural Resources A	307M	Exxon Mobil,Royal Dutch Shell PLC, Rio Tinto PLC
Dow Jones U.S. Oil Equipment & Services Index Fund	770M	Schlumberger NV, National Oilwell Drilling, Halliburton
Dow Jones U.S. Oil & Gas Exploration & Production Index Fund	286M	Occidental Petroleum Corporation, Apache Corp, Andarko Petroleum Corp.
S&P Global Clean Energy Index Fund	28M	China Everbright International, LTD, Covanta Energy, CEMIG SA -SPONS ADR
U.S. Global Investors Global Res	595M	SM Energy Company , EQT Corporation, Hess Corporation
S&P Global Nuclear Energy Index Fund	10M	Exelon. Corp, Cameco Corp., Mitsubishi Electric Corp.
Ivy Energy A	108M	National Oilwell Varco, Inc., Continental Resources Inc, Schlumberger NV
"Clean Energy" Mutual Funds		
Fidelity Select Envir Alt Energy	69.4M	Danaher Corporation, Emerson Electric Co.,
Republic Services Inc Class A		
New Alternatives	148.8M	American Water Works Co Inc, Koninklijke Philips Electronics, Johnson Controls Inc
Winslow Green Growth Inv	172.2M	Trimble Navigation Ltd., United Natural Foods, Inc.,Clean Harbors, Inc.
Gabelli SRI Green AAA	51.94M	Mead Johnson Nutrition Company, Danone, Dialight PLC
Leuthold Global Clean Tech Retail	11.7M	World Energy Solutions, Inc.,Pentair, Inc., DA-ES, Inc.
DWS Clean Technology S	18.23M	Danaher Corporation, BG Group PLC, ABB, Ltd.

Source: Morningstar

Table 29.26.   Energy research and development expenditures in selected nations[a], 1975-2010 (million USD, 2010 prices and exchange rates)

						Change
	1975	1985	1995	2005	2010	2005-10
Energy Efficiency	651	1,088	1,464	1,626	3,305	103%
Fossil fuels	1,088	2,086	1,263	1,433	2,089	46%
Renewable energy	471	1,197	953	1,271	3,304	160%
Nuclear Energy	9,207	10,932	5,614	5,558	5,400	-3%
Hydrogen & fuel cells	0	0	0	856	791	-8%
Other power & storage	293	406	453	476	881	85%
Cross-cutting research	1,076	1,044	1,468	1,438	444	-69%

[a]Australia, Austria, Belgium, Canada, Denmark, Finland, France, Germany, Greece, Hungary, Ireland, Italy, Japan, Korea, Luxembourg, the Netherlands, New Zealand, Norway, Portugal, Spain, Sweden, Switzerland, Turkey, the United Kingdom and the United States.
Source: International Energy Agency, Statistics database, <http://www.iea.org/stats/>.

Table 29.27.   Experience curve studies for energy demand technologies

Technology cluster	Technology	Dependent variable	Independent variable	Country	Time period	LR [%]	Error[a] [%]	N[b]
Automotive	Ford, model T	Price	cum. prod. [units]	USA	1910–1926	14	n. s.	10
	Ford, model T	Price [USD_{1958}/car]	cum. prod. [units]	USA	1909–1918	15	n. s.	~ 9
	Ford, model T	Costs [USD_{1978}/car]	cum. prod. [units]	USA	1909–1923	12	1	9
Building insulation and glazing	Selective window coatings	Prod. costs	cum. production [m^2]	n. s.	1992–2000	20	4	5
	Building façades insulation[c]	Costs [$CHF/kWh_{conserved}$]	cum. energy conserved [GWh]	CH	1975–2001	15	n. s.	~ 3.3
	Building façades insulation[d]	Costs [$CHF/kWh_{conserved}$]	cum. energy conserved [GWh]	CH	1975–2001	18	n. s.	~ 3.2
	Building façades insulation[c]	Costs [$CHF/kWh_{conserved}$]	cum. façade area [m^2]	CH	1975–2001	17	n. s.	~ 2.5
	Building façades insulation[d]	Costs [$CHF/kWh_{conserved}$]	cum. façade area [m^2]	CH	1975–2001	21	n. s.	~ 2.5
Residential heat pumps	Heat pumps	Investment costs [EUR_{2000}/kW_{th}]	inst. German capacity [MW_{th}]	NL	1980–2002	30	n. s.	~ 1.4
	Heat pumps	Price [EUR_{2006}/kW_{th}]	cum. Swiss sales [MW_{th}]	CH	1980–2004	35	1	3.5
Other residential heating technologies	Condensing gas boilers	Inv. costs [EUR_{2000}/kW_{th}]	cum. German capacity [MW_{th}]	GER	1992–1999	4	n. s.	~ 3.6
	Condensing gas boilers	Inv. costs [EUR_{2000}/kW_{th}]	cum. Dutch sales [MW_{th}]	NL	1983–1997	4	n. s.	~ 5
	Condensing gas space heating boilers	Price [EUR_{2006}/kW_{th}]	cum. Dutch sales [MW_{th}]	NL	1983–2006	6	1	6.8
	Condensing gas combi boilers	Price [EUR_{2006}/kW_{th}]	cum. Dutch sales [MW_{th}]	NL	1988–2006	14	1	5
	Gas water heaters	Price	cum. US shipments [units]	USA	1962–1993	25	2	n. s.
	Residential electrical water heaters	Costs [USD_{1995}/unit]	cum. US shipments [units]	USA	1982–1995	5	2	4.1

Technology cluster	Technology	Dependent variable	Independent variable	Country	Time period	LR [%]	Error[a] [%]	N[b]
Air conditioners	Room air conditioners	Price	cum. US industry sales [units]	USA	1946–1961	8	1	n. s.
	Room air conditioners	Price	cum. US industry sales [units]	USA	1946–1974	12	2	n. s.
	Air conditioners	Sales price [Yen/unit]	cum. sales [units]	JPN	1972–1997	10	n. s.	n. s.
	Room air conditioners	Price	cum. US shipments [units]	USA	1958–1993	23	1	n. s.
	Central air conditioners	Price	cum. US shipments [units]	USA	1967–1988	24	2	n. s.
	Room air conditioners	Unit costs	cum. US shipments [units]	USA	1980–1998	15[f]	n. s.	4.7
Washing machines	Washing machines	Unit costs	cum. US shipments [units]	USA	1980–1998	13[f]	n. s.	4.6
	Washing machines	Price [EUR$_{2006}$/kg l. c.]	cum. global prod. [units]	NL	1965–2008	33	9	2.5
Laundry dryers	Laundry dryers (electric)	Price	cum. US industry sales [units]	USA	1950–1961	6	3	n. s.
	Laundry dryers (gas)	Price	cum. US industry sales [units]	USA	1950–1974	12	2	n. s.
	Laundry dryers (electric)	Unit costs	cum. US shipments [units]	USA	1980–1998	12[f]	n. s.	4.6
	Laundry dryers (gas)	Unit costs	cum. US shipments [units]	USA	1980–1998	10[f]	n. s.	4.7
	Laundry dryers	Price [EUR$_{2006}$/kg l. c.]	cum. global prod. [units]	NL	1969–2003	28	7	2.3
Dishwashers	Dishwashers	Price	cum. US industry sales [units]	USA	1947–1968	10	2	n. s.
	Dishwashers	Price	cum. US industry sales [units]	USA	1947–1974	11	2	n. s.
	Dishwashers	Unit costs	cum. US shipments [units]	USA	1980–1998	16[f]	n. s.	4.7
	Dishwashers	Price [EUR$_{2006}$/SPS]	cum. global prod. [units]	NL	1968–2007	27	7	4.7
Refrigerators	Refrigerators	Price	cum. US industry sales [units]	USA	1922–1940	7	1	n. s.
	Refrigerators	Unit costs	cum. US shipments [units]	USA	1980–1998	12[f]	n. s.	4.6
	Refrigerators	Price [EUR$_{2006}$/hl]	cum. global prod. [units]	NL	1964–2008	9	4	5.7
Freezers	Freezers	Unit costs	cum. US shipments [units]	USA	1980–1998	22[f]	n. s.	3.9
	Upright Freezers	Price [EUR$_{2006}$/hl]	cum. global prod. [units]	NL	1970–2003	10	5	4.8
	Chest Freezers	Price [EUR$_{2006}$/hl]	cum. global prod. [units]	NL	1970–1998	8	2	4.4

Continued

Table 29.27. Continued

Technology cluster	Technology	Dependent variable	Independent variable	Country	Time period	LR [%]	Error[a] [%]	N[b]
Compact fluorescent light bulbs	Modular-electronic CFLs	Price [USD$_{1995}$/klm]	cum. global prod. [units]	USA	1992–1998	20	n. s.	~ 2.3
	Integral-electronic CLFs	Price [USD$_{1995}$/km]	cum. global prod. [units]	USA	1992–1998	16	n. s.	~ 2.8
	Modular-magnetic CFLs	Price [USD$_{1995}$/klm]	cum. global prod. [units]	USA	1992–1998	41	n. s.	~ 2.3
	CFLs	Price [USD$_{1995}$/km]	cum. global prod. [units]	USA	1992–1998	21^e	~ 6	~ 2.3
	CFLs	Price [USD$_{2004}$/unit]	cum. global sales [units]	USA	1990–2004	10	n. s.	~ 4.2
	CFLs	Price [EUR$_{2006}$/klm]	cum. global sales [Glm]	GER, NL	1988–2006	19	4	6.2
	CFLs	Price [EUR$_{2006}$/W$_e$]	cum. global sales [MW$_e$]	GER, NL	1988–2006	19	5	5.9
	CFLs	Price [EUR$_{2007}$]	cum. global sales [units]	INT	1985–2007	21	5	8.5
Lamp ballasts	Magnetic ballasts for CFLs	Price [USD$_{1995}$/unit]	cum. US prod. [units]	USA	1981–1988	16	n. s.	~ 3.9
	Magnetic ballasts for CFLs	Price [USD$_{1995}$/unit]	cum. US prod. [units]	USA	1990–1993	41	n. s.	~ 0.3
	Magnetic ballasts for FLs	Prod. costs [USD$_{1993}$/unit]	cum. shipments [units]	USA	1977–1993	3	3	4.1
	Electronic ballasts for FLs	Price [USD$_{1997}$/unit]	cum. prod. [units]	USA	1986–1997	11	2	~ 8.9
	Electronic ballasts for CFLs	Price [USD$_{1995}$/unit]	cum. US prod. [units]	USA	1986–1998	13	n. s.	~ 7.5
	Electronic ballasts for FLs	Prod. costs [USD$_{1996}$/unit]	cum. shipments [units]	USA	1986–2001	11	2	9.7
Television sets (cathode ray tube)	Black-and-white TV	Price	cum. US industry sales [units]	USA	1948–1960	13	4	n. s.
	Black-and-white TV	Price	cum. US industry sales [units]	USA	1948–1974	22	5	n. s.
	Color TV	Price	cum. US industry sales [units]	USA	1961–1971	5	1	n. s.
	Color TV	Price	Cum. US industry sales [units]	USA	1961–1974	7	2	n. s.

Technology cluster	Technology	Dependent variable	Independent variable	Country	Time period	LR [%]	Error[a] [%]	N[b]
Other consumer electronics	4-Function pocket calculators	Price [USD/unit]	cum. prod. [units]	USA	early 1970 s	30	n. s.	n. s.
	Digital watches	Av. factory price [const. USD/unit]	cum. prod. [units]	USA	1975–1978	26	n. s.	4
	Hand-held calculators	Av. factory price [const. USD/unit]	cum. prod. [units]	USA	1975–1978	26	n. s.	2
	Sony laser diodes	Prod. costs [Yen/unit]	cum. prod. [units]	prod. by Sony	1982–1994	23	n. s.	~ 17
Electronic components	Integrated circuits	Av. price [const. USD/unit]	cum. industry experience [units]	USA	1964–1972	25	n. s.	~ 10
	Integrated circuits	Av. price [const. USD/unit]	cum. prod. [units]	USA	1964–1972	28	n. s.	10
	Integrated circuits	Prod. costs [USD_{1993}/unit]	cum. prod. [USD]	USA	1962–1968	26	5	7.7
	MOS/LSI	Av. price [const. USD/unit]	cum. prod. [units]	USA	1970–1976	20	n. s.	10
	MOS dynamic RAM	Av. factory price [const. USD/unit]	cum. number [bits]	USA	1973–1978	32	n. s.	6
	Disk memory drives	Av. price [const. USD/bit]	cum. number [bits]	USA	1975–1978	24	n. s.	3
	4 kB - DRAM	Price [USD/MB]	cum. global shipments [units]	EU, JPN, USA	~ 1974–1986	20	2	n. s.
	16 kB - DRAM	Price [USD/MB]	cum. global shipments [units]	EU, JPN, USA	~ 1977–1922	24	2	n. s.
	16 kB-5 - DRAM	Price [USD/MB]	cum. global shipments [units]	EU, JPN, USA	~ 1977–1992	18	1	n. s.
	64 kB - DRAM	Price [USD/MB]	cum. global shipments [units]	EU, JPN, KR, USA	~ 1978–1992	23	1	n. s.
	256 kB - DRAM	Price [USD/MB]	cum. global shipments [units]	EU, JPN, KR, USA	~ 1982–1992	21	2	n. s.

Continued

Table 29.27. Continued

Technology cluster	Technology	Dependent variable	Independent variable	Country	Time period	LR [%]	Error[a] [%]	N[b]
	1 MB - DRAM	Price [USD/MB]	cum. global shipments [units]	EU, JPN, KR, USA	~ 1985–1992	16	2	n. s.
	4 MB - DRAM	Price [USD/MB]	cum. global shipments [units]	EU, JPN, KR, USA	~ 1988–1992	20	2	n. s.
	16 MB - DRAM	Price [USD/MB]	cum. global shipments [units]	JPN, KR, USA	n. s.	16	2	n. s.

Abbreviations, units, and symbols: LR: learning rate, the percentage decrease in wind power cost for each doubling of cumulative capacity or production.av. — average; CFLs — compact fluorescent light bulbs; const. — constant; cum. — cumulative; CH — Switzerland; DRAM — dynamic random access memory; EPROM — erasable programmable read only memory; EU — Europe; EUR — Euro (subscript numbers indicate the base year of currency deflation); FL — fluorescent light bulbs; GER — Germany; Glm — gigalumen; GWh — gigawatt hours; hl — hectolitre; inst. —installed; INT —international data; inv. —investment; JPN —Japan; kB —kilobyte; klm —kilolumen; KR—South Korea; kWhth —kilowatt hour thermal; LSI—large-scale integration; MB—megabyte; MOS— metal oxide semiconductor; MWe—gigawatt electric; NL—the Netherlands; n.s.—not specified; prod. —production, RAM—random access memory; SPS—standard place setting; TV—television, USD—United States dollars (subscript numbers indicate the base year of currency deflation).

[a]Representing the 95% confidence interval of learning rates.
[b]N — number of doublings of cumulative production.
[c]Building facades insulation of 1.0 W/m²K (Watt per square meter and degree Kelvin).
[d]Building facades insulation of 1.25 W/m²K.
[e]Weighted average of modular and integral CFLs.
[f]Learning rate refers to the cost difference between base year and final year of analysis.

Table 29.28. Demand and price elasticity of demand[a] for crude oil, selected nations,1971-2000[b]

Country	Oil consumption % growth per capita	Real GDP % growth per capita	Price elasticity of demand	
			Short-run	Long-run
Australia	−0.3	1.7	−0.034	−0.068
Austria	−0.7	3.1	−0.059	−0.092
Canada	−1.3	1.6	−0.041	−0.352
China	3.6	8.6	0.001	0.005
Denmark	−2.5	1.5	−0.026	−0.191
Finland	−1.2	2.1	−0.016	−0.033
France	−1.5	1.7	−0.069	−0.568
Germany	−1.4	1.2	−0.024	−0.279
Greece	2.2	1.5	−0.055	−0.126
Iceland	0.5	2.2	−0.109	−0.452
Ireland	0.2	3.9	−0.082	−0.196
Italy	−0.4	2.2	−0.035	−0.208
Japan	−1.0	8.1	−0.071	−0.357
Korea	8.3	6.4	−0.094	−0.178
Netherlands	−0.5	1.7	−0.057	−0.244
New Zealand	−0.4	1.4	−0.054	−0.326
Norway	0.2	2.9	−0.026	−0.036
Portugal	3	2.9	0.023	0.038
Spain	1.3	2.1	−0.087	−0.146
Sweden	1.3	2.8	−0.043	−0.289
Switzerland	−0.7	0.9	−0.030	−0.056
United	−1.1	2	−0.068	−0.182
Unites	−0.7	2	−0.061	−0.453

[a]The percentage change in the quantity demanded relative to a percentage change in its own price.
[b]The calculations for China and South Korea are based on the period 1979–2000.
Source: Adapted from Cooper, John C.B. 2003. Price elasticity of demand for crude oil: estimates for 23 countries, Volume 27, Issue 1, pages 1–8.

Table 29.29. Estimates of the price[a] and income[b] elasticity of demand for oil and motor gasoline

Study	Product	Method	Short run price elasticity	Long-run price elasticity	Long-run income elasticity
Dahl and Sterner (1991)	gasoline	literature survey	-0.26	-0.86	1.21
Espey (1998)	gasoline	literature survey	-0.26	-0.58	0.88
Graham and Glaister (2004)	gasoline	literature survey	-0.25	-0.77	0.93
Brons, et. al. (2008)	gasoline	literature survey	-0.34	-0.84	--
Dahl (1993)	oil (developing nations)	literature survey	-0.07	-0.30	1.32
Cooper (2003)	oil (average of 23 countries	annual time series regression	-0.05	-0.21	--

Brons, Martijn, Peter Nijkamp, Eric Pels, and Piet Rietveld. 2008. "A Meta- Analysis of the Price Elasticity of Gasoline Demand: A SUR Approach," Energy Economics 30, pp. 2105–2122.
Cooper, John C.B. 2003. "Price Elasticity of Demand for Crude Oil: Estimates for 23 Countries," OPEC Review 27(1), pp. 1-8.
Dahl, Carol A. 1993. "A Survey of Oil Demand Elasticities for Developing Countries," OPEC Review 17(Winter), pp. 399-419.
Dahl, Carol A. and Thomas Sterner. 1991. "Analysing Gasoline Demand Elasticities: A Survey," Energy Economics 13, pp. 203-210.
Espey, Molly. 1998. "Gasoline Demand Revisited: An International Meta-Analysis of Elasticities," Energy Economics 20, pp. 273-295.
Graham, Daniel J., and Stephen Glaister. 2004. "Road Traffic Demand Elasticity Estimates: A Review," Transport Reviews 24(3), pp. 261–274.
[a]The percentage change in the quantity demanded relative to a percentage change in its own price.
[b]The percentage change in the quantity demanded relative to a percentage change in income by consumers.
Source: Adapted from Hamilton, James D., 2008. Understanding crude oil prices, National Bureau of Economic Research, Working Paper 14492, <http://www.nber.org/papers/w14492>.

Table 29.30. Estimates of the long and short run price elasticity of demand for electricity[a]

Source	Type of model	Type of data	Long term	Short term
Al Faris (2002)	Error correction model	Annual time series, 1970–1997	– 0.82/– 3.39	– 0.04/– 0.18
Beenstock et al. (1999)	Error correction model	Quarterly time series, 1973–1994	Households: – 0.579	Households: – 0.124
			Industry: – 0.311	Industry: – 0.123
Bjørner and Jensen (2002)	Loglinear, fixed effects	Panel, 1983–1996	–	– 0.479
Boonekamp (2007)	Bottom-up	Annual time series, 1990–2000	Households: – 0.09/– 0.13	–
Brännlund et al. (2007)	AID-model	Quarterly time series, 1980–1997	–	Households: – 0.24
Caloghirou et al. (1997)	Translog	Panel, 1980–1991	Industry: – 0.77	Industry: – 0.51
Elkhafif (1992)	Loglinear	Annual time series, 1963–1990	– 0.697	– 0.147
Filippini and Pachuari (2002)	Loglinear	Monthly household panel, 1993–1994	–	Households: – 0.16/– 0.39
Hesse and Tarkka (1986)	Translog	Panel, 1973–1980	–	– 0.14/– 0.49
Holtedahl and Loutz (2004)	Long term: loglinear	Annual time series, 1955–1996	Households: – 0.16	Households: – 0.15
	Short term: error correction model			

Table 29.30. Continued

Source	Type of model	Type of data	Long term	Short term
Ilmakunnas and Törmä (1989)	Generalized Leontief	Annual time series, 1960–1981	–	– 0.73
Jones (1995)	Loglinear	Annual time series, 1960–1992	– 0.207	– 0.05
	Translog		– 0.201	– 0.276
Roy et al. (in press)	Translog	Pooled country panel, 1980–1993	Industrial: – 0.8/– 1.76	–
Taheri (1994), Urga and Walters (2003)	Translog	Panel, 1974–1981	– 0.845	– 0.888[b]
	Loglinear	Annual time series, 1960–1992[c]	– 0.2609	– 0.071
	Translog		– 0.1042	– 0.101
Woodland (1993)	Translog	Panel, 1977–1985	–	– 1.113
Zachariadis and Pashourtidou (2007)	Error correction model	Annual time series, 1960–2004	– 0.3/– 0.4	[b]

Al Faris, A.R.F., 2002. The demand for electricity in the GCC countries. Energy Policy 30, 117–124.

Beenstock, M., Goldin, E., Nabot, D., 1999. The demand for electricity in Israel. Energy Economics 21, 168–183.

Bjørner, T.B., Jensen, H.H., 2002. Interfuel substitution within industrial companies: an analysis based on panel data at company level. The Energy Journal 23 (2), 27–50.

Boisvert, R., Cappers, P., Neenan, B., Scott, B., 2004. Industrial and Commercial Customer Response to Real Time Electricity Prices. Neenan Associates.

Boonekamp, P.G.M., 2007. Price elasticities, policy measures and actual developments in household energy consumption—a bottom up analysis for the Netherlands. Energy Economics 29, 133–157.

Brännlund, R., Ghalwash T., Nordström, J., Increased energy efficiency and the rebound effect: effects on consumption and emissions. Energy Economics, Volume 29, Issue 1, January 2007.

Caloghirou, Y.D., Mourelatos, A.G., Thompson, H., 1997. Industrial energy substitution during the 1980s in the Greek economy. Energy Economics 19, 476–491.

Doorman, G., 2003. Capacity subscription and security of supply in deregulated electricity markets. Paper Presented at Research Symposium on European Electricity Markets. The Hague. http://www.electricitymarkets.info/symp03/doc/b1_1-paper.pdf.

Elkhafif, M.A.T., 1992. Estimating disaggregated price elasticities in industrial energy demand. The Energy Journal 13 (4), 209–217.

Filippini, M., Pachuari, S., 2002. Elasticities of electricity demand in urban Indian households. CEPE Working Paper, vol. 16. Centre for Energy Policy and Economics, Swiss Federal Institutes of Technology.

Hesse, D.M., Tarkka, H., 1986. The demand for capital, labor and energy in European manufacturing industry before and after the oil price shocks. Scandinavian Journal of Economics 88 (3), 529–546.

Holtedahl, P., Loutz, F.J., 2004. Residential electricity demand in Taiwan. Energy Economics 26, 201–224.

Ilmakunnas, P., Törmä, H., 1989. Structural change in factor substitution in Finnish manufacturing. Scandinavian Journal of Economics 91 (4), 705–721.

Jones, C.T., 1995. A dynamic analysis of interfuel substitution in U.S. industrial energy demand. Journal of Business and Economic Statistics 13 (4), 459–465.

Roy, J., Sanstad, A.H., Sathaye J.A., Khaddaria, R., in press. Substitution and price elasticity estimates using inter-country pooled data in a translog cost model. Energy Economics, vol. 28(5-6), pages 706-719,.

Taheri, A.A., 1994. Oil Shocks and the dynamics of substitution adjustments of industrial fuels in the US. Applied Economics 26 (8), 751–756.

Taylor, L.D., 1975. The demand for electricity: a survey. The Bell Journal of Economics 6 (1), 74–110.

Urga, G.,Walters, C., 2003. Dynamic translog and linear logit models: a factor demand analysis of interfuel substitution in US industrial energy demand. Energy Economics 25, 1–21.

Woodland, A.D., 1993. A micro-econometric analysis of the industrial demand for energy in NSW. The Energy Journal 14 (2), 57–89.

Zachariadis, T., Pashourtidou, N., 2007. An empirical analysis of electricity consumption in Cyprus. Energy Economics 29, 183–198.

[a]The percentage change in the quantity demanded relative to a percentage change in its own price.

[b]Not significantly different from zero.

[c]Same data as Jones (1995).

Source: Adapted from Lijesen, Mark G. 2007. The real-time price elasticity of electricity, Energy Economics, Volume 29, Issue 2, Pages 249-258.

Table 29.31. Impacts assessed in the ExternE project of the European Union

Impact Category	Pollutant/Burden	Effects
Human Health mortality	PM_{10}	Reduction in life expectancy due to short and long time exposure
	SO_2, O_3	Reduction in life expectancy due to short time exposure
	Benzene, BaP, 1,3-butad., Diesel part.	Reduction in life expectancy due to long time exposure
	Noise	Reduction in life expectancy due to long time exposure
	Accident risk	Fatality risk from traffic and workplace accidents
Human Health morbidity	PM_{10}, SO_2, O_3	Respiratory hospital admissions
	PM_{10}, O_3	Restricted activity days
	PM_{10}, CO	Congestive heart failure
	Benzene, BaP, 1,3-butad., Diesel part.	Cancer risk (non-fatal)
	PM_{10}	Cerebrovascular hospital admissions, cases of chronic bronchitis, cases of chronic cough in children, cough in asthmatics, lower respiratory symptoms
	O_3	Asthma attacks, symptom days
	Noise	Myocardial infarction, angina pectoris, hypertension, sleep disturbance
	Accident risk	Risk of injuries from traffic and workplace accidents
Building Material	SO_2, Acid deposition	Ageing of galvanised steel, limestone, mortar, sandstone, paint, rendering, and zinc for utilitarian buildings
	Combustion particles	Soiling of buildings
Crops	SO_2	Yield change for wheat, barley, rye, oats, potato, sugar beet
	O_3	Yield change for wheat, barley, rye, oats, potato, rice, tobacco, sunflower seed
	Acid deposition	Increased need for liming
	N, S	Fertilising effects
Global warming	CO_2, CH_4, N_2O	World-wide effects on mortality, morbidity, coastal impacts, agriculture, energy demand, and economic impacts due to temperature change and sea level rise
Amenity lossess	Noise	Amenity losses due to noise exposure
Ecosystems	SO_2, NO_x, NH_3	Eutrophication, Acidification

Source: Bickel, Peter and Rainer Friedrich (Editors). 2005. ExternE: Externalities of Energy, Methodology 2005 Update (Luxembourg, Office for Official Publications of the European Communities).

Table 29.32.   Value of energy depletion, top 20 nations, 1970-2008 (billion USD)

	1970	1980	1990	2000	2008	Change 2000-08
Russian Federation	0.0	0.0	65.0	72.8	333.7	358.2%
China	1.2	27.3	22.6	30.2	293.8	872.4%
United States	6.9	120.5	42.1	66.8	276.1	313.5%
Saudi Arabia	1.0	83.1	31.6	51.4	208.4	305.3%
Iran	1.0	12.1	14.6	24.6	105.7	329.8%
Mexico	0.2	18.5	12.9	22.1	88.0	297.6%
Canada	0.6	15.6	7.5	20.6	76.9	273.0%
Norway	0.0	5.6	9.2	25.4	72.0	183.7%
United Arab Emirates	0.2	13.6	8.6	14.9	60.9	309.0%
Kuwait	0.8	14.1	4.3	12.4	60.1	385.4%
Venezuela	1.1	19.8	11.0	20.5	58.7	186.9%
Indonesia	0.2	14.0	7.9	13.1	58.0	341.3%
United Kingdom	0.4	18.3	12.3	27.8	57.7	107.3%
India	0.3	3.4	6.1	8.6	56.1	554.0%
Algeria	0.2	9.2	5.3	12.2	48.9	301.4%
Iraq	0.4	21.4	9.2	14.5	47.1	224.5%
Nigeria	0.3	17.2	7.8	13.1	46.5	254.6%
Brazil	0.0	1.6	3.0	7.7	42.0	448.3%
Australia	0.1	4.7	3.5	5.7	39.9	603.2%
Angola	0.0	1.2	2.5	3.8	38.5	911.9%
World	18.7	500.9	355.0	584.6	2,591.1	343.2%

Energy depletion is equal to the ratio of present value of rents, discounted at 4%, to exhaustion time of the resource. It covers crude oil, natural gas, and coal. Exhaustion time = Min (25 years, Reserves/Production). Rents = Unit Rents * Production. Unit Rents = Unit Price – Unit Cost.

Source: World Bank, Changing Wealth of Nations, 2010 update, <http://data.worldbank.org/data-catalog/wealth-of-nations>.

Table 29.33. Value of human and natural capital by region, 2005 (billions of 2005 USD)

Region	Population	Total Wealth	Human capital	Total Natural Capital	Natural capital							
					Crop	Pastures	Forests	Protected Areas	Oil	Natural Gas	Coal	Minerals
East Asia & Pacific	1,796,063,936	36,115	28,275	7,841	4,318	479	788	481	630	440	956	186
Europe & Central Asia	408,064,333	29,684	23,429	6,256	876	662	309	506	1,734	1,954	2,084	68
Latin America & Caribbean	531,341,503	42,079	35,670	6,410	2,138	651	1,114	595	1,316	221	229	365
Middle East & North Africa	239,764,358	6,951	4,579	2,372	474	201	20	37	1,055	574	575	11
South Asia	1,439,901,500	15,040	11,243	3,797	1,854	954	271	232	118	154	310	54
Sub-Saharan Africa	700,684,434	9,709	6,976	2,733	897	320	319	125	876	69	149	47
Low income	715,963,412	4,670	3,012	1,658	738	260	261	119	94	173	175	12
Lower middle income	3,519,642,686	60,228	44,893	15,335	7,032	1,839	1,320	895	2,429	864	1,563	251
Upper middle income	880,213,965	74,681	62,266	12,415	2,787	1,168	1,241	963	3,206	2,376	2,566	468
High income: OECD	953,832,017	554,581	544,141	10,441	2,066	2,137	1,054	2,393	1,158	1,201	1,355	256
High income: non-OECD	58,676,250	13,566	9,788	3,778	47	37	4	237	2,993	460	460	0
World	6,128,328,330	707,726	664,099	43,627	12,669	5,441	3,880	4,606	9,881	5,073	6,118	987

Source: World Bank, Changing Wealth of Nations, 2010 update, <http://data.worldbank.org/data-catalog/wealth-of-nations>.

Table 29.34.   Monetary values used in the ExternE project of the European Union[a]

Health end-point	Recommended central unit values in € price year 2000
Value of a prevented Fatality	1,000,000
Year of Life Lost	50,000 / year lost
Hospital admissions	2,000 / admission
Emergency Room Visit for respiratory illness	670 / visit
General Practitioner visits:	
Asthma	53 / consultation
Lower respiratory symptoms	75 / consultation
Respiratory symptoms in asthmatics:	
Adults	130 / event
Children	280 / event
Respiratory medication use – adults and children	1 / day
Restricted activity days	130 / day
Cough day	38 / day
Symptom day	38 / day
Work loss day	82 / day
Minor restricted activity day	38 / day
Chronic bronchitis	190,000/ case

[a]<http://www.externe.info>
Source: Friedrich, Rainer, ExternE: Methodologyand Results, paper presented at the meeting "External costs of energy and their internalisation in Europe," 9 December 2005 at the European Commission, <http://www.externe.info/externe_d7/?q=node/60>.

Table 29.35. Estimates of the marginal abatement costs (MAC) of greenhouse gas emissions

| Model | Source | Type of model | | | MAC 25 | MAC50 | | | | | Stablization target ppm | CO2 | | |
		CGE	IDO	ENE	REG	CCS	MTG	ITC	IMC	USC		BAS	(€2005/ tCO2-eq.)	(€2005/ tCO2-eq.)
AIM	Fujino et al. (2006)	1	0	3	18	0	0	0	0	0	650	2.32	27.85	120.82
AIM		1	0	3	18	0	1	0	0	0	650	2.32	16.16	75.27
AMIGA	Hanson and Laitner (2006)	1	0	9	3	0	0	1	0	0	650	2.79	18.03	28.35
AMIGA		1	0	9	3	0	1	1	0	0	650	2.79	12.18	19.15
GTEM	Jakeman and Fisher (2006)	1	0	3	16	1	0	0	0	0	650	na	54.63	449.27
GTEM		1	0	3	16	1	1	0	0	0	650	na	29.74	159.64
GEMINI	Bernard et al. (2006)	1	0	3	21	0	0	0	0	0	650	na	22.11	75.02
GEMINI		1	0	3	21	0	1	0	0	0	650	na	7.83	28
PACE	Böhringer et al. (2006)	1	1	3	7	0	0	0	0	0	650	1.92	0.7	2.56
PACE		1	1	3	7	0	1	0	0	0	650	1.92	0.37	1.42
EPPA	Reilly et al. (2006)	1	0	9	16	0	0	0	0	0	650	3.92	27.53	129.87
EPPA		1	0	9	16	1	1	0	0	0	650	3.92	10.49	81.11
IPAC	Jiang et al. (2006)	1	0 na		9 na		0	0	0	0	650	2.28	21.76	44.76
IPAC		1	0 na		9 na		1	0	0	0	650	2.28	9.32	24.87
SGM	Fawcett and Sands (2006)	1	0	9	14	0	0	0	0	0	650	na	57.44	120.27
SGM		1	0	9	14	0	1	0	0	0	650	na	16.21	33.97
WIAGEM	Kemfert et al. (2006)	1	1	3	11	0	0	1	0	0	650	2.49	10.32	21.51
WIAGEM		1	1	3	11	0	0	1	0	0	650	2.49	4.03	9.25
COMBAT	Aaheim et al. (2006)	1	1	1	1	0	0	0	0	0	650	4.06	19.69	32.2
COMBAT		1	1	1	1	0	1	0	0	0	650	4.06	16.71	27.33

Model	Source	Type of model — CGE	IDO	ENE	MAC 25 REG	MAC50 CCS	MTG	ITC	IMC	USC	CO₂ Stablization target ppm	BAS	(€2005/ tCO₂-eq.)	(€2005/ tCO₂-eq.)
FUND	Tol (2006a)	1	1	1	16	0	0	1	0	0	650	4.8	119.9	52.34
FUND		1	1	1	16	0	1	1	0	0	650	4.8	97.97	79.69
GRAPE	Kurosawa (2006)	1	1	8	10	1	0	0	0	0	650	3.34	3.08	16.71
GRAPE		1	1	8	10	1	1	0	0	0	650	3.34	1.72	9.13
MERGE	Manne and Richels (2006)	1	1	9	9	1	0	0	0	0	650	3.2	5.67	20.24
MERGE		1	1	9	9	1	1	0	0	0	650	3.2	2.66	9.5
MERGE		1	1	9	9	1	1	0	0	0	535	3.2	12.43	48.49
IMAGE	van Vuuren et al. (2006b)	0	0	9	26	1	0	1	0	0	650	2.34	25.31	47.72
IMAGE		0	0	9	26	1	1	1	0	0	650	2.34	13.2	34.56
IMAGE		0	0	9	26	1	1	1	0	0	550	2.34	49.73	116.87
MESSAGE	Rao and Riahi (2006)	0	1	9	11	1	0	0	0	0	650	2.2	10.47	51.97
MESSAGE		0	1	9	11	1	1	0	1	0	650	2.2	3.26	18.65
MiniCAM	Smith and Wigley (2006)	0	0	9	14	0	0	0	0	0	650	2.94	6.24	36.4
MiniCAM		0	0	9	14	0	1	0	1	0	650	2.94	2.54	19.87
ENTICE-BR	Popp (2006)	1	1	1	1	1	0	1	1	0	575	na	0	17.77
ENTICE-BR		1	1	1	1	1	0	0	1	0	650	na	0	9.42
ENTICE-BR		1	1	1	1	1	0	1	1	0	650	na	0	6.94
DEMETER	Gerlagh (2006)	1	1	1	1	1	0	0	1	0	550	2	5.62	28.11
DEMETER		1	1	1	1	1	0	1	1	0	550	2	2.81	15.46
IMACLIM-R	Crassous et al. (2006)	1	na	na	na	na	0	0	1	0	550	na	69.01	155.26
IMACLIM-R		1	na	na	na	na	0	1	1	0	550	na	21.6	48.6

Continued

Table 29.35.　Continued

Model	Source	Type of model			MAC 25	MAC50					CO$_2$				
		CGE	IDO	ENE	REG	CCS	MTG	ITC	IMC	USC	Stablization target ppm	BAS	(€^{2005}/ tCO$_2$-eq).	(€^{2005}/ tCO$_2$-eq).
IMACLIM-R	Barker et al. (2006a)	1	na	na	na	na	0	0	1	0	650	na	22.73	51.13
IMACLIM-R		1	na	na	na	na	0	1	1	0	650	na	6.81	15.31
E3MG	Barker et al. (2006a)	0	0	9	20	0	0	0	1	0	550	1.88	75.9	208.02
E3MG		0	0	9	20	0	0	1	1	0	550	1.88	47.79	123.69
MESSAGE	Rao et al. (2006)	0	1	9	11	1	0	0	1	0	550	1.08	8.43	28.11
MESSAGE		0	1	9	11	1	0	1	1	0	550	1.08	8.15	18.83
DNE21+	Sano et al. (2006)	0	0	8	77	1	0	0	1	0	550	3.43	32.89	64.65
DNE21+		0	0	8	77	1	0	1	1	0	550	3.43	35.14	70.28
FEEM-RICE	Bosetti et al. (2006)	1	1	1	8	0	0	0	1	0	550	1.08	28.11	76.18
FEEM-RICE		1	1	1	8	0	0	1	1	0	550	1.08	25.86	64.37
GET-LFL	Hedenus et al. (2006)	0	0	7	1	1	0	0	1	0	550	3.13	32.89	28.11
GET-LFL		0	0	7	1	1	0	1	1	0	550	3.13	35.14	30.36
IGSM	Clarke et al. (2006)	1	0	8	16	1	1	0	0	1	525	3.43	79.94	209.37
IGSM		1	0	8	16	1	1	0	0	1	675	3.43	23.25	60.92
MiniCAM		1	0	9	14	1	1	0	0	1	525	3.21	32.7	115.87
MiniCAM		1	0	9	14	1	1	0	0	1	675	3.21	5.1	17.16
MERGE		1	1	9	9	1	1	0	0	1	525	3.43	37.42	142.73
MERGE		1	1	9	9	1	1	0	0	1	675	3.43	2.61	8.95
IMAGE	van Vuuren et al. (2006a)	0	0	9	26	0	1	1	0	0	450	2.64	98.28	196.57

CGE = computable general equilibrium (model); IDO = Intertemporal Dynamic Optimisation; ENE = Energy sources; REG = Regions; CCS = Carbon Capture and Storage; MTG = Multigas; ITC = Induced Technical Change; IMC = IMCP; USC = to USC = USCCCP; United States Climate Change Science Program; TAR = Target; BAS = Baseline; MAC25 = Marginal abatement cost in 2025; MAC50 = Marginal abatement cost in 2050. na = not available.

J. Fujino, R. Nair, M. Kainuma, T. Masui, Y. Matsuoka, Multi-gas mitigation analysis on stabilization scenarios using aim global model, The Energy Journal Special Issue on Multi-Greenhouse Gas Mitigation and Climate Policy (2006), pp. 343–353D.

A. Hanson, J.A. Laitner, Technology policy and world greenhouse gas emissions in the Amiga modeling system, The Energy Journal Special Issue on Multi-Greenhouse Gas Mitigation and Climate Policy (2006)

G. Jakeman, B.S. Fisher, Benefits of multi-gas mitigation: an application of the global trade and environment model (GTEM), The Energy Journal Special Issue on Multi-Greenhouse Gas Mitigation and Climate Policy (2006), pp. 323–342

A. Bernard, M. Vielle, L. Viguer, Burden sharing within a multi-gas strategy, The Energy Journal Special Issue on Multi-Greenhouse Gas Mitigation and Climate Policy (2006), pp. 289–304

C. Böhringer, A. Löschel, T.F. Rutherford, Efficiency gains from "what"-flexibility in climate policy, An Integrated CGE Assessment. The Energy Journal Special Issue on Multi-Greenhouse Gas Mitigation and Climate Policy (2006), pp. 405–424

D. Popp, Comparison of climate policies in the ENTICE-BR model, The Energy Journal Special Issue on Endogenous Technological Change and the Economics of Atmospheric Stabilisation (2006), pp. 163–174S. Rao, K. Riahi, The role of Non-CO2 greenhouse gases in climate change mitigation: long-term scenarios for the 21st century, The Energy Journal Special Issue on Multi-Greenhouse Gas Mitigation and Climate Policy (2006), pp. 177–200J. Reilly, M. Sarofim, S. Paltsev, R. Prinn, The role of non-CO2 GHGs in climate policy: analysis using the MIT IGSM, The Energy Journal Special Issue on Multi-Greenhouse Gas Mitigation and Climate Policy (2006), pp. 503–520

K. Jiang, X. Hu, Z. Songli, Multi-gas mitigation analysis by IPAC, The Energy Journal Special Issue on Multi-Greenhouse Gas Mitigation and Climate Policy (2006), pp. 425–440

A.A. Fawcett, R.D. Sands, Non-CO2 greenhouse gases in the second generation model, The Energy Journal Special Issue on Multi-Greenhouse Gas Mitigation and Climate Policy (2006), pp. 305–322

C. Kemfert, T.P. Truong, T. Bruckner, Economic impact assessment of climate change–a multi-gas investigation with WIAGEM-GTAPEL-ICM, The Energy Journal Special Issue on Multi-Greenhouse Gas Mitigation and Climate Policy (2006), pp. 441–460

A. Aaheim, J.S. Fuglestvedt, O. Godal, Costs savings of a flexible multi-gas climate policy, The Energy Journal Special Issue on Multi-Greenhouse Gas Mitigation and Climate Policy (2006), pp. 485–501

R.S.J. Tol, Multi-gas emission reduction for climate change policy: an application of FUND, The Energy Journal (2006), pp. 235–250

A. Kurosawa, Multigas mitigation: an economic analysis using GRAPE model, The Energy Journal Special Issue on Multi-Greenhouse Gas Mitigation and Climate Policy (2006), pp. 275–288

A. Manne, R.G. Richels, The role of Non-CO2 greenhouse gases and carbon sinks in meeting climate objectives, The Energy Journal Special Issue on Multi-Greenhouse Gas Mitigation and Climate Policy (2006), pp. 393–404

van Vuuren, M.G.J. den Elzen, P.L. Lucas, B. Eickhout, B.J. Strengers, B. van Ruijven, S. Wonink, R. van Houdt, Stabilizing greenhouse gas concentrations at low levels: an assessment of reduction strategies and costs, Climatic Change, 81 (2007), pp. 119–159

S.J. Smith, T.M.L. Wigley, Multi-gas forcing stabilization with minicam, The Energy Journal Special Issue on Multi-Greenhouse Gas Mitigation and Climate Policy (2006), pp. 373–392R.

Gerlagh, ITC in a global growth-climate model with CCS: the value of induced technical change for climate stabilization, The Energy Journal Special Issue on Endogenous Technological Change and the Economics of Atmospheric Stabilisation (2006), pp. 223–240

R. Crassous, J.-C. Hourcade, O. Sassi, Endogenous structural change and climate targets modeling experiments with Imaclim-R, The Energy Journal Special Issue on Endogenous Technological Change and the Economics of Atmospheric Stabilisation (2006), pp. 259–276

T. Barker, H. Pan, J. Köhler, R. Warren, S. Winne, Decarbonizing the global economy with induced technological change: scenarios to 2100 using E3MG, The Energy Journal Special Issue on Endogenous Technological Change and the Economics of Atmospheric Stabilisation (2006)

S. Rao, I. Keppo, K. Riahi, Importance of technological change and spillovers in long-term climate policy, The Energy Journal Special Issue on Endogenous Technological Change and the Economics of Atmospheric Stabilisation (2006), pp. 123–139

F. Sano, K. Akimoto, T. Homma, T. Tomoda, Analysis of technological portfolios for CO2 stabilization and effects on technological changes, The Energy Journal Special Issue on Endogenous Technological Change and the Economics of Atmospheric Stabilisation (2006), pp. 141–161

V. Bosetti, C. Carraro, M. Galeotti, The dynamics of carbon and energy intensity in a model of endogenous technical change, The Energy Journal Special Issue on Endogenous Technological Change and the Economics of Atmospheric Stabilisation (2006), pp. 191–205

F. Hedenus, C. Azar, K. Lindgren, Induced technical change in a limited foresight optimization model, The Energy Journal Special Issue on Endogenous Technological Change and the Economics of Atmospheric Stabilisation (2006), pp. 109–122

L.E. Clarke, J.P. Weyant, J.A. Edmonds, On the sources of technological change: what do the models assume?, Energy Economics, 30 (2008), pp. 409–424

D. van Vuuren, M.G.J. Den Elzen, P. Lucas, B. Eickhout, B.J. Strengers, V.R. B., M.M. Berk, H.J.M. De Vries, M. Hoogwijk, M. Meinshausen, S.J. Wonink, R. Van den Houdt, R. Oostenrijk, Stabilising Greenhouse Gas Concentrations at Low Levels: An Assessment of Options and Costs, Netherlands Environmental Assessment Agency, Bilthoven (2006) p. 273

Source: Adapted from: Kuik, Onno, Luke Brander, Richard S.J. Tol. 2009. Marginal abatement costs of greenhouse gas emissions: A meta-analysis. Energy Policy, Volume 37, Issue 4, Pages 1395–1403.

Table 29.36.   Global trends in renewable energy investment, by region 2004-2011 (billion USD)

	2004	2005	2006	2007	2008	2009	2010	2011	Change 2004-11
United States	7.4	11.2	27.2	28.5	37.7	22.5	32.5	50.8	586%
Brazil	0.4	1.9	4.3	9.3	12.7	7.3	6.9	7.5	1775%
AMER (excl. US & Brazil)	1.3	3.3	3.3	4.7	5.4	6.4	11	7	438%
Europe	18.6	27.7	37.4	57.8	67.1	67.9	92.3	101	443%
Middle East & Africa	0.3	0.4	1.6	1.9	3.7	3.1	6.7	5.5	1733%
China	2.2	5.4	10	14.9	24.3	37.4	44.5	52.2	2273%
India	2	2.9	4.7	5.6	4.7	4.2	7.6	12.3	515%
ASOC (excl. China & India)	7.2	8	8	10.1	11	12.1	18.4	21.1	193%

Source; Adapted from Frankfurt School of Finance and Management gGmbH. 2012. Global trends in renewable energy investment 2012, (Frankfurt, Frankfurt School of Finance and Management gGmbH).

Table 29.37.   Global trends in renewable energy investment, by sector, 2004-2011 (billion USD)

	2004	2005	2006	2007	2008	2009	2010	2011	Change 2004-11
Wind	13.3	22.9	32	51.1	67.7	74.6	95.5	83.8	530%
Solar	13.8	16.4	19.5	37.7	57.4	58	96.9	147.4	968%
Biofuels	3.5	8.2	26.6	24.5	19.2	9.1	8.5	6.8	94%
Biomass $ ETE	6.1	7.8	10.8	11.8	13.6	12.2	12	10.6	74%
Small hydro	1.4	4.4	5.4	5.5	6.6	4.7	3.6	5.8	314%
Geothermal	1.4	1	1.4	1.4	1.9	2	3.1	2.9	107%
Marine	0	0	0.9	0.7	0.2	0.3	0.3	0.2	

Source; Adapted from Frankfurt School of Finance and Management gGmbH. 2012. Global trends in renewable energy investment 2012, (Frankfurt, Frankfurt School of Finance and Management gGmbH).

Table 29.38. The Energy Development Index (EDI) for selected developing nations[a]

Country	EDI Rank (out of 64 nations)	EDI value	Commercial energy use per capita index	Electrification index	Per capita electricity consumption index	Modern fuels for cooking index
Libya	1	0.923	1.000	0.999	0.812	0.882
Iran	2	0.889	0.932	0.983	0.644	1.000
Lebanon	3	0.850	0.478	1.000	1.000	0.924
Venezuela	4	0.844	0.740	0.990	0.687	0.959
Argentina	5	0.798	0.570	0.970	0.675	0.976
Jordan	6	0.773	0.392	1.000	0.701	0.997
Malaysia	7	0.741	0.733	0.994	0.642	0.596
Algeria	8	0.706	0.353	0.993	0.485	0.993
Syria	9	0.703	0.333	0.919	0.560	1.000
Uruguay	10	0.692	0.343	0.982	0.845	0.599
Cambodia	55	0.081	0.023	0.145	0.037	0.118
Eritrea	56	0.075	0.000	0.235	0.008	0.055
Zambia	57	0.074	0.045	0.087	0.134	0.029
Togo	58	0.053	0.011	0.100	0.042	0.058
Kenya	59	0.037	0.029	0.056	0.024	0.038
Tanzania	60	0.022	0.022	0.032	0.021	0.014
Myanmar	61	0.018	0.020	0.021	0.021	0.008
Ethiopia	62	0.017	0.000	0.066	0.003	0.000
Mozambique	63	0.013	0.024	0.007	0.017	0.003
Dem. Rep. of Congo	64	0.010	0.018	0.000	0.019	0.002

[a]The EDI is composed of four indicators, each of which captures a specific aspect of potential energy poverty: (1) per capita commercial energy consumption: which serves as an indicator of the overall economic development of a country; (2) Per capita electricity consumption in the residential sector: which serves as an indicator of the reliability of, and consumer's ability to pay for, electricity services; (3) Share of modern fuels in total residential sector energy use: which serves as an indicator of the level of access to clean cooking facilities, and (4) Share of population with access to electricity. A separate index is created for each indicator, using the actual maximum and minimum values for the developing countries covered. Performance in each indicator is expressed as a value between 0 and 1, and the EDI is then calculated as the arithmetic mean of the four values for each country.
Source: International Energy Agency (IEA). 2011. World Energy Outlook, (Paris, IEA).

Table 29.39.   Access to electricity to electricity and reliance on biomass for various regions and nations

	Without access to electricity		Relying on the traditional use of biomass for cooking	
	Population (million)	Share of population	Population (million)	Share of population
Africa	587	58%	657	65%
Nigeria	76	49%	104	67%
Ethiopia	69	83%	77	93%
DR of Congo	59	89%	62	94%
Tanzania	38	86%	41	94%
Kenya	33	84%	33	83%
Other Sub-Saharan Africa	310	68%	335	74%
North Africa	2	1%	4	3%
Developing Asia	675	19%	1921	54%
India	289	25%	836	72%
Bangladesh	96	59%	143	88%
Indonesia	82	36%	124	54%
Pakistan	64	38%	122	72%
Myanmar	44	87%	48	95%
Rest of developing Asia	102	6%	648	36%
Latin America	31	7%	85	19%
Middle East	21	11%	0	0%
Developing Countries	1314	25%	2662	51%
World[a]	**1317**	**19%**	**2662**	**39%**

[a]World total includes OECD and Eastern Europe/Eurasia.
Source: Adapted from International Energy Agency. 2011. Energy for All: financing access for the poor, (Paris, IEA)

Table 29.40.   Important environmental aspects of materials: the case of aluminum alloys

Stage	Quantity
Raw material	
Global production, main component	37×10^6 metric ton/yr
Reserves	2.03×10^9 metric ton
Embodied energy, primary production	200-220 MJ/kg
CO_2 footprint, primary production	11-13 kg/kg
Water usage	495-1490 l/kg
Processing	
Casting energy	11 - 12.2 MJ/kg
Casting CO_2 footprint	0.82 - 0.91 kg/kg
Deformation processing energy	3.3 - 6.8 MJ/kg
Deformation processing CO_2 footprint	0.19 - 0.23 kg/kg
End of life	
Embodied energy, recycling	22 - 39 MJ/kg
CO_2 footprint, recycling	1.9 - 2.3 kg/kg
Recycle fraction in current supply	41-45%

Source: Adapted from Ashby, Michael F. 2013. Materials and the Environment (Second Edition), (Boston, Butterworth-Heinemann).

Communication

Charts

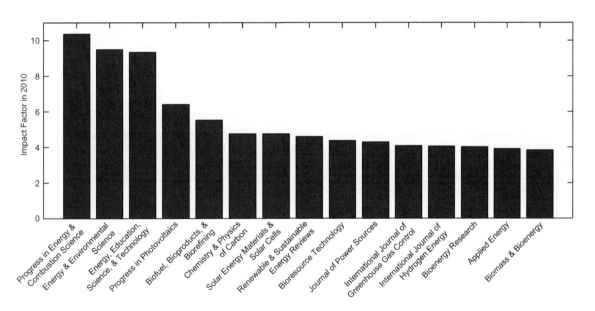

Chart 30.1. Impact factor for energy journals in natural sciences, 2010. In a given year, the impact factor of a journal is the average number of citations received per paper published in that journal during the two preceding years.
Source: Data from Thomson Reuters, Journal Citation Reports®, <http://thomsonreuters.com/ products_services/science/science_products/a-z/journal_citation_reports/>.

Handbook of Energy. http://dx.doi.org/10.1016/B978-0-08-046405-3.00030-9

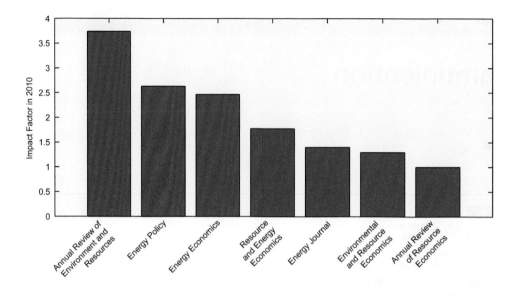

Chart 30.2. Impact factor for energy journals in social sciences, 2010. In a given year, the impact factor of a journal is the average number of citations received per paper published in that journal during the two preceding years.
Source: Data from Thomson Reuters, Journal Citation Reports®, <http://thomsonreuters.com/ products_services/science/science_products/a-z/journal_citation_reports/>.

Printed and bound by CPI Group (UK) Ltd, Croydon, CR0 4YY

08/05/2025

01864785-0001